Grundlehren der mathematischen Wissenschaften 261

A Series of Comprehensive Studies in Mathematics

S. Bosch · U. Güntzer · R. Remmert

Non-Archimedean Analysis

A Systematic Approach to Rigid Analytic Geometry

Springer-Verlag
Berlin Heidelberg NewYork Tokyo
1984

Prof. Dr. Siegfried Bosch
Mathematisches Institut der Universität Münster
Einsteinstraße 64, D-4400 Münster

Prof. Dr. Ulrich Güntzer
Institut für Informatik, Technische Universität München
Arcisstraße 21, D-8000 München 2

Prof. Dr. Reinhold Remmert
Mathematisches Institut der Universität Münster
Einsteinstraße 64, D-4400 Münster

AMS Subject Classification: 12Bxx, 12Jxx, 14G20, 30G05, 32K10, 46P05

ISBN 3-540-12546-9 Springer-Verlag Berlin Heidelberg New York Tokyo
ISBN 0-387-12546-9 Springer-Verlag New York Heidelberg Berlin Tokyo

Library of Congress Cataloging in Publication Data.
Bosch, S. (Siegfried), 1944.
Non-Archimedean analysis.
(Grundlehren der mathematischen Wissenschaften
A Series of comprehensive studies in mathematics; 261)
Bibliography: p. Includes index
1. Geometry, Analytic. 2. Functional analysis.
I. Güntzer, U. (Ulrich), 1941. II. Remmert, R. (Reinhold), 1930.
III. Title. IV. Series: Grundlehren der mathematischen Wissenschaften; 261.
QA551.B73 1983 516.3 83-10324

Printed in GDR

Bookbinding: Lüderitz & Bauer, Berlin
2141/3020-543210

Preface

So eine Arbeit wird eigentlich nie fertig,
man muß sie für fertig erklären,
wenn man nach Zeit und Umständen
das möglichste getan hat.

(Goethe, Caserta, 16. März 1787)

This book was planned in the late sixties by the second and third author after the revival of non-Archimedean function theory. Drafts of Parts A and B existed by 1970. They were mainly written while the authors were visiting at the University of Maryland. However, many other commitments delayed the completion. Fortunately in 1973 the first author joined the enterprise and gave a new impetus to the project. The bulk of Part C is his work.*

Courses on the material covered in this book have been given by us on several occasions. We were stimulated by our students and encouraged by colleagues asking for a systematic presentation of this topic which finds more and more applications in other fields of mathematics.

Parts of the manuscript were read in 1974 by J. Horvath, University of Maryland, and by Mrs. J. Snow during her stay at Münster in 1980. We thank them for useful suggestions. It is our special pleasure to express our gratitude to P. Ullrich, University of Münster, who critically read the text at the final stage and pointed out several flaws. He also set up the index and was of invaluable help in proof-reading. Last not least our thanks go to Springer-Verlag for the beautiful printing which is up to its old standards again.

München, Münster, December 1982 S. Bosch, U. Güntzer, R. Remmert

* We thank the following institutions for granting Forschungssemester to the authors (first author WS 77/78, WS 82/83; second author SS 73, WS 80/81; third author WS 79/80):

«Der Minister für Wissenschaft und Forschung des Landes Nordrhein-Westfalen»
«Der Präsident der Freien Universität Berlin»
«Das Bayerische Staatsministerium für Unterricht und Kultus»

Substantial work on the book was done during parts of these sabbaticals.

Contents

Part C. Rigid analytic geometry

Introduction

The discovery of p-adic numbers by K. HENSEL in 1905 led to the creation of the non-Archimedean completions $\mathbb{Q}_p$ of the field of rational numbers $\mathbb{Q}$. From a number theoretic point of view these fields are as natural as the Archimedean completion $\mathbb{R}$ of $\mathbb{Q}$. As the field $\mathbb{R}$ is the basis of classical analysis, so the fields $\mathbb{Q}_p$ are the fundamentals of non-Archimedean analysis. The fields $\mathbb{Q}_p$ are special examples of valued fields (bewertete Körper). J. KÜRSCHAK, 1913, and A. OSTROWSKI, 1918, considered such fields and gave a classification.

In the twenties R. STRASSMANN and others studied power series in one variable over the p-adic numbers. A very important paper dealing with function theoretic problems in a general non-Archimedean setting is the paper by W. SCHÖBE "Beiträge zur Funktionentheorie in nichtarchimedisch bewerteten Körpern", Helios Verlag, Münster 1930; unfortunately SCHÖBE's work never really became available to the mathematical community. In the forties M. KRASNER started to work systematically on problems in non-Archimedean function theory. In spite of the efforts of all these people, the topic was considered exotic by most contemporary mathematicians.

Modern non-Archimedean function theory was born in 1961 when J. TATE gave a seminar at Harvard entitled "Rigid Analytic Spaces". Motivated by the question how to characterize elliptic curves with bad reduction, he discovered a new category of analytic-algebraic objects with a structure rich enough to make possible the impossible: analytic continuation over totally disconnected ground fields. The notes by TATE were distributed in Paris in Spring 1962 with(out) his permission*. Incidently H. GRAUERT and the last named author of this book were able to obtain a copy. It soon turned out that, for complex analytically minded readers, the local part of TATE's new theory could be understood very well as a *partially global* analogue of the WEIERSTRASS approach to *local* complex function theory of several variables. This point of view is explained in detail in [15] where the non-Archimedean WEIERSTRASS Division and Preparation Theorems are stated and proved and used as point of departure. The models for the new spaces and their structure algebras were called *affinoid*; this word is to indicate that the models are a hybrid: they carry affine algebraic as well as analytic algebroid features.

The affinoid spaces are the local parts of TATE's global analytic spaces. He considered analytic spaces (defined analogously as in the classical complex case) and provided them with an extra "topological" structure which, from a

* Not till 9 years later were these notes published officially in the Inventiones, cf. [37]

higher point of view, amounts to a non-trivial notion of connectedness, although strictly speaking all these spaces are totally disconnected. The main idea is to carry out analytic continuation only with respect to certain *admissible open coverings*. Analytic functions are no longer considered on all open subsets of such a space; one has to restrict oneself to *admissible open sets**. TATE called his new spaces rigid analytic spaces; this is to distinguish them from ordinary analytic spaces which are only of very limited use over non-Archimedean ground fields.

The definition of rigid analytic spaces was rather clumsy at the beginning. Substantial simplifications are due to L. GERRITZEN and H. GRAUERT [12] in the late sixties. Their notion of *rational subdomains* led to the classification of affinoid subdomains of affinoid spaces. As a result, admissible open sets and coverings of rigid analytic spaces became easier to handle. Simultaneously R. KIEHL [23], [24] obtained some fundamental results on coherent modules: the rigid analytic analogues of GRAUERT's Direct Image Theorem and of Theorems A and B of CARTAN and SERRE. Thereby it became clear that rigid analytic geometry is related to algebraic geometry in the same way as is complex analysis in the classical complex case. Namely, the methods of SERRE's fundamental article "Géométrie Algébrique et Géométrie Analytique" carry over to the non-Archimedean case almost verbatim; the results remain unchanged [26].

Thus we see that, in the late sixties, TATE's ideas had been worked out to a considerable extent, and the new theory was ready to earn its first merits in applications. There have been several fields of interest stemming from complex analysis, algebraic geometry and number theory. All this is beyond the scope of this book. However we cannot refrain from mentioning a special subject which has attracted many rigid analysts. In the early seventies MUMFORD [27] succeeded in generalizing the uniformization of elliptic curves with bad reduction to curves of higher genus. The curves he considered had split degenerate reduction; they are nowadays called SCHOTTKY or MUMFORD curves. At the same time RAYNAUD [32] indicated how to obtain the uniformization for abelian varieties. Both MUMFORD and RAYNAUD were thinking in terms of formal algebraic geometry over discrete valuation rings; however the relationship to rigid analytic geometry was clear (see also the article [33] of RAYNAUD). Thus rigid analysts made efforts to set up a rigid analytic approach towards the uniformization of algebraic curves and varieties, cf. [11], [14] and [3]. Substantial progress was achieved. Today the subject is still of essential interest.

The aim of this book is to develop in a systematic way affinoid and rigid analytic geometry ab ovo. There are three parts; they are fairly independent of each other. In Part A we are concerned with certain preliminaries and fundamentals of ultrametric analysis and with non-Archimedean valuations. Since

* The framework of GROTHENDIECK topologies provides a useful formalism for this concept.

ultrametric spaces are totally disconnected, there are no continuous paths. Hence the power of line integrals is not at our disposal for the investigation of analytic functions. Other devices of algebraic or functional analytic nature have to be marshalled. In the literature, there is a variety of such methods which are applied with slight variations again and again. It is one of the main purposes of Part A to single out these tools and concepts once and for all. To enumerate a few of them: reduction functors, the smoothing of norms and semi-norms, spectral and supremum norms, BANACH function algebras, and stable fields will be most helpful. Within the framework of this systematization we have included a detailed treatment of valuation theory from our point of view.

In Part B the category of affinoid algebras over a given complete valued ground field k is at the center of our considerations. The TATE algebra T_n consisting of all strictly convergent power series in n variables (i.e., converging on the "closed" unit polydisc in k^n) is the prototype of such an algebra. It is a proper subalgebra of the algebra of *all* power series converging about the origin in k^n. The restriction to T_n (which is surprising for complex analysts), in conjunction with the concept of GROTHENDIECK topologies, will open the door again to the "paradise lost" of analytic continuation (see Part C). Using our WEIERSTRASS techniques, we show that RÜCKERT's classical results on the structure of the ring of germs of holomorphic functions at $0 \in \mathbb{C}^n$ carry over to TATE algebras:

T_n is a Noetherian factorial k-Banach algebra; all ideals in T_n are closed.

The same methods lead to the affinoid analogue of the classical NOETHER Normalization Lemma for finitely generated algebras. Thereby we obtain a key result which is of indispensible value for handling general affinoid algebras. It is needed throughout the rest of Part B for the discussion of the supremum norm and for the finiteness of the reduction functor $A \rightsquigarrow \tilde{A}$. Also we have included a complete presentation of the finiteness theorem of GRAUERT REMMERT GRUSON for the functor $A \rightsquigarrow \mathring{A}$.

In Part C we study rigid analytic spaces or *rigid analytic varieties* as we prefer to call them. (In the course of the book we say "analytic", which is meant as an abbreviation for "rigid analytic".) First we introduce model spaces, namely the affinoid varieties. They are just the spectra of affinoid algebras. In terms of algebraic geometry these varieties correspond to affine schemes; in terms of complex analysis they can be viewed as a special type of STEIN spaces: the Acyclicity Theorem of TATE is a first step towards the Theorem B of CARTAN and SERRE. We have paid particular attention to the discussion of affinoid subdomains of affinoid varieties. These subdomains are the rigid analytic analogues of the open affine subspaces of affine schemes; they can be constructed by "analytic localization", a process which corresponds to ordinary localization in algebraic geometry. The GERRITZEN GRAUERT Theorem on locally closed immersions gives a classification.

Global varieties are obtained by pasting model spaces in a certain way. Of course, one has to specify the overlappings. In our case, the model spaces are the affinoid varieties, and the overlapping is described by affinoid subdomains (more generally, by admissible open subsets of affinoid varieties). Thus, from its construction, a global rigid analytic variety is always equipped with a distinguished affinoid covering. This covering induces a GROTHENDIECK topology and thereby defines the *rigid* structure of the analytic variety. There are equivalent coverings, called *admissible coverings.* However, not all affinoid coverings are admissible; for an *affinoid variety,* the admissible coverings are just the finite coverings by affinoid subdomains.

The discussion of rigid analytic varieties requires some formal techniques which have to be developed judiciously; namely the theory of GROTHENDIECK topologies, of sheaves and their ČECH cohomology. We give a self-contained presentation which is adapted to our needs. Also we have included some sections dealing with KIEHL's results on coherent modules. Except for the Direct Image Theorem and the Theorem on Formal Functions, complete proofs are given.

At the end of our book, there is a section on elliptic curves. The problem to classify elliptic curves with bad reduction has attracted mathematicians since the epochal paper by M. DEURING [7]; it has initiated TATE's approach to non-Archimedean analysis. By establishing some simple facts of rigid analytic uniformization theory, we show that elliptic curves with bad reduction correspond bijectively to one-dimensional rigid analytic tori. In contrast to TATE's original proof (see [34]), the necessary facts from algebraic geometry are kept to a minimum. Although it is not the intention of the book to deal with applications, we hope to demonstrate thereby that rigid analytic geometry can act in algebraic geometry as a powerful analytical method which is more than just a "Glasperlenspiel".

Reading instructions

There are different ways to read a book like this one, depending on the experience and the interests of the reader. In order to get a feeling for the subject, a novice should begin with Part B, after picking up some basic definitions from Part A if necessary. One can go through sections (5.1), (5.2), (6.1), and maybe (6.2), and proceed then with Part C. After reading sections (7.1) and (7.2) ((7.1.5) and (7.2.6) can be skipped), it is possible to pass on directly to Chapter 9 if one is willing to accept the main result of (7.3) (Theorem 7.3.5/1) as well as Corollary 8.2.1/2 of TATE's Acyclicity Theorem. The sections (9.4) to (9.6) are not used for the discussion of elliptic curves in (9.7).

Prerequisites

We have tried to make our presentation self-contained; besides the Open Mapping Theorem for BANACH spaces, only some basic facts from commutative algebra are assumed. Among these are the Lemma of ARTIN-REES and KRULL's Intersection Theorem (see [28], Theorems 3.7 and 3.11). In section (3.8), we assume that the reader is familiar with the notion of integral dependence. In almost all other cases, we have incorporated proofs for auxiliary results going beyond the listed facts. For example, the Appendix to Part A contains a discussion of tame modules and Japanese rings.

References

There are only a few references to the literature in the text; however we give a detailed bibliography at the end of the book. Besides a list of articles referred to by numbers in square brackets, we have included a variety of several other publications in non-Archimedean analysis. Some of them are related to problems considered in this book, some go beyond it.

Most of the material we present has become standard by now. It goes without saying that our treatment of the whole subject is deeply influenced by TATE's original notes [37], although the investigation of affinoid algebras follows more or less the lines of the articles [15] and [16]. In addition to the references given in the introduction, we mention the following other articles whose ideas have served as guidelines: [20] (smoothing procedures), [2] (Lifting Theorem), [19] (function algebras), [13] (finite T_n-modules), [10] (supremum norm on affinoid algebras), [9] (Japaneseness of affinoid algebras), [17] (stability), [1] (GROTHENDIECK topologies).

Cross references in the book are subject to the following conventions: the different sections and subsections are referred to by their numbers in brackets; theorems, propositions and definitions are cited with their numbers and section numbers. Thus (7.3) means section (7.3), and Theorem 7.3.5/1 means Theorem 1 in subsection (7.3.5). For cross references within a certain subsection, the subsection number is not repeated.

PART A

Linear Ultrametric Analysis and Valuation Theory

CHAPTER 1

Norms and Valuations

In this chapter we are concerned with basic properties of non-Archimedean norms and semi-norms. We look at the particular phenomena caused by the non-Archimedean triangle inequality and give some important examples (strictly convergent power series rings, ideal-adic semi-norms, formal power series in countably many indeterminates). Norms and semi-norms are usually considered on rings or modules. In order to avoid repetitions, we first study semi-norms on groups. Then we add the multiplicative structure and continue with semi-norms (and valuations) on rings. The smoothing procedures in (1.3.2) as well as a first discussion of spectral values in (1.5.4) are of particular interest for our treatment of valuation theory in Chapter 3.

The sections (1.7) on bald B-rings and (1.8) on quasi-Noetherian rings have to be seen in the light of the articles [2], [16] and [21]. The theory of bald or quasi-Noetherian B-rings is the basis for an alternative approach to affinoid geometry, different from the one we have chosen in Part B. Except for the proof of the Lifting Theorem in (2.7), we do not pursue this possibility any further.

1.1. Semi-normed and normed groups

Let G denote an abelian group, which we shall write additively.

1.1.1. Ultrametric functions. — We start with the basic

Definition 1. *A function*

$$| \ | : G \to \mathbb{R}_+$$

on G with values in the set $\mathbb{R}_+$ of non-negative real numbers is said to be ultrametric if it has the following properties:

(a) $|0| = 0$,
(b) $|x - y| \leqq \max\{|x|, |y|\}$ *for all $x, y \in G$.*

Condition (b) is the crucial point for all non-Archimedean function theory. It implies the usual triangle inequality $|x - y| \leqq |x| + |y|$. As a matter of fact, (b) is much stronger than this inequality; e.g., (b) immediately implies

Proposition 2. *For each positive real number r, the sets*

$$G^0(r) := \{x \in G;\ |x| \leqq r\}, \qquad G^\vee(r) := \{x \in G;\ |x| < r\}$$

are subgroups of G.

Remark. We have $G^\vee(r) \subset G^0(r)$; therefore the "residue groups"

$$G^\sim(r) := G^0(r)/G^\vee(r), \qquad r > 0,$$

are well defined. These groups will turn out to be very important in our later considerations.

The set

$$\ker |\ \ | := \{x \in G;\ |x| = 0\}$$

is called the *kernel* of $|\ \ |$. Clearly,

$$\ker |\ \ | = \bigcap_{r>0} G^0(r) = \bigcap_{r>0} G^\vee(r).$$

In particular, $\ker |\ \ |$ is a subgroup of G.

Proposition 3. *For all elements x, y of an abelian group G with ultrametric function* $|\ \ |$, *we have*

(a) $|-y| = |y|$,
(b) $|x + y| \leqq \max\{|x|, |y|\}$,
(c) $|x + y| = \max\{|x|, |y|\}$, *if* $|x| \neq |y|$.

Proof. Inequality (b) of Definition 1 gives $|-y| \leqq \max\{|0|, |y|\}$; this yields $|-y| \leqq |y|$. In the same way, we get $|y| \leqq |-y|$. So we have (a).

By (a) and $|x - (-y)| \leqq \max\{|x|, |-y|\}$, we get (b).

In order to prove (c), we may assume $|x| < |y|$. Applying (b) of Definition 1, we find $|y| \leqq \max\{|x + y|, |x|\} \leqq \max\{|y|, |x|\}$. The assumption $|x| < |y|$ gives $\max\{|y|, |x|\} = |y|$. Hence $|y| = \max\{|x + y|, |x|\}$. From $|x| < |y|$, we deduce $|y| = |x + y|$. This gives (c). □

By induction we get from (c) the so-called Principle of Domination:

Proposition 4. *Let* $x_\nu \in G$, $\nu = 1, \ldots, n$, *such that* $|x_1| > |x_\nu|$ *for all* $\nu > 1$. *Then*

$$\left|\sum_{\nu=1}^{n} x_\nu\right| = |x_1|.$$

Corollary 5. *If* $\sum_{\nu=1}^{n} x_\nu = 0$, $x_\nu \in G$, $n \geqq 2$, *there exist two indices i, j,* $1 \leqq i < j \leqq n$, *such that*

$$|x_i| = |x_j| = \max_{1 \leqq \nu \leqq n} \{|x_\nu|\}.$$

We denote by $|G|$ the set $\{|x|;\ x \in G\} \subset \mathbb{R}_+$. The set $|G|$ can be bounded or even finite; e.g., $|G| = \{0, 1\}$ for the trivial ultrametric function defined by $|x| := 1$ for all $x \neq 0$. Obviously, for all $r > 0$, one has the equivalence: $r \in |G|$ if and only if $G^\sim(r) \neq 0$.

Proposition 6. *Let* $|\ \ |$ *be an ultrametric function on* G. *Let* $|\ \ |' \colon G \to \mathbb{R}_+$ *be a map with* $|0|' = 0$ *such that*

$$|x| \leqq |y| \text{ always implies } |x|' \leqq |y|', \qquad x, y \in G.$$

Then $|\ \ |'$ *is ultrametric.*

Proof. We have to show $|x - y|' \leqq \max\{|x|', |y|'\}$ for all $x, y \in G$. If $|x| \leqq |y|$, one deduces from $|x - y| \leqq \max\{|x|, |y|\} = |y|$ that $|x - y|' \leqq |y|' \leqq \max\{|x|', |y|'\}$. If $|y| \leqq |x|$ one proceeds in exactly the same way. □

1.1.2. Filtrations. — With each ultrametric function $|\ \ |$ on G, we associate the function

$$\nu(x) := \begin{cases} -\ln |x| & \text{if } |x| \neq 0, \\ \infty & \text{if } |x| = 0. \end{cases}$$

An easy computation shows that $\nu \colon G \to \mathbb{R} \cup \{\infty\}$ satisfies the following properties:

(a′) $\nu(0) = \infty$,
(b′) $\nu(x - y) \geqq \min\{\nu(x), \nu(y)\}$.

Let us call each map $\nu \colon G \to \mathbb{R} \cup \{\infty\}$ satisfying (a′) and (b′) a *filtration of* G. We have

If $|\ \ |$ *is an ultrametric function on* G, *then* $\nu := -\ln |\ \ |$ *is a filtration of* G.

Each filtration ν of G has the following properties (analogous to Proposition 1.1.1/3):

$$\nu(-y) = \nu(y),$$
$$\nu(x + y) \geqq \min\{\nu(x), \nu(y)\},$$
$$\nu(x + y) = \min\{\nu(x), \nu(y)\}, \quad \text{if } \nu(x) \neq \nu(y).$$

In applications it is sometimes convenient to use filtrations instead of ultrametric functions. However the concepts are of equal strength, since it is easy to pass back and forth from a filtration to an ultrametric function. Namely,

Let ν *be a filtration of* G *and let* $\varepsilon \in \mathbb{R}$, $0 < \varepsilon < 1$. *Then*

$$|x| := \varepsilon^{\nu(x)}, \qquad x \in G,$$

defines an ultrametric function $|\ \ |$ *on* G. *For* $\varepsilon := e^{-1}$ (*where* $e = \exp(1)$), *we have* $\nu = -\ln |\ \ |$.

In this book we shall always use ultrametric functions instead of filtrations. This means that we adopt the "multiplicative" and not the "additive" point of view.

1.1.3. Semi-normed and normed groups. Ultrametric topology. —

Definition 1. *A pair* $(G, |\ \ |)$ *consisting of an abelian group* G *and an ultrametric function* $|\ \ | \colon G \to \mathbb{R}_+$ *is called a semi-normed group. The pair* $(G, |\ \ |)$ *is called a normed group if* $\ker |\ \ | = \{0\}$.

We often write G instead of $(G, |\ \ |)$, when there can be no confusion. The function $|\ \ |$ is referred to as the semi-norm or the norm on G, respectively. For the rest of this subsection, G always denotes a semi-normed group.

With the definition

$$d(x, y) := |x - y|, \qquad x, y \in G,$$

we obtain a *pseudometric topology* on G which makes G into a topological group; i.e., the group operation $(x, y) \mapsto x - y$ is continuous. The group G admits a countable neighborhood basis at 0, consisting of subgroups of G. It is not hard to check that, conversely, every topological abelian group with such a fundamental system of neighborhoods at 0 posesses an ultrametric function defining the topology.

Proposition 2. *The map* $|\ \ |: G \to \mathbb{R}_+$ *is continuous, more precisely,*

$$\big||x| - |y|\big| \leqq |x - y| \quad \textit{for all} \quad x, y \in G.$$

Actually one has $|x| = |y|$, *whenever* $|x - y| < \max\{|x|, |y|\}$. *The subgroup* $\ker |\ \ |$ *is closed in* G.

Proposition 3. *G is Hausdorff if and only if G is a normed group.*

The proofs are easy and we omit them.

The following statements show essential differences between the ultrametric topologies and the familiar ones of the real or complex Euclidian spaces. These differences all come from condition (b) of Definition 1.1.1/1, which implies that the metric d is ultrametric; i.e., d satisfies the additional axiom:

$$d(x, z) \leqq \max\{d(x, y), d(y, z)\} \quad \text{for all} \quad x, y, z \in G.$$

This can be stated geometrically: *Each side of a triangle is at most as long as the longest one of the two other sides.* This implies that *each triangle is isosceles*, a geometrical interpretation of Proposition 1.1.1/3(c).

To further see how the topology is influenced by condition (b) of Definition 1.1.1/1, let $r > 0$ be given. Two elements $x, y \in G$ are called *r-near* if $d(x, y) < r$. In this case we write $x \underset{r}{\sim} y$. Obviously, $x \underset{r}{\sim} y$ if and only if $x - y \in G^{\vee}(r)$. Therefore,

The relation $\underset{r}{\sim}$ *is an equivalence relation on* G.

Let

$$B^+(a, r) := \{x \in G;\ |x - a| \leqq r\},$$
$$B^-(a, r) := \{x \in G;\ |x - a| < r\}$$

be the balls with center $a \in G$ and radius $r > 0$ (with or without circumference, respectively). Obviously, $B^-(a, r)$ is just the equivalence class of $\underset{r}{\sim}$ containing a. This implies $B^-(a', r) = B^-(a, r)$ for each $a' \in B^-(a, r)$. Similarly, one obtains $B^+(a', r) = B^+(a, r)$ for each $a' \in B^+(a, r)$. Therefore,

Each point of a ball is a center of this ball. In particular, all balls in G are open.

From this we derive

Two balls in G are either disjoint or one is contained in the other.

Indeed, if $a \in B_1 \cap B_2$, where B_1 and B_2 are balls, we may view a as the common center of both balls. Then, of course, one ball must be contained in the other.

Let

$$S(a, r) := \{x \in G; |x - a| = r\}$$

be the sphere with center $a \in G$ and radius $r > 0$, which is closed in G by definition. If $r \notin |G|$, this sphere is empty. We claim

Proposition 4. *$B^-(x, r) \subset S(a, r)$ for each $x \in S(a, r)$.*

Proof. Let $y \in B^-(x, r)$. Then $|x - y| < |x - a| = r$. Thus Proposition 1.1.1/3(c) implies: $|y - a| = |(y - x) + (x - a)| = \max\{|y - x|, |x - a|\} = r$; i.e., $y \in S(a, r)$. □

In particular,

Each sphere in G is open.

Moreover, since $B^-(a, r) = B^+(a, r) - S(a, r)$ and since $B^+(a, r)$ is closed, we see that

Each ball in G is closed.

This also follows from the fact that, in a topological group, open subgroups are always closed. Next we prove

Proposition 5. *A normed group G is totally disconnected (in the ultrametric topology).*

Proof. For each $a \in G$ the connected component T of $\{a\}$ is contained in every open and closed neighborhood of $\{a\}$. By our previous remarks, the balls $B^-(a, r)$ are such neighborhoods. Hence $T = \{a\}$. □

A group homomorphism $\varphi: G \to G'$ between semi-normed groups is called *bounded* if there is a real constant $M > 0$ such that $|\varphi(x)|' \leqq M\,|x|$ for all $x \in G$. The homomorphism φ is called *contractive* if $|\varphi(x)|' \leqq |x|$ for all $x \in G$.

As in the Archimedean case, one has the following connection between continuity and boundedness.

Proposition 6. *Let $\varphi: G \to G'$ be an (abstract) group homomorphism. Then:*

(i) *If φ is continuous at the origin $0 \in G$, then φ is continuous everywhere.*

(ii) *If φ is bounded, it is continuous.*

In particular if φ is bijective, it defines an isomorphism of topological groups between G and G' if there are positive real numbers ϱ, ϱ' such that

$$|\varphi(x)| \leqq \varrho\,|x| \leqq \varrho'\,|\varphi(x)| \quad \textit{for all } x \in G.$$

The proof is obvious.

Note that continuous homomorphisms need not be bounded (cf. (2.1.8)).

Let G and H be semi-normed groups and let $\varphi: G \to H$ be a contractive group homomorphism. For all $r > 0$, we have

$$\varphi(G^0(r)) \subset H^0(r) \quad \text{and} \quad \varphi(G^{\vee}(r)) \subset H^{\vee}(r).$$

Therefore φ induces a group homomorphism

$$\varphi^{\sim}(r): G^{\sim}(r) \to H^{\sim}(r).$$

It is easy to verify that $G \rightsquigarrow G^{\sim}(r)$ is a functor from the category of semi-normed groups (with contractive homomorphisms as morphisms) to the category of groups. The group $G^{\sim}(r)$ is a purely algebraic object, but nevertheless it allows us to derive information about the semi-norm of G. This functor (especially for $r = 1$) is an essential tool of non-Archimedean analysis.

1.1.4. Distance. — Let G be a semi-normed group, and let H be any subgroup of G.

Definition 1. *For each $a \in G$, the non-negative real number*

$$|a, H| := \inf_{y \in H} |a + y|$$

is called the distance from a to H.

Obviously, $|a + y, H| = |a, H|$ for all $y \in H$ and $|a, H| = \inf_{y \in H} d(a, y)$. Hence

$$\bar{H} = \{x \in G;\, |x, H| = 0\}.$$

Thus, if H is not dense in G, the function $x \mapsto |x, H|$ is not identically zero.

For later applications we prove

Proposition 2. *Let $(G, |\ \ |)$ be a normed group and let H be a subgroup of G which is "ε-dense" in G in the following sense: there is a real number $\varepsilon < 1$ such that for each $g \in G$ there exists an $h \in H$ with $|g + h| \leqq \varepsilon\, |g|$.*

Then H is dense in G; i.e., $\bar{H} = G$.

Proof. We may assume $\varepsilon > 0$. Since $\bar{H} = \{x \in G;\, |x, H| = 0\}$, it is enough to show that $|x, H| = 0$ for all $x \in G$. Assume there is a $g \in G$ such that $|g, H| > 0$. Choose an $h_1 \in H$ such that $|g + h_1| < \varepsilon^{-1}\, |g, H|$; this is possible since $\varepsilon^{-1} > 1$. By hypothesis there is an element $h_2 \in H$ such that $|(g + h_1) + h_2| \leqq \varepsilon\, |g + h_1|$. Since $h_1 + h_2 \in H$, the inequality $|g + (h_1 + h_2)| < |g, H|$ is impossible. Therefore, $|x, H| = 0$ for all $x \in G$. □

1.1.5. Strictly closed subgroups. — We choose this place to introduce an interesting notion for later use.

Definition 1. *A subgroup H of a normed group $(G, |\ \ |)$ is called strictly closed if for each $a \in G$ there exists an element $y_0 \in H$ such that $|a, H| = |a + y_0|$.*

Obviously, this condition can only be restrictive for elements $a \notin H$, since otherwise one may always choose $y_0 := -a$. Rephrasing the definition in the language of functional analysis, we may say: H is strictly closed in G if and only if to each $a \in G$ there exists a "best approximation" y_0 in H. Notice that y_0 is by no means uniquely determined.

Lemma 2. *For all $r > 0$, the ball groups $G^0(r)$, $G^\vee(r)$ are strictly closed in G.*

Proof. Let us restrict our attention to the case of a subgroup $G^0(r)$. Take $a \in G - G^0(r)$. Then $|a| > r$ and therefore $|a + y| = |a|$ for all $y \in G^0(r)$. Hence we may take $y_0 := 0$. □

Since $\bar{H} = \{x \in G;\ |x, H| = 0\}$, we have

Lemma 3. *Each strictly closed subgroup of a normed group is closed.*

The following converse holds:

Proposition 4. *Each closed subgroup H of a normed group G such that $|H - \{0\}|$ is discrete in $\mathbb{R}_+ - \{0\}$ is strictly closed. In particular, if $|G - \{0\}|$ is discrete in $\mathbb{R}_+ - \{0\}$, then every closed subgroup is strictly closed.*

Proof. Obviously, it is enough to show the first assertion. For each $a \in G$, we have

$$(*) \qquad \{|a + h|;\ h \in H \text{ and } |a + h| > |a, H|\} \subset |H|.$$

Indeed, for each $h \in H$ with $|a + h| > |a, H|$, we can find an element $h' \in H$ such that $|a + h| > |a + h'|$, and then we see

$$|a + h| = |(a + h) - (a + h')| = |h - h'| \in |H|,$$

which proves $(*)$. Assume now that $a \in G - H$. Since H is closed, this implies $|a, H| > 0$. Knowing that $|H - \{0\}|$ is discrete in $\mathbb{R}_+ - \{0\}$, we may deduce from $(*)$ that the set

$$\{|a + h|;\ h \in H \text{ and } |a, H| \leqq |a + h| < |a, H| + 1\}$$

is finite. Therefore, the infimum of $\{|a + h|; h \in H\}$ is actually assumed; i.e., H is strictly closed. □

If $|G - \{0\}|$ is not discrete, closedness does not necessarily imply strict closedness as we shall see later. In this case the question of which closed subgroups of a normed group G are strictly closed is difficult to handle (being of the type that an infimum has to be a minimum). As a matter of fact we shall be very much concerned with this question in (2.4.2) and (5.2.7).

Lemma 5. *Let $G \supset H_1 \supset H_2$ be normed groups such that H_1 is a strictly closed subgroup of G and H_2 is a strictly closed subgroup of H_1. Then H_2 is also strictly closed in G.*

Proof. Let $g \in G$ be given. Then there is an $h_1 \in H_1$ such that $|g + h_1| \leqq |g + x|$ for all $x \in H_1$. Furthermore, there is an $h_2 \in H_2$ such that $|-h_1 + h_2|$

$\leqq |-h_1 + y|$ for all $y \in H_2$. We want to show

$$(*) \qquad |g + h_2| \leqq |g + y| \quad \text{for all} \quad y \in H_2.$$

In order to do so, consider the inequalities

$$|g + h_2| \leqq \max\{|g + h_1|, |-h_1 + h_2|\} \leqq \max\{|g + y|, |-h_1 + y|\}$$
$$\text{for all} \quad y \in H_2.$$

But for all $y \in H_2$, one has $|-h_1 + y| \leqq \max\{|g + y|, |g + h_1|\} = |g + y|$. Thus, $(*)$, and hence Lemma 5, has been proved. □

1.1.6. Quotient groups. — Let G denote a semi-normed group and let H be a subgroup of G. The distance function fulfills the ultrametric inequality

$$|a + a', H| \leqq \max\{|a, H|, |a', H|\}, \qquad a, a' \in G,$$

since for all $y, y' \in H$ we have $|a + a' + (y + y')| \leqq \max\{|a + y|, |a' + y'|\}$. Thus, we see

Proposition 1. *Denote by $\pi\colon G \to G/H$ the residue epimorphism. Then $(G/H, |\ \ |_{\text{res}})$, where*

$$|\pi(x)|_{\text{res}} := |x, H| = \inf_{\pi(x') = \pi(x)} |x'|,$$

is a semi-normed group. The corresponding topology is the quotient topology. The epimorphism π is contractive. If H is closed in G, the function $|\ \ |_{\text{res}}$ is a norm.

We call $|\ \ |_{\text{res}}$ the *residue semi-norm* of G/H.

Corollary 2. *If $H := \ker |\ \ |$, then $(G/H, |\ \ |_{\text{res}})$ is a normed group. Moreover, π is an isometry; i.e., $|\pi(x)|_{\text{res}} = |x|$ for all $x \in G$.*

Proof of the corollary. The first statement follows from the second one. To prove the second one, it is enough to show $|x + y| = |x|$ for all $y \in \ker |\ \ |$, $x \in G$. But if $|x| > 0$, then $|x + y| = \max\{|x|, |y|\} = |x|$. If $|x| = 0$, then $x + y \in \ker |\ \ |$ and therefore $|x + y| = 0 = |x|$. □

The following lemma is an immediate consequence of the definition of strict closedness.

Lemma 3. *If H is a strictly closed subgroup of a normed group G, then $|G/H|_{\text{res}} \subset |G|$.*

For later reference we need

Lemma 4. *Let $G \supset H_1 \supset H_2$ be normed groups such that H_2 is a strictly closed subgroup of G and such that H_1/H_2 is a strictly closed subgroup of G/H_2, where both are provided with the residue norm. Then H_1 is also strictly closed in G.*

Proof. Let π denote the residue epimorphism $\pi: G \to G/H_2$. Let $g \in G$ be given. Because $\pi(H_1)$ is strictly closed in $\pi(G)$, there exists an $h_1 \in H_1$ such that $|\pi(g) + \pi(h_1)|_{\text{res}} \leqq |\pi(g) + \pi(x)|_{\text{res}} \leqq |g + x|$ for all $x \in H_1$. Furthermore, because H_2 is strictly closed in G, we can find an $h_2 \in H_2$ such that $|(g + h_1) + h_2| \leqq |(g + h_1) + x|$ for all $x \in H_2$ or, equivalently, such that $|\pi(g + h_1)|_{\text{res}} = |g + h_1 + h_2|$. Finally, we get

$$|g + (h_1 + h_2)| = |\pi(g) + \pi(h_1)|_{\text{res}} \leqq |g + x| \quad \text{for all} \quad x \in H_1. \qquad \square$$

1.1.7. Completions. — In every topological space X with pseudometric d one has the notion of a "Cauchy sequence". A sequence $(x_\nu)_{\nu\in\mathbf{N}}$, $x_\nu \in X$, is a Cauchy sequence if and only if $d(x_\nu, x_\mu)$ tends to 0 as $\nu, \mu \to \infty$.

In a semi-normed group G, one has, due to property (b) of Definition 1.1.1/1, the following characterization of Cauchy sequences:

Proposition 1. *The sequence $(a_\nu)_{\nu\in\mathbf{N}}$, $a_\nu \in G$, is a Cauchy sequence if and only if $|a_{\nu+1} - a_\nu|$ tends to 0 as $\nu \to \infty$.*

Note that this condition is not sufficient in the familiar metrics of real or complex Euclidian spaces.

Definition 2. *A semi-normed group G is called complete if every Cauchy sequence in G has a limit in G.*

Proposition 3. *If G is complete and if H is a subgroup of G, then G/H is complete.*

Proof. Let $\pi: G \to G/H$ be the projection; let (c_ν) be a Cauchy sequence in G/H. We shall construct a Cauchy sequence (a_ν), $a_\nu \in G$, such that $\pi(a_\nu) = c_\nu$. Let a_1 be an arbitrary point in $\pi^{-1}(c_1)$. If $a_1, \ldots, a_n$ are already determined, choose an element $a'_n \in \pi^{-1}(c_{n+1} - c_n)$ such that $|a'_n| \leqq |c_{n+1} - c_n|_{\text{res}} + \dfrac{1}{n}$. Define $a_{n+1} = a'_n + a_n$. Then $\pi(a_{n+1}) = c_{n+1}$ and $|a_{n+1} - a_n| \leqq |c_{n+1} - c_n|_{\text{res}} + \dfrac{1}{n}$. Hence (a_ν) is a Cauchy sequence. Since G is complete, (a_ν) tends to a limit $a \in G$. Because the map π is continuous, $(c_\nu) = (\pi(a_\nu))$ tends to $\pi(a) \in G/H$. $\square$

Definition 4. *Let G be a semi-normed group. A pair $(\hat{G}, i)$ is called a completion of G if the following conditions are fulfilled:*

(i) *$\hat{G}$ is a complete normed group.*
(ii) *$i: G \to \hat{G}$ is an isometric homomorphism.*
(iii) *$i(G)$ is dense in $\hat{G}$.*

Obviously, (i) and (ii) imply $\ker i = \ker |\ \ |$, where $|\ \ |$ is the semi-norm of G. If U is a subgroup of G and $(\hat{G}, i)$ is a completion of G, then clearly $(\overline{i(U)}, i|U)$ is a completion of U, where $\overline{i(U)}$ denotes the closure of $i(U)$ in $\hat{G}$.

Proposition 5. *Each semi-normed group G admits a completion.*

Proof. Let C be the subgroup of all Cauchy sequences of the group $G^{\mathbb{N}}$ of all sequences of elements of G. For $\mathfrak{a} = (a_\nu)_{\nu \in \mathbb{N}} \in C$, we define

$$|\mathfrak{a}|' = \lim_{\nu \to \infty} |a_\nu| .$$

This limit exists, since $\big||a_\nu| - |a_\mu|\big| \leqq |a_\nu - a_\mu|$ for all ν, μ by Proposition 1.1.3/2. Obviously, $(C, |\ \ |')$ is a semi-normed group. Let $\varphi: G \to C$ be the isometry defined by

$$a \mapsto (a, a, \ldots) \in C, \qquad a \in G.$$

If $\mathfrak{a} = (a_\nu)_{\nu \in \mathbb{N}} \in C$, then $\varphi(a_\nu)$ tends to $\mathfrak{a}$, since $|\varphi(a_\nu) - \mathfrak{a}|' = \lim_{\mu \to \infty} |a_\nu - a_\mu|$ tends to 0 as $\nu \to \infty$. Hence $\varphi(G)$ is dense in C.

We claim that C is complete. Let $(\mathfrak{a}_n)_{n \in \mathbb{N}}$, $\mathfrak{a}_n \in C$, be a Cauchy sequence. By looking at a subsequence, we may assume that $|\mathfrak{a}_{n+1} - \mathfrak{a}_n|' \leqq \frac{1}{n}$. Let $\mathfrak{a}_n = (a_{n\nu})_{\nu \in \mathbb{N}}$. There is a $\nu(n)$ such that $|\varphi(a_{n,\nu(n)}) - \mathfrak{a}_n|' \leqq \frac{1}{n}$ for all $n \in \mathbb{N}$. Define

$$b_n = a_{n,\nu(n)} .$$

We have the following estimate:

$$\begin{aligned} |b_{n+1} - b_n| &= |\varphi(b_{n+1}) - \varphi(b_n)|' \\ &\leqq \max \{|\varphi(b_{n+1}) - \mathfrak{a}_{n+1}|', |\mathfrak{a}_{n+1} - \mathfrak{a}_n|', |\mathfrak{a}_n - \varphi(b_n)|'\} \\ &\leqq \frac{1}{n} . \end{aligned}$$

Therefore, the diagonal sequence $\mathfrak{b} = (b_n)_{n \in \mathbb{N}}$ is a Cauchy sequence. Since $|\mathfrak{a}_n - \mathfrak{b}|' \leqq \max \{|\mathfrak{a}_n - \varphi(b_n)|', |\varphi(b_n) - \mathfrak{b}|'\}$ and since this expression tends to 0, the sequence $\mathfrak{a}_n$ has $\mathfrak{b}$ as limit in C. Thus, C is complete.

Set $B := \ker |\ \ |'$. Then $\hat{G} := C/B$, provided with the norm $|\ \ |^\wedge := |\ \ |'_{\text{res}}$, is a normed group and the projection $\pi: C \to \hat{G}$ is isometric (Corollary 1.1.6/2). Furthermore, G is complete by Proposition 3. The map $i := \pi \circ \varphi$: $G \to \hat{G}$ is isometric, and $i(G)$ is dense in $\hat{G}$, since $\varphi(G)$ is dense in C and since π is continuous. □

Proposition 6. *Let G and H be semi-normed groups, let $(\hat{G}, i)$ (resp. $(\hat{H}, j)$) be a completion of G (resp. H), and let $\varphi : G \to H$ be a continuous group homomorphism.*

(i) *Then there is a unique continuous group homomorphism $\hat{\varphi} : \hat{G} \to \hat{H}$ such that the diagram*

$$\begin{array}{ccc} G & \xrightarrow{\varphi} & H \\ {\scriptstyle i}\downarrow & & \downarrow{\scriptstyle j} \\ \hat{G} & \xrightarrow{\hat{\varphi}} & \hat{H} \end{array}$$

commutes.

(ii) *If φ is bounded, then $\hat{\varphi}$ is bounded with the same bound. If φ is an isometry, so is $\hat{\varphi}$.*

(iii) *If $G = H$, $(\hat{G}, i) = (\hat{H}, j)$ and if φ is the identity map, then also $\hat{\varphi}$ is the identity. If F is a semi-normed group with a completion $(\hat{F}, l)$ and $\psi : F \to G$ is a continuous group homomorphism, then $\widehat{\varphi \circ \psi} = \hat{\varphi} \circ \hat{\psi}$.*

Proof. First we have to define $\hat{\varphi}$. Let $\hat{g}$ be an arbitrary element of $\hat{G}$. Then there is a Cauchy sequence $(g_n)_{n\in\mathbf{N}}$ in G such that $i(g_n)$ converges to $\hat{g}$. Since φ is a continuous homomorphism, the image $\big(\varphi(g_{n+1}) - \varphi(g_n)\big)$ of the zero sequence $(g_{n+1} - g_n)$ is a zero sequence too. Since $\hat{H}$ is a complete normed group, the sequence $j\big(\varphi(g_n)\big)$ must converge to a unique element $\hat{h} \in \hat{H}$ which we define to be the image of $\hat{g}$ under $\hat{\varphi}$. It is easy to see that $\hat{h}$ does not depend on the choice of a particular sequence (g_n) approximating $\hat{g}$. Thus, we have defined a map $\hat{\varphi} : \hat{G} \to \hat{H}$. Clearly $\hat{\varphi}$ is a group homomorphism and makes the diagram commutative. Since φ is continuous, for every $\varepsilon > 0$, we can find a $\delta > 0$ such that $|g| \leqq \delta$ implies $|\varphi(g)| \leqq \varepsilon$. From the construction of $\hat{\varphi}$, one easily sees that $|\hat{g}| \leqq \delta$ implies $|\hat{\varphi}(\hat{g})| \leqq \varepsilon$. Thus, $\hat{\varphi}$ is continuous at the origin and hence everywhere. It is immediate that $\hat{\varphi}$ is the only continuous map from $\hat{G}$ to $\hat{H}$ making the diagram commutative. Part (i) of the proposition is proved. Ad (ii): For $\hat{g} \in \hat{G}$ choose a sequence (g_n) in G such that $i(g_n)$ converges to $\hat{g}$. Then we have $|\hat{\varphi}(\hat{g})| = \big|\lim_n j\big(\varphi(g_n)\big)\big| = \lim_n \big|j\big(\varphi(g_n)\big)\big| = \lim_n |\varphi(g_n)|$ and $|\hat{g}| = |\lim_n i(g_n)| = \lim_n |i(g_n)| = \lim_n |g_n|$, whence the assertion follows. Statement (iii) is a straightforward consequence of the construction of $\hat{\varphi}$ (resp. $\hat{\psi}$ and $\widehat{\varphi \circ \psi}$). □

It follows immediately from the proposition that completions are uniquely determined up to isometric isomorphisms. Thus, we may speak of *the* completion $\hat{G}$ of a semi-normed group G. (In most cases, we shall consider normed groups G. Then i is injective and G can be embedded into its completion $\hat{G}$.) Proposition 6 may be reformulated in the following way: $\hat{}$ is a covariant functor of the category of semi-normed groups with continuous (resp. bounded) homomorphisms into the category of complete normed groups with continuous (resp. bounded) group homomorphisms.

For later reference note that the injection $i : G \to \hat{G}$ induces a group isomorphism $i^\sim(r) : G^\sim(r) \to \hat{G}^\sim(r)$, for all $r > 0$ (for the notation see (1.1.1)). That is, the functor $^\sim$ does not distinguish between a group and its completion.

1.1.8. Convergent series. — In a metric space the notion of a convergent sequence is fundamental. In a normed group G, one also has the notion of a convergent series $\sum_{\nu=1}^{\infty} a_\nu$, $a_\nu \in G$, meaning that the sequence of partial sums $s_n := \sum_{\nu=1}^{n} a_\nu$ is convergent. The group G being Hausdorff, this limit is unique and will also be denoted by $\sum_{\nu=1}^{\infty} a_\nu$.

In the following G is always assumed to be a complete normed group.

Proposition 1. $\sum_{\nu=1}^{\infty} a_\nu$ *is convergent if and only if* $\lim |a_\nu| = 0$.

Proof. The assertion is clear by Proposition 1.1.7/1, because $s_{\nu+1} - s_\nu = a_{\nu+1}$. □

Concerning the rearrangement of the terms of a convergent series, one gets the following results:

Proposition 2. *Let* $\sum_{\nu=1}^{\infty} a_\nu$, $a_\nu \in G$, *be a convergent series and let* $\mathbb{N} = \bigcup_{\mu=1}^{\infty} I_\mu$ *be a decomposition of* $\mathbb{N}$ *into pairwise disjoint (finite or infinite) subsets. Then for all* $\mu \in \mathbb{N}$, *the series* $\sum_{\nu \in I_\mu} a_\nu$ *converges and we have* $\sum_{\nu=1}^{\infty} a_\nu = \sum_{\mu=1}^{\infty} (\sum_{\nu \in I_\mu} a_\nu)$.

Proof. For all $\mu \in \mathbb{N}$, fix a specific order for the indices $\nu \in I_\mu$. Taking the elements a_ν in that order, we get a sequence, which is a finite or a zero sequence. Hence $b_\mu := \sum_{\nu \in I_\mu} a_\nu$ (where the sum is computed in the given order) is a well-defined element of G. (In Corollary 4 below we shall see that these precautions are unnecessary, because the sum of the elements a_ν, $\nu \in I_\mu$, always has the same value regardless of the ordering.) We have to show $\sum_{\nu=1}^{\infty} a_\nu = \sum_{\mu=1}^{\infty} b_\mu$. For any $\varepsilon > 0$, we can find an $n \in \mathbb{N}$ such that $|a_\nu| < \varepsilon$ for all $\nu > n$. Choose an $m \in \mathbb{N}$ such that $I_1 \cup \cdots \cup I_m \supset \{1, \ldots, n\}$. Then we have $\left|\sum_{\mu=1}^{m} b_\mu - \sum_{\nu=1}^{n} a_\nu\right| = \left|\sum_{\substack{\nu \in I_1 \cup \cdots \cup I_m \\ \nu \notin \{1,\ldots,n\}}} a_\nu\right| < \varepsilon$ and $|b_\mu| < \varepsilon$ for $\mu > m$. Therefore, for all $m' \geqq m$, we can make the following estimate: $\left|\sum_{\mu=1}^{m'} b_\mu - \sum_{\nu=1}^{\infty} a_\nu\right| = \left|\left(\sum_{\mu=1}^{m} b_\mu - \sum_{\nu=1}^{n} a_\nu\right) + \sum_{\mu=m+1}^{m'} b_\mu - \sum_{\nu=n+1}^{\infty} a_\nu\right| < \varepsilon$, which yields $\sum_{\mu=1}^{\infty} b_\mu = \sum_{\nu=1}^{\infty} a_\nu$. □

Corollary 3. *Let* $a_{\nu\mu}$ *be elements of* G, *for* $\nu, \mu \in \mathbb{N}$, *such that for all* $\varepsilon > 0$, *one has* $|a_{\nu\mu}| < \varepsilon$ *for almost all pairs* (ν, μ). *Then one has*

$$\sum_{\nu=1}^{\infty}\left(\sum_{\mu=1}^{\infty} a_{\nu\mu}\right) = \sum_{\mu=1}^{\infty}\left(\sum_{\nu=1}^{\infty} a_{\nu\mu}\right).$$

Proof. The elements $a_{\nu\mu}$ can be arranged into a zero sequence c_i, e.g., by some diagonal procedure. Then the series $\sum c_i$ is convergent. Applying the proposition to this series, we get the corollary. □

Now one could introduce, as usual, the notion of an *absolutely convergent series* — i.e., of a series $\sum_{1}^{\infty} a_\nu$ such that $\sum_{1}^{\infty} |a_\nu|$ is convergent. But, although not all convergent series are absolutely convergent, they nevertheless behave as if they were. Namely, we can draw the following conclusion from Proposition 2:

Corollary 4. *Let* $\sum_{\nu=1}^{\infty} a_\nu$ *be a convergent series and let* $\pi : \mathbb{N} \to \mathbb{N}$ *be a bijection. Then* $\sum_{\nu=1}^{\infty} a_{\pi(\nu)}$ *is also convergent and*

$$\sum_{\nu=1}^{\infty} a_\nu = \sum_{\nu=1}^{\infty} a_{\pi(\nu)}.$$

Proof. Choose $I_\mu := \{\pi(\mu)\}$ for all $\mu \in \mathbb{N}$. Clearly this a disjoint decomposition of $\mathbb{N}$. Now Proposition 2 immediately yields the assertion. □

1.1.9. Strict homomorphisms and completions. — A map $\varphi : X \to Y$ between topological spaces is called strict if the quotient topology on $\varphi(X)$ coincides with the topology of $\varphi(X)$ inherited from Y. We need this concept for homomorphisms of semi-normed groups.

Definition 1. *Let* G *and* H *be semi-normed groups. A group homomorphism* $\varphi : G \to H$ *is said to be strict if the induced isomorphism*

$$\bar{\varphi} : G/\ker \varphi \to \varphi(G)$$

is a homeomorphism, where $G/\ker \varphi$ *is provided with the quotient topology and* $\varphi(G)$ *with the topology inherited from* H.

Strict group homomorphisms can be characterized as follows:

Lemma 2. *A group homomorphism* $\varphi : G \to H$ *between semi-normed groups is strict if and only if* φ *is continuous and for every* $\varepsilon > 0$ *there is a real number* $\delta > 0$ *such that, for every* $g \in G$, *the inequality* $|\varphi(g)| < \delta$ *implies* $|g + k| < \varepsilon$ *for some* $k \in \ker \varphi$.

Proof. Denote by $\pi : G \to G/\ker \varphi$ the canonical residue epimorphism. By definition we have $\iota \circ \bar{\varphi} \circ \pi = \varphi$, where $\iota : \varphi(G) \to H$ is the canonical injection. The map π is continuous, and $\bar{\varphi}$ is continuous if and only if φ is continuous. Furthermore, $\bar{\varphi}^{-1}$ is continuous if and only if it is continuous at the point $0 \in \varphi(G)$. This is equivalent to the condition: for every $\varepsilon > 0$ there exists a $\delta > 0$ such that, for every $g \in G$, the inequality $|\varphi(g)| < \delta$ implies $|\bar{\varphi}^{-1}(\varphi(g))|_{\text{res}} < \varepsilon$, where $|\ \ |_{\text{res}}$ denotes the residue semi-norm on $G/\ker \varphi$. Therefore, $\bar{\varphi}^{-1}$ is continuous if and only if for every $\varepsilon > 0$ there is a real number $\delta > 0$ such that $|\varphi(g)| < \delta$ implies the existence of an element $k \in \ker \varphi$ such that $|g + k| < \varepsilon$. □

Proposition 3. (i) *A monomorphism* $\varphi : G \to H$ *is strict if and only if* $\varphi : G \to \varphi(G)$ *is a homeomorphism.*

(ii) *An epimorphism* $\varphi : G \to H$ *is strict if and only if it is continuous and open.*

(iii) *An isomorphism* $\varphi : G \to H$ *is strict if and only if it is a homeomorphism.*

Proof. Assertion (i) is immediate from the definition. Assertion (iii) is a special case of (i). Thus we only have to show (ii). The map $\pi : G \to G/\ker \varphi$

is open, because, for any open $U \subset G$, the set $U + \ker \varphi$ is open in G. Therefore, if $\bar{\varphi}$ is open, so is $\varphi = \bar{\varphi} \circ \pi$. Conversely if φ is continuous and open, $\bar{\varphi}$ has these properties too and hence $\bar{\varphi}$ is a homeomorphism. □

We want to study the behavior of strict homomorphisms under completion.

Proposition 4. *If $\varphi : G \to H$ is a strict group homomorphism between seminormed groups, then $\hat{\varphi} : \hat{G} \to \hat{H}$ is strict too.*

Proof. The homomorphism $\hat{\varphi}$ is continuous by Proposition 1.1.7/6. Let $\varepsilon > 0$. By Lemma 2, there exists a $\delta > 0$ such that $|\varphi(g)| < \delta$, $g \in G$, implies $|g + k| < \varepsilon$ for some $k \in \ker \varphi$. We have only to show that $\hat{\varphi}$ has the corresponding property; i.e., that $|\hat{\varphi}(\hat{g})| < \delta$, $\hat{g} \in \hat{G}$, implies $|\hat{g} + \hat{k}| < \varepsilon$ for some $\hat{k} \in \ker \hat{\varphi}$. Assuming $|\hat{\varphi}(\hat{g})| < \delta$, we choose a sequence $(g_n) \subset G$ such that $i(g_n)$ converges to $\hat{g}$, where i denotes the isometric homomorphism from G into $\hat{G}$. According to the construction of $\hat{\varphi}$, we know that $j(\varphi(g_n))$ converges to $\hat{\varphi}(\hat{g})$, where j is the isometry $H \to \hat{H}$. For n large enough, we have simultaneously $|\hat{g} - i(g_n)| < \varepsilon$ and $|\varphi(g_n)| = |j(\varphi(g_n))| < \delta$. Hence we can find an element $k_n \in \ker \varphi$ such that $|g_n + k_n| < \varepsilon$. Then we get $i(k_n) \in \ker \hat{\varphi}$ and $|\hat{g} + i(k_n)| \leqq \max\{|\hat{g} - i(g_n)|, |i(g_n) + i(k_n)|\} < \varepsilon$. □

Proposition 5. *Let $\varphi : G \to H$ be a strict group homomorphism, and consider the commutative diagram*

$$\begin{array}{ccc} G & \xrightarrow{\varphi} & H \\ {\scriptstyle i}\downarrow & & \downarrow{\scriptstyle j} \\ \hat{G} & \xrightarrow{\hat{\varphi}} & \hat{H}. \end{array}$$

Then i restricts to a homomorphism $i' : \ker \varphi \to \ker \hat{\varphi}$ such that $(\ker \hat{\varphi}, i')$ is the completion of $\ker \varphi$. Similarly, j restricts to a homomorphism $j' : \varphi(G) \to \hat{\varphi}(\hat{G})$ such that $(\hat{\varphi}(\hat{G}), j')$ is the completion of $\varphi(G)$. In particular, we have $\widehat{\ker \varphi} = \ker \hat{\varphi}$ and $\widehat{\varphi(G)} = \hat{\varphi}(\hat{G})$.

Proof. We verify the conditions of Definition 1.1.7/4. The subgroup $\ker \hat{\varphi} \subset \hat{G}$ is closed, since the point 0 ist closed in $\hat{H}$. Hence $\ker \hat{\varphi}$ is complete with respect to the norm on $\hat{G}$. Furthermore, the homomorphism $i' : \ker \varphi \to \ker \hat{\varphi}$ is isometric, since $i : G \to \hat{G}$ is. Thus, we have only to show that $i(\ker \hat{\varphi})$ is dense in $\ker \hat{\varphi}$. Let $\hat{g}$ be an element of $\ker \hat{\varphi}$ and choose a sequence $(g_n) \subset G$ such that $i(g_n)$ converges to $\hat{g}$. Then we have $0 = \hat{\varphi}(\hat{g}) = \lim_n j(\varphi(g_n))$. Hence $\varphi(g_n)$ is a zero sequence. By Lemma 2 we can find a subsequence $(g_{n_\nu})_{\nu \in \mathbb{N}}$ of (g_n) and a sequence $(k_\nu) \subset \ker \varphi$ such that also $|g_{n_\nu} - k_\nu|$ tends to zero. Since $i(g_{n_\nu})$ tends to $\hat{g}$ and $i(g_{n_\nu}) - i(k_\nu)$ tends to zero, $i(k_\nu)$ converges to $\hat{g}$. To verify the remaining statement, we look at the subgroup $\hat{\varphi}(\hat{G})$ of the normed group

$\hat{H}$. Since $\hat{\varphi}$ is strict by Proposition 4, we see that $\hat{\varphi}(\hat{G}) \simeq \hat{G}/\ker \hat{\varphi}$ is complete (cf. Proposition 1.1.7/3). The homomorphism $j'\colon \varphi(G) \to \hat{\varphi}(\hat{G})$ is isometric, since j is. Thus, it remains to show that the image of j' is dense in $\hat{\varphi}(\hat{G})$. Let $\hat{h}$ be an element of $\hat{\varphi}(\hat{G})$, say $\hat{h} = \hat{\varphi}(\hat{g})$, where $\hat{g} \in \hat{G}$. There exists a sequence $(g_n) \subset G$ such that $\hat{g} = \lim_n i(g_n)$. Then we have

$$\hat{h} = \hat{\varphi}(\hat{g}) = \lim_n \hat{\varphi} \circ i(g_n) = \lim_n j \circ \varphi(g_n);$$

i.e., $\hat{h}$ can be approximated by elements in the image of j'. Consequently, the image of j' is dense in $\hat{\varphi}(\hat{G})$, and $(\hat{\varphi}(\hat{G}), j')$ is the completion of $\varphi(G)$. □

Corollary 6. *Let $F \xrightarrow{\psi} G \xrightarrow{\varphi} H$ be an exact sequence of strict homomorphisms between semi-normed groups. Then also the induced sequence $\hat{F} \xrightarrow{\hat{\psi}} \hat{G} \xrightarrow{\hat{\varphi}} \hat{H}$ is exact. If $\varphi\colon G \to H$ is a strict monomorphism (resp. epimorphism), so is $\hat{\varphi}\colon \hat{G} \to \hat{H}$.*

1.2. Semi-normed and normed rings

By a ring we mean always a commutative ring with identity 1. Let A be such a ring. The additive group of A is denoted by A^+. If $(A^+, |\ \ |)$ is a semi-normed group, it is natural to ask for conditions under which A, provided with the induced ultrametric topology, is a topological ring (i.e., the multiplication in A is continuous). Clearly this is the case if there is a real constant K such that

$$|x \cdot y| \leqq K\, |x| \cdot |y|, \qquad x, y \in A.$$

This condition plays a critical role because it is also necessary for the continuity of multiplication in many important cases.

1.2.1. Semi-normed and normed rings. —

Definition 1. *The pair $(A, |\ \ |)$ is called a semi-normed (resp. normed) ring if the following conditions are satisfied:*

(i) *$(A^+, |\ \ |)$ is a semi-normed (resp. normed) group.*
(ii) *$|xy| \leqq |x|\,|y|, \qquad x, y \in A$.*
(iii) *$|1| \leqq 1$.*

The function $|\ \ |$ is called a semi-norm (resp. a norm) on A.

As in (1.1), we often write A instead of $(A, |\ \ |)$ and call A semi-normed (resp. normed). Note that $\ker |\ \ |$ is an ideal in A. Furthermore, since $|1| \leqq |1|^2$, we have $|1| = 1$ or $|1| = 0$. In the latter case, the semi-norm $|\ \ |$ is identically zero.

The following proposition shows that the notion of a ring semi-norm is as general as the property "$|x \cdot y| \leqq K\,|x| \cdot |y|$ for a real constant K and for all $x, y \in A$".

Proposition 2. *Let* $|\ \ |_1$ *be a non-zero ultrametric function on the additive group of* A *such that* $|x \cdot y|_1 \leqq K\,|x|_1 \cdot |y|_1$ *for a fixed real* K *and for all* $x, y \in A$. *Then we have*

(i) *The function* $|\ \ |_2 \colon A \to \mathbb{R}$, *given by*

$$|x|_2 := \sup\{|xy|_1 \cdot |y|_1^{-1};\ y \in A,\ |y|_1 \neq 0\},$$

makes A *into a semi-normed ring. Furthermore, the functions* $|\ \ |_1$ *and* $|\ \ |_2$ *are equivalent* (*i.e. induce the same topology on* A).

(ii) *If* $a \in A$ *satisfies* $|ax|_1 = |a|_1 \cdot |x|_1$ *for all* $x \in A$, *then* $|a|_2 = |a|_1$ *and* $|ax|_2 = |a|_2 \cdot |x|_2$ *for all* $x \in A$.

(iii) *If* $a \in A$ *satisfies* $|ax|_1 \leqq |a|_1 \cdot |x|_1$ *for all* $x \in A$ *and if* $|1|_1 \leqq 1$, *then* $|a|_2 = |a|_1$.

(iv) *We have* $\ker |\ \ |_2 = \ker |\ \ |_1$; *in particular,* $|\ \ |_2$ *is a norm if and only if* $|\ \ |_1$ *is a norm.*

Proof. The definition of $|\ \ |_2$ makes sense, because $|\ \ |_1$ is a non-zero function (in particular, $|1|_1 \neq 0$) and because $|xy|_1 \leqq K\,|x|_1 \cdot |y|_1$ implies $|x|_2 \leqq K\,|x|_1$ for all $x \in A$. We have $|x|_2 \geqq |x|_1\,|1|_1^{-1}$ and, hence, $|x|_1 \leqq |1|_1\,|x|_2$ for all $x \in A$. It is easy to check that $(A^+, |\ \ |_2)$ is a semi-normed group. The semi-norms $|\ \ |_2$ and $|\ \ |_1$ define the same topology by Proposition 1.1.3/6.

We now check that conditions (ii) and (iii) of Definition 1 hold for $|\ \ |_2$. That $|1|_2 = 1$ is clear by the definition of $|\ \ |_2$. In order to prove $|xy|_2 \leqq |x|_2\,|y|_2$, it is enough to verify $|(xy)\,z|_1 \leqq |x|_2 \cdot |y|_2 \cdot |z|_1$ for all $z \in A$ with $|z|_1 \neq 0$. We may assume $|xyz|_1 \neq 0$; whence $|yz|_1 \neq 0$ and $|z|_1 \neq 0$ follow. Therefore we get $|(xy)\,z|_1 \leqq |x|_2 \cdot |yz|_1 \leqq |x|_2 \cdot |y|_2 \cdot |z|_1$. The remaining statements of the proposition follow directly from the definition of $|\ \ |_2$. □

Remark 1. *If* A *is a field, then every non-zero semi-norm* $|\ \ |$ *is a norm*, because the kernel of $|\ \ |$ must be the zero ideal.

Remark 2. Let $\mathfrak{a}$ be an ideal in A. Consider the quotient ring $A/\mathfrak{a}$ — thought of as an additive abelian group — with the residue semi-norm $|\ \ |_{\text{res}}$ (cf. (1.1.6)). A direct computation shows that $(A/\mathfrak{a}, |\ \ |_{\text{res}})$ *is a semi-normed ring.*

Remark 3. According to Proposition 1.1.7/5, each semi-normed ring A admits a completon $(\hat{A}, |\ \ |^\wedge)$ when viewed as a semi-normed group. It is easy to see that the ring multiplication on A induces a ring multiplication on $\hat{A}$ and that $(\hat{A}, |\ \ |^\wedge)$ *is a normed ring.*

We now give some useful estimates for semi-norms.

Proposition 3. *Let* $|\ \ |$ *be a semi-norm on* A. *Then*

$$|n| \leqq 1$$

(n denotes the n-fold sum of the identity of A). *For all* $a, b \in A$, *we have the inequality*

$$|(a-b)^n| \leqq \max_{\nu+\mu=n} \{|a^\nu|\,|b^\mu|\}.$$

Proof. The first assertion follows immediately from Definition 1 (iii), and the second one follows from the first by the binomial formula: $(a-b)^n = \sum_{\nu=0}^{n} (-1)^\nu \binom{n}{\nu} a^{n-\nu} b^\nu$. □

Because of the property "$|n| \leqq 1$ for all $n \in \mathbb{Z}$", semi-norms, as we have defined them, are often referred to as *non-Archimedean* semi-norms. (Recall that the Archimedean axiom of the real numbers says that, for each $\alpha \in \mathbb{R}$, there exists a natural number n, which is strictly greater than α.)

If $|\ \ |$ is a semi-norm on A, the corresponding filtration $\nu = -\ln |\ \ |$ of the additive group A^+ fulfills the conditions

$$\nu(1) \geqq 0, \quad \nu(xy) \geqq \nu(x) + \nu(y).$$

Let us call each filtration for which this holds a *filtration of the ring* A. We have

If ν is a filtration of the ring A, then for each (fixed) real ε, $0 < \varepsilon < 1$, the function $|\ \ | := \varepsilon^\nu$ is a semi-norm on A.

1.2.2. Power-multiplicative and multiplicative elements. — The pair $(A, |\ \ |)$ always denotes a semi-normed ring. For each $a \in A$ and for all $n \in \mathbb{N}$, we have $|a^n| \leqq |a|^n$.

Definition 1. *An element $a \in A$ is called power-multiplicative if*

$$|a^n| = |a|^n \text{ for all } n \in \mathbb{N}.$$

For filtrations, this condition reads $\nu(a^n) = n \cdot \nu(a)$. Obviously, all powers of a power-multiplicative element are power-multiplicative. If $a \in A$ is a root of unity and power-multiplicative, then $|a| = 1$ if $|\ \ |$ is not identically zero.

Proposition 2. *All power-multiplicative elements of the nilradical* rad A *of A are elements of* ker $|\ \ |$.

Proof. Let a be an element of rad A; i.e., $a^n = 0$ for some $n \in \mathbb{N}$. Suppose a is power-multiplicative. Then

$$0 = |0| = |a^n| = |a|^n; \quad \text{i.e.,} \quad |a| = 0; \quad \text{i.e.,} \quad a \in \ker |\ \ |. \qquad \square$$

Definition 3. *An element $a \in A$ is called multiplicative (with respect to $|\ \ |$) if $a \notin \ker |\ \ |$ and if*

$$|ax| = |a| \cdot |x| \quad \text{for all} \quad x \in A.$$

Each multiplicative element is power-multiplicative.

Proposition 4. *Let e be a unit in A. We suppose that $|\ \ |$ is not identically zero. Then $|e| \neq 0$ and $|e^{-1}| \geqq |e|^{-1}$. If $|e^{-1}| = |e|^{-1}$, then e is multiplicative.*

Proof. From $1 = |e \cdot e^{-1}| \leqq |e| \cdot |e^{-1}|$, we get the first statement. Under the assumption $|e^{-1}| = |e|^{-1}$, we have $|x| = |e^{-1}ex| \leqq |e|^{-1}\,|ex|$ for all $x \in A$; whence $|e| \cdot |x| = |e \cdot x|$. □

The multiplicative elements of A form a "multiplicatively closed set" in A; i.e., if s_1, s_2 are in this set, then $s_1 \cdot s_2$ is also in this set.

Proposition 5. *Let S be a multiplicatively closed subset of A consisting of multiplicative elements, and let A_S denote the ring of fractions with respect to S. Then the semi-norm $|\ \ |$ can be uniquely extended to A_S so that all elements of S remain multiplicative. One has $\left|\frac{a}{s}\right| = \frac{|a|}{|s|}$ for all $a \in A$ and all $s \in S$. If $|\ \ |$ is a norm, then the extension is also a norm.*

The *proof* is obvious.

1.2.3. The category $\mathfrak{N}$ and the functor $A \rightsquigarrow A^\sim$. — Let $\mathfrak{N}$ denote the category of semi-normed rings with contractive homomorphisms as morphisms. Often we write $A \in \mathfrak{N}$ instead of $(A, |\ \ |) \in \mathfrak{N}$.

Since every $A \in \mathfrak{N}$ is also a semi-normed group, the objects $A^\circ(r)$ and $A^\vee(r)$, $r > 0$, are well-defined semi-normed groups (cf. 1.1.1). We may ask the question: for which real numbers $r > 0$, are $A^\circ(r)$ and $A^\vee(r)$ objects of $\mathfrak{N}$. If $r > 1$, one cannot in general expect $A^\circ(r)$ or $A^\vee(r)$ to be closed under multiplication. If $r < 1$, the unit element is missing in $A^\circ(r)$, $A^\vee(r)$, and even in $A^\vee(1)$, unless the semi-norm on A is identically zero. Therefore, $A^\circ(1)$ is the only obvious candidate for a ring amongst the $A^\circ(r)$ and $A^\vee(r)$. Notice also that all groups $A^\circ(r)$ and $A^\vee(r)$, $r > 0$, may be viewed as $A^\circ(1)$-modules.

We set $A^\circ := A^\circ(1)$, $A^\vee := A^\vee(1)$, and call $A^\sim := A^\sim(1) = A^\circ/A^\vee$ the *residue ring* of the semi-normed ring A. If $\varphi\colon A \to B$ is a morphism in $\mathfrak{N}$, the map $\varphi^\sim(1) : A^\sim(1) \to B^\sim(1)$ of (1.1.3) is not only a group homomorphism but also a ring homomorphism. Thus, writing $\varphi^\sim := \varphi^\sim(1)$, we see that $A \rightsquigarrow A^\sim$ is a covariant functor from $\mathfrak{N}$ to the category of rings. As indicated already in (1.1.3), this functor is very important in non-Archimedean analysis. Namely, the rings $A^\sim$, carrying only an algebraic structure, are easier to handle than the rings A, and the functor $A \rightsquigarrow A^\sim$ can be used to derive valuable information about A if $A^\sim$ is known. Only for some special questions must one look at all the $A^\sim(r)$, $r \in |A - \{0\}|$.

Finally, one should remark that the functor $^\sim$ depends heavily on the given semi-norm. It may happen that one has two topologically equivalent norms on a ring A such that the residue ring with respect to the first norm is a transcendental extension of the residue ring with respect to the second one (cf. the example given at the end of (1.4.2)).

1.2.4. Topologically nilpotent elements and complete normed rings. — Let $A \in \mathfrak{N}$.

Definition 1. *An element $a \in A$ is called topologically nilpotent if* $\lim a^n = 0$. *The set of all topologically nilpotent elements of A is denoted by $\check{A}$.*

Obviously, $A^{\vee} \subset \check{A}$. Since $1 \notin \check{A}$ (unless $|A| = \{0\}$), we see that A° is not, in general, contained in $\check{A}$. The set $\check{A}$ depends only on the topology of A.

Proposition 2. *The set $\check{A}$ is a subgroup of A^+, which is multiplicatively closed. Furthermore, $\check{A}$ is open and closed with respect to the topology of A.*

Proof. Let $a, b \in \check{A}$. Obviously, $ab \in \check{A}$. To prove $a - b \in \check{A}$, let $\varepsilon > 0$ be arbitrary. Choose $M > 0$ such that $|a^\nu| \leqq M$, $|b^\nu| \leqq M$ for all ν. There is an integer $m \in \mathbb{N}$ such that $|a^n| \leqq \varepsilon M^{-1}$ and $|b^n| \leqq \varepsilon M^{-1}$ for $n \geqq m$. Hence for $n \geqq 2m$, we get

$$|(a-b)^n| \leqq \max_{0 \leqq \nu \leqq n} \{|a^\nu|\,|b^{n-\nu}|\} \leqq \varepsilon; \quad \text{i.e.,} \quad a - b \in \check{A}.$$

Since $A^{\vee} \subset \check{A}$ and since $\check{A}$ is a subgroup of the topological group A, we see that $\check{A}$ is open and closed in A. □

Corollary 3. *If A is complete, then $\check{A}$ is complete.*

In complete normed rings, the geometric series is a powerful tool.

Proposition 4. *If A is complete, each element of the form $e = 1 - y$, $y \in \check{A}$, is a unit in A. We have $e^{-1} = \sum_0^\infty y^n = 1 + z$, where $z \in \check{A}$.*

Proof. The ring A being complete, we can set

$$v := 1 + z, \quad z := \sum_1^\infty y^n.$$

Since $\check{A}$ is closed and since $y^n \in \check{A}$ for all $n \geqq 1$, we get $z \in \check{A}$. We have

$$ev = (1-y) \sum_0^\infty y^n = 1; \quad \text{i.e.,} \quad v = e^{-1}.$$ □

Note that Proposition 4 remains true if one replaces $\check{A}$ by $A^{\vee}$.

Corollary 5. *In a complete normed ring A, the multiplicative group $E(A)$ of units is open. Consequently, all maximal ideals of A are closed.*

Proof. For each $u \in E(A)$, we have $u + u\check{A} \subset E(A)$ by Proposition 4. Now $u\check{A}$ is a neighborhood of $0 \in A$, since the homothety $x \mapsto ux$ is topological. Hence $E(A)$ is open.

Let $\mathfrak{m}$ be a maximal ideal in A. From $\mathfrak{m} \subset A - E(A)$, we conclude $\overline{\mathfrak{m}} \subset A - E(A)$, since $A - E(A)$ is closed. Since $\overline{\mathfrak{m}}$ is an ideal, we get $\mathfrak{m} = \overline{\mathfrak{m}}$. □

Another important consequence of Proposition 4 is the following "Nakayama Lemma", which allows us to derive equations from congruences modulo topologically nilpotent elements.

Lemma 6. *Let A be complete and let M be an A-module. Let N be a submodule of M such that there are elements $x_1, \ldots, x_n$ in M with the property: $M \subset N + \sum_{\mu=1}^n \check{A}x_\mu$. Then $N = M$.*

Proof. By assumption there are elements $c_{\nu\mu} \in \check{A}$ and $y_\nu \in N$ such that

$$x_\nu = y_\nu + \sum_{\mu=1}^{n} c_{\nu\mu} x_\mu, \quad \nu = 1, \ldots, n.$$

If we denote by x (resp. y) the column vector with entries x_ν (resp. y_ν) and by I (resp. C) the $n \times n$-unit matrix (resp. the $n \times n$-matrix with entries $c_{\nu\mu}$), we have

$$y = (I - C)\, x.$$

If we can show that the matrix $I - C$ is invertible, we get $x = (I - C)^{-1}\, y$. Thus, $x_1, \ldots, x_n \in N$, and $M \subset N$.

Using CRAMER's rule, it is enough to show that $\det(I - C)$ is a unit in A. But clearly $\det(I - C)$ is of the form $1 - c$ with $c \in \check{A}$ (since $\check{A}$ is closed under the algebraic operations performed in computing the determinant). Hence Proposition 4 gives $\det(I - C) \in E(A)$. □

Remark. An analogue of Lemma 6 is the point of departure of TATE's approach to non-Archimedean function theory ([37], Proposition 3.1).

1.2.5. Power-bounded elements. — The set $\check{A}$ is not, in general, an ideal in A. However, it is easy to introduce a subring of A containing 1 and having $\check{A}$ as an ideal.

Definition 1. *An element $a \in A$ is called power-bounded if the set*

$$\{|a^n|;\, n \in \mathbb{N}\} \subset \mathbb{R}_+ \text{ is bounded.}$$

We denote by $\mathring{A}$ the set of all power-bounded elements of A.

Obviously $\check{A} \subsetneq \mathring{A}$, unless $|A| = \{0\}$. If $|A|$ is bounded, we have $A = \mathring{A}$. The inclusion $A^\circ \subset \mathring{A}$ is proper, in general.

Proposition 2. *The set $\mathring{A}$ is a subring of A and $\check{A}$ is an ideal in $\mathring{A}$. The subring $\mathring{A}$ is open and closed in A.*

Proof. Let $a, b \in \mathring{A}$. Choose $M > 0$ such that, for all $n \in \mathbb{N}$,

$$|a^n| \leqq M, \quad |b^n| \leqq M.$$

We conclude

$$|(ab)^n| \leqq |a^n|\, |b^n| \leqq M^2,$$

and

$$|(a - b)^n| \leqq \max_{0 \leqq \nu \leqq n} \{|a^\nu|\, |b^{n-\nu}|\} \leqq M^2.$$

Since $1 \in \mathring{A}$, we see that $\mathring{A}$ is a subring of A.

If $a \in \check{A}$, $b \in \mathring{A}$, then $|(ab)^n| \leqq |a^n|\, |b^n| \leqq |a^n|\, M \to 0$ — i.e., $ab \in \check{A}$. So $\check{A}$ is an ideal in A.

$\mathring{A}$ is a subgroup of A and contains $\check{A}$, which is open in A. Hence $\mathring{A}$ is open and closed in A. □

Corollary 3. *If A is complete, $\mathring{A}$ is also complete.*

The ring $\mathring{A}$ and its ideal $\check{A}$ give rise to another residue ring of A, namely,

$$\tilde{A} := \mathring{A}/\check{A}.$$

The inclusions $A^\circ \subset \mathring{A}$, $A^\vee \subset \check{A}$ induce a ring homomorphism $A^\sim \to \tilde{A}$, which, in general, is not a bijection.

Proposition 4. *Let $\varphi\colon A \to B$ be a bounded ring homomorphism; i.e., there exists a constant $M > 0$ such that $|\varphi(x)| \leqq M\,|x|$ for all $x \in A$. Then $\varphi(\mathring{A}) \subset \mathring{B}$ and $\varphi(\check{A}) \subset \check{B}$. Hence φ induces a ring homomorphism $\tilde{\varphi}\colon \tilde{A} \to \tilde{B}$.*

Proof. Let a be an element of $\mathring{A}$, and choose $L > 0$ such that $|a^n| \leqq L$ for all $n \in \mathbb{N}$. If $M > 0$ is a bound for φ, it follows that $|\varphi(a)^n| = |\varphi(a^n)| \leqq M\,|a^n| \leqq LM$; i.e., $\varphi(a) \in \mathring{B}$.

Let $a \in \check{A}$. The continuity of φ implies

$$0 = \varphi(0) = \varphi(\lim_n a^n) = \lim_n \left(\varphi(a)^n\right), \quad \text{i.e.} \quad \varphi(a) \in \check{B}. \qquad \square$$

Now it easily follows that

$A \rightsquigarrow \tilde{A}$, $A \in \mathfrak{R}$, is a covariant functor of $\mathfrak{R}$ into the category of rings.

Furthermore, we see by Proposition 4 that this is a functor even if one allows as morphisms of $\mathfrak{R}$ *all bounded* (and not only contractive) ring homomorphisms.

The rings $A^\sim$ and $\tilde{A}$ are not topological invariants of A. (The ring $A^\sim$ is invariant with respect to isometric automorphisms of A; whereas, $\tilde{A}$ is even invariant with respect to automorphisms of A which are bounded in both directions.) In order to attach a topologically invariant residue ring to each $A \in \mathfrak{R}$, we consider the set

$$\dot{A} := \{a \in A;\, a\check{A} \subset \check{A}\}.$$

This set is a subring of A containing 1 and $\check{A}$, which depends only on the topology, but not on the semi-norm of A. Obviously, $\dot{A}$ is *the largest subring of A containing $\check{A}$ as an ideal.* Therefore $\mathring{A} \subset \dot{A}$. In important cases we have equality.

Proposition 5. *Assume that $\check{A} = A^\vee$ and that $\check{A}$ contains a unit u of A. Then $\dot{A} = \mathring{A}$; i.e., $\mathring{A}$ is the largest subring of A containing $\check{A}$ as an ideal.*

Proof. We must show $\dot{A} \subset \mathring{A}$. Let a be an element of $\dot{A}$. As $a^n u \in \check{A}$ for all $n \in \mathbb{N}$, we conclude

$$|a^n| = |a^n u u^{-1}| \leqq |a^n u|\,|u^{-1}| < |u^{-1}|; \quad \text{i.e.,} \quad a \in \mathring{A}. \qquad \square$$

Remark. The assumption $\check{A} = A^\vee$ is always satisfied if $|\ \ |$ is power-multiplicative (cf. (1.3.1)). The assumption "$\check{A}$ contains a unit of A" is fulfilled, for example, if A is an algebra over a field with a non-trivial valuation (cf. (3.1.1)).

Definition 6. *The ring* $^\sim A := \dot{A}/\check{A}$ *is called the invariant residue ring of* A.

Since $\dot{A}$ and $\check{A}$ depend only on the topology and not on the semi-norm of A, the ring $^\sim A$ is indeed a topological invariant of A. The inclusion $\mathring{A} \subset \dot{A}$ induces a ring monomorphism $\tilde{A} \hookrightarrow {}^\sim A$, which, in general, is not a bijection. Proposition 5 gives a condition for $\tilde{A} = {}^\sim A$.

Now we shall prove some simple algebraic statements.

Proposition 7. *The rings* $\tilde{A}$ *and* $^\sim A$ *are reduced; i.e., they have no nilpotent elements* $\neq 0$.

Proof. Let b be an element of A. It is enough to prove that if $b^n \in \check{A}$ for some $n \geqq 1$, then $b \in \check{A}$. Now $b^n \in \check{A}$ means $\lim\limits_{\nu} |b^{n\nu}| = 0$. Choose $M > 0$ such that $|b^j| < M$ for $0 \leqq j < n$. Each $m \in \mathbb{N}$ can be written in the form $m = n\nu + j$, $0 \leqq j < n$. This implies $|b^m| \leqq |b^{n\nu}| \, M$, and hence $|b^m| \to 0$ as $m \to \infty$; i.e., $b \in \check{A}$. □

Proposition 8. *Let* $A \in \mathfrak{N}$ *be complete. Then an element* $a \in \mathring{A}$ *(resp.* $\dot{A}$*, resp.* A°*) is a unit in* $\mathring{A}$ *(resp.* $\dot{A}$*, resp.* A°*) if and only if its residue class* $\tilde{a} \in \tilde{A}$ *(resp.* $^\sim a \in {}^\sim A$*, resp.* $a^\sim \in A^\sim$*) is a unit in* $\tilde{A}$ *(resp.* $^\sim A$*, resp.* $A^\sim$*).*

Proof. If a is a unit, then $\tilde{a}$, $^\sim a$ and $a^\sim$ are obviously units. To show the converse, consider an element $b \in \mathring{A}$ with $\tilde{a}\tilde{b} = \tilde{1}$. This means $ab = 1 - x$, where $x \in \check{A}$. From Proposition 1.2.4/4, we conclude that ab and hence a is a unit in $\mathring{A}$. The case, where $a \in \dot{A}$ or $a \in A^\circ$, can be attacked in a similar way. □

1.3. Power-multiplicative semi-norms

1.3.1. Definition and elementary properties. — Let $(A, |\ \ |) \in \mathfrak{N}$ be given.

Definition 1. *The semi-norm* $|\ \ |$ *is called power-multiplicative or a pm-semi-norm if all elements of* A *are power-multiplicative. If in addition* $\ker |\ \ | = 0$, *we call* $|\ \ |$ *a pm-norm.*

It follows immediately from Proposition 1.2.2/2 that $\operatorname{rad} A \subset \ker |\ \ |$ for pm-semi-norms. In particular, all rings A with a pm-norm are *reduced*; i.e., $\operatorname{rad} A = 0$.

Proposition 2. *Let* $(A, |\ \ |)$ *and* $(A', |\ \ |')$ *be given, and let* $|\ \ |'$ *be power-multiplicative. Then every bounded ring homomorphism* $\varphi\colon A \to A'$ *is a contraction:*

$$|\varphi(a)|' \leqq |a|, \quad a \in A.$$

Proof. There exists a positive real number K with $|\varphi(a)|' \leqq K\,|a|$, $a \in A$. For all $n \in \mathbb{N}$, we have $|\varphi(a^n)|' \leqq K\,|a^n| \leqq K\,|a|^n$. The pm-property of $|\ \ |'$ implies $|\varphi(a^n)|' = (|\varphi(a)|')^n$. Hence $|\varphi(a)|'^n \leqq K\,|a|^n$ for all $n \in \mathbb{N}$; i.e., $|\varphi(a)|' \leqq \sqrt[n]{K}\,|a|$. Since $\lim\limits_{n\to\infty} \sqrt[n]{K} = 1$, we get $|\varphi(a)|' \leqq |a|$. □

Corollary 3. *Let* $|\ \ |$, $|\ \ |'$ *be pm-semi-norms on* A *such that there are real numbers* $\varrho, \varrho' > 0$ *with*

$$|\ \ |' \leqq \varrho |\ \ | \leqq \varrho' |\ \ |'.$$

Then these semi-norms are equal: $|\ \ | = |\ \ |'$.

Proof. The identity map id: $(A, |\ \ |) \to (A, |\ \ |')$ is an isometry by Proposition 2. □

For pm-semi-normed rings A, the associated rings $\mathring{A}$ and their ideals $\check{A}$ have a simple characterization.

Proposition 4. *If* $|\ \ |$ *is a pm-semi-norm on* A, *then* $\mathring{A} = \{a \in A; |a| \leqq 1\}$, $\check{A} = \{a \in A; |a| < 1\}$. *In particular,* $\mathring{A} = A^\circ$ and $\check{A} = A^{\vee}$.

The *proof* is straightforward.

Remark. If $|\ \ |$ is a pm-semi-norm on A and if $|A|$ is finite, then $|A| = \{0\}$ or $|A| = \{0, 1\}$, and $\tilde{A} = A/\ker |\ \ |$.

The following proposition shows how, within the context of pm-semi-norms, the usual triangle inequality $|x + y| \leqq |x| + |y|$ is related to the non-Archimedean triangle inequality $|x - y| \leqq \max\{|x|, |y|\}$.

Proposition 5. *A function* $|\ \ |: A \to \mathbb{R}_+$ *is a pm-semi-norm if and only if the following conditions are satisfied for all* $x, y \in A$:

(a) $|0| = 0$,
(b) $|xy| \leqq |x| \cdot |y|$, $\quad |x^n| = |x|^n$ *for all* $\quad n \geqq 1$,
(c) $|x + y| \leqq \max\{|x|, |y|\}$.

Furthermore, if $|\ \ |$ *satisfies* (b), *then condition* (c) *is equivalent to*

(c′) $|x + y| \leqq |x| + |y|$, *and* $|n| \leqq 1$ *for all* $n \in \mathbb{N}$.

Proof. If $|\ \ |$ is a semi-norm on A, then $|-x| = |x|$ for all $x \in A$ (cf. Proposition 1.1.1/3). Therefore, any pm-semi-norm $|\ \ |$ satisfies conditions (a), (b) and (c). Conversely, assume that $|\ \ |$ satisfies these conditions. Then $|1| = |1|^2 = |-1|^2$ by condition (b), hence $|1| \leqq 1$ and $|-1| = |1| \leqq 1$. In particular, we have $|x| \leqq |-x|\,|-1| \leqq |-x|$ and, similarly, $|-x| \leqq |x|$ so that $|x| = |-x|$ for all $x \in A$. Thus, from (c), one deduces $|x - y| \leqq \max\{|x|, |y|\}$, and it is clear that $|\ \ |$ is a pm-semi-norm. It remains only to show that condition (c′) implies (c) if (b) holds for $|\ \ |$. Using $(x + y)^n = \sum_{\nu=0}^{n} \binom{n}{\nu} x^\nu y^{n-\nu}$, we get

$$|x + y|^n = |(x + y)^n| \leqq \sum_{\nu=0}^{n} \left|\binom{n}{\nu}\right| |x^\nu|\,|y^{n-\nu}| \leqq \sum_{\nu=0}^{n} |x|^\nu\,|y|^{n-\nu}$$

since $\binom{n}{\nu} \in \mathbb{N}$ and therefore $\left|\binom{n}{\nu}\right| \leqq 1$. Assume $|x| \leqq |y|$. Then

$$|x + y|^n \leqq (n + 1)\,|y|^n; \quad \text{i.e.,} \quad |x + y| \leqq \sqrt[n]{n + 1}\,|y|.$$

As $\lim_{n\to\infty} \sqrt[n]{n+1} = 1$, we see that

$$|x + y| \leqq |y| = \max\{|x|, |y|\}. \qquad \square$$

1.3.2. Smoothing procedures for semi-norms. — First we describe a procedure that allows us to derive a power-multiplicative semi-norm from an arbitrarily given semi-norm.

If $|\ \ |$ is any semi-norm on A, we define a new function $|\ \ |' \colon A \to \mathbb{R}_+$ by setting

$$|x|' := \inf_{n\geqq 1} |x^n|^{1/n}, \qquad x \in A.$$

First we claim

$$|x|' = \lim_{n\to\infty} |x^n|^{1/n} \quad \textit{for all} \quad x \in A.$$

Proof. Fix $x \in A$ and set $\varrho := \inf_{n\geqq 1} |x^n|^{1/n}$. Clearly, $0 \leqq \varrho \leqq |x|$. For each $\varepsilon > 0$, we can find an integer m such that $|x^m|^{1/m} \leqq \varrho + \varepsilon$. Each $n \in \mathbb{N}$ can be written in the form $n = qm + r$, q, $r \in \mathbb{N}$, $0 \leqq r < m$. This implies

$$|x^n|^{1/n} \leqq (|x^m|^q\, |x^r|)^{1/n} \leqq (\varrho + \varepsilon)^{\left(1 - \frac{r}{n}\right)} |x^r|^{1/n}.$$

Now $|x^r|^{1/n}$ tends to 1 or 0 for all r, $0 \leqq r < m$. Since $\frac{r}{n} \to 0$, we get

$$\varrho \leqq |x^n|^{1/n} \leqq \varrho + 2\varepsilon \quad \text{for large } n\text{; i.e.,} \quad \lim_{n\to\infty} |x^n|^{1/n} = \varrho. \qquad \square$$

Now we prove

Proposition 1. *The function* $|\ \ |' \colon A \to \mathbb{R}_+$ *is a power-multiplicative semi-norm on* A. *We have* $|\ \ |' \leqq |\ \ |$. *The equation* $|a|' = |a|$ *holds whenever* a *is power-multiplicative with respect to* $|\ \ |$. *If* c *is multiplicative with respect to* $|\ \ |$, *then* c *is also multiplicative with respect to* $|\ \ |'$.

Proof. The equations $|0|' = 0$ and $|1|' \leqq 1$ are trivial. Furthermore, we have

$$|xy|' = \lim_{n\to\infty} |(xy)^n|^{1/n} \leqq \lim_{n\to\infty} (|x^n|^{1/n})\,(|y^n|^{1/n}) = (\lim_{n\to\infty} |x^n|^{1/n})\,(\lim_{n\to\infty} |y^n|^{1/n}) = |x|'\,|y|'$$

for all x, $y \in A$. Next we verify the triangle inequality (this is the only nontrivial point in the proof). Let x, $y \in A$ be given. Then

$$|x - y|' \leqq |(x - y)^n|^{1/n} \leqq \max_{\mu+\nu=n} \{|x^\mu|\, |y^\nu|\}^{1/n}.$$

For each n, we choose $\mu(n)$ and $\nu(n)$ such that

$$\mu(n) + \nu(n) = n \quad \text{and} \quad |(x - y)^n| \leqq |x^{\mu(n)}|\, |y^{\nu(n)}|.$$

Since $0 \leqq \frac{\mu(n)}{n} \leqq 1$, we can choose a sequence $(n_i) \subset \mathbb{N}$ such that

$\alpha := \lim_{i\to\infty} \frac{\mu_i}{n_i}$ exists, where μ_i stands for $\mu(n_i)$. The limit $\beta := \lim_{i\to\infty} \frac{\nu_i}{n_i}$, where $\nu_i := \nu(n_i)$, also exists, and we have $\alpha + \beta = 1$. Now it is enough to show that

$$\overline{\lim_{i\to\infty}}\, |x^{\mu_i}|^{1/n_i} \leqq |x|'^{\alpha} \quad \text{and} \quad \overline{\lim_{i\to\infty}}\, |y^{\nu_i}|^{1/n_i} \leqq |y|'^{\beta}.$$

Namely, if $\varepsilon > 0$ is given, this yields for $i \in \mathbb{N}$ big enough

$$|(x-y)^{n_i}|^{1/n_i} \leqq |x^{\mu_i}|^{1/n_i}\, |y^{\nu_i}|^{1/n_i} \leqq |x|'^{\alpha}\, |y|'^{\beta} + \varepsilon \leqq \max\{|x|', |y|'\} + \varepsilon.$$

To verify the stated estimates, first assume $\alpha \neq 0$. Then $\lim_{i\to\infty} \mu_i = \infty$ and

$$\lim_{i\to\infty} |x^{\mu_i}|^{1/n_i} = \lim_{i\to\infty} (|x^{\mu_i}|^{1/\mu_i})^{\mu_i/n_i} = |x|'^{\alpha}.$$

If $\alpha = 0$, we have

$$\overline{\lim_{i\to\infty}}\, |x^{\mu_i}|^{1/n_i} \leqq \overline{\lim_{i\to\infty}}\, |x|^{\mu_i/n_i} \leqq 1 = |x|'^{\alpha}.$$

Hence $\overline{\lim}_{i\to\infty} |x^{\mu_i}|^{1/n_i} \leqq |x|'^{\alpha}$ in all cases. The analogous inequality for y is proved in the same way. Thus, $|\ \ |'$ is a semi-norm. The inequality $|\ \ |' \leqq |\ \ |$ is clear by definition of $|\ \ |'$.

For each $x \in A$ and each exponent m, we have

$$|x^m|' = \lim_{n\to\infty} |x^{mn}|^{1/n} = \lim_{mn\to\infty} (|x^{mn}|^{1/mn})^m = |x|'^m;$$

i.e., $|\ \ |'$ is power-multiplicative.

If $|a^n| = |a|^n$ for all $n \geqq 1$, clearly $|a^n|^{1/n} = |a|$, and hence $|a|' = |a|$.

If c is multiplicative with respect to $|\ \ |$, we have $|(cx)^n| = |c|^n\, |x^n|$ for all $x \in A$. Therefore

$$|cx|' = \lim_{n\to\infty} |(cx)^n|^{1/n} = \lim_{n\to\infty} |c|\, |x^n|^{1/n} = |c| \lim_{n\to\infty} |x^n|^{1/n} = |c|' \cdot |x|'. \qquad \square$$

Remark. The topology induced by $|\ \ |$ is finer (and in most cases strictly finer) than the topology induced by $|\ \ |'$. The power-multiplicative semi-norm $|\ \ |'$ is sometimes referred to as the *spectral* semi-norm on A induced by $|\ \ |$.

Next we describe a device that enables us to produce multiplicative elements.

Proposition 2. *Let $|\ \ |$ be a pm-semi-norm on A and let c be an element of A such that $|c| \neq 0$. Then, for each $x \in A$, the limit*

$$|x|_c := \lim_{n\to\infty} \frac{|xc^n|}{|c|^n} \quad \textit{exists.}$$

The function $|\ \ |_c\colon A \to \mathbb{R}_+$ is a power-multiplicative semi-norm on A. We have $|\ \ |_c \leqq |\ \ |$. If a is multiplicative with respect to $|\ \ |$, then $|a| = |a|_c$, and a is multiplicative with respect to $|\ \ |_c$. Moreover, c is multiplicative with respect to $|\ \ |_c$ and we have $|c| = |c|_c$.

Proof. Since $|xc^n| \cdot |c|^{-n} \geqq |xc^{n+1}| \cdot |c|^{-(n+1)} \geqq 0$ for all $n \geqq 0$, the limit in question exists and we have $|x|_c \leqq |x|$. The equations $|0|_c = 0$ and $|1|_c \leqq 1$ are trivial. Furthermore

$$|xy|_c = \lim_{n\to\infty} \frac{|xyc^{2n}|}{|c|^{2n}} \leqq \lim_{n\to\infty} \left(\frac{|xc^n|}{|c|^n} \frac{|yc^n|}{|c|^n}\right) = |x|_c\,|y|_c \quad \text{for all} \quad x, y \in A.$$

The triangle inequality follows in the same way:

$$|x - y|_c = \lim_n \frac{|(x-y)\,c^n|}{|c|^n} \leqq \lim_n \left(\max\left\{\frac{|xc^n|}{|c|^n}, \frac{|yc^n|}{|c|^n}\right\}\right) = \max\{|x|_c, |y|_c\}.$$

For each $m \geqq 1$, we have

$$|x|_c^m = \left(\lim_{n\to\infty} \frac{|xc^n|}{|c|^n}\right)^m = \lim_{n\to\infty} \frac{|xc^n|^m}{|c|^{nm}} = \lim_{mn\to\infty} \frac{|x^m c^{mn}|}{|c|^{mn}} = |x^m|_c.$$

All further assertions follow in the same way by direct verification. □

1.3.3. Standard examples of norms and semi-norms. — In this section we describe some classical examples of norms. Let A be an arbitrary ring. We denote by $A[\![X]\!]$ the (commutative) *A-algebra of formal power series* $\sum_0^\infty a_\nu X^\nu$ over A in one indeterminate X. Recall that, for $f = \sum_0^\infty a_\nu X^\nu$, $g = \sum_0^\infty b_\mu X^\mu \in A[\![X]\!]$, one has

$$f \pm g = \sum_0^\infty (a_\nu \pm b_\nu)\, X^\nu, \qquad f \cdot g = \sum_{\lambda=0}^\infty \Big(\sum_{\mu+\nu=\lambda} a_\mu b_\nu \Big)\, X^\lambda.$$

The polynomials $\sum_0^{<\infty} a_\nu X^\nu$ form an A-subalgebra $A[X]$ of $A[\![X]\!]$.

Example (a) For each polynomial $p = \sum_0^{<\infty} a_\nu X^\nu \neq 0$, the *degree*

$$\deg p := \max\{\nu; a_\nu \neq 0\}$$

of p is a non-negative integer. If we set $\deg 0 := -\infty$, the following rules are clear:

(i) $\deg p = 0$ if and only if $p \in A - \{0\}$,
(ii) $\deg(pq) \leqq \deg p + \deg q$,
(iii) $\deg(p - q) \leqq \max\{\deg p, \deg q\}$.

From this we deduce that *the function* $-\deg$ *is a filtration of the ring* $A[X]$. Thus for any real α, $\alpha > 1$, the function $|\ \ | := \alpha^{\deg}$ is a *norm* on $A[X]$. We have $|p| = 1$ if and only if $p \in A - \{0\}$. Furthermore, $A[X]^\circ = A$, $A[X]^\vee = 0$, and hence $A[X]^\sim = A$. It is easy to see that $|\ \ |$ *is power-multiplicative if and only if A is reduced.*

Example (b) For each formal power series $f = \sum_0^\infty a_\nu X^\nu \neq 0$, the *order*

$$\operatorname{ord} f := \min\{\nu; a_\nu \neq 0\}$$

of f is a non-negative integer. If we set $\operatorname{ord} 0 := \infty$, it is clear that ord is a *filtration* of the ring $A[\![X]\!]$. Thus, for each real ε, $0 < \varepsilon < 1$, the function $|\ \ | := \varepsilon^{\mathrm{ord}}$ is a *norm* on $A[\![X]\!]$. We have $|f| = 1$ if $f \in A - \{0\}$. Furthermore, $A[\![X]\!]^\circ = A[\![X]\!]$, $A[\![X]\!]^\vee = \left\{\sum_1^\infty a_\nu X^\nu\right\}$, and hence $A[\![X]\!]^\sim = A$. As above, $|\ \ |$ *is power-multiplicative if and only if A is reduced.*

The last example admits an important generalization.

Proposition 1 ($\mathfrak{a}$-*adic semi-norm*). *Let A be a ring and $\mathfrak{a}$ an ideal in A. Define for $x \in A$*

$$v_\mathfrak{a}(x) := \begin{cases} \infty, & \text{if } x \in \mathfrak{a}^i \text{ for all } i \in \mathbb{N}, \\ \max\{i;\, x \in \mathfrak{a}^i\}, & \text{otherwise.} \end{cases}$$

Then $v_\mathfrak{a}$ is a filtration of the ring A, and any associated semi-norm $|\ \ |_\mathfrak{a} = \varepsilon^{v_\mathfrak{a}}$ (where $0 < \varepsilon < 1$) satisfies $\ker |\ \ |_\mathfrak{a} = \bigcap_{i=1}^\infty \mathfrak{a}^i$.

Proof. We have $v_\mathfrak{a}(1) \geqq 0$, and $v_\mathfrak{a}$ satisfies the inequalities

$$v_\mathfrak{a}(xy) \geqq v_\mathfrak{a}(x) + v_\mathfrak{a}(y), \quad v_\mathfrak{a}(x - y) \geqq \min\{v_\mathfrak{a}(x), v_\mathfrak{a}(y)\}.$$

Therefore $|\ \ |_\mathfrak{a}$ is a semi-norm on A. Obviously,

$$\ker |\ \ |_\mathfrak{a} = \{x \in A;\, v_\mathfrak{a}(x) = \infty\} = \bigcap_{i=1}^\infty \mathfrak{a}^i. \qquad \square$$

Remark. The trivial semi-norm on A (associating the value 1 to every non-zero element of A) is obtained by taking $\mathfrak{a} = (0)$.

The filtration $v_\mathfrak{a}$, as well as the semi-norm $|\ \ |_\mathfrak{a}$, are called $\mathfrak{a}$-*adic*. In example (b) we describe the (X)-*adic norm* on $A[\![X]\!]$, where (X) stands for the principal ideal $A[\![X]\!] \cdot X$.

In the next section we give an example of a norm which is basic for non-Archimedean function theory.

1.4. Strictly convergent power series

Let a semi-normed ring $(A, |\ \ |)$ be given.

1.4.1. Definition and structure of $A\langle X\rangle$. — In this section we want to look at convergent power series. For our purposes the following definition is fundamental.

Definition 1. *A formal power series $\sum_0^\infty a_\nu X^\nu \in A[\![X]\!]$ is called strictly convergent if*

$$\lim_\nu |a_\nu| = 0.$$

We denote by $A\langle X\rangle$ the set of all strictly convergent power series over A. For each $f = \sum_0^\infty a_\nu X^\nu \in A\langle X\rangle$ we set

$$|f|' := \max_\nu |a_\nu|.$$

Proposition 2. *The set $A\langle X\rangle$ is an A-subalgebra of $A[\![X]\!]$, and $A \subset A[X] \subset A\langle X\rangle$. The function $|\ |'$ is a semi-norm on $A\langle X\rangle$, which extends the semi-norm $|\ |$ on A. Furthermore, the polynomials are dense in $A\langle X\rangle$.*

Proof. Obviously, $A \subset A[X] \subset A\langle X\rangle$ and $|a|' = |a|$ for all $a \in A$. Let $f = \sum_0^\infty a_\mu X^\mu$, $g = \sum_0^\infty b_\nu X^\nu$ be elements of $A\langle X\rangle$. Then the inequalities $|a_\nu \pm b_\nu| \leqq \max\{|a_\nu|, |b_\nu|\} \leqq \max\{|f|', |g|'\}$ show that $\lim_\nu |a_\nu \pm b_\nu| = 0$, that $f \pm g \in A\langle X\rangle$ and that

$$|f \pm g|' = \max_\nu |a_\nu \pm b_\nu| \leqq \max(|f|', |g|').$$

Hence $A\langle X\rangle$ is an abelian group and $|\ |'$ is an ultrametric function on $A\langle X\rangle$.
Similarly, the inequalities

$$\Big|\sum_{\mu+\nu=\lambda} a_\mu b_\nu\Big| \leqq \max_{\mu+\nu=\lambda}(|a_\mu| \cdot |b_\nu|) \leqq |f|' \cdot |g|', \quad \lambda = 0, 1, 2, \ldots,$$

show that $\lim_\lambda \big|\sum_{\mu+\nu=\lambda} a_\mu b_\nu\big| = 0$, that $f \cdot g \in A\langle X\rangle$ and that

$$|f \cdot g|' = \max_\lambda \Big|\sum_{\mu+\nu=\lambda} a_\mu b_\nu\Big| \leqq |f|' \cdot |g|'.$$

Hence $A\langle X\rangle$ is closed under multiplication and $|\ |'$ is a semi-norm on $A\langle X\rangle$. From $|f - \sum_0^m a_\mu X^\mu| \leqq \max_{\mu > m} |a_\mu| \to 0$, we deduce that the polynomials are dense in $A\langle X\rangle$. □

The semi-norm $|\ |'$ on $A\langle X\rangle$ is called the *Gauss semi-norm on* $A\langle X\rangle$ (induced by $|\ |$). A motivation for this terminology will be given in (1.5.3). If $|\ |$ is trivial on A — i.e., if $|a| = 1$ for all $a \neq 0$ — then $A\langle X\rangle = A[X]$ and $|\ |'$ is trivial on $A\langle X\rangle$.

The kernel of the Gauss semi-norm $|\ |'$ consists of all power series of $A\langle X\rangle$ whose coefficients are in the kernel of the semi-norm $|\ |$ of A. Hence

$|\ |'$ is a norm on $A\langle X\rangle$ if and only if $|\ |$ is a norm on A.

Furthermore, $|A| = |A\langle X\rangle|'$; i.e., no new values are added to the value set.

Proposition 3. *If $(A, |\ |)$ is complete, then $(A\langle X\rangle, |\ |')$ is complete.*

Proof. Let $(f_i) = \left(\sum_{\nu=0}^\infty a_{i\nu} X^\nu\right)_{i \in \mathbf{N}}$ be a Cauchy sequence in $A\langle X\rangle$. Since $|a_{i+1,\nu} - a_{i\nu}| \leqq |f_{i+1} - f_i|'$ for fixed ν, each sequence $(a_{i\nu})_{i\in\mathbf{N}}$ is a Cauchy sequence in A. Let $a_\nu \in A$ be a limit of this sequence, $\nu = 0, 1, 2, \ldots$. Set $f := \sum_0^\infty a_\nu X^\nu \in A[\![X]\!]$. We have to prove that

$$f \in A\langle X\rangle \text{ and } \lim_{i\to\infty} |f - f_i|' = 0.$$

We may assume $|f_j - f_i|' \leqq \frac{1}{i}$ for all $j \geqq i$, $i = 1, 2, \ldots$. From $|a_{j\nu} - a_{i\nu}| \leqq \frac{1}{i}$

for all $j \geqq i$, we deduce by the continuity of $|\ \ |$ that $|a_\nu - a_{i\nu}| \leqq \frac{1}{i}$ for all $\nu \geqq 0$ and all $i \in \mathbb{N}$. For ν big enough, we have $|a_{i\nu}| < \frac{1}{i}$ since $f_i \in A\langle X\rangle$. Therefore $|a_\nu| \leqq \frac{1}{i}$ and we get $f \in A\langle X\rangle$. Furthermore, we have

$$|f - f_i|' = \max_\nu |a_\nu - a_{i\nu}| \leqq \frac{1}{i}; \quad \text{i.e.,} \quad f = \lim f_i. \qquad \square$$

From now on we consider always the Gauss semi-norm on $A\langle X\rangle$, unless specified otherwise. Instead of $|\ \ |'$, we simply write $|\ \ |$ for this norm.

By induction we define the A-algebra $A\langle X_1, \ldots, X_n\rangle$ of strictly convergent power series in n indeterminates $X_1, \ldots, X_n$. Namely for $n > 1$, we set

$$A\langle X_1, \ldots, X_n\rangle := A\langle X_1, \ldots, X_{n-1}\rangle \langle X_n\rangle.$$

Propositions 2 and 3 remain true mutatis mutandis.

1.4.2. Structure of $\widehat{A\langle X\rangle}$. — For any subset M of A, we set

$$M\langle X\rangle := \left\{\sum_0^\infty a_\nu X^\nu \in A\langle X\rangle;\, a_\nu \in M \text{ for all } \nu \geqq 0\right\}.$$

If M is a subring of A, obviously $M\langle X\rangle$ is a subring of $A\langle X\rangle$.

Proposition 1. $\mathring{A}\langle X\rangle = \overset{\circ}{\widehat{A\langle X\rangle}}$; $\check{A}\langle X\rangle = \overset{\vee}{\widehat{A\langle X\rangle}}$.

Proof. Let $f = \sum_0^\infty a_\nu X^\nu$ be an element of $A\langle X\rangle$, and assume that all coefficients a_ν are power-bounded in A. Then, since $|(a_\nu X^\nu)^i| = |a_\nu^i|$, all monomials $a_\nu X^\nu$ are power-bounded in $A\langle X\rangle$. Since $\overset{\circ}{\widehat{A\langle X\rangle}}$ is a closed subgroup of $A\langle X\rangle$ (cf. Proposition 1.2.5/2), it follows that the series f belongs to $\overset{\circ}{\widehat{A\langle X\rangle}}$. Thus $\mathring{A}\langle X\rangle \subset \overset{\circ}{\widehat{A\langle X\rangle}}$. Using Proposition 1.2.4/2, the inclusion $\check{A}\langle X\rangle \subset \overset{\vee}{\widehat{A\langle X\rangle}}$ is verified in the same way.

Now assume that f is power-bounded in $A\langle X\rangle$. If $f \notin \mathring{A}\langle X\rangle$, there exists an integer m such that $a_m \notin \mathring{A}$. Choose m minimal. From $\sum_0^{m-1} a_\nu X^\nu \in \mathring{A}\langle X\rangle \subset \overset{\circ}{\widehat{A\langle X\rangle}}$, we get

$$g := a_m X^m + a_{m+1} X^{m+1} + \cdots = f - \sum_0^{m-1} a_\nu X^\nu \in \overset{\circ}{\widehat{A\langle X\rangle}}.$$

Now g^i is of the form $a_m^i X^{mi}$ + higher terms. Since $|a_m^i| \leqq |g^i|$ and $|g^i|$ is bounded, we obtain $a_m \in \mathring{A}$, which is in contradiction with the choice of m. Thus we see that $\overset{\circ}{\widehat{A\langle X\rangle}} \subset \mathring{A}\langle X\rangle$. The inclusion $\overset{\vee}{\widehat{A\langle X\rangle}} \subset \check{A}\langle X\rangle$ is proved in the same way. $\square$

Let $\sim\colon \mathring{A} \to \tilde{A}$ denote the canonical residue epimorphism. If $\sum_0^\infty a_\nu X^\nu$ is a series in $\mathring{A}\langle X\rangle$, then we have $a_\nu \in \check{A}$ for almost all ν, and $\sum_0^\infty \tilde{a}_\nu X^\nu$ may be viewed as a polynomial in $\tilde{A}[X]$. Thus, the map

$$\tau\colon \mathring{A}\langle X\rangle \to \tilde{A}[X], \qquad \sum_0^\infty a_\nu X^\nu \mapsto \sum_0^\infty \tilde{a}_\nu X^\nu,$$

is a ring epimorphism having $\check{A}\langle X\rangle$ as kernel. Due to Proposition 1, we can interpret $\mathring{A}\langle X\rangle$ or $\check{A}\langle X\rangle$ as the set of power-bounded or topologically nilpotent elements in $A\langle X\rangle$, respectively. Therefore we get

Proposition 2. *The residue ring* $\widetilde{A\langle X\rangle} = \mathring{\widehat{A\langle X\rangle}}/\check{\widehat{A\langle X\rangle}}$ *of the ring of strictly convergent power series over* A *can be canonically identified with the polynomial ring* $\tilde{A}[X]$ *over the residue ring* $\tilde{A}$ *of* A. *The residue epimorphism* $\mathring{\widehat{A\langle X\rangle}} \to \widetilde{A\langle X\rangle}$ *corresponds to the extension* $\mathring{A}\langle X\rangle \to \tilde{A}[X]$ *obtained from the residue epimorphism* $\mathring{A} \to \tilde{A}$ *by mapping* X *to* X.

As an application we prove

Proposition 3. *Let* A *be complete. An element* $f = \sum_0^\infty a_\nu X^\nu \in \mathring{A}\langle X\rangle$ *is a unit in* $\mathring{A}\langle X\rangle$ *if and only if* a_0 *is a unit in* $\mathring{A}$ *and* $a_\nu \in \check{A}$ *for all* $\nu > 0$.

Proof. According to Corollary 1.2.5/3 and Proposition 1.4.1/3, the rings $\mathring{A}$ and $\mathring{A}\langle X\rangle = \mathring{\widehat{A\langle X\rangle}}$ are complete. Hence by Proposition 1.2.5/8, we see that $f = \sum_0^\infty a_\nu X^\nu$ is a unit in $\mathring{A}\langle X\rangle$ if and only if its residue class $\tilde{f} = \sum_0^{<\infty} \tilde{a}_\nu X^\nu$ is a unit in $\tilde{A}[X]$, and likewise that a_0 is a unit in $\mathring{A}$ if and only if $\tilde{a}_0$ is a unit in $\tilde{A}$. Therefore it is enough to show that any unit in $\tilde{A}[X]$ is a constant polynomial and hence a unit in $\tilde{A}$. Let r be a unit in $\tilde{A}[X]$ and denote by s its inverse so that $rs = 1$. If $\tilde{A}$ is an integral domain, we see from the equation $\deg r + \deg s = 0$ that $\deg r = \deg s = 0$. Hence r is a unit in $\tilde{A}$ in this case. Now let $\tilde{A}$ be arbitrary. If $\deg r > 0$ or $\deg s > 0$, then r or s contains a (non-zero) monomial αX^ν of degree $\nu > 0$. Choose a prime ideal $\mathfrak{p} \subset \tilde{A}$ which does not contain the coefficient α. This is possible since $\tilde{A}$ is reduced; just take for $\mathfrak{p}$ a maximal element in the set of all ideals which do not contain any power of α. Denoting by $\bar{r}$ and $\bar{s}$ the images of r and s in $\tilde{A}/\mathfrak{p}[X]$, we have $\bar{r}\bar{s} = 1$ and thus $\deg \bar{r} = \deg \bar{s} = 0$ because $\tilde{A}/\mathfrak{p}$ is an integral domain. However by our construction, the monomial αX^ν gives rise to a (non-zero) monomial of degree $\nu > 0$ in $\bar{r}$ or $\bar{s}$; hence a contradiction. Thus we must have $\deg r = \deg s = 0$, and r is a unit in $\tilde{A}$. □

Let us note that the proof of Proposition 2 also yields the result $A\langle X\rangle^\sim = A^\sim[X]$ if one replaces the functor $A \rightsquigarrow \tilde{A}$ by $A \rightsquigarrow A^\sim$. We shall apply this result to give an example which demonstrates that the functor $A \rightsquigarrow A^\sim$ discriminates rather strongly between two different norms on A, even if they are topologically equivalent. Let A be a normed ring such that there is a real $\varrho > 1$ with $\varrho^{-1} \notin |A|$. Such rings do exist; for example, all discrete valuation rings (cf. (1.6)) are of this type. On $A\langle X\rangle$ define $|\ \ |_1$ to be the usual Gauss norm and $|\sum_\nu a_\nu X^\nu|_2 := \max\{|a_0|, |a_1|\varrho, |a_2|\varrho, \ldots\}$. Then it is not hard to check that $|\ \ |_2$ is also a ring norm on $A\langle X\rangle$ extending the norm on A. Obviously one has $|f|_1 \leqq |f|_2 \leqq \varrho|f|_1$, for all $f \in A\langle X\rangle$. Therefore $|\ \ |_1$ and $|\ \ |_2$ induce the same topology on $A\langle X\rangle$. For $i = 1, 2$, set $A_i := (A\langle X\rangle, |\ \ |_i)$. We want to show that $A_1^\sim$ is a transcendental extension of $A_2^\sim$. In order to do so, we consider the following statements concerning $f = \sum a_\nu X^\nu \in A\langle X\rangle$:

$$f \in \check{A}_2 \Leftrightarrow (|a_0| < 1, |a_\nu| < \varrho^{-1} \text{ for } \nu \geqq 1) \quad \text{and}$$

$$f \in \mathring{A}_2 \Leftrightarrow (|a_0| \leqq 1, |a_\nu| \leqq \varrho^{-1} \text{ for } \nu \geqq 1) \Leftrightarrow (|a_0| \leqq 1, |a_\nu| < \varrho^{-1} \text{ for } \nu \geqq 1)$$

(recall that $\varrho^{-1} \notin |A|$). Hence an element $f = \sum a_\nu X^\nu \in \mathring{A}_2$ lies in $\check{A}_2$ if and only if $|a_0| < 1$. Therefore $A_2^\sim = A^\sim$. The identity map $A_2 \to A_1$ induces an embedding of $A_2^\sim = A^\sim$ into $A_1^\sim = A^\sim[X]$. Thus $A_1^\sim$ is indeed a transcendental extension of $A_2^\sim$. □

1.4.3. Bounded homomorphisms of $A\langle X\rangle$. — For semi-normed rings A and B, let $\mathrm{Hom}_b(A, B)$ denote the set of all bounded ring homomorphisms $\varphi\colon A \to B$. For $\Phi \in \mathrm{Hom}_b(A\langle X\rangle, B)$, the restriction $\Phi|A$ is an element of $\mathrm{Hom}_b(A, B)$, and, by Proposition 1.2.5/4, the image $\Phi(X)$ of the power-bounded element X must also be power-bounded. The two objects $\Phi|A$ and $\Phi(X)$ suffice to characterize Φ; more precisely, $A\langle X\rangle$ has the following universal mapping property:

Proposition 1. *Let A and B be normed rings, where B is complete. For every bounded ring homomorphism $\varphi\colon A \to B$ and every power-bounded element $b \in \mathring{B}$, there is a unique bounded ring homomorphism $\Phi\colon A\langle X\rangle \to B$ such that $\Phi|A = \varphi$ and $\Phi(X) = b$. In other words, the map* $\mathrm{Hom}_b(A\langle X\rangle, B) \to \mathrm{Hom}_b(A, B) \times \mathring{B}$ *defined by $\Phi \mapsto (\Phi|A, \Phi(X))$ is bijective.*

Proof. If $(a_\nu)_{\nu\geqq 0}$ is a zero sequence in A, then $(\varphi(a_\nu)\, b^\nu)$ is a zero sequence in B, because φ is bounded and b is power-bounded. Because B is complete, we may define $\Phi(\sum a_\nu X^\nu) := \sum \varphi(a_\nu)\, b^\nu$. Clearly Φ is a map from $A\langle X\rangle$ to B such that $\Phi|A = \varphi$ and $\Phi(X) = b$. Furthermore, for $f = \sum a_\nu X^\nu$, we have

$$|\Phi(f)| = |\Phi(\sum a_\nu X^\nu)| = |\sum \varphi(a_\nu)\, b^\nu| \leqq \max\{|\varphi(a_\nu)|\, |b^\nu|\}.$$

Choose $\varrho \in \mathbb{R}_+$ such that $|\varphi(a)| \leqq \varrho\, |a|$ for all $a \in A$ and such that $|b^\nu| \leqq \varrho$ for all $\nu \geqq 0$. Then one has

$$|\Phi(f)| \leqq \max\{|\varphi(a_\nu)|\, |b^\nu|\} \leqq \varrho^2 \max\{|a_\nu|\} = \varrho^2\, |f|,$$

and hence Φ is bounded. The restriction of Φ to $A[X]$ is a ring homomorphism. Thus, by continuity, Φ must be a ring homomorphism on $A\langle X\rangle$. If Ψ is another element of $\mathrm{Hom}_b(A\langle X\rangle, B)$ with $\Psi|A = \varphi$ and $\Psi(X) = b$, then

$$\Psi(f) = \Psi\left(\sum_{\nu=0}^{\infty} a_\nu X^\nu\right) = \Psi\left(\lim_{n\to\infty} \sum_{\nu=0}^{n} a_\nu X^\nu\right)$$

$$= \lim_{n\to\infty} \Psi\left(\sum_{\nu=0}^{n} a_\nu X^\nu\right) = \lim_{n\to\infty} \sum_{\nu=0}^{n} \varphi(a_\nu)\, b^\nu = \sum_{\nu=0}^{\infty} \varphi(a_\nu)\, b^\nu = \Phi(f). \quad \square$$

If in addition B is an A-algebra and Φ is an A-algebra homomorphism, then one has $(\Phi|A)(a) = a \cdot 1 \in B$ for all $a \in A$. So for this special case the proposition may be reformulated as:

Corollary 2. *Let A and B be normed rings, where B is complete. Assume that B is an A-algebra. Then, for every power-bounded element $b \in \mathring{B}$, there is a unique bounded A-algebra homomorphism $\Phi\colon A\langle X\rangle \to B$ such that $\Phi(X) = b$. In particular, if A is complete, the set of bounded A-algebra endomorphisms of $A\langle X\rangle$ may be identified with $\mathring{A}\langle X\rangle$.*

For the rest of this section, we want to study another class of bounded A-module endomorphisms of $A\langle X\rangle$, namely, the class of all continuous A-derivations of $A\langle X\rangle$. Recall that an A-module endomorphism D of an A-algebra B is called an "A-derivation" if $D(b_1b_2) = D(b_1)\, b_2 + b_1D(b_2)$ for all $b_1, b_2 \in B$. In particular, $D(1) = 0$ and hence $D(a \cdot 1) = 0$ for all $a \in A$. Let $\mathrm{Der}_A(B)$ denote the set of all continuous A-derivations of a normed A-algebra B. In the same way as the bounded A-algebra endomorphisms of $A\langle X\rangle$, the A-derivations of $A\langle X\rangle$ are already determined by their values at X. More precisely, one can describe $\mathrm{Der}_A(A\langle X\rangle)$ as follows.

Proposition 3. *The map $\delta\colon \mathrm{Der}_A(A\langle X\rangle) \to A\langle X\rangle$ defined by $\delta(D) := D(X)$ is bijective. All $D \in \mathrm{Der}_A(A\langle X\rangle)$ are bounded with* $\sup_{f\neq 0} |D(f)|/|f| = |D(X)|$.

Proof. For $f = \sum a_\nu X^\nu \in A\langle X\rangle$ define $\frac{\partial}{\partial X} f := \sum \nu a_\nu X^{\nu-1}$. One easily verifies that $\frac{\partial}{\partial X} \in \mathrm{Der}_A(A\langle X\rangle)$. Now let D be an arbitrary derivation in $\mathrm{Der}_A(A\langle X\rangle)$. By induction on ν, one gets $D(X^\nu) = \nu X^{\nu-1}D(X)$ for all $\nu \geqq 0$, and hence, by A-linearity and continuity, $D(f) = \sum_{\nu=0}^{\infty} a_\nu D(X^\nu) = \sum_{\nu=1}^{\infty} \nu a_\nu X^{\nu-1}D(X) = \left(\frac{\partial}{\partial X} f\right) D(X)$. Thus δ is injective. Furthermore, we see that $|D(f)| \leqq \left|\frac{\partial}{\partial X} f\right| |D(X)| \leqq |f|\,|D(X)|$, and therefore $\|D\| := \sup_{f\neq 0} |D(f)|/|f| \leqq |D(X)|$. On the other hand, choosing $f := X$, one sees $\|D\| \geqq |D(X)|$; whence $\|D\| = |D(X)|$. Finally, in order to show that δ is surjective, choose $g \in A\langle X\rangle$ arbitrarily and define $D(f) := \left(\frac{\partial}{\partial X} f\right) g$. Then D is a continuous A-derivation with $D(X) = g$. $\square$

1.5. Non-Archimedean valuations

Let A be a commutative ring with identity 1.

1.5.1. Valued rings. —

Definition 1. *A map* $| \ |: A \to \mathbb{R}$, *where* $A \neq 0$, *is called a (non-Archimedean) valuation on* A *if*

(a) $|0| = 0$ *and* $|x| > 0$ *for all* $x \neq 0$,
(b) $|x - y| \leqq \max\{|x|, |y|\}$,
(c) $|xy| = |x| \cdot |y|$.

From (c) one gets $|1| \leqq 1$; hence, *a valuation on A is a norm on A such that all elements $\neq 0$ are multiplicative.* The pair $(A, | \ |)$ will be called a *valued ring.* Condition (c) immediately implies the following:

A valued ring is an integral domain. The ideal $\check{A}$ is prime in $\mathring{A}$; hence $\tilde{A}$ is also an integral domain.

The sets $|A|$ and $|A - \{0\}|$ are semi-groups (with respect to multiplication).

Definition 2. *A valuation $| \ |$ is called bounded if $|A|$ is bounded. A valuation $| \ |$ is called degenerate if $|A| \subset \{0\} \cup \{r \in \mathbb{R}; r \geqq 1\}$ or $|A| \subset \{r \in \mathbb{R}_+; r \leqq 1\}$. A valuation $| \ |$ is called trivial if $|A - \{0\}| = \{1\}$.*

For fields these three notions are equivalent.

Proposition 3. *Let $(A, | \ |)$ be a valued ring. The valuation topology is linear (i.e., there exists a fundamental system of neighborhoods of 0 consisting of ideals) if and only if $| \ |$ is degenerate.*

Proof. If $| \ |$ is degenerate, then $| \ |$ is bounded by 1 or we have $\check{A} = \{0\}$. If $| \ |$ is bounded, we have $A = \mathring{A}$, and each ball $\{x \in A; |x| < \varepsilon\}$, $\varepsilon > 0$, around the origin is an ideal in A. If $\check{A} = \{0\}$, then $\{0\}$ is open. Thus, in both cases, there is a fundamental system of neighborhoods of 0 consisting of ideals.

To show the converse, we first remark that there is an ideal $\mathfrak{a}$ of A such that $\mathfrak{a}$ is a neighborhood of 0 and $\mathfrak{a} \subset \mathring{A}$. If $\mathfrak{a} = (0)$, there exists an $\varepsilon > 0$ such that $\{x \in A; |x| < \varepsilon\} = \{0\}$. Hence $|x| \geqq 1$ for all $x \neq 0$, because otherwise $0 < |x^n| < \varepsilon$ for some n. If $\mathfrak{a} \neq (0)$, choose $a \in \mathfrak{a}$, $a \neq 0$. From $A \cdot a \subset \mathfrak{a} \subset \mathring{A}$, we deduce that $|A| \leqq |a|^{-1}$; i.e., $| \ |$ is bounded. So in both cases, $| \ |$ is degenerate. □

Proposition 4. *Each valuation $| \ |$ on A can be uniquely extended to a valuation on the field of fractions Q of A.*

Proof. By definition, Q equals the ring of fractions $A_{A-\{0\}}$ of A with respect to the multiplicative system $A - \{0\}$. Since $A - \{0\}$ is the set of multiplicative elements of $(A, | \ |)$, the assertion follows from Proposition 1.2.2/5. □

The completion $(\hat{A}, | \ |^\wedge)$ of a valued ring (resp. valued field) is a valued ring (resp. valued field). The proof is obvious.

1.5.2. Examples. — If A is a finite integral domain, i.e., a finite field, there exists only the trivial valuation on A, as follows immediately from the remark following Proposition 1.3.1/4.

If ν is a filtration of a ring A and if $|\ \ | = \alpha^\nu$, where $0 < \alpha < 1$, is a corresponding semi-norm, then $|\ \ |$ is a valuation if and only if ν satisfies the following conditions:

$$(*)\ \nu(x) = \infty \Leftrightarrow x = 0, \quad x \in A,$$
$$(**)\ \nu(xy) = \nu(x) + \nu(y), \quad x, y \in A.$$

From this we deduce

Proposition 1. *Let A be any commutative ring. Let $|\ \ |$ denote the norm on $A[X]$ (resp. $A[\![X]\!]$) defined by the degree function* deg *(resp. by the order function* ord*), cf.* (1.3.3). *Then $|\ \ |$ is a valuation if and only if A is an integral domain.*

Proof. One has only to realize that the function $-$deg (resp. ord) fulfills equation $(**)$ if and only if A has no zero divisors $\neq 0$. □

Remark. Let $\alpha \in \mathbb{R}$, $0 < \alpha < 1$. The valuation $|\ \ | = \alpha^{-\deg}$ on $A[X]$ is degenerate but not bounded. The valuation $|\ \ | = \alpha^{\mathrm{ord}}$ on $A[\![X]\!]$ is bounded but not trivial.

Further examples of valuations are obtained by looking at special $\mathfrak{a}$-adic semi-norms, as introduced in Proposition 1.3.3/1.

Proposition 2. *Let $\mathfrak{p}$ be a principal prime ideal in an integral domain A such that $\bigcap_{\nu=1}^{\infty} \mathfrak{p}^\nu = 0$. (Note that this condition is always fulfilled if A is factorial or Noetherian.) Then the $\mathfrak{p}$-adic filtration $\nu_\mathfrak{p}$ of A induces a bounded valuation.*

Proof. We only have to verify $(**)$ for $\nu_\mathfrak{p}$. Let $\mathfrak{p} = Ap$ for some element $p \in A$. Write $x = x_0 p^{\nu_\mathfrak{p}(x)}$, $y = y_0 p^{\nu_\mathfrak{p}(y)}$, where $x_0, y_0 \notin \mathfrak{p}$. Then $xy = (x_0 y_0)\, p^{\nu_\mathfrak{p}(x)+\nu_\mathfrak{p}(y)}$ and $x_0 y_0 \notin \mathfrak{p}$. Hence $\nu_\mathfrak{p}(xy) = \nu_\mathfrak{p}(x) + \nu_\mathfrak{p}(y)$. □

The following converse holds for principal ideal domains.

Proposition 3. *Let A be a principal ideal domain. Then each bounded valuation of A is a $\mathfrak{p}$-adic one.*

Proof. Let $|\ \ |$ be a bounded valuation. Then $A = \mathring{A}$. Hence $\mathfrak{p} := \check{A}$ is a prime ideal in A. By assumption $\mathfrak{p}$ is principal: $\mathfrak{p} = Ap$. Set $\varepsilon := |p|$. Obviously $0 \leqq \varepsilon < 1$. Let $x \in A$ be arbitrary, $x \neq 0$. Write $x = x_0 p^{\nu_\mathfrak{p}(x)}$, where $x_0 \notin \mathfrak{p}$. Then $|x_0| = 1$ and $|x| = \varepsilon^{\nu_\mathfrak{p}(x)}$. Thus $|\ \ |$ is $\mathfrak{p}$-adic. Note that $|\ \ |$ is trivial if $\varepsilon = |p| = 0$. □

Looking at the ring $\mathbb{Z}$ of integers, the non-zero prime ideals are just the ideals $p\mathbb{Z}$, where $p \in \mathbb{N}$ is a prime. We say p-adic instead of $(p\mathbb{Z})$-adic.

Corollary 4. *The non-trivial (non-Archimedean) valuations on $\mathbb{Z}$ are exactly the p-adic ones, where p runs through all prime numbers.*

Proof. Since $|n| \leqq 1$ for each $n \in \mathbb{Z}$ by Proposition 1.2.1/3, we may apply Proposition 3. □

Corollary 5. *Let $k[X]$ be the polynomial ring in one indeterminate over a field k. The $\mathfrak{p}$-adic valuations on $k[X]$ are exactly those which are trivial on k and associate a value $\leqq 1$ to X.*

Proof. Since $k[X]$ is a principal ideal domain, one only has to remark that a valuation of $k[X]$ is bounded if and only if its restriction to k is trivial and if $|X| \leqq 1$. □

Remark. If k is algebraically closed, the non-zero prime ideals in $k[X]$ are of the form $(X - c)$, where c varies over k. Therefore, we deduce from Corollary 5 that all non-trivial bounded valuations on the polynomial ring $k[X]$ are of the form

$$|\ \ | = \alpha^{\text{vanishing order at } c}, \quad \text{where} \quad 0 < \alpha < 1 \quad \text{and where} \quad c \in k.$$

Of course there are unbounded valuations on $k[X]$ which are trivial on k, e.g., $\varrho^{\deg}$ for each $\varrho > 1$. If k carries a valuation, the definition

$$|f| := \max \{|a_\nu|\, \varrho^\nu\}, \quad \text{where} \quad f = \sum a_\nu X^\nu,$$

gives rise to a valuation on $k[X]$ for each real $\varrho > 0$, as we shall see later (this norm is called the generalized Gauss norm). If $|k| = \{0, 1\}$, the valuation equals $\varrho^{\deg}$ or ϱ^{ord} when $\varrho > 1$ or $\varrho < 1$, respectively.

1.5.3. The Gauss-Lemma. — First we write down a simple sufficient condition for a norm to be a valuation.

Proposition 1. *Let $(A, |\ \ |)$ be a normed ring with the following properties:*

(i) *for each $a \in A$, $a \neq 0$, there exists a multiplicative element $m \in A$ and an exponent $s \in \mathbb{N}$ such that $|ma^s| = |m|\, |a|^s = 1$,*
(ii) *$A^\sim = A^\circ / A^\vee$ is an integral domain.*

Then $|\ \ |$ is a valuation on A.

Proof. Assume there are elements $a_1, a_2 \in A$ such that $|a_1 a_2| < |a_1|\, |a_2|$. Clearly $a_1 \neq 0$, $a_2 \neq 0$. By (i) we may choose $m_\nu \in A$ and $s_\nu \geqq 1$ such that

$$|m_\nu|\, |a_\nu|^{s_\nu} = |m_\nu a_\nu^{s_\nu}| = 1 \quad \text{and hence,} \quad m_\nu a_\nu^{s_\nu} \in A^\circ - A^\vee, \quad \nu = 1, 2.$$

Assume $s_2 \geqq s_1$. Since m_ν is multiplicative, we get

$$\begin{aligned} |(m_1 a_1^{s_1})(m_2 a_2^{s_2})| &= |m_1| \cdot |m_2| \cdot |a_1^{s_1} a_2^{s_2}| \leqq |m_1|\, |m_2|\, |a_1 a_2|^{s_1}\, |a_2|^{s_2 - s_1} \\ &< |m_1| \cdot |m_2| \cdot (|a_1| \cdot |a_2|)^{s_1} \cdot |a_2|^{s_2 - s_1} \\ &= |m_1|\, |a_1|^{s_1} \cdot |m_2|\, |a_2|^{s_2} = 1. \end{aligned}$$

Thus, $(m_1 a_1^{s_1})(m_2 a_2^{s_2}) \in A^\vee$. However, this is in contradiction with the fact that, by condition (ii), the ideal $A^\vee$ is prime in A°. □

Corollary 2 (Gauss Lemma). *If* $(A, |\ \ |)$ *is a valued ring, the Gauss norm on* $A\langle X\rangle$ *is a valuation.*

Proof. Without loss of generality, we may assume that A is a field (use Proposition 1.5.1/4). Then condition (i) of Proposition 1 (where A must be replaced by $A\langle X\rangle$) is fulfilled because $|A\langle X\rangle - \{0\}| = |A^*|$ is a group. Since A is valued, we have $(A\langle X\rangle)^\circ = A^\circ\langle X\rangle = \mathring{A}\langle X\rangle$ and $(A\langle X\rangle)^\vee = A^\vee\langle X\rangle = \check{A}\langle X\rangle$. Furthermore, by Propositions 1.4.2/1 and 1.4.2/2, we have $(A\langle X\rangle)^\sim = \widetilde{A\langle X\rangle} = \tilde{A}[X]$. Thus, condition (ii) of Proposition 1 is fulfilled, since $\check{A}$ is a prime ideal in $\mathring{A}$ and, hence, $\tilde{A}[X]$ is an integral domain. □

In order to explain the connection with the classical Gauss Lemma, first we recall that a polynomial f with coefficients in a factorial domain A is called *primitive* if the greatest common divisor of all coefficients of f is a unit in A.

(Classical Gauss Lemma). *If* A *is factorial, the product of primitive polynomials is primitive.*

Proof (Reduction to Corollary 2). For each prime element $p \in A$, we denote by $|\ \ |_p$ the Gauss norm on $A[X] \subset A\langle X\rangle$ which extends the p-adic valuation of A (Proposition 1.5.2/2). Then $|\ \ |_p$ is a valuation by the corollary. Since $f \in A[X]$ is primitive if and only if $|f|_p = 1$ for all prime elements $p \in A$, the assertion follows. □

For the convenience of the reader, we also include the direct argument used in the classical proof of the Gauss Lemma. Let $f = \sum_0^\infty a_\mu X^\mu$, $g = \sum_0^\infty b_\nu X^\nu \in A\langle X\rangle$; let i (resp. j) be the smallest index such that $|f| = |a_i|$ (resp. $|g| = |b_j|$). If $g \cdot f = \sum_0^\infty c_\lambda X^\lambda$, we consider the coefficient c_{i+j} and write

$$c_{i+j} = a_i b_j + \sum_{\mu+\nu=i+j}' a_\mu b_\nu ,$$

where $\sum'$ means that the couple (i, j) is to be omitted. Since $|a_\mu b_\nu| < |a_i b_j|$ for all $(\mu, \nu) \neq (i, j)$ with $\mu + \nu = i + j$, it follows that

$$|f \cdot g| \geqq |c_{i+j}| = |a_i|\,|b_j| = |f| \cdot |g| . \qquad \square$$

Remark. If in Proposition 1 one weakens condition (ii) to "$A^\sim$ is reduced", the same type of proof shows that $|\ \ |$ is a *power-multiplicative norm.*

1.5.4. Spectral value of monic polynomials. — Let $(A, |\ \ |)$ be a seminormed ring. For each monic polynomial $p = X^m + a_1 X^{m-1} + \cdots + a_m \in A[X]$ of degree $m \geqq 1$, we set

$$\sigma(p) := \max_{1 \leqq \mu \leqq m} |a_\mu|^{1/\mu} ,$$

and call $\sigma(p)$ the *spectral value of* p. The use of the adjective "spectral" will be motivated later. Note that

$$\sigma(p) \leqq \max\{1, |a_1|, \ldots, |a_m|\} = \text{Gauss norm of } p$$

and that $\sigma(X^m) = 0$. The spectral function σ has the following fundamental property:

Proposition 1. *Let $p, q \in A[X]$ be monic. Then $\sigma(pq) \leqq \max\{\sigma(p), \sigma(q)\}$. If $\sigma(p) \neq \sigma(q)$ or if $|\ |$ is a valuation, the above inequality is, in fact, an equality.*

Proof. (1) We set $a_0 := 1$, $b_0 := 1$ and write $p = \sum_{\mu=0}^{m} a_\mu X^{m-\mu}$, $q = \sum_{\nu=0}^{n} b_\nu X^{n-\nu}$. Then

$$pq = \sum_{\lambda=0}^{m+n} c_\lambda X^{m+n-\lambda}, \quad \text{where} \quad c_\lambda = \sum_{\mu+\nu=\lambda} a_\mu b_\nu .$$

From $|a_\mu| \leqq \sigma(p)^\mu$, $|b_\nu| \leqq \sigma(q)^\nu$, $\mu = 0, \ldots, m$; $\nu = 0, \ldots, n$, (where $\sigma(p)^0 = \sigma(q)^0 = 1$), we conclude

$$|c_\lambda| \leqq \max_{\mu+\nu=\lambda} \{|a_\mu|\, |b_\nu|\} \leqq \max_{\mu+\nu=\lambda} \{\sigma(p)^\mu\, \sigma(q)^\nu\}, \quad \lambda = 1, \ldots, m+n.$$

Suppose $\sigma(p) \leqq \sigma(q)$. Then

$$|c_\lambda| \leqq \max_{\mu+\nu=\lambda} \{\sigma(q)^\mu\, \sigma(q)^\nu\} = \sigma(q)^\lambda .$$

Thus

$$\sigma(pq) = \max_{1 \leqq \lambda \leqq m+n} |c_\lambda|^{1/\lambda} \leqq \sigma(q) = \max\{\sigma(p), \sigma(q)\}.$$

(2) Assume now $\sigma(p) < \sigma(q)$. We choose j, $1 \leqq j \leqq n$, such that $|b_j| = \sigma(q)$ and consider the coefficient

$$c_j = b_j + a_1 b_{j-1} + \cdots + a_{j-1} b_1 + a_j .$$

From $|a_\mu| \leqq \sigma(p)^\mu < \sigma(q)^\mu$ for all $\mu \geqq 1$ and $|b_\nu| \leqq \sigma(q)^\nu$ for all $\nu \geqq 1$, we conclude that $|a_\mu b_{j-\mu}| \leqq |a_\mu|\, |b_{j-\mu}| < \sigma(q)^\mu\, \sigma(q)^{j-\mu} = \sigma(q)^j$ for $\mu \geqq 1$, and hence

$$|c_j| = |b_j| = \sigma(q)^j .$$

Since $\sigma(pq) \geqq |c_j|^{1/j}$, we see that $\sigma(pq) \geqq \sigma(q) = \max\{\sigma(p), \sigma(q)\}$.

(3) Assume now that $|\ |$ is a valuation. We only have to deal with the case $\sigma(p) = \sigma(q)$. Just as in the classical proof of the GAUSS Lemma, let i (resp. j) be the smallest index $\geqq 1$ such that $|a_i| = \sigma(p)^i$ (resp. $|b_j| = \sigma(q)^j$). We consider

$$c_{i+j} = a_i b_j + \sum_{\mu+\nu=i+j}{}' \, a_\mu b_\nu$$

where $\sum'$ means that the pair (i, j) is to be omitted. Now

$$|a_\mu b_\nu| < \sigma(p)^\mu\, \sigma(q)^\nu = \sigma(q)^{i+j} \quad \text{for all} \quad (\mu, \nu) \neq (i, j) \quad \text{with} \quad \mu + \nu = i + j .$$

Since $|a_i b_j| = |a_i|\, |b_j| = \sigma(q)^{i+j}$, we get $|c_{i+j}| = \sigma(q)^{i+j}$, and therefore

$$\sigma(pq) \geqq |c_{i+j}|^{1/(i+j)} = \sigma(q) = \max\{\sigma(p), \sigma(q)\}. \qquad \square$$

Remark. To our knowledge the spectral function first occurred implicitly in [15], p. 435. L. GRUSON introduced this function explicitly ([17], p. 56, 57).

1.5.5. Formal power series in countably many indeterminates. — We conclude section (1.5) with an example of a *non-Noetherian* valued ring, which often can serve as counter-example. Let us start with an arbitrary ring A; let $Y_1, Y_2, \ldots$ be an infinite sequence of indeterminates. The ring $F := A\,[\![Y_1, Y_2, \ldots]\!]$ of "formal power series in $Y_1, Y_2, \ldots$ over A" consists of *all* countable formal series whose terms are pairwise distinct monomials over A in finitely many Y_i's. To be more precise, let $p_1 < p_2 < \ldots$ be the sequence of all prime numbers in $\mathbb{N}$. Let $e\colon \mathbb{N} \to \bigoplus_1^\infty (\mathbb{N} \cup \{0\})$ denote the map which attaches to each natural number $\mu = p_1^{e_1} \cdot \ldots \cdot p_r^{e_r}$ the exponent system $e(\mu) := (e_1, \ldots, e_r, 0, \ldots)$ of its prime decomposition. The map e is a bijection and we have $e(\mu \cdot \nu) = e(\mu) + e(\nu)$. We write $Y^{e(\mu)}$ for $Y_1^{e_1} \cdot \ldots \cdot Y_r^{e_r}$ so that, in particular, $Y^{e(p_i)} = Y_i$. Now a formal power series f in $Y_1, Y_2, \ldots$ over A is nothing more than an expression of the form $\sum_{\mu=1}^\infty a_\mu Y^{e(\mu)}$, $a_\mu \in A$. Two such series are equal if and only if they have the same coefficients. Addition and multiplication are introduced in the obvious way. For

$$f = \sum_1^\infty a_\mu Y^{e(\mu)}, \qquad g = \sum_1^\infty b_\nu Y^{e(\nu)},$$

we set

$$f + g := \sum_1^\infty (a_\mu + b_\mu)\, Y^{e(\mu)},$$

$$f \cdot g := \sum_1^\infty c_\lambda Y^{e(\lambda)}, \quad \text{where} \quad c_\lambda := \sum_{\mu\nu=\lambda} a_\mu b_\nu .$$

This multiplication is similar to CAUCHY's multiplication since $e(\mu) + e(\nu) = e(\mu\nu)$.

Thus the set F of all these formal power series becomes a ring. As usual we identify the elements of A with the "constant" power series (meaning here that all $a_i = 0$, $i \geqq 2$). Viewed as an A-module, F is isomorphic to the direct product $\prod_1^\infty A$.

The ring F is not Noetherian; e.g., the ideal $\mathfrak{a}$ generated by all Y_ν, $\nu \geqq 1$, is not finitely generated.

Proof. Suppose the contrary. Then there is an index $r \geqq 1$ such that $\mathfrak{a} = \sum_{i=1}^r F Y_i$. In particular, we have an equation

$$Y_{r+1} = \sum_{i=1}^r f_i Y_i, \qquad f_i \in F .$$

Writing $f_i = \sum_{\mu=1}^\infty a_{i\mu} Y^{e(\mu)}$, $i = 1, \ldots, r$, and using $Y_\nu = Y^{e(p_\nu)}$, we conclude

$$Y^{e(p_{r+1})} = \sum_{\mu=1}^\infty a_{1\mu} Y^{e(p_1\mu)} + \cdots + \sum_{\mu=1}^\infty a_{r\mu} Y^{e(p_r\mu)} .$$

However this is impossible, since the injectivity of e gives $e(p_{r+1}) \neq e(p_i\mu)$ for all $i = 1, \ldots, r$ and all $\mu \geqq 1$. □

From now on we assume that A carries a bounded (possibly trivial) valuation $|\ \ |$. We extend this function from A to F by defining

$$|f| := \max_{\mu \geqq 1} \left\{\frac{1}{\mu} |a_\mu|\right\} \leqq 1 \quad \text{if} \quad f = \sum_1^\infty a_\mu Y^{e(\mu)}.$$

Obviously this is an ultrametric function on F. Let $f = \sum a_\mu Y^{e(\mu)}$ and $g = \sum b_\nu Y^{e(\nu)}$ be two series in F. Then, from the definition of the product $f \cdot g = \sum c_\lambda Y^{e(\lambda)}$, we conclude

$$\frac{1}{\lambda} |c_\lambda| \leqq \max_{\mu\nu=\lambda} \left\{\frac{1}{\mu} |a_\mu| \cdot \frac{1}{\nu} |b_\nu|\right\} \leqq |f| \cdot |g|;$$

i.e., $|fg| \leqq |f| \cdot |g|$. To get equality, we proceed as in the classical proof of the Gauss Lemma: let i (resp. j) be the smallest index such that

$$|f| = \frac{1}{i} |a_i| \quad \left(\text{resp. } |g| = \frac{1}{j} |b_j|\right).$$

Then $|c_{ij}| = |a_i| \cdot |b_j|$ and hence $|f \cdot g| \geqq \frac{1}{ij} |c_{ij}| = |f| \cdot |g|$. Thus we have proved

$|\ \ |$ is a bounded valuation on $F = A[\![Y_1, Y_2, \ldots]\!]$ which extends the valuation on A.

We have $|Y_\nu| = p_\nu^{-1}$; i.e., $\lim_\nu Y_\nu = 0$. Furthermore, each series $\sum_1^\infty a_\mu Y^{e(\mu)}$ is the limit of its partial sums in the topology induced by the valuation.

We have $F = \mathring{F}$ and $\check{F} = \left\{\sum_1^\infty a_\mu Y^{e(\mu)}; |a_1| < 1\right\}$. Therefore

$$\tilde{F} = \tilde{A}.$$

Moreover it can be shown (cf. Proposition 2.2.4/1)

If A is complete, F is complete.

Not all ideals of F are closed (see, for example, Proposition 2.2.4/2).

If A is a field (necessarily with trivial valuation), each $f \in F$, $|f| = 1$, can be written in the form $f = a(1 - g)$, $a \in A^*$, $g \in \check{F}$. The element $a^{-1} \sum_{i=0}^\infty g^i \in F$ is the inverse of f in F; i.e., f is a unit in F. Hence

If A is a trivially valued field, F is a local ring with the maximal ideal $\check{F}$. The value semi-group $|F - \{0\}|$ is discrete in $\mathbb{R}_+ - \{0\}$ and is not cyclic.

Furthermore, we have the following description of the elements of the maximal ideal of this local ring:

Each $g \in \check{F}$ can be written as a convergent series $g = \sum_1^\infty g_i Y_i$, $g_i \in F$.

Proof. The valuation being trivial on A, we have

$$g = \sum_1^\infty a_\mu Y^{e(\mu)} \in \check{F} \quad \text{if and only if} \quad a_1 = 0.$$

Set $N_i := \{\mu \in \{2, 3, 4, \ldots\}: p_1 \nmid \mu, \ldots, p_{i-1} \nmid \mu, p_i \mid \mu\}, i = 1, 2, \ldots$. This gives a partition of the set $\{2, 3, 4, \ldots\}$. Since, for $i = 1, 2, \ldots$,

$$g_i := \sum_{\mu \in N_i} a_\mu Y^{e(\mu/p_i)}$$

is a well-defined element in F, it is easily seen that $g = \sum_1^\infty g_i Y_i$. □

1.6. Discrete valuation rings

We denote by A an integral domain; Q denotes the field of fractions of A.

1.6.1. Definition. Elementary properties. — The most important class of valued rings is given by the class of so-called *valuation rings*. These rings first occurred in classical p-adic number theory as rings of "integers" of valued fields.

Definition 1. *A non-trivially valued ring* $(A, |\ \ |)$, *is called a valuation ring if* $A = \mathring{Q} = \{x \in Q; |x| \leqq 1\}$ *(where* Q *is provided with the extended valuation).*

Thus A is just the ring of *all* power-bounded elements of Q. The elements of A are often called the integers of the valued field Q, and we always have $A \neq Q$.

Each valuation ring A *is a local ring; i.e.,* A *contains a unique maximal ideal (which consists of all non-units of* A*). The units of* A *are all elements* $a \in A$ *such that* $|a| = 1$. *The maximal ideal is the set*

$$\check{A} = \{x \in A; |x| < 1\}.$$

In particular, the residue ring $\tilde{A} = A/\check{A}$ *is a field. It is called the residue field of* A.

Each ring A with a bounded valuation is contained in the valuation ring $\mathring{Q}$ of its field of fractions Q; in general, we have $A \neq \mathring{Q}$.

Proposition 2. *Let* A *be a valuation ring. Then each torsion-free, finitely generated* A*-module* M *is free, and each minimal system of generators of* M *is a basis.*

Proof. Let $\{x_1, \ldots, x_n\}$ be a system of generators of M such that n is minimal. Suppose there exists a non-trivial linear relation $\sum_1^n a_\nu x_\nu = 0$, $a_\nu \in A$. We may arrange the terms so that $a_1 \neq 0$ and $|a_1| \geqq |a_\nu|$, $\nu \geqq 2$. Then $b_\nu := -a_1^{-1} a_\nu \in A$, $\nu \geqq 2$, and $a_1 \left(x_1 - \sum_2^n b_\nu x_\nu\right) = 0$. Since M is torsion-free, we deduce that $x_1 = \sum_{2}^{n} b_\nu x_\nu$, in contradiction to the minimality of n. Thus $\{x_1, \ldots, x_n\}$ is a basis of M and M is free. □

Corollary 3. *Each Noetherian valuation ring A is a principal ideal domain.*

Proof. Since each ideal of A is torsion-free, all ideals of A are free by Proposition 2. However, a free ideal must be principal. □

A valuation ring A is called *discrete* if the value semi-group $|A - \{0\}|$ is discrete in $\mathbb{R}_+ - \{0\}$: this is the case if and only if the value group $|Q^*|$ is a discrete, and hence *cyclic*, subgroup of $\mathbb{R}_+ - \{0\}$. Then $\varepsilon := \max\{\alpha \in |Q^*|; \alpha < 1\}$ generates $|Q^*|$, and we have $|A - \{0\}| = \{\varepsilon^\nu; \nu \geqq 0\}$. Each $\pi \in A$ such that $|\pi| = \varepsilon$ is called a *uniformizing* element of A. (There is a close connection between this concept and the uniformization of a complex Riemann surface, but we cannot elaborate on this here.)

Proposition 4. *Let A be a discrete valuation ring, with maximal ideal $\mathfrak{m}$, and let $\pi \in A$ be a uniformising element. Then $\mathfrak{m} = A\pi$, and each ideal $\neq 0$ of A is of the form $\mathfrak{m}^n = A\pi^n$.*

Proof. Let $\mathfrak{a} \neq 0$ be an ideal in A. Choose $a \in \mathfrak{a}$ such that $|a| = \max_{x \in \mathfrak{a}} |x|$. We have $a \neq 0$, and $|a| = |\pi|^n$ for a suitable $n \geqq 0$. For each $x \in \mathfrak{a}$, write $x = e\pi^n$. Then $e := x\pi^{-n} \in Q$ belongs to A, since $|x| \leqq |\pi^n|$ and, hence, $|e| \leqq 1$. If $x = a$, we see that $e = a\pi^{-n}$ is a unit in A. Therefore we have $\mathfrak{a} \subset A\pi^n = Aa \subset \mathfrak{a}$; hence $\mathfrak{a} = A\pi^n$. In particular, $A\pi$ is the maximal ideal of A. □

Discrete valuation rings can be characterized in a purely algebraic way.

Proposition 5. *For a ring A the following two statements are equivalent:*

(i) *A is a local integral domain such that the maximal ideal $\mathfrak{m}$ of A is principal, non-zero, and satisfies $\bigcap_{1}^{\infty} \mathfrak{m}^\nu = 0$.*

(ii) *A can be provided with a valuation such that A becomes a discrete valuation ring.*

For such rings A each non-trivial bounded valuation $|\ \ |$ on A is $\mathfrak{m}$-adic; i.e., $|\ \ | = \varepsilon^{\nu_\mathfrak{m}}$, where $0 < \varepsilon < 1$.

Proof. As we have already seen, a discrete valuation ring possesses the algebraic properties listed under (i). Conversely, assume that A fulfills condition (i). Then by Proposition 1.5.2/2 the function $|\ \ |_\mathfrak{m} := \varepsilon^{\nu_\mathfrak{m}}$, where ε is fixed and $0 < \varepsilon < 1$, is a bounded non-trivial valuation on A. We extend this valuation to Q and claim $\mathring{Q} = A$. Let z be a generator of $\mathfrak{m}$. Each $q \in A - \{0\}$ can be uniquely written as $q = ez^n$ where $n := \nu_\mathfrak{m}(z)$ and where $e \in A - \mathfrak{m}$ is a unit, since A is local. From this we see that each $q \in Q^*$ is of the form $q = ez^n$ where $n \in \mathbb{Z}$ and where e is a unit in A. Thus $\mathring{Q} = \{0\} \cup \{q = ez^n; e \text{ unit in } A, n \geqq 0\} = A$. Since $\nu_\mathfrak{m}(Q^*) = \mathbb{Z}$, the value group $|Q^*|_\mathfrak{m}$ is discrete; i.e., $(A, |\ \ |_\mathfrak{m})$

is a discrete valuation ring. Therefore A satisfies condition (ii). From Proposition 4 we now deduce that all ideals $\neq 0$ of A are powers of $\mathfrak{m}$. Thus $\mathfrak{m}$ is the only prime ideal $\neq 0$ of A. Therefore each bounded non-trivial valuation on A is $\mathfrak{m}$-adic by Proposition 1.5.2/3. □

Remark. By KRULL's Intersection Theorem, any local integral domain A which is a principal ideal domain fulfills the assumptions of Proposition 5. Hence all these rings can be valued in such a way that they become discrete valuation rings. In particular, we immediately get from Corollary 3:

Proposition 6. *Every Noetherian valuation ring is a discrete valuation ring.*

1.6.2. The example of F. K. Schmidt. — We give an example due to F. K. SCHMIDT (cf. [35]) of a discrete valuation ring $A = \mathring{Q}$ which is not Japanese; i.e., Q permits finite algebraic extension fields Q' such that the integral closure A' of A in Q' is not a finite A-module (for the notion of Japaneseness and general facts see Chapter 4). We start with a general technique to construct valuations.

Each ring homomorphism $\Phi\colon R \to B$ of a ring R into a valued ring $(B, |\ \ |)$ induces a valuation $|\ \ | \circ \Phi$ on the residue ring $R/\ker \Phi$. We apply this remark to the case where R is a polynomial ring $k[X, Y]$ in two indeterminates over a field k and where B is the ring $k[\![T]\!]$ of formal power series in one indeterminate over k provided with a valuation $|\ \ |$ induced by the order function (cf. (1.3.3) and Proposition 1.5.2/1). For the homomorphism Φ, we want to choose a k-algebra substitution homomorphism which sends X into T and Y into a formal power series $f(T)$. If $f(T)$ is not algebraic over $k[T]$, i.e., if there exists no non-zero polynomial $q(T, W) \in k[T][W]$ such that $q(T, f(T)) = 0$, the homomorphism $\Phi\colon k[X, Y] \to k[\![T]\!]$ is injective and we get a valuation on $k[X, Y]$. We denote this valuation and its extension to the quotient field $K := k(X, Y)$ of rational functions in X, Y by $|\ \ |_f$, because it is uniquely determined by f. The value group $|K^*|_f$ is discrete; hence the ring $\mathring{K}$ is a discrete valuation ring. Note that

$$|X|_f < 1 \qquad \text{and} \qquad |Y|_f \leqq 1,$$

since $\operatorname{ord} T = 1$ and $\operatorname{ord} f(T) \geqq 0$. In particular, $k[X, Y] \subset \mathring{K}$. However, $\mathring{K} \neq k[X, Y]$. As a matter of fact, we have (if $f = \sum_0^\infty c_\nu T^\nu$)

$$Z_m := [Y - (c_0 + c_1 X^1 + \cdots + c_m X^m)] \cdot X^{-(m+1)} \in \mathring{K} \quad \text{for each} \quad m \geqq 1,$$

since

$$\operatorname{ord}\left(f(T) - \sum_0^m c_\nu T^\nu\right) = \operatorname{ord} \sum_{m+1}^\infty c_\nu T^\nu \geqq m + 1.$$

Since $|Y - c_0|_f < 1$, the residue field $\mathring{K}/\check{K}$ equals the field k.

Next we state a sufficient condition for a formal power series $f \in k[\![T]\!]$ not to be algebraic over the polynomial ring $k[T]$.

Let $f = \sum_0^\infty c_\nu T^\nu$ *be a series in* $k[\![T]\!]$ *such that the subfield* k' *of* k *generated by all coefficients* $c_0, c_1, \ldots$ *over the prime field of* k *has infinite transcendence degree. Then* f *is not algebraic over* $k[T]$.

Obviously it is enough to prove the following (contrapositive) statement:

Let $f = \sum_0^\infty c_\nu T^\nu$ *be a series in* $k[\![T]\!]$, *and let* $q(T, W) \in k[T, W]$ *be a non-zero polynomial such that* $q(T, f(T)) = 0$. *Denote by* k_0 *the subfield of* k *generated by all coefficients of* $q(T, W)$ *over the prime field of* k *(this field is of finite transcendence degree over the prime field). Then each coefficient* c_n *of* f *is algebraic over* k_0 *(and hence* $k_0(c_0, c_1, \ldots)$ *is of finite transcendence degree over the prime field of* k*).*

Proof. We proceed by induction on n. First let $n = 0$. We may assume that T does not divide $q(T, W)$. Then $q(0, W) \in k_0[W]$ is not zero; however, $q(0, f(0)) = q(0, c_0) = 0$. Hence c_0 is algebraic over k_0.

Now let $n \geqq 1$. We consider the polynomial

$$q'(T, W) := q(T, c_0 + c_1 T + \cdots + c_{n-1} T^{n-1} + W T^n) \in k_0(c_0, \ldots, c_{n-1})\,[T, W].$$

The polynomial q' is not the zero polynomial. In order to see this, write $q = \sum_0^r a_i(T)\, W^i$, where $a_r(T) \neq 0$. Then

$$q' = \sum_0^r a_i(T)\,(c_0 + c_1 T + \cdots + c_{n-1} T^{n-1} + W T^n)^i$$

contains the term $a_r(T) \cdot T^{rn} W^r$, which does not cancel with any other term.

Let T^s be the highest power of T which divides q', and write $q' = T^s q_1$, where $q_1 \in k_0(c_0, \ldots, c_{n-1})\,[T, W]$ and $q_1(0, W) \neq 0$. We have

$$q'(T, c_n + c_{n+1} T + \cdots) = q(T, f(T)) = 0,$$

and hence $q_1(T, c_n + c_{n+1} T + \cdots) = 0$. In particular, we see $q_1(0, c_n) = 0$. Thus c_n is algebraic over $k_0(c_0, \ldots, c_{n-1})$. Since $c_0, \ldots, c_{n-1}$ are algebraic over k_0 by the induction hypothesis, it follows that c_n is algebraic over k_0. □

We now come to the example of F. K. Schmidt. As field k we choose a field generated over $\mathbb{Z}/p\mathbb{Z}$ (where $p \neq 0$ is prime) by an infinite sequence $t_0, t_1, \ldots$ of indeterminates. We set

$$f(T) := t_0 + t_1 T + \cdots + t_n T^n + \cdots.$$

By the statement just proved, f is not algebraic over $k[T]$, and hence induces a valuation $|\ \ |_f$ on the field $K = k(X, Y)$. We consider the valued subfield $Q := k(X, Y^p)$ of K. Obviously $[K:Q] = p$ and the elements $1, Y, \ldots, Y^{p-1}$ form a Q-basis of K. The ring $A := \mathring{Q}$ is the discrete valuation ring for which we are looking. Namely,

$\mathring{K}$ is the integral closure of A in K. The A-module $\mathring{K}$ is not finitely generated.

Proof. Since any $z \in K$ is a p-th root of some element in Q, we see that all elements of $\mathring{K}$ are integral over $\mathring{Q}$. Furthermore, an integral equation

$$z^r + a_1 z^{r-1} + \cdots + a_r = 0$$

with coefficients $a_1, \ldots, a_r \in \mathring{Q}$ is impossible if $|z| > 1$. Thus $\mathring{K}$ is the integral closure of A in K. Now assume that $\mathring{K}$ is generated as an A-module by finitely many elements $e_1, \ldots, e_n \in \mathring{K}$. Write

$$e_\nu = \sum_{i=0}^{p-1} c_{\nu i} Y^i, \quad c_{\nu i} \in Q, \quad \nu = 1, \ldots, n.$$

Since $|X| < 1$, we can choose an exponent $s \geqq 1$ such that $|c_{\nu i} X^s| \leqq 1$ for all ν, i. Then $X^s e_1, \ldots, X^s e_n \in A + AY + \cdots + AY^{p-1}$, and hence

$$\mathring{K} \cdot X^s \subset A + AY + \cdots + AY^{p-1}.$$

Now consider the element

$$Z_s := (Y - t_0 - t_1 X - \cdots - t_s X^s) \cdot X^{-(s+1)} \in K.$$

We already saw that $Z_s \in \mathring{K}$, and hence

$$Z_s X^s = (Y - t_0)\, X^{-1} - (t_1 + \cdots + t_s X^{s-1}) \in A + AY + \cdots + AY^{p-1}.$$

Since $t_1 + \cdots + t_s X^{s-1}$ is an element of A, we must have $(Y - t_0)\, X^{-1} \in A + AY + \cdots + AY^{p-1}$. However this is impossible, since $X^{-1} \notin A$ and since the elements $1, Y, \ldots, Y^{p-1}$ form a Q-basis of K. Therefore $\mathring{K}$ is not finitely generated over A. □

Remark. The valuation on the field Q is *not* complete. We shall see later (Proposition 3.5.4/1) that the valuation ring A belonging to a field Q with a complete and discrete valuation is always *Japanese*. This statement is false if one drops the assumption of discreteness (cf. the remark following Proposition 6.3.4/1).

1.7. Bald and discrete *B*-rings

Discrete valuation rings enjoy the nice property that a series $\sum_0^\infty a_\nu$ is convergent if, say, $|a_0| \leqq 1$ and $|a_{\nu+1}| < |a_\nu|$. Then $|a_\nu| \leqq \varepsilon^\nu$, where ε is the biggest number < 1 in the value set. Such a criterion is necessary for the convergence of certain iteration techniques. For example, the usual proof of HENSEL's Lemma (see [38], § 144) relies heavily on the fact that the valuation in question is discrete. Unfortunately, in applications to k-affinoid geometry where one wants to "lift" theorems of $\tilde{k}$-affine geometry, we cannot restrict ourselves to the discrete case. So the question comes up as to how strongly the discreteness is really needed. The main point of the following considerations is to introduce certain "almost discrete" subrings of arbitrary valuation rings $\mathring{k}$ which will be as good as discrete valuation rings for the purposes mentioned above.

By K we always mean a (valued) field; A is a (valued) subring of K. We write ∂A for the multiplicatively closed subset of elements of value 1 in A.

1.7.1. *B*-rings. — We start with the simple

Definition 1. *A valued ring A is called a B-ring if*

(i) *$|A|$ is bounded, i.e., $|x| \leqq 1$ for all $x \in A$,*
(ii) *each $x \in \partial A$ is a unit in A.*

Each valuation ring is a B-ring. We have the trivial

Proposition 2. *If the valuation of A is bounded, then the ring of fractions $A_{\partial A}$ is the smallest B-ring containing A. The rings A and $A_{\partial A}$ have the same value semi-groups.*

For any B-ring A, we have $A = \mathring{A}$ by (i). Due to (ii), the ideal $\check{A} = \{x \in A;\ |x| < 1\}$ is the only maximal ideal in A. Hence *each B-ring is a local ring*, and we have a residue *field* $\tilde{A} = A/\check{A}$. The residue homomorphism $\tilde{\ }: A \to \tilde{A}$ maps the group ∂A onto $\tilde{A}^*$.

A ring A with a bounded valuation is a B-ring if and only if A is a local ring with maximal ideal $\check{A}$. In particular, the ring $F = k[\![Y_1, Y_2, \ldots]\!]$ of formal power series in countably many indeterminates over a trivially valued field k (cf. (1.5.5)) is a B-ring.

A B-subring of a B-ring A is a subring of A which is again a B-ring (with respect to the induced valuation). The intersection of any family of B-subrings of a B-ring is a B-subring. Hence for each subset M of a valuation ring A, there exists *the smallest B-subring of A containing M.*

Proposition 3. *The completion $\hat{A}$ of a B-ring A is a B-ring.*

Proof. Since $|\hat{A}| = |A|$, we only have to show that every $x \in \partial\hat{A}$ is a unit in $\hat{A}$. Choose $a \in A$ with $|x - a| < 1$. Then $|a| = 1$ and a is a unit in A. Write $x = a(1 - z)$ where $z := a^{-1}(a - x)$. Then $|z| < 1$ and $\sum_0^\infty z^\nu \in \hat{A}$ is the inverse of $1 - z$. Hence x is a unit in $\hat{A}$. □

The next technical lemma will turn out to be very helpful. Recall that K is a valued field.

Lemma 4. *Let $A \subset \mathring{K}$ be a B-ring and let y be a fixed element of $\mathring{K}$. Then there exists a polynomial $g \in A[X]$ with $|g(y)| < 1$ such that the following is true:*

Each polynomial $f \in A[X]$ with $|f(y)| < 1$ admits in $A[X]$ a decomposition

$$f = q \cdot g + r$$

where all coefficients of r are in $\check{A}$.

Proof. The residue field $\tilde{K}$ of K is an extension of the residue field $\tilde{A}$ of A. Let $\tilde{y}$ be the image of y in $\tilde{K}$. If $\tilde{y}$ is transcendental over $\tilde{A}$, we set $g := 0$, $q := 0$, $r := f$. Then the lemma holds trivially, because $|f(y)| < 1$ implies $\tilde{r}(\tilde{y}) = 0$, i.e., $\tilde{r} = 0$, which exactly means that all coefficients of r belong to $\check{A}$.

If $\tilde{y}$ is algebraic over $\tilde{A}$, we choose

$$g := X^n + a_1 X^{n-1} + \cdots + a_n \in A[X]$$

as an inverse image of the minimal polynomial $\tilde{g} = X^n + \tilde{a}_1 X^{n-1} + \cdots + \tilde{a}_n \in \tilde{A}[X]$ of $\tilde{y}$ over $\tilde{A}$ and choose $q, r \in A[X]$ according to Euclid's division theorem: $f = qg + r$. Then

$$\tilde{f} = \tilde{q}\tilde{g} + \tilde{r} \quad \text{for the image polynomials} \quad \tilde{f}, \tilde{q}, \tilde{g}, \tilde{r} \in \tilde{A}[X].$$

We have $\tilde{g}(\tilde{y}) = 0$ by definition; i.e., $|g(y)| < 1$. If $|f(y)| < 1$, we have $\tilde{f}(\tilde{y}) = 0$. Therefore $\tilde{r}(\tilde{y}) = 0$. From the fact that $\tilde{g}$ is the minimal polynomial of $\tilde{y}$ over $\tilde{A}$ and from $\deg \tilde{r} \leqq \deg r < \deg g = \deg \tilde{g}$ we conclude that $\tilde{r} = 0$, i.e., $r \in \check{A}[X]$. □

1.7.2. Bald rings. — Value semi-groups may contain numbers arbitrarily close to 1.

Definition 1. *A valued ring A is called bald if*

$$\sup_{x \in \check{A}} |x| < 1.$$

Each subring of a discrete valuation ring is bald. The completion $\hat{A}$ of a bald ring A remains bald.

Proposition 2. *Let $A \subset \mathring{K}$ be a bald ring and let M be a subset of $\mathring{K}$ such that $\sup_{y \in M} |y| < 1$. Then the polynomial ring $A[M]$ is bald; more precisely*

$$\sup_{z \in \widecheck{A[M]}} |z| \leqq \max \left\{ \sup_{a \in \check{A}} |a|, \sup_{y \in M} |y| \right\}.$$

Proof. Take $z \in \widecheck{A[M]}$, say $z = a_0 + z'$, where $a_0 \in A$ and $z' = \sum_{\nu_1 + \cdots + \nu_n > 0} a_{\nu_1 \ldots \nu_n} y_1^{\nu_1} \ldots y_n^{\nu_n}$ with $y_1, \ldots, y_n \in M$. We have

$$|z'| \leqq \max |a_{\nu_1 \ldots \nu_n}| \, |y_1|^{\nu_1} \ldots |y_n|^{\nu_n} \leqq \max |y_1|^{\nu_1} \ldots |y_n|^{\nu_n} \leqq \sup_{y \in M} |y|,$$

since the case $\nu_1 = \cdots = \nu_n = 0$ is excluded. From $|z| < 1$, we conclude that $|a_0| < 1$; i.e.,

$$|z| \leqq \max \left\{ \sup_{a \in \check{A}} |a|, \sup_{y \in M} |y| \right\}.$$ □

The preceding proposition remains true for arbitrary finite sets $M \subset \mathring{K}$. Namely,

Proposition 3. *If $A \subset \mathring{K}$ is bald, each polynomial ring $A[y]$, $y \in \mathring{K}$, is bald.*

Proof. Due to Proposition 1.7.1/2, we may assume without loss of generality that A is a B-ring. Set $\varepsilon := \sup_{x \in \check{A}} |x|$. Choose g as in Lemma 1.7.1/4 and set

$\varepsilon' := \max \{\varepsilon, |g(y)|\}$. We have $\varepsilon' < 1$. It is enough to show that $|z| \leqq \varepsilon'$ for each $z \in A[y]$, $|z| \neq 1$. Write $z = f(y)$ where $f \in A[X]$. By Lemma 1.7.1/4 we may write $f = qg + r$, where $q \in A[X]$ and all coefficients of r have absolute value $\leqq \varepsilon$. Hence we get (since $|q(y)| \leqq 1$)

$$|f(y)| = |q(y)\, g(y) + r(y)| \leqq \max \{|q(y)|\, |g(y)|, |r(y)|\}$$
$$\leqq \max \{|g(y)|, \varepsilon\} = \varepsilon'. \qquad \square$$

We now easily obtain

Theorem 4. *Let K be complete and let M be a subset of $\mathring{K}$ such that the set $M \cap \partial\mathring{K}$ is finite and such that* $\sup \{|y|; y \in M \cap \check{K}\} < 1$. *Then the smallest complete B-ring $B \subset \mathring{K}$ containing M is bald.*

Proof. Denote by I the intersection of all subrings of $\mathring{K}$. If char $K = p > 0$, we have $I \simeq \mathbb{Z}/p\mathbb{Z}$ and the valuation on I is trivial. If char $K = 0$, we have $I \simeq \mathbb{Z}$ and the valuation on I is either trivial or a p-adic one (cf. Corollary 1.5.2/4). Thus in any case I is a bald ring. From Propositions 2 and 3 we deduce that $I[M]$ is also bald. Proposition 1.7.1/2 gives the existence of a bald B-ring containing M. Now Proposition 1.7.1/3 gives the desired result since baldness is not destroyed by passing to a completion. $\square$

The most important case for applications of Theorem 4 is the following

Corollary 5. *Let $\{y_1, y_2, \ldots\}$ be a zero sequence in $\mathring{K}$. Then the smallest complete B-subring of K containing all y_ν, $\nu \geqq 1$, is bald.*

In (1.8.2) we shall see that much more than baldness holds for such a ring.

1.8. Quasi-Noetherian B-rings

Quasi-Noetherian rings were first introduced in [16]. They were used in order to ensure the convergence of certain approximation techniques. (Later it was discovered that the more general class of bald rings can do the same job.) In this book we include this interesting class of rings for the sake of completeness. However they will not be used in later applications (e.g., the Lifting Theorem 2.7.3/2). It will always suffice to use the property of baldness.

1.8.1. Definition and characterization. — In general, B-rings are not Noetherian; e.g., the B-ring $F = k[\![Y_1, Y_2, \ldots]\!]$ of formal power series in countably many indeterminates over a trivially valued field k is not Noetherian (cf. (1.5.5)). In many cases however all ideals of such rings are quasi-finitely generated.

Definition 1. *Let A be a B-ring and let $\mathfrak{a}$ be an ideal in A. A zero sequence $(a_\nu)_{\nu \geqq 1}$ of elements $a_\nu \in \mathfrak{a}$ is called a quasi-finite system of generators of $\mathfrak{a}$ if each $a \in \mathfrak{a}$ can be written in the form*

$$a = \sum_1^\infty c_\nu a_\nu, \qquad c_\nu \in A.$$

In this case the ideal $\mathfrak{a}$ is called quasi-finite.

A B-ring A is called quasi-Noetherian if each ideal in A is quasi-finite.

Each Noetherian B-ring is quasi-Noetherian.

Remark. If $\{a_1, a_2, \ldots\}$ is a quasi-finite system of generators of $\mathfrak{a}$, we by no means claim that each element $x \in A$ which can be written in the form $\sum_1^\infty c_\nu a_\nu$ belongs to $\mathfrak{a}$. Of course this holds if $\mathfrak{a}$ is closed.

Each quasi-Noetherian B-ring is bald. As a matter of fact the following is true:

Let A be a B-ring such that the maximal ideal $\check{A}$ of A possesses a quasi-finite system $\{x_1, x_2, \ldots\}$ of generators. Then

$$\sup_{x \in \check{A}} |x| = \max_{\nu \geqq 1} |x_\nu| < 1.$$

This is clear since each $x \in \check{A}$ is of the form $\sum_1^\infty c_\nu x_\nu$ where $|c_\nu| \leqq 1$ for all ν.

In order to characterize quasi-Noetherian rings, we associate to each bald ring A a sequence of vector spaces in the following way:

We set $\varrho := \sup_{x \in \check{A}} |x| < 1$ and consider the sequence

$$A_\nu := \{x \in A;\ |x| \leqq \varrho^\nu\}, \qquad \nu = 0, 1, \ldots$$

of closed ideals in A. We have $A_0 = A$, $A_1 = \check{A}$, $A_\nu \supset A_{\nu+1}$, $A_\mu A_\nu \subset A_{\mu+\nu}$, in particular $\check{A} A_\nu \subset A_{\nu+1}$. Therefore each A-residue module

$$\alpha_\nu := A_\nu / A_{\nu+1}, \qquad \nu = 0, 1, 2, \ldots$$

may be viewed in a canonical way as an α-vector space where $\alpha := \alpha_0 = A/\check{A}$ denotes the residue field of the B-ring A. Let $\varphi_\nu : A_\nu \to \alpha_\nu$ be the canonical residue epimorphism. The main result of this section will be

Theorem 2. *For a bald B-ring A, the following statements are equivalent:*

(1) *A is quasi-Noetherian.*
(2) *The maximal ideal $\check{A}$ of A is quasi-finite.*
(3) *All α-vector spaces α_ν, $\nu \geqq 1$, are of finite dimension.*

Proof. The implication (1) $\to$ (2) is trivial. In order to derive (3) from (2), we shall prove more; namely,

Each ideal A_ν possesses a quasi-finite system N_ν of generators, $\nu \geqq 1$.

Obviously this will imply (3), because the α-space α_ν is generated by the image of N_ν in α_ν (with respect to the residue map $\varphi_\nu : A_\nu \to \alpha_\nu$). Note that almost all of these image vectors are zero since almost all elements of N_ν belong to $A_{\nu+1}$.

We shall prove the existence of N_ν by induction on ν. The existence of N_1 is guaranteed by assumption (2); we write $N_1 = \{x_1, x_2, \ldots\}$. Assume that we have already constructed a quasi-finite system of generators $N_\nu = \{y_1, y_2, \ldots\}$

for A_ν. Since $\lim_j y_j = 0$, we can choose an index s such that $y_j \in A_{\nu+1}$ for all $j > s$. This means that $\varphi_\nu(y_j) = 0$ for all $j > s$.

The kernel $\ker\psi$ of the α-linear map $\psi\colon \alpha^s \to \alpha_\nu$ given by $(\xi_1, \ldots, \xi_s) \mapsto \sum_{j=1}^{s} \xi_j\varphi_\nu(y_j)$ is of finite dimension. Therefore we may select finitely many vectors

$$(a_{\mu 1}, \ldots, a_{\mu s}) \in A^s, \qquad \mu = 1, \ldots, m,$$

euch that their images

$$\big(\varphi_0(a_{\mu 1}), \ldots, \varphi_0(a_{\mu s})\big) \in \alpha^s, \qquad \mu = 1, \ldots, m,$$

in α^s (with respect to the residue map $\varphi_0\colon A \to \alpha$) generate $\ker\psi$ over α. We consider the m elements

$$z_\mu := \sum_{j=1}^{s} a_{\mu j} y_j \in A_\nu, \qquad \mu = 1, \ldots, m.$$

We have

$$\varphi_\nu(z_\mu) = \sum_{j=1}^{s} \varphi_0(a_{\mu j})\,\varphi_\nu(y_j) = 0; \quad \text{i.e.,} \quad z_1, \ldots, z_m \in A_{\nu+1}.$$

Now we claim

$$N_{\nu+1} := (N_\nu \cap A_{\nu+1}) \cup \{z_1, \ldots, z_m\} \cup \{x_i y_j;\, i \geqq 1, j = 1, \ldots, s\}$$

is a quasi-finite system of generators of $A_{\nu+1}$.

From $A_1 \cdot A_\nu \subset A_{\nu+1}$, we deduce $x_i y_j \in A_{\nu+1}$ for all i, j. Therefore $N_{\nu+1} \subset A_{\nu+1}$. Because the set $\{x \in N_{\nu+1};\, x \notin A_n\}$ is finite for each n, we see that the elements of $N_{\nu+1}$ form a zero sequence.

Now choose $v \in A_{\nu+1}$ arbitrarily. Since $A_{\nu+1} \subset A_\nu$, we may write $v = \sum_{j=1}^{\infty} c_j y_j$. Since $\sum_{j>s} c_j y_j \in A_{\nu+1}$, we see that $\sum_{j=1}^{s} c_j y_j \in A_{\nu+1}$. Hence it is enough to show that $\sum_{j=1}^{s} c_j y_j$ can be written as an infinite series in $\{z_1, \ldots, z_m\} \cup \bigcup_{j=1}^{s} (N_1 y_j)$. From $\sum_{j=1}^{s} c_j y_j \in A_{\nu+1}$ we get

$$\big(\varphi_0(c_1), \ldots, \varphi_0(c_s)\big) \in \ker\psi.$$

Therefore we can find elements $u_1, \ldots, u_m \in A$ such that

$$\big(\varphi_0(c_1), \ldots, \varphi_0(c_s)\big) = \sum_{\mu=1}^{m} \varphi_0(u_\mu)\big(\varphi_0(a_{\mu 1}), \ldots, \varphi_0(a_{\mu s})\big);$$

i.e.,

$$c'_j := c_j - \sum_{\mu=1}^{m} u_\mu a_{\mu j} \in \ker\varphi_0 = A_1, \qquad j = 1, \ldots, s.$$

Now

$$\sum_{j=1}^{s} c'_j y_j = \sum_{j=1}^{s} c_j y_j - \sum_{j=1}^{s}\sum_{\mu=1}^{m} u_\mu a_{\mu j} y_j;$$

i.e.,

$$\sum_{j=1}^{s} c_j y_j = \sum_{j=1}^{s} c_j' y_j + \sum_{\mu=1}^{m} u_\mu z_\mu .$$

Finally, each $c_j' \in A_1$ can be written as an infinite linear combination of elements from N_1. Thus we see that $N_{\nu+1}$ has the desired property.

It remains to deduce (1) from (3). Let $\mathfrak{a}$ be any ideal in A (we may assume $\mathfrak{a} \subset A_1$). For $\nu \geqq 1$, each set $\varphi_\nu(\mathfrak{a} \cap A_\nu) \subset \alpha_\nu$ is an α-vector space of finite dimension. Choose elements $a_{\nu 1}, \ldots, a_{\nu n_\nu} \in \mathfrak{a} \cap A_\nu$ such that their φ_ν-images generate $\varphi_\nu(\mathfrak{a} \cap A_\nu)$. Set

$$N := \bigcup_{\nu=1}^{\infty} \{a_{\nu 1}, \ldots, a_{\nu n_\nu}\} .$$

Obviously this is a zero sequence. Moreover it is clear (by induction) that each $a \in \mathfrak{a}$ can be written in the form

$$a = \sum_{i=1}^{n} (c_{i1} a_{i1} + \cdots + c_{i n_i} a_{i n_i}) + a_{n+1}, \quad \text{where} \quad a_{n+1} \in \mathfrak{a} \cap A_{n+1} .$$

Hence N is a quasi-finite system of generators of $\mathfrak{a}$. □

The theorem just proved together with the observations made at the end of (1.5.5) imply

The B-ring $F = k[\![Y_1, Y_2, \ldots]\!]$ of formal power series over a trivially valued field k is quasi-Noetherian.

We state a surprising corollary of Theorem 2.

Corollary 3. *The value semi-group $|A - \{0\}|$ of a quasi-Noetherian B-ring A is discrete in $\mathbb{R}_+ - \{0\}$.*

Proof. Using the same notations as above, we claim

(*) *Let $x_1, \ldots, x_m \in A_\nu$ such that $\varrho^\nu \geqq |x_1| > |x_2| > \cdots > |x_m| > \varrho^{\nu+1}$. Then the m vectors $\varphi_\nu(x_1), \ldots, \varphi_\nu(x_m) \in \alpha_\nu$ are linearly independent.*

If this were not the case, we could find (by lifting a non-trivial linear relation from α_ν to A_ν) elements $c_1, \ldots, c_m \in A$ such that

$$c_\mu = 0 \quad \text{or} \quad |c_\mu| = 1, \quad |c_1| + \cdots + |c_m| \neq 0, \quad \sum_{\mu=1}^{m} c_\mu x_\mu \in A_{\nu+1} .$$

Denoting by t the smallest index such that $c_t \neq 0$, we get the contradiction

$$\left| \sum_{1}^{m} c_\mu x_\mu \right| = |c_t|\,|x_t| > \varrho^{\nu+1} .$$

Hence (*) holds and implies (since all spaces α_ν are of finite dimension) that each set $\{|x| \in |A|; \varrho^{\nu+1} < |x| \leqq \varrho^\nu\}$, $\nu = 1, 2, \ldots$, is finite. Hence $|A - \{0\}|$ is discrete in $\mathbb{R}_+ - \{0\}$. □

Whenever we are given a B-ring with a discrete value set $|A - \{0\}|$, we can order the set $|A - \{0\}| = \{\varrho_0, \varrho_1, \varrho_2, \ldots\}$ in a unique way such that

$$1 = \varrho_0 > \varrho = \varrho_1 > \varrho_2 > \cdots > \varrho_n > \cdots .$$

We have $\varrho_\mu \varrho_\nu \leqq \varrho_{\mu+\nu}$ and $\lim \varrho_\nu = 0$. For such rings, it is natural to consider *all* ideals

$$A_\nu := \{x \in A\,;\, |x| \leqq \varrho_\nu\}, \qquad \nu = 0, 1, \ldots,$$

instead of only those where the radius ϱ_ν is replaced by $\varrho^\nu = \varrho_1^\nu$. We shall call the sequence $\{A_\nu\}$ the *natural filtration* of A; as before we have

$$A_0 = A, \quad A_1 = \check{A}, \quad A_\nu \supset A_{\nu+1}, \quad A_\mu A_\nu \subset A_{\mu+\nu}.$$

Furthermore, all residue modules $\alpha_\nu := A_\nu/A_{\nu+1}$ are α-vector spaces. The ring A is quasi-Noetherian if and only if all α_ν are of finite dimension (use Theorem 2).

We say that a quasi-finite system N of generators of an ideal $\mathfrak{a} \subset A$ is *filtered* if for all $\nu \geqq 1$ the set $N \cap A_\nu$ is a quasi-finite system of generators of the ideal $\mathfrak{a} \cap A_\nu$. Proceeding in the same way as in the proof of Theorem 2 (where we deduced (1) from (3)) one gets

Proposition 4. *For each ideal $\mathfrak{a}$ of a quasi-Noetherian ring A there exists a filtered quasi-finite system of generators.*

Proposition 5. *Let $\mathfrak{a}$ be an ideal in a quasi-Noetherian ring A, and let $N = \{a_1, a_2, \ldots\}$ be a filtered quasi-finite system of generators of $\mathfrak{a}$ such that*

$$\mathfrak{a} = \left\{x \in A\,;\, x = \sum_1^\infty c_\nu a_\nu,\ c_\nu \in A\right\}.$$

Then $\mathfrak{a}$ is closed in A.

Proof. Take a sequence $(x_\nu) \subset \mathfrak{a}$ and set $x := \lim x_\nu$. Since $x = x_1 + \sum_{\nu=2}^\infty (x_\nu - x_{\nu-1})$ and $\lim (x_\nu - x_{\nu-1}) = 0$, it is enough to show

For each zero sequence $(z_\nu) \subset \mathfrak{a}$ we have $\sum_1^\infty z_\nu \in \mathfrak{a}$.

We may assume $z_\nu \in \mathfrak{a} \cap A_\nu$. We have equations

$$z_\nu = \sum_{j=1}^\infty b_{\nu j} a_j, \quad \nu = 1, 2, \ldots, \quad \text{for suitable coefficients} \quad b_{\nu j} \in A.$$

Since N is filtered, we may assume $b_{\nu j} = 0$ whenever $a_j \notin A_\nu$. Thus for fixed j, almost all $b_{\nu j}$ vanish; hence the element

$$b_j := \sum_{\nu=1}^\infty b_{\nu j} \in A$$

is well-defined. Moreover, for all $\varepsilon > 0$, one has $|b_{\nu j} a_j| < \varepsilon$ for almost all (ν, j). Therefore (by Corollary 1.1.8/3) one gets

$$\sum_{\nu=1}^\infty z_\nu = \sum_{\nu=1}^\infty \left(\sum_{j=1}^\infty b_{\nu j} a_j\right) = \sum_{j=1}^\infty \left(\sum_{\nu=1}^\infty b_{\nu j}\right) a_j = \sum_{j=1}^\infty b_j a_j,$$

and by assumption this series is in $\mathfrak{a}$. □

Notice that there exist quasi-Noetherian rings having non-closed ideals, e.g., the ring $F = k[\![Y_1, Y_2, \ldots]\!]$ of formal power series over a trivially valued field k.

1.8.2. Construction of quasi-Noetherian rings. — We start with the obvious

Proposition 1. *The smallest B-subring A' of a B-ring A is quasi-Noetherian.*

Proof. If char $A = 0$, then $A' \supset \mathbb{Z}$. If $\mathbb{Z}$ carries the trivial valuation, one sees that $\mathbb{Q} \subset A'$, and therefore $A' = \mathbb{Q}$. If $\mathbb{Z}$ is not trivially valued, its valuation is a p-adic one for some prime number p (cf. Corollary 1.5.2/4). Then A' is the localization of $\mathbb{Z}$ with respect to (p). If $q := \text{char } A \neq 0$ then $A' = \mathbb{Z}/q\mathbb{Z}$, and A' carries the trivial valuation. In all these cases A' is quasi-Noetherian for obvious reasons. □

Next we prove

Proposition 2. *The completion $\hat{A}$ of a quasi-Noetherian B-ring A is quasi-Noetherian. Each filtered quasi-finite system N of generators of the maximal ideal A_1 of A is a quasi-finite system of generators of the maximal ideal $\hat{A}_1$ of $\hat{A}$.*

Proof. Each $\bar{y} \in \hat{A}_1$ can be written in the form

$$\bar{y} = \sum_{\nu=1}^{\infty} y_\nu, \quad y_\nu \in A_\nu.$$

Let $N = \{v_1, v_2, \ldots\}$. Since N is filtered, we have equations:

$$y_\nu = \sum_{j=1}^{\infty} b_{\nu j} v_j, \qquad \nu = 1, 2, \ldots,$$

for suitable $b_{\nu j} \in A$ such that $b_{\nu j} = 0$ whenever $v_j \notin A_\nu$. In the same way as in the proof of Proposition 1.8.1/5, one sees that $\bar{y} = \sum_{j=1}^{\infty} \left(\sum_{\nu=1}^{\infty} b_{\nu j} \right) v_j$, which proves that N is a quasi-finite system of generators of $\hat{A}_1$. This implies (by Theorem 1.8.1/2) that $\hat{A}$ is quasi-Noetherian. □

Proposition 3. *Let $A \subset \mathring{K}$ be quasi-Noetherian; let $y \in \mathring{K}$ be arbitrary. Then the smallest B-ring containing A and y, i.e., the ring*

$$L := A[y]_{\{x \in A[y];\, |x| = 1\}},$$

is quasi-Noetherian.

If $\{v_1, v_2, \ldots\}$ is a quasi-finite system of generators of A_1 and if $g \in A[X]$ denotes the polynomial belonging to $y \in \mathring{K}$ according to Lemma 1.7.1/4, then the sequence

$$\{v_0 := g(y), v_1, v_2, \ldots\}$$

is a quasi-finite system of generators of the maximal ideal L_1 of L.

Proof. We have $v_0 \in L_1$, since $|g(y)| < 1$, and $v_1, v_2, \ldots \in L_1$, since $A_1 \subset L_1$. Obviously $\{v_0, v_1, \ldots\}$ is a zero sequence. Choose an arbitrary element z in L_1, say

$$z = \frac{a}{b},\ a, b \in A[y],\quad |a| < 1,\quad |b| = 1.$$

Let $f \in A[X]$ such that $f(y) = a$. Since $|f(y)| < 1$ we can choose polynomials

$$q \in A[X],\quad r = r_0 + r_1 X + \cdots + r_s X^s,\quad r_0, r_1, \ldots, r_s \in A_1,$$

according to Lemma 1.7.1/4. We get

$$a = q(y)\, v_0 + \sum_0^s r_i y^i.$$

Since $\{v_1, v_2, \ldots\}$ is a quasi-finite system of generators of A_1, we have equations

$$r_i = \sum_{\nu=1}^{\infty} b_{i\nu} v_\nu,\quad b_{i\nu} \in A,\quad i = 0, 1, 2, \ldots, s.$$

We conclude that

$$z = \frac{q(y)}{b} v_0 + \sum_{\nu=1}^{\infty} \frac{b_{0\nu} + b_{1\nu} y + \cdots + b_{s\nu} y^s}{b} v_\nu .$$

All coefficients occurring here are in L. Hence L_1 is generated quasi-finitely by $\{v_0, v_1, \ldots\}$ and L is quasi-Noetherian. □

We now easily obtain

Proposition 4. *Let $A \subset \mathring{K}$ be quasi-Noetherian, let $\{y_1, y_2, \ldots\}$ be a zero sequence in K. Then the smallest B-ring containing A and all y_ν, $\nu \geqq 1$, i.e., the ring*

$$L := A[y_1, y_2, \ldots]_{\{x \in A[y_1, y_2, \ldots];\, |x| = 1\}}$$

is quasi-Noetherian.

Proof. Only finitely many of the y's — say $y_1, y_2, \ldots, y_d$ — have absolute value 1. Applying Proposition 3 d-times, we see that the B-ring

$$L' := A[y_1, y_2, \ldots, y_d]_{\{x \in A[y_1, \ldots, y_d];\, |x| = 1\}}$$

is quasi-Noetherian. Obviously

$$L = L'[y_{d+1}, \ldots]_{\{x \in L'[y_{d+1}, \ldots];\, |x| = 1\}}.$$

Hence it is sufficient to show

If $N' = \{v_1, v_2, \ldots\}$ is a quasi-finite system of generators of the maximal ideal L'_1 of L', the set

$$N := N' \cup \{\text{all monomials in finitely many of the } y_{d+1}, y_{d+2}, \ldots\}$$

is a quasi-finite system of generators of the maximal ideal L_1 of L.

Clearly N is a zero sequence (remember that $|y_{d+i}| < 1$ for all $i \geqq 1$ and that $|y_i| \to 0$). Choose $z \in L_1$ and write

$$z = \frac{a}{b},\quad a, b \in L'[y_{d+1}, \ldots],\quad |a| < 1,\quad |b| = 1.$$

For sufficiently large $n \in \mathbb{N}$, we have an equation

$$a = \sum_{0}^{<\infty} a_{\nu_1 \ldots \nu_n} y_{d+1}^{\nu_1} \ldots y_{d+n}^{\nu_n}, \qquad a_{\nu_1 \ldots \nu_n} \in L'.$$

From $|a| < 1$, $|y_{d+i}| < 1$, we deduce that $a_{0 \ldots 0} \in L'_1$. So we may write

$$a_{0 \ldots 0} = \sum_{1}^{\infty} c_j v_j, \qquad c_j \in L'.$$

This yields

$$z = \sum_{1}^{\infty} \frac{c_j}{b} v_j + \sum_{\nu_1 + \cdots + \nu_n > 0}^{<\infty} \frac{a_{\nu_1 \ldots \nu_n}}{b} y_{d+1}^{\nu_1} \cdots y_{d+n}^{\nu_n}.$$

Since all the coefficients are in L, the proposition is proved. □

As a corollary of Propositions 1, 2 and 4, we have the following significant improvement of Corollary 1.7.2/5.

Corollary 5. *Let $\{y_1, y_2, \ldots\}$ be a zero sequence in $\mathring{K}$. Then the smallest complete B-subring of $\mathring{K}$ containing all y_ν, $\nu \geqq 1$, is quasi-Noetherian.*

CHAPTER 2

Normed modules and normed vector spaces

The main subject of this chapter is the theory of normed vector spaces. We begin by discussing some generalities for the wider class of normed modules. Particular attention is paid to the construction of complete tensor products which are necessary for the investigation of affinoid subdomains in (7.2).

The first section dealing with true vector space problems is (2.3). Among other things, we show that normed vector spaces over complete non-Archimedean fields are weakly cartesian. This result implies the uniqueness of valuation extensions in the complete case (see (3.2.4)). Weakly cartesian vector spaces are then specialized to cartesian and strictly cartesian spaces. We discuss questions of orthogonal and orthonormal bases which are fundamental for the theory of stable fields (see (3.6) and (5.3)).

In the subsequent section on weakly cartesian spaces of countable dimension, we prove the existence of weakly cartesian bases (cf. [9] or [29]). This key result sometimes has to be used as a "last resort" in affinoid geometry, namely in the particular case of ground fields k satisfying $p := \operatorname{char} k > 0$ and $[k^{1/p}: k] = \infty$. See, for example, the proof of the Japaneseness of T_n in (5.3.1). Having considered weakly cartesian bases of infinite length, it is natural to look at SCHAUDER bases. This is done in (2.7). The main result of this section is the Lifting Theorem which can be used as point of departure for an alternative approach to affinoid geometry (cf. the introduction to Chapter 1). However, the Lifting Theorem will not be referred to elsewhere in the book.

The chapter is concluded by a short section on Banach spaces.

2.1. Normed and faithfully normed modules

By $A = (A, |\ \ |)$ we always mean a normed ring. All modules $L, M, N, \ldots,$ are A-modules.

2.1.1. Definition. — In classical analysis the notion of a normed vector space is basic. In the following we introduce the corresponding notion for non-Archimedean analysis.

Definition 1. *A pair* $(M, |\ \ |)$ *is called a normed* A*-module if the following hold:*

(i) $(M, |\ \ |)$ *is a normed group (with respect to addition in* M*),*
(ii) $|ax| \leqq |a|\,|x|$ *for all* $a \in A$, $x \in M$.

If, in addition, A *is a valued ring and if in* (ii) *we always have equality, we call* $(M, |\ \ |)$ *a faithfully normed* A*-module. The function* $|\ \ |$ *is called a (faithful)* A*-module norm.*

Remark. If $M \neq 0$ is a normed A-module satisfying the equality $|ax| = |a|\,|x|$ for all $a \in A$ and $x \in M$, then the norm on A automatically is a valuation, because

$$|a_1a_2|\,|x| = |a_1a_2x| = |a_1|\,|a_2x| = |a_1|\,|a_2|\,|x|, \qquad a_1, a_2 \in A, \quad x \in M.$$

Normed modules over normed rings which are not faithfully normed will not often occur in our applications, because most algebras we are interested in will turn out to be finite extensions of valued rings carrying a faithful norm.

Normed A-modules with bounded A-linear maps (resp. contractions) as morphisms form a category. If A is valued, the faithfully normed A-modules with bounded A-linear maps as morphisms form a subcategory which is of great importance. However, this subcategory need not be closed with respect to the operation of forming quotient modules, whereas the bigger category is (cf. (2.1.2)).

Lemma 2. *Each faithfully normed* A*-module* M *is torsion-free; i.e.,* $ax = 0$, $a \in A$, $x \in M - \{0\}$, *if and only if* $a = 0$.

The ring A itself can be viewed as a normed A-module. This module is faithfully normed if the norm on A is a valuation. Up to homeomorphism, there are no other faithfully normed cyclic A-modules; more precisely,

Proposition 3. *If* $M \neq 0$ *is a faithfully normed* A*-module generated by one element* x*, then* $\varphi\colon A \to M$ *defined by* $a \mapsto ax$ *is a bounded homeomorphism, whose inverse is also bounded. Therefore, every surjective* A*-linear map from* A *to* M *is a homeomorphism.*

Proof. The map φ (resp. φ^{-1}) is bounded with $|x|$ (resp. $|x|^{-1}$) as bound:

$$|\varphi(a)| = |x|\,|a|, \qquad |\varphi^{-1}(ax)| = |x|^{-1}\,|ax|. \qquad \square$$

For normed cyclic A-modules, Proposition 3 need not hold, even if the module is torsion-free. We give a simple example. We provide the polynomial ring $A := \mathbb{Z}[X]$ with a valuation $|\ \ |_{\deg} := \alpha^{-\deg}$ induced by the degree function, $0 < \alpha < 1$. Viewed as an A-module, we provide $M := \mathbb{Z}[X]$ with the norm $|\ \ |_{\mathrm{ord}} := \alpha^{\mathrm{ord}}$ induced by the order function, same α as above. Since $\operatorname{ord} a \geqq 0 \geqq -\deg a$, we have

$$|am|_{\mathrm{ord}} = |a|_{\mathrm{ord}}\,|m|_{\mathrm{ord}} \leqq |a|_{\deg}\,|m|_{\mathrm{ord}} \quad \text{for all} \quad a \in A, \quad m \in M.$$

Thus M is a normed A-module. It is clear that M is not faithfully normed and that the identity map $A \to M$ is not a homeomorphism (however it is a contraction).

In important cases A-module norms are always faithful.

Proposition 4. *If A is a valued field, each A-module norm is faithful.*

Proof. For each $a \in A^*$, $x \in M$, we have

$$|x| = |a^{-1}ax| \leqq |a^{-1}|\,|ax| = |a|^{-1}\,|ax|, \quad \text{i.e.,} \quad |ax| \geqq |a|\,|x|. \qquad \square$$

Normed modules over valued fields are called *normed vector spaces.*

In the next sections, we describe some fundamental constructions in the category of (faithfully) normed modules.

2.1.2. Submodules and quotient modules. — Each submodule N of a (faithfully) normed A-module M is a (faithfully) normed A-module with respect to the restricted norm. The *distance function*

$$x \mapsto |x, N| = \inf_{y \in N} |x + y| \quad \text{(as introduced in (1.1.4))}$$

satisfies the inequality

$$|ax, N| \leqq |a|\,|x, N| \quad \text{for all} \quad a \in A, \quad x \in M,$$

since

$$\inf_{y \in N} |ax + y| \leqq \inf_{y \in N} |ax + ay| \leqq \inf_{y \in N} |a|\,|x + y| = |a| \inf_{y \in N} |x + y|.$$

Hence using Proposition 1.1.6/1, we get

Proposition 1. *If N is closed in M, the residue ultrametric function $|\ \ |_{\mathrm{res}}$ on M/N is an A-module norm on M/N. If N is strictly closed in M (cf. 1.1.5), one has $|M/N|_{\mathrm{res}} \subset |M|$.*

Definition 2. *In the situation of the above proposition, the module M/N provided with the norm $|\ \ |_{\mathrm{res}}$ is called the normed quotient module of M by N.*

Warning. If M is faithfully normed, M/N need not be so (due to torsion: take $M = A$, $N \neq 0$ a closed ideal $\neq A$ in A). However, if A is a valued field, no complications arise (cf. Proposition 2.1.1/4).

From Proposition 1.1.7/3 we get

Proposition 3. *If M is complete, M/N is complete.*

2.1.3. Modules of fractions. Completions. — For each normed module M over a valued ring A, we introduce the set

$$S := \{a \in A - \{0\};\ |ax| = |a|\,|x| \text{ for all } x \in M\} \subset A$$

of *M-multiplicative elements.* Then S is a multiplicatively closed set containing 1; if M is faithfully normed, we have $S = A - \{0\}$.

The tensor product

$$M_S := M \otimes_A A_S = \left\{\frac{x}{a};\, x \in M,\, a \in S\right\}$$

is an A_S-module; by defining

$$\left|\frac{x}{a}\right| := \frac{|x|}{|a|},$$

we obviously provide M_S with an A_S-module norm.

If M and M' are normed A-modules and if S is a multiplicative set contained in the set of multiplicative, M-multiplicative, and M'-multiplicative elements of A, then each A-linear map $\varphi\colon M' \to M$ extends uniquely to an A_S-linear map $\varphi_S\colon M'_S \to M_S$ by setting $\varphi_S\left(\frac{x'}{a}\right) := \frac{\varphi(x')}{a}$. The map φ is bounded if and only if φ_S is bounded (with the same bound). For faithfully normed modules, these remarks imply (if we denote by $Q := A_{A-\{0\}}$ the field of fractions of A):

$-\otimes_A Q$ is a covariant functor from the category of faithfully normed A-modules with bounded linear maps (resp. contractions) as morphisms into the category of normed Q-vector spaces with bounded linear maps (resp. contractions) as morphisms.

For each normed A-module M, the completion $\hat{M}$ of M is a well-defined normed group (cf. (1.1.7)). A straightforward verification shows that $\hat{M}$ is a normed A-module and even a normed $\hat{A}$-module. If M is a faithfully normed A-module, $\hat{M}$ is a faithfully normed $\hat{A}$-module. We leave the details to the reader.

2.1.4. Ramification index. — For a given normed A-module M, it is natural to compare the value set $|M - \{0\}| \subset \mathbb{R}_+ - \{0\}$ with the value set $W := |A - \{0\}| \subset \mathbb{R}_+ - \{0\}$ as follows: one calls two elements $r_1, r_2 \in |M - \{0\}|$ "related with respect to W" if there exist elements $w_1, w_2 \in W$ such that $w_1 r_1 = w_2 r_2$. This relation obviously is reflexive and symmetric. In addition, it is transitive if W is multiplicatively closed (which is always the case if M is faithfully normed). Then $|M - \{0\}|$ decomposes into equivalence classes modulo W, and we may define

Definition 1. *Assume that W is multiplicatively closed. The number of equivalence classes of $|M - \{0\}|$ modulo W is called the ramification index of the normed A-module M. We denote this number by $e(M/A)$ or simply by $e(M)$ or e. (If it is not finite, we just write $e = \infty$).*

Remark. The letter e is chosen in analogy to classical valuation theory of field extensions (see (3.6)). If W and $|M - \{0\}|$ are multiplicative subgroups of $\mathbb{R}_+ - \{0\}$ and if $W \subset |M - \{0\}|$, obviously $e(M)$ is nothing more than the order of the residue group $|M - \{0\}|/W$.

For any module M over a commutative ring A, the rank of M (written $\mathrm{rk}_A M$ or just $\mathrm{rk}\, M$) is defined to be the maximal number of A-free elements. If A is an integral domain and if M is torsion-free, we have $\mathrm{rk}_A M = \dim_Q M \otimes_A Q$, where Q is the field of fractions of A.

Now we can prove

Proposition 2. *Let M be a faithfully normed A-module of finite rank. Then $e(M/A)$ is finite and we have* $e(M/A) \leq \mathrm{rk}_A M$.

Proof. Set $n := \mathrm{rk}_A M$. We have to show that among $n + 1$ elements $c_1, \ldots, c_{n+1} \in |M - \{0\}|$ at least two are W-equivalent. Choose $x_\nu \in M$ such that $|x_\nu| = c_\nu$, $\nu = 1, \ldots, n + 1$. By assumption there is a non-trivial linear relation

$$\sum_1^{n+1} a_\nu x_\nu = 0, \qquad a_\nu \in A.$$

From Corollary 1.1.1/5 (the Principle of Domination), we deduce the existence of two different indices i, j such that $|a_i x_i| = |a_j x_j| = \max |a_\nu x_\nu| \neq 0$. Hence we have $|a_i|\, c_i = |a_j|\, c_j$ with $|a_i|, |a_j| \in W$. Thus c_i and c_j are W-equivalent. □

2.1.5. Direct sum. Bounded and restricted direct product. — Let I be an index set, and let $(M_i, |\ \ |_i)_{i \in I}$ be a family of normed (resp. faithfully normed) A-modules. We provide the module $\bigoplus_i M_i$ with the function

$$|x| := \max_i \{|x_i|_i\} \quad \text{if} \quad x = \sum x_i, \quad x_i \in M_i.$$

It is easy to check that $(\bigoplus_i M_i, |\ \ |)$ is a normed (resp. faithfully normed) A-module.

Definition 1. *The pair $(\bigoplus M_i, |\ \ |)$ is called the (normed) direct sum of the modules $(M_i, |\ \ |_i)$. The completion of $\bigoplus M_i$ is denoted by $\widehat{\bigoplus} M_i$.*

The direct product $\prod_i M_i$ cannot be treated the same way because $\sup_i \{|x_i|_i\}$ need not be finite. But

$$b(\prod_i M_i) := \left\{(x_i)_{i \in I} \in \prod_i M_i;\ \sup_i \{|x_i|_i\} < \infty\right\}$$

is obviously an A-submodule of $\prod_i M_i$. On this module we introduce an A-module norm (resp. a faithful A-module norm) by

$$|(x_i)_{i \in I}| := \sup_i \{|x_i|_i\}.$$

Definition 2. *The pair $\left(b(\prod M_i), |\ \ |\right)$ is called the (normed) bounded direct product of the modules $(M_i, |\ \ |_i)$.*

We denote by $c(\prod_i M_i)$ the set of all elements $(x_i)_{i\in I}$ such that x_i converges to zero with respect to the filter of complements of finite subsets of I; i.e., for any $\varepsilon > 0$, we have $|x_i| < \varepsilon$ for almost all $i \in I$ (we simply write $\lim_{i\to\infty} x_i = 0$; note that the set of indices $i \in I$ where $x_i \neq 0$ is at most countable). Then

$$c(\prod_i M_i) = \left\{(x_i)_{i\in I} \in \prod_i M_i;\quad \lim_{i\to\infty} x_i = 0\right\}$$

is a normed A-submodule of $b(\prod_i M_i)$.

Definition 3. *The A-module $c(\prod M_i)$ is called the (normed) restricted direct product of the modules $(M_i, |\ \ |_i)$.*

For each $j \in I$, we have canonical injections

$$M_j \hookrightarrow \bigoplus_i M_i \hookrightarrow c(\prod_i M_i) \hookrightarrow b(\prod_i M_i),$$

which are isometries. Note also that the direct sum coincides with the restricted and the bounded product of the M_i if the index set I is finite.

Proposition 4. *Each module M_j is strictly closed in all modules $\bigoplus M_i$, $c(\prod M_i)$, $b(\prod M_i)$.*

Proof. It is enough to show that M_j is strictly closed in $b(\prod M_i)$. Choose $x = (x_i) \in b(\prod M_i)$ arbitrarily. Then for each $y \in M_j$, we have

$$|x + y| = \max\left\{\sup_{i\neq j} |x_i|, |x_j + y|\right\} \geqq \sup_{i\neq j} |x_i| = |x - x_j|. \qquad \square$$

Proposition 5. *The module $c(\prod M_i)$ is closed in $b(\prod M_i)$. The direct sum $\bigoplus M_i$ is dense in $c(\prod M_i)$.*

Proof. Let $y_\nu = (y_{\nu i})_{i\in I} \in c(\prod M_i)$ be a sequence which converges to an element $y = (y_i)_{i\in I} \in b(\prod M_i)$. In order to show y is an element of $c(\prod M_i)$, let $\varepsilon > 0$ be given. Choose an index μ such that $|y - y_\mu| < \varepsilon$; i.e.,

$$|y_i - y_{\mu i}| < \varepsilon \quad \text{for all} \quad i \in I.$$

We conclude that

$$|y_i| \leqq \max\{|y_i - y_{\mu i}|, |y_{\mu i}|\} \leqq \max\{\varepsilon, |y_{\mu i}|\} \quad \text{for all} \quad i \in I.$$

Since y_μ is an element of $c(\prod M_i)$, we can find a finite set $F \subset I$ such that $|y_{\mu i}| \leqq \varepsilon$ for all $i \notin F$. Thus,

$$|y_i| \leqq \varepsilon \quad \text{for all} \quad i \notin F; \quad \text{i.e.,} \quad y \in c(\prod M_i).$$

Let $z = (z_i)$ be an element of $c(\prod M_i)$; let $\varepsilon > 0$ be given. Choose a finite set $F \subset I$ such that $|z_i| < \varepsilon$ for all $i \notin F$. Define $x := (x_i) \in \bigoplus_i M_i$ by

$$x_i := z_i \quad \text{for} \quad i \in F, \qquad x_i := 0 \quad \text{for} \quad i \notin F.$$

Then

$$|x - z| = \max_i |x_i - z_i| < \varepsilon;$$

i.e., $\bigoplus M_i$ is dense in $c(\prod M_i)$. □

We now generalize Proposition 1.4.1/3.

Proposition 6. *If all M_i are complete, the modules $b(\prod M_i)$ and $c(\prod M_i)$ are complete. In particular, $\bigoplus M_i$ is complete if the index set I is finite.*

Proof. Since $c(\prod M_i)$ is closed in $b(\prod M_i)$, it is enough to show that $b(\prod M_i)$ is complete. Let $y_\nu = (y_{\nu i}) \in b(\prod M_i)$ be a Cauchy sequence. Let $\varepsilon > 0$ be given. Choose ν_0 such that $|y_\mu - y_\nu| \leqq \varepsilon$ for all $\mu, \nu \geqq \nu_0$; i.e.,

$$|y_{\mu i} - y_{\nu i}| \leqq \varepsilon \quad \text{for} \quad \mu, \nu \geqq \nu_0 \quad \text{and} \quad i \in I.$$

Then the sequence $(y_{\nu i}) \subset M_i$, $\nu = 1, 2, \ldots$, is a Cauchy sequence in M_i and hence has a limit $y_i \in M_i$. We claim $y := (y_i) \in b(\prod M_i)$ and $\lim_\nu |y - y_\nu| = 0$. The continuity of $|\ \ |$ implies that $\lim_{\mu \to \infty} |y_{\mu i} - y_{\nu i}| = |y_i - y_{\nu i}| \leqq \varepsilon$ for all $i \in I$, $\nu \geqq \nu_0$. Hence

$$|y - y_\nu| \leqq \varepsilon \quad \text{for all} \quad \nu \geqq \nu_0;$$

i.e.,

$$y = (y - y_\nu) + y_\nu \in b(\prod M_i) \quad \text{and} \quad \lim_{\nu \to \infty} |y - y_\nu| = 0.$$ □

Proposition 7. $c(\prod \hat{M}_i) = \widehat{\bigoplus M_i}$.

Proof. By the preceding proposition, $c(\prod \hat{M}_i)$ is complete. Since $\bigoplus M_i \subset c(\prod M_i) \subset c(\prod \hat{M}_i)$, we see that $\widehat{\bigoplus M_i} \subset c(\prod \hat{M}_i)$. To show the opposite inclusion, we prove that $\bigoplus M_i$ is dense in $c(\prod \hat{M}_i)$. Let $x = (\hat{x}_i)_{i \in I} \in c(\prod \hat{M}_i)$, and let $\varepsilon > 0$ be given. Then $|\hat{x}_i| < \varepsilon$ for $i \in I - F$, where F is a finite subset of I. For each $i \in F$, there exists $x_i \in M_i$ such that $|\hat{x}_i - x_i| < \varepsilon$. If we define $x_i := 0$ for $i \in I - F$ and $x := (x_i)_{i \in I}$, then we have $x \in \bigoplus M_i$ and $|\hat{x} - x| = \sup |\hat{x}_i - x_i| < \varepsilon$. □

2.1.6. The module $\mathscr{L}(L, M)$ of bounded A-linear maps. — If L, M are normed A-modules, we denote by $\mathscr{L}(L, M)$ the A-module of bounded A-linear maps $\varphi: L \to M$. For each such φ, we define the real number

$$|\varphi| := \begin{cases} 0 & \text{if} \quad L = 0 \\ \sup\limits_{x \neq 0} \dfrac{|\varphi(x)|}{|x|} = \text{infimum of all bounds of } \varphi & \text{if} \quad L \neq 0. \end{cases}$$

Then $|\varphi(x)| \leqq |\varphi|\,|x|$ for all $x \in L$.

Proposition 1. *The function $|\ \ |$ is an A-module norm on $\mathscr{L}(L, M)$. It is even a faithful A-module norm, if M is faithfully normed.*

Proof. First, $|\varphi| = 0$ if and only if $\varphi = 0$. Assuming $L \neq 0$, the ultrametric inequality

$$|\varphi + \varphi'| \leqq \max\{|\varphi|, |\varphi'|\}, \qquad \varphi, \varphi' \in \mathscr{L}(L, M),$$

follows easily from the fact that for all $x \in L$ we have

$$|(\varphi + \varphi')(x)| = |\varphi(x) + \varphi'(x)| \leqq \max\{|\varphi(x)|, |\varphi'(x)|\}$$
$$\leqq \max\{|\varphi|\,|x|, |\varphi'|\,|x|\} = |x| \max\{|\varphi|, |\varphi'|\}.$$

Furthermore, we have for $a \in A$

$$|a\varphi| = \sup_{x \neq 0} \frac{|a\varphi(x)|}{|x|} \leqq \sup_{x \neq 0} \left\{|a| \frac{|\varphi(x)|}{|x|}\right\} = |a|\,|\varphi|,$$

and equality holds if M is faithfully normed. □

Definition 2. *The module $\mathscr{L}(L, M)$, provided with the norm just introduced, is called the (normed) A-module of bounded A-linear maps $L \to M$.*

Obviously we have

$$|\varphi| \geqq \sup_{|y| \leqq 1} |\varphi(y)|.$$

In important cases this is an equality (as is always the case in real and complex analysis); e.g.,

Proposition 3. *Let M be a faithfully normed A-module. Assume that $|L|$ lies in the closure of $|A|$. Then if $|A - \{0\}|$ is a group, we have $|\varphi| = \sup_{|y| \leqq 1} |\varphi(y)|$ for all $\varphi \in \mathscr{L}(L, M)$.*

Proof. Set $r := \sup_{|y| \leqq 1} |\varphi(y)|$. It is enough to show

$$|\varphi(x)| \leqq r\,|x| \quad \text{for all} \quad x \in L, \quad x \neq 0.$$

By assumption for each such $x \in L$, there exists a sequence $a_\nu \in A - \{0\}$ such that $|a_\nu|^{-1}$ converges to $|x|$. Since $|A - \{0\}|$ is either discrete or dense in $\mathbb{R}_+ - \{0\}$ we can suppose $|a_\nu|^{-1} \geqq |x|$. Then we have $|a_\nu x| \leqq |a_\nu|\,|x| \leqq 1$ and hence

$$|\varphi(a_\nu x)| \leqq r.$$

Since M is faithfully normed, this implies

$$|\varphi(x)| \leqq r\,|a_\nu|^{-1} \quad \text{for all} \quad \nu = 1, 2, \ldots.$$

Thus $|\varphi(x)| \leqq r\,|x|$ in the limit. □

Proposition 4. *If M is complete, $\mathscr{L}(L, M)$ is complete.*

Proof. We may assume $L \neq 0$. Let $\varphi_\nu \in \mathscr{L}(L, M)$ be a Cauchy sequence, i.e., $|\varphi_\mu - \varphi_\nu| \leqq \varepsilon$ for $\mu, \nu \geqq n_0(\varepsilon)$, $\varepsilon > 0$ given arbitrarily. From

$$(*) \qquad |\varphi_\mu(x) - \varphi_\nu(x)| \leqq |\varphi_\mu - \varphi_\nu| \cdot |x| \leqq \varepsilon\,|x|, \qquad \mu, \nu \geqq n_0(\varepsilon), \quad x \in L,$$

we conclude that $\varphi_\nu(x)$ is a Cauchy sequence in M and hence (by assumption) has a well-defined limit. Define the map $\varphi: L \to M$ by

$$\varphi(x) := \lim_\nu \varphi_\nu(x), \qquad x \in L.$$

By reasons of continuity, φ is A-linear. From (*) we get (due to continuity of the norm)

$$|\varphi(x) - \varphi_\nu(x)| = \lim_\mu |\varphi_\mu(x) - \varphi_\nu(x)| \leqq \varepsilon\,|x|\,, \quad \nu \geqq n_0(\varepsilon)\,, \quad x \in L\,.$$

Hence $\varphi - \varphi_\nu$ and therefore φ are elements of $\mathscr{L}(L, M)$. Moreover we see

$$|\varphi - \varphi_\nu| = \sup_{x \neq 0} \left\{\frac{|\varphi(x) - \varphi_\nu(x)|}{|x|}\right\} \leqq \sup_{x \neq 0} \left\{\frac{\varepsilon\,|x|}{|x|}\right\} = \varepsilon\,, \quad \nu \geqq n_0(\varepsilon)\,.$$

Therefore φ is the limit of the sequence φ_ν. □

If N is a third normed A-module, the composition of homomorphisms yields a map $\mathscr{L}(L, M) \times \mathscr{L}(M, N) \to \mathscr{L}(L, N)$, $(\varphi, \psi) \mapsto \psi \circ \varphi$. Namely, we have

$$|\psi \circ \varphi(x)| \leqq |\psi|\,|\varphi(x)| \leqq |\psi|\,|\varphi|\,|x|\,, \qquad x \in L\,,$$

and hence $|\psi \circ \varphi| \leqq |\psi|\,|\varphi|$ for all $\psi \in \mathscr{L}(M, N)$, $\varphi \in \mathscr{L}(L, M)$. In particular, we see

The set $\mathscr{L}(M, M)$ is a normed "ring" (in general, not commutative).

For each $\varphi \in \operatorname{Aut} M$ (i.e, the group of units in $\mathscr{L}(M, M)$), we have

$$1 = |\varphi \circ \varphi^{-1}| \leqq |\varphi|\,|\varphi^{-1}|\,.$$

Therefore

Each contraction $\varphi \in \operatorname{Aut} M$ whose inverse is also a contraction is an isometry.

2.1.7. Complete tensor products. — As before let L and M denote normed A-modules. Considering the (ordinary) tensor product $L \otimes_A M$, we define a function $|\ \ |: L \otimes_A M \to \mathbb{R}_+$ in the following way. For $g \in L \otimes_A M$ let

$$|g| := \inf \left(\max_{1 \leqq i \leqq r} |x_i|\,|y_i|\right),$$

where the infimum runs over all possible representations

$$g = \sum_{i=1}^{r} x_i \otimes y_i\,, \qquad x_i \in L\,, \qquad y_i \in M\,,$$

of g. It is easily verified that $|\ \ |$ is an ultrametric function on the additive group of $L \otimes_A M$, which, in addition, satisfies condition (ii) of Definition 2.1.1/1. Hence $L \otimes_A M$, together with $|\ \ |$, could be called a semi-normed A-module. Applying Proposition 1.1.7/5, we construct the completion of $L \otimes_A M$ (as a semi-normed group), and it is not hard to see that the resulting normed group is canonically a normed A-module. We call this complete normed A-module the *complete tensor product of L and M over A* and use the notation $L \mathbin{\widehat{\otimes}}_A M$. Note that $L \mathbin{\widehat{\otimes}}_A M$ is also a normed $\hat{A}$-module with $\hat{A}$ denoting the completion of A (cf. the last paragraph of (2.1.3)).

The bilinear map $\tau'\colon L \times M \to L \otimes_A M$, $(x, y) \mapsto x \otimes y$, induces by composition with the canonical map $L \otimes_A M \to L \mathbin{\widehat{\otimes}}_A M$ a bilinear map $\tau\colon L \times M \to L \mathbin{\widehat{\otimes}}_A M$, $(x, y) \mapsto x \mathbin{\widehat{\otimes}} y$, where $x \mathbin{\widehat{\otimes}} y$ is defined as the image of $x \otimes y$ in $L \mathbin{\widehat{\otimes}}_A M$. Calling a bilinear map $\Phi\colon L \times M \to N$ into a normed (or semi-normed) A-module N ***bounded*** if there exists a constant $\varrho > 0$ such that $|\Phi(x, y)| \leqq \varrho\, |x|\, |y|$ for all $x \in L$, $y \in M$, it follows immediately from our definitions that the maps τ and τ' are bounded by $\varrho = 1$. Note that also in the case of bilinear maps boundedness implies continuity. We want to show that the bilinear map $\tau\colon L \times M \to L \mathbin{\widehat{\otimes}}_A M$ factors bounded bilinear maps from $L \times M$ into complete normed A-modules. (See (2.1.8) for situations, in which boundedness is equivalent to continuity.)

Proposition 1. *The bilinear map $\tau\colon L \times M \to L \mathbin{\widehat{\otimes}}_A M$ satisfies the following properties.*

(i) *Let $\Phi\colon L \times M \to N$ be a bounded bilinear map into a complete normed A-module N. Then there is a unique bounded A-linear map $\varphi\colon L \mathbin{\widehat{\otimes}}_A M \to N$ such that $\Phi = \varphi \circ \tau$.*

(ii) *If, in the situation of* (i), *the map Φ is bounded by $\varrho > 0$, then also φ is bounded by ϱ so that $|\varphi| \leqq \varrho$.*

Proof. Due to the universality of the (ordinary) tensor product, there is a unique A-linear map $\varphi'\colon L \otimes_A M \to N$ rendering commutative the diagram

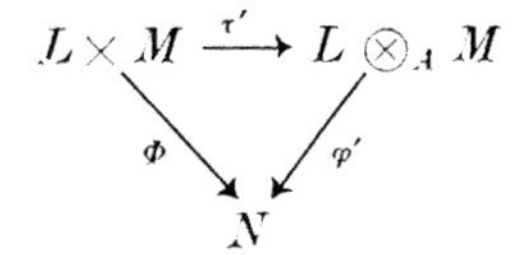

Let $\varrho > 0$ be a bound for Φ; we want to show that also φ' is bounded by ϱ. For an arbitrary element $g = \sum_{i=1}^{r} x_i \otimes y_i \in L \otimes_A M$, we have $\varphi'(g) = \sum_{i=1}^{r} \Phi(x_i, y_i)$, and therefore

$$|\varphi'(g)| \leqq \max_{1 \leqq i \leqq r} |\Phi(x_i, y_i)| \leqq \varrho \max_{1 \leqq i \leqq r} |x_i|\, |y_i|.$$

Then taking the infimum on the right-hand side over all possible representations of g, we get $|\varphi'(g)| \leqq \varrho\, |g|$; hence ϱ is a bound for φ'.

Now we can apply Proposition 1.1.7/6 to the map φ' and hence get a homomorphism $\varphi\colon L \mathbin{\widehat{\otimes}}_A M \to N$ such that the diagram

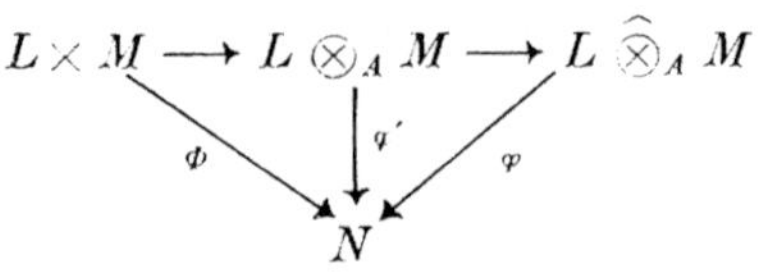

commutes. The map φ is bounded by ϱ and is easily checked to be A-linear. Thus it remains only to be shown that φ is the unique bounded A-linear map making the big triangle commutative. If $g = \sum_{i=1}^{r} x_i \mathbin{\widehat{\otimes}} y_i \in L \mathbin{\widehat{\otimes}}_A M$ is an element in the image of the map $L \otimes_A M \to L \mathbin{\widehat{\otimes}}_A M$, then due to the linearity of φ we must have $\varphi(g) = \sum_{i=1}^{r} \Phi(x_i, y_i)$. Hence φ is uniquely determined on a dense submodule of $L \mathbin{\widehat{\otimes}}_A M$, which means by Proposition 1.1.7/6 that φ must be unique. □

The A-module of bounded bilinear maps $\Phi: L \times M \to N$ can be interpreted as the A-module of bounded linear maps $\mathscr{L}(L, \mathscr{L}(M, N))$ for an arbitrary normed A-module N. This goes in the usual way by identifying a bounded bilinear map Φ with the linear map $L \to \mathscr{L}(M, N)$, $x \mapsto \Phi_x$, where Φ_x denotes the linear map $M \to N$, $y \mapsto \Phi(x, y)$. It is not hard to see that the norm of Φ as an element of $\mathscr{L}(L, \mathscr{L}(M, N))$ equals the infimum of all (positive) bounds for Φ as a bilinear map. Keeping this in mind, we define an A-module homomorphism

$$\iota: \mathscr{L}(L \mathbin{\widehat{\otimes}}_A M, N) \to \mathscr{L}(L, \mathscr{L}(M, N))$$

$$\varphi \mapsto \varphi \circ \tau$$

by composing bounded linear maps $\varphi: L \mathbin{\widehat{\otimes}}_A M \to N$ with the bilinear map $\tau: L \times M \to L \mathbin{\widehat{\otimes}}_A M$. As a consequence, we get from Proposition 1.

Corollary 2. *If N is a complete normed A-module, then $\iota: \mathscr{L}(L \mathbin{\widehat{\otimes}}_A M, N) \to \mathscr{L}(L, \mathscr{L}(M, N))$ is an isometric isomorphism of normed A-modules.*

Proof. The map ι is contractive, since τ admits 1 as a bound. Furthermore, ι is bijective due to assertion (i) of Proposition 1 and ι^{-1} is contractive due to assertion (ii) of Proposition 1. Hence ι must be an isometric isomorphism. □

For many applications it is useful to know that the complete tensor product $L \mathbin{\widehat{\otimes}}_A M$ is characterized by the properties of the bilinear map $\tau: L \times M \to L \mathbin{\widehat{\otimes}}_A M$ mentioned in Proposition 1. We state this explicitly.

Corollary 3. *Let P denote a complete normed A-module, and assume that $\sigma: L \times M \to P$ is a bounded bilinear map satisfying assertion* (i) *of Proposition* 1 *with P substituting for $L \mathbin{\widehat{\otimes}}_A M$ and σ substituting for τ. In particular, denote by $\psi: P \to L \mathbin{\widehat{\otimes}}_A M$ and $\psi': L \mathbin{\widehat{\otimes}}_A M \to P$ the unique bounded linear maps such that $\tau = \psi \circ \sigma$ and $\sigma = \psi' \circ \tau$. Then ψ and ψ' are inverse to each other; hence they are isomorphisms which are bounded in both directions. If additionally σ is bounded by* 1 *and satisfies assertion* (ii) *of Proposition* 1, *then ψ and ψ' are isometries.*

Proof. We have

$$\mathrm{id}_P \circ \sigma = \psi' \circ \tau = \psi' \circ \psi \circ \sigma \quad \text{and}$$

$$\mathrm{id}_{L \hat{\otimes}_A M} \circ \tau = \psi \circ \sigma = \psi \circ \psi' \circ \tau.$$

Hence the uniqueness assertion of (i) in Proposition 1 implies

$$\mathrm{id}_P = \psi' \circ \psi, \qquad \mathrm{id}_{L \hat{\otimes}_A M} = \psi \circ \psi',$$

and the first part of the corollary is clear. If σ is bounded by 1 and satifies assertion (ii) of Proposition 1, it follows that ψ and ψ' are contractive; thus they must be isometries. □

Just as for ordinary tensor products, one can derive various properties, including canonical isomorphisms, for complete tensor products. We state some of them below.

Proposition 4. *Let $i\colon L \to \hat{L}$ and $j\colon M \to \hat{M}$ denote the injections of the normed A-modules L and M into their completions. Then there is a unique isometric isomorphism $L \mathbin{\widehat{\otimes}}_A M \to \hat{L} \mathbin{\widehat{\otimes}}_{\hat{A}} \hat{M}$ such that $x \mathbin{\widehat{\otimes}} y \mapsto i(x) \mathbin{\widehat{\otimes}} j(y)$.*

Proof. The canonical bilinear map $L \times M \to L \mathbin{\widehat{\otimes}}_A M$ extends to a bilinear map $\hat{L} \times \hat{M} \to L \mathbin{\widehat{\otimes}}_A M$, which is obviously $\hat{A}$-bilinear, bounded by 1, and which furthermore satisfies assertions (i) and (ii) of Proposition 1. Thus we are done by Corollary 3. □

Proposition 5. *Let $\psi_i\colon L_i \to M_i$, $i = 1, 2$, be bounded A-linear maps between normed A-modules. Then there is a unique bounded A-linear map $L_1 \mathbin{\widehat{\otimes}}_A L_2 \to M_1 \mathbin{\widehat{\otimes}}_A M_2$, denoted by $\psi_1 \mathbin{\widehat{\otimes}} \psi_2$, such that $x \mathbin{\widehat{\otimes}} y \mapsto \psi_1(x) \mathbin{\widehat{\otimes}} \psi_2(y)$. Furthermore $|\psi_1 \mathbin{\widehat{\otimes}} \psi_2| \leqq |\psi_1|\,|\psi_2|$.*

Proof. The bilinear map $\Phi\colon L_1 \times L_2 \xrightarrow{\psi_1 \times \psi_2} M_1 \times M_2 \to M_1 \mathbin{\widehat{\otimes}}_A M_2$ is bounded by $|\psi_1|\,|\psi_2|$. Hence according to Proposition 1, the map Φ equals the composition of the canonical bilinear map $L_1 \times L_2 \to L_1 \mathbin{\widehat{\otimes}}_A L_2$ and of a unique A-linear map $L_1 \mathbin{\widehat{\otimes}}_A L_2 \to M_1 \mathbin{\widehat{\otimes}}_A M_2$ which is bounded by $|\psi_1|\,|\psi_2|$. □

Note that in the situation of Proposition 5 we have, in particular, a commutative diagram

$$\begin{array}{ccc}
L_1 \otimes_A L_2 & \xrightarrow{\psi_1 \otimes \psi_2} & M_1 \otimes_A M_2 \\
\downarrow & & \downarrow \\
L_1 \mathbin{\widehat{\otimes}}_A L_2 & \xrightarrow{\psi_1 \widehat{\otimes} \psi_2} & M_1 \mathbin{\widehat{\otimes}}_A M_2
\end{array}$$

of bounded maps; hence $\psi_1 \mathbin{\widehat{\otimes}} \psi_2$ may also be interpreted as the unique extension of $\psi_1 \otimes \psi_2$ in the sense of Proposition 1.1.7/6.

Proposition 6. *Let L, M, N be normed A-modules. Then there are isometric isomorphisms*

(i) $A \mathbin{\widehat{\otimes}}_A L \to \hat{L}$ *where $\hat{L}$ denotes the completion of L,*

(ii) $L \mathbin{\widehat{\otimes}}_A M \to M \mathbin{\widehat{\otimes}}_A L$,

(iii) $(L \mathbin{\widehat{\otimes}}_A M) \mathbin{\widehat{\otimes}}_A N \to L \mathbin{\widehat{\otimes}}_A (M \mathbin{\widehat{\otimes}}_A N)$,

(iv) $(L \oplus M) \mathbin{\widehat{\otimes}}_A N \to (L \mathbin{\widehat{\otimes}}_A N) \oplus (M \mathbin{\widehat{\otimes}}_A N)$

which are uniquely determined by requiring that, respectively,

$$a \mathbin{\widehat{\otimes}} x \mapsto ax,$$

$$x \mathbin{\widehat{\otimes}} y \mapsto y \mathbin{\widehat{\otimes}} x,$$

$$(x \mathbin{\widehat{\otimes}} y) \mathbin{\widehat{\otimes}} z \mapsto x \mathbin{\widehat{\otimes}} (y \mathbin{\widehat{\otimes}} z),$$

$$(x, y) \mathbin{\widehat{\otimes}} z \mapsto (x \mathbin{\widehat{\otimes}} z, y \mathbin{\widehat{\otimes}} z).$$

The *proof* goes exactly the same way as in the ordinary tensor product case; one has to rely on the universal property of the complete tensor product and eventually has to apply Corollary 3. We leave the details to the reader.

Now let $\sigma\colon A \to B$ be a contractive homomorphism of normed rings. Then B is a normed A-module via σ; hence, for any normed A-module L, the complete tensor product $L \mathbin{\widehat{\otimes}}_A B$ is defined and carries the structure of a normed B-module such that $b(x \mathbin{\widehat{\otimes}} b') = x \mathbin{\widehat{\otimes}} bb'$. We say that $L \mathbin{\widehat{\otimes}}_A B$ is derived from L by "extension of scalars".

Furthermore, if M is an arbitrary normed B-module, we get by "restriction of scalars" on M the structure of a normed A-module. In particular, the complete tensor product $L \mathbin{\widehat{\otimes}}_A M$ is defined, and just as above, $L \mathbin{\widehat{\otimes}}_A M$ is a normed B-module. With these preparations we can state the following generalized version of the associativity formula for complete tensor products:

Proposition 7. *Let $\sigma\colon A \to B$ be a contractive homomorphism of normed rings, and let L be a normed A-module, M, N be normed B-modules. Then there is a unique isometric isomorphism*

$$(L \mathbin{\widehat{\otimes}}_A M) \mathbin{\widehat{\otimes}}_B N \to L \mathbin{\widehat{\otimes}}_A (M \mathbin{\widehat{\otimes}}_B N)$$

such that $(x \mathbin{\widehat{\otimes}} y) \mathbin{\widehat{\otimes}} z \mapsto x \mathbin{\widehat{\otimes}} (y \mathbin{\widehat{\otimes}} z)$.

Proof. For an arbitrary $z \in N$, we consider the B-linear map

$$\psi_z\colon M \to M \mathbin{\widehat{\otimes}}_B N$$

$$y \mapsto y \mathbin{\widehat{\otimes}} z$$

which satisfies $|\psi_z| \leqq |z|$. Tensoring over A with the identical map $\mathrm{id}\colon L \to L$ yields by Proposition 5 an A-linear map

$$\mathrm{id} \mathbin{\widehat{\otimes}} \psi_z\colon L \mathbin{\widehat{\otimes}}_A M \to L \mathbin{\widehat{\otimes}}_A (M \mathbin{\widehat{\otimes}}_B N)$$

$$x \mathbin{\widehat{\otimes}} y \mapsto x \mathbin{\widehat{\otimes}} (y \mathbin{\widehat{\otimes}} z)$$

with $|\mathrm{id} \mathbin{\widehat{\otimes}} \psi_z| \leqq |z|$, which is obviously also B-linear. Since $\mathrm{id} \mathbin{\widehat{\otimes}} \psi_z$ depends linearly on z, we get a B-linear map

$$(L \mathbin{\widehat{\otimes}}_A M) \times N \to L \mathbin{\widehat{\otimes}}_A (M \mathbin{\widehat{\otimes}}_B N)$$

$$(x \mathbin{\widehat{\otimes}} y, z) \mapsto x \mathbin{\widehat{\otimes}} (y \mathbin{\widehat{\otimes}} z),$$

which is bounded by 1, and hence a B-linear map

$$\varphi\colon (L \mathbin{\widehat{\otimes}_A} M) \mathbin{\widehat{\otimes}_B} N \to L \mathbin{\widehat{\otimes}_A} (M \mathbin{\widehat{\otimes}_B} N)$$

$$(x \mathbin{\widehat{\otimes}} y) \mathbin{\widehat{\otimes}} z \mapsto x \mathbin{\widehat{\otimes}} (y \mathbin{\widehat{\otimes}} z)$$

with $|\varphi| \leqq 1$. Similarly one constructs a contractive B-linear map

$$\varphi'\colon L \mathbin{\widehat{\otimes}_A} (M \mathbin{\widehat{\otimes}_B} N) \to (L \mathbin{\widehat{\otimes}_A} M) \mathbin{\widehat{\otimes}_B} N$$

$$x \mathbin{\widehat{\otimes}} (y \mathbin{\widehat{\otimes}} z) \mapsto (x \mathbin{\widehat{\otimes}} y) \mathbin{\widehat{\otimes}} z .$$

Then it is clear that the contractive maps φ and φ' are inverse to each other; hence they must be isometric isomorphisms. □

We conclude by the following generalized version of assertion (iv) in Proposition 6.

Proposition 8. *Let $(L_i)_{i\in I}$ be a family of normed A-modules. Then for any normed A-module M, there are unique isometric isomorphisms*

$$\Big(\bigoplus_{i\in I} L_i\Big) \mathbin{\widehat{\otimes}_A} M \to c\Big(\prod_{i\in I} (L_i \mathbin{\widehat{\otimes}_A} M)\Big) \qquad \text{and}$$

$$c\Big(\prod_{i\in I} L_i\Big) \mathbin{\widehat{\otimes}_A} M \to c\Big(\prod_{i\in I} (L_i \mathbin{\widehat{\otimes}_A} M)\Big)$$

such that

$$(x_i)_{i\in I} \mathbin{\widehat{\otimes}} y \mapsto (x_i \mathbin{\widehat{\otimes}} y)_{i\in I} .$$

Proof. We begin by showing that the bilinear map

$$\sigma\colon \Big(\bigoplus_{i\in I} L_i\Big) \times M \to c\Big(\prod_{i\in I} (L_i \mathbin{\widehat{\otimes}_A} M)\Big)$$

$$\big((x_i)_{i\in I}, y\big) \mapsto (x_i \mathbin{\widehat{\otimes}} y)_{i\in I} ,$$

which is bounded by 1, satisfies assertions (i) and (ii) of Proposition 1. For this purpose let $\Phi\colon (\bigoplus L_i) \times M \to N$ be a bilinear map, bounded by $\varrho > 0$, into a complete normed A-module N. Then also the restrictions $\Phi_i\colon L_i \times M \to N$ are bounded by ϱ. Denoting by $\sigma_i\colon L_i \times M \to L_i \mathbin{\widehat{\otimes}_A} M$ the canonical bilinear map, we get according to Proposition 1 for each $i \in I$ a unique bounded A-linear map $\varphi_i\colon L_i \mathbin{\widehat{\otimes}_A} M \to N$ such that $\Phi_i = \varphi_i \circ \sigma_i$ and such that $|\varphi_i| \leqq \varrho$. The φ_i yield an A-linear map

$$\varphi\colon c\Big(\prod (L_i \mathbin{\widehat{\otimes}_A} M)\Big) \to N$$

$$(g_i)_{i\in I} \mapsto \sum_{i\in I} \varphi_i(g_i)$$

which is bounded by ϱ and obviously satisfies $\Phi = \varphi \circ \sigma$. Furthermore, it is easy to see that φ is unique since all φ_i are unique. Thus according to Corollary 3, the map

$$(\bigoplus_{i \in I} L_i) \hat{\otimes}_A M \to c\big(\prod_{i \in I} (L_i \hat{\otimes}_A M)\big)$$

induced by σ is an isometric isomorphism. This settles the first part of our proposition. The second one can be verified in literally the same way with $\bigoplus_{i \in I} L_i$ replaced by its completion $c(\prod_{i \in I} L_i)$ or more simply by relying on Proposition 4. □

2.1.8. Continuity and boundedness. — Each bounded A-module homomorphism $\varphi: L \to M$ is continuous. To prove a converse, we need the following observation (which is a substitute for the trivial fact that in a real (resp. complex) normed vector space each vector $\neq 0$ can be "normed to length 1 by multiplying with a scalar").

Proposition 1. *Let the valuation of A be non-degenerate; let a, b be elements of A such that $0 < |a| < 1 < |b|$. Set $\varrho := \max \{|a|^{-1}, |b|\}$.*

Let M be a faithfully normed A-module. Then for each $x \in M$, $x \neq 0$, there exists a scalar $c \in A$ such that $1 \leqq |cx| < \varrho$.

Proof. If $|x| \geqq 1$, choose $n \in \mathbb{N} \cup \{0\}$ such that $|a|^{-n} \leqq |x| < |a|^{-(n+1)}$, and define $c := a^n$. If $0 < |x| < 1$, take $n \in \mathbb{N} \cup \{0\}$ such that $|b|^{-(n+1)} \leqq |x| < |b|^{-n}$, and define $c := b^{n+1}$. In both cases we get $1 \leqq |cx| < \varrho$. □

Proposition 2. *Assume that the valuation of A is non-degenerate, and let L, M, N be faithfully normed A-modules. Then for any A-linear map $\varphi: L \to M$ and any A-bilinear map $\Phi: L \times M \to N$, continuity is equivalent to boundedness.*

Proof. It is only necessary to show that continuity implies boundedness. Assume that $\varphi: L \to M$ is not bounded. Then there exists a sequence $x_n \in L$ such that $|\varphi(x_n)| > |a|^{-n} |x_n|$ where $a \in A$, $0 < |a| < 1$. Due to the preceding proposition, we may assume $1 \leqq |x_n| < \varrho$. Set $x_n' := a^n x_n$. Then $|\varphi(x_n')| = |a^n| \, |\varphi(x_n)| > |x_n| \geqq 1$, but

$$|x_n'| = |a|^n \, |x_n| < |a|^n \varrho \to 0;$$

i.e., φ is not continuous.

If $\Phi: L \times M \to N$ is not bounded, there exist sequences $x_n \in L$ and $y_n \in M$ such that $|\Phi(x_n, y_n)| > |a|^{-2n} |x_n| \, |y_n|$, and a computation similar to the one above shows that Φ cannot be continuous. □

Corollary 3. *Let A be a normed ring which is an algebra over a field $k \subset A$. Assume that the norm of A induces on k a non-trivial valuation. Then for all A-linear maps between normed A-modules, continuity is equivalent to boundedness. The same holds for A-bilinear maps between normed A-modules.*

Proof. By restriction of scalars, we can view A-modules as normed k-modules which are even faithfully normed by Proposition 2.1.1/4. Since A-linearity implies, in particular, k-linearity we are done by Proposition 2. □

Hence if Proposition 2 or Corollary 3 is applicable, the space $\mathscr{L}(L, M)$ coincides with the space of continuous A-linear maps $L \to M$. Likewise, if A fulfills the condition of Corollary 3, the complete tensor product $L \mathbin{\widehat{\otimes}}_A M$ of the normed A-modules L and M is also characterized by the following universal property: For any continuous bilinear map $\Phi\colon L \times M \to N$ into a complete normed A-module N, there exists a unique continuous A-linear map $\varphi\colon L \mathbin{\widehat{\otimes}}_A M \to N$ such that

$$\begin{array}{ccc} L \times M & \xrightarrow{\tau} & L \mathbin{\widehat{\otimes}}_A M \\ & {\scriptstyle \Phi} \searrow \quad \swarrow {\scriptstyle \varphi} & \\ & N & \end{array}$$

commutes.

Two A-module norms $|\ \ |$ and $|\ \ |'$ on an A-module N are called *equivalent* if they induce the same topology on N or, in other words, if the identity maps

$$(N, |\ \ |) \to (N, |\ \ |'), \qquad (N, |\ \ |') \to (N, |\ \ |)$$

are continuous. As a consequence, we derive immediately from Proposition 2 and Corollary 3

Corollary 4. *Let $|\ \ |, |\ \ |'$ be A-module norms on an A-module N, and assume that one of the following conditions is satisfied.*

(i) *The norm of A is a non-degenerate valuation, and $|\ \ |, |\ \ |'$ are faithful A-module norms.*

(ii) *A contains a field k, and the norm of A induces on k a non-trivial valuation.*

Then $|\ \ |, |\ \ |'$ are equivalent norms on N if and only if there are constants $\varrho, \varrho' > 0$ such that

$$|\ \ |' \leq \varrho |\ \ | \leq \varrho' |\ \ |'.$$

It is natural to ask how the complete tensor product $L \mathbin{\widehat{\otimes}}_A M$ changes when the norms on L and M are replaced by equivalent ones. We can give an answer in the special case where Corollary 3 is applicable.

Proposition 5. *Let $\psi_j\colon L_j \to M_j$, $j = 1, 2$, be linear homeomorphisms of normed A-modules, and, as in Corollary 3, assume that A contains a field with a non-trivial valuation. Then the map $\psi_1 \otimes \psi_2\colon L_1 \otimes_A L_2 \to M_1 \otimes_A M_2$, as well as $\psi_1 \mathbin{\widehat{\otimes}} \psi_2\colon L_1 \mathbin{\widehat{\otimes}}_A L_2 \to M_1 \mathbin{\widehat{\otimes}}_A M_2$, is a homeomorphism.*

Proof. Due to Corollary 3 the maps ψ_1, ψ_2 are isomorphisms which are bounded in both directions. Then by Proposition 2.1.7/5 and the remark following it, the maps $\psi_1 \otimes \psi_2$ and $\psi_1^{-1} \otimes \psi_2^{-1}$, as well as their "completions" $\psi_1 \mathbin{\widehat{\otimes}} \psi_2$ and $\psi_1^{-1} \mathbin{\widehat{\otimes}} \psi_2^{-1}$, are bounded and inverse to each other. □

The assertion of Proposition 5 is, of course, also true if Proposition 2 is applicable — i.e., when L_j, M_j, $j = 1, 2$, are faithfully normed A-modules and the norm of A is a non-degenerate valuation. However, here one is restricted to faithful A-module norms, whereas under the assumption of Proposition 5, one concludes that any equivalent A-module norms on A-modules L_1 and L_2 lead to equivalent norms on $L_1 \hat{\otimes}_A L_2$. We give an application of this fact, which is a generalization of Proposition 5.

Proposition 6. *Let $\psi_j\colon L_j \to M_j$, $j = 1, 2$, be strict epimorphisms of normed A-modules, and assume that A contains a field with a non-trivial valuation. Then also $\psi_1 \otimes \psi_2$ and $\psi_1 \hat{\otimes} \psi_2$ are strict epimorphisms.*

Proof. According to Proposition 5, we may replace the norm on M_j by an equivalent one; hence, in particular, we may assume that M_j carries the residue norm via ψ_j, $j = 1, 2$. Applying Corollary 1.1.9/6, we only have to show that $\psi_1 \otimes \psi_2$, which is clearly surjective, is also strict. More specifically, we will see that $M_1 \otimes_A M_2$ carries the residue norm via $\psi_1 \otimes \psi_2$. Consider an element $g \in M_1 \otimes_A M_2$, and let δ be a real number with $|g| < \delta$. Then there is a representation $g = \sum_{i=1}^{r} x_i \otimes y_i$ with $x_i \in M_1$, $y_i \in M_2$ such that $\max_{1\leqq i\leqq r} |x_i|\,|y_i| < \delta$. Furthermore, the elements x_i and y_i admit inverse images $x_i' \in L_1$ and $y_i' \in L_2$, $i = 1, \ldots, r$, such that $\max_{1\leqq i\leqq r} |x_i'|\,|y_i'| < \delta$. The element

$$g' := \sum_{i=1}^{r} x_i' \otimes y_i' \in L_1 \otimes_A L_2$$

is mapped onto g by $\psi_1 \otimes \psi_2$ and satisfies $|g'| < \delta$. Furthermore by Proposition 2.1.7/5, the map $\psi_1 \hat{\otimes} \psi_2$ is contractive, and hence so is $\psi_1 \otimes \psi_2$. Thus, we have

$$|g| \leqq |g'| < \delta .$$

Since this holds for all $\delta > |g|$, it follows that $M_1 \otimes_A M_2$ carries the residue norm via $\psi_1 \otimes \psi_2$. □

We give an *example* which shows that, for rings A with degenerate valuations, Proposition 2 and Corollary 4 are not true in general.

Let $M_i := Ae_i$, $i = 1, 2, \ldots$, be a sequence of free cyclic A-modules and set $M := \bigoplus_{i=1}^{\infty} M_i$. Then, for each $\nu \in \mathbb{Z}$, we can define an A-module norm $|\ \ |_\nu$ on M by setting

$$\left|\sum_{1}^{n} a_i e_i\right|_\nu = \max_i \{i^{-\nu}|a_i|\}, \qquad a_i \in A .$$

Obviously $|\ \ |_\nu \leqq |\ \ |_{\nu-1}$; i.e., each identity map $(M, |\ \ |_{\nu-1}) \to (M, |\ \ |_\nu)$ is a contraction. The inverse map $\mathrm{id}_\nu\colon (M, |\ \ |_\nu) \to (M, |\ \ |_{\nu-1})$ is *not bounded*, because

$$|e_i|_{\nu-1} = i^{-(\nu-1)} = i\,|e_i|_\nu , \qquad i = 1, 2, \ldots .$$

However, id_ν *is continuous if the valuation of A is bounded and $\nu \geqq 2$ or if $|a| \geqq 1$ for all $a \in A - \{0\}$ and $\nu \leqq 0$.* To see this, first we consider the case where $|A|$ is bounded and where $\nu \geqq 2$. Let $\varepsilon > 0$ be given. Set $\delta := \varepsilon^\nu$, and choose $x = \sum_1^n a_i e_i \in M$ such that $|x|_\nu \leqq \delta$, i.e., $|a_i| \leqq i^\nu \varepsilon^\nu$ for $i = 1, \ldots, n$. Since $|a_i| \leqq 1$, we have $|a_i| \leqq |a_i|^{1/\nu}$, and hence

$$|a_i| \leqq i\varepsilon \leqq i^{\nu-1}\varepsilon; \quad \text{i.e.,} \quad |x|_{\nu-1} \leqq \varepsilon \quad \text{for} \quad \nu \geqq 2.$$

Next we look at the second case: $|a| \geqq 1$ for all $a \in A - \{0\}$ and $\nu \leqq 0$. Then one has $i^{-\nu} \geqq 1$ for all $i \geqq 1$. Therefore, $|x|_\nu \geqq 1$ for all $x \in M - \{0\}$. Hence $|\ \ |_\nu$ induces the discrete topology on M, and every map from $(M, |\ \ |_\nu)$ is continuous.

2.1.9. Density condition. — We give the following

Definition 1. *A normed A-module M is said to fulfill the density condition if $|M|$ is contained in the closure of $|A|$ in $\mathbb{R}_+$:*

$$|M| \subset \overline{|A|}.$$

Note that this condition, which already occurred in Proposition 2.1.6/3, is automatically fulfilled if $|A|$ is dense in $\mathbb{R}_+ - \{0\}$. Thus if $|A - \{0\}|$ is a group, the only interesting case is when $|A - \{0\}|$ is discrete in $\mathbb{R}_+ - \{0\}$.

One easily proves

Proposition 2. *If the normed A-modules L, M_i, where i varies in some index set I, fulfill the density condition, so do the modules*

$$\bigoplus_{i \in I} M_i, \qquad c(\prod_{i \in I} M_i), \qquad b(\prod_{i \in I} M_i).$$

If $|A - \{0\}|$ is a group, then $\mathcal{L}(L, M_i)$ fulfills the density condition.

We are going to describe a natural device for passing from an A-module norm on M to an equivalent one fulfilling the density condition. Let A be a valued ring and assume that the valuation on A is non-degenerate. Then for ach normed A-module $(M, |\ \ |)$ and for each $x \in M$, the set

$$R_x := \{r \in |A|;\, r \geqq |x|\} \subset \mathbb{R}_+, \qquad x \in M,$$

is non-empty. Hence the function $|\ \ |' : M \to \mathbb{R}_+$ given by

$$(*) \qquad |x|' := \inf R_x$$

is well-defined.

Proposition 3. *The pair $(M, |\ \ |')$ is a normed A-module fulfilling the density condition.*

Proof. Clearly $|x|' = 0$ implies $x = 0$, since $|\ \ | \leqq |\ \ |'$. In order to see that $|\ \ |'$ is ultrametric, it is sufficient to prove (see Proposition 1.1.1/6) that

$$|x| \leqq |y| \quad \text{implies} \quad |x|' \leqq |y|', \qquad x, y \in M.$$

However, $|x| \leqq |y|$ implies $R_y \subset R_x$, and hence $|y|' \geqq |x|'$.

To verify $|ax|' \leqq |a|\,|x|'$, $a \in A$, $x \in M$, it is enough to show that $R_{ax} \supset |a|\,R_x$. So take $r \in R_x$, i.e., $r \in |A|$ and $r \geqq |x|$. Obviously, $s := |a|r \in |A|$ and $s \geqq |a|\,|x| \geqq |ax|$; i.e., $s \in R_{ax}$. Thus $|\ \ |'$ is a norm on M which, by construction, fulfills the density condition. □

Remark. To guarantee that no set R_x, $x \neq 0$, is empty, it is actually enough to assume that A contains elements a with $|a| > 1$. Then the function $|\ \ |'$ can be defined as above, and the proposition remains true. If A has elements a with $0 < |a| < 1$, one can introduce the sets $S_x := \{s \in |A|;\ s \leqq |x|\}$, which are non-empty, and consider the map $x \mapsto \sup S_x$. Again one gets an A-module norm on M fulfilling the density condition.

Next we improve the last proposition (notations as above):

Proposition 4. *If the valuation on A is non-degenerate, the norms $|\ \ |$ and $|\ \ |'$ are equivalent.*

Proof. Since $|\ \ | \leqq |\ \ |'$, the equivalence of the norms will follow if we show that the identity map $(M, |\ \ |) \to (M, |\ \ |')$ is continuous — i.e., that for each sequence $x_\nu \in M$ with $|x_\nu| \to 0$, we also have $|x_\nu|' \to 0$. Take $\varepsilon > 0$ arbitrarily. Choose $r \in |A|$ such that $0 < r < \varepsilon$; this is possible by the assumption on $|A|$. Choose ν_0 such that $|x_\nu| \leqq r$ for $\nu \geqq \nu_0$. Then $r \in R_{x_\nu}$ for all $\nu \geqq \nu_0$, and hence $|x_\nu|' \leqq r < \varepsilon$ for $\nu \geqq \nu_0$. □

Finally, we state

Proposition 5. *Let $|A - \{0\}|$ be a group $\neq \{1\}$. If $(M, |\ \ |)$ is a faithfully normed A-module so is $(M, |\ \ |')$. (Observe that the assumption on A implies that A carries a non-degenerate valuation.)*

Proof. Assume that $|ax| = |a|\,|x|$ for all $a \in A$, $x \in M$. We only have to show that $|ax|' \geqq |a|\,|x|'$. For this it is enough to verify that $R_{ax} \subset |a|\,R_x$ for all $a \neq 0$. So take $s \in R_{ax}$, i.e., $s \in |A|$ and $s \geqq |ax| = |a|\,|x|$. Set $r := |a|^{-1} s$. Then $r \in |A|$ (since $|A - \{0\}|$ is a group) and $r \geqq |x|$; i.e., $r \in R_x$. So $s = |a|\,r \in |a|\,R_x$. □

2.1.10. The functor $M \rightsquigarrow M^\sim$. Residue degree. — To each normed A-module M, we attach the A°-modules

$$M^\circ := \{x \in M;\ |x| \leqq 1\} \quad \text{and} \quad M^\vee := \{x \in M^\circ;\ |x| < 1\}.$$

The quotient module $M^\sim := M^\circ/M^\vee$ is an $A^\sim$-module in a canonical way. Each A-linear contraction $\varphi\colon L \to M$ between normed A-modules induces an $A^\sim$-linear map $\varphi^\sim\colon L^\sim \to M^\sim$. Thus, we see that

$M \rightsquigarrow M^\sim$ is a covariant functor from the category of normed A-modules with contractions as morphisms into the category of $A^\sim$-modules.

Furthermore, recall that $A^\circ = \mathring{A}$, $A^\vee = \check{A}$ and $A^\sim = \tilde{A}$ if the norm on A is a valuation. In particular for faithfully normed A-modules M, we can talk about the $\tilde{A}$-module $M^\sim$.

Proposition 1. *If M is a faithfully normed A-module, then $M^\sim$ is a torsion-free $\tilde{A}$-module. (Observe that M is also torsion-free.)*

Proof. Assume $\tilde{a}x^\sim = 0$, $\tilde{a} \in \tilde{A}$, $x^\sim \in M^\sim$, $x^\sim \neq 0$. Choose inverse images $a \in \mathring{A}$, $x \in M^\circ$. Then $|a|\,|x| = |ax| < 1$. But $|x| = 1$ and therefore $|a| < 1$; i.e., $\tilde{a} = 0$. □

Definition 2. *For each faithfully normed A-module M, we define the residue degree of M over A by* $f(M/A) := \mathrm{rk}_{\tilde{A}}\, M^\sim$.

Remark. In classical valuation theory (where A and M are fields), $f(M/A)$ equals the degree of the field $\tilde{M}$ over the field $\tilde{A}$.

Proposition 3. *Let $|A - \{0\}|$ be a group. Then for each faithfully normed A-module M, we have $f(M/A) \leqq \mathrm{rk}_A\, M$. More precisely,*

If $x_1, \ldots, x_n \in M^\circ$ have $\tilde{A}$-linearly independent images in $M^\sim$, then $\left| \sum_{\nu=1}^{n} a_\nu x_\nu \right| = \max_{1 \leqq \nu \leqq n} |a_\nu|$ for all $a_1, \ldots, a_n \in A$. In particular, $x_1, \ldots, x_n$ are A-linearly independent.

Proof. Let $m := \sum_{\nu=1}^{n} a_\nu x_\nu$ be given. The inequality $|m| \leqq \max_{1 \leqq \nu \leqq n} |a_\nu|$ is obvious. In order to verify the opposite inequality, we may assume that $|a_1| \geqq |a_\nu|$ for $\nu = 1, \ldots, n$ and that $a_1 \neq 0$. We can choose $c \in A$ such that $|c| = |a_1|^{-1}$. Then $|ca_\nu| \leqq 1$ for all $\nu \geqq 1$. Now we proceed indirectly: Assume $|m| < \max_{1 \leqq \nu \leqq n} |a_\nu|$. Then we would get $\left| \sum_{\nu=1}^{n} ca_\nu x_\nu \right| < \max_{1 \leqq \nu \leqq n} |ca_\nu| = 1$, from where we could pass to $\sum_{\nu=1}^{n} \widetilde{ca_\nu} x_\nu^\sim = 0$. However, we have $\widetilde{ca_1} \neq 0$ in contradiction to the linear independence of the $x_\nu^\sim$. Thus, we must have $|m| = \max_{1 \leqq \nu \leqq n} |a_\nu|$. □

The functor $M \rightsquigarrow M^\sim$ is only useful under special assumptions on the norm on M (cf. the case of strictly cartesian vector spaces (2.5)). In general, $M^\sim$ does not inherit much structure from M. Indeed, if $M = Ax$ is cyclic and faithfully normed (A may even be a field) and if $|x|^{-1} \notin |A|$, we have $M^\circ = M^\vee$ and hence $M^\sim = 0$.

2.2. Examples of normed and faithfully normed A-modules

2.2.1. The module A^n. — As before, A denotes a normed ring. Viewing A as a module over itself, we get a normed A-module. For each integer $n \geqq 1$, we can consider the normed n-fold direct sum of n copies of this module; i.e., the A-module of all n-tuples $\{a = (a_1, \ldots, a_n);\, a_\nu \in A\}$ provided with the maximum norm

$$|a| = \max_{1 \leqq \nu \leqq n} \{|a_\nu|\}.$$

The module A^n is *free*; by $\{e_1, \ldots, e_n\}$ we always mean the canonical basis of A^n, i.e., e_ν has all coordinates 0 except 1 in the ν-th place. We remark that A^n is faithfully normed if A is a valued ring. Each A-linear map $A^m \to A^n$ is bounded; more generally,

If M is any normed A-module, $\mathscr{L}(A^m, M)$, $1 \leq m < \infty$, consists of all A-linear maps. For each such map φ, we have $|\varphi| = \max_{1 \leq \nu \leq m} |\varphi(e_\nu)|$.

The *proof* is obvious due to the definition of the norm on A^m.

Warning. Even if $\varphi: A^m \to M$ is bijective, φ^{-1} may fail to be continuous. The standard example which works even for valued fields is as follows:

Let K be a *field with a non-complete valuation*; let $\hat{K}$ be the completion of K. If we provide $\hat{K}$ with the extended valuation, $\hat{K}$ is a normed K-vector space. Choose $x \in \hat{K} - K$ and consider the 2-dimensional normed K-subspace $U := K + Kx$ of $\hat{K}$. Since K is not closed in U, the map $K^2 \to U$ given by $(a_1, a_2) \mapsto a_1 + a_2 x$ has no continuous inverse.

We state a sufficient condition for a normed A-module M to be homeomorphic to A^n:

Proposition 1. *An A-linear bijection $\psi: M \to A^n$, $1 \leq n < \infty$, is bounded (and hence a homeomorphism) if and only if each A-linear map $M \to A$ is bounded.*

Proof. First we show the if-part. For all $a \in A^n$, we have $|a| = \max_{1 \leq i \leq n} |\pi_i(a)|$, where $\pi_i: A^n \to A$ denotes the projection onto the i-th coordinate. Since by assumption all maps $\pi_i \circ \psi: M \to A$ are bounded, we deduce that

$$|\psi(x)| = \max_{1 \leq i \leq n} |(\pi_i \circ \psi)(x)| \leq (\max_{1 \leq i \leq n} |\pi_i \circ \psi|)\, |x| \quad \text{for all} \quad x \in M.$$

To show the converse, take the A-basis $m_1, \ldots, m_n$ of M such that $\psi(m_i) = e_i \in A^n$, $i = 1, \ldots, n$. Then one has for each A-linear map $\lambda: M \to A$ the estimates $\left|\lambda\left(\sum_{i=1}^n a_i m_i\right)\right| \leq \max_{1 \leq i \leq n} |a_i|\, |\lambda(m_i)| \leq (\max_{1 \leq i \leq n} |\lambda(m_i)|) \left|\psi\left(\sum_{i=1}^n a_i m_i\right)\right|$, and therefore λ is bounded, if ψ is bounded. □

It is a serious fault of the norm topology on A^n that, in general, A-submodules of A^n are not closed in A^n, even if A is faithfully normed and complete and quasi-Noetherian. We shall give an example in Proposition 2.2.4/2 showing that this can even happen in the case of ideals ($n = 1$). For valued fields K, all subspaces of K^n are closed (Proposition 2.3.1/1). One of our important later propositions (3.7.2/2) says that for normed Noetherian Banach algebras A (over a complete valued field k) — so, in particular, for strictly convergent power series algebras $k\langle X_1, \ldots, X_n\rangle$ over complete fields — all submodules of A^n, $1 \leq n < \infty$, are closed.

2.2.2. The modules $A^{(I)}$, $A^{(\infty)}$, $c(A)$ and $b(A)$. — Let I be a non-empty index set. Define $A^{(I)} := \bigoplus_{i \in I} M_i$, where $M_i = A$ for all $i \in I$ (for the definition of the (normed) direct sum $\bigoplus_{i\in I} M_i$ see (2.1.5)). Then $A^{(I)}$ is the A-module of all functions $f: I \to A$ which are zero for almost all $i \in I$, provided with the norm $|f| = \max_{i\in I} |f(i)|$. If e_i, $i \in I$, denotes the function defined by

$$e_i(i) = 1, \qquad e_i(j) = 0 \quad \text{for} \quad j \neq i,$$

each element $f \in A^{(I)}$ can be written uniquely in the form $f = \sum_{i\in I} f(i)\, e_i$. We call $\{e_i\}_{i\in I}$ the *canonical basis* of $A^{(I)}$.

If I is countable, we identify I with $\mathbb{N}$ (or sometimes with $\mathbb{N} \cup \{0\}$). Up to a canonical isometric isomorphism, $A^{(I)}$ is the normed A-module $A^{(\infty)}$ of all sequences

$$s = (a_1, a_2, \ldots), \quad \text{where} \quad a_i \in A \quad \text{for all} \quad i \in \mathbb{N} \quad \text{and} \quad a_i = 0 \quad \text{for almost all } i.$$

For all $n \in \mathbb{N}$, the A-module A^n can be embedded isometrically into $A^{(\infty)}$ in an obvious way. Thus, one has natural isometric injections

$$A^1 \hookrightarrow A^2 \hookrightarrow \cdots \hookrightarrow A^n \hookrightarrow A^{(\infty)}.$$

Note that $A^{(\infty)} = \bigcup_{n=1}^{\infty} A^n$ and that A^n is closed in $A^{(\infty)}$.

Proposition 1. *A linear map $\varphi: A^{(I)} \to M$ into an arbitrary normed A-module M is bounded if and only if the set $\{|\varphi(e_i)|;\, i \in I\}$ is bounded. We have*

$$|\varphi| = \sup_{i\in I} |\varphi(e_i)|$$

if φ is bounded.

Proof. Obvious. □

Still bigger normed A-modules than $A^{(I)}$ are the normed A-modules

$$c_I(A) := c(\prod_i M_i), \qquad b_I(A) := b(\prod_i M_i),$$

where $M_i := A^1$ for all $i \in I$. In the case $I = \mathbb{N}$, we just write $c(A)$, $b(A)$ instead of $c_{\mathbb{N}}(A)$, $b_{\mathbb{N}}(A)$. Thus

$$c(A) = \{a = (a_1, a_2, \ldots);\, a_\nu \in A, \lim a_\nu = 0\}, \quad \text{and}$$
$$b(A) = \{a = (a_1, a_2, \ldots);\, a_\nu \in A, \sup |a_\nu| < \infty\}$$

with the norm $|a| = \sup_\nu |a_\nu|$.

We have canonical isometric injections

$$A^{(I)} \hookrightarrow c_I(A) \hookrightarrow b_I(A).$$

If A is a valued ring, all modules considered here are faithfully normed A-modules.

Remark. The A-algebra $A[X]$ (resp. $A\langle X\rangle$) provided with the Gauss norm is a normed A-module which is isometrically isomorphic to $A^{(\infty)}$ (resp. $c(A)$). In these cases it is convenient to use $\mathbb{N} \cup \{0\}$ instead of $\mathbb{N}$ as index set.

2.2.3. Structure of $\mathscr{L}(c_I(A), M)$. — Let M be a normed A-module. In complete analogy to the definitions of the two preceding sections, we introduce the normed A-modules M^n, $M^{(\infty)}$, $c(M)$, $b(M)$ and $c_I(M)$, $b_I(M)$: e.g.,
$M^n := \bigoplus_1^n M_i$, $M^{(\infty)} := \bigoplus_1^\infty M_i$, $c(M) := c\left(\prod_1^\infty M_i\right)$, $b(M) := b\left(\prod_1^\infty M_i\right)$, and so on, where always $M_i := M$.

Again $M^n \subset M^{(\infty)} \subset c(M) \subset b(M)$ in a canonical way. From Proposition 2.1.5/6 we deduce

If M is complete, $b_I(M)$, $c_I(M)$ and the modules M^n, $n \geqq 0$, are complete.

We are going to determine the structure of $\mathscr{L}(c_I(A), M)$, i.e. of the A-module of bounded linear maps from $c_I(A)$ to M. Assigning to each $\varphi \in \mathscr{L}(c_I(A), M)$ the element $(\varphi(e_i))_{i\in I} \in b_I(M)$, yields an A-linear map

$$\pi\colon \mathscr{L}(c_I(A), M) \to b_I(M).$$

Since

$$|\pi(\varphi)| = \sup_{i\in I} |\varphi(e_i)| = |\varphi|,$$

we see that π is an isometry.

Proposition 1. *If M is complete, π is an (isometric) isomorphism. In particular, $\mathscr{L}(c(A), A)$ is isometrically isomorphic to $b(A)$ if A is complete.*

Proof. It is enough to show that π is onto (because then π^{-1} must also be isometric). Take (x_i) in $b_I(M)$. Since M is complete, for each $(a_i) \in c_I(A)$ the element $\sum_{i\in I} a_i x_i \in M$ is well-defined, and we have

$$|\sum_{i\in I} a_i x_i| \leqq \sup\{|a_i x_i|\} \leqq (\sup |x_i|)(\sup |a_i|) = |x| \cdot \max |a_i|$$

(note that the set of indices $i \in I$ where $a_i \neq 0$ is at most countable). Hence

$$(a_i) \mapsto \sum_{i\in I} a_i x_i$$

defines a linear map $\varphi\colon c_I(A) \to M$ with $|\varphi| \leqq |x|$; i.e., $\varphi \in \mathscr{L}(c_I(A), M)$. Since $\varphi(e_i) = x_i$, we have $\pi(\varphi) = x$; i.e., π is onto. □

2.2.4. The ring $F = A[[Y_1, Y_2, \ldots]]$ of formal power series. — In this section we assume that A has a bounded valuation. In (1.5.5) we provided the ring $F = A[[Y_1, Y_2, \ldots]]$ with a bounded valuation extending the valuation on A. Thus, F *is a faithfully normed A-module.* In order to describe this faithful A-module norm on F in a convenient way, we introduce the faithfully normed A-modules M_ν, $\nu = 1, 2, \ldots$, where M_ν equals A as A-module and where the norm $|\ \ |_\nu$ on M_ν is related to the valuation $|\ \ |$ on A by the equation

$$|\ \ |_\nu = \frac{1}{\nu} |\ \ |.$$

Then obviously

$$F = c\left(\prod_1^\infty M_\nu\right) = b\left(\prod_1^\infty M_\nu\right) = \prod_1^\infty M_\nu$$

as normed A-modules, since $|\ \ |$ is bounded. If A is complete, all M_ν are complete. We derive from Proposition 2.1.5/6

Proposition 1. *If A is complete, so is F.*

This was already announced in (1.5.5). Furthermore, we stated without proof that not all ideals of F are closed.

Proposition 2. *The ideal $\mathfrak{a}$ of F generated by $Y_1, Y_2, Y_3, \dots$ is not closed in F. For example,*

$$f := \sum_1^\infty Y_\nu \in \bar{\mathfrak{a}}, \quad \textit{but} \quad f \notin \mathfrak{a}.$$

Proof. We have $f_n := \sum_1^n Y_\nu \in \mathfrak{a}$. From $|Y_\nu| = p_\nu^{-1}$ (where $p_1 < p_2 < \cdots$ is the sequence of prime numbers), we conclude $|f - f_n| = |\sum_{n+1}^\infty Y_\nu| \leqq \max_{\nu>n} \{p_\nu^{-1}\} = p_{n+1}^{-1} \to 0$; i.e., $f \in \bar{\mathfrak{a}}$. In order to prove $f \notin \mathfrak{a}$, it is convenient to write the elements of F in their canonical unique way $\sum_{\mu=1}^\infty a_\mu Y^{e(\mu)}$ (notation as in (1.5.5): e.g., $Y_i = Y^{e(p_i)}$). Now assume $f \in \mathfrak{a}$; say $f = \sum_{\varrho=1}^r g_\varrho Y_\varrho$, $1 \leqq r < \infty$. If $g_\varrho = \sum_{\mu=1}^\infty a_{\mu\varrho} Y^{e(\mu)}$, this means, since $Y^{e(\mu)} Y^{e(\nu)} = Y^{e(\mu\nu)}$, that

$$f = \sum_{i=1}^\infty Y^{e(p_i)} = \sum_{\mu=1}^\infty \left(\sum_{\varrho=1}^r a_{\mu\varrho} Y^{e(\mu p_\varrho)}\right).$$

However, due to the bijectivity of e and the uniqueness of the power series expansion, this is impossible, since, on the right-hand side, the exponent $e(p_{r+1})$ never occurs. □

2.2.5. b-separable modules. — The following notion will turn out to be extremely useful.

Definition 1. *A normed A-module M is called separable with respect to bounded linear maps or simply b-separable if for each $x \neq 0$ in M there exists a bounded A-linear map $\lambda: M \to A$ such that $\lambda(x) \neq 0$.*

Normed submodules of b-separable modules are b-separable.

Proposition 2. *Let $\{M_i\}_{i \in I}$ be a family of b-separable A-modules. Then the modules*

$$b(\prod M_i) \supset c(\prod M_i) \supset \bigoplus M_i$$

are b-separable.

Proof. Let $x = (x_i) \in b(\prod M_i)$, $x \neq 0$. Choose j such that $x_j \neq 0$, and choose a bounded linear map $\lambda\colon M_j \to A$ such that $\lambda(x_j) \neq 0$. Define $\Lambda\colon b(\prod M_i) \to A$ by $(y_i) \mapsto \lambda(y_j)$. Then Λ is A-linear and $\Lambda(x) = \lambda(x_j) \neq 0$. Moreover $|\Lambda| = |\lambda|$; i.e., Λ is bounded. □

Corollary 3. *All the A-modules $b(A)$, $c(A)$, $A^{(\infty)}$, A^n, $A^{(I)}$ are b-separable. The module $F = A[\![Y_1, Y_2, \ldots]\!]$ is b-separable.*

As another corollary to Proposition 2, we get

Corollary 4. *Let $\{M_i; i \in I\}$ be a family of normed A-modules such that each finitely generated submodule of M_i is b-separable. Then each finitely generated submodule of $\bigoplus_{i\in I} M_i$ is b-separable.*

Proof. Let $N = \sum_{\varrho=1}^{r} An_\varrho$ be a finitely generated submodule of $\bigoplus_{i\in I} M_i$. Then there are elements $m_{\varrho i} \in M_i$ for $\varrho = 1, \ldots, r$ and $i \in I$ such that $n_\varrho = \sum_{i\in I} m_{\varrho i}$. Define $N_i := \sum_{\varrho=1}^{r} Am_{\varrho i}$. Then N_i is a finitely generated A-submodule of M_i and hence b-separable. According to Proposition 2, we know that $\bigoplus_{i\in I} N_i$ is b-separable. Because N is obviously contained in $\bigoplus_{i\in I} N_i$, N is b-separable. □

2.2.6. The functor $M \rightsquigarrow T(M)$. — Let $n \geqq 0$ be a given fixed integer. We write $T_n(A)$ or just $T(A)$ for the normed ring $A\langle X_1, \ldots, X_n\rangle$ provided with the Gauss norm as introduced in (1.4). ($T_0(A)$ is to be interpreted as A.)

For each normed A-module M, we denote by $T(M)$, or more explicitly by $T_n(M)$, the set of "strictly convergent power series" with coefficients in M:

$$T_n(M) = \left\{\sum_0^\infty x_{\nu_1\ldots\nu_n} X_1^{\nu_1} \ldots X_n^{\nu_n};\ x_{\nu_1\ldots\nu_n} \in M, \lim x_{\nu_1\ldots\nu_n} = 0\right\}.$$

Again, $T_0(M)$ stands for the module M itself. Obviously $T(M)$ is an A-module if addition and scalar multiplication are introduced in the usual way. By defining

$$\left|\sum_0^\infty x_{\nu_1\ldots\nu_n} X_1^{\nu_1} \ldots X_n^{\nu_n}\right| := \max |x_{\nu_1\ldots\nu_n}|,$$

we introduce an ultrametric function on $T(M)$ which again will be called the Gauss norm. Then we easily see

Lemma 1. *$T(M)$ is a normed A-module (isometrically isomorphic to $c(M)$).*

Obviously there is a canonical isometric A-isomorphism

$$T_m\big(T_n(M)\big) \xrightarrow{\sim} T_{m+n}(M) \quad \text{for all } m, n;$$

hence, it will often suffice to consider $T_1(M)$.

Now we are going to provide $T(M)$ with the structure of a $T(A)$-module. If

$$f = \sum_{\mu=0}^{\infty} a_\mu X^\mu, \qquad x = \sum_{\nu=0}^{\infty} x_\nu X^\nu$$

is the shorthand notation for elements $f \in T(A)$, $x \in T(M)$ (where μ stands for $(\mu_1, \ldots, \mu_n)$ and X^μ for $X_1^{\mu_1} \ldots X_n^{\mu_n}$), we define their "Cauchy product" $f \cdot x$ by

$$f \cdot x := \sum_{\lambda=0}^{\infty} \Big(\sum_{\mu+\nu=\lambda} a_\mu x_\nu \Big) X^\lambda .$$

As in the case of $T(A)$, one checks that $f \cdot x \in T(M)$ and proves that

$$|f \cdot x| \leqq |f| \cdot |x| .$$

Thus, we have

Proposition 2. *For each normed A-module M, the set $T(M)$ of strictly convergent power series over M is a normed $T(A)$-module.*

Next we prove a statement closely related to Proposition 1.4.3/1:

Proposition 3. *Let L, M be normed A-modules; let $\varphi \in \mathscr{L}(L, M)$. Then the map $T\varphi\colon T(L) \to T(M)$ defined by*

$$\sum_0^{\infty} y_\nu X^\nu \mapsto \sum_0^{\infty} \varphi(y_\nu)\, X^\nu$$

is a $T(A)$-linear bounded homomorphism with $|T\varphi| = |\varphi|$. The map $T\colon \mathscr{L}(L, M) \to \mathscr{L}\big(T(L), T(M)\big)$, $\varphi \mapsto T\varphi$, is an A-linear isometry.

Proof. We have $|\varphi(y_\nu)| \leqq |\varphi|\,|y_\nu| \to 0$; i.e., the power series on the right-hand side above actually is in $T(M)$. Obviously $T\varphi$ is additive. Moreover for $f = \sum_0^{\infty} a_\mu X^\mu \in T(A)$, $\quad y = \sum_0^{\infty} y_\nu X^\nu \in T(L)$, we get

$$(T\varphi)\,(f \cdot y) = (T\varphi) \sum_{\lambda=0}^{\infty} \Big(\sum_{\mu+\nu=\lambda} a_\mu y_\nu \Big) X^\lambda = \sum_{\lambda=0}^{\infty} \varphi \Big(\sum_{\mu+\nu=\lambda} a_\mu y_\nu \Big) X^\lambda$$

$$= \sum_{\lambda=0}^{\infty} \Big(\sum_{\mu+\nu=\lambda} a_\mu \varphi(y_\nu) \Big) X^\lambda = f \cdot (T\varphi)\,(y);$$

i.e., $T\varphi$ is a $T(A)$-module homomorphism. Finally

$$|(T\varphi)\,(y)| = \max_\nu |\varphi(y_\nu)| \leqq |\varphi| \max_\nu |y_\nu| = |\varphi| \cdot |y| ,$$

whence $|T\varphi| \leqq |\varphi|$. Since $M \subset T(M)$, we also have $|T\varphi| \geqq |\varphi|$ and therefore $|T\varphi| = |\varphi|$. Because the relations $T(a\varphi) = aT\varphi$ for all $a \in A$ and $T(\varphi + \psi) = T\varphi + T\psi$ are obvious, all assertions of the proposition are clear. □

The map $T\varphi$ is called the canonical extension of φ. An obvious consequence is

Proposition 4. *$T: M \rightsquigarrow T(M)$ is a covariant additive functor from the category of normed A-modules into the category of normed $T(A)$-modules (with bounded linear maps as morphisms in both cases).*

We state an important corollary of Proposition 3:

Corollary 5. *If M is a b-separable A-module, then $T(M)$ is a b-separable $T(A)$-module.*

Proof. Take $x = \sum_{0}^{\infty} x_\nu X^\nu \neq 0$ in $T(M)$, say $x_i \neq 0$. By assumption there exists a $\lambda \in \mathscr{L}(M, A)$ such that $\lambda(x_i) \neq 0$. Then $T\lambda \in \mathscr{L}\big(T(M), T(A)\big)$ by Proposition 3, and $(T\lambda)(x) \neq 0$ by definition of $T\lambda$. □

Proposition 6. *If M is a faithfully normed A-module, then $T(M)$ is a faithfully normed $T(A)$-module.*

The *proof* is the same as the classical proof of the Gauss Lemma in (1.5.3).

We conclude this section by listing some properties of T which will be used later.

Proposition 7. *Let $\varphi \in \mathscr{L}(L, M)$ be given. Then*

$$\ker T\varphi = T(\ker \varphi).$$

In particular, $T\varphi$ is injective if and only if φ is injective. If φ is open and surjective, $T\varphi$ is surjective.

The *proof* is left to the reader.

The functor T preserves finite direct sums. More precisely,

Proposition 8. *Let $M_1, \ldots, M_s$ be normed A-modules. There is a canonical isometric $T(A)$-module isomorphism*

$$T\left(\bigoplus_{1}^{s} M_\sigma\right) \cong \bigoplus_{1}^{s} T(M_\sigma).$$

The *proof* is straightforward. (Alternatively, one can use the fact that T is additive. Hint: consider the injections $T(M_j) \to T\left(\bigoplus_{1}^{s} M_\sigma\right)$ and projections $T\left(\bigoplus_{1}^{s} M_\sigma\right) \to T(M_i)$ induced by $M_j \to \bigoplus_{1}^{s} M_\sigma$ and $\bigoplus_{1}^{s} M_\sigma \to M_i$, respectively.) □

We remark that Proposition 8 fails to hold for infinite direct sums.

2.3. Weakly cartesian spaces

In the following, we always work over a field K with a non-trivial valuation. Let V denote a normed (hence faithfully normed, cf. Proposition 2.1.1/4) K-vector space. We often write "space" instead of "normed K-vector space". From Proposition 2.1.8/2, we get that, for K-linear maps between spaces, continuity and boundedness are equivalent properties.

2.3.1. Elementary properties of normed spaces. — In this section, we list important topological properties of K^n (the (normed) direct sum of n copies of K) which depend heavily on the fact that K is a field.

Proposition 1. *Let U be a subspace of K^n. Then U is closed, and there exists a linear homeomorphism $U \to K^r$ where $r := \dim_K U$.*

Proof. An automorphism of K^n is always a homeomorphism (cf. (2.2.1)). Because each subspace U may be transformed by an automorphism into the subspace $\{(c_1, \ldots, c_r, 0, \ldots, 0);\ c_i \in K\}$, where $r := \dim_K U$, the assertion is evident. □

Recall, however, that a (bounded) K-linear bijection $K^n \to V$ need not be a homeomorphism (cf. (2.2.1)), whereas each bounded K-linear bijection $V \to K^n$ is a homeomorphism.

The space V obviously is b-separable if each K-linear map $V \to K$ is bounded. We have the following converse for finite-dimensional spaces.

Proposition 2. *If a finite-dimensional normed space U is b-separable, then each K-linear map $U \to K$ is bounded.*

Proof. It is enough to construct $n := \dim_K U$ linearly independent bounded K-linear maps $\lambda_1, \ldots, \lambda_n$ of U into K, since each K-linear map is a linear combination of these and hence bounded. Choose at random a bounded K-linear map $\lambda_1 \neq 0$. Let $\lambda_1, \ldots, \lambda_{m-1}$, $m - 1 < n$, already be constructed. Choose $u \neq 0$ in $\bigcap_1^{m-1} \ker \lambda_\mu$, and take a bounded K-linear map $\lambda_m \colon U \to K$ such that $\lambda_m(u) = 1$. Since $\lambda_1(u) = \cdots = \lambda_{m-1}(u) = 0$, each linear relation $\sum_1^m a_\mu \lambda_\mu = 0$, $a_\mu \in K$, implies $a_m = 0$ and hence $a_1 = \cdots = a_{m-1} = 0$, since $\lambda_1, \ldots, \lambda_{m-1}$ are independent by assumption. □

Corollary 3. *A finite-dimensional normed space U is b-separable if and only if $\mathrm{Hom}_K(U, K) = \mathscr{L}(U, K)$.*

For each bounded K-linear map $\lambda \colon V \to K$, the kernel space $\ker \lambda$ is closed in V. We shall need the following converse.

Proposition 4. *Each K-linear map $\lambda \colon V \to K$ with a closed kernel is bounded.*

Proof. Assume $\lambda \neq 0$. Since $\ker \lambda$ is closed in V, the residue space $V/\ker \lambda$ provided with the residue norm is a 1-dimensional normed K-vector space. Therefore, it follows from Proposition 2.1.1/3 that the K-linear bijection $\bar{\lambda} \colon V/\ker \lambda \to K$ induced by λ is bounded ($V/\ker \lambda$ is a faithfully normed K-module since K is a field). Now the boundedness of λ follows since λ is the composition of the canonical contraction map $V \to V/\ker \lambda$ with $\bar{\lambda}$. □

2.3.2. Weakly cartesian spaces. — For each K-vector space V, we denote by $\mathfrak{F}(V)$ the family of all finite-dimensional K-subspaces. We have the following

Theorem 1. *The following statements over a normed K-vector space V are equivalent:*

(1) *For each $U \in \mathfrak{F}(V)$, there exists a linear homeomorphism $U \xrightarrow{\sim} K^n$, $n := \dim_K U$.*
(2) *Each $U \in \mathfrak{F}(V)$ is closed in V.*
(3) *Each $U \in \mathfrak{F}(V)$ is b-separable.*

Proof. (1) $\rightarrow$ (2): Take $U \in \mathfrak{F}(V)$ and $x \in \overline{U}$. Then $U' := U + Kx \in \mathfrak{F}(V)$. Hence by assumption U' is homeomorphic to a space K^n. Therefore $U \subset U' \simeq K^n$ is closed in U' by Proposition 2.3.1/1. Since $x \in U'$ is in the U'-closure of U, we deduce that $x \in U$. Thus $U = \overline{U}$.

(2) $\rightarrow$ (3): Take $U \in \mathfrak{F}(V)$ and $u \in U - \{0\}$. Then there exists a K-linear map λ: $U \rightarrow K$ such that $\lambda(u) \neq 0$. By assumption $\ker \lambda \in \mathfrak{F}(V)$ is closed in V and hence also closed in U. Thus, by Proposition 2.3.1/4, λ is bounded.

(3) $\rightarrow$ (1): Take $U \in \mathfrak{F}(V)$. Choose $n := \dim_K U$ linearly independent maps $\lambda_\nu: U \rightarrow K$, $1 \leqq \nu \leqq n$. By Proposition 2.3.1/2 these maps are continuous. Therefore the product map $\lambda_1 \times \cdots \times \lambda_n$: $U \rightarrow K^n$ is continuous. Since it is bijective, it is a homeomorphism. □

Definition 2. *A normed K-vector space V is called weakly cartesian (more precisely weakly K-cartesian) if the conditions of Theorem 1 are fulfilled.*

We say that a finite-dimensional K-vector space carries the *product topology* if there exists a linear homeomorphism $U \rightarrow K^n$, $n := \dim_K U$. Note that, in terms of topological spaces, the (normed) direct sum K^n is in fact the n-fold direct product of K with itself (provided with the product topology). Therefore,

Lemma 3. *V is weakly cartesian if and only if each finite-dimensional subspace of V carries the product topology.*

An immediate consequence of Proposition 2.3.1/1 is

Proposition 4. *An n-dimensional space V is weakly cartesian if and only if there exists a linear homeomorphism $\varphi: V \rightarrow K^n$. (Observe that φ need not be isometric.)*

Next we state some facts which are obvious from the definition of weakly cartesian spaces.

Lemma 5. *Each space V which can be exhausted by weakly cartesian spaces is weakly cartesian. Each subspace of a weakly cartesian space is weakly cartesian.*

Lemma 6. *If each finite-dimensional subspace of V is weakly cartesian, V itself is weakly cartesian.*

Proposition 7. *Each b-separable space V is weakly cartesian.*

So, in particular, the spaces K^n, $K^{(\infty)}$, $K^{(I)}$, $c(K)$ and $b(K)$ are weakly cartesian.

2.3.3. Properties of weakly cartesian spaces. — First we prove

Proposition 1. *The direct sum of weakly cartesian spaces is weakly cartesian.*

Proof. Since each finite-dimensional subspace of a direct sum $\bigoplus_{i\in I} V_i$ is contained in a *finite* direct sum of finite-dimensional subspaces of the V_i, it is enough to consider the direct sum of two finite-dimensional weakly cartesian spaces V_1, V_2. However, if V_i is homeomorphic to K^{n_i}, $i = 1, 2$, the sum $V_1 \oplus V_2$ is homeomorphic to $K^{n_1+n_2}$. □

The following application of Proposition 1 shows that weakly cartesian spaces behave well if one passes to smaller fields of coefficients.

Proposition 2. *Let K be a subfield of a valued field K' such that K' is weakly K-cartesian. Then each weakly K'-cartesian K'-vector space V is weakly K-cartesian.*

Proof. Let $U = \sum_1^m Ke_\mu$ be a finite-dimensional K-subspace of V. Then $U' := \sum_1^m K'e_\mu$ is a finite-dimensional K'-subspace, and hence by assumption homeomorphic to a space K'^s. Since K' is weakly K-cartesian, the direct sum K'^s is also weakly K-cartesian by Proposition 1. Thus the K-space U' is weakly K-cartesian. Hence the K-subspace $U \subset U'$ is also weakly K-cartesian. □

The following criterion will be useful in later applications.

Lemma 3. *Let A be a valued integral domain and K its valued field of fractions. Let M be a faithfully normed A-module such that each finitely generated A-submodule of M is b-separable.*

Then the K-vector space $V := M \otimes_A K$ (provided with the canonical norm extension) is weakly K-cartesian.

Proof. Let $U = \sum_1^m Ke_\mu \subset V$ be a finite-dimensional K-vector space. We may assume $e_\mu \in M$, since each e_μ is of the form $\frac{x_\mu}{a_\mu}$, $x_\mu \in M$, $a_\mu \in A - \{0\}$. By assumption the A-module $N := \sum_1^m Ae_\mu \subset M$ is b-separable. Take $u \neq 0$ in U. Choose $c \neq 0$ in A such that $cu \in N$. Since $cu \neq 0$, there is a bounded A-linear map $\lambda' : N \to A$ with $\lambda'(cu) \neq 0$. Now λ' extends uniquely to a bounded K-linear map $\lambda : U \to K$ (cf. (2.1.3), use that $U = N \otimes_A K$). From $c\lambda(u) = \lambda'(cu) \neq 0$, we conclude $\lambda(u) \neq 0$. Hence U is a b-separable K-space. Thus V is weakly K-cartesian. □

In important cases, normed vector spaces are always weakly cartesian.

Proposition 4. *If K is complete, each normed K-space V is weakly cartesian. In particular, V is complete if $\dim_K V < \infty$.*

Proof. We only have to show that each finite-dimensional K-space V is weakly cartesian. That such a space is complete follows then from Proposition 2.1.5/6. We proceed by induction on $n := \dim_K V$. The case $n = 0$ is clear. Suppose $n > 0$; let $U \subset V$ be a subspace. We want to show that U is closed in V. This is clear if $U = V$. Therefore, assume that $U \neq V$. By the induction hypothesis, U is weakly cartesian and hence complete. As a complete space, U is closed in V. □

Remark. Proposition 4 characterizes complete valued fields. Namely, if the completion $\hat{K}$ of K is weakly cartesian, K is closed in $\hat{K}$, i.e., $K = \hat{K}$.

A trivial consequence of Proposition 4 is

Corollary 5. *If K is complete, any two norms $|\ \ |_1$, $|\ \ |_2$ on a finite-dimensional K-vector space are equivalent, i.e., there exist real constants ϱ_1, ϱ_2 such that $|\ \ |_1 \leqq \varrho_2 |\ \ |_2 \leqq \varrho_1 |\ \ |_1$.*

Taking into account that the completion of a valued field is again a valued field, we have the following rank estimate:

Proposition 6. *Let V be a finite-dimensional normed K-vector space and $\hat{V}$ its completion. Then $\dim_K V \geqq \dim_{\hat{K}} \hat{V}$, and equality holds if and only if V is weakly K-cartesian. More precisely, every K-generating system of V is also a $\hat{K}$-generating system of $\hat{V}$. A K-basis of V is a $\hat{K}$-basis of $\hat{V}$ if and only if V is weakly K-cartesian.*

Proof. Choose a K-generating system $\{v_1, \ldots, v_n\}$ of V. Define $U := \sum_{i=1}^{n} \hat{K} v_i \subset \hat{V}$. According to Proposition 4, we know that U is complete. Since $V = \sum_{i=1}^{n} K v_i \subset \sum_{i=1}^{n} \hat{K} v_i = U$, we conclude $\hat{V} \subset \hat{U} = U \subset \hat{V}$ and furthermore $\hat{V} = U$. Therefore $\{v_1, \ldots, v_n\}$ is a generating system for $\hat{V}$ over $\hat{K}$, whence $\dim_K V \geqq \dim_{\hat{K}} \hat{V}$.

Now let $\{v_1, \ldots, v_n\}$ be a K-basis of V. Consider the isomorphism $\varphi \colon K^n \to V$ mapping the canonical basis of K^n onto $\{v_1, \ldots, v_n\}$. Then φ is continuous and extends to a continuous $\hat{K}$-linear map $\hat{\varphi} \colon \hat{K}^n \to \hat{V}$ which is surjective by what we have seen before. The $\hat{K}$-space $\hat{K}^n$ is complete by Proposition 2.1.5/6, and K^n is dense in $\hat{K}^n$. Therefore $\hat{K}^n$ can be interpreted as the completion of K^n. From this we see that $\hat{\varphi}$ is a homeomorphism if and only if φ is a homeomorphism. Using (2.2.1) and Proposition 4, we see that φ is a homeomorphism if and only if V is weakly K-cartesian and that $\hat{\varphi}$ is a homeomorphism if and only if $\hat{\varphi}$ is bijective. Thus, V is weakly K-cartesian if and only if $\{v_1, \ldots, v_n\}$ is a $\hat{K}$-basis of $\hat{V}$. □

2.3.4. Weakly cartesian spaces and tame modules. — We start with a simple remark:

If V is weakly K-cartesian, each $\mathring{K}$-submodule M of V° of finite rank is b-separable.

In order to see this, take $x \in M - \{0\}$. Since $\operatorname{rk} M < \infty$, the K-vector space $U := K \cdot M \subset V$ is finite-dimensional; hence, there exists a bounded K-linear map $\lambda\colon U \to K$ such that $\lambda(x) \neq 0$. Choose $c \in K^*$ such that $|\lambda(U^\circ)| \leq |c|$ and set $\Lambda := c^{-1}\lambda$. Since $M \subset U^\circ$, the map Λ induces a bounded $\mathring{K}$-linear map $\Lambda|M\colon M \to \mathring{K}$ with $\Lambda(x) \neq 0$. □

Next we prove

Proposition 1. *If V° is a tame $\mathring{K}$-module (i.e., if each $\mathring{K}$-submodule $M \subset V^\circ$ of finite rank is finitely generated), then V is weakly K-cartesian.*

Proof. Assume there is a finite-dimensional K-vector space $U \subset V$ admitting an unbounded K-linear map $\lambda\colon U \to K$. Then we can choose an infinite sequence $u_1, u_2, \ldots$ in U° such that $|\lambda(u_{n+1})| > \max_{1 \leq \nu \leq n} |\lambda(u_\nu)|$ for all n. The $\mathring{K}$-submodule M of V° generated by all u_n has finite rank since $M \subset U$. Thus, M is finitely generated over $\mathring{K}$, since V° is $\mathring{K}$-tame. Let $u_1, \ldots, u_m$ be generators. Then each $x \in M$ is of the form $x = \sum_1^m a_\mu u_\mu$, $a_\mu \in \mathring{K}$, whence

$$|\lambda(x)| \leq \max_{1 \leq \mu \leq m} |\lambda(u_\mu)| \quad \text{for all} \quad x \in M,$$

which is in contradiction to the choice of the sequence $u_1, u_2, \ldots$. Hence all finite-dimensional subspaces of V are b-separable. □

From the results of this section, we get in particular (since discrete valuation rings are Noetherian (Proposition 1.6.1/4) and since b-separable modules over such rings are tame (Corollary 4.1/5)):

Proposition 2. *If $\mathring{K}$ is a discrete valuation ring, a normed K-vector space V is weakly K-cartesian if and only if V° is a tame $\mathring{K}$-module.*

2.4. Cartesian spaces

As always let K be a field with a non-trivial valuation. We want to consider a subclass of the class of weakly K-cartesian vector spaces defined by the following additional requirement: that every finite-dimensional subspace not only carries the product topology but also admits an orthogonal basis (in the sense made precise in the following Definition 1). We shall proceed in a manner analogous to that used in (2.3) and study the finite-dimensional case first.

2.4.1. Cartesian spaces of finite dimension. —

Definition 1. *A finite-dimensional normed K-vector space V is called K-cartesian if there is a K-basis $\{v_1, \ldots, v_n\}$ of V such that, for all $c_1, \ldots, c_n \in K$, one has*

$$\left|\sum_{i=1}^n c_i v_i\right| = \max_{1 \leq i \leq n} |c_i|\,|v_i|.$$

A set $\{v_1, \ldots, v_n\}$ with this property is called K-orthogonal or K-cartesian.

Remark. Some results for weakly K-cartesian spaces might be derived from the corresponding results for K-cartesian spaces in the following way: the norm on a weakly K-cartesian space V of finite dimension can be modified (without affecting the vector space topology) in such a way that V becomes K-cartesian. Therefore all results which can be expressed in the language of topological vector spaces carry over.

The main result of this section is the assertion that every subspace of a finite-dimensional K-cartesian space is again K-cartesian. To show this, we need some technical lemmas and the following notion:

Definition 2. *Let $V_1, \ldots, V_n$ be subspaces of a normed space V (the dimensions need not be finite). We say that the subspace $V' := \sum_{i=1}^{n} V_i$ is the norm-direct sum of the spaces V_i if the canonical map from the (normed) direct sum $\bigoplus_{i=1}^{n} V_i$ onto V' is an isometry.*

Thus, *V' is the norm-direct sum of the V_i if and only if $|\sum_{i=1}^{n} v_i| = \max_{1 \leq i \leq n} |v_i|$ for all $v_i \in V_i$, $i = 1, \ldots, n$.* The connection to the notion "K-cartesian" is clear: V is K-cartesian if and only if there are a finite number of lines Kx_i, $i = 1, \ldots, n$, in V such that V is the norm-direct sum of these lines. This implies that the norm-direct sum of two K-cartesian spaces V_1, V_2 is again K-cartesian. More explicitly, if $\{v_1, \ldots, v_m\}$ (resp. $\{v_{m+1}, \ldots, v_l\}$) is a K-orthogonal basis of V_1 (resp. V_2), then $\{v_1, \ldots, v_m, v_{m+1}, \ldots, v_l\}$ is a K-orthogonal basis for the norm-direct sum $V_1 \oplus V_2$.

For simplicity we shall use the following definition:

Definition 3. *A subspace U of a normed K-vector space V admits a norm-direct supplement in V if there is a subspace $U^\perp$ such that $V = U + U^\perp$ is a norm-direct sum.*

Lemma 4. *If V is a finite-dimensional normed K-cartesian vector space and U a subspace of V of dimension 1, then U admits a K-cartesian norm-direct supplement. More precisely, if $\{v_1, \ldots, v_n\}$ is a K-orthogonal basis of V and $u \in U - \{0\}$, then (after a suitable renumbering of the v_i) $\{u, v_2, \ldots, v_n\}$ is a K-orthogonal basis for V.*

Proof. There are coefficients $c_1, \ldots, c_n \in K$ such that $u = \sum_{i=1}^{n} c_i v_i$ and $|u| = \max_{1 \leq i \leq n} |c_i v_i| > 0$. Renumber the v_i in such a way that $|c_1 v_1| = \max_{1 \leq i \leq n} |c_i v_i|$. Replacing u by $c_1^{-1} u$, we may assume $c_1 = 1$ and get $u = v_1 + \sum_{i=2}^{n} c_i v_i$ and $|c_i v_i| \leq |v_1| = |u|$ for $i = 2, \ldots, n$. Because $\{v_2, \ldots, v_n\}$ is a K-orthogonal set, we only have to show that $V' := \sum_{i=2}^{n} K v_i$ is a norm-direct supplement of $U = Ku$

in V. In order to verify this, take $v' = \sum_{i=2}^{m} a_i v_i \in V'$ and $au \in U$ with arbitrary coefficients $a, a_2, \ldots, a_n \in K$. Define $v := au + v'$. The relation $u = v_1 + \sum_{i=2}^{n} c_i v_i$ implies $v = av_1 + \sum_{i=2}^{n} b_i v_i$, where the b_i are suitable elements of K. The orthogonality of $\{v_1, \ldots, v_n\}$ yields $|au| = |av_1| \leqq |v|$ and hence $|v'| = |v - au| \leqq \max\{|v|, |au|\} = |v|$. □

Now it is rather straightforward to extend this orthogonal Steinitz exchange procedure for one-dimensional subspaces to an orthogonalization process for subspaces of arbitrary dimension. Thus we get the main result of this section.

Proposition 5. *If V is a finite-dimensional K-cartesian vector space, then every subspace U of V is K-cartesian and admits a norm-direct supplement. In particular, every orthogonal basis of U can be extended to an orthogonal basis of V.*

Proof. The last statement follows from the two previous ones. In order to prove these, we proceed by induction on $n := \dim V$. The cases $n = 0$ or $n = 1$ are clear. Therefore, assume $n \geqq 2$. Let $U' = Ku_1$ be a one-dimensional subspace of U. (If $U = 0$, there is nothing to show.) According to Lemma 4, we may assume that V is the norm-direct sum of U' and $V' := \sum_{i=2}^{n} Kv_i$, where $\{v_2, \ldots, v_n\}$ is an orthogonal basis of V'. Then it is clear that U is the norm-direct sum of U' and $V' \cap U$. Applying the induction hypothesis to the $(n-1)$-dimensional space V', we see that $V' \cap U$ admits a K-orthogonal basis $\{u_2, \ldots, u_m\}$ and a norm-direct supplement W in V'. Clearly $\{u_1, \ldots, u_m\}$ is a K-orthogonal basis of U, and W is a norm-direct supplement of U in V. □

The property that every subspace admits a norm-direct supplement is characteristic of K-cartesian spaces. Indeed, it is rather easy to show the following converse to Lemma 4 and Proposition 5:

Proposition 6. *If V is a finite-dimensional normed K-vector space such that every subspace of dimension 1 admits a norm-direct supplement, then V is K-cartesian.*

Proof. By induction on $n := \dim V$. For $n = 1$, the assertion is true. For $n > 1$, choose $u \in V - \{0\}$ and set $U := Ku$. There is a norm-direct supplement $U^\perp$ to U. Since every one-dimensional subspace W of $U^\perp$ admits a norm-direct supplement in V, hence a fortiori in $U^\perp$, we may apply the induction hypothesis to $U^\perp$ and see that $U^\perp$ is K-cartesian. Now V, being the norm-direct sum of the two K-cartesian spaces U and $U^\perp$, is K-cartesian itself. □

2.4.2. Finite-dimensional cartesian spaces and strictly closed subspaces. — In the last section, we have shown that a normed K-vector space is K-cartesian if and only if all its subspaces admit norm-direct supplements. In this section we want to derive another characterization of a K-cartesian space in terms of a more intrinsic property of its subspaces. In Definition **1.1.5/1**, we introduced the concept of a *strictly closed* subspace. In striking analogy to the fact that a normed K-vector space is weakly K-cartesian if and only if every finite-dimensional subspace is closed, we have

Proposition 1. *A finite-dimensional normed K-vector space is K-cartesian if and only if every subspace is strictly closed.*

Let us make a rather trivial remark before we prove the proposition.

Observation 2. *Let V be a normed K-vector space and let U be a strictly closed subspace of V. For all $v \in V$, there is an element $u_0 \in U$ such that $U + Kv = U + K(v - u_0)$, where the right-hand sum is norm-direct.*

Proof. Since U is strictly closed, we can find $u_0 \in U$ such that $|v - u_0| \leqq |v - u|$ for all $u \in U$. This element u_0 will do the job. We have to show that, for all $c \in K$ and all $u \in U$, the equality $|u + c(v - u_0)| = \max\{|u|, |c|\,|v - u_0|\}$ holds. For $c = 0$, there is nothing to prove. If $c \neq 0$, we set $u' := c^{-1}u$. It is enough to show that $|u' + (v - u_0)| = \max\{|u'|, |v - u_0|\}$. By the triangle inequality, we have $|u' + (v - u_0)| \leqq \max\{|u'|, |v - u_0|\}$. This is an equality if $|u'| \neq |v - u_0|$. Furthermore if $|u'| = |v - u_0|$, we have

$$|u' + (v - u_0)| = |v - (u_0 - u')| \geqq |v - u_0| = \max\{|u'|, |v - u_0|\}$$

by the choice of u_0. □

Now we are able to *prove* Proposition 1. By induction on $n := \dim V$, we want to show first that V is K-cartesian if all subspaces are strictly closed. For $n = 1$, there is nothing to show. Assume $n > 1$ and choose U to be an $(n - 1)$-dimensional subspace of V. Since every subspace of U is strictly closed in V, and hence a fortiori in U, it follows from the induction hypothesis that U is K-cartesian. We only have to exhibit an element $v_0 \in V - U$ such that V is the norm-direct sum of U and Kv_0. Choose an arbitrary $v \in V - U$, and, for that v, choose u_0 according to the preceding observation. Then it is clear that $v_0 := v - u_0$ has the required properties. (This direction of the proof consisted basically of the construction of an orthogonal basis by an orthogonalization process). The converse is a rather obvious consequence of Proposition 2.4.1/5. Indeed, every subspace U of V admits a supplement $U^{\perp}$ such that $V = U + U^{\perp}$ is a norm-direct sum, and this, of course, implies that U is strictly closed. □

For the important special case of a discretely valued ground field K, the notions of weakly K-cartesian and K-cartesian spaces coincide. Namely,

Proposition 3. *Let K be a discretely valued field and let V be a normed finite-dimensional K-vector space. If V is weakly K-cartesian, then V is also K-cartesian. In particular, $|V - \{0\}|$ is discrete.*

Proof. By induction on $n := \dim V$. For $n = 1$, there is nothing to show. For $n > 1$, let U be an arbitrary proper subspace of V. Of course, U is weakly

K-cartesian and of dimension $< n$, and therefore by the induction hypothesis, U is K-cartesian. Hence $|U - \{0\}|$ is discrete. Because U is closed in V, Proposition 1.1.5/4 tells us that U is strictly closed in V. From Proposition 1, we see that V is K-cartesian. □

Finally we need the following criterion:

Proposition 4. *A finite-dimensional normed K-vector space is K-cartesian if and only if it is weakly K-cartesian and its completion is $\hat{K}$-cartesian.*

Proof. Let V be a K-cartesian space, and let $\{v_1, \ldots, v_n\}$ be an orthogonal basis of V. Then, by Proposition 2.3.3/6, we know that $\{v_1, \ldots, v_n\}$ is also a basis of $\hat{V}$ over $\hat{K}$. All we have to show is that $\{v_1, \ldots, v_n\}$ is $\hat{K}$-orthogonal. Let $\hat{v} = \sum_{i=1}^{n} \hat{c}_i v_i$ be an element in $\hat{V}$, where $\hat{c}_1, \ldots, \hat{c}_n \in \hat{K}$. We have to show that $|\hat{v}| = \max_{1 \leq i \leq n} \{|\hat{c}_i| \, |v_i|\}$. Choose coefficients c_i in K such that $c_i = 0$ if $\hat{c}_i = 0$ and such that $|c_i - \hat{c}_i| < \min \{|\hat{v}| \, |v_i|^{-1}, |\hat{c}_i|\}$ if $\hat{c}_i \neq 0$ (in particular, $\hat{v} \neq 0$ if the latter case occurs). Set $v := \sum_{i=1}^{n} c_i v_i$. Then we have $|c_i| = |\hat{c}_i|$ and $|v - \hat{v}| = |\sum_{i=1}^{n} (c_i - \hat{c}_i) v_i| < |\hat{v}|$. Hence $|\hat{v}| = |v| = \max_{1 \leq i \leq n} \{|\hat{c}_i| \, |v_i|\}$. This proves the "only if" part of the proposition.

To show the converse, assume $\hat{V}$ to be $\hat{K}$-cartesian, and let $\{\hat{v}_1, \ldots, \hat{v}_n\}$ be a $\hat{K}$-orthogonal basis of $\hat{V}$. Since all $\hat{v}_i \neq 0$, there are $v_i \in V$ such that $|v_i - \hat{v}_i| < |\hat{v}_i|$ for $i = 1, \ldots, n$. We claim that $\{v_1, \ldots, v_n\}$ is a K-orthogonal set. For arbitrary coefficients $c_1, \ldots, c_n \in K$, we want to show that $|\sum c_i v_i| = \max \{|c_i| \, |v_i|\}$. Assuming that not all c_i are zero, we get

$$|\sum c_i (v_i - \hat{v}_i)| \leq \max \{|c_i| \, |v_i - \hat{v}_i|\} < \max \{|c_i| \, |\hat{v}_i|\} = |\sum c_i \hat{v}_i|$$

and hence

$$|\sum c_i v_i| = |\sum c_i (v_i - \hat{v}_i) + \sum c_i \hat{v}_i| = |\sum c_i \hat{v}_i| = \max \{|c_i| \, |\hat{v}_i|\} = \max \{|c_i| \, |v_i|\}.$$

Since V is weakly K-cartesian, we know $\dim_K V = \dim_{\hat{K}} \hat{V}$. Therefore $\{v_1, \ldots, v_n\}$ is an orthogonal K-basis of V. □

Let us remark that the computations in the two halves of this proof are very similar. The difference is this: in the first half one works with a little perturbation of the scalar coefficients, and in the second half with a perturbation of the basis vectors.

2.4.3. Cartesian spaces of arbitrary dimension. — In this section we no longer require the dimension of the given normed K-vector space to be finite. As in (2.3.2), let $\mathfrak{F}(V)$ denote the family of all finite-dimensional subspaces of V. Then we have

Proposition 1. *For a normed K-vector space V, the following conditions are equivalent:*

(1) *Every $U \in \mathfrak{F}(V)$ is strictly closed in V.*
(2) *Every $U \in \mathfrak{F}(V)$ is K-cartesian.*
(3) *For all $U, W \in \mathfrak{F}(V)$ with $W \subset U$ and* $\dim W = 1$, *there is a norm-direct supplement of W in U.*

We give a cyclic *proof*. To derive (2) from (1), we only have to observe that all subspaces of $U \in \mathfrak{F}(V)$ are strictly closed in V, hence a fortiori in U. Then U is K-cartesian by Proposition 2.4.2/1. That (2) implies (3) is an immediate consequence of Lemma 2.4.1/4. To go from (3) to (1) (via (2)), we just have to use Proposition 2.4.1/6 and Proposition 2.4.2/1. □

According to Proposition 2.4.1/5, a finite-dimensional K-cartesian vector space satisfies condition (2) of the preceding statement. Therefore the following definition is an extension of Definition 2.4.1/1.

Definition 2. *A normed K-vector space V (of arbitrary dimension) is called K-cartesian if it satisfies the (equivalent) conditions of Proposition* 1.

Examples of K-cartesian spaces are provided by the class of normed K-vector spaces admitting an orthonormal Schauder basis (cf. Proposition 2.7.5/1 in a later section). In particular, $T(K)$ is K-cartesian. Furthermore, its field of fractions $Q(T(K))$ is K-cartesian due to the following

Lemma 3. *Let A be a normed K-algebra, and let S be a multiplicative system in A consisting only of multiplicative elements (cf.* (2.1.3)). *If A is K-cartesian, then A_S (provided with the canonical extension of the norm of A) is also K-cartesian.*

Proof. Let U be a finite-dimensional subspace of A_S. Then there are elements $a_i \in A$, $s_i \in S$ $(i = 1, \ldots, n)$ such that all $s_i \neq 0$ and such that $U = \sum_{i=1}^{n} K \frac{a_i}{s_i}$. Without loss of generality, we may assume $s_1 = s_2 = \cdots = s_n =: s$. Define

$$V := sU = \sum_{i=1}^{n} K a_i \subset A .$$

According to the assumption, V has an orthogonal basis $x_1, \ldots, x_m$. Define $y_i := \frac{x_i}{s} \in U$ for $i = 1, \ldots, m$. Then obviously $\{y_1, \ldots, y_m\}$ is a K-generating system for U, but it is also an orthogonal basis, for if $c_1, \ldots, c_m \in K$, one has

$$\left|\sum_{i=1}^{m} c_i y_i\right| = |s|^{-1} \left|\sum_{i=1}^{m} c_i(s y_i)\right| = |s|^{-1} \max_{1 \leq i \leq m} \{|c_i|\,|x_i|\} = \max_{1 \leq i \leq m} \{|c_i|\,|y_i|\} .$$ □

Another procedure to build up K-cartesian spaces is given by

Lemma 4. *Let E be an extension field of K, where E carries a valuation such that E is a K-cartesian K-vector space. Then every E-cartesian space V is K-cartesian.*

Proof. Let U be a K-subspace of V such that $\dim_K U < \infty$. There exist $u_1, \ldots, u_n \in U$ such that $U = \sum_{i=1}^{n} Ku_i$. Define $U' := \sum_{i=1}^{n} Eu_i \subset V$. Then U' is a finite-dimensional E-subspace of the E-cartesian space V, and therefore U' is a norm-direct sum of finitely many "copies" of E, say $U' = \bigoplus_{j=1}^{m} E'_j$ where $E'_j = Ex_j$ for suitable elements $x_1, \ldots, x_m \in U'$. We have $U \subset U'$. Since $\dim_K U < \infty$, there exist finite-dimensional K-subspaces $E_j \subset E'_j$ such that $U \subset \bigoplus_{j=1}^{m} E_j$. Now E and hence all spaces E'_j and E_j are K-cartesian. Therefore $\bigoplus_{j=1}^{m} E_j$ is K-cartesian (see (2.4.1)), since this is a finite norm-direct sum of finite-dimensional K-cartesian spaces. In particular, we see that U, as a subspace of $\bigoplus_{j=1}^{m} E_j$, is K-cartesian. □

Furthermore, the argument used in the above proof shows that

Proposition 5. *The norm-direct sum of K-cartesian spaces (of arbitrary dimension) is K-cartesian.*

Of course, K-cartesian spaces are weakly K-cartesian. Furthermore, if V can be exhausted by a countable ascending family $\{V_i\}$ of K-cartesian spaces, V itself is K-cartesian. The field K itself is K-cartesian, whereas $\hat{K}$ can be K-cartesian only if $\hat{K}$ coincides with K. For later reference we need

Proposition 6. *If V is a normed K-vector space such that every one-dimensional subspace admits a norm-direct supplement in V, then V is K-cartesian and every finite-dimensional subspace admits a norm-direct supplement in V.*

Proof. Let $U, W \in \mathfrak{F}(V)$ with $W \subset U$ and assume $\dim W = 1$. Then W admits a norm-direct supplement $W^\perp$ in V by assumption. The intersection $U \cap W^\perp$ is then a norm-direct supplement of W in U, whence condition (3) of Proposition 1 is fulfilled and V is K-cartesian. In order to show that any $U \in \mathfrak{F}(V)$ admits a norm-direct supplement in V, we proceed by induction on $n := \dim U$. For $n \leqq 1$, there is nothing to show. If $n > 1$, choose a one-dimensional subspace $W \subset U$ and, as before, denote by $W^\perp$ the norm-direct supplement of W in V. Every one-dimensional subspace of $W^\perp$ admits a norm-direct supplement in V and a fortiori in $W^\perp$. Hence we may apply the induction hypothesis to the $(n-1)$-dimensional subspace $U \cap W^\perp$ of the space $W^\perp$ and get a norm-direct supplement S of $U \cap W^\perp$ in $W^\perp$. Clearly, S is also a norm-direct supplement of U in V. □

We want to derive further criteria for testing whether a given weakly K-cartesian space is K-cartesian and to establish connections between the notions of K-cartesian, weakly K-cartesian, and $\hat{K}$-cartesian spaces.

Proposition 7. *A normed K-vector space V is K-cartesian if and only if V is weakly K-cartesian and contains a dense K-cartesian subspace.*

Proof. We only have to show that if V is weakly K-cartesian and admits a dense K-cartesian subspace W, then every finite-dimensional subspace U of V is K-cartesian. In order to do so, it suffices to find a subspace U' of W which is isometric to U. Let $\{u_1, \ldots, u_n\}$ be an arbitrary basis of U. Since V is weakly K-cartesian, U carries the product topology, i.e., one can find a positive real constant α such that $|\sum_{i=1}^{n} c_i u_i| \geqq \alpha \max_{1\leqq i\leqq n} |c_i|$ for all $c_1, \ldots, c_n \in K$. For $i = 1, \ldots, n$, choose $w_i \in W$ such that $|u_i - w_i| < \alpha$ and define $U' := \sum_{i=1}^{n} K w_i$. If at least one of the coefficients c_i is non-zero, we have $|\sum_{i=1}^{n} c_i(u_i - w_i)| < \alpha \max_{1\leqq i\leqq n} |c_i|$. Since $|\sum_{i=1}^{n} c_i u_i| \geqq \alpha \max_{1\leqq i\leqq n} |c_i|$, we can conclude $|\sum_{i=1}^{n} c_i w_i| = \max\left\{|\sum_{i=1}^{n} c_i(u_i - w_i)|, |\sum_{i=1}^{n} c_i u_i|\right\} = |\sum_{i=1}^{n} c_i u_i|$ for all $c_1, \ldots, c_n \in K$. Therefore the map $\varphi\colon U \to U'$ defined by $\varphi\left(\sum_{i=1}^{n} c_i u_i\right) = \sum_{i=1}^{n} c_i w_i$ is an isometric isomorphism. □

For complete fields the preceding proposition can be sharpened in the following direction:

Proposition 8. *A normed $\hat{K}$-vector space V' is $\hat{K}$-cartesian if it contains a dense K-cartesian K-vector space.*

Proof. Let W be a dense K-cartesian K-subspace of V'. We want to show that the $\hat{K}$-space $W' := \hat{K}W$ (which is dense in V') is $\hat{K}$-cartesian. Then Proposition 7 can be applied. For every $U' \in \mathfrak{F}(W')$, we can find a $U \in \mathfrak{F}(W)$ such that $U' \subset \hat{K}U$. Then $\dim_{\hat{K}} \hat{K}U < \infty$ and $\hat{K}U$ is complete (use Proposition 2.3.3/4). Since U is dense in $\hat{K}U$, we may view $\hat{K}U$ as the completion of U. Thus from Proposition 2.4.2/4, we see that $\hat{K}U$ is $\hat{K}$-cartesian, since U is K-cartesian. Therefore, $U' \in \mathfrak{F}(\hat{K}U)$ is $\hat{K}$-cartesian and hence W' is $\hat{K}$-cartesian. □

The two preceding propositions allow us to conclude that a space is cartesian if we know that the property holds for a suitable subspace. Now we want to go the other way around.

Proposition 9. *A normed K-vector space V is K-cartesian if V is weakly K-cartesian and if V is contained in some $\hat{K}$-cartesian space V'.*

Proof. Let $U \in \mathfrak{F}(V)$. We have to show that U is K-cartesian. As in the preceding proof, we can conclude that $\hat{K}U$ equals the completion $\hat{U}$ of U.

Furthermore, $\hat{U} = \hat{K}U \in \mathfrak{F}(V')$ is $\hat{K}$-cartesian, and, since U is weakly K-cartesian, Proposition 2.4.2/4 yields that U is K-cartesian. □

As an immediate corollary to Propositions 8 and 9, we get the following generalization of Proposition 2.4.2/4:

Corollary 10. *A normed K-vector space V is K-cartesian if and only if V is weakly K-cartesian and $\hat{V}$ is $\hat{K}$-cartesian.*

Also Proposition 2.4.2/3 may be easily carried over to the infinite-dimensional case to yield

Corollary 11. *If the valuation on K is discrete, then a normed weakly K-cartesian vector space is K-cartesian. In particular, if K is discretely valued and complete, then every normed K-vector space is K-cartesian.*

2.4.4. Normed vector spaces over a spherically complete field. — In this section we shall consider a class of ground fields K with the property that all normed vector spaces over K are K-cartesian.

Definition 1. *A normed K-vector space V is called spherically complete if every descending sequence of balls $B^+(v_\nu, r_\nu) = \{x \in V;\ |x - v_\nu| \leqq r_\nu\}$, where $v_\nu \in V$ and $r_\nu > 0$, $\nu = 1, 2, \ldots$, has non-empty intersection.*

Of course "spherically complete" implies "complete", and the two notions are equivalent if $|V - \{0\}|$ is discrete.

Our goal for this section is the following

Proposition 2. *If K is spherically complete, then every normed K-vector space is K-cartesian.*

For its proof we need two lemmas.

Lemma 3. *If U is a spherically complete subspace of a normed K-vector space V, then U is strictly closed in V.*

Proof. For $v \in V - U$, define $d := |v, U|$, and, for $\nu \in \mathbb{N}$, choose (by induction on ν) $u_\nu \in U$ such that $d_\nu := |v - u_\nu| \leqq d + \nu^{-1}$ and $d_{\nu+1} \leqq d_\nu$. Define $B_\nu := \{u \in U;\ |u - u_\nu| \leqq d_\nu\}$. Then B_ν is a non-empty ball in U. We claim that $B_\nu \supset B_{\nu+1}$. In order to prove this claim, consider $u_{\nu+1} - u_\nu$. From $|u_{\nu+1} - u_\nu| \leqq \max\{|u_{\nu+1} - v|,\ |v - u_\nu|\} = \max\{d_{\nu+1}, d_\nu\} = d_\nu$, we deduce $u_{\nu+1} \in B_\nu$, and, because $d_{\nu+1} \leqq d_\nu$, this implies $B_{\nu+1} \subset B_\nu$. Since U is spherically complete by assumption, we can find an element $u \in \bigcap_{\nu \in \mathbb{N}} B_\nu$. Then one has $|u - v| \leqq \max\{|u - u_\nu|,\ |u_\nu - v|\} \leqq d_\nu \leqq d + \nu^{-1}$ for all $\nu \in \mathbb{N}$, and thus $|u - v| = d = |v, U|$. Hence U is strictly closed in V. □

Lemma 4. *If K is spherically complete and V is a finite-dimensional K-cartesian space, then V is spherically complete.*

Proof. Let $\{v_1, \ldots, v_n\}$ be a K-orthogonal basis of V. First we show that all balls in V can be written as n-fold norm-direct sums of suitable balls in K. More precisely, let $v = \sum_{i=1}^{n} c_i v_i \in V$ with $c_1, \ldots, c_n \in K$, and let $r > 0$. Then the following statements for $x = \sum_{i=1}^{n} a_i v_i \in V$ are equivalent:

$$|x - v| \leqq r \Leftrightarrow \max_{1 \leqq i \leqq n} |a_i - c_i|\, |v_i| \leqq r \Leftrightarrow |a_i - c_i| \leqq r\, |v_i|^{-1},\ i = 1, \ldots, n$$
$$\Leftrightarrow a_i \in B^+(c_i, r\, |v_i|^{-1}),\ i = 1, \ldots, n.$$

Hence $B^+(v, r) = \sum_{i=1}^{n} B^+(c_i, r\, |v_i|^{-1})\, v_i$. This equality allows us to transfer the property "spherically complete" from K to V; namely, let $B_\nu = B^+(w_\nu, r_\nu)$ (where $w_\nu = \sum_{i=1}^{n} c_{\nu i} v_i \in V$ and $r_\nu > 0$, $\nu = 1, 2, \ldots$) be an arbitrary descending sequence of balls in V. Define $B_{\nu i} := B^+(c_{\nu i}, r_\nu |v_i|^{-1}) \subset K$ for $\nu \in \mathbb{N}$ and $i = 1, \ldots, n$. Then $B_\nu = \sum_{i=1}^{n} B_{\nu i} \cdot v_i$. Therefore, for fixed i, $(B_{\nu i})_{\nu \in \mathbb{N}}$ is a descending sequence of balls in K. Take $a_i \in \bigcap_{\nu \in \mathbb{N}} B_{\nu i}$. Then $w := \sum_{i=1}^{n} a_i v_i \in \sum_{i=1}^{n} B_{\nu i} \cdot v_i = B_\nu$ for all $\nu \in \mathbb{N}$. Thus V is spherically complete. □

Proof of Proposition 2. We may assume that V is of finite dimension and proceed by induction on $n := \dim V$. For $n = 0$ or 1, there is nothing to show. Let U be a proper subspace of V and apply the induction hypothesis to U. Then U is K-cartesian and furthermore, due to Lemma 4, spherically complete. By Lemma 3, the subspace U is strictly closed in V. Since this holds for all proper subspaces U, Proposition 2.4.2/1 tells us that V is K-cartesian. □

The proposition and its proof strongly resemble Proposition 2.3.3/4, which asserts that the completeness of K implies that all normed K-vector spaces are weakly K-cartesian. Furthermore, the proposition characterizes spherically complete fields: i.e., if K is not spherically complete, there is always a normed vector space over K which is not K-cartesian. Indeed, if K is not spherically complete, it is not maximally complete, i.e., it admits a proper field extension L having the same residue field and the same value group (for the equivalence of spherical completeness and maximal completeness see, e.g., I. Kaplansky [22]). Take $l \in L - K$ and set $V := K + lK$. We claim that V does not admit an orthogonal basis. Assume the contrary, and let $\{v_1, v_2\}$ be a K-orthogonal basis of V. Because $|K^*| = |L^*|$, we may assume $|v_1| = |v_2| = 1$. Since $\tilde{L} = \tilde{K}$, we know that $\tilde{v}_2 = \tilde{c}\tilde{v}_1$ for a suitable $c \in \mathring{K}$. Therefore $|cv_1 - v_2| < 1$, a contradiction to the orthogonality of $\{v_1, v_2\}$. Hence V is not K-cartesian.

2.5. Strictly cartesian spaces

In (2.4), we studied normed K-vector spaces admitting orthogonal bases. Now we want to restrict our attention to the existence of *orthonormal* bases. As always, let K be a field with a non-trivial valuation, and let V be a normed K-vector space.

Definition 1. *A system $\{v_i; i \in I\}$ of elements of V is called orthonormal if, for every system $\{c_i; i \in I\}$ of elements in K such that $c_i = 0$ for almost all $i \in I$, one has*

$$|\sum_{i\in I} c_i v_i| = \max_{i\in I} |c_i| .$$

Obviously, an orthonormal system is linearly independent over K and all its elements must have a norm equal to 1.

As in (2.3) and (2.4), let us consider the finite-dimensional case first.

2.5.1. Finite-dimensional strictly cartesian spaces. — Throughout this section, we assume $\dim_K V < \infty$. We introduce the concept of a strictly K-cartesian vector space by

Definition 1. *V is called strictly K-cartesian if there is an orthonormal basis of V.*

Clearly V is strictly K-cartesian if and only if there is a generating system $\{v_1, \ldots, v_n\}$ of V such that

$$(*) \qquad |\sum_{i=1}^{n} c_i v_i| = \max_{1\leq i\leq n} |c_i| \quad \text{for all} \quad c_1, \ldots, c_n \in K.$$

The concepts "cartesian" and "strictly cartesian" are closely connected, as the following statement shows.

Observation 2. *V is strictly K-cartesian if and only if V is K-cartesian and $|V| \subset |K|$.*

Proof. Obviously a strictly K-cartesian space V is K-cartesian and fulfills $|V| \subset |K|$ due to the equality (*). On the other hand, if V is K-cartesian, then there is an orthogonal basis $\{w_1, \ldots, w_n\}$ of V, and since $|V| \subset |K|$, there exist $a_i \in K - \{0\}$ such that $|a_i| = |w_i|$, $i = 1, \ldots, n$. Define $v_i := a_i^{-1} w_i$. Clearly $\{v_1, \ldots, v_n\}$ is a basis of V over K, and, for all $c_1, \ldots, c_n \in K$, one gets $|\sum c_i v_i| = |\sum c_i a_i^{-1} w_i| = \max |c_i a_i^{-1}|\,|w_i| = \max |c_i|\,|a_i|^{-1}\,|w_i| = \max |c_i|$. □

We could use this observation in order to specialize our results on K-cartesian spaces to the strictly K-cartesian case. However strictly K-cartesian spaces enjoy particular properties which behave well with respect to the functor $\sim$ defined in (2.1.10). Therefore we will give direct proofs for the main results on strictly K-cartesian spaces. As we will see, there is substantial simplification due to the functor $\sim$.

A surprisingly simple remark regulates the going up and down between V and $V^\sim$:

Lemma 3. *Let $v_1, \ldots, v_n \in V^\circ$. Then $\{v_1, \ldots, v_n\}$ is orthonormal if and only if $\{v_1^\sim, \ldots, v_n^\sim\}$ is linearly independent over $\tilde{K}$.*

Proof. First we assume that $\{v_1^\sim, \ldots, v_n^\sim\}$ is linearly independent. We have to show that $|\sum c_i v_i| = \max |c_i|$ for all $(c_1, \ldots, c_n) \in K^n$. Assuming that $|c_1| = \max\limits_{1 \leqq i \leqq n} |c_i| > 0$, we define $d_i := c_1^{-1} c_i \in K$. Then one knows $|d_i| \leqq 1$ and $|d_1| = 1$. Therefore $\tilde{d}_1, \ldots, \tilde{d}_n \in \tilde{K}$ are well-defined, and $\tilde{d}_1 \neq 0$. Since $\{v_1^\sim, \ldots, v_n^\sim\}$ is linearly independent, we conclude that $(\sum d_i v_i)^\sim = \sum \tilde{d}_i v_i^\sim \neq 0$; i.e., $|\sum d_i v_i| = 1$. Hence $|\sum c_i v_i| = |c_1| \, |\sum d_i v_i| = |c_1| = \max |c_i|$. To show the converse, assume that $\{v_1, \ldots, v_n\}$ is orthonormal. Let $d_1, \ldots, d_n$ be coefficients in $\mathring{K}$ such that $\sum \tilde{d}_i v_i^\sim = 0$. Then we get $\max |d_i| = |\sum d_i v_i| < 1$, whence $\tilde{d}_i = 0$ for $i = 1, \ldots, n$. □

Now we can derive the following criterion:

Proposition 4. *Let V be strictly K-cartesian, and let $S = \{v_1, \ldots, v_n\}$ be contained in V°. Then S is an orthonormal basis of V over K if and only if $S^\sim$ is a basis of $V^\sim$ over $\tilde{K}$.*

Proof. First we assume that S is an orthonormal basis of V. Then $S^\sim$ is linearly independent according to Lemma 3; it remains to be shown that $S^\sim$ is a generating system. For all $\alpha \in V^\sim$, there is a $v \in V^\circ$ such that $v^\sim = \alpha$. One can find $c_1, \ldots, c_n \in K$ with $v = \sum\limits_{i=1}^{n} c_i v_i$. From $1 \geqq |v| = |\sum c_i v_i| = \max |c_i|$, we get $c_i \in \mathring{K}$ and hence $\alpha = v^\sim = \sum\limits_{i=1}^{n} \tilde{c}_i v_i^\sim$, which proves the contention. In particular, we see that $\dim_K V = \dim_{\tilde{K}} V^\sim$. Now assume that $S^\sim$ is a basis. Then S is orthonormal and a fortiori linearly independent over K. Since $\dim_K V = \dim_{\tilde{K}} V^\sim$, it follows that S is a basis. □

The assumption that V is strictly K-cartesian cannot be dropped. The property "strictly K-cartesian" is inherited by subspaces according to

Proposition 5. *Let V be a strictly K-cartesian space and U a subspace. Then U is strictly K-cartesian and every orthonormal basis of U can be enlarged to an orthonormal basis of V.*

Proof. To show the first assertion, we proceed as follows: choose $u_1, \ldots, u_t \in U^\circ$ such that $\{u_1^\sim, \ldots, u_t^\sim\}$ is a $\tilde{K}$-basis of $U^\sim$. Enlarge this set (by the usual procedure) to a basis of $V^\sim$ by adding $u_{t+1}^\sim, \ldots, u^\sim$, where $u_{t+1}, \ldots, u_n$ are suitable elements in V°. Then due to the preceding proposition $\{u_1, \ldots, u_n\}$ is an orthonormal basis of V, and therefore $\{u_1, \ldots, u_t\}$ is an orthonormal basis of $W := \sum\limits_{i=1}^{t} K u_i \subset U$. It remains to be shown that $W = U$. Take $u \in U$; then there are $c_1, \ldots, c_n \in K$ with $u = \sum\limits_{i=1}^{n} c_i u_i$. We are done if we can show $c_i = 0$ for $i = t+1, \ldots, n$. Otherwise, we may assume $\max\limits_{t+1 \leqq i \leqq n} |c_i| = |c_n| = 1$. Applying the functor $^\sim$, we get $\sum_{i=t+1}^{n} \tilde{c}_i u_i^\sim = \left(u - \sum_{i=1}^{t} c_i u_i\right)^\sim \in U^\sim$ and $\tilde{c}_n \neq 0$, a contradiction to the construction of $\{u_{t+1}^\sim, \ldots, u_n^\sim\}$. Hence $W = U$ and $\{u_1, \ldots, u_t\}$ is an orthonormal basis of U. To convince ourselves of the truth of the second assertion, we only have to start with an orthonormal basis of U, push it down to a basis of $U^\sim$, enlarge this basis to a basis of $V^\sim$ and lift the enlarged basis to an orthonormal basis of V. □

Corollary 6. *V is strictly K-cartesian if and only if* $\dim_K V = \dim_{\tilde{K}} V^\sim$.

Proof. It follows immediately from Proposition 4 that $\dim_K V = \dim_{\tilde{K}} V^\sim$ for a strictly K-cartesian space V. To show the converse, assume that $n = \dim_K V = \dim_{\tilde{K}} V^\sim$ and that $\{v_1, \ldots, v_n\} \subset V^\circ$ is such that $\{v_1^\sim, \ldots, v_n^\sim\}$ is a basis of $V^\sim$. Then, by Lemma 3, $\{v_1, \ldots, v_n\}$ is orthonormal over K and, since $n = \dim_K V$, an orthonormal basis of V. □

2.5.2. Strictly cartesian spaces of arbitrary dimension. — Similar to the case of weakly K-cartesian spaces and of K-cartesian spaces, Proposition 2.5.1/5 allows us to define:

Definition 1. *A normed K-vector space V is called strictly K-cartesian if every finite-dimensional subspace is strictly K-cartesian.*

We see that $K^{(\infty)}$ is an *example* of a strictly K-cartesian space, because every finite-dimensional subspace is contained in K^n for n large enough. Also the space $c(K)$ of all zero sequences in K is strictly K-cartesian. This is asserted by Proposition 2.7.5/1 in a later section or can be verified already at this stage as follows: since $|c(K)| = |K|$, we only have to show that $c(K)$ is K-cartesian. According to Proposition 2.3.2/7, the space $c(K)$ is at least weakly K-cartesian. Because $K^{(\infty)}$ is dense in $c(K)$, we may apply Proposition 2.4.3/7 to see that $c(K)$ is K-cartesian and so strictly K-cartesian.

One sees immediately that Observation 2.5.1/2 carries over verbatim to the infinite-dimensional case. According to Definition 2.5/1, a system $\{v_i; i \in I\}$ is orthonormal if and only if every finite subsystem is orthonormal. Therefore, Lemma 2.5.1/3 carries over mutatis mutandis; i.e., one has

Lemma 2. *Let $\{v_i; i \in I\}$ be a system of elements in V°. Then $\{v_i; i \in I\}$ is orthonormal if and only if $\{v_i^\sim; i \in I\}$ is linearly independent over $\tilde{K}$.*

We want to tackle the question of whether or not Propositions 2.5.1/4 and 2.5.1/5 can be generalized to the infinite-dimensional case. Let us first look at our standard examples. One verifies immediately that $(K^{(\infty)})^\sim = c(K)^\sim = \tilde{K}^{(\infty)}$, since $K^{(\infty)}$ is dense in $c(K)$, and since the canonical orthonormal basis of $K^{(\infty)}$ induces a $\tilde{K}$-linearly independent system of elements in $(K^{(\infty)})^\sim$ which generates this vector space. However, the canonical basis of $K^{(\infty)}$ does not generate $c(K)$. Furthermore, it can be shown that there are no orthogonal systems in $c(K)$ which generate $c(K)$ as K-vector space. If K is complete, this can be seen as follows. The length of such a basis would be countably infinite due to Lemma 2.

Hence $c(K)$ would be isometric to $K^{(\infty)}$. However $c(K)$ is complete, whereas $K^{(\infty)}$ is not. In particular, we see that any lifting of a $\tilde{K}$-basis of $\tilde{K}^{(\infty)}$ is an orthonormal system in $c(K)$ which does not generate $c(K)$. Therefore, Proposition 2.5.1/4 does not carry over to the infinite-dimensional case. Again, because $c(K)$ has no orthogonal basis, in spite of being strictly K-cartesian, Proposition 2.5.1/5 cannot be extended to the infinite-dimensional case. The picture changes completely if one considers Schauder bases, as shall be done in (2.7).

2.6. Weakly cartesian spaces of countable dimension

As always, let K denote a field with a non-trivial valuation. All vector spaces which occur are K-normed. For an arbitrary index set I, let $e_i := (\delta_{ij})_{j\in I}$, $i \in I$, denote the canonical K-basis of $K^{(I)}$. A vector space V is said to have *countable dimension* if there exists a K-linear bijection $K^{(\mathbb{N})} \to V$.

2.6.1. Weakly cartesian bases. — Each K-basis $\{y_i\}_{i\in I}$ of a vector space V induces a K-linear bijection $\Phi\colon K^{(I)} \to V$ given by $\sum_{i\in I} a_i e_i \mapsto \sum_{i\in I} a_i y_i$ (of course, $a_i = 0$ for almost all $i \in I$). If $\sup_i |y_i| < \infty$, then Φ is bounded with $|\Phi| = \sup_i |y_i|$ (cf. Proposition 2.2.2/1).

Fixing an element $\varrho \in |K|$, $\varrho > 1$, we see by Proposition 2.1.8/1 that for each vector $v \in V, v \neq 0$, there exists an element $c \in K^*$ such that $1 \leqq |cv| \leqq \varrho$. Thus, from any basis of V, we can pass (just multiply each basis vector by an appropriate constant $c \in K^*$) to a basis $\{y_i\}_{i\in I}$ such that $1 \leqq |y_i| \leqq \varrho$ for all $i \in I$. If we call such sets of V *bounded* (more precirely, *ϱ-bounded*), we have proved

Proposition 1. *Each space V admits a ϱ-bounded basis. Each such basis gives rise to a bounded K-linear bijection $\Phi\colon K^{(I)} \to V$ with $1 \leqq |\Phi| \leqq \varrho$.*

As in the finite-dimensional case, the Φ's occurring in Proposition 1 need not be homeomorphisms (e.g., assume that K is not complete and take for V the completion $\hat{K}$ of K). Next we derive a sufficient condition for Φ^{-1} to be bounded.

Proposition 2. *Let $\{y_i\}_{i\in I}$ be a ϱ-bounded basis of V, and let $\alpha > 0$ be a real number such that*

$$\max_{i\in I} \{|a_i y_i|\} \leqq \alpha \left|\sum_{i\in I} a_i y_i\right|$$

for all vectors $\sum_{i\in I} a_i y_i$ of V. Then $|\Phi^{-1}| \leqq \alpha$.

Proof. Since $|y_i| \geqq 1$ for all $i \in I$, we have

$$|\Phi^{-1}(\sum_{i\in I} a_i y_i)| = |\sum_{i\in I} a_i e_i| = \max_{i\in I} |a_i| \leqq \max_{i\in I} \{|a_i|\,|y_i|\} \leqq \alpha \left|\sum_{i\in I} a_i y_i\right|$$

for all vectors of V. □

The following definition is motivated by the last proposition.

Definition 3. *Let α be a positive real number. A ϱ-bounded family $\{y_i; i \in I\}$ of V with $y_i \neq 0$ is called α-cartesian if*

$$\max_{i\in I} \{|a_i y_i|\} \leqq \alpha \ |\sum_{i\in I} a_i y_i|$$

for every vector $v = \sum a_i y_i \in V$, where $a_i = 0$ for almost all $i \in I$.

A ϱ-bounded family $\{y_i; i \in I\}$ is called weakly K-cartesian (or simply weakly cartesian) if there exists a real number $\alpha > 0$ such that it is α-cartesian.

If the family $\{y_i; i \in I\}$ is α-cartesian, then $\alpha \geqq 1$. Proposition 2 and Definition 3 imply

Proposition 4. *For each normed vector space V admitting a weakly K-cartesian basis, there is a linear homeomorphism onto the space $K^{(I)}$. In particular, all these spaces are b-separable and weakly K-cartesian.*

2.6.2. Existence of weakly cartesian bases. Fundamental theorem. — The question is whether each weakly cartesian space possesses a weakly cartesian basis (and hence is b-separable). The answer is no, as we will point out at the end of this section. However, we shall prove that each weakly cartesian space of countable dimension has weakly cartesian bases. The proof depends on the following two rather technical observations, the first one being a criterion for a basis of V to be weakly cartesian. As before, we denote by ϱ a fixed element in $|K|$ such that $\varrho > 1$.

Observation 1. *Let $\alpha > 1$. A ϱ-bounded family $\{y_1, y_2, \ldots\}$ of V is α-cartesian if there exists a strictly increasing sequence $1 =: \alpha_1 < \alpha_2 < \cdots$ of real numbers converging to α such that, for $n = 1, 2, \ldots$, we have*

$$(*) \qquad \alpha_n \cdot \max\{|u|, |a y_{n+1}|\} \leqq \alpha_{n+1} |u + a y_{n+1}| \quad \text{for all} \quad a \in K, \quad u \in \sum_1^n K y_\nu.$$

Proof. It is enough to show by induction on n that

$$(\text{o}) \qquad \max_{1\leqq\nu\leqq n} \{|a_\nu y_\nu|\} \leqq \alpha_n \left|\sum_1^n a_\nu y_\nu\right|.$$

This is clear for $n = 1$, since $\alpha_1 = 1$. If we already know (o) for n, we conclude from $(*)$

$$\alpha_{n+1} \left|\sum_1^{n+1} a_\nu y_\nu\right| \geqq \alpha_n \max\left\{\left|\sum_1^n a_\nu y_\nu\right|, |a_{n+1} y_{n+1}|\right\}$$
$$\geqq \max\{|a_1 y_1|, \ldots, |a_n y_n|, \alpha_n |a_{n+1} y_{n+1}|\},$$

which gives (o) for $n + 1$ since $\alpha_n \geqq 1$. □

The next observation describes the crucial step for the construction of weakly cartesian bases.

Observation 2. *Let V be any K-space (not necessarily weakly cartesian and not necessarily of countable dimension). Let $U \subset V$ be a K-subspace, and let*

$x \in V$ be not in the closure of U. Then for each real number $\beta > 1$, there exists a vector $y \in U' := U + Kx$ such that $U' = U + Ky$ and such that

$$\max\{|u|, |ay|\} \leqq \beta\,|u + ay| \quad \text{for all} \quad u \in U, \quad a \in K.$$

Proof. Since $x \notin \overline{U}$, we have $|x, U| = \inf_{u \in U} |x + u| > 0$. Choose $u_0 \in U$ such that $|x + u_0| \leqq \beta\,|x, U|$. We claim that $y := x + u_0$ has the required properties. Since $y \notin U$, we have $U' = U + Ky$. If $|u| \neq |ay|$, we have (since $\beta > 1$)

$$\beta\,|u + ay| \geqq |u + ay| = \max\{|u|, |ay|\}.$$

So assume $|u| = |ay|$. It remains to show that $|ay| \leqq \beta\,|u + ay|$. We may assume $a \neq 0$. The condition $|y| \leqq \beta\,|x, U|$ implies (since $u_0 \in U$)

$$\beta\,|x + u_0 + a^{-1}u| \geqq |y|.$$

Multiplying by $|a|$ and using $x + u_0 = y$, we get $\beta\,|ay + u| \geqq |ay|$. □

Now the proof of the following proposition which may be regarded as a non-Archimedean analogue of the classical "Orthogonalisierungsprozess" of E. Schmidt is fairly simple:

Proposition 3. *Let V be weakly cartesian of at most countable dimension. Let $\{v_i;\ 1 \leqq i < d\}$ be any basis of V (where $d = \infty$ if V is infinite-dimensional). Then for each $\alpha > 1$, there is an α-cartesian basis $\{y_i;\ 1 \leqq i < d\}$ of V such that*

$$\sum_1^n K v_i = \sum_1^n K y_i \quad \text{for all } n, \quad 1 \leqq n < d.$$

Proof. Set $U_n := \sum_1^n K v_i$, and choose a strictly increasing sequence $1 =: \alpha_1 < \alpha_2 < \cdots$ of real numbers converging to α. Due to Observation 1, it is enough to construct a system $\{y_i;\ 1 \leqq i < d\}$ of vectors in V with the following properties:

$$(1) \qquad 1 \leqq |y_n| \leqq \varrho \quad \text{and} \quad U_n = \sum_1^n K y_i \quad \text{for all } n,$$

$$(2) \qquad \alpha_n \max\{|u|, |ay_{n+1}|\} \leqq \alpha_{n+1}\,|u + ay_{n+1}|, \quad a \in K, \quad u \in U_n \quad \text{for all } n.$$

We proceed by induction on n: choose $y_1 := c_1 v_1$, $c_1 \in K^*$, such that $1 \leqq |y_1| \leqq \varrho$. Let $y_1, \ldots, y_n$ be already constructed, $n \geqq 1$. Then $U_n = \sum_1^n K y_i$. Since V is weakly cartesian, U_n is closed in V, and hence v_{n+1} is not in the closure of U_n. Thus, we may apply Observation 2 (with $U := U_n$, $x := v_{n+1}$ and $\beta := \frac{\alpha_{n+1}}{\alpha_n}$). We get a vector $y \in U_{n+1}$ such that $U_{n+1} = U_n + Ky$ and

$$\max\{|u|, |ay|\} \leqq \frac{\alpha_{n+1}}{\alpha_n}\,|u + ay| \quad \text{for all} \quad a \in K, \quad u \in U_n = \sum_1^n K y_i.$$

Choose $c \in K^*$ such that $1 \leqq |cy| \leqq \varrho$, and set $y_{n+1} := cy$. Then (1) and (2) are fulfilled for $y_1, \ldots, y_n, y_{n+1}$. □

As the main conclusion of the results of this section, we can now state

Theorem 4. *For each weakly K-cartesian vector space V of countable (non-finite) dimension, there exists a linear homeomorphism onto $K^{(\infty)}$.*

In fact, we proved slightly more: *For each $\varrho \in |K|$, $\varrho > 1$, and each $\alpha > 1$, there is a linear homeomorphism $\Phi\colon K^{(\infty)} \to V$ such that $|\Phi| \leqq \varrho$, $|\Phi^{-1}| \leqq \alpha$.*

We explicitly state an easy but very important consequence of Theorem 4

Corollary 5. *Each weakly cartesian vector space of countable dimension is b-separable.*

The question arises whether this statement is true for each weakly cartesian vector space without any assumption on the dimension. The answer is no; more precisely,

There exists a complete valued field K and a weakly cartesian vector space $V \neq 0$ such that each continuous K-linear map $\lambda\colon V \to K$ is trivial.

For the proof see [31].

2.7. Normed vector spaces of countable type. The Lifting Theorem

In this section we always assume that K is complete and that its valuation is non-trivial.

2.7.1. Spaces of countable type. — By Proposition 2.3.3/4 all normed K-vector spaces are weakly cartesian (K is complete, as we said). For such spaces we now introduce a concept generalizing the notion of "weakly cartesian spaces of countable dimension".

Definition 1. *A normed K-vector space V is said to be of countable type if V contains a dense linear subspace of at most countable dimension.*

The space $c(K)$ of all zero sequences over K is of countable type, since $K^{(\infty)}$ is dense in $c(K)$. This is, in fact, the most general example of a space of countable type, as can be seen from

Proposition 2. *Each normed K-vector space V of countable type admits a linear homeomorphism onto some space K^n or onto a dense subspace of $c(K)$. In particular, V is b-separable.*

Because $c(K)$ is b-separable (see Corollary 2.2.5/3), we only have to *prove* the first assertion. If $\dim_K V < \infty$, the assertion follows from Proposition 2.3.3/4. If $\dim_K V = \infty$, we can apply Theorem 2.6.2/4 and the fact that $c(K)$ is the completion of $K^{(\infty)}$. □

However, we want to deal with the infinite-dimensional case more explicitly.

Proposition 3. *Let W be a dense linear subspace of V of countable dimension, and let $\{w_i; i \in \mathbb{N}\}$ be an α-cartesian (and ϱ-bounded) basis of W (cf. Proposition 2.6.2/3). Then the map $\psi: W \to K^{(\infty)}$, given by*

$$\sum c_i w_i \mapsto (c_i), \quad \textit{where} \quad c_i = 0 \quad \textit{for almost all} \quad i \in \mathbb{N},$$

extends uniquely to a strict K-linear injection $\Psi: V \to c(K)$. We have

$$\text{(o)} \qquad \alpha^{-1} |\Psi(v)| \leqq |v| \leqq \varrho\, |\Psi(v)| \quad \textit{for all} \quad v \in V.$$

The map Ψ is an epimorphism if and only if V is complete so that $V = \left\{ \sum_{i \in \mathbb{N}} c_i w_i; c_i \in K,\ c_i \to 0 \right\}$ in this case.

Proof. The map ψ is a homeomorphism. Namely, for each $w = \sum_{i \in \mathbb{N}} c_i w_i \in W$, where $c_i = 0$ for almost all $i \in \mathbb{N}$, we have

$$\alpha^{-1} \max_{i \in \mathbb{N}} |c_i|\, |w_i| \leqq |w| \leqq \max_{i \in \mathbb{N}} |c_i|\, |w_i|.$$

Since $1 \leqq |w_i| \leqq \varrho$ for all $i \in \mathbb{N}$ and since $\max_{i \in \mathbb{N}} |c_i| = |\psi(w)|$, we conclude that

$$\alpha^{-1} |\psi(w)| \leqq |w| \leqq \varrho\, |\psi(w)| \quad \text{for all} \quad w \in W.$$

Because W is dense in V and $c(K)$ is complete and contains $K^{(\infty)}$, the map ψ can be extended to a K-linear homeomorphism $\hat{V} \to c(K)$. Let $\Psi: V \to c(K)$ be its restriction to V. By continuity arguments, (o) holds. Hence Ψ is injective and bounded, and $\Psi^{-1}: \Psi(V) \to V$ is bounded. The last assertion is obvious. □

Proposition 4. *Let V be a complete space of countable type. Then every closed subspace U of V is also of countable type and admits a direct supplement $U^\perp$ in V such that the canonical map $U \oplus U^\perp \to V$ is a homeomorphism.*

Proof. We can exclude the trivial case $U = V$. Because U is closed, we may thus assume that $\overline{U} \neq V$. Then V/U provided with $|\ \ |_{\mathrm{res}}$ is a normed vector space. Let $\pi: V \to V/U$ denote the canonical residue epimorphism. The space V contains a dense subspace W of at most countable dimension. Then $\pi(W)$ is dense in V/U and also of at most countable dimension. By Proposition 2.6.2/3, for all $\alpha > 1$, we can find a (ϱ-bounded) α-cartesian basis $\{\pi(v_i); i \in I\}$ of $\pi(W)$, where I is at most countable and v_i is a suitable element of V for all $i \in I$. Furthermore, we may assume that $|v_i| \leqq 2\, |\pi(v_i)|_{\mathrm{res}}$. Define $U^\perp := \sum_{i \in I} K v_i \subset V$. We have to show that $U^\perp$ is a supplement of U in V. First we prove that $U \cap U^\perp = \{0\}$. In order to do so, take a sequence $x_\nu = \sum_{i \in I} \lambda_{\nu i} v_i \in \overline{\sum_{i \in I} K v_i}$, $\nu \in \mathbb{N}$, $\lambda_{\nu i} \in K$, such that x_ν converges to some element $u \in U \cap U^\perp$. Then $\pi(x_\nu) = \sum_{i \in I} \lambda_{\nu i} \pi(v_i)$ tends to zero, whence the elements $\lambda_{\nu i}$, $\nu \in \mathbb{N}$, $i \in I$, form a zero sequence. Therefore also $(x_\nu)_{\nu \in \mathbb{N}}$ is a zero sequence, and we have proved that $U \cap U^\perp = \{0\}$. It remains to be shown that $V = U + U^\perp$. Let $v \in V$. Then there are $\lambda_i \in K$,

$i \in I$, such that $|\lambda_i|$ tends to zero and $\pi(v) = \sum_{i \in I} \lambda_i \pi(v_i)$ (use the fact that $\pi(W)$ is dense in V/U). Since V is complete, $u^\perp := \sum_{i \in I} \lambda_i v_i$ exists and is clearly an element of $U^\perp$. Furthermore, $u := v - u^\perp \in \ker \pi = U$. Hence V is the direct sum of U and $U^\perp$ in the ordinary sense. Moreover, we have the estimates:

$$|u^\perp| = |\sum_{i \in I} \lambda_i v_i| \leqq \max_{i \in I} |\lambda_i|\ |v_i| \leqq 2 \max_{i \in I} |\lambda_i|\ |\pi(v_i)|_{\text{res}} \leqq 2\alpha \left| \sum_{i \in I} \lambda_i \pi(v_i) \right|_{\text{res}}$$

$$= 2\alpha\ |\pi(v)|_{\text{res}} \leqq 2\alpha\ |v|\,.$$

Since $u = v - u^\perp$, we get $|u| \leqq \max\{|v|, |u^\perp|\} \leqq (2\alpha)\ |v|$, and therefore $|v| \leqq \max\{|u|, |u^\perp|\} \leqq (2\alpha)\ |v|$, which shows that the canonical map $U \oplus U^\perp \to V$ is a homeomorphism. Hence the canonical isomorphism $\varphi\colon V/U^\perp \to U$ is in fact a homeomorphism. According to what we showed in the very beginning of the proof, the quotient space of V modulo a closed subspace is of countable type, and thus $V/U^\perp$ and U are of countable type. □

Proposition 5. *The space $b(K)$ is not of countable type. In particular, there is no linear homeomorphism of $b(K)$ onto $c(K)$.*

Before we can prove this, we need a few auxiliary statements.

Lemma 6. *Let L be a field. Then the dimension of the direct product $\prod_1^\infty L$ over L is uncountable.*

Proof. First we assume that the cardinality of L is at most countable. Since L contains at least two elements, Cantor's diagonal procedure shows that $\prod_1^\infty L$ is uncountable. Therefore, its dimension over L must be uncountable, because a space of countable dimension over a countable field is still countable. Next we consider the general case where L may be uncountable. Let P be the prime field of L. Then P is finite or equals $\mathbb{Q}$. In any case, the cardinality of P is at most countable. According to what we have proved already, we know that $\dim_P \prod_1^\infty P$ is uncountable. By tensoring with L, we find $\dim_L \left(\prod_1^\infty P\right) \otimes_P L = \dim_P \prod_1^\infty P$. If we can show that $\left(\prod_1^\infty P\right) \otimes_P L$ may be embedded into $\prod_1^\infty L$, then the proof of the lemma is complete. By defining $\Phi\big((p_i)_{i \in \mathbb{N}}, c\big) := (p_i c)_{i \in \mathbb{N}}$, for all $p_i \in P$ and $c \in L$, we get a P-bilinear map $\Phi\colon \left(\prod_1^\infty P\right) \times L \to \prod_1^\infty L$. It induces a P-linear map $\varphi\colon \left(\prod_1^\infty P\right) \otimes_P L \to \prod_1^\infty L$. We have to show that φ is injective. Let $x = \sum_{\nu=1}^n q_\nu \otimes c_\nu$ be an element of $\ker \varphi$, where $c_\nu \in L$ and $q_\nu = (p_{\nu i})_{i \in \mathbb{N}} \in \prod_1^\infty P$. We may assume that $(c_1, \ldots, c_n)$ is linearly independent over P (otherwise one can eliminate one of the c_ν and shorten the representation of x). Then $\varphi(x) = 0$

implies $\sum_{\nu=1}^{n} p_{\nu i} c_\nu = 0$ for all $i \in \mathbb{N}$. Since $(c_1, \ldots, c_n)$ are linearly independent, $p_{\nu i}$ must vanish for $\nu = 1, \ldots, n$ and all $i \in I$. Hence $q_\nu = 0$ for $\nu = 1, \ldots, n$, and therefore $x = 0$. □

Lemma 7. *The space $b(K)$ contains an uncountable orthonormal system.*

Proof. There is a well-defined K-homomorphism $\varphi\colon \big(b(K)\big)^\sim \to \prod_1^\infty \tilde{K}$ such that $\varphi\big((c_i)^\sim_{i\in\mathbb{N}}\big) = (\tilde{c}_i)_{i\in\mathbb{N}}$. Let $\tau\colon \big(b(K)\big)^\circ \to \big(b(K)\big)^\sim$ denote the canonical residue epimorphism, and set $\psi := \varphi \circ \tau$. Then we have the following diagram

$$\begin{array}{ccc} \big(b(K)\big)^\circ & & \\ {\scriptstyle\tau}\downarrow & \searrow^{\psi} & \\ \big(b(K)\big)^\sim & \xrightarrow{\varphi} & \prod_1^\infty \tilde{K}\,. \end{array}$$

Both φ and ψ are surjective, but φ is not, in general, injective. By Lemma 6, there exists an uncountable subset $\{q_j; j \in J\}$ of $\prod_1^\infty \tilde{K}$ which is linearly independent over $\tilde{K}$. Let $v_j \in \big(b(K)\big)^\circ$ be a pre-image of q_j. The set $\{\tau(v_j); j \in J\}$ is linearly independent over $\tilde{K}$, since its image under φ has this property. By Lemma 2.5.2/2, the set $\{y_j; j \in J\}$ is orthonormal and, of course, uncountable. □

By Lemma 7, we know already that $b(K)$ cannot be isometrically isomorphic to $c(K)$, since $\big(b(K)\big)^\sim$ has uncountable dimension over $\tilde{K}$, whereas the dimension of $\big(c(K)\big)^\sim = \tilde{K}^{(\infty)}$ is countable. But, in order to prove the stronger statement of Proposition 5, we need one more lemma.

Lemma 8. *Let $\{w_j; j \in J\}$ be a weakly K-cartesian system of $c(K)$. Then it must be of at most countable cardinality.*

Proof. For every $j \in J$, there is an element $c_j \in K^*$ such that $|c_j| = |w_j|$. By multiplying w_j by c_j^{-1}, we may assume $|w_j| = 1$ for all $j \in J$. Then the new set $\{w_j; j \in J\}$ is still weakly K-cartesian; in fact, it is α-cartesian with the same real constant $\alpha \geqq 1$ for which the original set was α-cartesian. Since $K^{(\infty)}$ is dense in $c(K)$, we can approximate the w_j by elements $v_j \in K^{(\infty)}$. Assume that $|w_j - v_j| < \alpha^{-1}$ for all $j \in J$. Then $\{v_j; j \in J\}$ is α-cartesian. Namely, let c_j be elements in K such that $c_j = 0$ for almost all $j \in J$. Then

$$\left|\sum c_j(w_j - v_j)\right| < \alpha^{-1} \max |c_j| \leqq \left|\sum c_j w_j\right|$$

(provided, not all c_j are zero), and we have

$$\alpha \left|\sum c_j v_j\right| = \alpha \max \left\{\left|\sum c_j w_j\right|, \left|\sum c_j(w_j - v_j)\right|\right\} \geqq \max |c_j|\,.$$

In particular, $\{v_j; j \in J\}$ is a linearly independent system in $K^{(\infty)}$. Since the dimension of $K^{(\infty)}$ is countable, the cardinality of J is at most countable. □

Now we can *prove* Proposition 5. We shall proceed indirectly and assume that $b(K)$ is of countable type. Then by Proposition 2, there is a strict monomorphism $\varphi: b(K) \to c(K)$. Lemma 7 guarantees the existence of an uncountable orthonormal system $\{v_j; j \in J\}$ in $b(K)$. Define $w_j := \varphi(v_j) \in c(K)$ for all $j \in J$. Let $\psi: \varphi(b(K)) \to b(K)$ be the inverse of φ. The map ψ is a continuous K-homomorphism and therefore bounded. It is straightforward to check that $\{w_j; j \in J\}$ is α-cartesian, where α is the product of the norms $|\varphi|$ and $|\psi|$. Thus, we have found an uncountable weakly cartesian system in $c(K)$, in contradiction to the preceding lemma. Hence our assumption was erroneous, and $b(K)$ is not of countable type nor, a fortiori, can there exist a linear homeomorphism onto $c(K)$. □

Remark. Even if there is no strict embedding of $b(K)$ into $c(K)$, we can at least exhibit a continuous embedding: choose $a \in K$ such that $0 < |a| < 1$ and define

$$\Phi: b(K) \to c(K)$$

by $\Phi((c_i)_{i \in \mathbb{N}}) := (a^i c_i)_{i \in \mathbb{N}}$. (Since (c_i) is bounded, the sequence $(a^i c_i)$ is in fact an element of $c(K)$.) Clearly Φ is a K-monomorphism. Furthermore Φ is bounded, with bound 1, and hence continuous.

2.7.2. Schauder bases. Orthogonality and orthonormality. — Let I be an index set of at most countable cardinality. A system $\{v_i; i \in I\}$ of vectors $v_i \in V$ is said to be a *topological generating system of* V if each $v \in V$ can be written as a convergent series

$$(*) \qquad v = \sum_{i \in I} c_i v_i, \qquad c_i \in K.$$

If $\{v_i; i \in I\}$ is any family of topological generators of V, we obviously may assume $v_i \neq 0$ for all $i \in I$. Moreover, if $\varrho \in |K^*|$ denotes one of the real numbers $\varrho > 1$ introduced by Proposition 2.1.8/1, we can always arrange that the given family is ϱ-bounded, i.e., $1 \leqq |v_i| \leqq \varrho$ for all $i \in I$. From now on, this will always be (tacitly) assumed for systems of topological generators. Then the sequence (c_i) occurring in $(*)$ must be a zero sequence, because $(c_i v_i)$ is a zero sequence.

Definition 1. *A system $\{v_i; i \in I\}$ of topological generators of V is called a Schauder basis of V if, for each $v \in V$, the sequence $\{c_i\}_{i \in I}$ of coefficients c_i in $(*)$ is uniquely determined by v.*

Example. The set $\{w_i; i \in \mathbb{N}\}$ occurring in Proposition 2.7.1/3 is a Schauder basis of V. The canonical basis $\{e_i = (\delta_{i\nu})_{\nu \in \mathbb{N}}; i \in \mathbb{N}\}$ of $K^{(\infty)}$ is a Schauder basis of $c(K)$.

In particular, there is a difference between bases in the ordinary sense and Schauder bases (unless we are working in the finite-dimensional case). Note that, for a Schauder basis $\{v_i; i \in \mathbb{N}\}$ of V, not all infinite sums $\sum c_i v_i$, where (c_i) is a zero sequence in K, need converge in V.

Next we derive a sufficient condition for a system of topological generators to be a Schauder basis.

Proposition 2. *Let $\{v_i; i \in I\}$ be a family of topological generators of V. Assume that this family is α-cartesian for a suitable $\alpha \geqq 1$. Then $\{v_i; i \in I\}$ is a Schauder basis of V, and, for each vector $v = \sum_{i\in I} c_i v_i \in V$, we have*

$$(**) \qquad \max_{i\in I} |c_i|\,|v_i| \leqq \alpha\,|v|\,.$$

Proof. We may assume $I = \{1, 2, \ldots\}$. Let $v = \sum_1^\infty c_i v_i$. Let $\varepsilon > 0$. Choose m so large that, for each vector $w_n := \sum_{i=1}^{n} c_i v_i$, $n \geqq m$, one has $|w_n| \leqq \varepsilon + |v|$. Since the family $\{v_i; i \in I\}$ is α-cartesian, we now derive

$$\max_{1\leqq i\leqq n} |c_i|\,|v_i| \leqq \alpha\,|w_n| \leqq \alpha\cdot\varepsilon + \alpha\,|v| \quad \text{for all} \quad n \geqq m.$$

Therefore $\max_{i\in I} |c_i|\,|v_i| \leqq \alpha\,|v| + \alpha\varepsilon$, and hence (**) holds.

For $v := 0$, we get $c_i = 0$ for all i, which proves the uniqueness of the coefficients. □

Remark. Let $\{v_i; i \in I\}$ be a ϱ-bounded set of V. According to Definition 2.6.1/3, the set $\{v_i; i \in I\}$ is α-cartesian if and only if the inequality (**) holds for finite sums, i.e., for all sums such that $c_i = 0$ for almost all $i \in I$. An important aspect of Proposition 2 is that the inequality (**) extends automatically to infinite convergent sums. To say it in other words, the inequality (**) holds for all infinite convergent sums if and only if every finite subset of $\{v_i; i \in I\}$ is α-cartesian.

Each space V admitting a Schauder basis $\{v_i; i \in I\}$ is of countable type, since the space W generated (in the ordinary sense) by all v_i is of at most countable dimension and is dense in V. The following converse holds:

Proposition 3. *Each normed vector space V of countable type admits an α-cartesian Schauder basis for each $\alpha > 1$.*

If W is a dense (linear) subspace of V of at most countable dimension and if $\{w_i; i \in I\}$ is an α-cartesian basis of W for some $\alpha \geqq 1$, then $\{w_i; i \in I\}$ is an α-cartesian Schauder basis of V.

Proof. Due to Proposition 2.6.2/3, we have only to verify the second assertion. By Proposition 2, it is enough to show that each $v \in V$ can be written as a convergent series $\sum c_i w_i$. We may assume $I = \mathbb{N}$. By Proposition 2.7.1/3, there exists a strict injection $\Psi\colon V \to c(K)$ such that $\Psi(\sum c_i w_i) = (c_i)$ holds for all finite sums $\sum c_i w_i \in W$.

Take $v \in V$ and set $(c_i) := \Psi(v) \in c(K)$. We have

$$\Psi(v) = \lim_n \Psi(v_n) \qquad \text{if} \qquad v_n := \sum_1^n c_i w_i \in W\,.$$

Since $\Psi^{-1}\colon \Psi(V) \to V$ is continuous, we conclude

$$v = \lim_{n\to\infty} v_n = \sum_1^\infty c_i w_i. \qquad \square$$

The topological structure of vector spaces admitting weakly K-cartesian Schauder bases is fairly simple due to the following:

Proposition 4. *Let $\{x_i; i \in I\}$ be an α-cartesian Schauder basis of V, $\alpha > 1$. Let $\{y_\nu := \sum_{i\in I} a_{i\nu} x_i; \nu \in \mathbb{N}\}$ be a sequence in V converging to a vector $y \in V$. Then all limits $a_i := \lim_{\nu\to\infty} a_{i\nu}$, $i \in I$, exist in K, and we have $y = \sum_{i\in I} a_i x_i$.*

Proof. We adopt the same notations as in the proof of the preceding proposition. Then $\Psi(y_\nu) = (a_{i\nu})_{i\in I}$ for all $\nu \in \mathbb{N}$. Since Ψ is continuous, we get $\Psi(y) = \Psi(\lim y_\nu) = \lim_\nu \Psi(y_\nu) = \lim_\nu (a_{i\nu})_{i\in I} = (\lim a_{i\nu})_{i\in I} = (a_i)_{i\in I} \in c(K)$. Therefore $y = \sum_{i\in I} a_i x_i$. (By means of the strict monomorphism Ψ, we have thus transferred the convergence problem from V to $c(K)$. The convergence in $c(K)$ is characterized by Proposition 2.1.5/6 and its proof.) $\square$

In some applications one must consider series $\sum a_i v_i$ where the coefficients take values only in some restricted subset R of K. Proposition 4 yields a useful result for this case. More precisely, for each subring R of K and each bounded family $\{v_i; i \in I\}$ of vectors of V, we consider the set

$$V_R(v_i; i \in I) := \Big\{v = \sum_{i\in I} a_i v_i \in V; a_i \in R, \lim a_i = 0\Big\}.$$

Obviously this is an R-module. We have the following

Corollary 5. *If R is closed in K and if the family $\{v_i; i \in I\}$ is α-cartesian, the R-module $V_R(v_i; i \in I)$ is the topological closure in V of the R-module generated (over R) by $\{v_i; i \in I\}$.*

Proof. Since each series $v = \sum_{i\in I} a_i v_i \in V_R(v_i; i \in I)$ is the limit of the sequence of its partial sums, we only have to show that the set $V_R(v_i; i \in I)$ is closed in V. This follows immediately from Proposition 4 since the limit of each sequence $(a_{i\nu}) \subset R$ converging in K belongs to R. $\square$

Now we want to restrict ourselves to the case of an α-cartesian Schauder basis $\{v_i; i \in I\}$, where $\alpha = 1$. Then for all zero sequences $(c_i)_{i\in I}$ of K such that $v := \sum_{i\in I} c_i v_i$ exists in V (note that V need not be complete), one has $\max_{i\in I} |c_i|\,|v_i| \leq |v|$ and, as always, $|v| \leq \max_{i\in I} |c_i|\,|v_i|$. Thus, one gets $|\sum_{i\in I} c_i v_i| = \max_{i\in I} |c_i|\,|v_i|$. Thus, specializing from α-cartesian Schauder bases to 1-cartesian ones corresponds in the finite-dimensional case to going over from weakly cartesian bases to orthogonal ones. Hence it is reasonable to make the following definition:

Definition 6. *A set $\{v_i; i \in I\}$ of vectors of V is called orthogonal if* $|\sum_{i\in I} c_i v_i| = \max_{i\in I} |c_i|\,|v_i|$ *for all convergent sums* $\sum_{i\in I} c_i v_i \in V$.

According to the remark following Proposition 2, the set $\{v_i\}$ is orthogonal if and only if every finite subset is orthogonal.

The following is an infinite-dimensional extension of Proposition 2.4.1/5.

Proposition 7. *Let V be a normed K-vector space admitting an orthogonal Schauder basis $\{v_i; i \in I\}$. Then V is K-cartesian, and every orthogonal basis of every finite-dimensional subspace can be extended to an orthogonal Schauder basis of V.*

Proof. First we claim that every one-dimensional subspace W of V admits a norm-direct supplement $W^\perp$ in V which has an orthogonal Schauder basis. Take $w = \sum_{i\in I} c_i v_i \in W - \{0\}$. Then $|w| = \max_{i\in I} |c_i|\,|v_i|$. Choose $j \in I$ such that $|w| = |c_j|\,|v_j|$. Exactly as in the proof of Lemma 2.4.1/4, one proves that $\{w\} \cup \{v_i; i \in I - \{j\}\}$ is an orthogonal basis of V. Thus our claim is justified. Using Proposition 2.4.3/6, we see that V is K-cartesian.

In order to verify the remaining assertion of the proposition, we have to show that each finite-dimensional subspace $U \subset V$ admits a norm-direct supplement $U^\perp$ which has an orthogonal Schauder basis. We use induction on $n := \dim U$. The case where $n = 1$ is settled above. If $n > 1$, choose a one-dimensional subspace $W \subset U$, and denote by $W^\perp$ a norm-direct supplement of W in V which has an orthogonal Schauder basis. Applying the induction hypothesis to the $(n-1)$-dimensional subspace $U \cap W^\perp$ of $W^\perp$, we get a norm-direct supplement S of $U \cap W^\perp$ in $W^\perp$ which has an orthogonal Schauder basis. Clearly, S is a norm-direct supplement of U in V. □

Remark. If one is only interested in the result that V is K-cartesian under the assumptions of Proposition 7, then one can argue as follows: $V' := \bigoplus_{i\in I} K v_i$ is a dense subspace of V, and V' is K-cartesian according to Proposition 2.4.3/5. Since V is of countable type and therefore weakly K-cartesian, Proposition 2.4.3/7 gives the desired result.

An orthogonal set $\{v_i; i \in I\}$ is orthonormal in the sense of Definition 2.5/1 if and only if $|v_i| = 1$ for all $i \in I$. We want to show that after a suitable change of norms — not affecting the topology — every space of countable type admits an orthonormal Schauder basis.

Proposition 8. *Each normed vector space V of countable type carries a K-norm equivalent to the given one such that V provided with this new norm admits an orthonormal Schauder basis.*

Proof. Choose $\alpha > 1$ and a (ϱ-bounded) α-cartesian Schauder basis $\{w_i; i \in I\}$ (e.g., by Proposition 3). Define a new norm $|\ \ |'$ by

$$|\sum c_i w_i|' := \max_{i\in I} |c_i|.$$

Then obviously $\alpha^{-1}|v|' \leqq |v| \leqq \varrho\,|v|'$ for all $v \in V$, which shows that $|\ \ |'$ is equivalent to $|\ \ |$. Clearly $\{w_i;\, i \in I\}$ is orthonormal with respect to $|\ \ |'$. □

Proposition 8 enables us to use orthonormal Schauder bases instead of weakly cartesian Schauder bases whenever we are dealing with topological properties of the space V. For applications in affinoid geometry, we need a criterion allowing us to decide when a given subset $\{x_i;\, i \in I\}$ of a normed vector space is an orthonormal Schauder basis. In [2] the first author gave such a criterion, and we shall discuss and prove that fundamental theorem in the next sections.

2.7.3. The Lifting Theorem. — Again K is complete. Let V be a normed K-vector space. Let us repeat for the particular case at hand the definitions of (2.1.10). There we considered the set $V^\circ := \{v \in V;\, |v| \leqq 1\}$ which is a faithfully normed $\mathring{K}$-module; $V^\vee := \{v \in V^\circ;\, |v| < 1\}$ is a $\mathring{K}$-submodule of V°. The residue module $V^\sim := V^\circ/V^\vee$ is a $\tilde{K}$-vector space. The canonical residue epimorphism $V^\circ \to V^\sim$ is denoted by $^\sim$.

The following remark is a partial generalization of Proposition 2.5.1/4.

Remark 1. *If $\{v_i;\, i \in I\}$ is an orthonormal Schauder basis of V, the family $\{v_i^\sim;\, i \in I\}$ is a $\tilde{K}$-basis of $V^\sim$.*

We already know that the family $\{v_i^\sim;\, i \in I\}$ is linearly independent over $\tilde{K}$ (cf. Lemma 2.5.1/3). In order to show that each $v^\sim \in V^\sim$ is a (finite) linear combination of the $v_i^\sim$, take an inverse image $v \in V$ of $v^\sim$ and write $v = \sum c_i v_i$. Then $\max |c_i| = |v| \leqq 1$, since we are working with an orthonormal Schauder basis. Decompose $v = \sum c_\alpha v_\alpha + \sum c_\beta v_\beta$ such that the first sum contains *all* summands with $|c_\alpha| = 1$. Then $|\sum c_\beta v_\beta| < 1$, the first sum contains at most a finite number of terms, and therefore $v^\sim = \sum \tilde{c}_\alpha v_\alpha^\sim$. □

A family $\{v_i;\, i \in I\}$ in V° such that $\{v_i^\sim;\, i \in I\}$ is a $\tilde{K}$-basis of $V^\sim$ is always orthonormal (cf. again Lemma 2.5.1/3); however it is not, in general, a Schauder basis of V. This is shown by the following

Example. Let the valuation on K be non-discrete, and let $\{x_1, x_2, \ldots\}$ be an orthonormal Schauder basis of $V := c(K)$. Choose a sequence $\{\lambda_2, \lambda_3, \ldots\}$ in K such that $0 < |\lambda_i| < 1$ and $\lim \lambda_i \lambda_{i-1} \cdots \lambda_2 \neq 0$. Set

$$y_i := x_i - \lambda_{i+1}\, x_{i+1}, \qquad i = 1, 2, \ldots.$$

Then $y_i^\sim = x_i^\sim$ for all $i \geqq 1$; i.e., $\{y_i^\sim;\, i \geqq 1\}$ is a $\tilde{K}$-basis of $V^\sim$. However x_1 cannot be written as a series $x_1 = \sum_1^\infty a_\nu y_\nu$, because this would imply

$$x_1 = \sum_1^\infty a_\nu(x_\nu - \lambda_{\nu+1} x_{\nu+1}) = a_1 x_1 + \sum_{\nu=2}^\infty (a_\nu - a_{\nu-1}\lambda_\nu)\, x_\nu,$$

which means

$$a_1 = 1 \qquad \text{and} \qquad a_\nu = \lambda_\nu a_{\nu-1} = \lambda_\nu \lambda_{\nu-1} \cdots \lambda_2 .$$

Thus, we would have $\lim a_\nu \neq 0$.

We now come to the crucial theorem of this section.

Theorem 2 (Lifting Theorem). *Let K be complete, and let V be a normed K-vector space admitting an orthonormal Schauder basis $\{v_i; i \in I\}$. Let*

$$\{w_j = \sum_i c_{ij} v_i; j \in J\} \subset V^\circ, \qquad c_{ij} \in K,$$

be a family of vectors such that the following two conditions are fulfilled:

(B 1) *the family $\{w_j^\sim; j \in J\} \subset V^\sim$ is a $\tilde{K}$-basis of $V^\sim$,*

(B 2) *there exists a bald subring B of $\mathring{K}$ containing all scalars c_{ij}, $i \in I$, $j \in J$.*

Then $\{w_j; j \in J\}$ is an orthonormal Schauder basis of V. There exist elements $b_{ij} \in B$ such that

$$v_i = \sum_{j \in J} b_{ij} w_j, \qquad i \in I.$$

Remark. We do not suppose that V is complete, or equivalently, we do not suppose that all series $\sum_{i \in I} c_i v_i$, where $(c_i)_{i \in I}$ is a zero sequence in K, converge in V. The condition (B 2), which is trivially fulfilled for discrete valuations (just take $B := \mathring{K}$), is the essential point of the theorem; the example just given above shows that such a condition cannot be avoided.

2.7.4. Proof of the Lifting Theorem. — Due to (B 1), the sets I and J have the same cardinality which implies that J is at most countable. Moreover, the family $\{w_j; j \in J\}$ is orthonormal by (B 1). We have to show that each $v \in V$ can be written as a convergent series $\sum_{j \in J} a_j w_j$ or, using the notation of (2.7.2), that $V = V_K(w_j; j \in J)$. Since $V = V_K(v_i; i \in I)$ by assumption, it will be sufficient to prove that each v_i, $i \in I$, is of the form $\sum_{j \in J} a_j w_j$, because then Corollary 2.7.2/5 yields $V = V_K(v_i; i \in I) \subset V_K(w_j; j \in J)$.

The proof that v_i is an element of $V_K(w_j; j \in J)$, $i \in I$, will rely heavily on (B 2). We may assume that the bald ring B occurring in (B 2) is a complete B-ring. (If not, just replace B by the completion of its localization with respect to the family $\{b \in B; |b| = 1\}$; this new ring is again bald (see (1.7.1)). Then B is closed in K. We shall write V_B (resp. V'_B) for the B-module $V_B(v_i; i \in I)$ (resp. $V_B(w_j; j \in J)$). By Corollary 2.7.2/5, the module V_B (resp. V'_B) is the closure in V of the B-module generated (over B) by $\{y_i; i \in I\}$ (resp. $\{w_j; j \in J\}$). By assumption, we have $w_j \in V_B$ for all $j \in J$, and therefore $V'_B \subset V_B$. If we can show that $V'_B = V_B$, we will have $v_i \in V'_B \subset V_K(w_j; j \in J)$ for all $i \in I$, and the proof will be finished. In order to prove the equation $V'_B = V_B$, we consider the situation in $V^\sim$. We know that $\{v_i^\sim; i \in I\}$ and $\{w_j^\sim; j \in J\}$ are

$\tilde{K}$-bases of the vector space $V^{\sim}$. Since $B \subset \mathring{K}$ is a B-ring, $\tilde{B}$ is a subfield of $\tilde{K}$. Thus

$$V'^{\sim}_{\tilde{B}} = \sum_{j \in J} \tilde{B} w_j^{\sim} \qquad \text{and} \qquad V^{\sim}_{\tilde{B}} = \sum_{i \in I} \tilde{B} v_i^{\sim}$$

are $\tilde{B}$-vector spaces. We have $w_j^{\sim} \in V^{\sim}_{\tilde{B}}$ for all $j \in J$; i.e., $V'^{\sim}_{\tilde{B}} \subset V^{\sim}_{\tilde{B}}$. As a first step in the direction of proving $V'_B = V_B$, we shall show $V'^{\sim}_{\tilde{B}} = V^{\sim}_{\tilde{B}}$ by using the following lemma for vector spaces.

Lemma 1. *Let F be a subfield of a field G, and let U be a G-vector space. Let $\{x_i; i \in I\}$ and $\{y_j; j \in J\}$ be G-bases of U. Define*

$$U_F := \sum_{i \in I} F x_i \qquad \textit{and} \qquad U'_F := \sum_{j \in J} F y_j .$$

Then $U'_F \subset U_F$ implies $U'_F = U_F$.

The lemma can be verified without difficulty by using tensor products (note that we have $U'_F \otimes_F G = U$ and $U_F \otimes_F G = U$). For completeness we give the following elementary *proof*. The lemma being trivial for finite-dimensional spaces, we want to reduce the problem to this case. It suffices to show that $x_i \in U'_F$ for all $i \in I$. Fix an element $x_{i_1} \in \{x_i; i \in I\}$. Since $\{y_j; j \in J\}$ generates U over G, we can choose elements $y_{j_1}, \ldots, y_{j_m} \in \{y_j; j \in J\}$ such that

$$x_{i_1} \in \sum_{\mu=1}^{m} G y_{j_\mu} .$$

Since all y_j belong to U_F, we can pick finitely many vectors $x_{i_2}, \ldots, x_{i_n} \in \{x_i; i \in I\}$ such that the F-vector space $W_F := \sum_{\nu=1}^{n} F x_{i_\nu}$ contains all vectors y_{j_μ}, $\mu = 1, \ldots, m$. Since $y_{j_1}, \ldots, y_{j_m}$ are linearly independent over F (in fact over G), we can choose vectors $z_1, \ldots, z_s \in W_F$ such that $\{y_{j_1}, \ldots, y_{j_m}, z_1, \ldots, z_s\}$ is an F-basis of W_F. These vectors also generate the G-vector space $W_G := \sum_{\nu=1}^{n} G x_{i_\nu}$. They are, in fact, a G-basis of this space, since $m + s = \dim_F W_F = n = \dim_G W_G$. From

$$x_{i_1} \in \sum_{\mu=1}^{m} G y_{j_\mu} \qquad \text{and} \qquad x_{i_1} \in \sum_{\mu=1}^{m} F y_{j_\mu} + \sum_{\sigma=1}^{s} F z_\sigma$$

and the linear independence of all vectors $y_{j_1}, \ldots, y_{j_m}, z_1, \ldots, z_s$ over G, we now conclude

$$x_{i_1} \in \sum_{\mu=1}^{m} F y_{j_\mu} \subset U'_F ,$$

which finishes the proof of the lemma. □

Let us resume the proof of the Lifting Theorem. By setting $F := \tilde{B}$, $G := \tilde{K}$, $U := V^{\sim}$, $x_i := v_i^{\sim}$ and $y_j := w_j^{\sim}$, we derive $V'^{\sim}_{\tilde{B}} = V^{\sim}_{\tilde{B}}$ from the lemma. Therefore, for each $i \in I$, there exist elements $b_{ij} \in B$, $j \in J$, such that

$$z_i := \sum_{j \in J} b_{ij} w_j \in V'_B \qquad \text{and} \qquad z_i^{\sim} = -v_i^{\sim}; \qquad \text{i.e.,} \qquad |z_i + v_i| < 1 .$$

Set $\varepsilon := \sup\{|b|;\, b \in B,\, |b| < 1\}$. Since B is bald, we have $\varepsilon < 1$. Obviously $\varepsilon = \sup\{|v|;\, v \in V_B,\, |v| < 1\}$. From $z_i + v_i \in V_B$, we now conclude

$$|z_i + v_i| \leqq \varepsilon \qquad \text{for all} \qquad i \in I.$$

Next we claim

For each $x \in V_B$, there exists an element $y \in V_B'$ such that $|x + y| \leqq \varepsilon\, |x|$.

In order to see this, write $x = \sum_{i \in I} b_i v_i$. Choose a finite subset I^* of I such that $|x - \sum_{i \in I^*} b_i v_i| \leqq \varepsilon |x|$. Now set $y := \sum_{i \in I^*} b_i z_i$. We have $y \in V_B'$. From $x + y = (x - \sum_{i \in I^*} b_i v_i) + \sum_{i \in I^*} b_i (v_i + z_i)$, we derive

$$\begin{aligned} |x + y| &\leqq \max\left\{|x - \sum_{i \in I^*} b_i v_i|,\, \max_{i \in I^*} |b_i|\, |v_i + z_i|\right\} \\ &\leqq \max\{\varepsilon\, |x|,\, \max_{i \in I^*} |b_i|\, \varepsilon\} = \varepsilon\, |x|. \end{aligned}$$

Thus V_B' is "ε-dense" in V_B in the sense of Proposition 1.1.4/2. Since V_B' is also closed in V_B, we get $V_B' = V_B$ from Proposition 1.1.4/2, which concludes the proof of the Lifting Theorem. □

2.7.5. Applications. — We state some immediate consequences of the Lifting Theorem. First we have in complete analogy to Proposition 2.7.2/7

Proposition 1. *If V admits an orthonormal Schauder basis, then V is strictly K-cartesian.*

Each orthonormal basis of a finite-dimensional subspace can be extended to an orthonormal Schauder basis of V.

Proof. Let $\{v_i;\, i \in I\}$ be an orthonormal Schauder basis of V, and let $U \subset V$ be a subspace of finite dimension. Set $n := \dim U^\sim$ and choose $u_j = \sum_{i \in I} c_{ji} v_i \in U^\circ$, $j = 1, \ldots, n$, such that $\{u_1^\sim, \ldots, u_n^\sim\}$ is a $\tilde{K}$-basis of $U^\sim$. Since $(c_{ji})_{i \in I}$, $j = 1, \ldots, n$, comprise only *finitely many* zero sequences, there exists a bald subring B of K containing all c_{ji} by Corollary 1.7.2/5.

We can choose a subset $I' \subset I$ such that the family

$$\{u_1^\sim, \ldots, u_n^\sim\} \cup \{v_i^\sim;\, i \in I'\}$$

is a $\tilde{K}$-basis of $V^\sim$. Then the family

$$\{u_1, \ldots, u_n\} \cup \{v_i;\, i \in I'\},$$

together with the bald ring B chosen above, fulfills conditions (B 1) and (B 2) of the Lifting Theorem and therefore is an orthonormal Schauder basis of V.

We claim that $\{u_1, \ldots, u_n\}$ is an orthonormal basis of U. We need only show that $U = \sum_{i=1}^{n} K u_i$. Take $u \in U$ and write $u = \sum_{1}^{n} c_\nu u_\nu + z$, where z is of the form $z = \sum_{i \in I'} d_i v_i$. Then $z \in U$. If $z \neq 0$, we can normalize $|z|$ to 1, which leads to the contradiction

$$0 \neq z^{\sim} = \sum_{i \in I'} \tilde{d}_i v_i^{\sim} \in U^{\sim}.$$

It remains to be shown that each orthonormal basis $\{y_1, \ldots, y_n\}$ of U can be extended. We have $|y_\nu| = 1$; moreover, $\{y_1^{\sim}, \ldots, y_n^{\sim}\}$ is a $\tilde{K}$-basis of $U^{\sim}$. Therefore, we can proceed exactly in the same way we did above with the vectors $u_1, \ldots, u_n$. In this way, we get an orthonormal Schauder basis of V extending the given basis of U. □

For spaces of countable type, one has the following improvement of Proposition 1:

Proposition 2. *The following statements concerning a normed K-vector space V of countable type are equivalent:*

(1) *V admits an orthonormal Schauder basis.*
(2) *V is strictly K-cartesian.*
(3) $\dim_{\tilde{K}} U^{\sim} = \dim_K U$ *whenever U is a finite-dimensional subspace of V.*

Proof. The equivalence of (2) and (3) is clear (use Corollary 2.5.1/6). The implication (1) → (2) is shown in Proposition 1. In order to deduce (1) from (2), we choose a dense subspace $W \subset V$ of countable dimension and exhaust W by a sequence W_n, $n \in \mathbb{N}$, of vector spaces such that

$$W_n \subset W_{n+1}, \qquad W = \bigcup_{n \in \mathbb{N}} W_n, \qquad \dim_K W_n = n.$$

Using Proposition 2.5.1/5, we can extend each orthonormal basis $\{w_1, \ldots, w_n\}$ of W_n to an orthonormal basis $\{w_1, \ldots, w_n, w_{n+1}\}$ of W_{n+1}. Proceeding in this way, we get an orthonormal K-basis $\{w_\nu; \nu \in \mathbb{N}\}$ for W and therefore, by Proposition 2.7.2/3, an orthonormal Schauder basis for V. □

2.8. Banach spaces

In this section, the valuation on the base field K is always complete and non-trivial.

2.8.1. Definition. Fundamental theorem. — We define the concept of a K-Banach space as in real and complex analysis:

Definition 1. *A complete normed K-vector space V is called a Banach space.*

Each finite-dimensional normed K-space is a Banach space. Each closed subspace of a Banach space is a Banach space. The direct sum of finitely many Banach spaces is a Banach space; more generally, it follows from Proposition 2.1.5/6

If $\{V_i\}_{i \in I}$ is an arbitrary family of K-Banach spaces, their bounded direct product $b(\prod_i V_i)$ and their restricted direct product $c(\prod_i V_i)$ are K-Banach spaces.

Thus, the simplest examples of Banach spaces which are not of finite dimension are the spaces

$$b(K) := b\left(\prod_1^\infty K^1\right) \quad \text{and} \quad c(K) := c\left(\prod_1^\infty K^1\right),$$

where one takes countably many copies of the line K^1. Recall that $b(K)$ (resp. $c(K)$) consists of all bounded sequences (resp. all zero sequences) $a = (a_1, a_2, \ldots)$, $a_\nu \in K$, provided with norm

$$|a| = \sup_\nu |a_\nu| \quad (\text{resp. } |a| = \max_\nu |a_\nu|).$$

We state without proof a fundamental theorem, which we shall use frequently later on.

Theorem of Banach (Open Mapping Theorem). *Let V, W be Banach spaces, and let $\Phi: V \to W$ be a bounded and surjective K-linear map. Then Φ is open, and W carries the quotient topology with respect to Φ. (In particular, Φ is a homeomorphism if Φ is bijective.)*

For a *proof* see [5].

An important corollary is the

Closed Graph Theorem. *Let $\Phi: V \to W$ be a K-linear map between Banach spaces. The map Φ is bounded if and only if* $\text{Graph}\,\Phi := \{(v, \Phi(v)) \in V \times W\}$ *is closed with respect to the product topology on $V \times W$.*

As an illustration of the type of reasoning used in dealing with Banach spaces, we deduce this statement from BANACH's Theorem. Obviously, the continuity of Φ implies the closedness of Graph Φ. So assume $\overline{\text{Graph}\,\Phi} = \text{Graph}\,\Phi$. Then $\text{Graph}\,\Phi \subset V \times W$ is itself a Banach space. Hence the K-linear projection $\pi: \text{Graph}\,\Phi \to V$ given by $(v, \Phi(v)) \mapsto v$ is open (since it is continuous). As the projection $\pi': \text{Graph}\,\Phi \to W$ given by $(v, \Phi(v)) \mapsto \Phi(v)$ is also continuous, the continuity of Φ follows from the commutative diagram

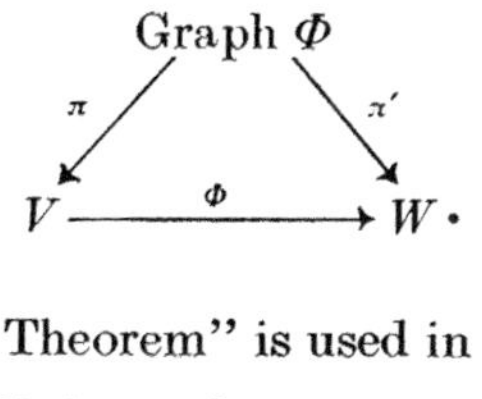

$\square$

Often the "Closed Graph Theorem" is used in the following version:

$\Phi: V \to W$ is continuous if, for each sequence $v_\nu \in V$ with

$$\lim v_\nu = 0 \quad \textit{and} \quad \lim \Phi(v_\nu) =: w \in W,$$

we have $w = 0$.

2.8.2. Banach spaces of countable type. — Here we specialize the results of (2.7) to the case of Banach spaces. We start by recalling Definition 2.7.1/1.

Definition 1. *A Banach space V is said to be of countable type if V contains a dense subspace V' of at most countable dimension.*

The space $c(K)$ of zero sequences is of countable type, since $K^{(\infty)}$ is obviously dense in $c(K)$. In fact, $c(K)$ is the only infinite-dimensional example of such a space: namely,

Theorem 2. *Each Banach space V of countable type admits a K-linear homeomorphism onto K^n if $n := \dim_K V < \infty$ and onto $c(K)$ if $\dim_K V = \infty$.*

Proof. Due to Proposition 2.3.3/4 we may assume that V is infinite-dimensional. Choose a dense subspace V' of countable dimension in V. By Theorem 2.6.2/4, there exists a K-linear homeomorphism $K^{(\infty)} \xrightarrow{\sim} V'$. Thus V may be viewed as the completion of $K^{(\infty)}$. Since $c(K)$ is also a completion of $K^{(\infty)}$, the theorem follows. □

Corollary 3. *Each Banach space of countable type is b-separable.*

CHAPTER 3

Extensions of norms and valuations

In this chapter, valuation theory — in its widest sense — is at the center of our interest. We start with some general facts on normed algebras. These include auxiliary results from commutative algebra as well as a further discussion of spectral values (continuation of (1.5.4)). The spectral value is used in (3.2) in order to extend power-multiplicative norms on fields. The procedure is totally elementary; it only requires the smoothing techniques of (1.3.2). Without investing the power of HENSEL's Lemma, we establish the extension theorem for valuations on fields in the complete case. In fact, HENSEL's Lemma is an easy consequence of our extension theorems which are generalized to the non-complete case in (3.3).

In (3.4), we prove the lemma on continuity of roots. One of its consequences is the fact that the completion of an algebraically closed field remains algebraically closed. Thus each valued field has a smallest extension which is both complete and algebraically closed. For HENSEL's field of p-adic numbers $\mathbb{Q}_p$, this extension is denoted by $\mathbb{C}_p$; its Archimedean analogue is the field $\mathbb{C}$ of complex numbers. Using KRASNER's Lemma, we show that the extension $\mathbb{C}_p/\mathbb{Q}_p$ is much more complicated than the extension $\mathbb{C}/\mathbb{R}$.

In the remaining sections we deal with some special topics. First there is the discussion of stable and weakly stable fields; we derive certain criteria which are necessary for the proof of the Stability Theorem for $Q(T_n)$ in (5.3). Then there are two sections on Banach algebras and function algebras; these sections should be seen from the viewpoint of affinoid algebras. They anticipate some technicalities which otherwise would have been dealt with in Chapter 6. (Note that integral torsion-free monomorphisms of the type considered throughout section (3.8) are furnished by NOETHER's Normalization Lemma 6.1.2/2.) The main result is Proposition 3.8.1/7 which relates the supremum norm on Banach algebras to the spectral norm with respect to integral extensions.

3.1. Normed and faithfully normed algebras

Let $A = (A, |\ \ |)$ be a normed ring. If $|\ \ |$ is a valuation, we denote by $Q(A)$ the valued field of fractions of A (provided with the extended valuation). The symbol B always denotes an A-algebra; i.e., there is given a ring homomorphism $\varphi: A \to B$ which by the definition $a \cdot b := \varphi(a)\, b$, $a \in A$, $b \in B$, equips B with the structure of an A-module.

3.1.1. A-algebra norms. — We want to look at A-module norms on B for which the multiplication in B is continuous. If A carries a non-degenerate valuation, then one knows due to Proposition **2.1.8/2** that

The ring multiplication in a faithfully normed A-module $(B, |\ \ |')$ is continuous if and only if there exists a real constant $\varrho > 0$ such that $|xy|' \leqq \varrho\, |x|'\, |y|'$ for all $x, y \in B$.

Applying Proposition **1.2.1/2** if $B \neq 0$, we can derive from $|\ \ |'$ an equivalent A-module norm $|\ \ |$ (which is again a faithful A-module norm) such that $|xy| \leqq |x|\, |y|$ and $|1| = 1$. Hence the following definition which combines the two concepts of "module norm" and "ring norm" is reasonable. It will actually cover all interesting cases.

Definition 1. *A ring norm $|\ \ |$ on an A-algebra B is called a (faithful) A-algebra norm if $|\ \ |$ is a (faithful) A-module norm on B.*

Let $B = (B, \|\ \ \|)$ be a normed ring, and consider a ring homomorphism $\varphi: A \to B$ making B into an A-algebra. Then $\|\ \ \|$ is an A-algebra norm on B if and only if φ is contractive. Namely, any A-algebra norm $\|\ \ \|$ on B must satisfy

$$\|\varphi(a)\| = \|\varphi(a) \cdot 1\| \leqq |a|\, \|1\| \leqq |a|$$

for all $a \in A$. Conversely if φ is contractive, we have

$$\|ab\| = \|\varphi(a)\, b\| \leqq \|\varphi(a)\|\, \|b\| \leqq |a|\, \|b\|$$

for all $a \in A$, $b \in B$. In particular, if A is a normed subring of B (i.e., provided with the ring norm inherited from B) and if $\varphi = \mathrm{id}$, then B is a normed A-algebra.

Let B_1 and B_2 denote normed A-algebras. Then the complete tensor product $B := B_1 \widehat{\otimes}_A B_2$ is a well-defined A-module. We want to show that B is even a normed A-algebra with a unique multiplication such that

$$(b_1 \widehat{\otimes} b_2)\, (b_1' \widehat{\otimes} b_2') = b_1 b_1' \widehat{\otimes} b_2 b_2'.$$

In order to verify this, we consider for $b_1 \in B_1$, $b_2 \in B_2$ the A-linear maps

$$\lambda_{b_i}: B_i \to B_i, \quad x \mapsto b_i x, \qquad i = 1, 2.$$

By Proposition **2.1.7/5**, we get an A-linear map

$$\lambda_{b_1} \widehat{\otimes} \lambda_{b_2}: B \to B$$

such that $b_1' \widehat{\otimes} b_2' \mapsto b_1 b_1' \widehat{\otimes} b_2 b_2'$ and satisfying $|\lambda_{b_1} \widehat{\otimes} \lambda_{b_2}| \leqq |\lambda_{b_1}|\, |\lambda_{b_2}| = |b_1|\, |b_2|$. Obviously $\lambda_{b_1} \widehat{\otimes} \lambda_{b_2}$ depends A-linearly on b_1 and b_2 so that we get an A-bilinear map

$$B_1 \times B_2 \to \mathscr{L}(B, B), \quad (b_1, b_2) \mapsto \lambda_{b_1} \widehat{\otimes} \lambda_{b_2},$$

which is bounded by 1. Since $\mathscr{L}(B, B)$ is complete (see Proposition 2.1.6/4), this map corresponds by Proposition 2.1.7/1 to a contractive A-linear map $B \to \mathscr{L}(B, B)$ such that $b_1 \mathbin{\widehat{\otimes}} b_2 \mapsto \lambda_{b_1} \mathbin{\widehat{\otimes}} \lambda_{b_2}$. This map, in turn, can be interpreted as an A-bilinear map $\mu: B \times B \to B$, which is bounded by 1. Thus B is a normed A-algebra with multiplication μ, and it is true that $\mu(b_1 \mathbin{\widehat{\otimes}} b_2, b_1' \mathbin{\widehat{\otimes}} b_2') = b_1b_1' \mathbin{\widehat{\otimes}} b_2b_2'$. Furthermore it is clear that μ, as a continuous map, is uniquely determined by this property, since by linearity it is determined on a dense subset of $B \times B$.

It follows from the definition of the multiplication on $B_1 \mathbin{\widehat{\otimes}}_A B_2$ that the maps

$$\tau_1: B_1 \to B_1 \mathbin{\widehat{\otimes}}_A B_2, \qquad b_1 \mapsto b_1 \mathbin{\widehat{\otimes}} 1,$$

$$\tau_2: B_2 \to B_1 \mathbin{\widehat{\otimes}}_A B_2, \qquad b_2 \mapsto 1 \mathbin{\widehat{\otimes}} b_2,$$

are contractive A-algebra homomorphisms. These maps characterize $B_1 \mathbin{\widehat{\otimes}}_A B_2$ as an A-algebra in the following way.

Proposition 2. *The canonical maps $\tau_i: B_i \to B_1 \mathbin{\widehat{\otimes}}_A B_2$, $i = 1, 2$, satisfy the following universal property:*

If $\varphi_i: B_i \to D$, $i = 1, 2$, are bounded A-algebra homomorphisms into a complete A-algebra D, then there is a unique bounded A-algebra homomorphism $\psi: B_1 \mathbin{\widehat{\otimes}}_A B_2 \to D$ such that the diagram

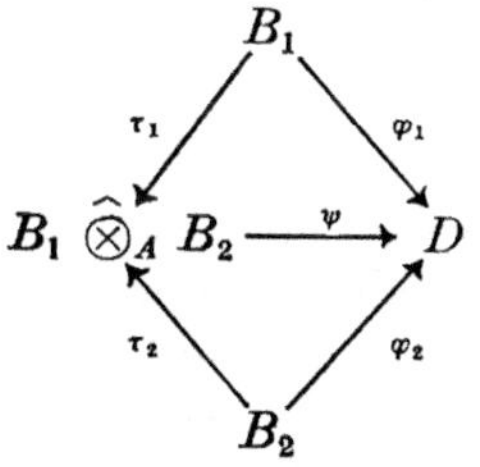

commutes. One has $|\psi| \leqq |\varphi_1|\,|\varphi_2|$.

Furthermore, if $\sigma_i: B_i \to B'$, $i = 1, 2$, are bounded A-algebra homomorphisms into a complete normed A-algebra B' satisfying the above universal property with τ_1, τ_2 replaced by σ_1, σ_2, then the unique homomorphism $\sigma: B_1 \mathbin{\widehat{\otimes}}_A B_2 \to B'$ satisfying $\sigma_i = \sigma \circ \tau_i$ for $i = 1, 2$, is an isomorphism and bounded in both directions.

Proof. In order to verify the universal property for $B_1 \mathbin{\widehat{\otimes}}_A B_2$, let $\varphi_i: B_i \to D$, $i = 1, 2$, denote two bounded A-algebra homomorphisms into a complete normed A-algebra D. The A-bilinear map $\Phi: B_1 \times B_2 \to D$, $(b_1, b_2) \mapsto \varphi_1(b_1)\,\varphi_2(b_2)$, which is bounded by $|\varphi_1|\,|\varphi_2|$, corresponds by Proposition 2.1.7/1 to an A-linear map $\psi: B_1 \mathbin{\widehat{\otimes}}_A B_2 \to D$ such that $|\psi| \leqq |\varphi_1|\,|\varphi_2|$ and such that the diagram

$$\begin{array}{ccc} B_1 \times B_2 & \xrightarrow{\tau} & B_1 \mathbin{\widehat{\otimes}}_A B_2 \\ & {\scriptstyle \Phi}\searrow \quad \swarrow{\scriptstyle \psi} & \\ & D & \end{array}$$

commutes. For $b_1 \in B_1$, $b_2 \in B_2$, the following holds:

$$\psi\big(\tau_1(b_1)\,\tau_2(b_2)\big) = \psi(b_1 \mathbin{\widehat{\otimes}} b_2) = \Phi(b_1, b_2) = \varphi_1(b_1)\,\varphi_2(b_2).$$

In particular, this equation is true for $b_1 = 1$ and for $b_2 = 1$, and one concludes that ψ is multiplicative on the dense image of $B_1 \otimes_A B_2$ in $B_1 \mathbin{\widehat{\otimes}}_A B_2$. Hence ψ is an A-algebra homomorphism which obviously satisfies $\varphi_i = \psi \circ \tau_i$ for $i = 1, 2$. Furthermore one easily derives from the uniqueness assertion in Proposition 2.1.7/1 that ψ is the unique algebra homomorphism with this property.

Thus we proved the universal property for $B_1 \mathbin{\widehat{\otimes}}_A B_2$, and it is now a formal argument to verify that any complete A-algebra with the same universal property is canonically isomorphic to $B_1 \mathbin{\widehat{\otimes}}_A B_2$. □

If B is any A-algebra, the group $G(B/A)$ of all A-algebra automorphisms of B is of great interest. For later reference we prove

Proposition 3. *Let B be a normed A-algebra, and let H be a subgroup of $G(B/A)$ such that all $h \in H$ are contractions. Then there is a canonical group homomorphism ${}^\sim: H \to G(B^\sim/A^\sim)$ such that $h^\sim(b^\sim) = h(b)^\sim$ for all $h \in H$, $b \in B^\circ$.*

Proof. Each $h \in H$ maps B° (resp. $B^\vee$) into itself and hence gives rise to a ring homomorphism $h^\sim: B^\sim \to B^\sim$ such that $h^\sim(b^\sim) = \big(h(b)\big)^\sim$ for all $b \in B^\circ$. Because h is an A-algebra homomorphism, we have for all $a \in A^\circ$ the equalities: $h^\sim(a^\sim b^\sim) = h^\sim\big((ab)^\sim\big) = \big(h(ab)\big)^\sim = \big(ah(b)\big)^\sim = a^\sim\big(h(b)\big)^\sim = a^\sim h^\sim(b^\sim)$. Therefore, $h^\sim$ is an $A^\sim$-homomorphism. It is straightforward to check $(hg)^\sim = h^\sim g^\sim$ for all $h, g \in H$ and $\mathrm{id}_B^\sim = \mathrm{id}_{B^\sim}$. Therefore $(h^{-1})^\sim = (h^\sim)^{-1}$ for all $h \in H$, and hence $h^\sim$ is indeed an $A^\sim$-automorphism of $B^\sim$.

Thus we proved that ${}^\sim: H \to G(B^\sim/A^\sim)$ is a group homomorphism. □

A subgroup H of $G(B/A)$ consisting only of contractions is actually a subgroup of isometries. If, more generally, one looks at bounded A-automorphisms of B, one can prove a result very similar to Proposition 3.

Proposition 4. *Let B be a normed A-algebra, and let H be a subgroup of $G(B/A)$ such that all $h \in H$ are bounded. Then there is a canonical group homomorphism ${}^\sim: H \to G(\tilde{B}/\tilde{A})$ such that $\tilde{h}(\tilde{b}) = \widetilde{h(b)}$ for all $h \in H$, $b \in \mathring{B}$.*

Proof. In the proof of Proposition 3, one has to replace everywhere the objects $B^\sim$, B°, $B^\vee$, etc., by the invariant objects $\tilde{B}$, $\mathring{B}$, $\check{B}$, etc. Then literally the same proof yields the proposition. □

If B is a non-zero faithfully normed A-algebra, the given homomorphism $A \to B$ is an isometry and hence injective. Thus B becomes a normed overring of $(A, |\ \ |)$, and the set $A - \{0\}$ is contained in the set of multiplicative elements of $(B, |\ \ |)$. Conversely, if $(A, |\ \ |)$ is a normed subring of a normed ring $(B, |\ \ |) \neq 0$ such that all elements of $A - \{0\}$ are multiplicative (in B), then $(A, |\ \ |)$ is a valued integral domain and $(B, |\ \ |)$ is a faithfully normed A-algebra. For example, each valuation on an overring B of A extending the valuation on A provides B with the structure of a faithfully normed A-algebra. Note, however, that a faithful A-algebra norm on B is by no means necessarily a valuation on B.

If A is a valued field, each A-algebra norm is faithful (cf. Proposition 2.1.1/4). Hence for every overring B of A with a ring norm $|\ \ |$, we have that $|\ \ |$ is an extension of the valuation on A if and only if $|\ \ |$ is an A-algebra norm.

3.1.2. Spectral values and power-multiplicative norms. — In this section we do not suppose that B is faithfully normed or that A is valued. In (1.5.4) we introduced for each monic polynomial $p = X^m + a_1X^{m-1} + \cdots + a_m \in A[X]$ its *spectral value* $\sigma(p) = \max\limits_{1 \leqq \mu \leqq m} |a_\mu|^{1/\mu}$. The following proposition indicates the importance of this concept.

Proposition 1. *Let B be a normed A-algebra with a power-multiplicative norm $|\ \ |$. Let $q = X^m + a_1X^{m-1} + \cdots + a_m \in A[X]$. Then we have*

$$|b| \leqq \sigma(q)$$

for each root $b \in B$ of q.

If in addition A is a normed subring of B (more precisely, if $\varphi: A \to B$ is an isometry) and if $q \in A[X]$ splits into linear factors $X - b_i$ over B (i.e., if $q = \prod\limits_{i=1}^{m} (X - b_i)$, $b_i \in B$), then we have

$$\max_{1 \leqq i \leqq m} |b_i| = \sigma(q).$$

Proof. (1) Let $b \in B$ be any root of q. Assume $|b| > \sigma(q) = \max\limits_{1 \leqq \mu \leqq m} |a_\mu|^{1/\mu}$ so that $|a_\mu| < |b|^\mu$ for all $\mu = 1, \ldots, m$. We conclude that

$$|a_\mu b^{m-\mu}| \leqq |a_\mu|\, |b|^{m-\mu} < |b|^\mu\, |b|^{m-\mu} = |b|^m = |b^m|, \quad \mu = 1, \ldots, m.$$

(The last equality is the only place in the whole proof where the assumption that $|\ \ |$ is power-multiplicative is used.) Thus $\left|\sum\limits_{\mu=1}^{m} a_\mu b^{m-\mu}\right| < |b^m|$, in contradiction to $q(b) = b^m + \sum\limits_{\mu=1}^{m} a_\mu b^{m-\mu} = 0$. Hence $|b| \leqq \sigma(q)$.

(2) Now assume that A is a normed subring of B. Since $a_\mu = \pm \sum\limits_{1 \leqq i_1 < \cdots < i_\mu \leqq m} b_{i_1} \ldots b_{i_\mu}$, we get

$$|a_\mu| \leqq \max_{i_1, \ldots, i_\mu} |b_{i_1} \ldots b_{i_\mu}| \leqq \max_{i_1, \ldots, i_\mu} (|b_{i_1}| \ldots |b_{i_\mu}|) \leqq (\max_{1 \leqq i \leqq m} |b_i|)^\mu.$$

Therefore $\sigma(q) \leqq \max\limits_{1 \leqq i \leqq m} |b_i|$, which together with (1) yields $\sigma(q) = \max\limits_{1 \leqq i \leqq m} |b_i|$. □

3.1.3. Residue degree and ramification index. — In (2.1.4) and (2.1.10) we introduced the notions of ramification index $e(M/A)$ and residue degree $f(M/A)$ for faithfully normed A-modules M. These notions are especially meaningful in the case of faithfully normed A-algebras B. The inequalities $e(B/A) \leqq \operatorname{rk}_A B$, $f(B/A) \leqq \operatorname{rk}_A B$ (Propositions 2.1.4/2 and 2.1.10/3) can then be improved considerably. We start with a simple lemma that combines the arguments of the proofs of the propositions just mentioned.

Lemma 1. *Let B be a faithfully normed A-algebra. Assume that $|A - \{0\}|$ is a group. Let $x_1, \ldots, x_r \in B^\circ$ be elements such that $x_1^\sim, \ldots, x_r^\sim \in B^\sim$ are linearly independent over $\tilde{A}$. Let $y_1, \ldots, y_s$ be elements in $B - \{0\}$ such that*

(i) $|xy_j| = |x|\,|y_j|$ *for all* $x \in \sum_{i=1}^{r} Ax_i$, $j = 1, \ldots, s$,

(ii) *the values* $|y_1|, \ldots, |y_s| \in |B - \{0\}|$ *represent different equivalence classes of* $|B - \{0\}|$ *modulo* $|A - \{0\}|$.

Then for all $a_{ij} \in A$, $i = 1, \ldots, r$ *and* $j = 1, \ldots, s$, *one has* $\left|\sum_{\substack{1 \leqq i \leqq r \\ 1 \leqq j \leqq s}} a_{ij}x_iy_j\right| = \max_{\substack{1 \leqq i \leqq r \\ 1 \leqq j \leqq s}} |a_{ij}|\,|x_iy_j|$. *In particular, the rs elements $x_iy_j \in B$, $1 \leqq i \leqq r$, $1 \leqq j \leqq s$, are linearly independent over A, and hence $rs \leqq \operatorname{rk}_A B$.*

Proof. For the given $a_{ij} \in A$, consider the element

$$b := \sum_{\substack{1 \leqq i \leqq r \\ 1 \leqq j \leqq s}} a_{ij}x_iy_j \in B.$$

Write it in the form

$$(*) \qquad b = \sum_{1 \leqq j \leqq s} (a_{1j}x_1 + \cdots + a_{rj}x_r)\, y_j .$$

Because $|A - \{0\}|$ is a group, we have $|a_{1j}x_1 + \cdots + a_{rj}x_r| = \max_{1 \leqq i \leqq r} |a_{ij}|$ by Proposition 2.1.10/3. Thus $|a_{1j}x_1 + \cdots + a_{rj}x_r| \in |A|$ for each j. Since $|(a_{1j}x_1 + \cdots + a_{rj}x_r)\, y_j| = |a_{1j}x_1 + \cdots + a_{rj}x_r|\,|y_j|$, the absolute values of two non-zero terms in the sum $(*)$ can never be equal, because otherwise $|y_1|, \ldots, |y_s| \in |B - \{0\}|$ would not represent different equivalence classes of $|B - \{0\}|$ modulo $|A - \{0\}|$.

The Principle of Domination yields

$$|b| = \max_{1 \leqq j \leqq s} |a_{1j}x_1 + \cdots + a_{rj}x_r|\,|y_j| .$$

Because $|x_iy_j| = |x_i|\,|y_j| = |y_j|$, we have

$$|b| = \max_{1 \leqq j \leqq s} \left\{\max_{1 \leqq i \leqq r} |a_{ij}|\,|y_j|\right\} = \max_{\substack{1 \leqq i \leqq r \\ 1 \leqq j \leqq s}} |a_{ij}|\,|x_iy_j| . \qquad \square$$

If $r := 1$, $x_1 := 1$, the lemma implies $e(B/A) \leqq \operatorname{rk}_A B$. Similarly, for $s := 1$, $y_1 := 1$ we get $f(B/A) \leqq \operatorname{rk}_A B$. A further consequence of the lemma is

Proposition 2. *For each valued ring B containing a valued subring A such that $|A - \{0\}|$ is a group, we have*

$$e(B/A)\, f(B/A) \leqq \mathrm{rk}_A\, B.$$

Let us finish this section with some general remarks concerning the residue degree. For any faithfully normed A-algebra B (where, by definition, A must be a valued ring), the ring $\tilde{A}$ is an integral domain and the $\tilde{A}$-algebra $B^{\sim}$ is torsion-free as an $\tilde{A}$-module. Hence

$$f(B/A) = \mathrm{rk}_{\tilde{A}}\, B^{\sim} = \dim_{Q(\tilde{A})} (B^{\sim})_{\tilde{A}-\{0\}}.$$

The $Q(\tilde{A})$-vector space $(B^{\sim})_{\tilde{A}-\{0\}}$ is contained in the full ring of fractions $Q(B^{\sim})$; therefore $f(B/A) \leqq \dim_{Q(\tilde{A})} Q(B^{\sim})$. In important cases we have equality; namely,

Proposition 3. *If B is a faithfully normed A-algebra and if B° is integral over $\mathring{A}$, then*

$$f(B/A) = \dim_{Q(\tilde{A})} Q(B^{\sim}).$$

This proposition is a direct consequence of the following purely algebraic statement:

If R' is an overring of an integral domain R such that each $r' \in R'$ is integral over R and such that no element of $R - \{0\}$ is a zero divisor in R', then $R'_{R-\{0\}} = Q(R')$ and therefore $\mathrm{rk}_R\, R' = \dim_{Q(R)} Q(R')$.

The ring $Q(R')$ is an integral extension of the field $Q(R)$.

Proof. Let $q = cd^{-1} \in Q(R')$, $c, d \in R'$, be given. By assumption, there is an equation

$$d^n + b_1 d^{n-1} + \cdots + b_n = 0, \quad b_1, \ldots, b_n \in R, \quad n \geqq 1.$$

Since d is not a zero divisor in R', we can assume that $b_n \neq 0$. Then

$$d^{-1} = -b_n^{-1}(d^{n-1} + b_1 d^{n-2} + \cdots + b_{n-1}),$$

and hence $q = cd^{-1}$ is an element of $R'_{R-\{0\}}$. This proves $Q(R') = R'_{R-\{0\}}$. The element $r := b_n q$ belongs to R' and satisfies an integral equation over R, say

$$r^m + a_1 r^{m-1} + \cdots + a_m = 0.$$

Then

$$q^m + (a_1 b_n^{-1})\, q^{m-1} + \cdots + a_m b_n^{-m} = 0,$$

and we see that $Q(R')$ is an integral extension of $Q(R)$. □

3.1.4. Dedekind's Lemma and a Finiteness Lemma. — In this section we shall supply some auxiliary results about reduced algebras which are of particular interest for valuation theory (see the following sections) as well as for the theory of k-affinoid algebras. First we prove a classical lemma by DEDEKIND describing the structure of finite-dimensional reduced algebras.

Proposition 1 (DEDEKIND's Lemma). *Every reduced finite-dimensional algebra L over a field K is the (ring-theoretic) direct sum of finitely many field extensions of K.*

Proof. First we claim that every prime ideal $\mathfrak{p}$ of L is maximal. Consider the integral domain $L/\mathfrak{p}$. For each $y \in L/\mathfrak{p}$, $y \neq 0$, the K-algebra endomorphism φ_y of $L/\mathfrak{p}$ defined by $\varphi_y(x) := xy$ is injective. Moreover, it is surjective, because $L/\mathfrak{p}$ has finite dimension over K. Hence, for every $y' \in L/\mathfrak{p}$, the equation $xy = y'$ has a solution $x \in L/\mathfrak{p}$; i.e., $L/\mathfrak{p}$ is a field, or equivalently $\mathfrak{p}$ is maximal, as claimed. Let $\mathfrak{p}_1, \ldots, \mathfrak{p}_n$ be different prime ideals in L. Since all $\mathfrak{p}_i$ are maximal, one has

$$(*) \qquad \bigcap_{\substack{i=1 \\ i \neq j}}^{n} \mathfrak{p}_i \not\subset \mathfrak{p}_j \quad \text{for} \quad j = 1, \ldots, n.$$

The relation (*) implies that the chain $\mathfrak{p}_1 \supset \mathfrak{p}_1 \cap \mathfrak{p}_2 \supset \cdots \supset \mathfrak{p}_1 \cap \cdots \cap \mathfrak{p}_n$ is strictly decreasing, whence $n \leqq \dim_K L$. Thus L admits only a finite number of prime ideals, say $\mathfrak{p}_1, \ldots, \mathfrak{p}_t$. Let $\pi_i : L \to L/\mathfrak{p}_i$ denote the residue epimorphism. Because L is reduced, one knows that $\bigcap_{i=1}^{t} \mathfrak{p}_i = 0$. (Namely for any $a \neq 0$ in L, there exists an ideal $\mathfrak{p} \subset L$ which does not contain any power a^n of a. Choosing $\mathfrak{p}$ maximal with this property, one gets a prime ideal not containing a.) Therefore the K-algebra homomorphism $\pi := (\pi_1, \ldots, \pi_t) : L \to \bigoplus_{i=1}^{t} L/\mathfrak{p}_i$ is injective. Since all $L/\mathfrak{p}_i$ are fields, only the surjectivity of π remains to be shown. Define $L_i := \pi(L) \cap L/\mathfrak{p}_i$ for $i = 1, \ldots, t$, where $\pi(L)$ and the fields $L/\mathfrak{p}_i$ are viewed as L-submodules of $\bigoplus_{i=1}^{t} L/\mathfrak{p}_i$. Then each L_i is an L-submodule of $L/\mathfrak{p}_i$ or, in other terms, an ideal in the field $L/\mathfrak{p}_i$. Using relation (*) again, we see that $L_i \neq 0$. Therefore $L_i = L/\mathfrak{p}_i$ for all i, and π is surjective. □

Recall that a ring homomorphism $\psi : R \to R'$ is called *finite* if R' is a finite R-module via ψ. Furthermore, ψ is called *integral* if each element in R' satisfies an integral equation over R (more precisely, over $\psi(R)$). Each finite homomorphism is integral. However the converse is not true. We use DEDEKIND's Lemma in order to give sufficient conditions for an integral ring homomorphism to be finite. (See Chapter 4 for the notion of Japaneseness.)

Lemma 2 (Finiteness Lemma). *Let R be a Noetherian and Japanese integral domain, and let $\psi : R \to R'$ be an integral ring homomorphism such that the following conditions are fulfilled:*

(i) *the ring R' is reduced; the R-module R' is torsion-free,*

(ii) $\mathrm{rk}_R R' < \infty$.

Then ψ is finite.

Proof. Because ψ is injective by (i), we may assume $R \subset R'$ and $\psi = \mathrm{id}$. The ring $R'_{R-\{0\}}$ (which coincides with $Q(R')$) is a finite-dimensional reduced

$Q(R)$-algebra. Hence by DEDEKIND's Lemma, $R'_{R-\{0\}}$ is the ring-theoretic direct sum of a finite number of finite field extensions of $Q(R)$:

$$R'_{R-\{0\}} = K_1 \oplus \cdots \oplus K_t.$$

Since R is Japanese, the integral closure R_i of R in K_i is a finite R-module, $i = 1, \ldots, t$. Because R' is integral over R, we have

$$R' \subset R_1 \oplus \cdots \oplus R_t.$$

Since R is Noetherian, R' is a finite R-module. □

3.1.5. Power-multiplicative and faithful A-algebra norms. — In some situations we shall have to deal with power-multiplicative A-algebra norms on B which are faithful as A-module norms, but which are, in general, not valuations. We list some useful properties of such norms.

Proposition 1. *Let the valuation on A be non-degenerate. Let $|\ \ |_i$, $i = 1, 2$, be power-multiplicative faithful A-algebra norms on B which are equivalent on each subring $A[y]$, $y \in B$. Then $|\ \ |_1 = |\ \ |_2$.*

The *proof* follows immediately from Proposition 2.1.8/2 and Corollary 1.3.1/3. □

Next we show

Proposition 2. *Let $|\ \ |$ be a power-multiplicative faithful A-algebra norm on B. Assume that there exists an integer $n \in \mathbb{N}$ such that each element of B annihilates a polynomial $\neq 0$ over A of degree $\leqq n$ (for example, this is the case if B is a finite A-module). Then, for each $b \in B$, there exists an integer m, $1 \leqq m \leqq n$, such that $|b|^m \in |Q(A)|$.*

Proof. Let $b \in B$. By assumption we can find elements $a_0, a_1, \ldots, a_n \in A$, not all zero, such that

$$a_0 b^n + a_1 b^{n-1} + \cdots + a_n = 0.$$

We can choose indices i, j, $0 \leqq i < j \leqq n$, such that $0 < |a_{n-i} b^i| = |a_{n-j} b^j|$. Since $|a_{n-\nu} b^\nu| = |a_{n-\nu}|\ |b|^\nu$ for all ν by the assumption on $|\ \ |$, we get $|b|^{j-i} = |a_{n-i}|\ |a_{n-j}|^{-1} \in |Q(A)|$. This verifies the assertion because $j - i \in \{1, \ldots, n\}$. □

Corollary 3. *Under the assumptions of Proposition 2, we have $|B|^{n!} \subset |Q(A)|$.*

In particular, we get $e(B/A) = 1$ if the (multiplicative) value group $|Q(A)^*|$ is *divisible*, i.e., closed under the operation of taking roots.

As an application of the Finiteness Lemma 3.1.4/2, we can prove

Proposition 4. *Let A be a valued integral domain, and assume that $\tilde{A}$ is Noetherian and Japanese and that $|A - \{0\}|$ is a group. Let B be a faithfully normed A-algebra such that*

(i) *The norm on B is power-multiplicative,*

(ii) $\mathrm{rk}_A B < \infty$,

(iii) *$\mathring{B}$ is integral over $\mathring{A}$.*

Then $\tilde{B}$ is a finite $\tilde{A}$-module.

Proof. We want to apply Lemma 3.1.4/2 with $R := \tilde{A}$, $R' := \tilde{B}$, and $\psi: \tilde{A} \to \tilde{B}$ the canonical injection map. Since $\mathring{B}$ is integral over $\mathring{A}$, the map ψ is integral. It remains to verify conditions (i) and (ii) of Lemma 3.1.4/2. The first one is fulfilled, since the norm on B is power-multiplicative and faithful (cf. Propositions 1.2.5/7 and 2.1.10/1). The second one follows from Proposition 2.1.10/3, which gives

$$\mathrm{rk}_{\tilde{A}} \tilde{B} \leqq \mathrm{rk}_A B < \infty. \qquad \square$$

3.2. Algebraic field extensions. Spectral norm and valuations

We denote by K a field with a power-multiplicative norm. It is not assumed that K is complete nor that K is valued. Let L be a K-algebra. Without loss of generality, we may assume that $K \subset L$ (unless $L = 0$). By $G = G(L/K)$ we mean the group of all K-algebra automorphisms of L.

3.2.1. Spectral norm on algebraic field extensions. — Let L be an algebraic extension of K. We are dealing with the problem of extending the norm on K to a power-multiplicative K-algebra norm and later on even to a valuation on L. (Of course, the latter is feasible only if the given norm on K is a valuation.) The following definition and theorem are important for our considerations.

Definition 1. *Let L be an algebraic extension of K. For each element $y \in L$, we set*

$$|y|_{\mathrm{sp}} := \text{spectral value } \sigma(q) \text{ of the minimal polynomial } q \in K[X] \text{ of } y \text{ over } K$$

and call the function $|\ \ |_{\mathrm{sp}}: L \to \mathbb{R}_+$ the spectral norm on L (induced by the norm on K).

This definition is motivated by

Theorem 2. *The function $|\ \ |_{\mathrm{sp}}$ is a power-multiplicative K-algebra norm on L extending the given norm on K. All K-algebra automorphisms of L are isometries with respect to the spectral norm. For any power-multiplicative K-algebra norm $|\ \ |$ on L, we have $|\ \ | \leqq |\ \ |_{\mathrm{sp}}$.*

If L is a finite quasi-Galois ($=$ normal) extension of K, the spectral norm $|\ \ |_{\mathrm{sp}}$ is the only power-multiplicative K-algebra norm on L extending the norm on K for which all $g \in G(L/K)$ are isometries. If $|\ \ |$ is an arbitrary power-multiplicative K-algebra norm on L extending the norm on K, we have

$$|y|_{\mathrm{sp}} = \max_{g \in G(L/K)} |g(y)| \quad \text{for all} \quad y \in L.$$

The proof will be based on

Lemma 3. *Each finite extension L of K carries at least one power-multiplicative K-algebra norm extending the given norm on K.*

Proof. Choose a basis $\{e_1 = 1, e_2, \ldots, e_n\}$ of the K-vector space L, and denote by $|\ \ |$ the norm given by $|\sum_{\nu=1}^{n} a_\nu e_\nu| := \max_{1\leq\nu\leq n} |a_\nu|$. In this way we get a K-module norm on L, extending the norm on K. Set $M := \max_{1\leq\mu,\nu\leq n} |e_\mu e_\nu|$. Then it follows immediately that

$$|yy'| \leq M\,|y|\,|y'| \quad \text{for all} \quad y, y' \in L.$$

Therefore we can "smooth" this norm to a ring norm on L according to Proposition 1.2.1/2, and by assertion (iii) of that proposition, the new norm is an extension of the given norm on K. From Proposition 1.3.2/1, we now deduce the existence of a power-multiplicative semi-norm on L which must, in fact, be a norm (L being a field). Since the norms of power-multiplicative elements remain unchanged throughout this last procedure, we see that L carries a power-multiplicative K-algebra norm extending the norm on K. □

Remark. In the proof just given, the norm $|\ \ |$ with which we started defines the product topology on L. So does the K-algebra norm we derived from this norm. However, the final norm we get by applying Proposition 1.3.2/1 may fail to provide L with the product topology.

Now we come to the *proof* of Theorem 2. Let y be an element in L and let $q \in K[X]$ be its minimal polynomial. We have $|y|_{sp} = 0$ if and only if $\sigma(q) = 0$, i.e., if and only if $q = X^m$, i.e., if and only if $y = 0$.

For each automorphism $g \in G := G(L/K)$, the element $g(y)$ is a root of q; hence the polynomial q is also the minimal polynomial of $g(y)$. Therefore $|g(y)|_{sp} = |y|_{sp}$; i.e., each map $g \in G$ leaves the spectral function invariant. From Proposition 3.1.2/1, we get $|\ \ | \leq |\ \ |_{sp}$ for each power-multiplicative K-algebra norm on L.

In order to show that $|\ \ |_{sp}$ is really a power-multiplicative K-algebra norm on L extending the norm on K, we have to verify the inequalities $|yy'|_{sp} \leq |y|_{sp}\,|y'|_{sp}$, $|y + y'|_{sp} \leq \max\{|y|_{sp}, |y'|_{sp}\}$ and the equations $|a|_{sp} = |a|$, $|y^\nu|_{sp} = |y|_{sp}^\nu$ for $y, y' \in L$, $a \in K$, $\nu \in \mathbb{N}$. We may restrict all considerations to the field $K(y, y') \subset L$, which is finite over K. From $K(y, y')$, we can pass to a smallest quasi-Galois extension K' over K (not necessarily contained in L) and work in this field which is still finite over K. Thus we may assume from the beginning that L itself is a finite quasi-Galois extension of K.

Let $|\ \ |$ be a power-multiplicative K-algebra norm on L extending the norm on K. By Lemma 3, we know that such norms exist. For each $g \in G$, we set

$$|y|_g := |g(y)|, \qquad y \in L.$$

Obviously $|\ \ |_g$ is a power-multiplicative K-algebra norm on L for each $g \in G$, and it extends the given norm on K. Since L is finite over K, the group G is finite; therefore, we can apply a further smoothing procedure:

$$|y|_G := \max_{g\in G}\{|y|_g\}.$$

It is clear that $|\ \ |_G$ is also a power-multiplicative K-algebra norm on L and that it extends the given norm. By construction, all $g \in G$ are isometries with respect to this norm. Since L is quasi-Galois over K, the minimal polynomial of $y \in L$ is of the form

$$q = \prod_{g \in G} (X - g(y))^{p^e},$$

where the product is taken over certain $g \in G$, where $p := \operatorname{char} K$ and where $e \geqq 0$ is the exponent of inseparability of y. Since $|g(y)|_G = |y|_G$, we conclude from Proposition 3.1.2/1 that

$$|y|_G = \sigma(q) = |y|_{\mathrm{sp}}.$$

Therefore, $|\ \ |_{\mathrm{sp}}$ is actually a power-multiplicative K-algebra norm extending the norm on K. Moreover, for any other such norm $|\ \ |$, we found
$|y|_{\mathrm{sp}} = \max\limits_{g \in G(L/K)} |g(y)|$ for all $y \in L$. □

From Proposition 3.1.1/3 and the preceding theorem, we immediately deduce

Proposition 4. *If L carries the spectral norm, there exists a canonical group homomorphism* $\tilde{}: G(L/K) \to G(\tilde{L}/\tilde{K})$ *such that* $\widetilde{g(y)} = \tilde{g}(\tilde{y})$ *for all* $g \in G(L/K)$, $y \in \mathring{L}$.

By definition of the spectral norm, the set $|L^*|$ is contained in the smallest divisible subgroup of $\mathbb{R}_+ - \{0\}$ containing $|K^*|$. Hence one gets

Proposition 5. *If K is a valued field, then for every $y \in L^*$ there exist an exponent $s \geqq 1$ and an element $c \in K^*$ such that $|cy^s|_{\mathrm{sp}} = 1$.*

In our preceding considerations, we may take for L an algebraic closure of K. Thus we may talk without any ambiguity about the spectral norm on an algebraic closure of K. For later reference we note the following corollary of Theorem 2, which improves Proposition 1.5.4/1.

Corollary 6. *If u and v are monic polynomials in $K[X]$, then one has*

$$\sigma(uv) = \max\{\sigma(u), \sigma(v)\}.$$

Proof. Let L be a finite extension of K such that u(resp. v) splits into linear factors $X - b_i$, $b_i \in L$, for $i = 1, \ldots, r$ (resp. for $i = r + 1, \ldots, r + s$). By the theorem we can provide L with the spectral norm $|\ \ |_{\mathrm{sp}}$ induced by the given norm on K. Then we can apply Proposition 3.1.2/1 to see that $\sigma(u) = \max\limits_{1 \leqq i \leqq r} |b_i|_{\mathrm{sp}}$, $\sigma(v) = \max\limits_{r+1 \leqq i \leqq r+s} |b_i|_{\mathrm{sp}}$ and $\sigma(uv) = \max\limits_{1 \leqq i \leqq r+s} |b_i|_{\mathrm{sp}}$, whence the assertion follows. □

3.2.2. Spectral norm on reduced integral K-algebras. — The concept of the spectral norm shall be used not only for algebraic field extensions, but also for reduced integral K-algebras L. Recall that a K-algebra L is called integral if each $y \in L$ is a zero of a monic polynomial $q \in K[X]$. Since the ideal $\{f \in K[X];$

$f(y) = 0\}$ is principal in $K[X]$ (K is a field), there must exist a unique monic polynomial $q \in K[X]$ of minimal degree such that $q(y) = 0$. This polynomial is called the minimal polynomial of y over K. It divides any other polynomial in $K[X]$ annihilating y. (Of course, if L is not a field, q may be reducible.) We want to extend Definition 3.2.1/1 to the case of reduced integral K-algebras.

Definition 1. *Let L be a reduced integral K-algebra. For each element $y \in L$, we set $|y|_{sp} := \sigma(q) = \max_{1 \leq i \leq n} |a_i|^{1/i}$, where $q = X^n + a_1 X^{n-1} + \cdots + a_n \in K[X]$ is the minimal polynomial of y. The function $|\ \ |_{sp}: L \to \mathbb{R}_+$ is called the spectral norm on L (induced by the norm on K).*

Similarly as in (3.2.1), we want to show that the spectral norm is, in fact, a norm.

Theorem 2. *Let L be a reduced integral K-algebra.*

(i) *The function $|\ \ |_{sp}$ is a power-multiplicative K-algebra norm on L extending the norm on K.*

(ii) *If $\mathfrak{M}$ denotes the set of all prime ideals $\mathfrak{p} \subset L$ and if $\pi_{\mathfrak{p}}: L \to L/\mathfrak{p}$ is the residue map, then one has for all $y \in L$*

$$|y|_{sp} = \max_{\mathfrak{p} \in \mathfrak{M}} |\pi_{\mathfrak{p}}(y)|_{sp}.$$

If L is of finite dimension over K and if $L = \bigoplus_{i=1}^{t} L_i$ is a decomposition of L into a ring-theoretic direct sum of fields (Dedekind's Lemma) then, for all $y \in L$, one has $|y|_{sp} = \max_{1 \leq i \leq t} |\pi_i(y)|_{sp}$, where $\pi_i: L \to L_i$ is the projection onto L_i.

(iii) *$|\ \ |_{sp}$ dominates every other power-multiplicative K-algebra norm on L.*

(iv) *$|\ \ |_{sp}$ is invariant under all K-algebra automorphisms of L.*

Proof. Without loss of generality, we may assume in proving (i) that L is of finite dimension over K (cf. the considerations in the proof of Theorem 3.2.1/2). Then according to Proposition 3.1.4/1, there are finite field extensions $L_1, \ldots, L_t$ of K such that $L = \bigoplus_{i=1}^{t} L_i$.

Since we already know the behavior of the spectral norm on the components L_i and since we want information about $|\ \ |_{sp}$ on the direct sum, we must try to describe norms on L in terms of norms on the components and vice versa. Here the following proposition is useful:

Proposition 3. *Let $L = \bigoplus_{i=1}^{t} L_i$ be the ring-theoretic direct sum of K-subalgebras $L_1, \ldots, L_t$, and, for $i = 1, \ldots, t$, let $\pi_i: L \to L_i$ denote the residue epimorphism. Then one has*

(i) *If $|\ \ |$ is a power-multiplicative K-algebra norm on L, then its restriction to L_i is also a power-multiplicative K-algebra norm and $|y| = \max_{1 \leq i \leq t} |\pi_i(y)|$ for all $y \in L$.*

(ii) *Conversely, if, for $i = 1, \ldots, t$, there is a power-multiplicative norm* $| \ |_i$ *on* L_i, *then* $|y| := \max_{1 \leq i \leq t} |\pi_i(y)|_i$ *defines a power-multiplicative K-algebra norm on L, whose restriction to* L_i *coincides with* $| \ |_i$.

Proof of Proposition 3. Ad (i): For $i = 1, \ldots, t$, let e_i denote the unit element of L_i. Then $e_i^2 = e_i \neq 0$, and hence $|e_i| = 1$. Therefore the restriction of $| \ |$ to L_i is a power-multiplicative K-algebra norm. Because $y = \pi_1(y) + \cdots + \pi_t(y)$, we deduce immediately that $|y| \leq \max_{1 \leq i \leq t} |\pi_i(y)|$. On the other hand, we know $\pi_i(y) = ye_i$, and therefore $|\pi_i(y)| = |ye_i| \leq |y| \, |e_i| = |y|$, whence $\max_{1 \leq i \leq t} |\pi_i(y)| = |y|$. Thus we have proved (i). The proof of (ii) consists of a straightforward checking of the axioms. Thus Proposition 3 is proved. □

Continuation of the *proof* of Theorem 2. Using assertion (ii) of Proposition 3, we see that statement (i) of Theorem 2 is an immediate consequence of statement (ii) of the theorem. In order to prove (ii), we proceed as follows: for $\mathfrak{p} \in \mathfrak{M}$, let $q_\mathfrak{p} \in K[X]$ be the minimal polynomial of $\pi_\mathfrak{p}(y)$ over K. Then $q_\mathfrak{p}$ is a prime polynomial. Furthermore, let $q \in K[X]$ be the minimal polynomial of y over K. Then $q(\pi_\mathfrak{p}(y)) = \pi_\mathfrak{p}(q(y)) = 0$, and therefore $q_\mathfrak{p}$ divides q for all $\mathfrak{p} \in \mathfrak{M}$. Hence there are only finitely many different polynomials, say $q_1, \ldots, q_r$, amongst the polynomials $q_\mathfrak{p}$, $\mathfrak{p} \in \mathfrak{M}$. Define $q' := \prod_{j=1}^{r} q_j \in K[X]$. Then one has the following equivalences for a polynomial $f \in K[X]$:

$$f(y) = 0 \Leftrightarrow \pi_\mathfrak{p}(f(y)) = 0 \text{ for all } \mathfrak{p} \in \mathfrak{M} \Leftrightarrow f(\pi_\mathfrak{p}(y)) = 0 \text{ for all } \mathfrak{p} \in \mathfrak{M}$$
$$\Leftrightarrow f \text{ is a common multiple of } q_1, \ldots, q_r \Leftrightarrow f \text{ is a multiple of } q'.$$

Hence the monic polynomial q' must coincide with the minimal polynomial q of y over K. Due to Corollary 3.2.1/6, we have

$$|y|_{\mathrm{sp}} = \sigma(q) = \sigma(q_1 \cdots q_r) = \max_{1 \leq i \leq r} \sigma(q_i) = \max_{\mathfrak{p} \in \mathfrak{M}} \sigma(q_\mathfrak{p}) = \max_{\mathfrak{p} \in \mathfrak{M}} |\pi_\mathfrak{p}(y)|_{\mathrm{sp}},$$

which finishes the proof of the first assertion of (ii). If L is finite over K and if $L = \bigoplus_{i=1}^{t} L_i$, then $\mathfrak{p}_i := \bigoplus_{\substack{1 \leq \nu \leq t \\ \nu \neq i}} L_\nu$ is a prime ideal in L for $i = 1, \ldots, t$ and there are no others. Since $L/\mathfrak{p}_i = L_i$ for $i = 1, \ldots, t$, the second part of (ii) follows from the first part. Assertion (iii) follows immediately from Proposition 3.1.2/1. Since the minimal polynomial q of y is invariant under K-algebra automorphisms, assertion (iv) is obvious. Thus Theorem 2 is proved. □

For later reference we need the following transitivity statement for spectral norms.

Proposition 4. *Let* K' *be an algebraic extension of* K *and let* L *be a reduced integral* K'*-algebra. Denote by* $| \ |_{K',K}$ *the spectral norm on* K' *induced by the norm on* K, *and by* $| \ |_{L,K'}$ *the spectral norm on* L *induced by the norm* $| \ |_{K',K}$ *on* K' *and by* $| \ |_{L,K}$ *the spectral norm on* L *induced by the norm on* K. *Then we have*

$$| \ |_{L,K} = | \ |_{L,K'}.$$

Proof. $|\ |_{L,K'}$ is a K'-algebra norm and $|\ |_{K',K}$ is a K-algebra norm; hence a fortiori $|\ |_{L,K'}$ is a K-algebra norm. Therefore, by Theorem 3.2.2/2 (iii), the norm $|\ |_{L,K'}$ is dominated by $|\ |_{L,K}$. The opposite inequality can be shown quite similarly: $|\ |_{L,K}$ is an extension of $|\ |_{K',K}$, because, for elements of K', both norms are defined via the same minimal polynomials. If we apply Theorem 3.2.2/2 (iii) again (this time to the extension L over K'), we see that $|\ |_{L,K'}$ dominates $|\ |_{L,K}$. □

3.2.3. Spectral norm and field polynomials. — We describe another way of computing the spectral norm.

Let L be a finite field extension of K. For each $y \in L$, the *field polynomial* of y over K is defined to be the characteristic polynomial of the K-linear map $L \to L$ defined by $x \mapsto yx$, $x \in L$. This polynomial, which is monic, depends not only on y but also on the field L. Nevertheless, we have

Proposition 1. *Let L be a finite extension of K. Then for each $y \in L$, the spectral norm $|y|_{\mathrm{sp}}$ equals the spectral value $\sigma(\xi)$ of the field polynomial ξ of y over K.*

Proof. As is well known, the field polynomial ξ of y is a power of the minimal polynomial q of y, say $\xi = q^m$. Hence $\sigma(\xi) = \sigma(q)$ by Corollary 3.2.1/6. □

If $\xi = X^n + a_1X^{n-1} + \cdots + a_n \in K[X]$ is the field polynomial of $y \in L$ over K, the element $-a_1 \in K$ is called the *trace* of $y \in L$ over K:

$$\mathrm{Tr}_{L/K}\, y = -a_1 .$$

The map $\mathrm{Tr}_{L/K}\colon L \to K$ is K-linear.

Corollary 2. *The K-linear trace map $\mathrm{Tr}_{L/K}\colon L \to K$ is a contraction (and hence continuous) if L is provided with the spectral norm.*

Proof. We have

$$|\mathrm{Tr}_{L/K}\, y| = |a_1| \leqq \max_{1 \leqq \nu \leqq n} \sqrt[\nu]{|a_\nu|} = \sigma(\xi) = |y|_{\mathrm{sp}} .$$ □

3.2.4. Spectral norm and valuations. — In important cases the spectral norm is not only a K-algebra norm, but also a valuation on L so that we can talk about the *spectral valuation* on L over K.

Proposition 1. *The spectral norm on L is a valuation on L if $L^\sim = L^\circ/L^{\vee}$ is an integral domain.*

Proof. The assertion follows immediately from Proposition 1.5.3/1, since, for each $y \in L^*$, there exist an $s \geqq 1$ and an element $c \in K^*$ such that $|cy^s| = 1$. □

Next we prove the important

Theorem 2. *Let K be complete with respect to the given valuation $|\ |$, and let L be an algebraic extension of K. Then the spectral norm on L is a valuation, and* *each power-multiplicative K-algebra norm on L coincides with this valuation. In particular, the spectral valuation is the unique valuation on L extending the valuation $|\ \ |$ from K. Furthermore, L is complete if $[L:K] < \infty$.*

Proof. By Proposition 2.3.3/4, each normed K-vector space V over a complete field K is weakly cartesian (and complete if $\dim_K V < \infty$). Therefore it is clear (see Proposition 3.1.5/1) that each power-multiplicative K-algebra norm on L coincides with the spectral norm $|\ \ |$ on L. It remains to be shown that each $y \neq 0$ in L is multiplicative for $|\ \ |$. By Proposition 1.3.2/2, we can smooth $|\ \ |$ to a power-multiplicative semi-norm $|\ \ |_y$ such that y is multiplicative for $|\ \ |_y$. Since L is a field, $|\ \ |_y$ is a norm. The proposition just mentioned tells us furthermore that $|\ \ |_y$ is a K-algebra norm. Hence $|\ \ |_y = |\ \ |$; i.e., y is multiplicative for $|\ \ |$. □

Since the question of extending a valuation on a field K to a valuation on a finite algebraic extension L is of great interest in classical valuation theory, let us repeat here briefly how it was done for K complete. We applied four smoothing procedures: we started with any K-vector space norm on L and first smoothed it to a K-algebra norm. In a second step, we passed to a power-multiplicative K-algebra norm $|\ \ |'$ which then was smoothed to *the* spectral norm by taking the maximum over all power-multiplicative norms $|\ \ |' \circ g$, $g \in G(L/K)$. This norm turned out to be a valuation due to a fourth smoothing device which made a given element $\neq 0$ multiplicative. As an important consequence of Theorem 2, we get

Proposition 3. *Let K be complete and let $q = X^m + a_1X^{m-1} + \cdots + a_m \in K[X]$ be irreducible. Then*

$$\sigma(q) = |a_m|^{1/m}; \quad i.e., \quad |a_\mu| \leqq |a_m|^{\mu/m} \quad \text{for all} \quad \mu = 1, \ldots, m.$$

Proof. Let L be a splitting field of q over K. We have an equation

$$q = \prod_{\mu=1}^{m} (X - \theta_\mu), \quad \text{where} \quad \theta_1, \ldots, \theta_m \in L.$$

Since q is the minimal polynomial of $\theta_1, \ldots, \theta_m$ over K, we have $|\theta_\mu| = \sigma(q)$ for all $\mu = 1, \ldots, m$ by definition of the spectral norm. Since the spectral norm is a valuation on L by Theorem 2, the equation $a_m = (-1)^m \prod_{\mu=1}^{m} \theta_\mu$ yields

$$|a_m| = \prod_{\mu=1}^{m} |\theta_\mu| = \sigma(q)^m. \qquad \square$$

The equation just proved $\sigma(q) = |a_m|^{1/m}$ is often used in classical valuation theory as the definition for the extension of the given valuation. Then it is rather obvious that one gets for free the multiplicativity of this function. However it is not clear at all that the triangle inequality is fulfilled. In order to show this, one needs the inequalities $|a_\mu|^{1/\mu} \leqq |a_m|^{1/m}$, $\mu = 1, \ldots, m$, which one usually proves by applying the classical Lemma of HENSEL. In (3.3.4) we shall show how HENSEL's Lemma can easily be deduced from Theorem 2 and Proposition 3.

3.3. Classical valuation theory

As before, let K be a field with a (not necessarily complete) valuation. Let L be a finite extension of K. In the following sections, we will deal with the problem of extending the valuation $|\ |$ on K to a power-multiplicative K-algebra norm on L. We will see that, besides the spectral norm, there are at most finitely many other such norms on L.

3.3.1. Spectral norm and completions. — We provide L with the spectral norm and consider the completion $\hat{L}$ with respect to this norm on L. For simplicity, we write $|\ |$ for the norm on L as well as for its canonical extension to $\hat{L}$. Then $|\ |$ is a power-multiplicative K-algebra norm on $\hat{L}$, and $\hat{L}$ is a reduced K-algebra (however not, in general, a field). Since $\hat{L}$ is complete and contains K, it must contain the completion $\hat{K}$ of K. Note that $|\ |$ is a valuation on $\hat{K}$ and that, due to the continuity of the map $K^* \to K^*$, $x \mapsto x^{-1}$ (K is a valued field), $\hat{K}$ is again a field. In particular, $|\ |$ is a power-multiplicative faithful $\hat{K}$-algebra norm on $\hat{L}$. Since $\dim_{\hat{K}} \hat{L} \leq \dim_K L < \infty$ (see Proposition 2.3.3/6), we get

Proposition 1. *The $\hat{K}$-algebra $\hat{L}$ is a finite ring-theoretic direct sum of finite extensions $\hat{L}_1, \ldots, \hat{L}_t$ of $\hat{K}$:*

$$\hat{L} = \hat{L}_1 \oplus \cdots \oplus \hat{L}_t .$$

If $|\ |_i$ denotes the spectral norm on $\hat{L}_i$ over $\hat{K}$ (which is a valuation by Theorem 3.2.4/2) and if $\pi_i\colon \hat{L} \to \hat{L}_i$ is the projection onto the i-th component, we have

$$|\hat{x}| = \max_{1 \leq i \leq t} |\pi_i(\hat{x})|_i , \qquad \hat{x} \in \hat{L} .$$

Proof. The first assertion follows from Dedekind's Lemma (Proposition 3.1.4/1). To verify the second one, consider the function $|\hat{x}|' := \max_{1 \leq i \leq t} |\pi_i(\hat{x})|_i$ for $\hat{x} \in \hat{L}$. By construction, $|\ |'$ is a power-multiplicative $\hat{K}$-algebra norm on $\hat{L}$. Thus by Corollaries 1.3.1/3 and 2.3.3/5, we see that $|\ | = |\ |'$. □

3.3.2. Construction of inequivalent valuations. — In the situation of the preceding proposition, it is easy to construct valuations on the field L which extend the given valuation on K. The maps $\pi_i|_L\colon L \to \hat{L}_i$ are contractive K-algebra homomorphisms. Each $\pi_i|_L$ is injective, because otherwise (since L is a field) we would have $\pi_i(L) = 0$, and hence $\pi_i(\hat{L}) = 0$. Thus for $i = 1, \ldots, t$, the function $|\ |_i'$, defined by

$$|x|_i' := |\pi_i(x)|_i , \qquad x \in L ,$$

is a valuation on L which extends the given valuation on K. Since $\pi_i(L)$ is dense in $\hat{L}_i$ with respect to the valuation $|\ |_i$, the field $\hat{L}_i$ equals the completion of $L \simeq \pi_i(L)$ with respect to the valuation $|\ |_i'$. For simplicity, we will write $|\ |_i$ instead of $|\ |_i'$. Then we have the following

Approximation Theorem. *Let* $x_1, \ldots, x_t$ *be elements in* L. *Then, for any real* $\varepsilon > 0$, *there exists an element* $x \in L$ *such that*

$$|x - x_i|_i < \varepsilon \qquad \textit{for} \qquad i = 1, \ldots, t.$$

Proof. Consider the element $\hat{x} := \big(\pi_1(x_1), \ldots, \pi_t(x_t)\big) \in \bigoplus_{i=1}^{t} \hat{L}_i = \hat{L}$. Since L is dense in $\hat{L}$, we can find an element $x \in L$ such that $|x - \hat{x}| < \varepsilon$. But $|x - \hat{x}| = \max_{1 \leqq i \leqq t} |\pi_i(x) - \pi_i(x_i)|_i = \max_{1 \leqq i \leqq t} |x - x_i|_i$. □

The Approximation Theorem has the following consequence:

The valuations $|\ \ |_1, \ldots, |\ \ |_t$ *are inequivalent.*

Proof. Suppose two valuations are equivalent, say $|\ \ |_1$ and $|\ \ |_2$. Then the t elements $1, 0, \ldots, 0 \in L$ cannot simultaneously be approximated if $\varepsilon < 1$, because the inequalities $|x - 1|_1 < \varepsilon$ and $|x|_2 < \varepsilon$ say that x is topologically nilpotent with respect to $|\ \ |_2$, however not with respect to $|\ \ |_1$. □

3.3.3. Construction of power-multiplicative algebra norms. — Using the valuations $|\ \ |_1, \ldots, |\ \ |_t$ on L, we now construct power-multiplicative K-algebra norms on L as follows: denote by $\mathfrak{T}$ the index set $\{1, \ldots, t\}$. For each non-empty subset $\mathfrak{S} \subset \mathfrak{T}$, the function

$$|\ \ |_{\mathfrak{S}} := \max_{i \in \mathfrak{S}} |\ \ |_i$$

is a power-multiplicative K-algebra norm on L. If $\mathfrak{S}, \mathfrak{S}'$ are different subsets of $\mathfrak{T}$, these norms are *inequivalent*, as is easily seen by the following argument: if j belongs to $\mathfrak{S} - \mathfrak{S}'$, the Approximation Theorem yields the existence of an element $x \in L$ such that $|x|_j \geqq 1$ and $|x|_i < 1$ for all $i \neq j$. Hence $|x|_{\mathfrak{S}} \geqq 1$ and $|x|_{\mathfrak{S}'} < 1$; i.e., $|\ \ |_{\mathfrak{S}}$ and $|\ \ |_{\mathfrak{S}'}$ are inequivalent.

Altogether we found $2^t - 1$ power-multiplicative K-algebra norms on L. Among them are t valuations. We shall prove now that there are no further power-multiplicative K-algebra norms or valuations on L extending the given valuation on K.

Proposition 1. *If* $\|\ \ \|$ *is any power-multiplicative* K*-algebra norm on* L, *then there exists a non-empty subset* $\mathfrak{S}$ *of* $\mathfrak{T}$ *such that* $\|\ \ \| = |\ \ |_{\mathfrak{S}}$. *For* $\mathfrak{S} = \mathfrak{T}$ *we get the spectral norm. The norm* $\|\ \ \|$ *is a valuation if and only if* $\mathfrak{S} = \{j\}$, *i.e., if and only if* $\|\ \ \| = |\ \ |_j$ *for some* $j \in \mathfrak{T}$.

Proof. Denote by L' the completion of L with respect to $\|\ \ \|$. We know by Theorem 3.2.2/2 (iii) that $\|\ \ \| \leqq |\ \ |$. Hence the identity map $(L, |\ \ |) \to (L, \|\ \ \|)$ is a contraction. Since L is dense in $\hat{L}$, this map extends uniquely to a contraction $\psi\colon \hat{L} \to L'$. It is easily verified that ψ is a $\hat{K}$-algebra homomorphism. Since L' is a finite-dimensional $\hat{K}$-vector space, $\psi(\hat{L})$ is a closed subspace. Since $L \subset \psi(\hat{L})$ is dense in L', we conclude $\psi(\hat{L}) = L'$. Then L' is isomorphic to $\hat{L}/\ker\psi$ as $\hat{K}$-algebra. Now $\ker\psi$ is necessarily a sum of certain of the components $\hat{L}_i$ occurring in the decomposition $\hat{L} = \hat{L}_1 \oplus \cdots \oplus \hat{L}_t$. Denote by $\mathfrak{S}$ the subset of all $j \in \mathfrak{T}$ such that $\hat{L}_j \not\subset \ker \psi$. We have $\mathfrak{S} \neq \emptyset$, because otherwise $L' = 0$. It follows that

$$L' = \bigoplus_{j \in \mathfrak{S}} \hat{L}_j,$$

and hence $\|x'\| = \max_{j \in \mathfrak{S}} |\pi_j(x')|_j$ for all $x' \in L'$ by Corollaries 1.3.1/3 and 2.3.3/5. Thus $\| \ \| = | \ |_{\mathfrak{S}}$.

If $\| \ \|$ is a valuation, L' is a field — i.e., $\mathfrak{S}$ can contain only one index. □

The preceding proposition may be extended to reduced finite-dimensional K-algebras L if one uses the results of (3.2.2). Roughly speaking, every power-multiplicative K-algebra norm on such an algebra L is the maximum of a finite number of valuations; more precisely,

Corollary 2. *Let $L = \bigoplus_{i=1}^{n} L_i$ be a finite-dimensional reduced K-algebra, where $L_1, \ldots, L_n$ are finite extensions of K and where $\pi_i \colon L \to L_i$ denotes the projection onto the i-th component. Let $| \ |_{i,1}, \ldots, | \ |_{i,t_i}$ be the different valuations on L_i extending the valuation on K.*

Then for every power-multiplicative K-algebra norm $| \ |$ on L, there are subsets $\mathfrak{S}_i \subset \mathfrak{T}_i := \{1, \ldots, t_i\}$ such that $|x| = \max_{1 \leq i \leq n} \max_{j \in \mathfrak{S}_i} |\pi_i(x)|_{i,j}$ for all $x \in L$.

Proof. Let $| \ |_i$ denote the restriction of $| \ |$ to L_i. By Proposition 3.2.2/3, we know that $|x| = \max_{1 \leq i \leq n} |\pi_i(x)|_i$ for $x \in L$ and that $| \ |_i$ is a power-multiplicative K-algebra norm on L_i. Now Proposition 1 yields the existence of subsets $\mathfrak{S}_i$ of $\mathfrak{T}_i$ such that $|y|_i = \max_{j \in \mathfrak{S}_i} |y|_{i,j}$ for $y \in L_i$. □

3.3.4. Hensel's Lemma. — We start with a simple

Observation. *Let $f = X^n + c_1 X^{n-1} + \cdots + c_n$, $n \geq 1$, be a monic polynomial over a (not necessarily valued) field K, and let L be an extension of K such that f splits into linear factors over L:*

$$f = (X - \theta_1) \ldots (X - \theta_n), \quad \theta_1, \ldots, \theta_n \in L.$$

Assume that the group $G(L/K)$ of K-algebra automorphisms of L operates transitively on the set of roots $\{\theta_1, \ldots, \theta_n\}$. Then f is a power of an irreducible polynomial over K.

Proof. Denote by q the minimal (irreducible) polynomial of θ_1 over K. Since $f(\theta_1) = 0$, we have an equation $f = q^s \cdot h$, where $h \in K[X]$ is monic and not divisible by q. Hence $h(\theta_1) \neq 0$. If h had a positive degree, we would have $h(\theta_i) = 0$ for some i, $1 < i \leq n$. Choose $g \in G(L/K)$ such that $g(\theta_i) = \theta_1$. We get the contradiction

$$0 = g\big(h(\theta_i)\big) = h\big(g(\theta_i)\big) = h(\theta_1).$$

Hence $h = 1$; i.e., $f = q^s$. □

From now on let K be provided with a *complete* valuation $|\ \ |$. We extend $|\ \ |$ to the Gauss norm on $K[X]$ and consider the residue map $\tilde{}: \mathring{K}[X] \to \tilde{K}[X]$.

Proposition 1. *Let $f \in K[X]$ be monic and irreducible such that $|f| = 1$. Then $\tilde{f} \in \tilde{K}[X]$ is a power of a monic irreducible polynomial $\pi \in \tilde{K}[X]$.*

Proof. Choose a splitting field L of K and write $f = (X - \theta_1) \dots (X - \theta_n)$, where $\theta_1, \dots, \theta_n \in L$. Provide L with the spectral valuation $|\ \ |$ (cf. Theorem 3.2.4/2). We have

$$|\theta_\nu| = \sigma(f) \leqq |f| = 1, \qquad \nu = 1, \dots, n.$$

Consider the image elements $\tilde{\theta}_1, \dots, \tilde{\theta}_n$ in the residue field $\tilde{L} = \mathring{L}/\check{L}$ and the image polynomial $\tilde{f}$ in $\tilde{K}[X]$. We have the splitting

$$\tilde{f} = (X - \tilde{\theta}_1) \dots (X - \tilde{\theta}_n).$$

Now we apply the group homomorphism $\tilde{}: G(L/K) \to G(\tilde{L}/\tilde{K})$ (cf. Proposition 3.1.1/4), where $\tilde{g}(\tilde{y}) = \widetilde{g(y)}$ for all $y \in \mathring{L}$. Since $G(L/K)$ operates transitively on the set $\{\theta_1, \dots, \theta_n\}$ (because L is a splitting field of f over K), we conclude that $G(\tilde{L}/\tilde{K})$ operates transitively on the set $\{\tilde{\theta}_1, \dots, \tilde{\theta}_n\}$ of all roots of $\tilde{f}$. Hence $\tilde{f}$ is a power of an irreducible polynomial $\pi \in \tilde{K}[X]$ by the observation made above. □

Proposition 2. *Let $f = a_0X^n + a_1X^{n-1} + \dots + a_n \in K[X]$, $a_0 \neq 0$, be irreducible, $n \geqq 1$. Assume $|f| = 1$. Then*

(i) $\deg \tilde{f} = \deg f$ *if and only if* $|a_0| = 1$,
(ii) $\tilde{f} \in \tilde{K}^*$ *if and only if* $|a_0| < 1$.

Proof. It is clear that $\deg \tilde{f} = \deg f$ if and only if $|a_0| = 1$. Furthermore it is clear that $\tilde{f} \in \tilde{K}^*$ implies $|a_0| < 1$.

Now assume $|a_0| < 1$. By applying Proposition 3.2.4/3 to $f^* := a_0^{-1}f$, we get

$$|a_\nu a_0^{-1}| \leqq |a_n a_0^{-1}|^{\nu/n}, \quad \text{i.e.,} \quad |a_\nu| \leqq |a_n|^{\nu/n}\, |a_0|^{1-\nu/n}.$$

From $|a_0| < 1$, we conclude that $|a_0|^{1-\nu/n} < 1$ for all $\nu < n$, and hence $|a_\nu| < |a_n|^{\nu/n} \leqq 1$ for all $\nu < n$. Thus $\tilde{f} = \tilde{a}_n \in \tilde{K}$, and in fact $\tilde{f} \in \tilde{K}^*$, since $\tilde{f} \neq 0$. □

Propositions 1 and 2 now easily give

Proposition 3 (Hensel's Lemma). *Let K be a complete valued field. Let f be a polynomial in $K[X]$, $|f| = 1$, such that $\tilde{f} \in \tilde{K}[X]$ is the product $\tilde{f} = \gamma \cdot \chi$ of two coprime polynomials $\gamma, \chi \in \tilde{K}[X]$ (i.e., the greatest common divisor of γ and χ is 1). Then there exist polynomials $g, h \in \mathring{K}[X]$ with the following properties:*

$$f = g \cdot h, \quad \tilde{g} = \gamma, \quad \tilde{h} = \chi \quad \textit{and} \quad \deg g = \deg \gamma.$$

Proof. Let $f = p_1 \dots p_n$ be a factorization of f into irreducible polynomials. We may assume $|p_\nu| = 1$ for all ν. Furthermore, we can label the polynomials $p_1, \dots, p_n$ in such a way that the leading coefficients of the first m polynomials

$p_1, \ldots, p_m$ have norm 1, while the leading coefficients of $p_{m+1}, \ldots, p_n$ have norm < 1. (Then $0 \leqq m \leqq n$.) Multiplying by suitable constants of norm 1 in K, we can assume that $p_1, \ldots, p_m$ are monic. By Proposition 1, we have equations $\tilde{p}_\mu = \pi_\mu^{s_\mu}$, where $\pi_\mu \in \tilde{K}[X]$ is irreducible and where $s_\mu \geqq 1$, $\mu = 1, \ldots, m$. By Proposition 2, the elements $\tilde{p}_{m+1}, \ldots, \tilde{p}_n$ are units in $\tilde{K}[X]$. Since γ and χ are coprime in the factorial ring $\tilde{K}[X]$, we may assume $\gamma = \tilde{c}\pi_1^{s_1} \cdots \pi_r^{s_r}$ where $c \in K$ has norm 1 and where $r \leqq m$. Then $g := cp_1 \ldots p_r$ and $h := c^{-1}p_{r+1} \ldots p_n$ have the desired properties. □

3.4. Properties of the spectral valuation

By K_a we always mean the algebraic closure of K provided with the spectral norm $|\ \ |$. All roots of polynomials $f \in K[X]$ are elements of K_a. For each $f \in K[X]$, we denote by $|f|$ its Gauss norm. If f is monic, we have $\sigma(f) \leqq |f|$ for the spectral value $\sigma(f)$ of f. Hence, in particular, $|\alpha| \leqq |f|$ for each root $\alpha \in K_a$ of f due to Proposition 3.1.2/1.

3.4.1. Continuity of roots. — Let $f, g \in K[K]$ be monic polynomials of the same degree n. Let $\alpha \in K_a$ be a root of f. We have the crucial inequality:

$$(*) \qquad |g(\alpha)| \leqq |f - g| \cdot |f|^{n-1}.$$

Proof. Write $f = X^n + \sum_{\nu=1}^{n} f_\nu X^{n-\nu}$, $g = X^n + \sum_{\nu=1}^{n} g_\nu X^{n-\nu}$. Then

$$g(\alpha) = g(\alpha) - f(\alpha) = \sum_{\nu=1}^{n} (g_\nu - f_\nu)\,\alpha^{n-\nu},$$

and hence

$$|g(\alpha)| \leqq \max_{1 \leqq \nu \leqq n} |f_\nu - g_\nu| \cdot |\alpha|^{n-\nu}.$$

Since $|f_\nu - g_\nu| \leqq |f - g|$, $|\alpha| \leqq |f|$ and $|f| \geqq 1$, we deduce the desired inequality. □

Proposition 1 (Continuity of roots). *Let K be complete; let $f, g \in K[X]$ be monic polynomials of the same degree n. Then for each root $\alpha \in K_a$ of f, there exists a root $\beta \in K_a$ of g such that*

$$|\alpha - \beta| \leqq \sqrt[n]{|f - g|} \cdot |f|.$$

Proof. Write $g = (X - \beta_1) \ldots (X - \beta_n)$, where $\beta_1, \ldots, \beta_n \in K_a$ are all roots of g. Assume $|\alpha - \beta_\nu| > \sqrt[n]{|f - g|}\,|f|$ for all ν. Then we get (since the spectral norm is a valuation)

$$|g(\alpha)| = |\alpha - \beta_1| \ldots |\alpha - \beta_n| > |f - g| \cdot |f|^n,$$

which contradicts (*) since $|f| \geqq 1$. □

Corollary 2. *Let K be complete, and let $f_i \in K[X]$, $i = 1, 2, \ldots$, be a sequence of monic polynomials of the same degree n which converge (with respect to the*

Gauss norm) to a polynomial $g \in K[X]$. Let $\alpha_i \in K_a$ be a root of f_i, $i \geqq 1$. Then the sequence $(\alpha_i)_{i\geq 1}$ contains a subsequence which converges to a root of g.

Proof. The polynomial g is monic of degree n. Hence, by applying Proposition 1 for each α_i, there exists a root β_i of g such that $|\alpha_i - \beta_i| \leqq \sqrt[n]{|f_i - g|}\,|f_i|$. Since there are at most n different roots of g, there exist a root β of g and an infinite subset S of $\mathbb{N}$ such that

$$|\alpha_i - \beta| \leqq \sqrt[n]{|f_i - g|} \cdot |f_i|, \qquad i \in S.$$

Since $|f_i| = |g|$ for large i, the sequence $(\alpha_i)_{i\in S}$ converges to β. □

For the convenience of the reader, we rephrase Proposition 1 to show that it implies the classical lemma on continuity of roots.

Let K be complete, let $f \in K[X]$ be monic of degree n, and let α be a root of f of multiplicity t. Choose $\varepsilon_0 > 0$ such that all roots $\neq \alpha$ of f have distance $\geqq \varepsilon_0$ from α. Then for each ε, $0 < \varepsilon < \varepsilon_0$, there exists a $\delta > 0$ such that each monic polynomial $g \in K[X]$ of degree n with $|g - f| < \delta$ has exactly t roots (counted with multiplicities) in the ball $B^-(\alpha, \varepsilon) = \{x \in K_a;\ |x - \alpha| < \varepsilon\}$.

Proof. If the assertion were not true, there would exist an ε_1, $0 < \varepsilon_1 < \varepsilon_0$, and a sequence $g_i \in K[X]$, $i \geqq 1$, of monic polynomials of degree n converging to f such that no g_i has exactly t roots in $B^-(\alpha, \varepsilon_1)$. Write $g_i = (X - \beta_{1i}) \ldots (X - \beta_{ni})$, $b_{\nu i} \in K_a$. By Corollary 2, we can choose a subset $S_1 \subset \mathbb{N}$ such that the sequence $(\beta_{1i})_{i\in S_1}$ converges to a root α_1 of f. From S_1 we extract a subset S_2 such that $(\beta_{2i})_{i\in S_2}$ converges to a root α_2 of f. After n steps we thus get a subset $S \subset \mathbb{N}$ such that, for each $\nu = 1, \ldots, n$, the sequence $(\beta_{\nu i})_{i\in S}$ has a limit α_ν, which is a root of f. Since $f = \lim_{i\in S} g_i = (X - \alpha_1) \ldots (X - \alpha_n)$, we see that the elements $\alpha_1, \ldots, \alpha_n$ are, in fact, all roots of f, counted with their multiplicities. Choose $i_0 \in S$ such that

$$|\beta_{\nu i} - \alpha_\nu| < \varepsilon_1 \quad \text{for all} \quad \nu = 1, \ldots, n \quad \text{and all} \quad i \in S, i \geqq i_0.$$

We label the roots of f (and the roots of the polynomials g_i) in such a way that $\alpha = \alpha_1 = \cdots = \alpha_t$. Then we have

$$|\beta_{\nu i} - \alpha| < \varepsilon_1 \quad \text{for all} \quad \nu = 1, \ldots, t \quad \text{and all} \quad i \in S, i \geqq i_0,$$

and (since $\alpha_\nu \neq \alpha$ and therefore $|\alpha_\nu - \alpha| \geqq \varepsilon_0$ for all $\nu > t$)

$$|\beta_{\nu i} - \alpha| = |(\beta_{\nu i} - \alpha_\nu) + (\alpha_\nu - \alpha)| \geqq \varepsilon_0 > \varepsilon_1$$

$$\text{for all} \quad \nu > t \quad \text{and all} \quad i \in S, i \geqq i_0.$$

Hence each g_i, $i \in S$, $i \geqq i_0$, has exactly t roots in the ball $B^-(\alpha, \varepsilon_1)$, in contradiction to our assumption. □

We now draw some conclusions from Proposition 1.

Proposition 3. *If K is algebraically closed, the completion $\hat{K}$ of K is algebraically closed.*

Proof. Let $\hat{f} = X^n + \hat{a}_1 X^{n-1} + \cdots + \hat{a}_n$ be a polynomial in $\hat{K}[X]$, $n \geqq 1$. We have to find a root $\hat{\alpha}$ of $\hat{f}$ in $\hat{K}$. Since K is dense in $\hat{K}$, there are n sequences $(a_{\nu i})_{i \in \mathbb{N}}$ in K with $\lim_i a_{\nu i} = \hat{a}_\nu$, $\nu = 1, \ldots, n$. The field K is algebraically closed. Therefore each polynomial $f_i = X^n + a_{1i} X^{n-1} + \cdots + a_{ni} \in K[X]$ has a root α_i in K, $i \in \mathbb{N}$. Denote by $\hat{K}_{\mathrm{a}}$ the algebraic closure of $\hat{K}$. Since $K \subset \hat{K}_{\mathrm{a}}$ and since the sequence f_i converges to $\hat{f}$ (with respect to the Gauss norm on $\hat{K}[X]$), a subsequence of the sequence $(\alpha_i)_{i \in \mathbb{N}}$ converges in $\hat{K}_{\mathrm{a}}$ (with respect to the spectral norm) to a root $\hat{\alpha}$ of $\hat{f}$ (cf. Corollary 2). However $\alpha_i \in K$ implies $\hat{\alpha} \in \hat{K}$. □

The proposition just proved enables us to construct for each valued field K a smallest algebraically closed extension K' which is complete with respect to a valuation extending the valuation on K. First we pass to a completion $\hat{K}$ of K and provide $\hat{K}$ with the extended valuation. In general, $\hat{K}$ will not be algebraically closed. So next we take an algebraic closure $\hat{K}_{\mathrm{a}}$ of $\hat{K}$ and provide K_{a} with the spectral valuation. In general, $\hat{K}_{\mathrm{a}}$ will not be complete, so we take a completion $K' = (\hat{K}_{\mathrm{a}})^\wedge$ of $\hat{K}_{\mathrm{a}}$ and extend the valuation from $\hat{K}_{\mathrm{a}}$ to K'. By Proposition 3, the field K' is algebraically closed and complete. It is obvious that up to (non-canonical) isometric K-algebra isomorphisms the field K' is the *smallest* extension of K with the mentioned properties. This implies that the algebraic closure of K in K' is dense in K'. Namely, the completion of this field (i.e., its topological closure in K') is a complete algebraically closed extension of K contained in K'.

As a by-product of these considerations, we want to determine the residue field of K' in terms of the residue field of K. First we show

Lemma 4. *If K is complete, then $\widetilde{K_{\mathrm{a}}} = (\tilde{K})_{\mathrm{a}}$ (where K_{a} is provided with the spectral valuation, i.e., with the unique extension of the valuation on K).*

Proof. The field $\widetilde{K_{\mathrm{a}}}$ is algebraically closed. In order to verify this, let $p \in \mathring{K}_{\mathrm{a}}[X]$ be a monic polynomial; we have to find a zero of $\tilde{p}$ in $\widetilde{K_{\mathrm{a}}}$. Since K_{a} is algebraically closed, we can find a zero c of p in K_{a}, where $c \in \mathring{K}_{\mathrm{a}}$ by Proposition 3.1.2/1. Then $\tilde{c}$ is the desired zero of $\tilde{p}$ in $\widetilde{K_{\mathrm{a}}}$, and hence $\widetilde{K_{\mathrm{a}}}$ is algebraically closed. In order to show that $\widetilde{K_{\mathrm{a}}}$ is indeed the algebraic closure of $\tilde{K}$, it remains to prove that $\widetilde{K_{\mathrm{a}}}$ is algebraic over $\tilde{K}$. For every $c \in \mathring{K}_{\mathrm{a}}$, we have to find a monic polynomial $p \in \mathring{K}[X]$ such that $\tilde{p}(\tilde{c}) = 0$. Let p be the minimal polynomial of c over K. Then $\sigma(p) = |c| \leqq 1$. Hence we have $p \in \mathring{K}[X]$ so that $p(c) = 0$ implies $\tilde{p}(\tilde{c}) = 0$. □

Remark. If K is not complete, this proof nevertheless makes sense if K_{a} is provided with the spectral norm. One gets that the reduced $\tilde{K}$-algebra $\widetilde{K_{\mathrm{a}}}$ is integral over $\tilde{K}$ and "algebraically closed". However, $\widetilde{K_{\mathrm{a}}}$ is not, in general, a field.

Using the lemma, we easily get the desired description of $\tilde{K}'$.

Proposition 5. *Let K be a valued field, and let $K' = (\hat{K}_a)^\wedge$ be the smallest complete algebraically closed extension of K. Then $\tilde{K}' = (\tilde{K})_a$; i.e., the residue field of K' equals the algebraic closure of the residue field of K.*

Proof. The residue field of $\hat{K}_a$ equals the algebraic closure of the residue field of $\hat{K}$ by the preceding lemma, i.e., $\widetilde{\hat{K}_a} = (\tilde{\hat{K}})_a$. Since the residue field of any valued field L remains unchanged if L is replaced by its completion, the assertion is obvious. □

Proposition 6. *Let K_a be the algebraic closure of the complete field K. Then the subfield K_{sep} of K_a consisting of all elements which are separable over K is dense in K_a. More precisely,*

If $\alpha \in K_a$ is of degree n over K, then, for each $\varepsilon > 0$, there exists a separable element $\beta \in K_a$ of degree $\leqq n$ such that $|\beta - \alpha| < \varepsilon$.

Proof. Let $f = X^n + a_1X^{n-1} + \cdots + a_n \in K[X]$ be the minimal polynomial of α over K. Set $\delta := (\varepsilon\, |f|^{-1})^n$. According to Proposition 1, it suffices to construct a polynomial $g = X^n + z_1X^{n-1} + \cdots + z_n \in K[X]$ which satisfies $|f - g| < \delta$ and which has only simple roots in K_a, i.e., a polynomial whose discriminant $\Delta \in K$ is different from zero. If $z_1, \ldots, z_n$ are considered as indeterminates, the discriminant Δ of g is a non-zero polynomial in $z_1, \ldots, z_n$. The zero set $Z \subset K^n$ of any such polynomial is closed in K^n and nowhere dense if we consider the product topology on K^n. (A subset of a topological space is called nowhere dense if its closure has no interior points.) Thus in our case, we see that the polynomial is not identically zero on any open ball in K^n. In particular, we can choose a point $(z_1, \ldots, z_n) \in K^n$ with $|z_\nu - a_\nu| < \delta$, $\nu = 1, \ldots, n$, such that $\Delta(z_1, \ldots, z_n) \neq 0$. □

3.4.2. Krasner's Lemma. — In this section the field K is always complete. For elements $\alpha \in K_a$ such that the minimal polynomial p of α over K has at least two different roots, we define $r(\alpha) := \min\limits_{\gamma} |\alpha - \gamma|$, where γ ranges over all roots $\neq \alpha$ of p. We have $r(\alpha) > 0$. The following proposition describes an important property of the elements $\beta \in K_a$ which lie in the "open" ball $B^-(\alpha, r(\alpha))$ of radius $r(\alpha)$ around α.

Proposition 1. *For each $\beta \in B^-(\alpha, r(\alpha))$, the minimal polynomial h of α with respect to the field $K(\beta)$ has no roots $\neq \alpha$.*

Proof. Let $\gamma \in K_a$ be a root of h. Then $\gamma - \beta$ and $\alpha - \beta$ are conjugate over $K(\beta)$. Therefore, $|\gamma - \beta| = |\alpha - \beta|$, and hence $|\alpha - \gamma| = |(\alpha - \beta) - (\gamma - \beta)| \leqq |\alpha - \beta| < r(\alpha)$. Since h is a factor (in $K(\beta)\,[X]$) of the minimal polynomial f of α over K, we have $f(\gamma) = 0$. Since α is the only root of f in the ball $B^-(\alpha, r(\alpha))$ (by the choice of $r(\alpha)$), we conclude $\gamma = \alpha$. □

Corollary 2 (Krasner's Lemma). *If α is separable of degree > 1 over K, we have $K(\alpha) \subset K(\beta)$ for each $\beta \in B^-(\alpha, r(\alpha))$.*

Proof. The element α is also separable over $K(\beta)$. Hence h has only simple roots; i.e., $h = X - \alpha \in K(\beta)[X]$ due to Proposition 1. Thus $\alpha \in K(\beta)$. $\square$

If α is of degree n over K, KRASNER's Lemma tells us, in particular, that each $\beta \in B^-(\alpha, r(\alpha))$ is of degree $\geqq n$ over K. If the degree of β is n, we must have $K(\alpha) = K(\beta)$. The next proposition shows that there are plenty of elements in $B^-(\alpha, r(\alpha))$ with this property.

Proposition 3. *Let α be separable over K of degree $n > 1$, and let $f \in K[X]$ be the minimal polynomial of α over K. Set $\varepsilon(\alpha) := (|f|^{-1} r(\alpha))^n$. Then each monic polynomial $g \in K[X]$ of degree n such that $|f - g| < \varepsilon(\alpha)$ has a root $\beta \in B^-(\alpha, r(\alpha))$. Furthermore, each such root β satisfies $K(\beta) = K(\alpha)$.*

Proof. Proposition 3.4.1/1 yields the existence of a root β of g such that

$$|\beta - \alpha| \leqq \sqrt[n]{|f - g|} \cdot |f| < \sqrt[n]{\varepsilon(\alpha)}\, |f| = r(\alpha).$$

Hence $\beta \in B^-(\alpha, r(\alpha))$. For any such β, we have $K(\alpha) \subset K(\beta)$ by KRASNER's Lemma. Now $\dim_K K(\beta) \leqq n$, and $\dim_K K(\alpha) = n$. Therefore $K(\alpha) = K(\beta)$. $\square$

Corollary 4. *Let f be a polynomial in $K[X]$ of degree $n > 1$, which is monic, irreducible and separable. Then any monic polynomial $g \in K[X]$ of the same degree n which is sufficiently close to f is also irreducible and separable.*

For later application we state another consequence of Proposition 3:

Proposition 5. *Let K' be a dense subfield of K, and let $\alpha \in K_a$ be separable of degree $n > 1$ over K. Then there exist elements $\alpha' \in K(\alpha)$ arbitrarily close to α and algebraic over K' such that $K(\alpha) = K(\alpha')$.*

Proof. Let $f \in K[X]$ be the minimal polynomial of α. Since K' is dense in K, there exists a sequence $f_i \in K'[X]$ of monic polynomials of degree n converging to f (with respect to the Gauss norm on $K[X]$). Then each f_i has a root $\alpha_i \in K_a$ such that $|\alpha_i - \alpha| \leqq \sqrt[n]{|f - f_i|}\, |f|$. For large i we have $\alpha_i \in B^-(\alpha, r(\alpha))$ and thus $K(\alpha_i) = K(\alpha)$ by Proposition 3. Now α_i is algebraic over K', and $|\alpha_i - \alpha| \leqq \sqrt[n]{|f - f_i|}\, |f| \to 0$. $\square$

3.4.3. Example. *p*-adic numbers. — Let p be a fixed prime number. We consider the p-adic valuation on $\mathbb{Z}$ (cf. (1.3.3) and (1.5.2)) and extend it to a valuation on the field of fractions $\mathbb{Q}$ (use Proposition 1.2.2/5). Thereby we get the so-called *p-adic valuation on $\mathbb{Q}$*, which is denoted by $|\ \ |_p$. Usually $|\ \ |_p$ is normalized in such a way that $|p|_p = p^{-1}$ (i.e., one sets $\varepsilon := p^{-1}$ in Proposition 1.3.3/1). Then the p-adic value of an element $x \in \mathbb{Q}$ is given by

$$|x|_p := \begin{cases} 0 & \text{if } x = 0, \\ p^{-n} & \text{if } x = p^n \dfrac{a}{b} \quad \text{with integers } n, a, b \in \mathbb{Z},\ p \nmid ab. \end{cases}$$

For simplicity, one may view this as a definition for $|\ \ |_p$. (The properties of a valuation follow easily by direct computation.)

In order to determine the residue field of $\mathbb{Q}$ (with respect to $|\ |_p$), one has to realize that the valuation ring $\mathring{\mathbb{Q}}$ equals the localization of $\mathbb{Z}$ with respect to the prime ideal (p), i.e., that $\mathring{\mathbb{Q}} = \mathbb{Z}_{(p)}$. The maximal ideal of $\mathring{\mathbb{Q}}$ is, of course, the ideal $p\mathbb{Z}_{(p)}$, and one obtains

$$\tilde{\mathbb{Q}} = \mathbb{Z}_{(p)}/p\mathbb{Z}_{(p)} = \mathbb{Z}/p\mathbb{Z}.$$

Thus, the residue field of $\mathbb{Q}$ (and that of its completion) is the finite field $\mathbb{F}_p = \mathbb{Z}/p\mathbb{Z}$.

The completion of $\mathbb{Q}$ with respect to $|\ |_p$ is referred to as *Hensel's field of p-adic numbers* $\mathbb{Q}_p$. It corresponds to the field $\mathbb{R}$ of real numbers if $|\ |_p$ is replaced by the ordinary absolute value on $\mathbb{Q}$. Using (3.4.1), we can construct an extension $\mathbb{C}_p$ of $\mathbb{Q}_p$ which is analogous to the field $\mathbb{C}$ of complex numbers Namely, let $\mathbb{C}_p$ be the completion of the algebraic closure of $\mathbb{Q}_p$. This is the smallest complete extension of $\mathbb{Q}_p$ which is algebraically closed. By abuse of language, $\mathbb{C}_p$ is often referred to as the *(algebraically closed) field of p-adic numbers*.

We will see that the extension $\mathbb{C}_p$ over $\mathbb{Q}_p$ is much more complicated than the extension $\mathbb{C}$ over $\mathbb{R}$. Let $\mathbb{Q}_p^{\mathrm{alg}}$ denote the algebraic closure of $\mathbb{Q}_p$ in $\mathbb{C}_p$. First we want to show that $\mathbb{Q}_p^{\mathrm{alg}}$ (hence also $\mathbb{C}_p$) is of infinite degree over $\mathbb{Q}_p$. Using Proposition 2.1.10/3 we have only to verify that the residue degree of the extension $\mathbb{Q}_p^{\mathrm{alg}}/\mathbb{Q}_p$ is infinite. However this is clear, since the residue field of $\mathbb{Q}_p$ is the finite field $\mathbb{F}_p$ and since the residue field of $\mathbb{Q}_p^{\mathrm{alg}}$ equals the algebraic closure of $\mathbb{F}_p$ (Lemma 3.4.1/4).

Next we claim that the field $\mathbb{Q}_p^{\mathrm{alg}}$ is not complete. (Note that $\mathbb{Q}_p^{\mathrm{alg}}$ would be complete if $[\mathbb{Q}_p^{\mathrm{alg}}: \mathbb{Q}_p]$ were finite.) Namely, we show that the following general fact is true.

Lemma 1. *Let K be a field with a complete non-trivial valuation. Assume that the algebraic closure K_{a} of K is of infinite degree over K. Then K_{a} (provided with the unique valuation extending the valuation on K) is not complete.*

Proof. First we want to show that $[K_{\mathrm{a}} : K] = \infty$ implies $[K_{\mathrm{sep}} : K] = \infty$ for the separable algebraic closure K_{sep} of K. Namely if $[K_{\mathrm{sep}} : K] < \infty$, then K_{sep} is complete (Proposition 2.3.3/4) and $K_{\mathrm{sep}} = K_{\mathrm{a}}$ (since K_{sep} is dense in K_{a}, see Proposition 3.4.1/6). Therefore $[K_{\mathrm{a}} : K] = \infty$ can only be true if $[K_{\mathrm{sep}} : K] = \infty$.

Now choose a sequence $x_0 = 1, x_1, x_2, \ldots$ of elements in K_{sep} which are linearly independent over K. Then one can find a sequence $c_1, c_2, \ldots$ of non-zero constants in K such that, for all $i \in \mathbb{N}$,

(i) $|c_{i+1}x_{i+1}| \leq |c_i x_i|$, $\lim\limits_{i\to\infty} |c_i x_i| = 0$,

(ii) $|c_{i+1}x_{i+1}| < r\left(\sum\limits_{\nu=1}^{i} c_\nu x_\nu\right)$,

where the function r is defined as in (3.4.2). Notice that any non-trivial linear combination $y = \sum_{\nu=1}^{i} d_\nu x_\nu \in K_{\text{sep}}$ with coefficients $d_\nu \in K$ is not contained in K so that $r(y)$ is well-defined.

We want to show that the infinite series $\sum_{i=1}^{\infty} c_i x_i$ has no limit in K_a. Assuming the contrary, let $x \in K_a$ be the limit of this series. Since

$$\left| x - \sum_{\nu=1}^{i} c_\nu x_\nu \right| = \left| \sum_{\nu=i+1}^{\infty} c_\nu x_\nu \right| \leqq |c_{i+1} x_{i+1}| < r\left(\sum_{\nu=1}^{i} c_\nu x_\nu\right),$$

we see that $x \in B^-\left(\sum_{\nu=1}^{i} c_\nu x_\nu,\ r\left(\sum_{\nu=1}^{i} c_\nu x_\nu\right)\right)$ for all $i \in \mathbb{N}$. Then KRASNER's Lemma (Corollary 3.4.2/2) says that

$$\sum_{\nu=1}^{i} c_\nu x_\nu \in K(x)$$

for all $i \in \mathbb{N}$. All coefficients c_i are non-zero. Hence we get $x_1, x_2, \ldots \in K(x)$ and, in particular, $[K(x) : K] = \infty$. However this is impossible, because x is algebraic over K. Consequently, the series $\sum_{i=1}^{\infty} c_i x_i$ has no limit in K_a, and K_a is not complete. □

The above lemma implies that the construction of the smallest complete and algebraically closed extension of a valued field K is a complicated process which is comparable to the construction of the classical extension $\mathbb{C}$ over $\mathbb{R}$ only if $[K_a : K] < \infty$. The general construction method given in (3.4.1) cannot be simplified as the example of p-adic numbers shows.

3.5. Weakly stable fields

We want to combine the notions of weakly cartesian vector space and spectral norm. As always, K denotes a valued field and K_a the algebraic closure of K. Let L be an algebraic extension of K provided with a K-algebra norm (usually the spectral norm).

3.5.1. Weakly cartesian fields. — We start with a general lemma concerning the connection between the product topology and the spectral norm.

Lemma 1. *Let L be a finite extension of K with a K-algebra norm $|\ \ |$ such that L is weakly K-cartesian. Then the semi-norm $|\ \ |'$ defined by $|x|' := \inf_{\nu\to\infty} |x^\nu|^{1/\nu}$ is the spectral norm. In particular, if the product topology on L can be induced by a power-multiplicative K-algebra norm, then this norm must be the spectral norm.*

Proof. Because L provided with $|\ \ |$ carries the product topology, the identity map $(L, |\ \ |) \to (L, |\ \ |_{\text{sp}})$ is continuous. Therefore, there is a constant $\varrho > 0$ such that $|x|_{\text{sp}} \leqq \varrho |x|$ for all $x \in L$, whence $|x|_{\text{sp}} \leqq |x|'$ for all $x \in L$. Thus we see by Proposition 1.3.2/1 that $|\ \ |'$ is a power-multiplicative

K-algebra norm dominating the spectral norm. According to Proposition 3.1.2/1, the two norms $|\ \ |$ and $|\ \ |_{sp}$ coincide. The second assertion of the lemma is obvious. □

In this section we shall study extensions of K whose product topology can be given by the spectral norm. We start with the simple

Observation 2. *The K-vector space L is weakly K-cartesian (resp. K-cartesian) if each finite extension $L' \subset L$ of K is weakly K-cartesian (resp. K-cartesian).*

Proof. Let W be any K-subspace of L of finite dimension. Since L is algebraic over K, the field $L' := K[W]$ generated over K by all elements of W is a finite extension of K. By assumption L' is weakly K-cartesian (resp. K-cartesian). Hence $W \subset L'$ is weakly K-cartesian (resp. K-cartesian) as well, cf. Lemma 2.3.2/5 (resp. Proposition 2.4.1/5). □

Next we prove

Proposition 3. *Let L be a finite separable extension of K such that the trace function $T := \mathrm{Tr}_{L/K}: L \to K$ is continuous. Then L is weakly K-cartesian.*

Proof. By assumption the K-bilinear map $L \times L \to K$ defined by $(x, y) \mapsto T(xy)$ is continuous. Since L is a separable extension, $T(xy)$ is non-degenerate, i.e., for given $x_0 \neq 0$ in L, there always exists a $y_0 \in L$ such that $T(x_0y_0) \neq 0$. Hence $x \mapsto T(xy_0)$ is a continuous K-linear map $\lambda: L \to K$ such that $\lambda(x_0) \neq 0$. Thus L is a b-separable K-vector space and therefore weakly K-cartesian (Proposition 2.3.2/7). □

From Proposition 3 we easily deduce

Proposition 4. *If K is perfect (in particular, if K is of characteristic 0) the algebraic closure K_a of K provided with the spectral norm is weakly K-cartesian.*

Proof. The field K being perfect, each finite extension $L \subset K_a$ is separable. The trace function $T: L \to K$ is a contraction with respect to the spectral norm on L (see Corollary 3.2.3/2). Hence L is weakly K-cartesian by Proposition 3. The observation made above concludes the proof. □

3.5.2. Weakly stable fields. — The proposition just proved is not true in general if K is not perfect. Namely if M is a complete non-perfect field, Proposition 3.4.1/6 says that the algebraic closure M_a of M cannot be weakly cartesian over the separable algebraic closure $K := M_{sep}$ of M in M_a, since this field is dense in M_a. That there exist complete non-perfect fields can be seen as follows. Start with a field k of characteristic $p > 0$. Consider on the polynomial ring $k[X]$ a valuation defined by the degree function (see (1.3.3)). Let M be the completion of the field of fractions of $k[X]$. Then M does not contain a p-th root of the element X, because $|X|^{1/p} \notin |M|$. Thus M is not perfect, and we have seen that

There exist valued fields K admitting purely inseparable extensions $L \neq K$

such that K is dense in L if L carries the spectral valuation. (For purely inseparable extensions, the spectral norm is always a valuation.)

In order to give a slightly more explicit example, we fix a field k of characteristic $p > 0$ with a complete non-trivial valuation such that the field $k^{p^{-1}}$ of p-th roots is of infinite degree over k. One can take for k the completion of the field of fractions of the formal power series ring $\mathbb{F}_p[[Y_1, Y_2, \ldots]]$ (as defined in (1.5.5)) where $\mathbb{F}_p$ is the (trivially valued) prime field of characteristic p. We consider the valued k-algebra $R := k^{p^{-1}}\langle X\rangle$ of strictly convergent power series over $k^{p^{-1}}$ in one indeterminate X (where $k^{p^{-1}}$ carries the spectral valuation). Let $A \subset R$ be the subset of all series $f = \sum_0^\infty f_\nu X^\nu$ such that the coefficients f_ν generate a finite extension of k. Obviously A is a k-algebra. Since each $\sum_0^\infty a_\nu X^\nu \in R$ is the limit of the sequence of its partial polynomials $\sum_0^n a_\nu X^\nu \in A$, the ring A is dense in R. Hence the field of fractions K of A is dense in the field of fractions L of R (with respect to the extended valuations).

It remains to be shown that $L \neq K$. Since $\dim_k k^{p^{-1}} = \infty$, we can choose a sequence $h_\nu \in k^{p^{-1}}$ converging to 0 and generating an extension of infinite degree over k. Set $h := \sum_0^\infty h_\nu X^\nu \in R$ and suppose h is an element of K. Then there exist series $f = \sum f_\lambda X^\lambda$, $g = \sum g_\mu X^\mu$ in A, $g \neq 0$, such that $gh = f$. By the definition of A, there exists a finite extension $L' \subset k^{p^{-1}}$ of k containing all f_λ and all g_μ. Let i be the smallest index such that $h_i \notin L'$, and assume that $g_0 \neq 0$. Then in the equation

$$f_i = g_0 h_i + g_1 h_{i-1} + \cdots + g_i h_0,$$

the left-hand side and all terms of the right-hand side except for the first one are in L'. Hence we conclude $g_0 h_i \in L'$, and therefore $h_i \in L'$, since $g_0 \neq 0$ and L' is a field. However, this is in contradiction to the choice of the index i so that we must have $L \neq K$. □

We now introduce the following

Definition 1. *A valued field K is called weakly stable if each finite extension L of K provided with the spectral norm is weakly K-cartesian.*

The example just given shows the existence of fields which are not weakly stable (for further examples, see (3.5.4)). Obviously our definition can be rephrased as follows:

The field K is weakly stable if and only if K_a is weakly K-cartesian with respect to the spectral norm.

By Proposition 3.5.1/4, *each perfect field is weakly stable.* Furthermore, it follows from Proposition 2.3.3/4 that *each complete field K is weakly stable.*

Remark. Another way of expressing the fact that K is weakly stable is to say that the completion $\hat{K}$ of K is a separable extension of K, i.e., that $\hat{K} \otimes_K K^{p^{-1}}$ is reduced (if $p := \operatorname{char} K \neq 0$). However we shall never use this characterization.

3.5.3. Criterion for weak stability. — We only have to consider the case where $p := \operatorname{char} K \neq 0$. The essential role is played by the field $K^{p^{-1}} = \{x \in K_a; x^p \in K\}$ of all p-th roots. The spectral norm on $K^{p^{-1}}$ is a valuation.

Theorem 1. *A valued field K is weakly stable if and only if $K^{p^{-1}}$ (provided with the spectral valuation) is weakly K-cartesian.*

Proof. We only have to show that K_a is weakly K-cartesian with respect to the spectral norm if $K^{p^{-1}}$ is weakly K-cartesian. First we show by induction on n

Each field $K_n := \{x \in K_a; x^{p^n} \in K\}$ is weakly K-cartesian, $n \geqq 1$.

Assume K_n is weakly K-cartesian (true by assumption for $n = 1$). In order to prove that K_{n+1} is weakly K-cartesian, it is enough to prove (use Proposition 2.3.3/2 with $V := K_{n+1}$ and $K' := K_n$ and the fact that the spectral norm on K_n is a valuation extending the valuation on K)

Each finite-dimensional K_n-vector space $U \subset K_{n+1}$ is closed in K_{n+1}.

The Frobenius homomorphism $x \mapsto x^{p^n}$ is a homeomorphism mapping the K_n-algebra K_{n+1} onto the K-algebra K_1. Thereby U is mapped onto a finite-dimensional K-subspace of K_1 which is closed in K_1 by assumption. Hence, U must be closed in K_{n+1}.

Next we consider the "perfect closure" $K_\infty := \bigcup_{n=1}^{\infty} K_n$ which is a valued subfield of K_a. From what we just proved, we conclude that K_∞ is weakly K-cartesian. Thus, again by Proposition 2.3.3/2, all that remains to be shown is that K_a is weakly K_∞-cartesian. This follows from Proposition 3.5.1/4, since K_∞ is perfect and since the spectral norm on K_a over K coincides with the spectral norm over K_∞ (see Proposition 3.2.2/4). □

In applications K is often given as the field of fractions of a valued ring A. Then

$$A_1 = A^{p^{-1}} = \{x \in K_1; x^p \in A\}$$

is a valued ring having K_1 as field of fractions. More precisely, $K_1 = A_1 \otimes_A K$; i.e., each $z \in K_1$ is of the form $z = \frac{x}{a}$, $x \in A_1$, $a \in A - \{0\}$. (In order to see this, just write $z^p = \frac{b}{a}$, $b \in A$, $a \in A - \{0\}$, and set $x := az$. Then $z = \frac{x}{a}$ and $x^p = a^{p-1}b \in A$; i.e., $x \in A_1$.)

Now Lemma 2.3.3/3 (with $M = A_1$, $V = K_1$) and Theorem 1 directly imply

Lemma 2. *Let K be the field of fractions of the ring A. Assume that each finitely generated A-submodule of $A^{p^{-1}}$ is b-separable. Then K is weakly stable.*

For later reference and as an illustration of Lemma 2, we want to show

Proposition 3. *Let $K = k(X_1, \ldots, X_n)$ be the field of rational functions over some field k. Provide K with the valuation induced by the total degree, i.e.,* $\left|\frac{f}{g}\right| := \exp(\deg f - \deg g)$ *for all polynomials $f, g \in k[X_1, \ldots, X_n]$, $g \neq 0$. Then K is weakly stable.*

Proof. Let $A := k[X_1, \ldots, X_n]$ be provided with the valuation induced by the total degree. Then K is the field of fractions of A, and the valuation on K extends the valuation on A. In order to apply the preceding lemma, we must show that $A^{p^{-1}}$ is a b-separable A-module. Viewing $k[X_1, \ldots, X_n]$ as a $k[X_1^p, \ldots, X_n^p]$-module, we have a canonical decomposition as a norm-direct sum

$$k[X_1, \ldots, X_n] = \bigoplus_{0 \leq \nu_i < p} k[X_1^p, \ldots, X_n^p]\, X_1^{\nu_1} \ldots X_n^{\nu_n}.$$

Then we can use the Frobenius homomorphism $f \mapsto f^p$ (or more precisely, its inverse) to see that $A^{p^{-1}}$ is a norm-direct sum of finitely many A-submodules linearly homeomorphic to $k^{p^{-1}}[X_1, \ldots, X_n]$. Thus by Proposition 2.2.5/2, we have only to show that $k^{p^{-1}}[X_1, \ldots, X_n]$ is a b-separable A-module. However this is clear, since $k^{p^{-1}}$ (carrying the trivial valuation) is b-separable over k. □

3.5.4. Weak stability and Japaneseness. — In this section we assume the reader is familiar with the notions and results of the appendix "Tame modules and Japanese rings". Lemma 3.5.3/2 and Proposition 4.4/3 indicate a close connection between the notions of Japaneseness and weak stability. Here we prove

Proposition 1. *A discrete valuation ring A is Japanese if and only if its field of fractions K is weakly stable.*

Proof. Set $p := \operatorname{char} A$. If $p = 0$, our statement is true due to Propositions 3.5.1/4 and 4.3/2, since $A = \mathring{K}$ is a principal ideal domain by Proposition 1.6.1/4 and, in particular, both Noetherian and normal.

Assume $p \neq 0$. Then it follows from Theorem 3.5.3/1 and Proposition 2.3.4/2 that K is weakly stable if and only if $\widehat{(K^{p^{-1}})}^{\circ}$ is a tame $\mathring{K}$-module. Now $A^{p^{-1}} = \widehat{(K^{p^{-1}})}^{\circ}$, and Proposition 4.4/2 says that A (being Noetherian and normal) is Japanese if and only if $A^{p^{-1}}$ is a tame A-module. Hence the assertion follows. □

In (1.6.2) we gave an example of a discrete valuation ring which is not Japanese. Thus we conclude

Proposition 2. *There exist fields K with a discrete valuation which are not weakly stable.*

3.6. Stable fields

We want to combine the notions of cartesian vector space and spectral norm. As in (3.5), let K be a valued field and let L be an algebraic extension of K. As always, we denote by K_a the algebraic closure of K and by $\hat{K}$ the completion of K.

3.6.1. Definition. — We strengthen Definition 3.5.2/1 in the following way.

Definition 1. *K is said to be stable if each finite extension L of K provided with the spectral norm is a K-cartesian vector space.*

By Observation 3.5.1/2, a field K is stable if and only if K_a is a K-cartesian vector space with respect to the spectral norm; in particular, algebraically closed fields are stable. Naturally, stable fields are weakly stable. However, there exist weakly stable fields, even fields of characteristic 0 with a complete valuation, which are not stable. We give the following

Example. Consider the field $\mathbb{Q}_2$ of 2-adic numbers, and let $\alpha_0 = 2, \alpha_1, \alpha_2, \ldots$ be a sequence of elements algebraic over $\mathbb{Q}_2$ such that α_{i+1} is always a square root of α_i. Writing $k_i := \mathbb{Q}_2(\alpha_i) = \mathbb{Q}_2(\alpha_0, \ldots, \alpha_i)$, we get a sequence of extensions of $\mathbb{Q}_2$ as follows:

$$k_0 = \mathbb{Q}_2 \subset k_1 \subset k_2 \subset \cdots .$$

Let K be the completion of the field $\bigcup_{i=0}^{\infty} k_i$. We claim that K is not stable. In order to verify this, we have to realize that, by construction, $[k_{i+1} : k_i] = 2$ and $e(k_{i+1}/k_i) \geqq 2$. This implies $e(k_{i+1}/k_i) = 2$ and $f(k_{i+1}/k_i) = 1$ by Proposition 3.1.3/2. Therefore if the valuation on $\mathbb{Q}_2$ is normalized in the usual way, we get

$$|k_i^*| = \left(\sqrt[2^i]{2}\right)^{\mathbb{Z}}, \qquad |K^*| = \bigcup_{i=0}^{\infty} |k_i^*|$$

and

$$\tilde{K} = \tilde{k}_i = \tilde{\mathbb{Q}}_2 = \mathbb{F}_2 .$$

In particular, the value group $|K^*|$ is closed under the process of taking square roots, and any quadratic extension L of K must satisfy $e(L/K) = 1$.

Let n be one of the numbers $-3, -1, 3$. We want to show that K does not contain a square root of n. Assuming the contrary, let $\alpha \in K$ satisfy $\alpha^2 = n$. Then there are elements in $\bigcup_{i=0}^{\infty} k_i$, which are arbitrarily close to α. Hence KRASNER's Lemma (Corollary 3.4.2/2) implies $\alpha \in k_i$ for some $i \geqq 0$. From this we can derive $\alpha \in k_{i-1}$ if $i > 0$. Namely, write $\alpha = a + b\alpha_i$ with coefficients $a, b \in k_{i-1}$. We have $|\alpha| = \max\{|a|, |b\alpha_i|\}$, since $|\alpha_i| \notin |k_{i-1}|$ implies $|a| \neq |b\alpha_i|$. Furthermore, $|b\alpha_i| < |a| = 1$ since $|\alpha| = |n|^{1/2} = 1$ and $|b\alpha_i| \neq 1$. Now the equation

$$n = (a + b\alpha_i)^2 = a^2 + 2ab\alpha_i + b^2\alpha_{i-1}$$

shows that we must have $b = 0$ (because otherwise $\alpha_i \in k_{i-1}$). Therefore $\alpha \in k_{i-1}$. Repeating this process gives $\alpha \in k_0 = \mathbb{Q}_2$. We want to derive a contradiction by showing that the polynomial $X^2 - n$ is irreducible over $\mathbb{Q}_2$. Replacing X by $X + 1$, we have to consider the polynomial $p = X^2 + 2X - (n - 1)$. If $n = -1, 3$, the spectral value $\sigma(p)$ equals $|2|^{1/2}$ so that $\sigma(p) \notin |\mathbb{Q}_2|$. Consequently, p is irreducible in this case (use a direct argument or apply Proposition 3.1.2/1). If $n = -3$, we can consider the polynomial $\frac{1}{4}p(2X) = X^2 + X + 1$, which is irreducible over the residue field $\mathbb{F}_2$ of $\mathbb{Q}_2$. Therefore $X^2 - n$ is irreducible also if $n = -3$, and we see that K does not contain a square root of n.

Next we claim that the quadratic extension $L := K(\sqrt{3})$ of K satisfies $f(L/K) = 1$. Again, we assume the contrary. Then $f(L/K) = 2$ and the only possibility for the residue field of L is

$$\tilde{L} = \mathbb{F}_4 = \mathbb{F}_2[X]/(X^2 + X + 1).$$

Since $X^2 + X + 1$ splits into different linear factors over $\tilde{L}$, the same must be true over L by HENSEL's Lemma (Proposition 3.3.4/3). Therefore the roots

$$-\frac{1}{2} + \frac{1}{2}\sqrt{-3}, \quad -\frac{1}{2} - \sqrt{-3},$$

and hence $\sqrt{-3}$ must belong to L. We can write $\sqrt{-3} = x + y\sqrt{3}$ with coefficients $x, y \in K$ so that $-3 = x^2 + 2xy\sqrt{3} + 3y^2$. However this is impossible, since $xy \neq 0$ implies $\sqrt{3} \in K$, since $x = 0$ implies $\sqrt{-1} \in K$, and since $y = 0$ implies $\sqrt{-3} \in K$.

Thus it is clear that $L = K(\sqrt{3})$ is a quadratic extension of K which satisfies $e(L/K) = 1 = f(L/K)$. Then L cannot be K-cartesian, since otherwise L would be strictly cartesian (due to $e(L/K) = 1$, see Observation 2.5.1/2), and we would have $f(L/K) = [L:K] = 2$ (Corollary 2.5.1/6). Consequently, K is a complete field of characteristic 0, which is not stable.

3.6.2. Criteria for stability. — We start with some simple criteria which are obtained by specializing some of our results on cartesian vector spaces.

Proposition 1. *Let the valuation on K be discrete. Then K is stable if and only if K is weakly stable. In particular, any field with a complete discrete valuation is stable.*

Proof. Apply Corollary 2.4.3/11. □

Proposition 2. *A complete field K is stable if and only if every finite separable extension of K is K-cartesian.*

Proof. Due to Proposition 3.4.1/6, the separable algebraic closure K_{sep} of K is dense in the algebraic closure K_a of K if one provides K_a with the spectral valuation. Applying Proposition 2.4.3/8, we see that K is stable if and only if K_{sep} is K-cartesian, whence the assertion follows. □

Next we want to reduce the question of stability to the case of complete fields, where it is easier to handle since here all spectral norms are valuations.

Proposition 3. *The field K is stable if and only if the following two conditions are fulfilled:*

(i) *K is weakly stable,*
(ii) *$\hat{K}$ is stable.*

Proof. Let K be stable. Then K is weakly stable, and we have to show that $\hat{K}$ is stable. In order to do this, consider a finite extension F of $\hat{K}$ (provided with the spectral valuation, see Theorem 3.2.4/2). We may assume by Proposition 2 (and by Proposition 2.3.3/4) that F is separable over $\hat{K}$, say $F = \hat{K}(\alpha)$ with a separable element $\alpha \in F$. Using Proposition 3.4.2/5 we may even assume that α is algebraic over K. Then $L := K(\alpha)$ is a dense subfield of F which is finite algebraic over K. Because K is stable we know that L is K-cartesian if provided with the spectral norm of L over K (this norm may be different from the valuation induced from F on L). Let $\hat{L}$ denote the completion of L with respect to the spectral norm. Then $\hat{L}$ is the norm-direct sum

$$\hat{L} = \hat{L}_1 \oplus \cdots \oplus \hat{L}_t$$

of finite extensions $\hat{L}_i$ of $\hat{K}$ (see Proposition 3.3.1/1). We may view L as a subfield of each $\hat{L}_i$. If $|\ \ |_i$ denotes the spectral valuation on $\hat{L}_i$ over $\hat{K}$, we know by Proposition 3.3.3/1 that any valuation on L extending the given valuation on K coincides with one of the valuations $|\ \ |_1, \ldots, |\ \ |_t$ on L. In particular, there exists an index i, say $i = 1$, such that the valuation induced from F on L equals $|\ \ |_1$ on L. Since L is dense in $\hat{L}_1$ as well as in F, we see that F is isometrically isomorphic to $\hat{L}_1$. Therefore we can view F as a $\hat{K}$-subspace of $\hat{L}$. Now K is stable; hence L is K-cartesian and, by Proposition 2.4.3/8, its completion $\hat{L}$ is $\hat{K}$-cartesian. So F, as a subspace of $\hat{L}$, must be $\hat{K}$-cartesian. This verifies the only if part of the assertion.

To verify the if part, assume that K is weakly stable and that $\hat{K}$ is stable. Let L be a finite extension of K (provided with the spectral norm). As above, $\hat{L}$ is a norm-direct sum

$$\hat{L} = \hat{L}_1 \oplus \cdots \oplus \hat{L}_t$$

of finite extensions $\hat{L}_i$ of $\hat{K}$. Each $\hat{L}_i$ is $\hat{K}$-cartesian; hence $\hat{L}$ is $\hat{K}$-cartesian. Thus we see by Proposition 2.4.3/9 that L is K-cartesian. □

In order to determine whether or not a field K is stable, it is sometimes useful to look at the ramification index and the residue degree of finite extensions of K. We want to use the remainder of this section to show how these concepts are related to the stability of K.

Proposition 4. *Let L be a finite extension of K (provided with a valuation extending the valuation on K). Write $n := [L:K]$, $e := e(L/K)$, $f := f(L/K)$, and choose $x_1, \ldots, x_f \in \mathring{L}$ and $y_1, \ldots, y_e \in L^*$ such that $\tilde{x}_1, \ldots, \tilde{x}_f$ is a $\tilde{K}$-basis of $\tilde{L}$ and such that $|y_1|, \ldots, |y_e|$ represent all equivalence classes of $|L^*|$ modulo $|K^*|$. Then the following conditions are equivalent:*

(i) *The family $F := \{x_iy_j;\ 1 \leqq i \leqq f,\ 1 \leqq j \leqq e\}$ is an orthogonal K-basis of L.*

(ii) $e \cdot f = n$.

(iii) *L is K-cartesian.*

If the above conditions are fulfilled, the valuation on L is the only extension to L of the given valuation on K and therefore coincides with the spectral norm.

Proof. By Lemma 3.1.3/1, the set F is orthogonal. This shows that (i) is equivalent to (ii). Clearly (i) implies (iii). It remains to be shown that (iii) implies (ii). Assume that F is K-cartesian and that $ef \neq n$, i.e., that $ef < n$. Then by Proposition 2.4.1/5, we can enlarge F to an orthogonal basis of L. In particular, there exists an element $z \in L^*$ such that $F \cup \{z\}$ is K-orthogonal. Arrange $y_1, \ldots, y_e$ such that $|z|$ and $|y_1|$ are in the same $|K^*|$-equivalence class of $|L^*|$. Choose $a \in K^*$ such that $|z| = |a|\ |y_1|$. The vector

$$w := a^{-1}y_1^{-1}z \in L$$

has absolute value 1. We have an equation

$$\tilde{w} = \sum_1^f \tilde{a}_i\tilde{x}_i, \quad \tilde{a}_i \in \tilde{K}.$$

Let $a_i \in \mathring{K}$ denote an inverse image of $\tilde{a}_i$, and consider the element

$$v := w - \sum_1^f a_ix_i \in L.$$

We have $|v| < 1$ since $\tilde{v} = 0$. Multiplying v by ay_1, we get

$$ay_1v = z - \sum_1^f aa_i(x_iy_1) \quad \text{and} \quad |ay_1v| < |ay_1|.$$

But the orthogonality of the family $\{z, x_iy_1;\ i = 1, \ldots, f\}$ yields

$$|ay_1v| = \max\{|z|, |aa_i|\ |x_iy_1|\} \geqq |z| = |ay_1|.$$

This is a contradiction; hence $e \cdot f = n$. In order to verify the remaining assertion, consider condition (iii), which implies that the given valuation induces the product topology on L. By Lemma 3.5.1/1, the valuation must be the spectral norm. □

If there is more than one extension to L of the valuation on K, the single term ef in the formula "$ef = n$" has to be replaced by the sum of the ramification indices times residue degrees of the different valuation extensions. More precisely, as in (3.3), let us denote by $|\ \ |_i$, $i = 1, \ldots, t$, the t different valuations on L extending the valuation on K and by L_i the normed space one gets if one provides L with the norm $|\ \ |_i$. Define $e_i := e(L_i/K)$ and $f_i := f(L_i/K)$. As always let $n = [L:K]$. Then we can formulate the following criterion:

Proposition 5. *We have* $\sum_{i=1}^{t} e_i f_i \leqq n$, *and L is K-cartesian if and only if this is an equality.*

Proof. As in (3.3), denote by $\hat{L}$ (resp. $\hat{L}_i$) the completion of L with respect to the spectral norm (resp. the valuation $|\ \ |_i$ for $i = 1, \ldots, t$). From (3.3.2) and Proposition 3.3.1/1, we know

(1) $\hat{L}$ is the norm-direct sum $\hat{L}_1 \oplus \cdots \oplus \hat{L}_t$.

Using Proposition 2.3.3/6, we get

(2) $n = [L:K] \geqq [\hat{L}:\hat{K}] = \sum_{i=1}^{t} [\hat{L}_i:\hat{K}]$.

(In spite of the fact that $\hat{L}$ is not in general a field, nevertheless we use the symbol $[\hat{L}:\hat{K}]$ as an abbreviation for $\dim_{\hat{K}} \hat{L}$.) In order to be able to apply Proposition 4 to $\hat{K}$ and $\hat{L}_i$, we compute

(3) $e(\hat{L}_i/\hat{K}) = \operatorname{card}(|\hat{L}_i^*|/|\hat{K}^*|) = e(L_i/K) = e_i$, for $i = 1, \ldots t$, and

(4) $f(\hat{L}_i/\hat{K}) = [\tilde{\hat{L}}_i:\tilde{\hat{K}}] = [\tilde{L}_i:\tilde{K}] = f(L_i/K) = f_i$, for $i = 1, \ldots, t$.

Then from Proposition 3.1.3/2 we get

(5) $[\hat{L}_i:\hat{K}] \geqq e(\hat{L}_i/\hat{K})\, f(\hat{L}_i/\hat{K}) = e_i f_i$ for $i = 1, \ldots, t$.

If one combines (2) and (5), one sees

(6) $n = [L:K] \geqq [\hat{L}:\hat{K}] = \sum_{i=1}^{t} [\hat{L}_i:\hat{K}] \geqq \sum_{i=1}^{t} e(\hat{L}_i/\hat{K})\, f(\hat{L}_i/\hat{K}) = \sum_{i=1}^{t} e_i f_i$,

where equality holds if and only if

(A) $[L:K] = [\hat{L}:\hat{K}]$ and

(B) $[\hat{L}_i:\hat{K}] = e(\hat{L}_i/\hat{K})\, f(\hat{L}_i/\hat{K})$ for $i = 1, \ldots, t$.

Using the equivalence of conditions (ii) and (iii) in Proposition 4, we see that (B) is equivalent to the condition "$\hat{L}_i$ is $\hat{K}$-cartesian for $i = 1, \ldots, t$", which in turn is equivalent by (1) to

(B′) $\hat{L}$ is $\hat{K}$-cartesian.

Due to Proposition 2.3.3/6, condition (A) can be reformulated as

(A′) L is weakly K-cartesian.

Hence it remains to be verified that L is K-cartesian if and only if L is weakly K-cartesian and $\hat{L}$ is $\hat{K}$-cartesian. But this follows from Corollary 2.4.3/10. □

Now the desired criterion for stability is an easy corollary.

Proposition 6. *K is stable if and only if for every finite extension L of K the formula* $\sum_{i=1}^{t} e_i f_i = n$ *holds.*

The proposition shows that the classical formula $\sum_{i=1}^{t} e_i f_i = n$ (valid, e.g., for local fields) characterizes a wider class of fields. This class of stable fields will be of special interest in (5.3) where we show that the field of fractions of the power series ring $K\langle X_1, \ldots, X_n\rangle$ is stable if K is stable. Furthermere, stability is inherited under the following circumstances:

Corollary 7. *If K is complete and stable, then every finite extension L is stable (and complete).*

Proof. Let E be a finite extension of L. Then there is exactly one extension of the valuation on K to L and to E. Proposition 6 gives us $[L:K] = e(L/K)\,f(L/K)$ and $[E:K] = e(E/K) \cdot f(E/K)$. Then we know $\big(e(E/L) \cdot e(L/K)\big)\big(f(E/L) \cdot f(L/K)\big) = e(E/K) \cdot f(E/K) = [E:K] = [E:L] \cdot [L:K] = [E:L]\big(e(L/K) \cdot f(L/K)\big)$. By cancelling, we see $e(E/L) \cdot f(E/L) = [E:L]$ which, by Proposition 6, yields the assertion. (For the completeness of L see Proposition 2.3.3/4.) □

In the special case where the value set of the spectral norm on L is contained in $|K|$, we can easily see that there is no ramification, i.e., that $e_i = 1$ for $i = 1, \ldots, t$. The formula $\sum_{i=1}^{t} e_i f_i = n$ then reduces to $\sum_{i=1}^{t} f_i = n$. Since, on the other hand, $[\tilde{L}:\tilde{K}] = [\tilde{\hat{L}}:\tilde{\hat{K}}] = \sum_{i=1}^{t} [\tilde{\hat{L}}_i:\tilde{\hat{K}}] = \sum_{i=1}^{t} f(\hat{L}_i/\hat{K}) = \sum_{i=1}^{t} f_i$, we get as a consequence of Proposition 5

Proposition 8. *Let L be a finite extension of K. Then the following two conditions are equivalent:*

(i) *L is K-cartesian and $|L| = |K|$; i.e., L is strictly K-cartesian.*
(ii) $[\tilde{L}:\tilde{K}] = [L:K]$.

Proof. Instead of deducing this corollary from Proposition 5 as indicated above, one can use Corollary 2.5.1/6. □

Now we look for a criterion which will guarantee that the equivalent conditions of the preceding proposition are fulfilled for *all* finite extensions of K. We need an additional condition on the value group of K.

Definition 9. *A multiplicative subgroup G of $\mathbb{R}_+ - \{0\}$ is said to be divisible if, for all $\varrho \in G$ and all $n \in \mathbb{N}$, the number $\varrho^{1/n} \in \mathbb{R}_+ - \{0\}$ belongs to G.*

Observation 10. *The value group $|K^*|$ is divisible if and only if it equals the value set $|K_a^*|$ of the algebraic closure K_a (provided with the spectral norm).*

Proof. First we assume that $|K^*| = |K_a^*|$. Let $\varrho \in |K^*|$ and $n \in \mathbb{N}$ be given. Then there are an element $a \in K^*$ with $|a| = \varrho$ and an $\alpha \in K_a$ with $\alpha^n = a$. Then $\varrho^{1/n} = |\alpha| \in |K_a^*| = |K^*|$, and therefore $|K^*|$ is divisible. In order to show the converse, we assume that $|K^*|$ is divisible. Because the spectral norm on K_a is computed by taking roots of elements in $|K|$, we know $|K_a| \subset |K|$ and hence $|K_a^*| = |K^*|$. □

This enables us to state the desired criterion.

Proposition 11. *The following three conditions on K are equivalent:*

(i) *K_a is strictly K-cartesian.*

(ii) *For all finite extensions L of K (provided with the spectral norm), one has $[L:K] = [\tilde{L}:\tilde{K}]$.*

(iii) *K is stable and $|K^*|$ is divisible.*

Proof. Apply Proposition 8 and Observation 10. □

Using Proposition 11, we can show how, within the context of stable fields, the algebraic closedness of K is related to the algebraic closedness of the residue field $\tilde{K}$.

Proposition 12. *Let K be complete. Then K is algebraically closed if and only if*

(i) *K is stable,*

(ii) *$|K^*|$ is divisible,*

(iii) *$\tilde{K}$ is algebraically closed.*

Proof. First we show that an algebraically closed field K fulfills conditions (i) to (iii). Condition (i) is clear; (ii) follows from Observation 10, and (iii) from Lemma 3.4.1/4. Next we suppose that conditions (i) to (iii) are satisfied. Let L be a finite extension of K. We have to show that $L = K$. By Proposition 11, we get $[L:K] = [\tilde{L}:\tilde{K}]$ from conditions (i) and (ii). Because K is complete, the spectral norm on L is a valuation, and hence $\tilde{L}$ is a field. By condition (iii), this implies that $[\tilde{L}:\tilde{K}] = 1$, whence $[L:K] = 1$. □

Remark. Instead of assuming the completeness of K, it suffices to suppose that for all finite extensions L of K there is exactly one valuation extension (namely, the spectral norm).

We conclude this section by listing some sufficient conditions for K to be stable.

Proposition 13. *K is stable if* char $\tilde{K} = 0$.

Proof. Use the remark following the proof of Theorem 19 in section VI.11 of Zariski Samuel [39] and the corollary to Theorem 24 in section VI.12 of the same volume. One gets $\sum_{i=1}^{t} e_i f_i = n$ for all finite extensions of K. Then Proposition 6 yields the assertion. □

If we combine the two preceding propositions, we get the following result:

Corollary 14. *Let K be complete and assume* char $\tilde{K} = 0$. *Then K is algebraically closed if and only if*

(i) *$|K^*|$ is divisible,*

(ii) *$\tilde{K}$ is algebraically closed.*

An immediate consequence of Proposition 2.4.4/2 is

Proposition 15. *K is stable (and complete) if K is spherically complete.*

As already mentioned in (2.4.4), a field K is spherically complete if and only if it is maximally complete. Thus maximal completeness is another sufficient condition for stability.

3.7. Banach algebras

3.7.1. Definition and examples. — For every field k with a complete non-Archimedean non-trivial valuation, we define as in the complex case

Definition 1. *A k-algebra with a complete k-algebra norm is called a k-Banach algebra.*

Any homomorphism of k-Banach algebras $\varphi: B \to A$ is, in particular, k-linear. Therefore by Proposition 2.1.8/2, such a map φ is continuous if and only if it is bounded. If $\mathfrak{a}$ is a closed ideal in a k-Banach algebra A, it is easy to see that the residue algebra $A/\mathfrak{a}$ provided with the residue norm is again a k-Banach algebra (cf. Proposition 1.1.6/1, Remark 2 of (1.2.1) and Proposition 2.1.2/3).

Obviously k itself is a k-Banach algebra. Applying Propositions 1.4.1/2 and 1.4.1/3, we see that, for every k-Banach algebra A, the algebra $A\langle X\rangle$ of strictly convergent power series with coefficients in A is a k-Banach algebra. Therefore the algebra $k\langle X_1, \ldots, X_n\rangle$ (the main object in the beginning of the following part B) is a Banach algebra. Further simple examples are provided by finite extensions of k provided with the spectral valuation. We already know that these extensions carry always the product topology and that all subspaces are closed. In the next sections, we shall get similar results for infinite-dimensional algebras under certain conditions.

3.7.2. Finiteness and completeness of modules over a Banach algebra. — The underlying vector space of a k-Banach algebra is a k-Banach space and the underlying ring is a complete normed ring. Therefore we may apply the results of (1.2.4) and of (2.8). This leads to the following proposition.

Proposition 1. *Let A be a k-Banach algebra and let M be a normed A-module such that the completion $\hat{M}$ of M is a finite A-module. Then M is complete.*

Proof. There are elements $x_1, \ldots, x_n \in \hat{M}$ such that the homomorphism $\pi: A^n \to \hat{M}$ defined by $\pi(a_1, \ldots, a_n) := \sum_{i=1}^{n} a_i x_i$ is surjective. By BANACH's Theorem, π is open, and therefore $\sum_{i=1}^{n} \check{A} x_i = \pi(\check{A}^n)$ is a neighborhood of 0 in $\hat{M}$. Since M is dense in $\hat{M}$, we have

$$x_\nu \in M + \sum_{\mu=1}^{n} \check{A} x_\mu \quad \text{for } \nu = 1, \ldots, n.$$

Now NAKAYAMA's Lemma 1.2.4/6 yields $M = \hat{M}$. □

As an immediate consequence of this proposition, we have that all submodules of a Noetherian complete normed module over a Banach algebra A are closed. This property characterizes the Noetherian complete normed modules over A; more precisely

Proposition 2. *Let A be a k-Banach algebra and M a complete normed A-module. Then M is Noetherian if and only if all submodules of M are closed. In particular, the ring A is Noetherian if and only if all ideals in A are closed.*

Proof. We only have to show that M is Noetherian if all submodules are closed. Let $M_1 \subset M_2 \subset \ldots$ be an ascending chain of submodules. Let $M' := \bigcup_{i=1}^{\infty} M_i$. Then M' being a closed submodule of the complete module M is a Baire space. Since all M_i are closed, we have by BAIRE's Theorem (cf. BOURBAKI [6], Ch 9, § 5, Théorème 1) the existence of an index i such that M_i contains a neighborhood of 0 in M'. This implies $M_i = M'$; hence the chain becomes stationary. □

3.7.3 The category $\mathfrak{M}_A$. — In this section, A always denotes a *Noetherian* k-Banach algebra. We denote by $\mathfrak{M}_A$ the category of all finite complete normed A-modules with continuous A-linear maps as morphisms. Note that, by Corollary 2.1.8/3, such an A-linear map is continuous if and only if it is bounded. Since each $M \in \mathfrak{M}_A$ is a Noetherian A-module, we conclude from Proposition 3.7.2/2

Proposition 1. *Every submodule M' of a module $M \in \mathfrak{M}_A$ is closed. Therefore $M' \in \mathfrak{M}_A$ and $M/M' \in \mathfrak{M}_A$ if M' carries the induced norm and M/M' the residue norm. Furthermore $M_1 \oplus M_2 \in \mathfrak{M}_A$ if $M_1, M_2 \in \mathfrak{M}_A$.*

Next we claim that the category $\mathfrak{M}_A$ is essentially the same as the category of all finite A-modules with A-linear maps as morphisms. This is made more precise by the following two statements.

Proposition 2. *If M, M' are objects of $\mathfrak{M}_A$, each A-linear map $\varphi: M \to M'$ is continuous.*

Proof. Choose an epimorphism $\pi: A^n \to M$ for a suitable $n \in \mathbb{N}$. Define $\varphi': A^n \to M'$ by $\varphi' := \varphi \circ \pi$. Since addition and scalar multiplication are continuous operations in normed modules, both maps π and φ' are continuous. Furthermore π is open (by BANACH's Theorem). Hence φ is continuous. □

If we consider the special case where φ is bijective, we get from Proposition 2 the uniqueness part of the following proposition.

Proposition 3. *Each finite A-module M can be provided with a complete A-module norm. All such norms are equivalent.*

Proof. We only have to prove the existence of such a norm. Take any A-linear epimorphism $\pi: A^n \to M$. Since $A^n \in \mathfrak{M}_A$, the kernel $\ker \pi$ is closed. The residue norm on $A^n/\ker \pi$ gives rise to a complete A-module norm on M. □

Recall that a continuous map $\xi\colon X \to Y$ between topological spaces X, Y is called *strict* if the topology of the subspace $\xi(X) \subset Y$ (i.e., the topology on $\xi(X)$ induced from Y) equals the quotient topology with respect to the map $X \xrightarrow{\xi} \xi(X)$. In particular, ξ is strict if ξ is open. Therefore one easily verifies (using BANACH's Theorem and Proposition 2.1.2/3)

Proposition 4. *A continuous k-linear map $\varphi\colon X \to Y$ between k-Banach spaces is strict if and only if $\varphi(X)$ is closed in Y.*

From this we immediately conclude

Corollary 5. *Each A-module homomorphism $\varphi\colon M \to M'$, where $M, M' \in \mathfrak{M}_A$, is strict.*

We want to give an application of the preceding results to complete tensor products (which were defined in (2.1.7)). In the following let B denote a normed A-algebra via a homomorphism $A \to B$, and assume that B itself is a Noetherian Banach algebra over k.

Proposition 6. *For finite A-modules $M, N \in \mathfrak{M}_A$, the canonical maps*

$$i\colon M \otimes_A N \to M \mathbin{\widehat{\otimes}}_A N, \quad x \otimes y \mapsto x \mathbin{\widehat{\otimes}} y$$

$$j\colon M \otimes_A B \to M \mathbin{\widehat{\otimes}}_A B, \quad x \otimes b \mapsto x \mathbin{\widehat{\otimes}} b$$

are bijective. In particular, $M \mathbin{\widehat{\otimes}}_A N \in \mathfrak{M}_A$ and $M \mathbin{\widehat{\otimes}}_A B \in \mathfrak{M}_B$.

Proof. If $M = A^n$ is a free A-module, the assertion follows directly from Proposition 2.1.7/6 by using the isomorphisms (i) and (iv), as well as their counterparts for ordinary tensor products. In the general case, we choose a free resolution of M, i.e., an exact sequence

$$A^m \xrightarrow{\varphi} A^n \xrightarrow{\psi} M \longrightarrow 0.$$

Tensoring with the identity map $\mathrm{id}\colon N \to N$, we get a commutative diagram of continuous maps

$$\begin{array}{ccccccc}
A^m \otimes_A N & \xrightarrow{\varphi\otimes\mathrm{id}} & A^n \otimes_A N & \xrightarrow{\psi\otimes\mathrm{id}} & M \otimes_A N & \longrightarrow & 0 \\
\downarrow{\scriptstyle i''} & & \downarrow{\scriptstyle i'} & & \downarrow{\scriptstyle i} & & \\
A^m \mathbin{\widehat{\otimes}}_A N & \xrightarrow{\varphi\widehat{\otimes}\mathrm{id}} & A^n \mathbin{\widehat{\otimes}}_A N & \xrightarrow{\psi\widehat{\otimes}\mathrm{id}} & M \mathbin{\widehat{\otimes}}_A N & \longrightarrow & 0
\end{array}$$

with i' and i'', as well as i, denoting the canonical maps from the finite A-modules in the first row (provided with a semi-norm as in (2.1.7)) into their completions. The first row is exact due to the right exactness of the ordinary tensor product, and it follows from what we have proved already that i' and i'' are bijective. Therefore the map $\varphi \otimes \mathrm{id}$ is strict, since $\varphi \mathbin{\widehat{\otimes}} \mathrm{id}$ is strict by Corollary 5. Furthermore, the map ψ, which is surjective, is strict by Corollary 5; hence, Proposition 2.1.8/6 applies and $\psi \otimes \mathrm{id}$ is strict. Thus, the first row in the above diagram is an exact sequence of strict homomorphisms.

But then according to Corollary 1.1.9/6, also the second row must be exact, and the bijectivity of i' and i'' implies the bijectivity of i.

The bijectivity of $j: M \otimes_A B \to M \hat{\otimes}_A B$ is proved in literally the same way with N being replaced by B. The only crucial thing is to see that the corresponding map $\varphi \hat{\otimes} \mathrm{id}: A^m \hat{\otimes}_A B \to A^n \hat{\otimes}_A B$ is strict. But as above, this is settled by Corollary 5 since $\varphi \hat{\otimes} \mathrm{id}$ is a B-linear map between finite B-modules. □

3.7.4. Finite homomorphisms. — In this section we study A-algebras B which — when viewed as A-modules — are finite. More precisely, we consider *finite* k-algebra homomorphisms $\varphi: A \to B$. If A is a Noetherian k-Banach algebra as in the last section, we can provide B with a complete A-module norm. Actually, more is true:

Proposition 1. *Let A be a Noetherian k-Banach algebra, and let $\varphi: A \to B$ be a finite k-algebra homomorphism from A into a k-algebra B. Then B is Noetherian and can be provided with a complete A-algebra norm (which by Definition 3.7.1/1 makes B a k-Banach algebra) such that φ is continuous and strict. All complete k-algebra norms on B such that φ is continuous are equivalent.*

Proof. Since φ is finite, B is a finite A-module. By Proposition 3.7.3/3, we can provide B with a complete A-module norm $|\ \ |'$. Then the map φ is contractive. In general $|\ \ |'$ will fail to be a ring norm on B. However, the following is true:

There exists a constant $\varrho > 0$ such that $|xy|' \leqq \varrho\, |x|'\, |y|'$ for all $x, y \in B$.

In order to see this, let $\{b_1, \ldots, b_n\} \subset B$ be a system of A-generators of B. Set $C := \max\limits_{1 \leqq \mu,\nu \leqq n} |b_\mu b_\nu|'$, and choose $\eta \geqq 1$ such that, for each $x \in B$, there is an equation $x = \sum\limits_1^n \varphi(a_\nu)\, b_\nu$ with $\max\limits_{1 \leqq \nu \leqq n} |a_\nu| \leqq \eta\, |x|'$ (such η exists since th norm $|\ \ |'$ is equivalent to the residue norm defined by the map $A^n \to B$, $(a_1, \ldots, a_n) \mapsto \sum\limits_1^n \varphi(a_\nu)\, b_\nu$, see Proposition 3.7.3/3). Setting $\varrho := \eta^2 C$, we have for all vectors

$$x_i = \sum_1^n \varphi(a_{i\nu})\, b_\nu \in B$$

$$\begin{aligned} |x_1 x_2|' &= \Big|\sum_{\mu,\nu=1}^n \varphi(a_{1\mu})\, \varphi(a_{2\nu})\, b_\mu b_\nu\Big|' \leqq \max_{1 \leqq \mu,\nu \leqq n} |\varphi(a_{1\mu})\, \varphi(a_{2\nu})\, b_\mu b_\nu|' \\ &\leqq \big(\max_{1 \leqq \mu,\nu \leqq n} |a_{1\mu} a_{2\nu}|\big)\ \big(\max_{1 \leqq \mu,\nu \leqq n} |b_\mu b_\nu|'\big) \leqq \big(\max_{1 \leqq \mu \leqq n} |a_{1\mu}|\big)\ \big(\max_{1 \leqq \nu \leqq n} |a_{2\nu}|\big) \cdot C \\ &\leqq \eta\, |x_1|' \cdot \eta\, |x_2|' \cdot C = \varrho\, |x_1|'\, |x_2|'. \end{aligned}$$

Now we define for $x \in B$

$$|x| := \sup_{y \in B - \{0\}} \frac{|xy|'}{|y|'}.$$

Applying Proposition 1.2.1/2, we see that $|\ \ |$ is a ring norm on B inducing the same topology as $|\ \ |'$. It is easy to verify that $|\ \ |$ is also an A-module norm. In particular, φ is continuous and hence strict by the results of (3.7.3).

It remains to be shown that each complete k-algebra norm $|\ \ |^*$ on B such that φ is continuous is equivalent to $|\ \ |$. The k-linear map $\pi\colon A^n \to B$ given by $(a_1, \ldots, a_n) \mapsto \sum\limits_{1}^{n} \varphi(a_\nu)\, b_\nu$ is continuous with respect to any such norm $|\ \ |^*$. Therefore, due to BANACH's Theorem, we see that $|\ \ |^*$ must induce the quotient topology with respect to π on B. □

Corollary 2. *Each finite continuous k-algebra homomorphism between Noetherian k-Banach algebras is strict.*

3.7.5. Continuity of homomorphisms. — In this section we consider a class of k-Banach algebras with the property that all k-algebra homomorphisms are automatically continuous.

Proposition 1. *Let A, B be k-Banach algebras, and let $\Phi\colon A \to B$ be a k-algebra homomorphism. Assume that there is a family $\mathfrak{B}$ of ideals in B such that*

(i) *each $\mathfrak{b} \in \mathfrak{B}$ is closed in B and each inverse image $\Phi^{-1}(\mathfrak{b})$ is closed in A,*
(ii) *for each $\mathfrak{b} \in \mathfrak{B}$ one has* $\dim_k B/\mathfrak{b} < \infty$,
(iii) $\bigcap\limits_{\mathfrak{b}\in\mathfrak{B}} \mathfrak{b} = (0)$.

Then Φ is continuous.

Proof. Fix $\mathfrak{b} \in \mathfrak{B}$ and denote by β the residue epimorphism $B \to B/\mathfrak{b}$. Define $\psi\colon A \to B/\mathfrak{b}$ by $\psi := \beta \circ \Phi$. Let $\overline{\psi}\colon A/\ker\psi \to B/\mathfrak{b}$ be the injection induced by ψ. Then we have the commutative diagram

$$\begin{array}{ccc} A & \xrightarrow{\ \Phi\ } & B \\ \downarrow & \searrow^{\psi} & \downarrow{\scriptstyle\beta} \\ A/\ker\psi & \xrightarrow{\ \overline{\psi}\ } & B/\mathfrak{b}\,. \end{array}$$

Obviously $\ker\psi = \Phi^{-1}(\mathfrak{b})$. According to (i) and (ii), the residue spaces $A/\ker\psi$ and $B/\mathfrak{b}$ provided with the residue norms are finite-dimensional weakly cartesian k-vector spaces (see Proposition 2.3.3/4). Therefore $\overline{\psi}$ and hence ψ are continuous. Now we get the continuity of Φ from the Closed Graph Theorem; namely, assume there is given a sequence $a_n \in A$ with $\lim a_n = 0$ and $\lim \Phi(a_n) = b$. Using the continuity of ψ and β, we get $\beta(b) = \beta(\lim \Phi(a_n)) = \lim (\beta \circ \Phi)(a_n) = \lim \psi(a_n) = \psi(\lim a_n) = \psi(0) = 0$, i.e., $b \in \mathfrak{b}$. Since this holds for all $\mathfrak{b} \in \mathfrak{B}$, we deduce $b = 0$ from (iii). This implies the continuity of Φ. □

From Propositions 1 and 3.7.2/2, we easily derive

Proposition 2. *Let B be a Noetherian k-Banach algebra with a family $\mathfrak{B}$ of ideals of B such that*

(i) $\dim_k B/\mathfrak{b} < \infty$ *for all* $\mathfrak{b} \in \mathfrak{B}$,
(ii) $\bigcap\limits_{\mathfrak{b}\in\mathfrak{B}} \mathfrak{b} = (0)$.

Then each k-algebra homomorphism of a Noetherian k-Banach algebra A into B is continuous.

Proposition 3. *All complete k-algebra norms (if there exist any) on a Noetherian k-algebra B with a family $\mathfrak{B}$ of ideals satisfying conditions* (i) *and* (ii) *of the preceding proposition are equivalent, i.e., all Banach algebra structures on B have the same underlying topological space. In particular, B admits at most one power-multiplicative complete norm.*

The above results are somewhat amazing, insofar as purely algebraic conditions have topological implications. A special case of Proposition 3 is of course the earlier result that a finite extension of k has at most one valuation extending the valuation on k.

3.8. Function algebras

Let k be a (not necessarily complete) valued field and let X be a set. Then the set of all bounded functions $f: X \to k$ forms a k-algebra. Defining $|f| := \sup_{x \in X} |f(x)|$, one gets a k-algebra semi-norm on this algebra, the so-called supremum semi-norm. In the next section we shall construct for a wide class of k-algebras A spaces X such that A may be interpreted as an algebra of bounded functions from X to the algebraic closure of k. In later sections, we shall consider the following questions: When is the supremum semi-norm a complete norm on A? If A is not only a k-algebra, but in fact a k-Banach algebra, how are the given Banach norm and the supremum semi-norm related?

We assume that the reader is familiar with the notion of integral dependence.

3.8.1 The supremum semi-norm on k-algebras. — For the purposes of this section, we do not suppose that our ground field k is complete. Let A be a k-algebra. We want to derive a semi-norm on A from the given valuation on k. In order to do so, we use the following

Definition 1. *Spectrum of k-algebraic maximal ideals of A:*

$\mathrm{Max}_k\, A := \{x;\ x$ maximal ideal in A and A/x algebraic over $k\}$.

Of course, $\mathrm{Max}_k\, A$ may be empty; e.g., this is the case if A is a field which is transcendent over k. But in many cases, the spectrum of k-algebraic maximal ideals of A yields substantial information about A.

For $x \in \mathrm{Max}_k\, A$ and $f \in A$, denote by $f(x)$ the image of f under the canonical residue epimorphism $\pi_x: A \to A/x$. Since A/x is an algebraic extension of k, it can be provided with the spectral norm derived from the given valuation on k (cf. (3.2)). Writing $|f(x)|$ for the spectral norm of the element $f(x) \in A/x$, we are able to introduce

Definition 2. *Semi-norm of uniform convergence on* $\mathrm{Max}_k A$ *or supremum semi-norm on* $\mathrm{Max}_k A$:

$$|f|_{\sup} := \begin{cases} 0 & \text{if } \mathrm{Max}_k A = \emptyset, \\ \sup \{|f(x)|;\, x \in \mathrm{Max}_k A\} & \text{if } \mathrm{Max}_k A \neq \emptyset \text{ and } f(\mathrm{Max}_k A) \text{ bounded}, \\ \infty & \text{otherwise.} \end{cases}$$

This definition generalizes the concept of the spectral norm. Namely, if A is an algebraic extension of k, this definition obviously yields the spectral norm on A. Furthermore, we will show (Proposition 7) that under suitable circumstances $|\ \ |_{\sup}$ may be interpreted as a spectral norm, of course, over some bigger ground field. However, although we have used the term semi-norm, it is not in general true that the function $|\ \ |_{\sup}$ defines a semi-norm on A. Namely, the third case occurring in Definition 2 is possible: e.g., take $A = k[X]$. Then $\mathrm{Max}_k A \supset \{(X - c)\, k[X];\, c \in k\}$. If one takes $f := X \in A$, then $f(\mathrm{Max}_k A) \supset k$, which is clearly not bounded (unless k carries the trivial valuation). We shall say that the supremum semi-norm on A is finite if $f(\mathrm{Max}_k A)$ is bounded for all $f \in A$. For the special applications we have in mind, it will be easy to verify that $|\ \ |_{\sup}$ is finite. First let us collect some rather obvious properties of $|\ \ |_{\sup}$.

Lemma 3. *If* $|\ \ |_{\sup}$ *is finite, it is a power-multiplicative "k-algebra semi-norm" on* A; *i.e., one has for all* $f, g \in A$, $c \in k$, $n \in \mathbb{N}$

(a) $|f|_{\sup} \in \mathbb{R}$, $|f|_{\sup} \geqq 0$, $|0|_{\sup} = 0$,

(b) $|f + g|_{\sup} \leqq \max \{|f|_{\sup}, |g|_{\sup}\}$,

(c) $|cf|_{\sup} = |c|\, |f|_{\sup}$,

(d) $|fg|_{\sup} \leqq |f|_{\sup}\, |g|_{\sup}$,

(e) $|1|_{\sup} \leqq 1$,

(f) $|f^n|_{\sup} = |f|_{\sup}^n$.

All these formulas remain true also in the non-finite case if one extends the usual operations on $\mathbb{R}$ to $\mathbb{R} \cup \{\infty\}$ in an obvious way.

The supremum semi-norm is compatible with k-algebra homomorphisms in the following sense.

Lemma 4. *Let* $\varphi\colon B \to A$ *be a k-algebra homomorphism between two k-algebras* A *and* B. *Then* φ *is a contraction with respect to* $|\ \ |_{\sup}$, *i.e.*, $|\varphi(g)|_{\sup} \leqq |g|_{\sup}$ *for all* $g \in B$.

Proof. If $\mathrm{Max}_k A = \emptyset$, there is nothing to show. If $x \in \mathrm{Max}_k A$, then φ induces a k-algebra monomorphism $B/\varphi^{-1}(x) \to A/x$. Because A/x is an algebraic extension of k, its subring $B/\varphi^{-1}(x)$ must also be an algebraic extension of k. Hence $\varphi^{-1}(x) \in \mathrm{Max}_k B$. Then one has

$$(*) \qquad |\varphi(g)|_{\sup} = \sup_{x \in \mathrm{Max}_k A} |\varphi(g)\,(x)| = \sup_{x \in \mathrm{Max}_k A} \left|g\big(\varphi^{-1}(x)\big)\right| \leqq \sup_{y \in \mathrm{Max}_k B} |g(y)| = |g|_{\sup}. \qquad \square$$

We apply this lemma to show that the supremum semi-norm of A can easily be derived from the supremum semi-norm on its prime components.

Lemma 5. *Let $\mathfrak{M}$ be the set of all minimal prime ideals in A, and let $\pi_{\mathfrak{p}}: A \to A/\mathfrak{p}$ denote the canonical residue map for all $\mathfrak{p} \in \mathfrak{M}$. Then we have*

$$|f|_{\sup} = \sup_{\mathfrak{p}\in\mathfrak{M}} |\pi_{\mathfrak{p}}(f)|_{\sup} \quad \textit{for all} \quad f \in A\,.$$

Remark. The above lemma can be used to show that $|\ \ |_{\sup}$ coincides with the function $|\ \ |_{sp}$ as defined in (3.2.2) if A is reduced and integral over k. Namely, then all prime ideals of A are minimal (as well as maximal), and the assertion follows from Theorem 3.2.2/2.

Proof of Lemma 5. From Lemma 4, we know $\sup_{\mathfrak{p}\in\mathfrak{M}} |\pi_{\mathfrak{p}}(f)|_{\sup} \leqq |f|_{\sup}$. In order to show the opposite inequality, take $x \in \operatorname{Max}_k A$. Then one can find $\mathfrak{p} \in \mathfrak{M}$ such that $x \supset \mathfrak{p}$. From $A/x = (A/\mathfrak{p})/(x/\mathfrak{p})$, we get $\pi_{\mathfrak{p}}(x) = x/\mathfrak{p} \in \operatorname{Max}_k A/\mathfrak{p}$ and $|f(x)| = \left|\left(\pi_{\mathfrak{p}}(f)\right)\left(\pi_{\mathfrak{p}}(x)\right)\right|$, whence $|f(x)| \leqq |\pi_{\mathfrak{p}}(f)|_{\sup} \leqq \sup_{\mathfrak{p}\in\mathfrak{M}} |\pi_{\mathfrak{p}}(f)|_{\sup}$. Since this holds for all $x \in \operatorname{Max}_k A$, we have found that $|f|_{\sup} \leqq \sup_{\mathfrak{p}\in\mathfrak{M}} |\pi_{\mathfrak{p}}(f)|_{\sup}$. □

For the case of integral monomorphisms, Lemma 4 can be improved considerably.

Lemma 6. *Let $\varphi: B \to A$ be an integral k-algebra monomorphism. Then one has*

(a) *φ is an isometry with respect to $|\ \ |_{\sup}$,*

(b) *$|f|_{\sup} \leqq \max_{1\leqq i\leqq n} |b_i|_{\sup}^{1/i}$ for $f \in A$, where $f^n + \varphi(b_1) f^{n-1} + \cdots + \varphi(b_n) = 0$ is an equation of integral dependence for f over $\varphi(B)$,*

(c) *$|\ \ |_{\sup}$ is finite on A if and only if it is finite on B.*

Proof. Ad(a): If $\operatorname{Max}_k B = \emptyset$, then by Lemma 4 we have $|\varphi(g)|_{\sup} \leqq |g|_{\sup} = 0$ for all $g \in B$. Therefore we may assume $\operatorname{Max}_k B \neq \emptyset$. Take $y \in \operatorname{Max}_k B$. Since φ is integral and injective, there is a maximal ideal x of A lying over y, i.e., $\varphi^{-1}(x) = y$. Now φ induces an integral monomorphism from B/y into A/x. The field B/y is an algebraic extension of k by our assumption, and A/x is integral over B/y. Therefore A/x is an algebraic extension of k. In other words, the map $x \mapsto \varphi^{-1}(x)$ from $\operatorname{Max}_k A$ to $\operatorname{Max}_k B$ is surjective. Therefore, one has equality in the formula (*) occurring in the proof of lemma 4, and so φ is an isometry.

Ad(b): For all $x \in \operatorname{Max}_k A$, one has $0 = f(x)^n + \left(\varphi(b_1)\right)(x) f(x)^{n-1} + \cdots + \varphi(b_n)(x) = f(x)^n + b_1\left(\varphi^{-1}(x)\right) f(x)^{n-1} + \cdots + b_n\left(\varphi^{-1}(x)\right)$. Due to Proposition 3.1.2/1, this equation implies $|f(x)| \leqq \max_{1\leqq i\leqq n} \left|b_i\left(\varphi^{-1}(x)\right)\right|^{1/i} \leqq \max_{1\leqq i\leqq n} |b_i|_{\sup}^{1/i}$. Since this holds for all $x \in \operatorname{Max}_k A$, we have $|f|_{\sup} \leqq \max_{1\leqq i\leqq n} |b_i|_{\sup}^{1/i}$.

Ad(c): If $g(\operatorname{Max}_k B)$ is bounded for all $g \in B$, then $f(\operatorname{Max}_k A)$ is also bounded for all $f \in A$ due to (b). The converse is true due to (a). □

In order to be able to transform inequality (b) into an equality — which would allow us to compute $| \ |_{\sup}$ on A in terms of $| \ |_{\sup}$ on the smaller algebra B — we must impose some additional assumptions on A and B. We say that the *Maximum Modulus Principle* holds for a k-algebra A if, for all $f \in A$, there exists an $x \in \operatorname{Max}_k A$ such that $|f(x)| = |f|_{\sup}$.

Proposition 7. *Let $\varphi\colon B \to A$ be an integral torsion-free k-algebra monomorphism between two k-algebras A and B, where B is an integrally closed (integral) domain. Then one has*

(a) $|f|_{\sup} = \max\limits_{1 \leq i \leq n} |b_i|_{\sup}^{1/i}$ *for $f \in A$, where $f^n + \varphi(b_1) f^{n-1} + \cdots + \varphi(b_n) = 0$ is the (unique) integral equation of minimal degree for f over $\varphi(B)$.*

(b) *The Maximum Modulus Principle holds for A if and only if it holds for B.*

(c) $| \ |_{\sup}$ *is a norm on A if and only if $| \ |_{\sup}$ is a norm on B and A is reduced.*

(d) *We have $|\varphi(b) f|_{\sup} = |b|_{\sup} |f|_{\sup}$ for all $b \in B$ and all $f \in A$ if and only if $|bb'|_{\sup} = |b|_{\sup} |b'|_{\sup}$ for all $b, b' \in B$.*

Before proving the proposition, let us recall briefly how, for a given element $f \in A$, its integral equation of minimal degree over $\varphi(B)$ is obtained. We may assume that B is a subalgebra of A and that $A = B[f]$. Let $Q(B)$ be the field of fractions of B and define $Q(A) := A_{B-\{0\}}$. Since A is torsion-free over B, we have a commutative diagram of inclusions

$$\begin{array}{ccc} B & \subset & A \\ \cap & & \cap \\ Q(B) & \subset & Q(A). \end{array}$$

The canonical epimorphism $\tau\colon B[X] \to A$, $X \mapsto f$, extends to an epimorphism $\tau'\colon Q(B)[X] \to Q(A)$, and since $Q(B)$ is a field, there is a unique monic polynomial $q \in Q(B)[X]$ generating the ideal $\ker \tau'$. We claim that q has, in fact, coefficients in B. Assuming this for the moment, we can easily see that $q(f) = 0$ is the unique integral equation of minimal degree for f over B. Namely, we show that the ideal $\ker \tau = B[X] \cap \ker \tau'$ is generated by q. Let p be any polynomial in $\ker \tau$. Using EUCLID's division for $q \in B[X]$, we have $p = q_1 q + r$, where $q_1, r \in B[X]$ and $\deg r < \deg q$ unless $r = 0$. Since $q \in \ker \tau$, we must have $r \in \ker \tau \subset \ker \tau'$. However this is only possible if $r = 0$. Thus we see that $\ker \tau$ is generated by q and that τ induces an isomorphism $B[X]/(q) \xrightarrow{\sim} A$.

It remains to be shown that the polynomial q has coefficients in B. Since q generates the kernel of the epimorphism $\tau'\colon Q(B)[X] \to Q(A)$, $X \mapsto f$, we have an isomorphism $Q(B)[X]/(q) \xrightarrow{\sim} Q(A)$. Let us first consider the case where q is irreducible in $Q(B)[X]$. Then $Q(A)$ is a field, namely the field generated by f over $Q(B)$, and q is the corresponding minimal polynomial. Let K be the splitting field of q over $Q(B)$ (where $Q(B) \subset Q(A) \subset K$). Then K contains all conjugates of f over $Q(B)$. Since f is integral over B so are its conjugates. Thus the coefficients of q (being the elementary symmetric polynomials in the conjugates of f) must be integral over B so that $q \in B[X]$ because B is integrally closed in its field of fractions $Q(B)$.

If q is not irreducible, we consider a decomposition into prime factors in $Q(B)[X]$, say $q = p_1 \dots p_r$. Then, for each i, one can consider $Q(B)[X]/(p_i)$ as a finite extension of $Q(B)$. Concluding as before, one sees that $p_i \in B[X]$ for all i. Thus we have $q \in B[X]$ also in this case.

Now we come to the *proof* of Proposition 7. Ad(a): We may assume that B is a subalgebra of A and that $A = B[f]$. Let $q(f) = 0$ be the integral equation of minimal degree for f over B, where $q = X^n + b_1X^{n-1} + \dots + b_n \in B[X]$. Then we have $A = B[f] \cong B[X]/(q)$ by the above considerations (where f corresponds to the residue class of X in $B[X]/(q)$). For $y \in \mathrm{Max}_k B$, denote by f_y the residue class of f in A/yA and by $\bar{f}_y$ its residue class in red A/yA (the quotient of A/yA modulo its nilradical). Similarly, let q_y be the residue class of q in $(B/y)[X]$. Since each $y \in \mathrm{Max}_k B$ is contained in some ideal $x \in \mathrm{Max}_k A$, we see that

$$|f|_{\sup} = \sup_{y \in \mathrm{Max}_k B} |f_y|_{\sup} = \sup_{y \in \mathrm{Max}_k B} |\bar{f}_y|_{\sup}.$$

For a fixed $y \in \mathrm{Max}_k B$, let $q_y = q_1^{n_1} \dots q_r^{n_r}$ be the decomposition of q_y into prime factors, where $q_1, \dots, q_r$ are pairwise different monic polynomials in $(B/y)[X]$. Then

$$A/yA = (B/y)[X]/(q_y)$$

and

$$\mathrm{red}\,(A/yA) = (B/y)[X]/(q_1 \dots q_r) = \bigoplus_{\nu=1}^{r} (B/y)[X]/(q_\nu).$$

Provide B/y with the spectral norm over k and denote by $|\ \ |_\nu$ the spectral norm on $(B/y)[X]/(q_\nu)$ over B/y (which equals the spectral norm over k by Proposition 3.2.2/4). Then if $\bar{f}_\nu$ is the residue class of $\bar{f}_y$ in $(B/y)[X]/(q_\nu)$, we have by Corollary 3.2.1/6 (cf. also the remark following Lemma 3.8.1/5)

$$|\bar{f}_y|_{\sup} = \max_{1 \le \nu \le r} |\bar{f}_\nu|_\nu = \max_{1 \le \nu \le r} \sigma(q_\nu) = \sigma(q_y).$$

Therefore

$$|f|_{\sup} = \sup_{y \in \mathrm{Max}_k B} \sigma(q_y) = \sup_{y \in \mathrm{Max}_k B} \max_{1 \le i \le n} |b_i(y)|^{1/i} = \max_{1 \le i \le n} |b_i|_{\sup}^{1/i},$$

which verifies assertion (a).

Ad(b): If the Maximum Modulus Principle holds for B, then (using the same notations as before) there is a $y \in \mathrm{Max}_k B$ such that $|\bar{f}_y|_{\sup} = \sigma(q_y) = |f|_{\sup}$. Since A/yA contains only finitely many maximal ideals, there must exist an $x \in \mathrm{Max}_k A$ such that $|\bar{f}_y|_{\sup} = |f(x)|$. Hence the Maximum Modulus Principle holds for A. The converse follows from Lemma 6 (a).

Ad(c): Assume that A is reduced and that $|\ \ |_{\sup}$ is a norm on B, i.e., $|\ \ |_{\sup}$ is finite and $|b|_{\sup} = 0$ implies $b = 0$ for all $b \in B$. We have to show that $|\ \ |_{\sup}$ is a norm on A. Take $f \in A$ with $f \neq 0$, and let $f^n + b_1 f^{n-1} + \dots + b_n = 0$ be the integral equation of minimal degree for f over B. Because A is reduced, there exists an index m with $1 \le m \le n$ such that $b_m \neq 0$. By assumption we have

$|b_m|_{\sup} \neq 0$; hence

$$|f|_{\sup} = \max_{1 \leq i \leq n} |b_i|_{\sup}^{1/i} \geqq |b_m|_{\sup}^{1/m} > 0.$$

Using assertion (c) of Lemma 6, we conclude that $|\ \ |_{\sup}$ is a norm on A. Since any ring with a power-multiplicative norm is reduced, the converse follows from (a) and (c) of Lemma 6.

Ad(d): Consider an element $f \in A$, and let $f^n + b_1 f^{n-1} + \cdots + b_n = 0$ be the corresponding integral equation of minimal degree over B. Then f is of degree n over the fraction field $Q(B)$, and so is any product bf where $b \in B - \{0\}$. Therefore

$$(bf)^n + bb_1(bf)^{n-1} + \cdots + b^n b_n = 0$$

is the integral equation of minimal degree for any such product bf. Now assume that $|\ \ |_{\sup}$ is multiplicative on B, i.e., that $|bb'|_{\sup} = |b|_{\sup}\, |b'|_{\sup}$ for all $b, b' \in B$. Then we can compute $|bf|_{\sup}$ according to (a):

$$|bf|_{\sup} = \max_{1 \leq i \leq n} |b^i b_i|_{\sup}^{1/i} = |b|_{\sup} \max_{1 \leq i \leq n} |b_i|_{\sup}^{1/i} = |b|_{\sup}\, |f|_{\sup}.$$

Since this equation is trivially true for $b = 0$, we have verified the if part of assertion (d). The only if part follows immediately from assertion (a) of Lemma 6. □

Corollary 8. *In addition to the assumptions of Proposition 7, assume that the Maximum Modulus Principle holds for B. Then for every $f \in A$ with $|f| \neq 0$, there exist $c \in k$ and $m \in \mathbb{N}$ such that $|cf^m|_{\sup} = 1$.*

Proof. For all $b \in B$, there exists $y \in \operatorname{Max}_k B$ such that $|b|_{\sup} = |b(y)|$. Hence $|B|_{\sup} \subset |k_a|$. The set $|k_a|$ is invariant under the operation of taking roots. Therefore $\{|b|_{\sup}^{1/i}; b \in B - \{0\}$ and $i \in \mathbb{N}\} \subset |k_a|$. According to assertion (a) of the preceding proposition, $|f|_{\sup} \in |k_a|$ for all $f \in A$. Hence there are an element $d \in k$ and an integer m such that $|d|^{1/m} = |f|_{\sup}$. Define $c := d^{-1} \in k$. Then $1 = |cf^m|_{\sup}$. □

For later reference we state the following trivial fact:

Lemma 9. *If $|\ \ |_{\sup}$ is a norm on a k-algebra A, then $\bigcap_{\mathfrak{m} \in \operatorname{Max}_k A} \mathfrak{m} = (0)$. In particular, A is reduced, and the Jacobson radical $\bigcap_{\mathfrak{m} \in \operatorname{Max} A} \mathfrak{m}$ vanishes.*

Remark. Assume that $|\ \ |_{\sup}$ is a valuation on B. Then it can be extended to a valuation on the field of fractions $Q(B)$. If we assume furthermore that A is reduced, then $Q(A) := A_{\varphi(B)-\{0\}}$ is a reduced integral $Q(B)$-algebra. According to assertion (d) of Proposition 7, the function $|\ \ |_{\sup}$ defines a faithful B-module norm on A; hence it can be extended to a $Q(B)$-algebra norm on $Q(A)$. According to assertion (a) of Proposition 7, this norm is nothing more than the spectral norm on the integral $Q(B)$-algebra $Q(A)$, as defined in (3.2.2). To put it in another way: *The spectral norm on $Q(A)$ considered as a*

$Q(B)$-algebra yields — if restricted to A — the supremum norm on A considered as a k-algebra.

This remark is used in showing that, under certain assumptions (including the stability of k), the algebras A and $Q(A)$ are k-cartesian vector spaces — a result which shall be needed in (5.3) for proving the stability of the fields of fractions $Q(T_n)$.

Proposition 10. *If k is stable, then every k-algebra A such that $|\ |_{\sup}$ is a norm on A fulfilling the Maximum Modulus Principle is k-cartesian.*

Proof. By Proposition 2.4.3/6, we only have to show that every one-dimensional subspace $k \cdot f$, $f \in A$, of a finite-dimensional subspace U of A admits a norm-direct supplement in U. We can find $x \in \operatorname{Max}_k A$ such that $|f(x)| = |f|_{\sup}$. Now $U/x \cap U$ is a finite-dimensional subspace of A/x, which in turn is a finite extension of k. Because k is stable, we know that A/x provided with the spectral norm is k-cartesian. Hence the one-dimensional space $k \cdot f(x)$ has a norm-direct supplement W in $U/x \cap U$ by Lemma 2.4.1/4. Let $V = \pi^{-1}(W)$, where $\pi: U \to U/x \cap U$ denotes the quotient map. The proposition is proved if we can show that V is a norm-direct supplement to $k \cdot f$ in U. In order to verify this claim, take $u \in U$; then one has $\pi(u) = cf(x) + w$ for suitable $c \in k$ and $w \in W$. Defining $v := u - cf$, we obtain that $u = cf + v$ and that $v \in V$. Furthermore, we have $|\pi(u)| = \max\{|c|\,|f(x)|, |\pi(v)|\}$. Then one gets the equalities $|cf|_{\sup} = |c|\,|f|_{\sup} = |c|\,|f(x)| \leqq |\pi(u)| \leqq |u|_{\sup}$, which imply that $|v|_{\sup} \leqq |u|_{\sup}$. Therefore $|u|_{\sup} = \max\{|cf|_{\sup}, |v|_{\sup}\}$. □

Remark. The assumption "k is stable" cannot be omitted. Indeed, choose $A = k_a$. Then $|\ |_{\sup}$ coincides with the spectral norm on A, since $\operatorname{Max}_k k_a = \{(0)\}$. But k_a can be cartesian only if k is stable.

Proposition 11. *Assume that k is stable. Let B be an integrally closed domain such that $|\ |_{\sup}$ is a valuation on B fulfilling the Maximum Modulus Principle, and let $\varphi: B \to A$ be an integral torsion-free k-algebra monomorphism, where A is reduced. Then $Q(A) := A_{\varphi(B)-\{0\}}$ is k-cartesian under the spectral norm relative to the field of fractions $Q(B)$.*

Proof. We see by Proposition 7 (b) and (c) that $|\ |_{\sup}$ is a norm on A fulfilling the Maximum Modulus Principle. Hence A is k-cartesian under $|\ |_{\sup}$. Let $|\ |_{sp}$ be the spectral norm on the reduced integral $Q(B)$-algebra $Q(A)$. According to the remark following Lemma 9, we know that $|\ |_{\sup}$ equals the restriction of $|\ |_{sp}$ to A. By Lemma 2.4.3/3, we conclude that $Q(A)$ is k-cartesian under $|\ |_{sp}$. □

3.8.2. The supremum semi-norm on k-Banach algebras. — So far we have studied k-algebras as purely algebraic objects. The norm or semi-norm we constructed on such an algebra A was only a paraphrasing of the structure of the set of all k-algebraic maximal ideals of A. From now on we assume that we are given a norm $|\ \ |$ on A and ask how are $|\ \ |$ and $|\ \ |_{\sup}$ interrelated. For simplicity, we restrict ourselves to the case where the ground field k is complete and A is a k-Banach algebra. Even in this case, the maximal ideals $\mathfrak{m}$ of A need not be algebraic over k (i.e., $A/\mathfrak{m}$ need not be algebraic over k). For example, just take A to be the completion of the field of fractions of $k\langle X\rangle$ provided with the Gauss valuation. One has the following preliminary result.

Lemma 1. *Let A be a k-Banach algebra with norm $|\ \ |$ and let $x \in \operatorname{Max}_k A$ be a k-algebraic maximal ideal. Then x is closed in A and A/x provided with the residue norm $|\ \ |_{\mathrm{res}}$ is a k-Banach algebra. Moreover, one has*

$$|f(x)| = \inf_{i\in\mathbb{N}} |f(x)^i|_{\mathrm{res}}^{1/i} \leqq |f(x)|_{\mathrm{res}} \leqq |f|\,.$$

Proof. Assume that x is not closed in A. Then its completion $\hat{x}$ is an ideal in A such that $x \subsetneqq \hat{x}$. Therefore $\hat{x} = A$; i.e., x is dense in A. In particular, x contains elements which are arbitrarily close to the unit element $1 \in A$. Hence by Proposition 1.2.4/4, the ideal x must contain units itself. However this is impossible so that x must be closed in A. Therefore the function $|\ \ |_{\mathrm{res}}$, given by

$$|f(x)|_{\mathrm{res}} = \inf_{f(x)=g(x)} |g| \quad \text{for} \quad f \in A\,,$$

is a norm on A/x (cf. (1.1.6)). Using Proposition 1.1.7/3, we see that $|\ \ |_{\mathrm{res}}$ is actually a complete k-algebra norm on A/x with $|f(x)|_{\mathrm{res}} \leqq |f|$.

We want to define another norm $|\ \ |'$ on A/x by $|f(x)|' := \inf_{i\in\mathbb{N}} |f(x)^i|_{\mathrm{res}}^{1/i}$. Due to Proposition 1.3.2/1, the map $|\ \ |'$ is a power-multiplicative "k-algebra seminorm" on A/x such that $|f(x)|' \leqq |f(x)|_{\mathrm{res}}$ for all $f \in A$. Because A/x is a field, $|\ \ |'$ is in fact a norm. The field k is complete, and A/x is an algebraic extension of k. Hence the spectral norm is the only power-multiplicative k-algebra norm on A/x, and therefore it coincides with $|\ \ |'$. Thus we have $|f(x)| = \inf_{i\in\mathbb{N}} |f(x)^i|_{\mathrm{res}}^{1/i}$. □

Applying this lemma to all $x \in \operatorname{Max}_k A$, we get

Corollary 2. *If A is a k-Banach algebra with norm $|\ \ |$, then for all $f \in A$ one has*

$$|f|_{\sup} \leqq |f|\,.$$

If A is not complete, this statement may fail to be true. For example, take $A = k[X]$ provided with the Gauss norm and $f := X$; then $|X| = 1$, whereas $|f|_{\sup} = \infty$, as we have seen already in (3.8.1).

The following result is closely related to Proposition 3.7.5/2.

Proposition 3. *Let A be a k-Banach algebra. Assume that $|\ \ |_{\sup}$ is a norm on A. Then every k-algebra homomorphism φ from an arbitrary k-Banach algebra B into A is continuous.*

Proof. For all $x \in \operatorname{Max}_k A$, consider the commutative diagram

$$\begin{array}{ccc} B & \xrightarrow{\varphi} & A \\ {\scriptstyle\beta}\downarrow & \searrow^{\psi} & \downarrow{\scriptstyle\alpha} \\ B/\varphi^{-1}(x) & \xrightarrow{\overline{\psi}} & A/x \end{array}$$

where α and β are the canonical epimorphisms. Because A/x is algebraic over k, so is $B/\varphi^{-1}(x)$. Hence $\varphi^{-1}(x) \in \operatorname{Max}_k B$. Provide A/x and $B/\varphi^{-1}(x)$ with the spectral norm. The map $\overline{\psi}$ is then an isometry. The epimorphisms α and β are contractions according to Lemma 1. Thus we know that $\psi = \overline{\psi} \circ \beta$ is continuous. Then by Lemma 3.8.1/9, we may finish the proof by using the Closed Graph Theorem (in almost literally the same way as in the proof of Proposition 3.7.5/1). □

Corollary 4. *If A is a k-Banach algebra such that $|\ \ |_{\sup}$ is a norm on A, then all complete k-algebra norms on A are equivalent.*

In Corollary 2 we only have an estimate for $|\ \ |_{\sup}$. Now we want to compute $|\ \ |_{\sup}$ in terms of $|\ \ |$. This can be done if we add topological conditions to the assumptions of Proposition 3.8.1/7.

Proposition 5. *Let $\varphi\colon B \to A$ be an integral torsion-free k-algebra monomorphism between two k-algebras A and B, where B is an integrally closed domain. Assume furthermore that A is a k-Banach algebra with norm $|\ \ |$ and that φ is continuous if B is provided with the topology induced by $|\ \ |_{\sup}$. Then one has for all $f \in A$*

$$\inf_{i \in \mathbb{N}} |f^i|^{1/i} = |f|_{\sup} = \max_{1 \leq i \leq n} |b_i|_{\sup}^{1/i},$$

where $f^n + \varphi(b_1) f^{n-1} + \cdots + \varphi(b_n) = 0$ is the integral equation of minimal degree for f over $\varphi(B)$.

Proof. We only have to show that $|f|_r = |f|_{\sup}$, where $|f|_r := \inf\limits_{i \in \mathbb{N}} |f^i|^{1/i}$. From Corollary 2 one deduces immediately that $|f|_{\sup} \leq |f|_r$. In order to show the opposite inequality, we shall use Proposition 3.8.1/7. Since $|\ \ |_r$ is a power-multiplicative semi-norm on A, one can apply Proposition 3.1.2/1 to $A/\ker|\ \ |_r$ (viewed as a normed algebra over itself), and one gets

$$|f|_r \leq \max_{1 \leq i \leq n} |\varphi(b_i)|_r^{1/i} \leq \max_{1 \leq i \leq n} |\varphi(b_i)|^{1/i}.$$

Because φ is continuous, there is a real constant $C > 1$ such that $|\varphi(b)| \leq C|b|_{\sup}$ for all $b \in B$, and a fortiori $|\varphi(b)|^{1/i} \leq C|b|_{\sup}^{1/i}$ for all $i \in \mathbb{N}$. Thus by Proposition 3.8.1/7, we have shown that $|f|_r \leq C \max\limits_{1 \leq i \leq n} |b_i|_{\sup}^{1/i} = C|f|_{\sup}$. This implies $|f|_r \leq |f|_{\sup}$, since both $|\ \ |_r$ and $|\ \ |_{\sup}$ are power-multiplicative. □

Remark 1. In spite of the fact that $|\ \ |_{\sup}$ depends only on the algebraic structure of the k-algebra A (and, of course, on the valuation on k), $|\ \ |_{\sup}$ nevertheless coincides with the norm $|\ \ |_r$ which is derived from the given

Banach norm on A. This is a result similar to Corollary 4, asserting that (roughly speaking) the topological structure of the k-Banach algebra A is already determined by the underlying algebraic structure.

Remark 2. The assumption "φ is continuous" of Proposition 5 is automatically fulfilled if A is reduced and if $|\ \ |_{\sup}$ is a complete norm on B. Namely, B provided with $|\ \ |_{\sup}$ is then a k-Banach algebra and $|\ \ |_{\sup}$ is a norm on A by Proposition 3.8.1/7 (c). Applying Proposition 3, we see that φ is continuous.

We want to conclude this section by looking at topologically nilpotent and power-bounded elements of a k-Banach algebra A. Since

$$\inf_{i\in\mathbb{N}} |f^i|^{1/i} = \lim_{i\to\infty} |f^i|^{1/i}$$

for any semi-norm (cf. (1.3.2)), it is easily seen that an element f is topologically nilpotent in A if and only if $\inf_{i\in\mathbb{N}} |f^i|^{1/i} < 1$. Furthermore, we have $\inf_{i\in\mathbb{N}} |f^i|^{1/i} \leqq 1$ if f is power-bounded. In the special situation described in Proposition 5, the latter implication is, in fact, an equivalence. Namely,

Corollary 6. *Under the hypotheses of Proposition* 5, *the following statements are equivalent for all* $f \in A$:

(a) *f is topologically nilpotent,*

(b) $\inf_{i\in\mathbb{N}} |f^i|^{1/i} < 1$,

(c) $|f|_{\sup} < 1$.

If the Maximum Modulus Principle holds for A or B, then (a), (b) *and* (c) *are equivalent to*

(d) $|f(x)| < 1$ *for all* $x \in \operatorname{Max}_k A$.

Furthermore, the following statements are equivalent for all $f \in A$:

(a′) *f is power-bounded,*

(b′) $\inf_{i\in\mathbb{N}} |f^i|^{1/i} \leqq 1$,

(c′) $|f|_{\sup} \leqq 1$.

Proof. It has already been mentioned that (a) and (b) are equivalent. The preceding proposition yields the equivalence of (b) and (c). Clearly (c) and (d) are equivalent if the Maximum Modulus Principle holds for A. Furthermore by Proposition 3.8.1/7 (b), if the Maximum Modulus Principle holds for B, it holds also for A.

As already indicated, (a′) implies (b′). The equivalence of (b′) and (c′) follows from Proposition 5. Hence it suffices to show that (c′) implies (a′). Use the notations of Proposition 5. Then from (c′) we get $|b_i|_{\sup} \leqq 1$ for $i = 1, \ldots, n$, and therefore $|b|_{\sup} \leqq 1$ for all $b \in P[b_1, \ldots, b_n]$, where P is the prime ring of B, i.e., the smallest subring of B containing 1. Since φ is continuous, $R := \varphi(P)\,[\varphi(b_1), \ldots, \varphi(b_n)]$ is bounded under $|\ \ |$. From the integral equation for f, one easily derives $f^j \in \sum_{i=0}^{n-1} Rf^i$ for all $j \in \mathbb{N}$ (use induction on j). Hence f is power-bounded with respect to $| \; |$. □

3.8.3. Banach function algebras. — The main result of the preceding section (Proposition 3.8.2/5) allows us to compute $| \; |_{\sup}$ on A in terms of the given complete norm on A, but it does not answer the question of whether or not $| \; |_{\sup}$ also induces the Banach topology on A. We shall attack this problem from a somewhat more general point of view and start with the following definition:

Definition 1. *A k-algebra A is called a Banach function algebra if $| \; |_{\sup}$ is a complete norm on A.*

For any k-algebra A, we may interpret the elements of A as functions on $\operatorname{Max}_k A$ with values in the algebraic closure k_a of k. The special case where the semi-norm of uniform convergence on $\operatorname{Max}_k A$ is, in fact, a norm making A a k-Banach algebra is singled out by the above definition. The question of whether or not A is a Banach function algebra depends only on the algebraic structure of A. If A is a k-Banach algebra with some given norm, we may ask: Does the fact that A is also a Banach function algebra influence the given norm?

Lemma 2. *A k-Banach algebra A is a Banach function algebra if and only if $| \; |_{\sup}$ is equivalent to the given norm on A.*

Proof. The if part is obvious. The only if part follows from Corollary 3.8.2/4. □

Lemma 3. *If A is a Banach function algebra, then $| \; |_{\sup}$ is the only power-multiplicative complete k-algebra norm on A.*

Proof. Since any two power-multiplicative k-algebra norms inducing the same topology must coincide (cf. Proposition 3.1.5/1), there is at most one power-multiplicative complete k-algebra norm on A by Lemma 2. On the other hand, $| \; |_{\sup}$ is power-multiplicative and complete if A is a Banach function algebra. □

Given a k-algebra homomorphism $B \to A$, we are interested as before in deriving information about $| \; |_{\sup}$ on A from properties of $| \; |_{\sup}$ on B, and vice versa. Here, where we are concerned with Banach function algebras, it is natural to ask: If B is a Banach function algebra, does A inherit this property from B, and vice versa? As one would expect, "going down" is easier than "going up". Thus we treat the easy case first.

Lemma 4. *Let $\varphi: B \to A$ be a k-algebra monomorphism between two k-algebras A and B. Assume that A is a Banach function algebra. Then B is a Banach function algebra if $\varphi(B)$ is closed in A. In particular, a closed subalgebra of a Banach function algebra is again a Banach function algebra.*

Proof. Let $|\ \ |_{\sup}^{(A)}$ (resp. $|\ \ |_{\sup}^{(B)}$) denote the supremum semi-norm on A (resp. B). The algebra $\varphi(B)$, provided with the restriction of $|\ \ |_{\sup}^{(A)}$, is a k-Banach subalgebra of A, because A is a Banach function algebra and because $\varphi(B)$ is closed in A. Lifting this Banach algebra structure back to B by means of φ (i.e., by defining $|b| := |\varphi(b)|_{\sup}^{(A)}$ for $b \in B$), we get a k-Banach algebra structure on B. Applying Corollary 3.8.2/2, we see that the new norm dominates the supremum semi-norm on B — i.e., $|b|_{\sup}^{(B)} \leqq |b|$ for all $b \in B$. On the other hand, by Lemma 3.8.1/4, the monomorphism φ is a contraction with respect to the supremum semi-norms, i.e., $|b| = |\varphi(b)|_{\sup}^{(A)} \leqq |b|_{\sup}^{(B)}$ for all $b \in B$. Putting these inequalities together, we see that $|\ \ |_{\sup}^{(B)}$ coincides with the complete norm $|\ \ |$. □

Remark. If, in addition to the assumptions of Lemma 4, the monomorphism φ is supposed to be integral, then the closedness of $\varphi(B)$ is also necessary for B to be a Banach function algebra. This follows from Lemma 3.8.1/6 (a).

We want to replace the condition "$\varphi(B)$ is closed" by other assumptions which are easier to handle.

Corollary 5. *Let $\varphi\colon B \to A$ be a finite k-algebra monomorphism, where B is a Noetherian k-Banach algebra and A is a Banach function algebra. Then B is a Banach function algebra.*

Proof. Provide A with the norm $|\ \ |_{\sup}$. By Proposition 3.8.2/3, the map φ is continuous. For all $b \in B$, we have $|\varphi(b)|_{\sup} \leqq |b|_{\sup}$ by Lemma 3.8.1/4 and $|b|_{\sup} \leqq |b|$ by Corollary 3.8.2/2. Therefore we can view A as a finite normed B-module. Since B is a Noetherian k-Banach algebra, all B-submodules of A are closed (see Proposition 3.7.2/2); in particular, $\varphi(B)$ is closed in A. Now the assertion follows from the preceding lemma. □

In the remainder of this section, we consider a k-algebra homomorphism $\varphi\colon B \to A$; we want to look at conditions which imply that A is a Banach function algebra. We start with a criterion which can be used if char $k = 0$.

Proposition 6. *Let* char $k = 0$, *and let $\varphi\colon B \to A$ be a finite torsion-free k-algebra monomorphism, where B is a Noetherian integrally closed domain and where A is reduced. Then A is a Banach function algebra if B is a Banach function algebra.*

Proof. First we want to show that $\mathring{B} = \{b \in B;\ |b|_{\sup} \leqq 1\}$ is integrally closed in its field of fractions. Let f, g be elements in $\mathring{B}$ with $g \neq 0$ such that $u := \dfrac{f}{g}$ is integral over $\mathring{B}$. Since $\mathring{B}$ and B have the same field of fractions $Q(B)$ and since B is integrally closed in $Q(B)$, we know that, in fact, $u \in B$. Let $u^m + b_1u^{m-1} + \cdots + b_m = 0$ be an integral equation for u over $\mathring{B}$. Then we have by Proposition 3.1.2/1

$$|u|_{\sup} \leqq \max |b_i|_{\sup}^{1/i} \leqq 1 .$$

Hence $u \in \mathring{B}$, and $\mathring{B}$ is integrally closed. Similarly, using Proposition 3.8.1/7 (a), we see that $\mathring{A}$ is integral over $\mathring{B}$.

Let us now look at the ring of fractions $Q(A) := A_{\varphi(B)-\{0\}}$, which is a finite $Q(B)$-algebra. We want to show that there exists a $Q(B)$-basis $v_1, \ldots, v_r$ of $Q(A)$ over $Q(B)$ such that $\mathring{A} \subset \sum_{i=1}^{r} \mathring{B} v_i$. First we consider the case, where A is an integral domain, or equivalently, where $Q(A)$ is a field. Since we are working in characteristic 0, we know that $Q(A)$ is a separable extension of $Q(B)$ of degree $r < \infty$, say generated by an element $a \in Q(A)$. We may assume that the minimal polynomial of a has coefficients in $\mathring{B}$ so that a is integral over $\mathring{B}$. Let $a = a_1, \ldots, a_r$ be the conjugates of a over $Q(B)$ and set $K := Q(B)(a_1, \ldots, a_r)$. Then any $f \in Q(A)$ can be written as $f = \sum_{i=0}^{r-1} b_i a^i$ with coefficients $b_i \in Q(B)$. Using Galois automorphisms over $Q(B)$ on K, we get equations

$$f_j = \sum_{i=0}^{r-1} b_i a_j^i, \qquad j = 1, \ldots, r,$$

where f_j is the image of f under the automorphism mapping a to a_j. If one views these as a system of linear equations in the unknown b_i, the determinant of the coefficient matrix is VANDERMONDE's determinant

$$d = \det(a_j^i) = \prod_{\nu > \mu} (a_\nu - a_\mu).$$

Since d^2 (the discriminant of the minimal polynomial of a) is invariant under Galois automorphisms, we have $d^2 \in Q(B)$. (In fact, $d^2 \in \mathring{B}$ since all a_j and hence d^2 are integral over $\mathring{B}$.) If f is integral over $\mathring{B}$, then all f_j are integral over $\mathring{B}$. Therefore CRAMER's rule shows that in this case all elements $d^2 b_i$ are integral over $\mathring{B}$; hence they are contained in $\mathring{B}$, because, as we saw, $\mathring{B}$ is integrally closed. Thus we have verified that the integral closure of $\mathring{B}$ in $Q(A)$ (and, in particular, $\mathring{A}$) is contained in $\sum_{i=0}^{r-1} \mathring{B}(d^{-2}a^i)$.

If $Q(A)$ is not a field, it is a finite direct sum of finite extensions $L_1, \ldots, L_t$ of $Q(B)$ (use DEDEKIND's Lemma 3.1.4/1). Then the above consideration can be carried out for each extension L_i separately, and the desired result follows.

Let $v_1 \ldots, v_r$ be a $Q(B)$-basis of $Q(A)$ such that $\mathring{A} \subset \sum_{i=1}^{r} \mathring{B} v_i$, and choose an element $b \neq 0$ in $\mathring{B}$ such that all products $v_{i_1} v_{i_2}$ are contained in $\sum_{i=1}^{r} \mathring{B} b^{-1} v_i$. Then

$$\mathring{A} \subset \sum_{i=1}^{r} \mathring{B} v_i \subset \sum_{i=1}^{r} \mathring{B} b^{-1} v_i$$

and, in particular,

$$A \subset \sum_{i=1}^{r} B v_i \subset F := \sum_{i=1}^{r} B b^{-1} v_i .$$

We assume that B is a k-Banach algebra under $| \;\; |_{\sup}$ and consider a norm $| \;\; |$ on the finite B-module F, which is given by

$$\left| \sum_{i=1}^{r} b_i \frac{v_i}{b} \right| := \max_{1 \leq i \leq r} |b_i|_{\sup} .$$

Then F is a complete Noetherian B-module. By Proposition 3.7.2/2, all submodules of F ars closed. In particular, A is closed in F and therefore a k-Banach space under $|\ \ |$. An easy verification shows that the multiplication is continuous in A (due to the choice of the element $b \in \mathring{B}$). Hence by Proposition 1.2.1/2, we can define a k-algebra norm $\|\ \ \|$ on A by

$$\|f\| := \sup_{g \in A - \{0\}} \frac{|fg|}{|g|},$$

which is equivalent to $|\ \ |$.

We want to conclude the proof by showing that $|\ \ |_{\sup}$ is equivalent to $\|\ \ \|$ and hence to $|\ \ |$ on A. Then it follows that A is a Banach function algebra. We have $|\ \ |_{\sup} \leqq \|\ \ \|$ by Corollary 3.8.2/2. Thus it is enough to find a constant $C > 0$ such that $|f| \leqq C|f|_{\sup}$ for all $f \in A$. Choose an element $c \in k$ with $|c| > 1$. We claim that $|f| \leqq |c|\ |f|_{\sup}$ for all $f \in A$. In order to verify this, consider a fixed element $f \neq 0$ in A. Then $|f|_{\sup} \neq 0$ by Proposition 3.8.1/7 (c). Hence there exists an $m \in \mathbb{Z}$ such that

$$|c|^{m-1} < |f|_{\sup} \leqq |c|^m.$$

Then $|c^{-m}f|_{\sup} \leqq 1$ and $|c|^m < |c|\ |f|_{\sup}$. Set $g := c^{-m}f$ so that $g \in \mathring{A}$. Since $\mathring{A} \subset \sum_{i=1}^{r} \mathring{B}b^{-1}v_i$, one can find elements $b_1, \ldots, b_r \in \mathring{B}$ such that $g = \sum_{i=1}^{r} b_i(b^{-1}v_i)$. Then one has

$$|f| = |c^m g| = \left|\sum_{i=1}^{r} (c^m b_i)\,(b^{-1}v_i)\right| = \max_{1 \leqq i \leqq r} |c^m b_i|_{\sup} \leqq |c|^m \leqq |c|\ |f|_{\sup},$$

which finishes our proof. □

For char $k = p > 0$, we do not know whether or not Proposition 6 holds without additional assumptions. At any rate, it remains true for char $k = p$ if A is finite over the k^p-algebra A^p. Instead of pursuing this further, we shall give another criterion for A to be a Banach function algebra, which is valid in all characteristics and which is quite adequate for the applications to k-affinoid algebras in (6.2.2).

Theorem 7. *Let $\varphi\colon B \to A$ be a finite torsion-free k-algebra monomorphism, where A is reduced and where B is a valued integrally closed Noetherian k-Banach algebra such that the field of fractions $Q(B)$ is weakly stable* (*cf.* (3.5)). *Then A is a Banach function algebra, and $|f|_{\sup} = |f|_{sp}$ for all $f \in A$, where $|\ \ |_{sp}$ denotes the spectral norm on the finite-dimensional reduced $Q(B)$-algebra $Q(A) := A_{\varphi(B)-\{0\}}$.*

Proof. Clearly $Q(A)$ is a finite-dimensional reduced $Q(B)$-algebra and hence a ring-theoretic direct sum of finitely many finite extensions of $Q(B)$ (see Dedekind's Lemma 3.1.4/1). Since $Q(B)$ is weakly stable, all these extensions are weakly $Q(B)$-cartesian under their spectral norm. Then $Q(A)$, provided with its spectral norm, is weakly $Q(B)$-cartesian (use Theorem 3.2.2/2). Let $a_1, \ldots, a_n$ be a $Q(B)$-basis of $Q(A)$. Because φ is finite, there is a universal denominator $b \in B - \{0\}$ such that $A \subset A' := \sum_{i=1}^{n} B \frac{a_i}{b} \subset Q(A)$. Since $| \ |_{sp}$ induces the $Q(B)$-product topology on $Q(A)$, and since B is complete, we get a complete finite (and hence Noetherian) B-module if we restrict $| \ |_{sp}$ to A'. By Proposition 3.7.2/2, the B-submodule A of A' is closed with respect to the restriction of $| \ |_{sp}$. Because A' is complete, A is complete and hence a k-Banach algebra. From Proposition 3.8.1/7 (a), we derive $|f|_{sup} = |f|_{sp}$ for all $f \in A$. Hence $| \ |_{sup}$ is a complete norm on A, and therefore A is a Banach function algebra. □

Remark. If A is assumed to be a k-Banach algebra with given norm $| \ |$, then by Corollary 3.8.2/4 and Proposition 3.8.2/5, $| \ |_{sup}$ is equivalent to $| \ |$ and $|f|_{sup} = \inf_{i \in \mathbf{N}} |f^i|^{1/i}$.

Appendix to Part A

CHAPTER 4

Tame modules and Japanese rings

The purpose of this chapter is to establish some auxiliary results from commutative algebra. By A we always mean an integral domain (no norm or valuation is required). Let $K = Q(A)$ be the field of fractions of A. For any A-module M, we denote by $Q(M)$ the K-vector space $M \otimes_A K$. This is the localization of M with respect to the multiplicative system $A - \{0\}$. By $\mathrm{rk}_A M$ we always mean the *rank* of M, i.e., the maximal number of A-linearly independent elements of M. We have $\mathrm{rk}_A M = \dim_K Q(M)$.

4.1. Tame modules

The results of this section are needed with others to derive certain criteria for Japaneseness of Noetherian integral domains.

Each finitely generated A-module M is of finite rank. The converse is not in general true if $A \neq K$; namely in this case, K itself is an A-module of rank 1 which is not finitely generated.

Definition 1. *An A-module M is called tame if each A-submodule $N \subset M$ of finite rank is finitely generated.*

Obviously, each such submodule N is then Noetherian (i.e., all submodules of N are finitely generated). *Thus M is tame if and only if each submodule $N \subset M$ with $\mathrm{rk}_A N < \infty$ is Noetherian.*

Submodules of tame modules are again tame. Furthermore,

Proposition 2. *The direct sum of finitely many tame A-modules is a tame A-module.*

Proof. Let M_1, M_2 be tame A-modules, and let $N \subset M_1 \oplus M_2$ be a submodule of finite rank. We have $N \subset \pi_1(N) \oplus \pi_2(N)$, where π_i denotes the projection from $M_1 \oplus M_2$ to M_i, $i = 1, 2$. Clearly, $\mathrm{rk}_A \pi_i(N) \leqq \mathrm{rk}_A N$, $i = 1, 2$. Hence $\pi_i(N) \subset M_i$ is Noetherian, $i = 1, 2$. Therefore, $\pi_1(N) \oplus \pi_2(N)$ and so N are Noetherian. □

Proposition 3. *Let $A \subset A'$ be a pair of integral domains such that A' is a tame A-module. Then each torsion-free tame A'-module M' (viewed as an A-module) is a tame A-module.*

Proof. Let N be an A-submodule of M' of finite rank. Denote by N' the A'-submodule of M' generated by N. Obviously, $\mathrm{rk}_{A'} N' \leqq \mathrm{rk}_A N$. Hence N' is a finitely generated A'-module. Since M' is torsion-free, we can embed N' into a finitely generated free A'-module F', say $F' = \bigoplus_1^s A'$. Then F' is a tame A-module, because A' is a tame A-module. Thus N, being an A-submodule of finite rank in F', is finitely generated. □

Proposition 4. *Let A be Noetherian. Let M be an A-module such that, for each submodule $N \neq 0$ of finite rank, there exists an A-linear map $\Phi\colon N \to A$ with* $\ker \Phi \neq N$ *(i.e., $\Phi \neq 0$). Then M is tame.*

Proof. First, M has no torsion. Let $N \subset M$ be a submodule of finite rank. We set $t := \mathrm{rk}_A N$, and we shall prove by induction on t that N is Noetherian. For $t = 0$, we have $N = 0$ (since M is torsion-free), and the assertion is trivial. Let $t > 0$. Then $N \neq 0$, and there exists an A-linear map $\Phi\colon N \to A$ with $\ker \Phi \neq N$. We extend Φ to a $Q(A)$-homomorphism $\Phi^*\colon Q(N) \to Q(A)$. Since $\Phi^* \neq 0$, we have

$$\mathrm{rk}_A \ker \Phi \leqq \dim_{Q(A)} \ker \Phi^* < \dim_{Q(A)} Q(N) = \mathrm{rk}_A N = t.$$

By the induction hypothesis, $\ker \Phi$ is Noetherian. Because A is Noetherian, the ideal $\Phi(N) \subset A$ is also a Noetherian A-module. From the exact sequence $0 \to \ker \Phi \to N \to \Phi(N) \to 0$, we now conclude that N itself is a Noetherian A-module. □

An A-module M is called *linearly separable* if, for each element $x \neq 0$ in M, there is an A-linear map $\Phi\colon M \to A$ such that $\Phi(x) \neq 0$. Since all submodules of a linearly separable module are again linearly separable, we can deduce from Proposition 4 the following result:

Corollary 5. *Let A be Noetherian. Then each linearly separable A-module — in particular each free A-module — is tame.*

4.2. A Theorem of Dedekind

Let K' be an algebraic extension of $K = Q(A)$. The *integral closure A' of A in K'* consists of all elements of K' which are *integral over* A, i.e., which satisfy an integral equation with coefficients in A. It is well known that A' is an A-algebra with K' as field of fractions; more precisely, $K' = A'_{A-\{0\}}$. The integral domain A is called *normal* or *integrally closed* if A is integrally closed in its field of fractions K. We are interested in the following question:

If K' is a finite K-module (i.e., if $\dim_K K' < \infty$), is it true that A' is a finite A-module?

First, we state (and repeat the proof of) a well-known theorem which goes back to Dedekind.

Theorem 1. *Let A be a normal Noetherian integral domain; let K' be a finite and separable extension of K. Then A' is a finite A-module (and, in particular, a Noetherian ring).*

Proof. Since K' is separable over K, the bilinear form $K' \times K' \to K$, $(x, y) \mapsto T(xy)$, is non-degenerate ($T := \mathrm{Tr}_{K'/K}: K' \to K$ denotes the trace function). Hence, for each basis $\{x_1, \ldots, x_n\}$ of the K-vector space K', there exists a K-basis $\{y_1, \ldots, y_n\}$ of K' such that

$$T(x_\mu y_\nu) = \delta_{\mu\nu}; \quad \mu, \nu = 1, \ldots, n \quad (\delta_{\mu\nu} := \text{KRONECKER's delta}).$$

We may assume $x_1, \ldots, x_n \in A'$. Then we claim

$$(*) \qquad A' \subset \sum_{\nu=1}^{n} A y_\nu,$$

from which the assertion follows, because A is Noetherian. In order to verify $(*)$, take $a' \in A'$, and write $a' = \sum_1^n a_\nu y_\nu$ where $a_\nu \in K$. We derive

$$T(x_\mu a') = \sum_{\nu=1}^{n} a_\nu T(x_\mu y_\nu) = a_\mu, \qquad \mu = 1, \ldots, n.$$

Since $x_\mu a' \in A'$ is integral over A, it follows that $a_\mu \in K$ is integral over A. Because A is normal, we conclude $a_\mu \in A$, $\mu = 1, \ldots, n$. □

4.3. Japanese rings. First criterion for Japaneseness

The Theorem of DEDEKIND is not, in general, true for inseparable extensions. Therefore, we introduce

Definition 1. *An integral domain A is called Japanese if the integral closure of A in any finite extension of K is always a finite A-module.*

Remark. A Noetherian integral domain A is Japanese if the integral closure of A in each finite quasi-Galois (= normal) extension of K is a finite A-module. This is clear, since an arbitrary finite extension of K can always be embedded into its quasi-Galois closure, which is also finite over K.

Since each algebraic extension of a perfect field is separable, we immediately deduce from the result of DEDEKIND (Theorem 4.2/1)

Proposition 2. *Each normal Noetherian integral domain A, whose field of fractions is perfect, is Japanese.*

The last proposition fails if we drop the assumption of perfectness. A counterexample is given by the discrete valuation ring of F. K. SCHMIDT (cf. (1.6.2)). In this example the condition of the following criterion is violated:

Proposition 3 (First criterion for Japaneseness). *A normal Noetherian integral domain A is Japanese if and only if the integral closure of A in each finite purely inseparable extension of K is a finite A-module.*

Proof. By Proposition 2, we only have to show that the condition of this criterion is sufficient for A to be Japanese when char $A =: p \neq 0$. Let L be an arbitrary finite quasi-Galois extension of K. Then there exists a field L_i between K and L such that L_i is purely inseparable over K and L is separable over L_i. By assumption, the integral closure A_i of A in L_i is a finite A-module. Since the integral closure A' of A in L equals the integral closure of A_i in L, it is enough to show that A' is a finite A_i-module. Because A_i is normal and Noetherian, this follows from DEDEKIND's Theorem 4.2/1. □

4.4. Tameness and Japaneseness

Let K_a be an algebraic closure of the field K. We denote by A_a the integral closure of A in K_a. The A-algebra A_a (which contains A) will be called a (universal) integral closure of A; this algebra A_a is uniquely determined up to (non-canonical) A-algebra isomorphisms.

If K' is any finite extension of K, we may always assume $K' \subset K_a$. Then $A' := A_a \cap K'$ is the integral closure of A in K'.

Proposition 1 (Second criterion for Japaneseness). *A Noetherian integral domain A is Japanese if and only if A_a is a tame A-module.*

Proof. (1) Let A be Japanese. Take any A-module $N \subset A_a$ of finite rank. Then $Q(N) \subset K_a$ is a finite K-vector space. Therefore the field $K' \subset K_a$ generated by $Q(N)$ is a finite extension of K. Thus $A_a \cap K'$ is a finite A-module by assumption. Hence $N \subset A_a \cap K'$ is a finite A-module, since A is Noetherian.

(2) Let A_a be a tame A-module. Take any finite extension K' of K. Since

$$\mathrm{rk}_A(A_a \cap K') = [K' : K] < \infty,$$

the integral closure $A_a \cap K'$ of A in K' is a finite A-module. □

The proposition just proved can be strengthened considerably if, in addition, A is assumed to be normal. Set $p :=$ char A. If $p = 0$, the ring A is always Japanese by Proposition 4.3/2. So assume $p \neq 0$. Then an important role is played by the subring

$$A^{p^{-1}} := \{x \in A_a;\, x^p \in A\}$$

of A_a, which we may view as an A-submodule of A_a.

Proposition 2 (Third criterion for Japaneseness). *A Noetherian normal integral domain A of characteristic $p \neq 0$ is Japanese if and only if $A^{p^{-1}}$ is a tame A-module.*

Proof. (1) If A is Japanese, A_a is a tame A-module by Proposition 1. Hence $A^{p^{-1}}$ is also tame, as a submodule of A_a.

(2) Let $A^{p^{-1}}$ be a tame A-module. As a first step, we prove by induction on n:

Each A-module $A_n := \{x \in A_a;\, x^{p^n} \in A\}$ is tame, $n = 1, 2, \ldots$.

Assume A_n is tame (true by assumption for $n = 1$). In order to see that A_{n+1} is a tame A-module, it is enough to prove (due to Proposition 4.1/3 with

$A' := A_n$ and $M' := A_{n+1}$) that $A_{n+1} \supset A_n$ is a tame A_n-module. But this is clear, since the Frobenius homomorphism $x \mapsto x^{p^n}$ maps the A_n-module A_{n+1} bijectively onto the A-module A_1.

Now it can be easily seen by using Proposition 4.3/3 that A is Japanese. Let $L \supset K$ be a finite purely inseparable extension of K. Choose $n \geqq 1$ such that $L \subset K_n := \{x \in K_a; x^{p^n} \in K\}$. The integral closure $L \cap A_a$ of A in L is an A-submodule of $K_n \cap A_a$ of finite rank (namely, $\mathrm{rk}_A(L \cap A_a) = \dim_K L$). If we can prove $K_n \cap A_a = A_n$, the tameness of A_n will imply the finiteness of the A-module $L \cap A_a$.

The inclusion $A_n \subset K_n \cap A_a$ is obvious. Now take $q \in K_n \cap A_a$. Then $q^{p^n} \in K \cap A_a$, and $K \cap A_a = A$, since A is normal. Thus we have $q \in A_n$ and hence $K_n \cap A_a \subset A_n$. □

We get directly from Proposition 2 and Corollary 4.1/5

Proposition 3. *A Noetherian normal integral domain A of characteristic $p \neq 0$ is Japanese if $A^{p^{-1}}$ is a linearly separable A-module.*

We apply this last proposition in order to obtain the (well-known) fact that polynomial rings and rings of formal power series are Japanese.

Proposition 4. *For each field k, the polynomial ring $k[X_1, \ldots, X_n]$, as well as the ring $k[\![X_1, \ldots, X_n]\!]$ of formal power series in n indeterminates, is Japanese.*

Proof. We discuss the case $A := k[\![X_1, \ldots, X_n]\!]$ and use the fact that A is Noetherian and factorial, and hence normal. Therefore, it is enough to show (by Proposition 3) that, for $p := \mathrm{char}\, k \neq 0$, the A-module $A^{p^{-1}}$ is linearly separable. We have

$$A^{p^{-1}} = k^{p^{-1}}[\![X_1^{1/p}, \ldots, X_n^{1/p}]\!],$$

and this last ring is, as an A-module, isomorphic to the direct sum of finitely many copies of the A-module

$$A' := k^{p^{-1}}[\![X_1, \ldots, X_n]\!]$$

(the monomials $X_1^{\mu_1/p} \ldots X_n^{\mu_n/p}$, $0 \leqq \mu_i < p$, form a basis). Hence it is sufficient to prove that A' is linearly separable. Take $h \in A'$, $h \neq 0$, say

$$h = \sum_0^\infty a_{\nu_1 \ldots \nu_n} X_1^{\nu_1} \ldots X_n^{\nu_n}, \qquad a_{\nu_1 \ldots \nu_n} \in k^{p^{-1}},$$

where $(\bar\nu_1, \ldots, \bar\nu_n)$ is an index tuple such that $a_{\bar\nu_1 \ldots \bar\nu_n} \neq 0$. Since $k^{p^{-1}}$ is linearly separable as a k-vector space, we can choose a k-linear map $\lambda' \colon k^{p^{-1}} \to k$ such that $\lambda'(a_{\bar\nu_1 \ldots \bar\nu_n}) \neq 0$. By

$$\lambda\left(\sum_0^\infty b_{\nu_1 \ldots \nu_n} X_1^{\nu_1} \ldots X_n^{\nu_n}\right) := \sum_0^\infty \lambda'(b_{\nu_1 \ldots \nu_n}) X_1^{\nu_1} \ldots X_n^{\nu_n} \in A$$

we extend λ' to an A-linear map $\lambda \colon A' \to A$. Since $\lambda(h) \neq 0$ by the choice of λ', we see that A' is linearly separable.

It is clear that the proof just given works for polynomial rings $k[X_1, \ldots, X_n]$ as well. □

PART B
Affinoid algebras

CHAPTER 5

Strictly convergent power series

In classical function theory of several complex variables, a function $f: U \to \mathbb{C}$, where U is open in $\mathbb{C}^n$, is called analytic if it has a convergent power series expansion around every point $z \in U$. The same definition makes sense for arbitrary complete fields k substituting for $\mathbb{C}$. Due to OSTROWSKI, one knows that a complete field k is either isomorphic to $\mathbb{R}$ or $\mathbb{C}$ or the valuation on k is non-Archimedean. Here we are only concerned with the second case. Since all non-Archimedean fields are totally disconnected, it is clear that functions, which are analytic on k in the sense mentioned above (or in any other *local* sense), cannot satisfy the classical Identity Theorem. For example, the function $f: k \to k$ defined by

$$f(z) := \begin{cases} 0 & \text{for} \quad |z| \leqq 1 \\ 1 & \text{for} \quad |z| > 1 \end{cases}$$

has a convergent power series representation (either $\equiv 0$ or $\equiv 1$) around each $z \in k$, since the sets $\{z; |z| \leqq 1\}$ and $\{z; |z| > 1\}$ are open in k. However f is not identically 0 or identically 1.

In order to exclude this phenomenon, we require an analytic function to have a globally convergent power series representation if it is defined on a polycylinder. In the following chapters, we will show how to extend this definition to more general domains. Before this can be done, we have to study in detail analytic functions on polycylinders and, in particular, on the unit ball in k^n (polycylinder of polyradius 1). It is easily seen that a power series converges on the ("closed") unit ball if and only if it is a strictly convergent power series (characterized by the fact that its coefficients form a zero sequence). Therefore this chapter is devoted to the investigation of strictly convergent power series. Such series have already been considered in (1.4) and (2.2.6). We start by repeating some properties of strictly convergent power series.

The heart of the present chapter is section (5.2). It contains a presentation of the WEIERSTRASS techniques which allow the application of RÜCKERT's fundamental method. Thereby we obtain important results on the TATE algebra T_n of strictly convergent power series in n variables. Also, the WEIERSTRASS Finiteness Theorem smoothes the way for the proof of the NOETHER Normalization Lemma in (6.1.2).

Section (5.3) is devoted to the Stability Theorem for the field of fractions

$Q(T_n)$. We give a proof which is a simplified version of GRUSON's approach in [17]. Together with the results of (5.2.7), we need the stability of $Q(T_n)$ as a key ingredient for the proof of the GRAUERT REMMERT GRUSON theorem on the finiteness of the functor $A \rightsquigarrow \mathring{A}$; see (6.4).

5.1. Definition and elementary properties of T_n and $\tilde{T}_n$

5.1.1. Description of T_n. — Let k be a commutative field with a complete non-trivial non-Archimedean valuation. For $n = 1, 2, \ldots$, define the following subalgebra of the k-algebra $k[\![X_1, \ldots, X_n]\!]$ of formal power series in n indeterminates over k (cf. (1.4.1)):

$$T_n(k) := k\langle X_1, \ldots, X_n\rangle := \Big\{ \sum_{\nu_1,\ldots,\nu_n \geqq 0} a_{\nu_1\ldots\nu_n} X_1^{\nu_1} \ldots X_n^{\nu_n};$$
$$a_{\nu_1\ldots\nu_n} \in k \text{ and } |a_{\nu_1\ldots\nu_n}| \to 0 \text{ for } \nu_1 + \cdots + \nu_n \to \infty\Big\}.$$

We call $T_n(k)$ the (free) *Tate algebra in n indeterminates over k.* The elements of $T_n(k)$ are called *strictly convergent power series.* It is easily checked that the inductive definition of strictly convergent power series in several variables, as given in (1.4.1), is equivalent to the one given here. In particular, we have $T_n(k) = T_{n-1}(k)\langle X_n\rangle$. If the ground field is clear from the context, we write T_n instead of $T_n(k)$; furthermore, we write $T_0(k) := k$. For simplicity we adopt the following notation:

$$X = (X_1, \ldots, X_n),\ \nu = (\nu_1, \ldots, \nu_n),\ X^\nu = X_1^{\nu_1} \ldots X_n^{\nu_n} \text{ and } |\nu| := \nu_1 + \cdots + \nu_n.$$

For $f = \sum_\nu a_\nu X^\nu \in T_n$, the real number

$$|f| := \max_\nu |a_\nu|$$

is well-defined. Similarly as in (1.4.1), we call $|\ \ |$ the Gauss norm on T_n. The results of (1.4.1) immediately give us the following

Proposition 1. *$T_n(k)$ is a k-subalgebra of the algebra of formal power series $k[\![X_1, \ldots, X_n]\!]$. The Gauss norm is a k-algebra norm on $T_n(k)$ making it into a k-Banach algebra containing the polynomial algebra $k[X_1, \ldots, X_n]$ as a dense k-subalgebra.*

Using the fact that $|T_n| = |k|$, we see that every non-zero series can be normed to length 1 by multiplication with a scalar from k:

Observation 2. *For every $f \in T_n - \{0\}$, there exists $c \in k$ such that $|cf| = 1$.*

For later reference we add two more remarks.

Remark 3. *T_n is a field if and only if $n = 0$.*

Remark 4. *As a k-vector space, each $T_n(k)$, $n \geqq 1$, is isometrically isomorphic to the space $c(k)$ of all zero sequences over k.*

The first remark follows simply from the fact that X_1 is not a unit. To prove the second one, just convert the multiple zero sequences $a_{\nu_1 \ldots \nu_n}$ into simple zero sequences by CANTOR's diagonal procedure. (Remark 4 is a special case of the general fact that every complete k-vector space of countable type and of infinite dimension admits a linear homeomorphism onto $c(k)$; cf. Proposition 2.7.1/2.)

5.1.2. The Gauss norm is a valuation and $\tilde{T}_n$ is a polynomial ring over $\tilde{k}$. — As in (1.2.4) and (1.2.5) we set

$$\check{T}_n := \{f \in T_n ; f \text{ topologically nilpotent}\},$$
$$\mathring{T}_n := \{f \in T_n ; f \text{ power-bounded}\}.$$

Then $\mathring{T}_n$ is a subring of T_n, and $\check{T}_n$ is a $\mathring{T}_n$-ideal. The residue ring $\mathring{T}_n/\check{T}_n$ is a $\tilde{k}$-algebra; it is denoted by $\tilde{T}_n$.

Proposition 1. *The Gauss norm is a valuation on T_n.*

Proposition 2. $\tilde{T}_n = \tilde{k}[X]$.

We give a combined *proof* for both assertions. Following (1.2.3), we use the Gauss norm in order to define the objects

$$T_n^\circ := \{f \in T_n ; |f| \leqq 1\},$$
$$T_n^\vee := \{f \in T_n ; |f| < 1\},$$
$$T_n^\sim := T_n^\circ / T_n^\vee .$$

We can extend the canonical epimorphism $^\sim : \mathring{k} \to \tilde{k}$ (where $\mathring{k}$ is the valuation ring of k and $\tilde{k}$ is the residue field of k) to a map $^\sim : T_n^\circ \to \tilde{k}[X]$ by setting

$$(\sum a_\nu X^\nu)^\sim := \sum \tilde{a}_\nu X^\nu \in \tilde{k}[X].$$

Obviously the kernel of this map is $T_n^\vee$, and the map is surjective. Therefore we get $T_n^\sim = \tilde{k}[X]$. In particular, the residue algebra $T_n^\sim$ is an integral domain. Since the Gauss norm of any non-zero element in T_n can be adjusted to 1 by scalar multiplication, it is easily verified that $|\ \ |$ is a valuation (see Proposition 1.5.3/1). But then we must have $\mathring{T}_n = T_n^\circ$, $\check{T}_n = T_n^\vee$, and hence $\tilde{T}_n = T_n^\sim = \tilde{k}[X]$. □

Alternatively, the above results can be deduced from Corollary 1.5.3/2 and Proposition 1.4.2/2; use induction on n.

5.1.3. Going up and down between T_n and $\tilde{T}_n$. — By reducing mod $\check{T}_n$, we move from power series to polynomials, thereby simplifying the problems at hand in many cases, as the following results will show.

Proposition 1. *A series $f \in T_n$ with $|f| = 1$ is a unit in T_n if and only if $|f(0)| = 1$ and $|f - f(0)| < 1$. Thus f is a unit if and only if $\tilde{f}$ is a unit (i.e., a constant) in $\tilde{T}_n$.*

Proof. Since $|\ \ |$ is a valuation, $f = \sum a_\nu X^\nu \in T_n$ with $|f| = 1$ is a unit in T_n if and only if it is a unit in $\mathring{T}_n$. Due to Proposition 1.4.2/3 (use induction on n and the fact that T_n is complete), this is equivalent to $a_{(0,\ldots,0)}$ being a unit in $\mathring{k}$ and a_ν belonging to $\check{k}$ for all ν, $|\nu| > 0$. This is the assertion. □

Using the above characterization of units, we get the following technical lemma:

Lemma 2. *For each $f \in T_n$ with $|f| = 1$, there is an element $c \in k$ with $|c| = 1$ such that $c + f$ is not a unit in T_n.*

Proof. We shall treat the two cases $|f(0)| < 1$ and $|f(0)| = 1$ separately. If $|f(0)| < 1$, then $|f| = 1$ implies $|f - f(0)| = 1$. For $g := 1 + f \in T_n$, we have $|g| = 1$ and $|g - g(0)| = |f - f(0)| = 1$. According to the preceding proposition, g is not a unit in T_n. If $|f(0)| = 1$, define $g := f - f(0)$. Then $g(0) = 0$, and hence g cannot be a unit in T_n. □

This lemma has two important consequences.

Proposition 3. $\bigcap_{\mathfrak{m} \in \operatorname{Max} T_n} \mathfrak{m} = (0)$, *where* $\operatorname{Max} T_n$ *denotes the set of all maximal ideals of T_n.*

Proof. Assume that there is a non-zero series f contained in all maximal ideals of T_n. We may assume that $|f| = 1$. Choose $c \in k$ with $|c| = 1$ such that $c + f$ is a non-unit. Then we can find a maximal ideal $\mathfrak{m}$ such that $c + f \in \mathfrak{m}$. By assumption, f also is an element of $\mathfrak{m}$. This implies $c \in \mathfrak{m}$, which is impossible since $c \in k^*$. □

Theorem 4. *Every k-algebra homomorphism $\phi: T_n \to T_m$ is a contraction, i.e., $|\phi(f)| \leqq |f|$ for all $f \in T_n$.*

Proof. Again we proceed indirectly. Assume that there is an $f \in T_n$ such that $|\phi(f)| > |f|$. Without loss of generality, we may assume that $|\phi(f)| = 1$. Choose $c \in k$ with $|c| = 1$ such that $c + \phi(f)$ is not a unit in T_m (Lemma 2). On the other hand, $g := c + f$ is a unit in T_n according to Proposition 1, since $|g| = 1$ and $|g - g(0)| = |f - f(0)| < 1$ (because $|f| < 1$). Hence $\phi(g) = c + \phi(f)$ must be a unit in T_m, which is a contradiction. □

The proof just given is similar to the proof that a k-algebra homomorphism between analytical local algebras is local.

We draw some conclusions from the above theorem.

Corollary 5. *Every k-algebra homomorphism $\phi: T_n \to T_m$ is a continuous substitution homomorphism; i.e., for any such map ϕ, there are $f_1, \ldots, f_n \in \mathring{T}_m$ such that*

$$\phi(\sum a_\nu X^\nu) = \sum a_{\nu_1 \ldots \nu_n} f_1^{\nu_1} \ldots f_n^{\nu_n}$$

for all series $\sum a_\nu X^\nu \in T_n$. More precisely, the map $\phi \mapsto (\phi(X_1), \ldots, \phi(X_n))$ defines a bijection between $\operatorname{Hom}(T_n, T_m)$ *and* $(\mathring{T}_m)^n$.

Proof. Let $\phi: T_n \to T_m$ be a k-algebra homomorphism. Define $f_i := \phi(X_i)$ for $i = 1, \dots, n$. According to the theorem, we have $|f_i| \leq |X_i| = 1$, whence $f_i \in \mathring{T}_m$. For $\sum a_\nu X^\nu \in k[X]$, one clearly has $\phi(\sum a_\nu X^\nu) = \sum a_\nu f_1^{\nu_1} \dots f_n^{\nu_n}$. Due to the theorem, ϕ is continuous, and therefore $\phi(\sum a_\nu X^\nu) = \sum a_\nu f_1^{\nu_1} \dots f_n^{\nu_n}$ for all $\sum a_\nu X^\nu \in T_n$. Thus ϕ is uniquely determined by the tuple $(\phi(X_1), \dots, \phi(X_n)) \in (\mathring{T}_m)^n$. Now let $f_1, \dots, f_n \in \mathring{T}_m$ be given. Then it is easy to verify that $\phi: T_n \to T_m$ defined by $\phi(\sum a_\nu X^\nu) = \sum a_\nu f_1^{\nu_1} \dots f_n^{\nu_n}$ is a k-algebra homomorphism with $\phi(X_i) = f_i$. Therefore it is clear that the map $\operatorname{Hom}(T_n, T_m) \to (\mathring{T}_m)^n$ given by $\phi \mapsto (\phi(X_1), \dots, \phi(X_n))$ is bijective. □

Corollary 6. *Every k-algebra isomorphism $\phi: T_n \to T_m$ is an isometry.*

This result follows immediately from Theorem 4. It can be improved as follows:

Corollary 7. *If $\phi: T_n \to T_m$ is a k-algebra isomorphism, then $n = m$ and ϕ is an isometric automorphism of T_n.*

Proof. The k-algebra homomorphisms ϕ and ϕ^{-1} induce $\tilde{k}$-algebra homomorphisms $\tilde{\phi}: \tilde{k}[X_1, \dots, X_n] \to \tilde{k}[X_1, \dots, X_m]$ and $\widetilde{\phi^{-1}}: \tilde{k}[X_1, \dots, X_m] \to \tilde{k}[X_1, \dots, X_n]$. Obviously, $\tilde{\phi}$ and $\widetilde{\phi^{-1}}$ are inverse to each other. Hence $\tilde{\phi}$ is a $\tilde{k}$-algebra isomorphism. Then $\tilde{\phi}$ extends to an isomorphism $\tilde{k}(X_1, \dots, X_n) \to \tilde{k}(X_1, \dots, X_m)$ between the fields of fractions, and by looking at transcendence degrees over $\tilde{k}$, we get $n = m$. □

The proof of Corollary 7 depends on the fact that the bijectivity of ϕ implies the bijectivity of $\tilde{\phi}$. The converse of this fact is also true.

Corollary 8. *A k-algebra endomorphism ϕ of T_n is bijective if and only if $\tilde{\phi}$ is bijective.*

Proof. We only have to show that "$\tilde{\phi}$ bijective" implies "ϕ bijective". It is easily seen that ϕ is an isometry if $\tilde{\phi}$ is injective. It remains to show that ϕ is surjective. Therefore assume that $\tilde{\phi}$ is bijective. There are elements $f_1, \dots, f_n \in \mathring{T}_n$ such that $\varepsilon := \max\limits_{1 \leq i \leq n} |X_i - \phi(f_i)| < 1$. An easy computation shows that, for any $g = \sum a_\nu X^\nu \in T_n$, one can find an $h \in T_n$ such that $|g - \phi(h)| \leq \varepsilon\, |g|$. Namely, using the standard estimate

$$|u_1 \dots u_r - v_1 \dots v_r| = \left| \sum_{i=1}^{r} u_1 \dots u_i v_{i+1} \dots v_r - u_1 \dots u_{i-1} v_i \dots v_r \right|$$
$$\leq (\max_{1 \leq i \leq r} |u_i - v_i|)\, (\max_{1 \leq i \leq r} |u_i|, |v_i|)^{r-1},$$

we see that the series $h := \sum a_\nu f_1^{\nu_1} \dots f_n^{\nu_n}$ is as desired. Thus $\phi(T_n)$ is "ε-dense" in T_n, and Proposition 1.1.4/2 shows that $\phi(T_n)$ is, in fact, dense in T_n. But ϕ is an isometry, and hence $\phi(T_n)$ is closed in T_n. So we must have $\phi(T_n) = T_n$. □

Alternatively, the above result can be obtained by applying the Lifting Theorem 2.7.3/2. — We conclude this section by considering a special class of automorphisms of T_n which will play an important role in the applications of the WEIERSTRASS Preparation Theorem.

Example. *Let* $c_1, \ldots, c_{n-1} \in \mathbb{N}$ *be given. Define* $\phi\colon T_n \to T_n$ *by* $\phi(X_\nu) := X_\nu + X_n^{c_\nu}$ *for* $\nu = 1, \ldots, n-1$ *and* $\phi(X_n) := X_n$. *Then* ϕ *is an isometric automorphism of* T_n.

Proof. Let $\psi\colon T_n \to T_n$ be defined by $\psi(X_\nu) := X_\nu - X_n^{c_\nu}$ for $\nu = 1, \ldots, n-1$ and $\psi(X_n) := X_n$. One easily checks that ψ is an inverse of ϕ. Applying Corollary 6, we see that ϕ is an isometric automorphism of T_n. □

5.1.4. T_n as a function algebra. — As before, we consider strictly convergent power series with coefficients in the complete valued field k. Let k_a be the algebraic closure of k provided with the spectral valuation (which is the unique valuation extending the valuation from k; see Theorem 3.2.4/2). For any valued field K, we denote by

$$B^n(K) := \left\{(x_1, \ldots, x_n) \in K^n;\ \max_{1 \leqq \nu \leqq n} |x_\nu| \leqq 1\right\}$$

the n-dimensional unit ball (or polydisc) around the origin.

We want to show that each power series $f = \sum a_\nu X^\nu \in T_n$ defines a map

$$B^n(k_a) \to k_a\,,$$
$$x \mapsto f(x) := \sum a_{\nu_1 \ldots \nu_n} x_1^{\nu_1} \ldots x_n^{\nu_n}\,,$$

which also shall be denoted by f. Namely, consider a point $x \in B^n(k_a)$, and let $L \subset k_a$ be a finite field extension of k containing all coordinates $x_1, \ldots, x_n$ of x. Then L is complete by Theorem 3.2.4/2. Since $a_\nu x^\nu$ is a zero-sequence in L, the series $\sum a_\nu x^\nu$ must converge to some element in L. Thus we see that, for all $x \in B^n(k_a)$, the element $f(x)$ is well-defined in k_a. In particular, $f(x) \in k$ for all $x \in B^n(k)$.

Conversely, every k_a-valued function $f\colon B^n(k_a) \to k_a$ admitting a power series expansion (with coefficients in k_a) converging for all $x \in B^n(k_a)$ and satisfying the additional requirement $f(B^n(k)) \subset k$ comes from a strictly convergent power series in the manner described above. Namely, if one starts with a power series $f = \sum c_\nu X^\nu \in k_a[\![X_1, \ldots, X_n]\!]$, the requirement that it must converge for $x = (1, \ldots, 1)$ immediately yields $|c_\nu| \to 0$ for $\nu_1 + \cdots + \nu_n \to \infty$. It remains to show that the second condition "$f(B^n(k)) \subset k$" implies $c_\nu \in k$ for all ν. Unfortunately, since k may have characteristic $p \neq 0$, the argument that $c_\nu = \dfrac{1}{\nu!}\dfrac{\partial^\nu f}{\partial x^\nu}(0) \in k$ is not conclusive. However, we can get the desired implication by a direct computation. First write $f = f_1 + f_2$, where all non-zero coefficients of f_1 do not lie in k, whereas those of f_2 do. It is clear that $f_1(B^n(k)) \subset k$. Therefore, we may assume that no non-zero coefficient of f lies in k, and we have to show $f = 0$. Assuming the contrary, we can write $f = X_n^s \sum_{\nu=s}^{\infty} g_\nu(X_1, \ldots, X_{n-1})\, X_n^{\nu-s}$

for some $s \in \mathbb{N} \cup \{0\}$, where $g_\nu \in T_{n-1}(k_{\mathrm{a}})$ and $g_s \neq 0$. Obviously no non-zero coefficient of g_s is in k. Since $f(B^n(k)) \subset k$, we derive for all $x_1, \ldots, x_{n-1} \in \mathring{k}$ that $g_s(x_1, \ldots, x_{n-1}) = \lim\limits_{x_n \to 0,\, x_n \in \mathring{k} - \{0\}} f(x_1, \ldots, x_n)\, x_n^{-s} \in k$. Therefore, g_s is a series in $n - 1$ indeterminates having the same properties as f. Applying this reduction n times, we end up with a non-zero constant in k which, on the other hand, is a coefficient of f and therefore cannot lie in k. This contradiction gives the desired result: a series $f \in T_n(k_{\mathrm{a}})$ which maps $B^n(k)$ into k is already an element of $T_n(k)$.

We summarize the preceding considerations in the following

Proposition 1. *The series of $T_n(k)$ give rise to exactly those functions $f\colon B^n(k_{\mathrm{a}}) \to k_{\mathrm{a}}$ which*

(i) *have a power series expansion over k_{a} converging on the whole unit ball $B^n(k_{\mathrm{a}})$ and*

(ii) *map $B^n(k)$ into k.*

If $L \subset k_{\mathrm{a}}$ is any finite algebraic extension of k, then $f(B^n(L)) \subset L$ for all $f \in T_n(k)$.

Later on (cf. Corollary 5), we shall see that two power series of $T_n(k)$ induce the same function $B^n(k_{\mathrm{a}}) \to k_{\mathrm{a}}$ if and only if they coincide. To prove this fact, we have to look at the norm of uniform convergence on $B^n(k_{\mathrm{a}})$.

Proposition 2. *Let f be a series in T_n. Then* $\sup\limits_{x \in B^n(k_{\mathrm{a}})} |f(x)| \leq |f|$, *and f gives rise to a continuous function on $B^n(k_{\mathrm{a}})$.*

Proof. For all $x \in B^n(k_{\mathrm{a}})$ and all ν, we have $|a_\nu x^\nu| \leq |a_\nu| \leq |f|$ if $f = \sum a_\nu X^\nu$. Therefore, $|f(x)| \leq \max |a_\nu x^\nu| \leq |f|$, whence the first assertion follows. Furthermore, f is a uniform limit of polynomials and hence continuous. □

Let $\tilde{}\colon B^n(k_{\mathrm{a}}) = \mathring{k}_{\mathrm{a}}^n \to \tilde{k}_{\mathrm{a}}^n$ denote the obvious extension of the residue map $\tilde{}\colon \mathring{k} \to \tilde{k}$. It is easy to see that the following diagram

$$\begin{array}{ccc} \mathring{k}_{\mathrm{a}}^n & \xrightarrow{\;f\;} & \mathring{k}_{\mathrm{a}} \\ {\scriptstyle\sim}\downarrow & & \downarrow{\scriptstyle\sim} \\ \tilde{k}_{\mathrm{a}}^n & \xrightarrow{\;\tilde{f}\;} & \tilde{k}_{\mathrm{a}} \end{array}$$

is commutative for all $f \in \mathring{T}_n$. (In the diagram, $\tilde{f}$ stands for the map induced by the polynomial $\tilde{f} \in \tilde{k}_{\mathrm{a}}[X_1, \ldots, X_n]$.) We shall use this connection between the functions in T_n and the polynomials over $\tilde{k}$ associated to them to show that the inequality in Proposition 2 is actually an equality.

Proposition 3 (Maximum Modulus Principle). *For all $f \in T_n$, there is an $x \in B^n(k_{\mathrm{a}})$ such that $|f(x)| = |f|$. If $|f(x)| = |f|$ for some $x \in \check{k}_{\mathrm{a}}^n$, then $|f(x)| = |f|$ for all $x \in \check{k}_{\mathrm{a}}^n$. These assertions remain valid if k_{a} is replaced by any field extension $L \subset k_{\mathrm{a}}$ of k, provided $\tilde{L}$ is infinite.*

Proof. We may assume $|f| = 1$. Since the residue field $\tilde{k}_a$ of k_a equals the algebraic closure of $\tilde{k}$ (see Lemma 3.4.1/4), it has infinitely many elements. Then there must be a point $x = (x_1, \ldots, x_n) \in B^n(k_a)$ such that $\tilde{f}(\tilde{x}_1, \ldots, \tilde{x}_n) \neq 0$, i.e., $\widetilde{f(x)} \neq 0$, which is equivalent to $|f(x)| = 1$. So the first assertion is proved.

From $|f(x)| = 1$ for some $x \in \check{k}_a^n$, we can conclude $\tilde{f}(0, \ldots, 0) \neq 0$, which again is equivalent to $\widetilde{f(x)} \neq 0$ or $|f(x)| = 1$ for all $x \in \check{k}_a^n$. The only property of k_a (besides being valued) needed for the proof was the fact that $\tilde{k}_a$ had to be infinite. Hence the last remark of the proposition is also justified. □

The preceding proposition can be strengthened in the following way:

Proposition 4. *The maximum of the values taken by a strictly convergent power series f is assumed on the subset $\{(x_1, \ldots, x_n) \in B^n(k_a);\ |x_1| = \cdots = |x_n| = 1\}$ of the unit ball $B^n(k_a)$.*

Proof. Use the fact that the Gauss norm is a valuation on T_n (Proposition 5.1.2/1), and apply Proposition 3 to the series $X_1 \ldots X_n f$. □

Corollary 5 (Identity Theorem). *If $f \in T_n$ vanishes for all $x \in B^n(k_a)$, then $f = 0$. Therefore the map associating to a series $f \in T_n$ its corresponding function from $B^n(k_a)$ to k_a is an injection.*

Proof. If f induces the zero function, then $|f| = \sup_{x \in B^n(k_a)} |f(x)| = 0$ and hence $f = 0$. □

This is a rather weak version of the Identity Theorem. Using elementary methods, one can show the following much better statement: The zero set of a strictly convergent series $f \neq 0$ is nowhere dense in $B^n(k_a)$. In the one variable case, the WEIERSTRASS Preparation Theorem will tell us that a non-zero series has only a finite number of zeros.

Corollary 6. *T_n is a Banach function algebra satisfying the Maximum Modulus Principle. The Gauss norm $|\ |$ and the supremum norm $|\ |_{\sup}$ coincide on T_n.*

Proof. For the definition of $|\ |_{\sup}$ and of Banach function algebras see (3.8). Since $|\ |_{\sup} \leqq |\ |$ by Corollary 3.8.2/2, we have only to show that, for each $f \in T_n$, there exists a k-algebraic maximal ideal $\mathfrak{m} \subset T_n$ such that $|f(\mathfrak{m})| = |f|$. In order to do this, consider a point $x = (x_1, \ldots, x_n) \in B^n(k_a)$ such that $|f(x)| = |f|$ (Proposition 3). Denote by $L := k(x_1, \ldots, x_n)$ the extension of k generated by the components of x. Then L is finite over k, and due to the last assertion of Proposition 1, there is an evaluation homomorphism

$$h_x \colon T_n \to L, \qquad g \mapsto h_x(g) := g(x).$$

Since the image of h_x contains k and the elements $x_1, \ldots, x_n$, it follows that h_x is surjective. Thus $\mathfrak{m}_x := \ker h_x$ is a k-algebraic maximal ideal in T_n, and $T_n/\mathfrak{m}_x$ is isomorphic to L over k. Corresponding elements in $T_n/\mathfrak{m}_x$ and L must have the same spectral norm over k so that $|g(\mathfrak{m}_x)| = |g(x)|$ for all $g \in T_n$. In particular, we have $|f(\mathfrak{m}_x)| = |f(x)| = |f|$. □

The proof relies on the fact that the sets $B^n(k_a)$, $\text{Hom}_k\,(T_n, k_a)$, and $\text{Max}_k\, T_n$ are essentially the same. This point of view shall be elaborated on in more detail in (7.1.1).

We now give a geometric interpretation of algebra homomorphisms $T_n \to T_m$. Let $\phi: T_n \to T_m$ be such a homomorphism, and consider the elements $f_i := \phi(X_i)$ for $i = 1, \ldots, n$. By Proposition 2 and Theorem 5.1.3/4, we have $|f_i(x)| \leqq |f_i| \leqq |X_i| = 1$ for all $x \in B^m(k_a)$, $i = 1, \ldots, n$. Therefore by defining $\phi'(x_1, \ldots, x_m) := \big(f_1(x_1, \ldots, x_m), \ldots, f_n(x_1, \ldots, x_m)\big)$, one gets a map ϕ': $B^m(k_a) \to B^n(k_a)$. If we call a mapping ψ: $B^m(k_a) \to B^n(k_a)$ affinoid whenever its coordinate mappings ψ_i: $B^m(k_a) \to k_a$ are given by elements of T_m, then $\phi \mapsto \phi'$ is a contravariant functor from the category $\{T_n; n \in \mathbb{N}\}$ with k-algebra homomorphisms as morphisms into the category $\{B^n(k_a); n \in \mathbb{N}\}$ with affinoid mappings as morphisms.

Using the Identity Theorem, one easily deduces the following

Proposition 7. *Let ϕ be a k-algebra endomorphism of T_n. Then ϕ is bijective if and only if the corresponding map ϕ': $B^n(k_a) \to B^n(k_a)$ is bi-affinoid (i.e., ϕ' is bijective, and ϕ' as well as ϕ'^{-1} are affinoid).*

When considering automorphisms of T_n, it is convenient to look at special topological generating systems of T_n.

Definition 8. *A system $\{f_1, \ldots, f_n\} \subset T_n$ is called an affinoid chart of T_n if there is a k-algebra automorphism ϕ of T_n with $\phi(X_i) = f_i$ for $i = 1, \ldots, n$, i.e., if every $f \in T_n$ can be written uniquely as $f = \sum a_{\nu_1 \ldots \nu_n} f_1^{\nu_1} \ldots f_n^{\nu_n}$ with $a_\nu \in k$ and $|a_\nu| \to 0$.*

Remark. It can be shown that $\{f_1, \ldots, f_n\} \subset \mathring{T}_n$ is already a chart if the map defined by $X_i \mapsto f_i$, $i = 1, \ldots, n$, is surjective. Loosely speaking, we could rephrase this in the following way: if a generating system has minimal length, the representation of any series by it is uniquely determined.

Using Corollary 5.1.3/8, we find the following characterization of charts:

Proposition 9. *The system $\{f_1, \ldots, f_n\} \subset \mathring{T}_n$ is an affinoid chart of T_n if and only if $\{\tilde{f}_1, \ldots, \tilde{f}_n\}$ generates $\tilde{T}_n$ as a $\tilde{k}$-algebra.*

Proof. Define ϕ: $T_n \to T_n$ by $\phi(X_i) = f_i$, $i = 1, \ldots, n$. The map ϕ is an automorphism if and only if $\tilde{\phi}$ is an automorphism of $\tilde{T}_n$. The latter is equivalent to $\tilde{\phi}$ being surjective. Namely if $\tilde{\phi}$ is surjective, consider the isomorphism $\tilde{T}_n/\ker \tilde{\phi} \to \tilde{T}_n$ and extend it to an isomorphism $Q(\tilde{T}_n/\ker \tilde{\phi}) \to Q(\tilde{T}_n)$ between the fields of fractions. By looking at transcendence degrees over $\tilde{k}$, we see that $Q(\tilde{T}_n/\ker \tilde{\phi})$ must have transcendence degree n. However this can only be true if $\ker \tilde{\phi} = 0$. □

Specializing to the case of one variable, we get the following description of the "group of automorphisms of the unit disc".

Corollary 10. *The series* $f = \sum_{\nu=0}^{\infty} a_\nu X \in T_1$ *defines a bi-affinoid map of the unit disc onto itself if and only if* $|a_0| \leqq 1$, $|a_1| = 1$ *and* $|a_\nu| < 1$ *for all* $\nu > 1$.

Proof. The system $\{f\}$ is a chart of T_1 if and only if $|f| \leqq 1$ and $\tilde{f} = \sum \tilde{a}_\nu X^\nu$ generates $\tilde{k}[X]$. This is equivalent to $|a_\nu| \leqq 1$ for all ν, $\tilde{a}_1 \neq 0$ and $\tilde{a}_\nu = 0$ for $\nu > 1$. □

A remark one should add is that, contrary to the classical complex case, the automorphism group of the unit disc has infinitely many parameters.

5.2. Weierstrass-Rückert theory for T_n

There are basically two ways to get further information on T_n and on finite T_n-modules. One can prove the Weierstrass Preparation Theorem and then follow rather closely the classical method of Rückert, or one can use the Lifting Theorem 2.7.3/2 and derive the desired results from well-known facts about the polynomial algebra $\tilde{T}_n$. Here we shall follow the first approach; for the second one, refer to [2].

5.2.1. Weierstrass Division Theorem. — Let us start with

Definition 1. *A strictly convergent power series* $g = \sum_{\nu=0}^{\infty} g_\nu(X_1, \ldots, X_{n-1}) X_n^\nu$ *is* X_n*-distinguished of degree* s *if*

(1) g_s *is a unit in* T_{n-1} *and*

(2) $|g_s| = |g|$ *and* $|g_s| > |g_\nu|$ *for all* $\nu > s$.

It is easy to see that a power series $g \in T_n$ with $|g| = 1$ is X_n-distinguished of degree s if and only if $\tilde{g} \in \tilde{T}_n$ is a unitary polynomial of degree s in the polynomial ring $\tilde{k}[X_1, \ldots, X_{n-1}][X_n]$. (Recall that a polynomial is called unitary if its highest coefficient is a unit, and use Proposition 5.1.3/1.) This remark already gives an idea of how to proceed if one wishes to carry out a division by a distinguished element g with $|g| = 1$. Namely, just use Euclid's division in $\tilde{k}[X_1, \ldots, X_{n-1}][X_n]$ and then pull back the results to T_n. To describe this procedure precisely, let us state the so-called Weierstrass Division Theorem.

Theorem 2. *Let* $g \in T_n$ *be* X_n*-distinguished of degree* s. *Then for each* $f \in T_n$, *there exist uniquely determined elements* $q \in T_n$ *and* $r \in T_{n-1}[X_n]$ *with* $\deg r < s$ *such that*

$$f = qg + r.$$

One has the following estimates

$$|f| = \max\{|q|\,|g|, |r|\}; \quad i.e.,$$

$$|q| \leqq |g|^{-1}\,|f| \quad \text{and} \quad |r| \leqq |f|.$$

If, in addition, f *and* g *are polynomials in* $T_{n-1}[X_n]$ *and if* g *has degree* s, *then also* q *is a polynomial in* $T_{n-1}[X_n]$.

Proof. Without loss of generality, we may assume $|g| = 1$. First we show that the existence of a representation

$$(*) \qquad f = qg + r, \quad r \in T_{n-1}[X_n], \quad \deg r < s, \quad q \in T_n$$

implies the estimates $|q| \leqq |f|$ and $|r| \leqq |f|$. This can be seen as follows. By multiplying $(*)$ with a scalar from k, we may assume

$$(**) \qquad \max\{|q|, |r|\} = 1,$$

which implies $|f| \leqq 1$. We have to show $|f| = 1$. Assume the contrary. Then we would have $0 = \tilde{f} = \tilde{q}\tilde{g} + \tilde{r}$. Since $\deg \tilde{g} = s > \deg r \geqq \deg \tilde{r}$, this would imply $\tilde{q} = \tilde{r} = 0$, in contradiction to $(**)$. So we have verified the estimates. Now it is trivial to show uniqueness. Namely if one has a representation $0 = qg + r$, $r \in T_{n-1}[X_n]$, $\deg r < s$, $q \in T_n$, the estimates just verified yield $q = r = 0$.

Next we want to show the existence of the representation $(*)$. Define $B := \{qg + r; r \in T_{n-1}[X_n], \deg r < s, q \in T_n\}$. It follows from what we have shown above that B is a closed subgroup of T_n. We claim $B = T_n$. Writing $g = \sum_{\nu=0}^{\infty} g_\nu(X_1, \ldots, X_{n-1}) X_n^\nu$, we define $\varepsilon := \max_{\nu > s} \{|g_\nu|\}$, where $\varepsilon < 1$. Furthermore, set $k_\varepsilon := \{x \in k; |x| \leqq \varepsilon\}$ and $\tilde{k}_\varepsilon := \mathring{k}/k_\varepsilon$. Then there is a natural ring epimorphism $\tau_\varepsilon\colon \mathring{T}_n \to \tilde{k}_\varepsilon[X_1, \ldots, X_n]$ with $\ker \tau_\varepsilon = \{f \in T_n; |f| \leqq \varepsilon\}$, and $\tau_\varepsilon(g)$ is a unitary polynomial in X_n of degree s. Therefore, Euclid's division with respect to $\tau_\varepsilon(g)$ is possible in the ring $(\tilde{k}_\varepsilon[X_1, \ldots, X_{n-1}])[X_n]$. So for all $f \in \mathring{T}_n$, we can find $q \in \mathring{T}_n$ and $r \in \mathring{T}_{n-1}[X_n]$ with $\deg r < s$ such that $\tau_\varepsilon(f) = \tau_\varepsilon(q)\,\tau_\varepsilon(g) + \tau_\varepsilon(r)$, or equivalently $|f - (qg + r)| \leqq \varepsilon$. Hence, for all $f \in T_n$, there is an element $b \in B$ such that $|f - b| \leqq \varepsilon\,|f|$. Therefore B is ε-dense in T_n, and Proposition 1.1.4/2 says that B, in fact, is dense in T_n. Since B is closed in T_n, we get $B = T_n$. Hence every $f \in T_n$ admits a representation $(*)$.

Only the last statement of the theorem remains to be shown. If $g \in T_{n-1}[X_n]$ and $\deg g = s$, then g is a unitary polynomial, and Euclid's division with respect to g can be applied in $T_{n-1}[X_n]$. For every $f \in T_{n-1}[X_n]$, one can find polynomials $q, r \in T_{n-1}[X_n]$ with $\deg r < s$ such that $f = qg + r$. Due to the uniqueness of q and r, the last assertion is clear. □

5.2.2. Weierstrass Preparation Theorem. — As an easy application we deduce the Preparation Theorem.

Theorem 1. *Let $g \in T_n$ be X_n-distinguished of degree s. Then there are a unique monic polynomial $\omega \in T_{n-1}[X_n]$ of degree s and a unique unit $e \in T_n$ such that $g = e \cdot \omega$. One has: $|\omega| = 1$ so that ω is X_n-distinguished of degree s. If $g \in T_{n-1}[X_n]$, then also $e \in T_{n-1}[X_n]$.*

Proof. By the Weierstrass Division Theorem, there exist $e' \in T_n$ and $r' \in T_{n-1}[X_n]$ with $\deg r' < s$ such that $X_n^s = e'g + r'$. Define $\omega := X_n^s - r'$. Then ω is a monic polynomial in $T_{n-1}[X_n]$ of degree s and $\omega = e'g$. To complete the existence part of the theorem, we only have to show that e' is a unit. Since

$|r'| \leqq |X_n^s| = 1$, we see that $|\omega| = 1$ and that ω is X_n-distinguished of degree s. We may assume $|g| = 1$. From $\tilde{\omega} = \tilde{e}'\tilde{g}$, we conclude that $\tilde{e}'$ is a unit in $\tilde{T}_{n-1}$, because $\tilde{\omega}$ and $\tilde{g}$ are unitary polynomials of the same degree. Then $\tilde{e}'$ is a fortiori a unit in $\tilde{T}_n$, and therefore e' is a unit in T_n (see Proposition 5.1.3/1). This proves the existence part of the theorem. Now let $\omega \in T_{n-1}[X_n]$ be a monic polynomial of degree s and e be a unit in T_n such that $g = e \cdot \omega$. Define $r := X_n^s - \omega$. Then one has $X_n^s = e^{-1}g + r$. The series g being given, this relation uniquely determines e and r, and therefore also ω. If $g \in T_{n-1}[X_n]$, then according to the last assertion of the Division Theorem, also e must be a polynomial in $T_{n-1}[X_n]$. □

5.2.3. Weierstrass polynomials and Weierstrass Finiteness Theorem. — The polynomials $\omega \in T_{n-1}[X_n]$ appearing in the preceding theorem will play an important role later on. Therefore we introduce a special name for them.

Definition 1. *A Weierstrass polynomial (in X_n) is a monic polynomial $\omega \in T_{n-1}[X_n]$ with $|\omega| = 1$.*

For later reference we mention the following simple fact:

Lemma 2. *Let ω_1 and ω_2 be monic polynomials in $T_{n-1}[X_n]$. If $\omega_1 \cdot \omega_2$ is a Weierstrass polynomial, then ω_1 and ω_2 are Weierstrass polynomials.*

Proof. Since ω_1 and ω_2 are monic, we have $|\omega_i| \geqq 1$ for $i = 1, 2$. On the other hand, $|\omega_1|\,|\omega_2| = |\omega_1 \cdot \omega_2| = 1$, and therefore we get $|\omega_i| = 1$. □

The importance of the concept of Weierstrass polynomials is shown by the fact that, for every X_n-distinguished power series g, there is a Weierstrass polynomial ω with $\omega T_n = gT_n$. Moreover, we have the following

Proposition 3. *Let ω be a Weierstrass polynomial of degree s in X_n. Then*

(i) *$T_n/\omega T_n$ is a finite free T_{n-1}-module;*

(ii) *$T_{n-1}[X_n]/\omega T_{n-1}[X_n] \cong T_n/\omega T_n$.*

More explicitly, the sequence

$$T_{n-1}^s \xrightarrow{j} T_{n-1}[X_n] \xrightarrow{i} T_n,$$

where T_{n-1}^s is the s-fold normed direct sum of copies of T_{n-1}, where j is given by $j(t_0, \ldots, t_{s-1}) := \sum_{\nu=0}^{s-1} t_\nu X_n^\nu$, and where i is the natural injection, induces a sequence of isometric T_{n-1}-module isomorphisms

$$T_{n-1}^s \xrightarrow{\bar{j}} T_{n-1}[X_n]/\omega T_{n-1}[X_n] \xrightarrow{\bar{i}} T_n/\omega T_n.$$

The map $\bar{i}$ is the k-algebra isomorphism mentioned in (ii).

Proof. We consider the following commutative diagram of T_{n-1}-module homomorphisms:

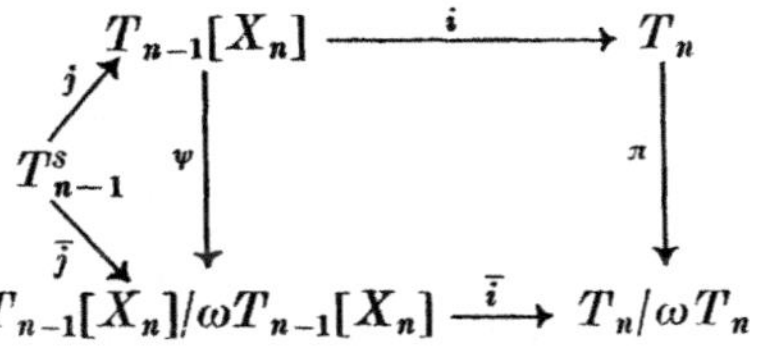

where ψ and π are the canonical residue epimorphisms and $\bar{i}$ and $\bar{j}$ are induced by i and j, respectively. The existence statement of the WEIERSTRASS Division Theorem tells us that $\pi \circ i \circ j$ and $\psi \circ j$ are surjective. Hence $\bar{i}$ and $\bar{j}$ must be surjective. Furthermore, the uniqueness part of the Division Theorem shows that $\pi \circ i \circ j$ is injective, whence the injectivity of $\bar{j}$ and $\bar{i}$ follows. Thus, $\bar{i}$ and $\bar{j}$ are bijections. Obviously, $\bar{i}$ is not only a T_{n-1}-module isomorphism, but also a k-algebra isomorphism, because i, ψ and π are k-algebra homomorphisms. It remains to be shown that $\bar{i}$ and $\bar{j}$ are isometries if one provides $T_{n-1}[X_n]/\omega T_{n-1}[X_n]$ and $T_n/\omega T_n$ with the residue norm derived from the Gauss norm on T_n. Since $\omega T_{n-1}[X_n]$ is dense in ωT_n, we see that $\bar{i}$ is an isometry. Furthermore, the map $\bar{j}$ is contractive. If $\bar{j}$ is not an isometry, there must exist a tuple $(t_0, \ldots, t_{s-1}) \in T_{n-1}^s$ and a polynomial $q \in T_{n-1}[X_n]$ such that $f := q\omega + \sum_{\nu=0}^{s-1} t_\nu X_n^\nu$ satisfies

$$|f| < \max_{0 \leq \nu \leq s-1} |t_\nu| = \left| \sum_{\nu=0}^{s-1} t_\nu X_n^\nu \right|.$$

However this is impossible by the WEIERSTRASS Division Theorem. Thus also $\bar{j}$ must be an isometry. □

The essence of the preceding proposition is rephrased in the following theorem which will turn out to be a useful tool for proofs by induction on the number of indeterminates.

Theorem 4 (WEIERSTRASS Finiteness Theorem). *Let A be a k-Banach algebra, let $\phi: T_n \to A$ be a finite k-algebra homomorphism and let $\omega \in T_{n-1}[X_n]$ be a Weierstrass polynomial contained in* ker ϕ. *Then the map $\phi': T_{n-1} \to A$ defined by $\phi' := \phi \mid T_{n-1}$ is also finite. In particular, the k-algebra monomorphism $T_{n-1} \hookrightarrow T_n/\omega T_n$ induced by the natural injection $T_{n-1} \hookrightarrow T_n$ is finite for every Weierstrass polynomial ω.*

Proof. Let us consider the following commutative diagram

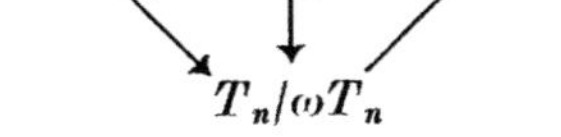

where ε denotes the natural embedding of T_{n-1} into T_n and π the canonical residue epimorphism. The maps $\bar{\varepsilon}$ and $\bar{\phi}$ are induced by ε and ϕ, respectively, in an obvious manner. Since ϕ is finite, so is $\bar{\phi}$. By the preceding proposition, $T_n/\omega T_n$ is a finite T_{n-1}-module via $\bar{\varepsilon}$; i.e., the map $\bar{\varepsilon}$ is finite. Then $\bar{\phi} \circ \bar{\varepsilon}$ is also finite. Since $\phi' = \phi \circ \varepsilon = \bar{\phi} \circ \bar{\varepsilon}$, the proof is finished. □

At this point we would like to mention without proof a stronger WEIERSTRASS Finiteness Theorem, which is not needed now and which will be a later consequence of more general facts about affinoid algebras. Namely,

If $g \in T_n$ is X_n-distinguished of degree $s > 0$, then the endomorphism ϕ of T_n defined by $\phi(X_n) := g$ and $\phi(X_i) := X_i$ for $i = 1, \ldots, n-1$ is finite with $1, X_n, \ldots, X_n^{s-1}$ as a free generating system.

5.2.4. Generation of distinguished power series. — The two preceding results show that Weierstrass polynomials are extremely useful in reducing problems to similar problems in a lower dimension. But, in order to exploit this fact for T_n, one has to make sure that there are "enough" of these Weierstrass polynomials. For the applications we have in mind, "enough" means that every $f \in T_n - \{0\}$ can be transformed by a suitable automorphism σ into an X_n-distinguished series $\sigma(f)$ which then is associated to some Weierstrass polynomial. That this is feasible is asserted by the following

Proposition 1. *For every $f \in T_n$, $f \neq 0$, there is a k-algebra automorphism σ of T_n such that $\sigma(f)$ is X_n-distinguished.*

Proof. We may assume $|f| = 1$. Let $f = \sum_\mu a_\mu X^\mu$. Let $m = (m_1, \ldots, m_n)$ be the maximal n-tuple (with respect to lexicographical ordering) such that $|a_m| = 1$. Let t be a natural number such that $t \geq \max_{1 \leq i \leq n} \mu_i$ for all indices $\mu = (\mu_1, \ldots, \mu_n)$ with $\tilde{a}_\mu \neq 0$; e.g., take t equal to the total degree of $\tilde{f}$. The automorphism σ for which we are looking will be one of the class we considered at the end of (5.1.3). Namely, set $\sigma(X_i) := X_i + X_n^{c_i}$ for $i = 1, \ldots, n-1$ and $\sigma(X_n) := X_n$ where, starting with an additional number $c_n := 1$, the exponents $c_{n-1}, \ldots, c_1$ are defined recursively by

$$c_{n-j} := 1 + t \sum_{d=0}^{j-1} c_{n-d} \quad \text{for } j = 1, \ldots, n-1 .$$

(The formula remains true for $j = 0$; it reproduces the definition of c_n.) We claim that then $\sigma(f)$ is X_n-distinguished of order $s := \sum_{i=1}^n c_i m_i$. First we observe that, for all $\mu = (\mu_1, \ldots, \mu_n)$ with $\tilde{a}_\mu \neq 0$ and $\mu \neq m$, we have $\sum_{i=1}^n c_i \mu_i < s$. Namely, there is an index p, $1 \leq p \leq n$, such that $\mu_1 = m_1, \ldots, \mu_{p-1} = m_{p-1}$ and $\mu_p < m_p$. Then

$$\sum_{i=1}^n c_i \mu_i \leq \sum_{i=1}^{p-1} c_i m_i + c_p(m_p - 1) + \sum_{i=p+1}^n c_i t = \sum_{i=1}^p c_i m_i - 1 < \sum_{i=1}^n c_i m_i = s ,$$

whence our claim is justified. Now let us compute $\widetilde{\sigma(f)}$. We have

$$\widetilde{\sigma(f)} = \tilde{\sigma}(\tilde{f}) = \sum_{\mu} \tilde{a}_\mu (X_1 + X_n^{c_1})^{\mu_1} \ldots (X_{n-1} + X_n^{c_{n-1}})^{\mu_{n-1}} X_n^{\mu_n}$$

$$= \sum_{\substack{\mu \\ \tilde{a}_\mu \neq 0}} \tilde{a}_\mu \sum_{\substack{\lambda_1,\ldots,\lambda_{n-1} \\ 0 \leqq \lambda_i \leqq \mu_i}} \binom{\mu_1}{\lambda_1} \cdots \binom{\mu_{n-1}}{\lambda_{n-1}} X_1^{\mu_1-\lambda_1} \ldots X_{n-1}^{\mu_{n-1}-\lambda_{n-1}} X_n^{c_1\lambda_1+\cdots+c_{n-1}\lambda_{n-1}+\mu_n} = \sum p_i X_n^i,$$

where the p_i are suitable elements of $\tilde{k}[X_1, \ldots, X_{n-1}]$. Using the above observation, we see that $\widetilde{\sigma(f)}$ is a polynomial in X_n of degree $\leqq s$. Furthermore, a power $X_n^{c_1\lambda_1+\cdots+c_{n-1}\lambda_{n-1}+\mu_n}$ occurring in the above representation for $\widetilde{\sigma(f)}$ equals X_n^s if and only if $\mu_n = m_n$ and $\lambda_i = \mu_i = m_i$ for $i = 1, \ldots, n-1$. Thus we have $p_s = \tilde{a}_m \in \tilde{k} - \{0\}$. In particular, $\widetilde{\sigma(f)}$ is a unitary polynomial of degree s in $(\tilde{k}[X_1, \ldots, X_{n-1}])[X_n]$, and hence $\sigma(f)$ is X_n-distinguished of that degree. □

For later reference we state what we have actually proved.

Proposition 2. *Let $f = \sum_{\mu} a_\mu X^\mu \in T_n$, $f \neq 0$, and let $t \in \mathbb{N} \cup \{0\}$ such that $t \geqq \max_{1 \leqq i \leqq n} \mu_i$ for all indices $\mu = (\mu_1, \ldots, \mu_n)$ with $|a_\mu| = |f|$. Define an automorphism $\sigma\colon T_n \to T_n$ by $\sigma(X_n) := X_n$ and $\sigma(X_i) := X_i + X_n^{c_i}$ for $i = 1, \ldots, n-1$, where, starting with $c_n := 1$, the coefficients c_i are determined recursively by $c_{n-j} := 1 + t \sum_{d=0}^{j-1} c_{n-d}$, $j = 1, \ldots, n-1$. Then $\sigma(f)$ is X_n-distinguished of order $s := \sum_{i=1}^{n} c_i m_i$ where $m = (m_1, \ldots, m_n)$ is the maximal index (with respect to lexicographical ordering) such that $|a_m| = |f|$.*

5.2.5. Rückert's theory. — Following the classical method of Rückert in the complex case, we want to establish some results about the ring structure of the algebra T_n. For clarity, we axiomatize the situation by introducing the following concept.

Definition 1. *Let I be a ring (commutative with identity element). An overring I' of $I[X]$ is called Rückert over I if there is a family W of monic polynomials in $I[X]$ such that the following three axioms are fulfilled:*

(1) *If the product of two monic polynomials lies in W, so do the factors.*

(2) *For all $\omega \in W$, there is an isomorphism of I-algebras $I'/\omega I' \cong I[X]/\omega I[X]$. In particular, the canonical map $I \to I'/\omega I'$ is finite.*

(3) *For all $f \in I' - \{0\}$, there is an automorphism σ of I' and a unit e of I' such that $e \cdot \sigma(f) \in W$.*

According to the results of (5.2.2), (5.2.3) and (5.2.4), the algebra T_n is Rückert over T_{n-1} if one takes W to be the family of Weierstrass polynomials in X_n. If one replaces the strictly convergent power series by the formal, or simply the convergent series, the same statement holds mutatis mutandis. With respect to many aspects, a Rückert overring of I behaves as $I[X]$ does. In particular, some ring properties of I are inherited by I', as the following three propositions show.

Proposition 2. *A Rückert overring I' of a Noetherian ring I is Noetherian.*

Proof. We have to show that every ideal $\mathfrak{a} \neq (0)$ in I' is finitely generated. According to axiom (3), we may assume that $\mathfrak{a}$ contains a polynomial $\omega \in W$. Since I is Noetherian by assumption, so is $I[X]$ by HILBERT's Basis Theorem. Because $I'/\omega I'$ is isomorphic to $I[X]/\omega I[X]$ due to axiom (2), the image of $\mathfrak{a}$ in $I'/\omega I'$ has a finite generating system. Pulling back that system to $\mathfrak{a}$ and adding ω, we get a finite generating system for $\mathfrak{a}$. □

Recall that a ring I is said to be a *Jacobson ring* if for every ideal $\mathfrak{a} \subset I$ the nilradical rad $\mathfrak{a}$ equals the Jacobson radical $\mathfrak{j}(\mathfrak{a})$ (which is the intersection of all maximal ideals in I containing $\mathfrak{a}$). Obviously any field is a Jacobson ring, whereas a local ring I is not Jacobson unless $I/\text{rad}\, I$ is a field. So one cannot expect that every ring I' which is Rückert over a Jacobson ring I is itself a Jacobson ring, because $I := k$ and $I' := k[\![X]\!]$ provide a counterexample. But at least one can show the following

Proposition 3. *Let I be a Jacobson ring, and let I' be a Rückert overring of I. Then* rad $\mathfrak{a} = \mathfrak{j}(\mathfrak{a})$ *for any non-zero ideal* $\mathfrak{a} \subset I'$.

Proof. Since the nilradical of any ideal $\mathfrak{a} \subset I'$ equals the intersection of all prime ideals containing $\mathfrak{a}$ (see the argument used in the proof of DEDEKIND's Lemma 3.1.4/1), we have only to show $\mathfrak{j}(\mathfrak{p}') = \mathfrak{p}'$ for any non-zero prime ideal $\mathfrak{p}' \subset I'$. This will be done by showing that the Jacobson radical $\mathfrak{j}(I'/\mathfrak{p}')$ of (the zero ideal in) $I'/\mathfrak{p}'$ vanishes for all such $\mathfrak{p}'$. Therefore let $\mathfrak{p}'$ be a non-zero prime ideal in I', and set $\mathfrak{p} := \mathfrak{p}' \cap I$. We may assume that $\mathfrak{p}'$ contains an element $\omega \in W$ so that by axiom (2) the canonical injection $I/\mathfrak{p} \hookrightarrow I'/\mathfrak{p}'$ is finite. For each $b \in \mathfrak{j}(I'/\mathfrak{p}')$ we consider an integral equation

$$b^n + a_1 b^{n-1} + \cdots + a_n = 0$$

of b over $I/\mathfrak{p}$ of minimal degree n. Then

$$a_n = -(b^n + a_1 b^{n-1} + \cdots + a_{n-1} b) \in \mathfrak{j}(I'/\mathfrak{p}') \cap I/\mathfrak{p}.$$

Since for each maximal ideal $\mathfrak{m} \subset I/\mathfrak{p}$ there exists a maximal ideal $\mathfrak{m}' \subset I'/\mathfrak{p}'$ lying over $\mathfrak{m}$, we see that

$$\mathfrak{j}(I'/\mathfrak{p}') \cap I/\mathfrak{p} = \mathfrak{j}(I/\mathfrak{p}) = 0.$$

Then $a_n = 0$, and due to the minimality of n, we must have $n = 1$ and therefore $b = 0$. This shows that $\mathfrak{j}(I'/\mathfrak{p}') = 0$. □

Recall that a factorial ring is an integral domain I such that each non-unit $f \in I - \{0\}$ can be written as a finite product of prime elements in I. (An element $p \in I - \{0\}$ is called a prime element if it generates a prime ideal in I.) Any such product decomposition of f is unique up to units.

Proposition 4. *Every integral domain I', which is Rückert over a factorial ring I, is factorial itself.*

Proof. The assertion is an easy consequence of the fact that $I[X]$ is factorial if I is factorial. However, since this result is not needed in full generality, we include here a direct proof which is based on the Classical GAUSS Lemma (1.5.3).

We have to factor every non-unit $f \in I' - \{0\}$ into prime elements. Since automorphisms and units do not matter for that task, we may assume $f \in W \subset I[X]$. The polynomial ring $Q(I)[X]$ over the field of fractions $Q(I)$ is factorial. Hence there is a factorization $f = p_1 \dots p_r$ into monic polynomials $p_1, \dots, p_r \in Q(I)[X]$. We can choose elements $c_1, \dots, c_r \in I$ such that the polynomials $c_1 p_1, \dots, c_r p_r$ are primitive in $I[X]$. (A polynomial $p \in I[X]$ is called primitive if there is no prime element in I dividing all coefficients of p.) Then we see by the Classical GAUSS Lemma (1.5.3) that

$$\left(\prod_{i=1}^{r} c_i\right) f = \prod_{i=1}^{r} (c_i p_i)$$

is a primitive polynomial in $I[X]$. But this can only be true if $\prod_{i=1}^{r} c_i$ and hence all c_i are units in I. Consequently, $f = p_1 \dots p_r$ is a factorization of f in $I[X]$. It remains to be shown that all p_i are prime elements in I'. Since $p_1, \dots, p_r \in W$ (axiom (1)), and since $I[X]/p_i I[X] \cong I'/p_i I'$ for all i (axiom (2)), it is enough to show that each p_i is a prime element in $I[X]$. However this follows from the equations

$$I[X] \cap p_i Q(I)[X] = p_i I[X], \qquad i = 1, \dots, r,$$

which are easily obtained from the Classical GAUSS Lemma by an argument similar to the one used above. □

5.2.6. Applications of Rückert's theory for T_n. — As we have already observed, Theorem 5.2.2/1, Lemma 5.2.3/2 and Propositions 5.2.3/3 and 5.2.4/1 guarantee that T_n is Rückert over T_{n-1}. Furthermore $T_0 = k$ is a Noetherian factorial ring. Thus using induction on n, we get from Propositions 5.2.5/2 and 5.2.5/4

Theorem 1. *The ring T_n is Noetherian and factorial.*

It is a well-known fact that any factorial ring I is normal (i.e., integrally closed in its field of fractions $Q(I)$). Namely if

$$\left(\frac{a}{b}\right)^r + c_1 \left(\frac{a}{b}\right)^{r-1} + \dots + c_r = 0$$

is an integral equation of some element $\frac{a}{b} \in Q(I)$ over I, we may assume that a and b have no common prime factor. However since

$$a^r + c_1 a^{r-1} b + \dots + c_r b^r = 0,$$

we see that any prime factor p of b must divide a^r and hence a. Thus b can only be a unit in I, and hence $\frac{a}{b}$ belongs to I. In particular, we see that

Theorem 2. *T_n is normal.*

Finally, we get

Theorem 3. *T_n is a Jacobson ring.*

Proof. Proposition 5.1.3/3 tells us that $\mathrm{j}(T_n) = 0$. Therefore we can conclude from Proposition 5.2.5/3 that T_n is a Jacobson ring if T_{n-1} is. Since $T_0 = k$ is a Jacobson ring, the assertion follows by induction on n. □

5.2.7. Finite T_n-modules. — Because T_n is Noetherian, all finite T_n-modules are Noetherian. In particular, all submodules of the s-fold normed direct sum T_n^s are finitely generated over T_n. We want to improve this result and derive finiteness theorems with estimates, comparable to CARTAN's Theorem in the classical case.

Proposition 1. *Let M be a submodule of a finite complete T_n-module. Then M itself is a finite complete T_n-module. Furthermore for every T_n-generating system $\{m_1, \ldots, m_s\}$ of M, there exists a real constant ϱ such that every $m \in M$ admits a representation $m = \sum_{i=1}^{s} t_i m_i$ with $\max_{1 \leqq i \leqq s} |t_i| \leqq \varrho\, |m|$.*

Proof. The first assertion follows from Proposition 3.7.3/1. In order to verify the second assertion, consider the T_n-epimorphism $\varphi\colon T_n^s \to M$ defined by $\varphi(t_1, \ldots, t_s) := \sum_{i=1}^{s} t_i m_i$. Due to Proposition 3.7.3/1, we know that $T_n^s/\ker \varphi$ provided with the residue norm is a finite complete T_n-module. Therefore the induced T_n-isomorphism $\bar{\varphi}\colon T_n^s/\ker \varphi \xrightarrow{\sim} M$ and its inverse are both continuous (see Proposition 3.7.3/2) and hence bounded (see Proposition 2.1.8/2). Let ϱ' be a bound for $\bar{\varphi}^{-1}$, and set $\varrho := \varrho' + 1$. Then the inverse image $\bar{\varphi}^{-1}(m)$ of any element $m \in M$ has norm $\leqq \varrho'\, |m|$ and can be represented by a tuple $(t_1, \ldots, t_s) \in T_n^s$ satisfying $\max_{1 \leqq i \leqq s} |t_i| \leqq \varrho\, |m|$. Since $m = \varphi(t_1, \ldots, t_s) = \sum_{i=1}^{s} t_i m_i$, the second assertion follows. □

Corollary 2. *All ideals of T_n are closed.*

We want to improve Proposition 1 by looking for generating systems admitting the bound $\varrho = 1$. For vector spaces (instead of modules), we have studied in detail such questions as the existence of orthonormal bases (cf. Chapter 2). Now we are going to handle analogous questions for A-modules, where A is a normed ring (thought to be equal to T_n for some n). We want to make precise what we mean by "generating system admitting the bound $\varrho = 1$".

Definition 3. *A finite generating system $\{m_1, \ldots, m_s\}$ of a normed A-module M*

is called pseudo-cartesian if every $m \in M$ *admits a representation* $m = \sum_{i=1}^{s} a_i m_i$ *with* $a_1, \ldots, a_s \in A$ *and* $|m| = \max_{1 \leq i \leq s} |a_i| \, |m_i|$. *If this equation is true for all possible representations of each* $m \in M$, *the system* $\{m_1, \ldots, m_s\}$ *is called cartesian. If such generating systems exist for* M, *we say that* M *is a pseudo-cartesian or a cartesian* A*-module, respectively.*

This is a generalization of Definition 2.4.1/1, where we defined finite k-cartesian spaces and bases. A pseudo-cartesian generating system is cartesian if and only if it is free:

Remark. A finite-dimensional normed k-vector space V is pseudo-cartesian if and only if it is cartesian.

Namely, let $\{v_1, \ldots, v_s\}$ be a pseudo-cartesian generating system of V, where $v_i \neq 0$ for all i. Consider the epimorphism $\phi: k^s \to V$, $(c_1, \ldots, c_s) \mapsto \sum_{i=1}^{s} c_i v_i$, and provide k^s with the norm given by $|(c_1, \ldots, c_s)| := \max_{1 \leq i \leq s} |c_i| \, |v_i|$. Then k^s is a cartesian space, and the norm on V equals the residue norm with respect to the map ϕ. The subspace $\ker \phi$ admits a norm-direct supplement U in V, and U is cartesian (see Proposition 2.4.1/5). Since ϕ induces an isometric isomorphism $U \xrightarrow{\sim} V$, we see that V is cartesian. □

The above remark is not true for general A-modules, since one can easily find pseudo-cartesian modules which are not free. For example, let $\mathfrak{m}$ denote the maximal ideal in T_n which is generated by the indeterminates. Then $k = T_n/\mathfrak{m}$ is a pseudo-cartesian T_n-module which is not cartesian (unless $n = 0$). We state some elementary properties of pseudo-cartesian and cartesian modules.

Lemma 4. (a) *The normed direct sum of finitely many pseudo-cartesian* A*-modules is pseudo-cartesian. The same is true if pseudo-cartesian is replaced by cartesian.*

(b) *Let* M *be a pseudo-cartesian* A*-module, and let* N *be a strictly closed submodule of* M. *Then* M/N *(provided with the residue norm) is a pseudo-cartesian* A*-module.*

Lemma 5. *Let* M *be a normed* A*-module over a normed* k*-algebra* A. *Suppose that* $|M| = |k|$. *Then* M *is a pseudo-cartesian* A*-module if and only if* M° *is a finite* A°*-module.*

The assumption $|M| = |k|$ occurring in Lemma 5 cannot in general be avoided. However it can be weakened (as far as the only if part of Lemma 5 is concerned) if the valuation on k is discrete.

Lemma 6. *Let* M *be a normed* A*-module over a normed* k*-algebra* A. *Suppose that* $|A| = |k|$ *and that the valuation on* k *is discrete. Then* M° *is a finite* A°*-module if* M *is a pseudo-cartesian* A*-module.*

Proof. Let $\{m_1, \ldots, m_s\}$ be a pseudo-cartesian generating system for M, where $m_i \neq 0$ for all i, and define $\eta := \max\{\alpha \in |k|; \alpha < 1\}$. Then $\eta < 1$. By multiplying $m_1, \ldots, m_s$ with suitable coefficients from k, we may assume that $\eta < |m_i| \leqq 1$ for all i. Given $m \in M^\circ$, $m \neq 0$, we can find $a_1, \ldots, a_s \in A$ such that $m = \sum_{i=1}^{s} a_i m_i$ and $|m| = \max_{1 \leqq i \leqq s} |a_i|\,|m_i|$. Then we get

$$\max |a_i| \leqq \frac{\max |a_i|\,|m_i|}{\min |m_i|} < \frac{|m|}{\eta} \leqq \frac{1}{\eta}.$$

Since $|a_i| \in |A| = |k|$, this is only possible if $\max_{1 \leqq i \leqq s} |a_i| \leqq 1$. Hence $\{m_1, \ldots, m_s\}$ is a finite generating system for M° over A°. □

For the remainder of this section, we restrict ourselves to the case $A = T_n$. We want to show that submodules of cartesian T_n-modules are always pseudo-cartesian.

Theorem 7. *Let M be a submodule of a cartesian T_n-module F. Then M is pseudo-cartesian and strictly closed in F. In particular if $|M| = |k|$, then M° is a finite $\mathring{T}_n$-module.*

We can apply the theorem in the special case, where F equals T_n and where M is an ideal in T_n. Thereby we obtain

Corollary 8. *Each ideal $\mathfrak{a} \subset T_n$ is strictly closed in T_n, and the residue norm on $T_n/\mathfrak{a}$ satisfies $|T_n/\mathfrak{a}| = |T_n| = |k|$.*

For the proof of Theorem 7, we need some preparations. We denote by $Q := Q(T_n)$ the field of fractions of T_n. If F is a normed T_n-module, the norm on F induces a (semi-) norm on the Q-vector space $F \otimes_{T_n} Q$ (see (2.1.7) for general facts). The process is very simple if F is a cartesian T_n-module. Namely, we may view F as a T_n-submodule of $F \otimes_{T_n} Q$, and any cartesian generating system of F gives rise to an orthogonal basis of $F \otimes_{T_n} Q$, all norms being preserved. In particular, $F \otimes_{T_n} Q$ is a cartesian Q-vector space.

Lemma 9. *Let F be a cartesian T_n-module and let ω be a non-zero element in T_n. Then ωF is a strictly closed submodule of F. Furthermore, F is strictly closed in $F \otimes_{T_n} Q$.*

Proof. First we show that ωT_n is strictly closed in T_n. Applying a suitable automorphism of T_n, we may assume that ω is a Weierstrass polynomial in T_n of some degree $s \geqq 0$. Then the WEIERSTRASS Division Theorem 5.2.1/2 says that T_n (viewed as a T_{n-1}-module) is the norm-direct sum of ωT_n and $\sum_{\nu=0}^{s-1} T_{n-1} X_n^\nu$. In particular, ωT_n is strictly closed in T_n. That ωF is strictly closed in F is an easy consequence of this fact.

Viewing F as a T_n-submodule of $F \otimes_{T_n} Q$, we can also say that F is strictly closed in $\frac{1}{\omega} F$. Since $F \otimes_{T_n} Q$ is the union of all T_n-submodules $\frac{1}{\omega} F$, $\omega \in T_n - \{0\}$, we see that F is strictly closed in $F \otimes_{T_n} Q$. □

Lemma 10. *Let M be a submodule of a cartesian T_n-module F. Then there exist an element $\omega \in T_n - \{0\}$ and a cartesian T_n-submodule $F' \subset F \otimes_{T_n} Q$ such that $\omega F' \subset M \subset F'$.*

Proof. We consider $V' := M \otimes_{T_n} Q$ as a Q-subspace of $V := F \otimes_{T_n} Q$. Then V' is cartesian, since V is cartesian (see Proposition 2.4.1/5). Let $\{v_1, \ldots, v_r\}$ denote an orthogonal basis for V'. We may assume $v_1, \ldots, v_r \in M$. Namely if $v_i = \frac{m_i}{t_i}$ with elements $m_i \in M$, $t_i \in T_n - \{0\}$, then $\{m_1, \ldots, m_r\}$ is an orthogonal basis of the desired type. Since M is finitely generated, there is a universal denominator $\omega \in T_n - \{0\}$ such that

$$M \subset \frac{1}{\omega} \sum_{i=1}^{r} T_n v_i .$$

Define $F' := \sum_{i=1}^{r} T_n \frac{v_i}{\omega}$. Then F' is a cartesian T_n-submodule of V satisfying $\omega F' \subset M \subset F'$. □

Proof of Theorem 7. Due to Lemma 5, we have only to show that M is pseudo-cartesian and strictly closed in F. We use induction on n. For $n = 0$, the assertion follows from Propositions 2.4.1/5 and 2.4.2/1. Therefore, let $n \geqq 1$. Choose a non-zero element $\omega \in T_n$ and a cartesian T_n-submodule $F' \subset F \otimes_{T_n} Q$ such that $\omega F' \subset M \subset F'$ (Lemma 10). There is a chart $\{X_1, \ldots, X_n\} \subset T_n$ such that ω is X_n-distinguished of some degree $s \geqq 0$ (use Proposition 5.2.4/1). Hence, by the Weierstrass Preparation Theorem 5.2.2/1, we may assume that ω is a Weierstrass polynomial in X_n. Writing $T_{n-1} := k\langle X_1, \ldots, X_{n-1}\rangle$, we see that $F'/\omega F'$ is a cartesian T_{n-1}-module. Namely, an easy computation verifies that $F'/\omega F'$ is a cartesian $T_n/\omega T_n$-module; furthermore, $T_n/\omega T_n$ is a cartesian T_{n-1}-module by Proposition 5.2.3/3. If we view $M/\omega F'$ as a T_{n-1}-submodule of $F'/\omega F'$, we can apply the induction hypothesis and see that $M/\omega F'$ is pseudo-cartesian and strictly closed in $F'/\omega F'$.

In order to construct a pseudo-cartesian generating system for M, we consider a pseudo-cartesian generating system for the T_{n-1}-module $M/\omega F'$. Since $\omega F'$ is strictly closed in F' (Lemma 9) and hence strictly closed in M, this system can be lifted to M without changing norms. Adding a cartesian generating system for $\omega F'$, we get a pseudo-cartesian generating system for the T_n-module M. It remains to show that M is strictly closed in F. Since $M/\omega F'$ is strictly closed in $F'/\omega F'$ and since $\omega F'$ is strictly closed in F' (Lemma 9), we can apply Lemma 1.1.6/4 and thereby see that M is strictly closed in F'. Now F' is strictly closed in $F' \otimes_{T_n} Q$ (Lemma 9), and $F' \otimes_{T_n} Q$ is strictly closed in $F \otimes_{T_n} Q$ because any subspace of a finite-dimensional cartesian vector space is strictly closed (Proposition 2.4.2/1). Thus M is strictly closed in $F \otimes_{T_n} Q$ by Lemma 1.1.5/5, and hence strictly closed in F. □

We derive some consequences from Theorem 7.

Corollary 11. *Every submodule N of a pseudo-cartesian T_n-module M is pseudo-cartesian.*

Proof. There exists an epimorphism $\phi: F \to M$ from a cartesian T_n-module F onto M such that the norm on M equals the residue norm with respect to ϕ. Namely, choose a pseudo-cartesian generating system $\{m_1, \ldots, m_s\}$ for M and consider the canonical epimorphism $F \to M$ where F is the "free cartesian T_n-module generated by $m_1, \ldots, m_s$"; i.e., $F := T_n^s$, the norm being defined by $|(t_1, \ldots, t_s)| := \max_{1 \leq i \leq s} |t_i|\, |m_i|$ for all tuples $(t_1, \ldots, t_s) \in T_n^s$. It follows easily from our construction that $\ker \phi$ is strictly closed in F (for general ϕ, this follows from Theorem 7). In particular for any $m \in M$, one can find an inverse image $m' \in F$ such that $|m| = |m'|$. Keeping this in mind, we see that the ϕ-image of any pseudo-cartesian generating system for $\phi^{-1}(N)$ is a pseudo-cartesian generating system for N. Thus it follows that N is pseudo-cartesian because $\phi^{-1}(N)$ is pseudo-cartesian by Theorem 7. □

Applying Lemma 5, we get

Corollary 12. *Let M be a normed T_n-module, and suppose that $|M| = |k|$. Let N be a submodule of M. Then N° is a finite $\mathring{T}_n$-module if M° is a finite $\mathring{T}_n$-module.*

5.3. Stability of $Q(T_n)$

5.3.1. Weak stability. — In the following sections we will show that the field of fractions $Q(T_n)$ is stable if the ground field k is stable. As a first step towards this result, we show in this section for arbitrary ground field k that

Theorem 1. *The field of fractions $Q(T_n)$ is weakly stable.*

Proof. All valued fields of characteristic 0 are weakly stable (see Proposition 3.5.1/4). Therefore we have only to consider the case where char $k = p > 0$. In this case, we apply the criterion given in Lemma 3.5.3/2 which says that $Q(T_n)$ is weakly stable if each finitely generated T_n-submodule of $T_n^{p^{-1}}$ is b-separable. We invest a little bit of extra work and show that

Lemma 2. *Each T_n-submodule of finite rank in $T_n^{p^{-1}}$ is b-separable.*

Proof. To give an interpretation of the pair of integral domains $T_n \subset T_n^{p^{-1}}$, consider the injection

$$k\langle X \rangle = k\langle X_1, \ldots, X_n \rangle \hookrightarrow k^{p^{-1}}\langle Y \rangle = k^{p^{-1}}\langle Y_1, \ldots, Y_n \rangle$$

extending the inclusion map $k \hookrightarrow k^{p^{-1}}$ by mapping each X_i onto Y_i^p, $i = 1, \ldots, n$. Note that $k^{p^{-1}}$ is complete because k is complete. The p-th power of any series in $k^{p^{-1}}\langle Y\rangle$ is a series in $k\langle X\rangle$, and conversely, any series in $k\langle X\rangle$ has a p-th root in $k^{p^{-1}}\langle Y\rangle$. Thus we have $(k\langle X\rangle)^{p^{-1}} = k^{p^{-1}}\langle Y\rangle$.

Now let M be a $k\langle X\rangle$-submodule of $k^{p^{-1}}\langle Y\rangle$, and let $\{m_1, \ldots, m_s\}$ be a maximal set of $k\langle X\rangle$-linearly independent elements of M. Writing

$$m_\varrho = \sum_\nu c_{\varrho\nu} Y^\nu, \qquad \varrho = 1, \ldots, r,$$

we denote by k' the smallest complete subfield of $k^{p^{-1}}$ containing all coefficients $c_{\varrho\nu}$ and k. Then k' is a k-vector space of countable type. Any such vector space is b-separable by Proposition 2.7.1/2. Therefore $k'\langle Y\rangle$ is a b-separable $k\langle Y\rangle$-module by Corollary 2.2.6/5. Since the direct sum decomposition

$$k\langle Y\rangle = \bigoplus_{\substack{\nu \\ 0 \leqq \nu_i < p}} k\langle X\rangle\, Y^\nu$$

is norm-direct, we see that $k\langle Y\rangle$ is a b-separable $k\langle X\rangle$-module (use Proposition 2.2.5/2). Thus $k'\langle Y\rangle$ can be viewed as a b-separable $k\langle X\rangle$-module, and we know that M is a b-separable $k\langle X\rangle$-module if we can show $M \subset k'\langle Y\rangle$.

In order to achieve this, consider an element $m \in M$. Then $m^p \in k\langle X\rangle \subset k'\langle Y\rangle$. By our construction, we have

$$M \subset k'\langle Y\rangle \otimes_{k\langle X\rangle} Q(k\langle X\rangle) \subset Q(k'\langle Y\rangle).$$

Thus m is an element of the field of fractions of $k'\langle Y\rangle$, and m is integral over $k'\langle Y\rangle$. Since $k'\langle Y\rangle$ is normal (Theorem 5.2.6/2), we have $m \in k'\langle Y\rangle$ and hence $M \subset k'\langle Y\rangle$. This concludes the proof of Lemma 2 and thereby also the proof of Theorem 1. □

The T_n-modules considered in the above lemma are, in fact, finite T_n-modules. Namely, each such M is a tame T_n-module by Proposition 4.1/4 so that $T_n^{p^{-1}}$ is a tame T_n-module. In particular, we get the following additional result:

Theorem 3. *T_n is Japanese.*

Proof. T_n is a normal Noetherian integral domain (see (5.2.6)). Therefore the assertion follows from Proposition 4.3/2 if char $k = 0$ and from Proposition 4.4/2 if char $k = p > 0$. □

5.3.2. The Stability Theorem. Reductions. — As indicated before, we want to prove the following Stability Theorem.

Theorem 1. *The field of fractions $Q(T_n)$ is stable if k is stable.*

In the next chapter, we will apply this result to affinoid algebras. We will look at finite homomorphisms $\varphi: A \to B$ between affinoid algebras and discuss the cases in which the induced homomorphism $\mathring{\varphi}: \mathring{A} \to \mathring{B}$ between the subrings of power-bounded elements is finite also. As we will see, this question is closely related to the stability of $Q(T_n)$.

Before starting with the proof of the Stability Theorem, we want to show the following converse.

Proposition 2. *If there is an $n \geqq 0$ such that $Q(T_n)$ is stable, then k is stable.*

Proof. Let $(Q(T_n))_a$ denote the algebraic closure of $Q(T_n)$, and assume that $Q(T_n)$ is stable for some $n \in \mathbb{N}$. Then $(Q(T_n))_a$ is $Q(T_n)$-cartesian. Since the algebraic closure of k is contained in $(Q(T_n))_a$, it is enough to show that $(Q(T_n))_a$ is k-cartesian. Using Lemma 2.4.3/4, this amounts to showing that $Q(T_n)$ is k-cartesian. However the latter follows from Lemma 2.4.3/3 because $T_n \cong c(k)$ is k-cartesian (Proposition 2.7.2/7). □

Thus we can conclude from Theorem 1 and Proposition 2 that $Q(T_n)$ is stable for all $n \geqq 0$ if the stability is known for at least one $n \geqq 0$. In order to attack the assertion of the Stability Theorem, we view the field of rational functions $k(X) = k(X_1, \ldots, X_n)$ as a valued subfield of $Q(T_n)$. Then $k(X)$ is dense in $Q(T_n)$, and the stability of $Q(T_n)$ can be derived from the stability of $k(X)$.

Proposition 3. *The field $Q(T_n)$ is stable if its subfield of rational functions $k(X)$ is stable.*

Proof. If $k(X)$ is stable, its completion $\widehat{k(X)}$ is stable (Proposition 3.6.2/3). Since $\widehat{k(X)}$ can also be interpreted as the completion of $Q(T_n)$ and since $Q(T_n)$ is weakly stable (Theorem 5.3.1/1), we see (again by Proposition 3.6.2/3) that $Q(T_n)$ is stable. □

5.3.3. Stability of $k(X)$ if $|k^*|$ is divisible. — Due to the preceding proposition, the burden in the proof of the Stability Theorem lies in showing that $k(X)$ (provided with the valuation induced by the Gauss norm) is stable. Things become easier if we consider the valuation induced by the total degree, which is defined by $\left|\frac{f}{g}\right|_t := \exp(\deg f - \deg g)$ for all $f, g \in k[X]$, $g \neq 0$.

Proposition 1. *If $k(X)$ is provided with the valuation induced by the total degree, then $k(X)$ is stable.*

Proof. The valuation induced by the total degree is discrete, and $k(X)$ is weakly stable by Proposition 3.5.3/3. Hence $k(X)$ is stable by Proposition 3.6.2/1. □

This is only an auxiliary result, which does not take into account the given valuation on k. Nevertheless, together with the following proposition it will play a central role in the final proof of the Stability Theorem. Instead of showing that all finite extensions L of $k(X)$ are $k(X)$-cartesian with respect to the Gauss norm on $k(X)$ (if k is stable), we shall check first that they are at least k-cartesian.

Proposition 2. *Suppose that k is stable. Let L be a finite extension of $k(X)$, and* *provide L with the spectral norm derived from the Gauss norm on $k(X)$. Then L is k-cartesian.*

Proof. Let $l_1, \ldots, l_r \in L$ be elements generating the extension L over $k(X)$. We may assume that $l_1, \ldots, l_r$ are integral over the polynomial ring $k[X]$. Let $\hat{L}$ be the completion of L with respect to the spectral norm on L, and let $\hat{Q}$ be the completion of $Q := Q(T_n)$. Then $\hat{Q}$ is also the completion of $k(X)$, and we have the following commutative diagram of inclusions

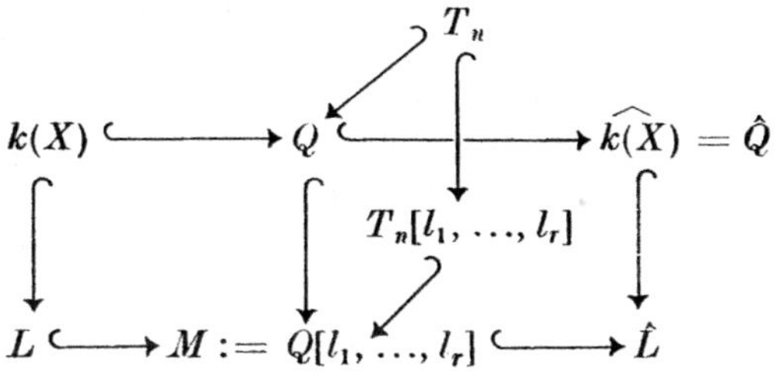

We claim that the spectral norm $|\ |_{M,Q}$ on M over Q restricts to the spectral norm $|\ |_{L,k(X)}$ on L over $k(X)$. Namely, the norm $|\ |_{M,Q}$ induces a power-multiplicative $k(X)$-algebra norm on L, and therefore we have $|l|_{M,Q} \leqq |l|_{L,k(X)}$ for all $l \in L$ (Theorem 3.2.1/2). On the other hand, the norm on $\hat{L}$ (which is the canonical extension of $|\ |_{L,k(X)}$) induces a power-multiplicative Q-algebra norm $|\ |$ on M so that $|l| \leqq |l|_{M,Q}$ for all $l \in M$ (Theorem 3.2.2/2) and hence $|l|_{L,k(X)} = |l| \leqq |l|_{M,Q}$ for all $l \in L$. Thus we see that $|\ |_{M,Q}$ restricts to $|\ |_{L,k(X)}$ on L, and in order to show that L is k-cartesian with respect to $|\ |_{L,k(X)}$ it is enough to show that M is k-cartesian with respect to $|\ |_{M,Q}$. However this follows from Proposition 3.8.1/11 applied to the inclusion map $T_n \hookrightarrow T_n[l_1, \ldots, l_r]$. Namely, T_n is normal (Theorem 5.2.6/2), and the Maximum Modulus Principle holds for T_n and the supremum norm $|\ |_{\sup}$ on T_n (Corollary 5.1.4/6). Furthermore, the Gauss norm equals the supremum norm on T_n (Corollary 5.1.4/6), and, by our construction, we have $M = Q[l_1, \ldots, l_r] = (T_n[l_1, \ldots, l_r])_{T_n - \{0\}}$. □

As indicated before, the Stability Theorem is settled by the following proposition (for the special case where $|k^*|$ is divisible).

Proposition 3. *Suppose that k is stable and that $|k^*|$ is divisible. Then $k(X)$ is stable.*

In order to verify this, it suffices to show (according to Proposition 3.6.2/8) that $[L : k(X)] = [\tilde{L} : \tilde{k}(X)]$ for every finite-dimensional extension L of $k(X)$ where L is provided with the spectral norm. What we know already from Proposition 2 is that at least $\dim_k V = \dim_{\tilde{k}} \tilde{V}$ for all finite-dimensional k-vector spaces V in L. Therefore, we have to establish a connection between k-dimensions and $k(X)$-dimensions of vector spaces in L and between $\tilde{k}$-dimensions and $\tilde{k}(X)$-dimensions of vector spaces in $\tilde{L}$. We shall do this by considering the following abstract situation:

Lemma 4. *Let F be a field, and provide $F(X) := F(X_1, \ldots, X_n)$ with the valuation defined by $|f/g|_t := \exp(\deg f - \deg g)$ for all $f, g \in F[X]$, $g \neq 0$, where* deg *denotes the total degree. Let M be a finite-dimensional reduced algebra over $F(X)$. Denote by $|\ \ |_t$ the spectral norm on M derived from the valuation $|\ \ |_t$ on $F(X)$. Furthermore, let A_M denote the integral closure of $F[X]$ in M. For $\nu \in \mathbb{N}$, define*

$$A_M(\nu) := \{a \in A_M;\ |a|_t \leq \exp \nu\}.$$

Then

$$\dim_{F(X)} M = n! \lim_{\nu\to\infty} \nu^{-n} \dim_F A_M(\nu).$$

Before proving Lemma 4, we shall show how to derive Proposition 3 from it. Apply the lemma to two different situations: first with $F := k$ and $M := L$ and second with $F := \tilde{k}$ and $M := \tilde{L}$, where the $\tilde{\ }$-functor is constructed with respect to the spectral norm on L derived from the Gauss norm on $k(X)$. In order to avoid confusion, this spectral norm shall be denoted by $|\ \ |_{\text{Gauss}}$ (for the purposes of this proof only). We then get

$$(1) \qquad [L : k(X)] = n! \lim_{\nu\to\infty} \nu^{-n} \dim_k A_L(\nu) \quad \text{and,}$$

since $[\tilde{L} : \tilde{k}(X)] \leq [L : k(X)] < \infty$ (see, e.g., Proposition 2.1.10/3) and $\tilde{L}$ is reduced,

$$(2) \qquad [\tilde{L} : \tilde{k}(X)] = n! \lim_{\nu\to\infty} \nu^{-n} \dim_{\tilde{k}} A_{\tilde{L}}(\nu).$$

According to Proposition 2, the field L is k-cartesian. Since $|k^*|$ is divisible, it follows that L is, in fact, strictly k-cartesian. Namely, we have $|L^*|_{\text{Gauss}} = |k(X)^*|_{\text{Gauss}}$ because $|k(X)^*|_{\text{Gauss}} = |k^*|$ is divisible. Thus considering $A_L(\nu)$ as a k-subspace of L, we see by Corollary 2.5.1/6 that

$$(3) \qquad \dim_k A_L(\nu) = \dim_{\tilde{k}} \big(A_L(\nu)\big)^{\sim}.$$

We claim

$$(4) \qquad \big(A_L(\nu)\big)^{\sim} \subset A_{\tilde{L}}(\nu).$$

In order to prove this, let l be an element in A_L with $|l|_t \leq \exp \nu$ and $|l|_{\text{Gauss}} \leq 1$. We must show that $\tilde{l}$ is integral over $\tilde{k}[X]$ and that $|\tilde{l}|_t \leq \exp \nu$. Because l is integral over $k[X]$, there is a polynomial $p = Y^m + p_1 Y^{m-1} + \cdots + p_m \in k[X][Y]$ such that $p(l) = l^m + p_1 l^{m-1} + \cdots + p_m = 0$. Furthermore, we may assume that p is the minimal polynomial of l over $k(X)$, because $k[X]$ is integrally closed in $k(X)$ (see the considerations preceding the proof of Proposition 3.8.1/7). From $\max_{1\leq i\leq m} |p_i|_{\text{Gauss}}^{1/i} = |l|_{\text{Gauss}} \leq 1$, we derive $|p_i|_{\text{Gauss}} \leq 1$, and therefore $\tilde{p}(\tilde{l}) = 0$ for the polynomial $\tilde{p} = Y^m + \tilde{p}_1 Y^{m-1} + \cdots + \tilde{p}_m \in \tilde{k}[X][Y]$. Hence $\tilde{l}$ is integral over $\tilde{k}[X]$. It remains to be shown that $|\tilde{l}|_t \leq \exp \nu$. We know that $\max_{1\leq i\leq m} |p_i|_t^{1/i} = |l|_t \leq \exp \nu$. Since $|\tilde{p}_i|_t = \exp \deg \tilde{p}_i \leq \exp \deg p_i = |p_i|_t$, we get $\sigma(\tilde{p}) = \max_{1\leq i\leq m} |\tilde{p}_i|_t^{1/i} \leq \max_{1\leq i\leq m} |p_i|_t^{1/i} = |l|_t \leq \exp \nu$ for the spectral value of the polynomial $\tilde{p}$. Applying Proposition 3.1.2/1, we find $|\tilde{l}|_t \leq \sigma(\tilde{p})$, and therefore $|\tilde{l}|_t \leq \exp \nu$. Thus the inclusion relation (4) is true as claimed. The relations (3) and (4) together yield $\dim_k A_L(\nu) \leq \dim_{\tilde{k}} A_{\tilde{L}}(\nu)$. Combining this with (1) and (2), we get $[L : k(X)] \leq [\tilde{L} : \tilde{k}(X)]$. Since $[\tilde{L} : \tilde{k}(X)] \leq [L : k(X)]$ always holds, we finally get the desired equality $[L : k(X)] = [\tilde{L} : \tilde{k}(X)]$. This shows that Proposition 3 can be derived from the assertion of Lemma 4. □

Proof of Lemma 4. First we want to reduce the problem to the case where M is a field. In general, M is a finite sum of finite field extensions $M_1, \ldots, M_s$ of $F(X)$ (Dedekind's Lemma 3.1.4/1). An easy computation shows that $A_M = \bigoplus_{i=1}^{s} A_{M_i}$; i.e., the integral closure of $F[X]$ in M equals the direct sum of the integral closures of $F[X]$ in the components M_i. If $|\ \ |_{t,i}$ denotes the spectral norm on M_i with respect to $|\ \ |_t$ on $F(X)$, then $|(m_1, \ldots, m_s)|_t = \max_{1 \leq i \leq s} |m_i|_{t,i}$. Therefore $A_M(\nu) = \bigoplus_{i=1}^{s} A_{M_i}(\nu)$. If we had already proved the lemma for the case where M is a field, we could also handle the general case, because

$$\dim_F A_M(\nu) = \sum_{i=1}^{s} \dim_F A_{M_i}(\nu)$$

and

$$\dim_{F(X)} M = \sum_{i=1}^{s} \dim_{F(X)} M_i .$$

Therefore we may assume that M is a field. Choose a basis $a_1, \ldots, a_r$ of M over $F(X)$ such that $a_i \in A_M$ for $i = 1, \ldots, r$, where $r := [M : F(X)]$. Choose $\nu_0 \in \mathbb{N}$ such that $\max_{1 \leq i \leq r} |a_i|_t \leq \exp \nu_0$. We claim

(1) $$\dim_F A_M(\nu) \geq rN(\nu - \nu_0) \text{ for all } \nu \in \mathbb{N} \text{ with } \nu > \nu_0,$$

where $N(s)$ denotes the number of monomials $X^\nu = X_1^{\nu_1} \ldots X_n^{\nu_n}$ of degree $\leq s$. In order to justify this claim, we take a monomial $m \in F[X]$ with $\deg m \leq \nu - \nu_0$ and estimate $|ma_i|_t$ in the following way:

$$|ma_i|_t \leq |m|_t\, |a_i|_t = \exp(\deg m) \cdot |a_i|_t \leq \exp(\nu - \nu_0) \exp \nu_0 = \exp \nu.$$

Hence $ma_i \in A_M(\nu)$ for $i = 1, \ldots, r$. Define

$$B := \{ma_i;\ i = 1, \ldots, r \text{ and } m \in F[X] \text{ is a monomial with } \deg m \leq \nu - \nu_0\}.$$

We just proved that $B \subset A_M(\nu)$. As one can easily check, B is linearly independent over F and B contains $rN(\nu - \nu_0)$ elements, whence (1) is proved.

Due to Proposition 1, we know that M admits an $F(X)$-orthogonal basis if M is provided with $|\ \ |_t$. Since A_M is finite over $F[X]$ (indeed, $F[X]$ is Japanese; see Proposition 4.4/4), we may choose an orthogonal basis $m_1, \ldots, m_r$ of M such that $A_M \subset \sum_{i=1} F[X]\, m_i$. Choose $\nu_1 \in \mathbb{N}$ large enough so that $\exp(-\nu_1)$

$\leqq \min_{1 \leqq i \leqq r} |m_i|_t$. In order to get a bound for dim $A_M(\nu)$ in the other direction, we want to show that

(2) $$\dim_F A_M(\nu) \leqq rN(\nu + \nu_1) \quad \text{for all} \quad \nu \in \mathbb{N}.$$

Namely, let a be an arbitrary element in $A_M(\nu)$ (so that $|a|_t \leqq \exp \nu$). Then there are $f_1, \ldots, f_r \in F[X]$ such that $a = \sum_{i=1}^{r} f_i m_i$ and $|a|_t = \max_{1 \leqq i \leqq r} |f_i|_t \, |m_i|_t$. From $|a|_t \leqq \exp \nu$, we derive

$$\exp(\deg f_i) = |f_i|_t \leqq |a|_t \, |m_i|_t^{-1} \leqq \exp \nu \exp \nu_1 = \exp(\nu + \nu_1),$$

whence $\deg f_i \leqq \nu + \nu_1$. If we denote by S the set of all polynomials in $F[X]$ of total degree $\leqq \nu + \nu_1$, then we just proved $A_M(\nu) \subset \sum_{i=1}^{r} S m_i$. Since $\dim_F \sum_{i=1}^{r} S m_i = rN(\nu + \nu_1)$, also (2) has been verified.

Now we want to determine $N(s)$ for $s \in \mathbb{N}$. To each monomial $X_1^{i_1} \ldots X_n^{i_n}$ with $i_1 + \cdots + i_n \leqq s$, we associate a sequence of s ones and n zeros in the following way:

$$X_1^{i_1} \ldots X_n^{i_n} \mapsto \underbrace{1 \ldots 1}_{i_1} 0 \underbrace{1 \ldots 1}_{i_2} 0 \ldots 0 \underbrace{1 \ldots 1}_{i_n} 0 \; \underbrace{1 \ldots 1}_{s-(i_1+\cdots+i_n)}.$$

Obviously this is a bijection. On the other hand, it is clear that there are exactly $\binom{s+n}{n} = \frac{(s+n)!}{s!\,n!}$ such sequences. Combining this with (1) and (2), we see

(3) $$r\binom{\nu - \nu_0 + n}{n} \leqq \dim_F A_M(\nu) \leqq r\binom{\nu + \nu_1 + n}{n} \quad \text{for all} \quad \nu \in \mathbb{N}, \; \nu > \nu_0.$$

Since the two binomial coefficients are both asymptotically equal to $\frac{\nu^n}{n!}$ as $\nu \to \infty$, inequality (3) implies $r = \lim_{\nu \to \infty} \frac{n!}{\nu^n} \dim_F A_M(\nu)$, which finishes the proof of Lemma 4 and also of Proposition 3. □

5.3.4. Completion of the proof for arbitrary $|k^*|$. — Let us return to the general situation; i.e., we no longer assume that $|k^*|$ is divisible. Define k' to be the completion of the algebraic closure k_a of k. According to Proposition 3.4.1/3, we know that k' is algebraically closed. In particular, k' is stable, and $|k'^*|$ is divisible (Observation 3.6.2/10). Therefore we are allowed to apply Proposition 5.3.3/3 to k', and we get $k'(X) = k'(X_1, \ldots, X_n)$ is stable. What we really want to show is that $k(X) = k(X_1, \ldots, X_n)$ is stable. Therefore, suppose we are given a finite extension L of $k(X)$. Then one can construct a finite extension L' of $k'(X)$ containing L. If we provide L' with the spectral norm derived from the Gauss norm on $k'(X)$, we know that L' is $k'(X)$-cartesian because $k'(X)$ is stable. We want to show that L' is also $k(X)$-cartesian. In order to do so, we need the following

Lemma 1. *If k is complete and stable, then $k'(X)$ is $k(X)$-cartesian.*

Proof. Let U be a $k(X)$-subspace of $k'(X)$ of dimension $m < \infty$. We have to show that U is $k(X)$-cartesian. There are elements $u_1, \ldots, u_m, u \in k'[X]$ such that $U = \sum_{i=1}^{m} k(X) \frac{u_i}{u}$. If we can find a $k(X)$-orthogonal basis for $u \cdot U = \sum_{i=1}^{m} k(X)\, u_i$, we immediately get one for U. Therefore we may assume $u = 1$. Let V be the k-subspace of k' generated by the finitely many coefficients of the polynomials $u_1, \ldots, u_m$. Then V is a finite-dimensional k-cartesian subspace of k'. Namely, k_a is k-cartesian and dense in k' so that k' is k-cartesian by Corollary 2.4.3/10. Hence V admits a k-orthogonal basis $v_1, \ldots, v_l$. We claim that

(1) $\{v_1, \ldots, v_l\}$ is also orthogonal over $k(X)$

and that

(2) $$U \subset \sum_{\lambda=1}^{l} k(X)\, v_\lambda .$$

If both (1) and (2) are verified, then $\sum_{\lambda=1}^{l} k(X)\, v_\lambda$ is a $k(X)$-cartesian space and therefore also U is $k(X)$-cartesian. Because $k(X)$ is the field of fractions of $k[X]$, in order to show the validity of (1), it suffices to prove

$$\left|\sum_{\lambda=1}^{l} f_\lambda v_\lambda\right| = \max_{1 \leq \lambda \leq l} |f_\lambda|\, |v_\lambda| \quad \text{for all} \quad f_1, \ldots, f_l \in k[X].$$

To prove this equality, we write $f_\lambda = \sum_{\nu} f_{\lambda,\nu} X^\nu$ with $f_{\lambda,\nu} \in k$ for $\lambda = 1, \ldots, l$. We know $|f_\lambda| = \max_{\nu} |f_{\lambda,\nu}|$. Furthermore, we have

$$\left|\sum_{\lambda=1}^{l} f_\lambda v_\lambda\right| = \left|\sum_{\lambda=1}^{l} \left(\sum_{\nu} f_{\lambda,\nu} X^\nu\right) v_\lambda\right| = \left|\sum_{\nu} \left(\sum_{\lambda=1}^{l} f_{\lambda,\nu} v_\lambda\right) X^\nu\right| = \max_{\nu} \left|\sum_{\lambda=1}^{l} f_{\lambda,\nu} v_\lambda\right| .$$

Using the fact that $\{v_\lambda\}$ is k-orthogonal, we may continue as follows:

$$\max_{\nu} \left|\sum_{\lambda=1}^{l} f_{\lambda,\nu} v_\lambda\right| = \max_{\nu} \left(\max_{\lambda} |f_{\lambda,\nu}|\, |v_\lambda|\right) = \max_{\lambda} \left(\max_{\nu} |f_{\lambda,\nu}|\right) |v_\lambda| = \max_{\lambda} |f_\lambda|\, |v_\lambda| .$$

So we have verified assertion (1). To justify also the second claim, we only have to show $u_i \in \sum_{\lambda=1}^{l} k(X)\, v_\lambda$ for $i = 1, \ldots, m$, because $U = \sum_{i=1}^{m} k(X)\, u_i$. According to our construction, we know that

$$u_i \in V[X] = \sum_{\lambda=1}^{l} k[X]\, v_\lambda \subset \sum_{\lambda=1}^{l} k(X)\, v_\lambda .$$

Thus assertion (2) and consequently Lemma 1 are proved. □

Now we can continue the proof of the Stability Theorem. First we claim that L' is $k(X)$-cartesian. Namely, $k'(X)$ is $k(X)$-cartesian due to Lemma 1, and L' is $k'(X)$-cartesian according to our considerations preceding that lemma. Lemma 2.4.3/4 tells us that L' is $k(X)$-cartesian. Because L is a $k(X)$-subspace of L', we see that L is also $k(X)$-cartesian, but only if L is provided with the restriction $|\ \ |'$ of the spectral norm on L' belonging to the Gauss norm on $k'(X)$. Since L provided with this norm is $k(X)$-cartesian, this norm induces the product topology on L. But then Lemma 3.5.1/1 gives us that $|\ \ |'$ must coincide with the spectral norm on L belonging to the Gauss norm on $k(X)$. Thus we finally have shown that an arbitrary finite extension L of $k(X)$ provided with the "right" spectral norm is $k(X)$-cartesian. This implies that $k(X)$ is stable. Together with Proposition 5.3.2/3 this finishes the proof of the Stability Theorem. □

CHAPTER 6

Affinoid algebras and Finiteness Theorems

In this chapter we study a more general type of TATE algebras, namely affinoid algebras. First we establish the NOETHER Normalization Lemma which is fundamental for almost all further investigations in this chapter. It says that each affinoid algebra A contains a subalgebra isomorphic to some algebra of strictly convergent power series T_d such that A is a finite T_d-module. This result makes it possible to reduce certain problems on affinoid algebras to problems on algebras of strictly convergent power series. For example, the procedure works well for the discussion of the supremum norm and for the proof of the Maximum Modulus Principle; the technical details have already been dealt with in (3.8).

In the second part of this chapter, we consider homomorphisms of affinoid algebras. We show that the reduction functor $A \rightsquigarrow \tilde{A}$ preserves finite homomorphisms. The same question for the functor $A \rightsquigarrow \mathring{A}$ is fairly complicated. It was first solved by GRAUERT and the third author of this book in [16] and by GRUSON in [17]. Our discussion of the problem uses the results of (5.2.7) on finite T_n-modules and the Stability Theorem of (5.3).

We denote by k a field with a complete valuation $|\ \ |$. All homomorphisms are k-algebra homomorphisms.

6.1. Elementary properties of affinoid algebras

6.1.1. The category $\mathfrak{A}$ of k-affinoid algebras. — Each residue algebra $T_n/\mathfrak{a}$ of T_n by a (closed) ideal $\mathfrak{a} \subset T_n$ becomes a k-Banach algebra if one defines the *residue norm* of the residue class $\bar{f}$ of an element $f \in T_n$ by

$$|\bar{f}| := |f, \mathfrak{a}| := \inf \{|h|;\, h \in \bar{f}\}.$$

The residue epimorphism $T_n \to T_n/\mathfrak{a}$ is contractive (hence continuous) and open. In particular, the residue norm induces the quotient topology on $T_n/\mathfrak{a}$. Notice that the residue norm is not in general power-multiplicative. For example, $T_n/\mathfrak{a}$ can have nilpotent elements $\neq 0$.

Definition 1. *A k-Banach algebra A is called affinoid (more precisely, k-affinoid) if there exists an integer $n \geq 0$ and a continuous epimorphism $\alpha: T_n \to A$.*

By BANACH's Theorem, α is open; hence A is isomorphic as a k-Banach algebra to the residue algebra $T_n/\ker \alpha$. In particular, the residue norm, which from now on will be denoted by $|\ \ |_\alpha$, induces the given Banach topology on A.

The residue norm $|\ \ |_\alpha$ depends heavily on the choice of the epimorphism α. However all norms $|\ \ |_\alpha$ are equivalent, since they induce the given Banach topology on A.

Proposition 2. *Let A be k-affinoid and let $|\ \ |_\alpha$ be a residue norm on A. Then*

$$|A|_\alpha \subset |k|.$$

In particular, each vector $\neq 0$ in A can be normed to length 1 *by multiplication with a scalar.*

The assertion follows directly from Corollary 5.2.7/8.

Remark. We shall see later that a *reduced* affinoid algebra can be provided in a natural way with a complete power-multiplicative norm, the so-called *spectral norm*. For this norm, all values are roots of elements of $|k|$.

Proposition 3. *Let A be a k-affinoid algebra. Then A is a Noetherian Jacobson ring. Each ideal $\mathfrak{a} \subset A$ is closed, and each quotient $A/\mathfrak{a}$ (provided with the residue norm) is k-affinoid.*

Proof. Let $\alpha\colon T_n \to A$ be a continuous epimorphism. Then $A \cong T_n/\ker \alpha$ is a Noetherian Jacobson ring since T_n is such a ring (Theorems 5.2.6/1 and 5.2.6/3). The closedness of any ideal $\mathfrak{a} \subset A$ follows from Proposition 3.7.2/2 (or simply from the closedness of ideals in T_n), and $A/\mathfrak{a}$ is k-affinoid, since $T_n \xrightarrow{\alpha} A \longrightarrow A/\mathfrak{a}$ is a continuous epimorphism. □

The k-affinoid algebras form the objects of a category $\mathfrak{A}$; the morphisms of this category are the continuous k-algebra homomorphisms. (We shall see later that every k-algebra homomorphism between k-affinoid algebras is continuous.) For our purposes, the category $\mathfrak{A}$ will play the same fundamental role as does the category of affine algebras in algebraic geometry.

As in (1.2.5), we use the notation $\mathring{A}$ for the subring of power-bounded elements in a k-affinoid algebra A. Morphisms of $\mathfrak{A}$ map power-bounded elements into power-bounded elements.

Proposition 4. *Let $\varphi\colon B \to A$ be a continuous homomorphism between k-Banach algebras A, B. Let $f_1, \ldots, f_n$ be power-bounded elements in A; let $X_1, \ldots, X_n$ be indeterminates. Then there exists a unique continuous homomorphism $\Phi\colon B\langle X_1, \ldots, X_n\rangle \to A$ such that*

$$\Phi \mid B = \varphi, \qquad \Phi(X_i) = f_i, \qquad i = 1, \ldots, n.$$

Proof. For $h = \sum_0^\infty a_{\nu_1 \ldots \nu_n} X_1^{\nu_1} \ldots X_n^{\nu_n} \in B\langle X_1, \ldots, X_n\rangle$, we set

$$\Phi(h) := \sum_0^\infty \varphi(a_{\nu_1 \ldots \nu_n})\, f_1^{\nu_1} \ldots f_n^{\nu_n}.$$

Since $f_i \in \mathring{A}$ and $\lim a_{\nu_1 \ldots \nu_n} = 0$, the series on the right-hand side represents a well-defined element of A. It is clear that this is the only way to extend φ to the polynomial algebra $B[X_1, \ldots, X_n]$ and that Φ is a homomorphism of $B[X_1, \ldots, X_n]$ into A. As $B[X_1, \ldots, X_n]$ is dense in $B\langle X_1, \ldots, X_n\rangle$, it follows that Φ is the unique continuous extension of φ to a homomorphism of $B\langle X_1, \ldots, X_n\rangle$ into A. □

In the case $B = k$, Proposition 4 says that, for each set $f_1, \ldots, f_n \in \mathring{A}$ and each chart $\{X_1, \ldots, X_n\}$ of T_n, there exists exactly one continuous homomorphism $\Phi\colon T_n \to A$ such that $\Phi(X_i) = f_i$, $i = 1, \ldots, n$. If Φ is surjective, we call the elements $f_1, \ldots, f_n$ a *system of affinoid generators of* A. In particular, A is a k-affinoid algebra, and we write suggestively $A = k\langle f_1, \ldots, f_n\rangle$.

We shall prove now that the category $\mathfrak{A}$ is closed under finite extensions, i.e., that $\mathfrak{A}$ contains all finite overalgebras of a given k-affinoid algebra. Recall that a ring homomorphism $\varrho\colon R \to S$ is called *finite* if S provided with the induced R-module structure (i.e., $r \cdot s := \varrho(r)\, s$) is a finite R-module. Recall further that the composition of two finite ring homomorphisms is finite. Each epimorphism is finite; hence each $A \in \mathfrak{A}$ admits finite homomorphisms $T_n \to A$.

Proposition 5. *Let B be an object of $\mathfrak{A}$, and let $\varphi\colon B \to A$ be a continuous finite homomorphism into a k-Banach algebra A. Then $A \in \mathfrak{A}$.*

Proof. We may assume $B = T_n$ for some n. By assumption there are elements $a_1, \ldots, a_m \in A$ such that $A = \sum_{i=1}^{m} \varphi(T_n)\, a_i$. We may assume $a_i \in \mathring{A}$. By Proposition 4, the map φ extends to a continuous homomorphism $\Phi\colon T_n\langle Y_1, \ldots, Y_m\rangle \to A$ such that $\Phi(Y_i) = a_i$. Then Φ is surjective, and hence $A \in \mathfrak{A}$. □

If φ is not assumed to be continuous and A is not assumed to be a Banach algebra, we still have

Proposition 6. *Let B be an object of $\mathfrak{A}$, and let $\varphi\colon B \to A$ be a finite homomorphism into a k-algebra A. Then A can be provided with a topology such that φ is continuous and strict and such that $A \in \mathfrak{A}$.*

Proof. By Proposition 3.7.4/1, we can provide A with a topology such that A becomes a k-Banach algebra and φ becomes continuous and strict. We have $A \in \mathfrak{A}$ by Proposition 5. □

Later we shall see that the topology on A is uniquely determined.

The category $\mathfrak{A}$ is closed with respect to the operation of forming direct sums; i.e., *if A, B are objects of $\mathfrak{A}$, then the ring-theoretic direct sum $A \oplus B$ belongs to $\mathfrak{A}$.*

Proof. Let $\alpha\colon T_n \to A$ and $\beta\colon T_m \to B$ be continuous epimorphisms. We may assume $m = n$. Define $\alpha \oplus \beta\colon T_n \oplus T_n \to A \oplus B$ by $(\alpha \oplus \beta)(f \oplus g) := \alpha(f) \oplus \beta(g)$. Then $\alpha \oplus \beta$ is a continuous k-algebra epimorphism. Therefore we only have to show that the k-Banach algebra $T_n \oplus T_n$ (viewed as the normed direct sum of T_n with itself) belongs to $\mathfrak{A}$. The map $\varphi\colon T_n \to T_n \oplus T_n$ given by $x \mapsto (x, x)$ is a continuous finite homomorphism. Hence $T_n \oplus T_n \in \mathfrak{A}$ by Proposition 5. $\square$

The statement $T_n \oplus T_n \in \mathfrak{A}$ can also be verified by checking directly that the map $\Phi\colon T_n\langle Y\rangle \to T_n \oplus T_n$ given by $\Phi\left(\sum_0^\infty a_\nu Y^\nu\right) := \left(a_0, \sum_0^\infty a_\nu\right)$ is a continuous epimorphism.

We want to see that $\mathfrak{A}$ is also closed with respect to complete tensor products. Beginning with a slightly more general situation, we let A and B denote k-Banach algebras. Then $A \mathbin{\hat{\otimes}}_k B$ is a complete normed k-algebra (cf. (3.1.1)), hence a k-Banach algebra. If in particular $B = k\langle X_1, \ldots, X_n\rangle$, there exists by Proposition 4 a unique continuous homomorphism $\sigma\colon A\langle X_1, \ldots, X_n\rangle \to A \mathbin{\hat{\otimes}}_k k\langle X_1, \ldots, X_n\rangle$ such that

$$X_i \mapsto 1 \mathbin{\hat{\otimes}} X_i, \qquad i = 1, \ldots, n,$$

$$a \mapsto a \mathbin{\hat{\otimes}} 1, \qquad a \in A.$$

More generally, Proposition 4 says that the inclusions $\sigma_1\colon A \hookrightarrow A\langle X_1, \ldots, X_n\rangle$ and $\sigma_2\colon k\langle X_1, \ldots, X_n\rangle \hookrightarrow A\langle X_1, \ldots, X_n\rangle$ satisfy the universal property stated in Proposition 3.1.1/2 which characterizes the complete tensor product $A \mathbin{\hat{\otimes}}_k k\langle X_1, \ldots, X_n\rangle$. Thus one concludes that σ is an isomorphism. Furthermore, σ is obviously contractive. Also σ^{-1} is contractive by Proposition 3.1.1/2, since σ_1 and σ_2 are contractive. Hence we get

Proposition 7. *Let A denote a k-Banach algebra. Then the canonical k-algebra homomorphism $\sigma\colon A\langle X_1, \ldots, X_n\rangle \to A \mathbin{\hat{\otimes}}_k k\langle X_1, \ldots, X_n\rangle$ is an isometric isomorphism.*

The proposition applies in particular to the cases where A is a Tate algebra $T_n(k)$ or where A is an extension field k' of k with a complete valuation on k' extending the valuation on k. Thus, we have

Corollary 8. *There are canonical isometric isomorphisms $T_m \mathbin{\hat{\otimes}}_k T_n \cong T_{m+n}$ and $k' \mathbin{\hat{\otimes}}_k T_n(k) \cong T_n(k')$.*

Corollary 9. *Let A and B denote k-affinoid algebras, and as above let k' be a complete valued extension field of k. Then $A \mathbin{\hat{\otimes}}_k B$ is k-affinoid and $k' \mathbin{\hat{\otimes}}_k B$ is k'-affinoid. The canonical homomorphism $B \to k' \mathbin{\hat{\otimes}}_k B$ is a strict monomorphism.*

Proof. Let $\varphi\colon k\langle X_1, \ldots, X_n\rangle \to B$ denote a continuous epimorphism. By Banach's Theorem, φ is open and hence strict by Proposition 1.1.9/3. Applying

Proposition 2.1.8/6, we get continuous epimorphisms

$$\mathrm{id}_A \mathbin{\widehat{\otimes}} \varphi\colon\; A \mathbin{\widehat{\otimes}}_k k\langle X_1, \ldots, X_n\rangle \to A \mathbin{\widehat{\otimes}}_k B$$

$$\mathrm{id}_{k'} \mathbin{\widehat{\otimes}} \varphi\colon\; k' \mathbin{\widehat{\otimes}}_k k\langle X_1, \ldots, X_n\rangle \to k' \mathbin{\widehat{\otimes}}_k B$$

showing that $A \mathbin{\widehat{\otimes}}_k B$ is k-affinoid and that $k' \mathbin{\widehat{\otimes}}_k B$ is k'-affinoid, since the corresponding facts obviously hold for the algebras $A \mathbin{\widehat{\otimes}}_k k\langle X_1, \ldots, X_n\rangle$ and $k' \mathbin{\widehat{\otimes}}_k k\langle X_1, \ldots, X_n\rangle$ by Proposition 7. To verify the remaining assertion, we view B as a k-Banach space. If $\{f_1, \ldots, f_n\}$ is a system of affinoid generators of B, then B contains $k[f_1, \ldots, f_n]$ as a dense subspace. Thus we see that B is a Banach space of countable type and that there is a linear homeomorphism of B onto $c(k)$ or onto k^r if $r := \dim_k B < \infty$ (Theorem 2.8.2/2). Since the inclusion map $k \hookrightarrow k'$ is strict and since the restricted direct product of normed vector spaces commutes with the complete tensor product (Proposition 2.1.7/8), it follows that the canonical homomorphism $B \to B \mathbin{\widehat{\otimes}}_k k'$ is a strict monomorphism. □

The category $\mathfrak{A}$ admits also complete tensor products of a more general type; namely the following holds:

Proposition 10. *Let $B_1, B_2 \in \mathfrak{A}$ be normed algebras over some algebra $A \in \mathfrak{A}$ via contractive homomorphisms $A \to B_i$, $i = 1, 2$. Then also $B_1 \mathbin{\widehat{\otimes}}_A B_2$, viewed as a k-algebra, belongs to $\mathfrak{A}$. If $A' \to A$ is a contractive homomorphism of k-affinoid algebras, the canonical homomorphism $B_1 \mathbin{\widehat{\otimes}}_{A'} B_2 \to B_1 \mathbin{\widehat{\otimes}}_A B_2$ is surjective.*

Proof. We start with the second assertion. According to Proposition 3.1.1/2, the canonical maps from B_1 and B_2 into $B_1 \mathbin{\widehat{\otimes}}_{A'} B_2$, respectively $B_1 \mathbin{\widehat{\otimes}}_A B_2$, induce a commutative diagram of contractive homomorphisms

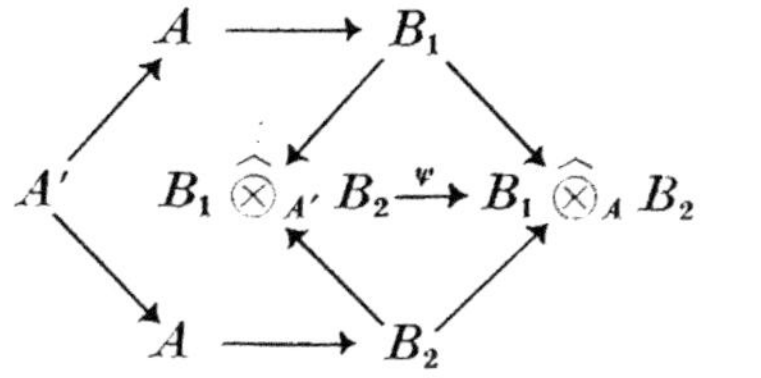

and furthermore a commutative diagram of contractive homomorphisms

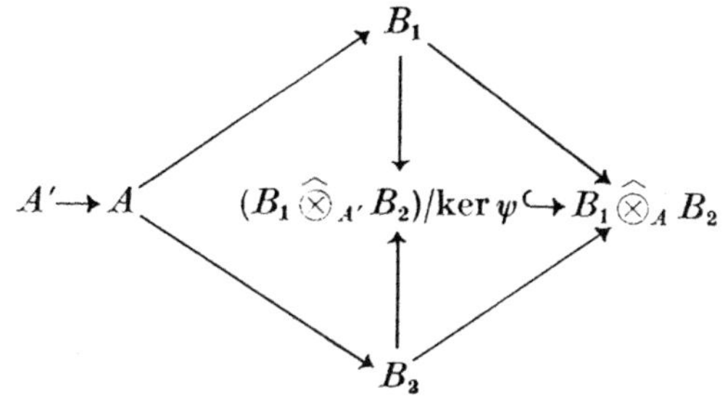

where $B_1 \hat{\otimes}_{A'} B_2/\ker\psi$ is provided with the canonical residue norm ($\ker\psi$ is a closed ideal in $B_1 \hat{\otimes}_{A'} B_2$). It is a straightforward verification to see that the maps $B_i \to (B_1 \hat{\otimes}_{A'} B_2)/\ker\psi$, $i = 1, 2$, satisfy the universal property stated in Proposition 3.1.1/2 which characterizes the complete tensor product $B_1 \hat{\otimes}_A B_2$. Hence $(B_1 \hat{\otimes}_{A'} B_2)/\ker\psi \to B_1 \hat{\otimes}_A B_2$ is an isomorphism showing that $B_1 \hat{\otimes}_{A'} B_2 \to B_1 \hat{\otimes}_A B_2$ is surjective. In particular, if A' equals k and if $A' \to A$ is the canonical map $k \to A$, it follows from Proposition 3 that $B_1 \hat{\otimes}_A B_2$ is k-affinoid since $B_1 \hat{\otimes}_k B_2$ is k-affinoid. □

Proposition 11. *In the situation of Proposition* 10, *let* $\mathfrak{b}_i \subset B_i$, $i = 1, 2$, *be ideals, and denote by* $(\mathfrak{b}_1, \mathfrak{b}_2) \subset B_1 \hat{\otimes}_A B_2$ *the ideal generated by the images of* $\mathfrak{b}_1$ *and* $\mathfrak{b}_2$ *in* $B_1 \hat{\otimes}_A B_2$. *Then the canonical map* $\pi: B_1 \hat{\otimes}_A B_2 \to B_1/\mathfrak{b}_1 \hat{\otimes}_A B_2/\mathfrak{b}_2$ *is surjective and satisfies* $\ker \pi = (\mathfrak{b}_1, \mathfrak{b}_2)$; *hence,* π *induces a strict isomorphism* $B_1 \hat{\otimes}_A B_2/(\mathfrak{b}_1, \mathfrak{b}_2) \xrightarrow{\sim} B_1/\mathfrak{b}_1 \hat{\otimes}_A B_2/\mathfrak{b}_2$.

Proof. The map π is surjective by Proposition 2.1.8/6, and obviously $(\mathfrak{b}_1, \mathfrak{b}_2) \subset \ker \pi$. Hence π induces a continuous homomorphism

$$\pi': (B_1 \hat{\otimes}_A B_2)/(\mathfrak{b}_1, \mathfrak{b}_2) \to B_1/\mathfrak{b}_1 \hat{\otimes}_A B_2/\mathfrak{b}_2 .$$

Furthermore, the canonical maps $B_i \to B_1 \hat{\otimes}_A B_2$ induce maps $B_i/\mathfrak{b}_i \to B_1 \hat{\otimes}_A B_2/(\mathfrak{b}_1, \mathfrak{b}_2)$, $i = 1, 2$. Just as in the preceding proof, it is not hard to see that these induced maps satisfy the universal property characterizing the complete tensor product $B_1/\mathfrak{b}_1 \hat{\otimes}_A B_2/\mathfrak{b}_2$. Thus $(B_1 \hat{\otimes}_A B_2)/(\mathfrak{b}_1, \mathfrak{b}_2)$ is k-affinoid and π' is a strict isomorphism by Proposition 3.1.1/2. □

As a consequence, we now have an explicit description of the complete tensor product over k in $\mathfrak{A}$. Namely for $T_m/\mathfrak{a}, T_n/\mathfrak{b} \in \mathfrak{A}$, it follows that

$$T_m/\mathfrak{a} \hat{\otimes}_k T_n/\mathfrak{b} = T_{m+n}/(\mathfrak{a}, \mathfrak{b}).$$

Using the same technique as in Proposition 11, one shows that

Proposition 12. *Let* A *be a* k*-affinoid algebra, and let* $\mathfrak{a}$ *be an ideal in* A. *If* k' *is a complete field extending* k, *the canonical homomorphism of* k'*-affinoid algebras* $\pi: A \hat{\otimes}_k k' \to (A/\mathfrak{a}) \hat{\otimes}_k k'$ *is surjective, and* $\ker \pi$ *equals the ideal* $\mathfrak{a}'$ *generated by* $\mathfrak{a}$ *in* $A \hat{\otimes}_k k'$. *Hence* π *induces a strict isomorphism* $(A \hat{\otimes}_k k')/\mathfrak{a}' \xrightarrow{\sim} A/\mathfrak{a} \hat{\otimes}_k k'$.

The category $\mathfrak{A}$ is not closed with respect to the operation of passing to rings of fractions. Set $A := T_1 = k\langle X\rangle$, $S := \{1, X, X^2, \ldots\}$, and consider the ring

$$A_S = \left\{h = \sum_{\nu > -\infty}^{\infty} a_\nu X^\nu;\ \lim a_\nu = 0\right\}$$

of *strictly convergent Laurent series with finite principal part.* Then A_S provided with the norm $|h| := \max\{|a_\nu|\}$ is not complete. In (6.1.4), we shall see that the completion of A_S is again k-affinoid.

Remark. A closed k-Banach subalgebra of a k-affinoid algebra is not necessarily Noetherian and hence not necessarily k-affinoid.

To give an *example*, we take $T_2 = k\langle X_1, X_2\rangle$ and set

$$A := \{f \in T_2; f(0, X_2) \in k\}.$$

Obviously A is a closed k-Banach subalgebra of T_2. We consider the ideal $\mathfrak{a}$ in A generated by the elements $X_1 \cdot X_2^i \in A$, $i \geqq 0$. If $\mathfrak{a}$ could be finitely generated, some of the elements $X_1 \cdot X_2^i$, say $X_1, X_1 \cdot X_2, \ldots, X_1 \cdot X_2^s$, would do it. However, an equation

$$X_1 X_2^{s+1} = \sum_{i=0}^{s} f_i X_1 X_2^i, \qquad f_i \in A,$$

is impossible because it would imply the equation

$$X_2^{s+1} = \sum_{i=0}^{s} f_i(0, X_2)\, X_2^i, \qquad f_i(0, X_2) \in k,$$

of linear dependence over k.

For each $A \in \mathfrak{A}$, we denote by $\mathfrak{M}_A$ the category of all finite complete normed A-modules with continuous A-module homomorphisms as morphisms. Since A is Noetherian, all results of (3.7.3) hold for this category. Thus, we know that each submodule of a module $M \in \mathfrak{M}_A$ is closed, that up to equivalence each finite A-module can be uniquely provided with a complete A-module norm, and that each A-linear homomorphism $\varphi: M \to M'$, $M, M' \in \mathfrak{M}_A$, is automatically continuous and strict.

6.1.2. Noether normalization. — By definition each $A \in \mathfrak{A}$ is the residue algebra of a free algebra T_n. In this section, we prove that, on the other hand, each $A \in \mathfrak{A}$ contains a free algebra T_d as subalgebra — more precisely, that each $A \in \mathfrak{A}$ is a finite overalgebra of a free algebra T_d.

Recall that a ring homomorphism $\varrho: R \to S$ is called *integral* if S is integral over the subring $\varrho(R)$. We will see later that all integral homomorphisms of k-affinoid algebras are already finite, but for the time being we are not yet able to use this result.

Theorem 1. (i) *Let A be a non-zero k-affinoid algebra. For every finite (resp. integral) homomorphism $\alpha: T_n \to A$, there exist a chart $\{X_1, \ldots, X_n\}$ of T_n and an integer $d \geqq 0$ such that $\alpha \mid k\langle X_1, \ldots, X_d\rangle$ is finite (resp. integral) and injective.*

(ii) *Let $\varphi: B \to A$ be a finite (resp. integral) homomorphism between non-zero k-affinoid algebras. Then there exists a homomorphism $\psi: T_d \to B$ for some $d \geqq 0$ such that $\varphi \circ \psi: T_d \to B \to A$ is finite (resp. integral) and injective.*

Proof. First let us reduce the second statement to the first one. By choosing an epimorphism $\alpha: T_n \to B$ and applying the first statement to the map $\varphi \circ \alpha: T_n \to A$, one gets a subalgebra $T_d \subset T_n$ such that $(\varphi \circ \alpha) \mid T_d: T_d \to A$

is finite (resp. integral) and injective. Then $\psi := \alpha \mid T_d$ has the required properties.

Now let us prove the first assertion. We proceed by induction on n. The case $n = 0$ is trivial. Let $n \geqq 1$. If $\ker \alpha = 0$, there is nothing to prove. Otherwise we can find a chart $\{X_1, \ldots, X_n\}$ of T_n and a Weierstrass polynomial

$$\omega \in T_{n-1}[X_n], \qquad T_{n-1} = k\langle X_1, \ldots, X_{n-1}\rangle,$$

such that $\omega \in \ker \alpha$. Then α induces a finite (resp. integral) homomorphism $\bar{\alpha}: T_n/\omega T_n \to A$. By the WEIERSTRASS Finiteness Theorem 5.2.3/4, the natural injection $T_{n-1} \to T_n$ induces a finite monomorphism $\beta: T_{n-1} \to T_n/\omega T_n$. Obviously, $\bar{\alpha} \circ \beta$ is finite (resp. integral), and one has $\bar{\alpha} \circ \beta = \alpha \mid k\langle X_1, \ldots, X_{n-1}\rangle$. Applying the induction hypothesis to T_{n-1} and $\bar{\alpha} \circ \beta$, we get the theorem. □

As a special consequence of the theorem, one obtains the so-called NOETHER Normalization Lemma (in analogy to the corresponding theorem for finitely generated k-algebras).

Corollary 2. *For every k-affinoid algebra $A \neq 0$, there exists a finite monomorphism $\varphi: T_d \to A$ for some $d \geqq 0$.*

Remark. Using some facts about Krull dimensions of rings, one can show that the integer d above is uniquely determined by the algebra A. Namely, $\dim A = \dim T_d$ (for example, use NAGATA [28] Corollary 10.10), and $\dim T_d = d$. The latter assertion is easily verified. Namely, the chain of prime ideals

$$0 \subset (X_1) \subset (X_1, X_2) \subset \ldots \subset (X_1, \ldots, X_d)$$

shows that $\dim T_d \geqq d$. Since each maximal ideal in T_d can be generated by d elements (Proposition 7.1.1/3), we must have $\dim T_d = d$.

Corollary 3. *Let A be a k-affinoid algebra; let $\mathfrak{q}$ be an ideal in A such that its nilradical* $\operatorname{rad} \mathfrak{q}$ *is a maximal ideal in A. Then $A/\mathfrak{q}$ is finite over k.*

Proof. By the theorem, there exists a finite monomorphism $\varphi: T_d \to A/\mathfrak{q}$ for some $d \geqq 0$. We claim $d = 0$, i.e., $T_d = k$. The composition of φ with the canonical epimorphism $\varrho: A/\mathfrak{q} \to A/\operatorname{rad} \mathfrak{q}$ is finite and injective (because T_d is reduced). The ideal $\operatorname{rad} \mathfrak{q}$ being maximal, we see that T_d has a finite extension which is a field. But then T_d must itself be a field so that $d = 0$ and $T_d = k$. □

Another consequence of Theorem 1 is

Proposition 4. *Every k-affinoid algebra A which is an integral domain is Japanese.*

Proof. Let A' be an integral extension of A such that its field of fractions $Q(A')$ is a finite extension of $Q(A)$. We have to show A' is a finite A-module. By the theorem, there is a normalization map $\varphi: T_d \to A$ for a suitable $d \geqq 0$.

Then A' is an integral extension of T_d, and $Q(A')$ is finite over $Q(T_d)$. Since T_d is Japanese by Theorem 5.3.1/3, we see that A' is a finite T_d-module and a fortiori a finite A-module. □

6.1.3. Continuity of homomorphisms. — We now apply the general results on k-Banach algebras of (3.7.5) to k-affinoid algebras. The following remark is crucial:

For each k-affinoid algebra $B \in \mathfrak{A}$, *the set*

$$\mathfrak{B} := \{\mathfrak{m}^\nu;\ \mathfrak{m} \text{ maximal ideal in } B,\ \nu \in \mathbb{N}\}$$

fulfills conditions (i) *and* (ii) *of Proposition* 3.7.5/2. *Namely,*

(i) $\dim_k B/\mathfrak{b} < \infty$ *for each* $\mathfrak{b} \in \mathfrak{B}$,

(ii) $\bigcap_{\mathfrak{b}\in\mathfrak{B}} \mathfrak{b} = (0)$.

Proof. We have $\dim_k B/\mathfrak{b} < \infty$ for all $\mathfrak{b} \in \mathfrak{B}$ by Corollary 6.1.2/3. In order to show $\bigcap_{\mathfrak{b}\in\mathfrak{B}} \mathfrak{b} = (0)$, take any $f \in B$ such that $f \in \bigcap_{\nu\geqq 1} \mathfrak{m}^\nu$ for all maximal ideals $\mathfrak{m} \subset B$. Krull's Intersection Theorem implies that for each $\mathfrak{m}$ there is an element $m \in \mathfrak{m}$ such that $(1 - m) f = 0$. Hence the annihilator of f is contained in no maximal ideal in B. Therefore, $f = 0$ and (ii) holds. □

Since each $B \in \mathfrak{A}$ is Noetherian, we derive from Proposition 3.7.5/2

Theorem 1. *Each k-algebra homomorphism of a Noetherian k-Banach algebra into a k-affinoid algebra is continuous.*

This theorem tells us for the category $\mathfrak{A}$ of k-affinoid algebras that (similarly as in the category of finite modules over a k-affinoid algebra A) one need not bother about questions of continuity. Each morphism is automatically continuous. Furthermore, a k-algebra can carry at most one k-affinoid structure, since the identity map must be continuous in both directions. Actually a stronger result holds:

Proposition 2. *If A is k-affinoid, then any k-Banach algebra topology on the k-algebra A coincides with the k-affinoid topology of A.*

Proof. The assertion is a direct consequence of Proposition 3.7.5/3. □

We can strengthen Theorem 1 also in the following way.

Proposition 3. *Let* $\varphi\colon B \to A$ *be a homomorphism of k-affinoid algebras. Then the algebra norm on A can be replaced by an equivalent one such that* φ *becomes contractive, and thus A becomes a normed B-algebra.*

Proof. Let $a_1, \ldots, a_n \in A$ denote affinoid generators of A. Then according to Proposition 6.1.1/4, the map φ extends uniquely to a continuous homomorphism $\psi\colon B\langle X_1, \ldots, X_n\rangle \to A$ such that $\psi(X_i) = a_i$, $i = 1, \ldots, n$. The map ψ

is obviously surjective and hence open by BANACH's Theorem. Therefore, the residue norm via ψ is equivalent to the original norm on A; thus ψ and, in particular, φ are contractive with respect to this norm on A. □

Consequently, for arbitrary homomorphisms $B \to A_1$ and $B \to A_2$ of k-affinoid algebras, the complete tensor product $A_1 \hat{\otimes}_B A_2$ can be constructed by taking suitable norms on A_1 and A_2. The B-algebra $A_1 \hat{\otimes}_B A_2$ is then k-affinoid by Proposition 6.1.1/10. Moreover, according to Proposition 2.1.8/5, equivalent norms on A_1 and A_2 (and even on B as a simple computation shows) lead to equivalent norms on $A_1 \hat{\otimes}_B A_2$. Thus, in our situation, $A_1 \hat{\otimes}_B A_2$ is a well-defined k-affinoid algebra, uniquely determined up to isomorphism.

Proposition 4. *Let $\varphi: B \to A$ be a finite homomorphism of a k-affinoid algebra B into a k-Banach algebra A. Then φ is strict and A is k-affinoid.*

Proof. According to Proposition 6.1.1/6 the k-algebra A can be provided with a complete k-algebra norm such that φ is strict and A is k-affinoid with respect to this new norm. From Proposition 2, we conclude that the corresponding k-affinoid topology of A coincides with the given Banach topology. □

Remark. We see that the category $\mathfrak{A}$ could also have been defined as the category of all k-Banach algebras A permitting a finite (not necessarily surjective and continuous) homomorphism from some T_n into A. Furthermore, this category is equivalent to the (purely algebraic) category of all k-algebras which are finite over some T_n.

6.1.4. Examples. Generalized rings of fractions. — If A is a k-affinoid algebra with norm $|\ \ |$ and if $X = (X_1, \ldots, X_m)$ denotes a system of indeterminates, then the ring $A\langle X\rangle$ of strictly convergent power series over A is k-affinoid. Similarly one can consider strictly convergent Laurent series; define

$$A\langle X, X^{-1}\rangle := \Big\{\sum_{\nu_i \in \mathbb{Z}} a_{\nu_1 \ldots \nu_m} X_1^{\nu_1} \ldots X_m^{\nu_m};\ a_\nu \in A \text{ and } |a_\nu| \to 0 \quad \text{for } |\nu_1| + \cdots + |\nu_m| \to \infty\Big\}.$$

If $\sum a_\nu X^\nu$ and $\sum b_\mu X^\mu$ are elements of $A\langle X, X^{-1}\rangle$, then $c_\lambda := \sum_{\nu+\mu=\lambda} a_\nu b_\mu$ converges for all $\lambda \in \mathbb{Z}$ and $\sum_\lambda c_\lambda X^\lambda$ is again an element of $A\langle X, X^{-1}\rangle$. Thus, it is easily verified that $A\langle X, X^{-1}\rangle$ is a k-Banach algebra with the norm given by

$$|\sum a_\nu X^\nu| := \max |a_\nu|.$$

Furthermore, $A\langle X, X^{-1}\rangle$ is even k-affinoid. Namely, let $Y = (Y_1, \ldots, Y_m)$ denote a second system of indeterminates. Then, according to Proposition 6.1.1/4, the injection $A\langle X\rangle \to A\langle X, X^{-1}\rangle$ extends to a homomorphism $A\langle X, Y\rangle \to A\langle X, X^{-1}\rangle$, $Y \mapsto X^{-1}$, which is obviously surjective.

The k-affinoid algebra $A\langle X, X^{-1}\rangle$ contains, in particular, the ring $A\langle X\rangle\,[X^{-1}]$ which stands for the localization of $A\langle X\rangle$ by X, and it is clear that $A\langle X\rangle\,[X^{-1}]$ is dense in $A\langle X, X^{-1}\rangle$. Hence, in some sense, $A\langle X, X^{-1}\rangle$ is the "smallest" k-affinoid algebra over $A\langle X\rangle$ such that $X_1, \ldots, X_m$ become units. We want to carry out similar constructions in a more general situation.

As before, let A denote a k-affinoid algebra, and let $f = (f_1, \ldots, f_m)$ and $g = (g_1, \ldots, g_n)$ be systems of elements in A. We are looking for a k-Banach algebra A' over A such that the g_j become units and the f_i, g_j^{-1} are power-bounded in A'. In order to give a construction for A', we start with the ring of fractions $A[g^{-1}]$. Any element $a \in A[g^{-1}]$ can be written as a finite sum

$$a = \sum_{\mu_i,\nu_j \geq 0} a_{\mu\nu} f^\mu g^{-\nu}, \qquad a_{\mu\nu} \in A,$$

and we define a semi-norm on $A[g^{-1}]$ by

$$|a| = \inf\,(\max_{\mu,\nu} |a_{\mu\nu}|),$$

where the infimum runs over all possible representations of a. Note that this semi-norm on $A[g^{-1}]$ depends not only on the system g but also on the system f. It is a natural semi-norm such that the elements of f and g^{-1} become power-bounded. Hence $A\langle f, g^{-1}\rangle$, the completion of $A[g^{-1}]$, is a k-Banach algebra over A, which has the properties we are looking for. We will see below that $A\langle f, g^{-1}\rangle$ is even k-affinoid; however, first note that, by construction, the canonical map $A \to A\langle f, g^{-1}\rangle$ satisfies the following universal property.

Proposition 1. *Let $\varphi\colon A \to B$ denote a continuous homomorphism from the k-affinoid algebra A into a k-Banach algebra B such that the elements $\varphi(g_j)$ are units and the elements $\varphi(f_i)$, $\varphi(g_j)^{-1}$ are power-bounded. Then there is a unique continuous homomorphism $\varphi'\colon A\langle f, g^{-1}\rangle \to B$ such that the diagram*

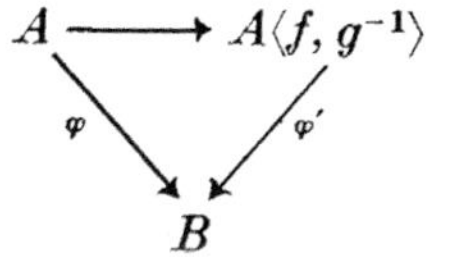

commutes.

In particular, if A is replaced by the algebra $A\langle X\rangle$ of strictly convergent power series in $X = (X_1, \ldots, X_m)$, and if $f := \emptyset$ and $g := X$, we see that the algebra $A\langle X\rangle\,\langle X^{-1}\rangle$ is canonically isomorphic to the algebra $A\langle X, X^{-1}\rangle$ of strictly convergent Laurent series in X. Namely, both algebras satisfy the same universal property. (The isomorphism can also be obtained by a direct argument.) We always use the notation $A\langle X, X^{-1}\rangle$ instead of $A\langle X\rangle\,\langle X^{-1}\rangle$.

Returning to the general case, there is another possible way to construct a k-Banach algebra A' over A satisfying the required properties. Let $X = (X_1, \ldots, X_m)$ and $Y = (Y_1, \ldots, Y_n)$ denote systems of indeterminates. Then

$A' := A\langle X, Y\rangle/(X - f,\ gY - 1)$ is obviously a k-Banach and even a k-affinoid algebra over A such that the g_j become units and the f_i, g_j^{-1} are power-bounded in A'. Furthermore, the canonical map $A \to A\langle X, Y\rangle/(X - f,\ gY - 1)$ also satisfies the universal property stated in Proposition 1. Namely, let $\varphi: A \to B$ be a continuous homomorphism as in Proposition 1. Then φ extends to a continuous homomorphism $\varphi'': A\langle X, Y\rangle \to B$, $X \mapsto \varphi(f)$, $Y \mapsto \varphi(g)^{-1}$, with

$$(X - f,\ gY - 1) \subset \ker \varphi''.$$

Thus φ'' gives rise to a continuous homomorphism φ': $A\langle X, Y\rangle/(X - f,$ $gY - 1) \to B$ such that the diagram

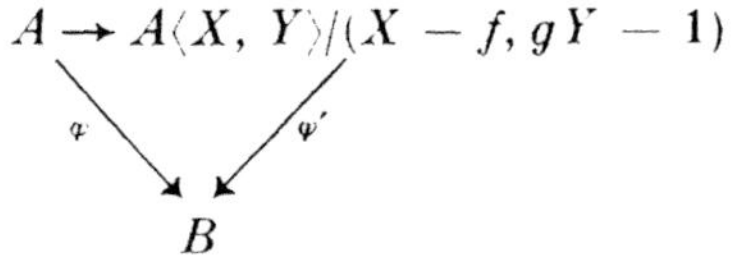

commutes, and φ' is uniquely determined by this diagram since the residue classes of X and Y must be mapped by φ' onto $\varphi(f)$ and $\varphi(g)^{-1}$ respectively. In particular, taking $B := A\langle f, g^{-1}\rangle$, we get

Proposition 2. *The continuous homomorphism* $A\langle X, Y\rangle \to A\langle f, g^{-1}\rangle$, $X \mapsto f$, $Y \mapsto g^{-1}$, *is surjective and gives rise to a strict isomorphism* $A\langle X, Y\rangle/(X - f, gY - 1) \xrightarrow{\sim} A\langle f, g^{-1}\rangle$.

Providing $A\langle X, Y\rangle/(X - f,\ gY - 1)$ with the canonical residue norm, it is not hard to see that the above isomorphism is in fact isometric. In particular, it is now clear that $A\langle f, g^{-1}\rangle$ is k-affinoid.

A few remarks concerning the notation $A\langle f, g^{-1}\rangle = A\langle f_1, \ldots, f_m, g_1^{-1}, \ldots, g_n^{-1}\rangle$ seem to be necessary. In the case where $g = \emptyset$, we simply write $A\langle f\rangle$ instead of $A\langle f, g^{-1}\rangle$; likewise we write $A\langle g^{-1}\rangle$ if $f = \emptyset$. Note also that $A\langle h^{-1}\rangle$ is defined in two ways when $h \in A$ is a unit. However, no difficulties will arise from that, since both definitions coincide in this special case. Finally it follows from Proposition 2 that we have associativity in the following sense:

$$A\langle f_1, \ldots, f_{m-1}, g_1^{-1}, \ldots, g_{n-1}^{-1}\rangle\langle f_m, g_n^{-1}\rangle = A\langle f_1, \ldots, f_m, g_1^{-1}, \ldots, g_n^{-1}\rangle.$$

There is another procedure which partially generalizes the above one. Let $g, f_1, \ldots, f_m \in A$ be elements generating the unit ideal in A; i.e., there are elements $a, a_1, \ldots, a_m \in A$ such that

$$ag + \sum_{i=1}^{m} a_i f_i = 1.$$

We are looking for a k-Banach algebra A' over A such that g becomes a unit and such that the fractions $\frac{f_i}{g}$ are power-bounded in A'. Since, in the ring of fractions $A[g^{-1}]$, we have

$$g^{-1} = a + \sum_{i=1}^{m} a_i \frac{f_i}{g},$$

it is clear that any element $b \in A[g^{-1}]$ can be written as a finite sum

$$b = \sum b_\nu \left(\frac{f}{g}\right)^\nu, \qquad b_\nu \in A,$$

where $\frac{f}{g}$ stands for the system $\left(\frac{f_1}{g}, \ldots, \frac{f_m}{g}\right)$. Similarly as before, one defines a semi-norm on $A[g^{-1}]$ by

$$|b| = \inf(\max |b_\nu|),$$

where the infimum runs over all possible representations of b. The completion of $A[g^{-1}]$ is denoted by $A\left\langle\frac{f}{g}\right\rangle$; it is a k-Banach algebra having the properties we are looking for. Furthermore, the canonical map $A \to A\left\langle\frac{f}{g}\right\rangle$ satisfies the following universal property:

Proposition 3. *Let $\varphi: A \to B$ denote a continuous homomorphism from the k-affinoid algebra A into the k-Banach algebra B such that $\varphi(g)$ is a unit and the elements $\frac{\varphi(f_i)}{\varphi(g)}$ are power-bounded. Then there is a unique continuous homomorphism $\varphi': A\left\langle\frac{f}{g}\right\rangle \to B$ such that*

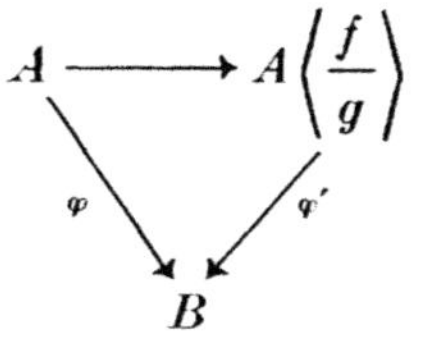

commutes.

Also in this case, we want to have an explicit description of $A\left\langle\frac{f}{g}\right\rangle$ which shows that it is k-affinoid. Let $X = (X_1, \ldots, X_m)$ be a system of indeterminates, and consider the k-affinoid algebra $A' = A\langle X\rangle/(gX - f)$. With $\overline{X}_i$ denoting the residue class of X_i in A', we get

$$\left(a + \sum_{i=1}^{m} a_i \overline{X}_i\right) g = ag + \sum_{i=1}^{m} a_i f_i = 1$$

which shows that g is a unit in A'. Moreover, $\overline{X}_i = \frac{f_i}{g}$ in A'; hence, the elements $\frac{f_i}{g}$ must be power-bounded in A'. It is now a straightforward verification to see that also $A' = A\langle X\rangle/(gX - f)$ satisfies the universal property stated in Proposition 3. Thus we get

Proposition 4. *The continuous homomorphism $A\langle X\rangle \to A\left\langle\frac{f}{g}\right\rangle$, $X \mapsto \frac{f}{g}$, is surjective and gives rise to a strict isomorphism $A\langle X\rangle/(gX - f) \to A\left\langle\frac{f}{g}\right\rangle$.*

Again, providing $A\langle X\rangle/(gX - f)$ with the canonical residue norm, one shows that the above isomorphism is isometric. Note also that our definition of $A\left\langle\frac{f}{g}\right\rangle$ is compatible with the one given before, if the f_i or g are units.

6.1.5. Further examples. Convergent power series on general polydiscs. — If the ground field k were equal to $\mathbb{R}$ or $\mathbb{C}$, then all polydiscs $P_\varrho(k) := P_\varrho := \{x \in k^n;\ |x_i| \leqq \varrho_i \text{ for } i = 1, \ldots, n\}$, with radii $\varrho_1, \ldots, \varrho_n > 0$ would be essentially the same, because $(x_1, \ldots, x_n) \mapsto (\varrho_1 x_1, \ldots, \varrho_n x_n)$ would map the unit polydisc bijectively (and bi-analytically) onto P_ϱ. In our case, i.e., with a non-Archimedean k, such a mapping is by no means always available, and in fact the polydiscs P_ϱ (and the algebras of power series convergent on P_ϱ) behave rather differently depending on ϱ. This shall be elaborated on in the present section.

Let $X = (X_1, \ldots, X_n)$ be a set of indeterminates, and let ϱ be an n-tuple of positive real numbers. It is clear that a formal power series $f = \sum a_\nu X^\nu \in k[\![X]\!]$ is convergent on the polydisc P_ϱ if $\lim |a_\varrho|\, \varrho^\nu = 0$. Conversely, if the components of ϱ belong to $|k^*|$, any series converging on P_ϱ must satisfy this condition. Therefore we define

$$T_{n,\varrho} = \left\{\sum a_\nu X^\nu \in k[\![X]\!];\ \lim |a_\nu|\, \varrho^\nu = 0\right\}.$$

In particular, $T_{n,\varrho} = T_n$ if $\varrho = (1, \ldots, 1)$. Generalizing the Gauss norm on T_n, we set for any $f = \sum a_\nu X^\nu \in T_{n,\varrho}$

$$|f|_\varrho := \max |a_\nu|\, \varrho^\nu .$$

Then, similarly as in the case $\varrho = (1, \ldots, 1)$, one shows that

Proposition 1. *The series in $T_{n,\varrho}$ form a k-subalgebra of $k[\![X]\!]$. The map $|\ \ |_\varrho$ is a k-algebra norm on $T_{n,\varrho}$, making it a k-Banach algebra, which contains $k[X]$ as a dense subalgebra.*

Furthermore, an argument similar to the one used in the classical proof of the Gauss Lemma (see (1.5.3)) shows that, in fact,

Proposition 2. *The norm $|\ \ |_\varrho$ is a valuation on $T_{n,\varrho}$.*

If ϱ consists of a tuple of numbers in $|k^*|$, then by definition, $T_{n,\varrho}$ consists of precisely those power series $f \in k[\![X]\!]$ which converge on the polydisc $P_\varrho(k)$. This assertion cannot be maintained if not all components of ϱ belong to $|k^*|$. For example if $n = 1$ and $\varrho \notin |k^*|$, then convergence on $P_\varrho(k) = B^+(0, \varrho)$ is the same as convergence on $B^-(0, \varrho)$. However there are power series $f = \sum a_\nu X^\nu$ converging on $B^-(0, \varrho)$, which do not satisfy the condition $\lim |a_\nu|\, \varrho^\nu = 0$, so that $f \notin T_{1,\varrho}$ in this case.

As a by-product of Proposition 2, it follows that there are valued fields k' extending k such that $|k'|$ contains arbitrary prescribed values $\varrho_1, \ldots, \varrho_n > 0$. Just take for k' the field of fractions of $T_{n,\varrho}$. Thereby we see that

Proposition 3. *The algebra $T_{n,\varrho}$ consists of precisely those series $f \in k[\![X]\!]$ such that f converges on $P_\varrho(k')$ for all complete fields k' extending k.*

We want to characterize the tuples ϱ, for which $T_{n,\varrho}$ is k-affinoid. Of course if $\varrho = (|c_1|, \ldots, |c_n|)$, where $c_1, \ldots, c_n \in k^*$, then $T_n \to T_{n,\varrho}$, $X_i \mapsto c_i^{-1}X_i$, defines an isometric isomorphism between T_n and $T_{n,\varrho}$ so that $T_{n,\varrho}$ is k-affinoid in this case. In order to deal with the general case, let k_a be the algebraic closure of k. Then a positive real α belongs to $|k_a^*|$ if and only if $\alpha^s \in |k^*|$ for some $s \in \mathbb{N}$.

Theorem 4. *The algebra $T_{n,\varrho}$ is k-affinoid if and only if all components ϱ_i of ϱ belong to $|k_a^*|$.*

Proof. First assume that there are $s_1, \ldots, s_n \in \mathbb{N}$ and $c_1, \ldots, c_n \in k^*$ such that $\varrho_i^{s_i} = |c_i^{-1}|$ for $i = 1, \ldots, n$. Then $|c_i X_i^{s_i}|_\varrho = 1$, and we can define a monomorphism φ: $T_n \to T_{n,\varrho}$ by setting $\varphi(X_i) := c_i X_i^{s_i}$, $i = 1, \ldots, n$ (see Proposition 6.1.1/4). We claim that φ is finite. Take $f = \sum a_\nu X^\nu \in T_{n,\varrho}$ and write

$$f = \sum_{\substack{\lambda \\ 0 \leqq \lambda_i < s_i}} X^\lambda \Big(\sum_{\substack{\mu \\ 0 \leqq \mu_i}} a_{\mu s + \lambda} c^{-\mu} (cX^s)^\mu \Big),$$

where $\lambda = (\lambda_1, \ldots, \lambda_n)$, $\mu = (\mu_1, \ldots, \mu_n)$, $s = (s_1, \ldots, s_n)$, and $c = (c_1, \ldots, c_n)$. For each λ with $0 \leqq \lambda_i < s_i$, define

$$g_\lambda := \sum_\mu (a_{\mu s + \lambda} c^{-\mu}) X^\mu \in k[\![X]\!].$$

Then $g_\lambda \in T_n$, since

$$|a_{\mu s + \lambda} c^{-\mu}| = |a_{\mu s + \lambda}| \varrho^{\mu s} = |a_{\mu s + \lambda}| \varrho^{\mu s + \lambda} \varrho^{-\lambda} \to 0$$

as $|\mu| \to \infty$. We have

$$f = \sum_{\substack{\lambda \\ 0 \leqq \lambda_i < s_i}} \varphi(g_\lambda) X^\lambda$$

so that $T_{n,\varrho}$ is a finite T_n-module via φ; the monomials X^λ, $0 \leqq \lambda_i < s_i$, $i = 1, \ldots, n$, are generators. Thus, $T_{n,\varrho}$ is k-affinoid by Proposition 6.1.1/5, and half of the theorem is proved.

To show the other half, assume that $T_{n,\varrho}$ is k-affinoid. Choose a finite normalization monomorphism φ: $T_d \to T_{n,\varrho}$. Then φ is strict by Proposition 6.1.3/4. Furthermore, φ must be an isometry, because both norms $|\ \ |$ (on T_d) and $|\ \ |_\varrho$ (on $T_{n,\varrho}$) are power-multiplicative (use Proposition 3.1.5/1). Thus we can apply Proposition 3.1.5/2 and thereby find integers $s_1, \ldots, s_n \in \mathbb{N}$ satisfying $\varrho_i^{s_i} = |X_i|^{s_i} \in |Q(T_d)| = |k|$ for $i = 1, \ldots, n$. □

Remark. If one uses the results of (6.2), the above ad hoc proof for the second part of the theorem can be replaced by the following consideration. If $T_{n,\varrho}$ is k-affinoid, the power-multiplicative norm $|\ \ |_\varrho$ must coincide with the supremum norm $|\ \ |_{\sup}$ on $T_{n,\varrho}$; the supremum norm takes values only in the value group of the algebraic closure of k.

Proposition 5. *If all components* ϱ_i *of* ϱ *belong to* $|k_a^*|$, *then the norm* $|\ |_\varrho$ *coincides with the supremum norm* $|\ |_{\sup}$ *on* $T_{n,\varrho}$, *and* $T_{n,\varrho}$ *is a Banach function algebra.*

Proof. Choose a finite algebraic extension k' of k such that $\varrho_1, \ldots, \varrho_n \in |k'|$ and consider $T_{n,\varrho}(k)$ as a k-subalgebra of $T_{n,\varrho}(k')$ (we add k, respectively k', in brackets in order to specify the ground field). Then $T_{n,\varrho}(k')$, considered as a k'-algebra, is a Banach function algebra, because there is an isometric k'-isomorphism $T_n(k') \to T_{n,\varrho}(k')$, and because $T_n(k')$ is a Banach function algebra (Corollary 5.1.4/6). Since the k-algebraic and the k'-algebraic maximal ideals coincide in $T_n(k')$ (k' is finite over k), we see that $T_{n,\varrho}(k')$ is also a Banach function algebra over k. Since $T_{n,\varrho}(k)$ is closed in $T_{n,\varrho}(k')$, it follows from Lemma 3.8.3/4 that $T_{n,\varrho}(k)$ is a Banach function algebra. □

Actually, the assumption of Proposition 5 is superflous. This relies on the fact that one has

$$|f|_\varrho = \sup \{|f(x)|\, ; x \in P_\varrho(k_a)\}$$

for all $f \in T_{n,\varrho}$ (where k_a is the algebraic closure of k). Knowing this, one can conclude along the lines of the proof of Corollary 5.1.4/6 in order to see that $T_{n,\varrho}$ is a Banach function algebra. Fruthermore, $T_{n,\varrho}$ satisfies the Maximum Modulus Principle if and only if the above supremum is always assumed. This is equivalent to the fact that $T_{n,\varrho}$ is k-affinoid.

See VAN DER PUT [30] and GÜNTZER [18] for further results on the algebras $T_{n,\varrho}$.

6.2. The spectrum of a k-affinoid algebra and the supremum semi-norm

6.2.1. The supremum semi-norm. — Let A denote an arbitrary k-affinoid algebra. We want to introduce an intrinsic semi-norm on every such algebra A, not depending on the incidental representation of A as a residue algebra of some algebra T_n. Such an intrinsic semi-norm is already at our disposal: we take the supremum semi-norm $|\ |_{\sup}$ as defined and studied in (3.8). For the convenience of the reader, we repeat some of the relevant material given there and adapt it to our special situation.

According to Corollary 6.1.2/3, all maximal ideals in a k-affinoid algebra A are k-algebraic. Therefore the supremum semi-norm is computed using the *whole* spectrum of maximal ideals $\operatorname{Max} A$. Furthermore, because A is a k-Banach algebra, Corollary 3.8.2/2 yields that $|\ |_{\sup}$ is finite; more precisely, $|f|_{\sup} \leq |f|_\alpha$ for all $f \in A$ and all epimorphisms $\alpha\colon T_n \to A$. Thus using Lemma 3.8.1/3, we see that

Lemma 1. *The function* $|\ |_{\sup}\colon A \to \mathbb{R}_+$ *defined by*

$$|f|_{\sup} := \sup_{x \in \operatorname{Max} A} |f(x)|, \quad f \in A,$$

is a power-multiplicative k-algebra semi-norm.

Definition 2. *This semi-norm is called the supremum (or the spectral) semi-norm on A. It is also referred to as the semi-norm of uniform convergence on* Max A.

Let red $A = A/\text{rad}\,A$ denote the nilreduction of A, and let red: $A \to \text{red}\,A$ be the canonical residue map. Since A is Noetherian, there are only finitely many minimal prime ideals $\mathfrak{p}_1, \ldots, \mathfrak{p}_r$ in A. Let $\pi_i\colon A \to A/\mathfrak{p}_i$, $i = 1, \ldots, r$, denote the canonical residue maps. With these notations we state the following lemma which allows us to reduce problems concerning affinoid algebras with zero divisors to integral domains.

Lemma 3. *For each $f \in A$, one has*

$$|f|_{\text{sup}} = \max_{1 \leq i \leq r} |\pi_i(f)|_{\text{sup}} \quad \textit{and} \quad |f|_{\text{sup}} = |\text{red}\,f|_{\text{sup}}.$$

Proof. The first equation is the assertion of Lemma 3.8.1/5. The second one follows from the fact that the nilradical rad A is contained in each maximal ideal of A. □

According to Corollary 5.1.4/6, the spectral semi-norm coincides with the Gauss norm on T_n. This fact allows us — by means of Proposition 3.8.1/7 — to extend some of our results on free Tate algebras to general affinoid algebras. Let k_a be the algebraic closure of k.

Proposition 4. (i) *The Maximum Modulus Principle holds for $|\ \ |_{\text{sup}}$ on each k-affinoid algebra A.*

(ii) *For all $f \in A$ such that $|f|_{\text{sup}} \neq 0$, there are $c \in k$ and $m \in \mathbb{N}$ such that $|cf^m|_{\text{sup}} = 1$. Consequently, $|A|_{\text{sup}} \subset |k_a|$.*

(iii) *An element $f \in A$ is nilpotent if and only if $|\ \ |_{\text{sup}} = 0$. In particular, $|\ \ |_{\text{sup}}$ is a norm on A if and only if A is reduced.*

Proof. Let us first consider the special case, where A is an integral domain. Applying the Noether Normalization Lemma (Corollary 6.1.2/2), we find a finite monomorphism $\varphi\colon T_d \to A$ for some $d \geq 0$. Since T_d is integrally closed (Theorem 5.2.6/2) and since the Maximum Modulus Principle holds for T_d (Corollary 5.1.4/6), assertions (i) and (ii) follow from Proposition 3.8.1/7. Furthermore, this proposition shows that $|\ \ |_{\text{sup}}$ is a norm on A (since $|\ \ |_{\text{sup}}$ is a norm on T_d).

Now let A be arbitrary. Denote by $\mathfrak{p}_1, \ldots, \mathfrak{p}_r$ the minimal prime ideals in A. By what we have just seen, assertions (i) and (ii) are true for the algebras $A/\mathfrak{p}_i$, $i = 1, \ldots, r$. Thus by Lemma 3, they must also be true for A. Furthermore it follows that $|\ \ |_{\text{sup}}$ is a norm on red $A = A/\text{rad}\,A$, since rad $A = \bigcap_{i=1}^{r} \mathfrak{p}_i$. Therefore assertion (iii) is clear by Lemma 3. □

Combining assertion (iii) of the above proposition with the assertion of Proposition 3.8.2/3, we get the following improvement of Theorem 6.1.3/1 for reduced k-affinoid algebras.

Corollary 5. *If A is a reduced k-affinoid algebra, then each homomorphism of a (not necessarily Noetherian) k-Banach algebra into A is continuous.*

Remark. The assertion (iii) of Proposition 4 simply says that

$$\operatorname{rad} A = \bigcap_{\mathfrak{m} \in \operatorname{Max} A} \mathfrak{m}.$$

If $A \cong T_n/\mathfrak{a}$, where $\mathfrak{a}$ is an ideal in the free TATE algebra T_n, this is equivalent to

$$\operatorname{rad} \mathfrak{a} = \bigcap_{\substack{\mathfrak{m} \in \operatorname{Max} T_n \\ \mathfrak{a} \subset \mathfrak{m}}} \mathfrak{m}.$$

Thus assertion (iii) of Proposition 4 is equivalent to the fact that each T_n is a Jacobson ring. (The latter has already been obtained in Theorem 5.2.6/3. The proof of this theorem can be viewed as a refinement of the detailed considerations in (3.8) which led to the proof of Proposition 4.)

6.2.2. Integral homomorphisms. — Some facts from (3.8) about integral homomorphisms have already implicitly been used in the preceding section. Here we want to write down explicitly some properties of such homomorphisms. First, Lemmata 3.8.1/4 and 3.8.1/6 yield

Proposition 1. *Every homomorphism of k-affinoid algebras $\varphi\colon B \to A$ is a contraction with respect to the supremum semi-norm. If φ is an integral monomorphism, it is an isometry.*

Since T_d is a valued integrally closed domain, we can derive from Proposition 3.8.1/7 (a) and (d) the following result.

Proposition 2. *Let $\varphi\colon T_d \to A$ be an integral torsion-free monomorphism into some k-affinoid algebra A. Then $|\ \ |_{\sup}$ is a faithful T_d-algebra norm on A (i.e., $|\varphi(t)\, f|_{\sup} = |t|\, |f|_{\sup}$ for all $f \in A$ and all $t \in T_d$). If*

$$f^n + \varphi(t_1)\, f^{n-1} + \cdots + \varphi(t_n) = 0$$

is the integral equation of minimal degree for f over T_d, then one has

$$|f|_{\sup} = \max_{1 \leq i \leq n} |t_i|^{1/i}.$$

We want to extend the equation $|f|_{\sup} = \max\limits_{1 \leq i \leq n} |t_i|^{1/i}$ to the case where $\varphi\colon B \to A$ is an arbitrary integral homomorphism of k-affinoid algebras. We need a simple lemma, which is closely related to Proposition 3.1.2/1.

Lemma 3. *Let $\varphi\colon B \to A$ be a homomorphism of k-affinoid algebras. Let $f \in A$, $b_1, \ldots, b_n \in B$ such that*

$$f^n + \varphi(b_1)\, f^{n-1} + \cdots + \varphi(b_n) = 0.$$

Then

$$|f|_{\sup} \leqq \max_{1\leqq i\leqq n} |b_i|_{\sup}^{1/i}.$$

Proof. There exists an index j, $1 \leqq j \leqq n$, such that

$$|f|_{\sup}^n = |f^n|_{\sup} \leqq |\varphi(b_j)\, f^{n-j}|_{\sup} \leqq |b_j|_{\sup}\, |f|_{\sup}^{n-j}.$$

Then $|f|_{\sup} \leqq |b_j|_{\sup}^{1/j}$. □

Now we are able to extend Proposition 2.

Proposition 4. *Let $\varphi: B \to A$ be an integral homomorphism of affinoid algebras. Then for each $f \in A$, there exists a monic polynomial $q = X^n + b_1X^{n-1} + \cdots + b_n \in B[X]$ such that $q(f) = 0$ and*

$$|f|_{\sup} = \sigma(q) = \max_{1\leqq i\leqq n} |b_i|_{\sup}^{1/i}.$$

Proof. According to the preceding lemma, it suffices to show $|f|_{\sup} \geqq \sigma(q)$ in order to get equality. First we treat the special case where A is an integral domain. Theorem 6.1.2/1 provides us with a homomorphism $\psi: T_d \to B$ such that $\varphi \circ \psi: T_d \to B \to A$ is an integral monomorphism. The map $\varphi \circ \psi$ is torsion-free. Hence we may apply Proposition 2 to find that, for all $f \in A$, we have $|f|_{\sup} = \sigma(p)$, where $p \in T_d[X]$ is the minimal polynomial of f over T_d with respect to $\varphi \circ \psi$. This means in particular that $p(f) = 0$. Consider the polynomial $q \in B[X]$, which is obtained from p by replacing all its coefficients by their ψ-images in B. Clearly, $q(f) = p(f) = 0$, and $|f|_{\sup} = \sigma(p) \geqq \sigma(q)$. Thus we have proved the proposition in the case where A is an integral domain.

In order to take care of the general case, let $\mathfrak{p}_1, \ldots, \mathfrak{p}_r$ be the minimal prime ideals in A and, for $i = 1, \ldots, r$, denote by π_i the residue epimorphism $A \to A/\mathfrak{p}_i$. According to what we already proved, there are monic polynomials $q_i \in B[X]$, $i = 1, \ldots, r$, such that $q_i(\pi_i(f)) = 0$ and $|\pi_i(f)|_{\sup} = \sigma(q_i)$. The first relation can be rephrased as $q_i(f) \in \mathfrak{p}_i$ for $i = 1, \ldots, r$. If one defines $q^* := \prod_{i=1}^{r} q_i$, one gets a monic polynomial in $B[X]$ such that

$$q^*(f) \in \bigcap_{i=1}^{r} \mathfrak{p}_i = \operatorname{rad} A.$$

Then there is an exponent $e \in \mathbb{N}$ such that $(q^*(f))^e = 0$. Setting $q := q^{*e}$, we get a monic polynomial in $B[X]$ such that $q(f) = 0$. Furthermore, Proposition 1.5.4/1 gives us

$$\sigma(q) \leqq \max_{1\leqq i\leqq r} \sigma(q_i) = \max_{1\leqq i\leqq r} |\pi_i(f)|_{\sup} = |f|_{\sup},$$

where the last equality follows from Lemma 6.2.1/3. Thus the proof is finished. □

6.2.3. Power-bounded and topologically nilpotent elements. — Let A be a k-affinoid algebra. Then each norm $|\ |$, inducing the given Banach topology on A, satisfies $|\ |_{\sup} \leqq |\ |$ (Corollary 3.8.2/2). In particular, all power-bounded elements $f \in A$ must satisfy $|f|_{\sup} \leqq 1$, since $|\ |_{\sup}$ is power-multiplicative. The converse is also true.

Proposition 1. *For each $f \in A$, the following statements are equivalent:*

(i) *f is power-bounded.*

(ii) $|f|_{\sup} \leqq 1$.

Proof. We have only to show that f is power-bounded if $|f|_{\sup} \leqq 1$. Choose a finite homomorphism $\varphi\colon T_d \to A$ (for example, an epimorphism). Then due to Proposition 6.2.2/4, there is an integral equation

$$f^n + t_1 f^{n-1} + \cdots + t_n = 0$$

of f over T_d such that

$$|f|_{\sup} = \max_{1 \leqq i \leqq n} |t_i|^{1/i}.$$

We have $t_1, \ldots, t_n \in \mathring{T}_d$ if $|f|_{\sup} \leqq 1$. Induction on ν gives then

$$f^{n+\nu} \in \sum_{i=0}^{n-1} \varphi(\mathring{T}_d)\, f^i, \qquad \nu = 0, 1, 2, \ldots .$$

Since $\varphi(\mathring{T}_d)$ is bounded in A, we see that $\sum_{i=0}^{n-1} \varphi(\mathring{T}_d)\, f^i$ is bounded. □

From this proposition we may derive the following characterization of topologically nilpotent elements.

Proposition 2. *For each $f \in A$, the following statements are equivalent:*

(i) *f is topologically nilpotent,*

(ii) $|f(x)| < 1$ *for all* $x \in \operatorname{Max} A$,

(iii) $|f|_{\sup} < 1$.

Proof. Statements (ii) and (iii) are equivalent due to the Maximum Modulus Principle (Proposition 6.2.1/4). Furthermore, statement (i) implies statement (iii), since any Banach norm on A dominates $|\ |_{\sup}$ and since $|\ |_{\sup}$ is power-multiplicative. In order to verify the opposite direction, assume $|f|_{\sup} < 1$. Then there exist a constant $c \in k$, $|c| > 1$, and an integer $m > 0$ such that $|cf^m|_{\sup} \leqq 1$. This follows from Proposition 6.2.1/4 (ii) if $|f|_{\sup} \neq 0$ and is trivial if $|f|_{\sup} = 0$. We have $cf^m \in \mathring{A}$ by Proposition 1. Therefore $f^m \in c^{-1}\mathring{A} \subset \check{A}$, and we see that f^m and hence also f are topologically nilpotent. □

Furthermore, Proposition 1 allows to compute $|\ |_{\sup}$ in terms of an arbitrary complete k-algebra norm on A.

Proposition 3. *Let* $|\ \ |$ *be a complete* k*-algebra norm on* A. *Then* $|f|_{\sup} = \inf_{i \in \mathbb{N}} |f^i|^{1/i}$ *for all* $f \in A$.

Proof. Define $|f|' := \inf_{i \in \mathbb{N}} |f^i|^{1/i}$ for all $f \in A$. Then $|\ \ |'$ is a power-multiplicative k-algebra semi-norm according to Proposition 1.3.2/1, and it follows from Corollary 3.8.2/2 that $|f|_{\sup} \leqq |f|'$ for all $f \in A$. The opposite inequality shall be shown indirectly. Assume that $|f|_{\sup} < |f|'$ for some $f \in A$. Proposition 6.2.1/4 (ii) allows us to assume $|f|_{\sup} = 1$, and hence $|f|' > 1$. This implies $|f^i| \geqq |f|'^i \to \infty$, and therefore f cannot be power-bounded, in contradiction to Proposition 1. □

Remark. For affinoid algebras without zero divisors, the preceding propositions are direct consequences of Proposition 3.8.2/5 and Corollary 3.8.2/6.

Using Propositions 1 and 2, we can give the following description of the residue algebra $\tilde{A} = \mathring{A}/\check{A}$ (as defined in (1.2.5)).

Proposition 4. $\tilde{A} = \{f \in A;\ |f|_{\sup} \leqq 1\}/\{f \in A;\ |f|_{\sup} < 1\}$.

We want to finish this section by giving a criterion for $|\ \ |_{\sup}$ to be a valuation on A. Namely, using Propositions 1.5.3/1, 6.2.1/4, and the above Proposition 4, we conclude that

Proposition 5. *The supremum semi-norm is a valuation on* A *if and only if* A *is reduced and* $\tilde{A}$ *is an integral domain.*

Even if A is an integral domain, the supremum semi-norm $|\ \ |_{\sup}$ is not in general a valuation on A. This can be seen from the following example.

Consider the k-algebra $k\langle X, X^{-1}\rangle$ of strictly convergent Laurent series in one variable X over k and look at the subalgebra A of all series which converge on the annulus $\{x \in k;\ |c| \leqq |x| \leqq 1\}$, where $c \in k$, $0 < |c| < 1$, is fixed. Then

$$A = \left\{f = \sum_{\nu=-\infty}^{+\infty} a_\nu X^\nu;\ \lim_{\nu\to\infty} a_\nu = 0,\ \lim_{\nu\to-\infty} c^\nu a_\nu = 0\right\},$$

and A is an integral domain, since $k\langle X, X^{-1}\rangle$ is an integral domain. A direct computation shows that there is a canonical isomorphism

$$A \simeq k\langle X, Y\rangle/(XY - c).$$

Identifying A with $k\langle X, Y\rangle/(XY - c)$, the algebra A becomes k-affinoid. Let $f_1, f_2 \in A$ denote the residue classes of $X, Y \in k\langle X, Y\rangle$. The ideals $(X - 1, Y - c)$ and $(X - c, Y - 1)$ are maximal ideals in $k\langle X, Y\rangle$ containing the ideal $(XY - c)$, because

$$XY - c = (X - 1)\,Y + (Y - c) = X(Y - 1) + (X - c).$$

Therefore

$$x_1 := (X - 1,\ Y - c)/(XY - c),$$

and

$$x_2 := (X - c, Y - 1)/(XY - c)$$

are maximal ideals in A. Since $f_1(x_1) = 1 = f_2(x_2)$, we see that $|f_1|_{\sup} = 1 = |f_2|_{\sup}$. However $|f_1 f_2|_{\sup} = |c|_{\sup} = |c| < 1$. Consequently, $|\ \ |_{\sup}$ cannot be a valuation on A.

6.2.4. Reduced k-affinoid algebras are Banach function algebras. — In this section we study the relationship between the supremum semi-norm $|\ \ |_{\sup}$ and the Banach topology on a k-affinoid algebra A.

Theorem 1. *Every reduced k-affinoid algebra A is a Banach function algebra; i.e.,* $|\ \ |_{\sup}$ *is a complete norm on A. It is equivalent to every other complete k-algebra norm on A.*

Proof. Let us first consider the case, where A is an integral domain. Choose a finite normalization monomorphism $\varphi: T_d \to A$ for a suitable $d \geqq 0$. Then φ is torsion-free, and the assertion follows immediately from Theorem 3.8.3/7, because the field of fractions $Q(T_d)$ is weakly stable (Theorem 5.3.1/1).

Now consider an arbitrary reduced k-affinoid algebra A. Let $\mathfrak{p}_1, \ldots, \mathfrak{p}_r$ denote the minimal prime ideals in A. Then the canonical homomorphism

$$\pi: A \to A' := \bigoplus_{i=1}^{r} A/\mathfrak{p}_i$$

is injective, and $|\ \ |_{\sup}$ is a complete norm on each algebra $A/\mathfrak{p}_i$. Provide A' with the maximum norm, i.e. $|(a_1, \ldots, a_r)| := \max_{1 \leqq i \leqq r} |a_i|_{\sup}$. Then A' is complete under $|\ \ |$, and due to Lemma 6.2.1/3, the norm $|\ \ |$ induces the supremum norm on A. Viewing A as a submodule of the finite A-module A', we see by Proposition 3.7.3/1 that A is closed in A'. Hence $|\ \ |_{\sup}$ is complete on A, and A is a Banach function algebra. That all complete k-algebra norms on A are equivalent to $|\ \ |_{\sup}$ follows from Proposition 6.1.3/2. □

It should be noted that the characterization of power-bounded elements and of topologically nilpotent elements as given in Propositions 6.2.3/1 and 6.2.3/2 is an easy consequence of the above theorem, at least in the case where A is reduced. However, the direct proofs given in (6.2.3) do not need the weak stability of $Q(T_d)$. This fact had to be used in the proof of Theorem 1.

6.3. The reduction functor $A \rightsquigarrow \tilde{A}$

In the following sections we consider homomorphisms $\varphi: B \to A$ between k-affinoid algebras A and B. As we already saw in (1.2.5), each such φ maps power-bounded elements into power-bounded elements and topologically nilpotent elements into topologically nilpotent elements. Thus φ gives rise to a homomorphism $\mathring{\varphi}: \mathring{B} \to \mathring{A}$ and furthermore, by reducing modulo topologically nilpotent elements, to a homomorphism $\tilde{\varphi}\colon \tilde{B} = \mathring{B}/\check{B} \to \tilde{A} = \mathring{A}/\check{A}$. This is described by the following commutative diagram

$$\begin{array}{ccc} B & \xrightarrow{\varphi} & A \\ \uparrow & & \uparrow \\ \mathring{B} & \xrightarrow{\mathring{\varphi}} & \mathring{A} \\ \downarrow{\scriptstyle \tau_B} & & \downarrow{\scriptstyle \tau_A} \\ \tilde{B} & \xrightarrow{\tilde{\varphi}} & \tilde{A}, \end{array}$$

where τ_A and τ_B denote the canonical reduction epimorphisms modulo topologically nilpotent elements. If there is no confusion possible, we just write τ instead of τ_A or τ_B.

We are interested in studying properties of the map φ which are inherited by $\mathring{\varphi}$ or $\tilde{\varphi}$ and vice versa. Of particular interest will be the case of integral, respectively finite homomorphisms. The characterization of power-bounded elements and of topologically nilpotent elements by means of the supremum semi-norm $|\ \ |_{\sup}$ (see Propositions 6.2.3/1 and 6.2.3/2) is important for our considerations. We will make use of the mentioned results without giving a further reference.

6.3.1. Monomorphisms, isometries and epimorphisms. — We begin with injectivity properties.

Lemma 1. *The homomorphism $\varphi\colon B \to A$ is an isometry with respect to $|\ \ |_{\sup}$ if and only if $|\varphi(f)|_{\sup} = 1$ for all $f \in B$ satisfying $|f|_{\sup} = 1$.*

Proof. We have only to verify the if part of the assertion. Therefore assume $|\varphi(f)|_{\sup} = 1$ for all $f \in B$ with $|f|_{\sup} = 1$. Consider an arbitrary element $g \in B$. If $|g|_{\sup} = 0$, we conclude $|\varphi(g)|_{\sup} = 0$ from $|\varphi(g)|_{\sup} \leqq |g|_{\sup}$. If $|g|_{\sup} \neq 0$, choose $c \in k^*$ and $m \in \mathbb{N}$ such that $|cg^m|_{\sup} = 1$ (Proposition 6.2.1/4). Then

$$|c|\ |\varphi(g)|_{\sup}^m = |\varphi(cg^m)|_{\sup} = 1 = |cg^m|_{\sup} = |c|\ |g|_{\sup}^m,$$

and hence $|\varphi(g)|_{\sup} = |g|_{\sup}$. □

An immediate consequence of the lemma is

Proposition 2. *The map $\tilde{\varphi}\colon \tilde{B} \to \tilde{A}$ is injective if and only if $\varphi\colon B \to A$ is an isometry.*

Corollary 3. *If $\tilde{\varphi}\colon \tilde{B} \to \tilde{A}$ is injective, then* ker φ *is contained in the nilradical* rad B.

Proof. We have $\operatorname{rad} B = \{g \in B;\ |g|_{\sup} = 0\}$ by Proposition 6.2.1/4. Consequently, the kernel of any isometry $\varphi\colon B \to A$ is contained in rad B. □

Under the additional hypothesis that φ is strict, the converse of Corollary 3 is true. In order to prove this, we look more closely at the ideal $\tau^{-1}(\ker \tilde{\varphi})$

$= \mathring{\varphi}^{-1}(\check{A})$ in $\mathring{B}$, which is reduced (i.e., equal to its nilradical), since $\check{A}$ is reduced. Obviously, we have $\check{B} + \ker \mathring{\varphi} \subset \tau^{-1}(\ker \tilde{\varphi})$ and therefore also

$$\operatorname{rad}(\check{B} + \ker \mathring{\varphi}) \subset \tau^{-1}(\ker \tilde{\varphi}).$$

If φ is strict, this inclusion relation is just an equality; namely,

Observation 4. *If φ is strict, then we have*

$$\tau^{-1}(\ker \tilde{\varphi}) = \operatorname{rad}(\check{B} + \ker \mathring{\varphi}) \quad \textit{and} \quad \ker \tilde{\varphi} = \operatorname{rad}\big(\tau(\ker \mathring{\varphi})\big).$$

Proof. In order to verify the first equation, assume that φ is strict. Then $\varphi(\check{B})$ is open in $\varphi(B)$. Consider an arbitrary element $g \in \tau^{-1}(\ker \tilde{\varphi}) = \mathring{\varphi}^{-1}(\check{A})$. From $\lim_{n\to\infty} \varphi(g)^n = 0$, we conclude $\varphi(g)^n \in \varphi(\check{B})$ and hence $g^n \in \check{B} + \ker \mathring{\varphi}$ for n big enough. This verifies the first equation. The second equation is a consequence of the first one, since the formation of the nilradical commutes with the map $\tau\colon \mathring{B} \to \tilde{B}$ for those ideals in $\mathring{B}$ which contain $\check{B} = \ker \tau$. □

Now we can give the converse to Corollary 3.

Proposition 5. *If $\varphi\colon B \to A$ is strict and $\ker \varphi \subset \operatorname{rad} B$, then $\tilde{\varphi}$ is injective.*

Proof. If $\ker \varphi \subset \operatorname{rad} B$, then a fortiori $\ker \mathring{\varphi} \subset \operatorname{rad} \mathring{B}$. Since $\check{B}$ is a reduced ideal, one has $\operatorname{rad}(\check{B} + \ker \mathring{\varphi}) = \check{B}$. Now the preceding observation implies $\ker \tilde{\varphi} = 0$. □

Summarizing the above results, we obtain

Theorem 6. *Let B be reduced. Then the following statements are equivalent:*

(i) *$\varphi\colon B \to A$ is injective and strict.*

(ii) *$\varphi\colon B \to A$ is an isometry with respect to $|\ |_{\sup}$.*

(iii) *$\tilde{\varphi}\colon \tilde{B} \to \tilde{A}$ is injective.*

Proof. The equivalence of statements (ii) and (iii) is asserted by Proposition 2. Furthermore, statement (i) implies (iii) by Proposition 5. So far we did not use the fact that B is reduced. However this assumption is necessary in order to verify the remaining implication, say from (ii) to (i). We know from Theorem 6.2.4/1 that B is a Banach function algebra, i.e., that $|\ |_{\sup}$ is a complete norm on B. Therefore, any isometry $\varphi\colon B \to A$ with respect to $|\ |_{\sup}$ is injective. It remains to verify that φ is also strict. Fix a Banach norm $|\ |$ on A. Then $|\ |$ dominates $|\ |_{\sup}$ on A and $|g|_{\sup} = |\varphi(g)|_{\sup} \leqq |\varphi(g)|$ for all $g \in B$. Since the supremum norm $|\ |_{\sup}$ induces the given Banach topology on B and since φ is continuous anyway, we see that φ is strict. □

There are no good criteria relating the surjectivity of φ and $\tilde{\varphi}$. We give two examples. The first one shows that $\tilde{\varphi}$ may be surjective, even bijective, without φ being surjective. The second one shows that φ may be surjective without $\tilde{\varphi}$ being surjective.

Example 1. Consider a ground field k, which admits a finite extension K of degree $n > 1$ such that $e(K/k) = n$. (The field $\mathbb{Q}_2$ of 2-adic numbers is such a field; one can set $K := \mathbb{Q}_2(\sqrt{2})$.) Then $f(K/k) = 1$ by Proposition 3.1.3/2. Thus, viewing k and K as k-affinoid algebras, the injection $\varphi\colon k \hookrightarrow K$ is a homomorphism of k-affinoid algebras which is not surjective. However, the residue homomorphism $\tilde{\varphi}\colon \tilde{k} \to \tilde{K}$ is bijective, since $f(K/k) = 1$.

Remark. As we will see later, in the case of a stable ground field k with divisible value group $|k^*|$, the surjectivity of $\tilde{\varphi}$ implies the surjectivity of $\varphi\colon B \to A$ if A is reduced. Therefore, for reduced k-affinoid algebras over such a field (e.g., over an algebraically closed field), φ is bijective if and only if $\tilde{\varphi}$ is bijective, see Corollary 6.4.2/2.

Example 2. Set $B := T_1 = k\langle X\rangle$ and $A := k \oplus k$ (ring-theoretic normed direct sum of two copies of k). Choose a constant $c \in k$, $0 < |c| < 1$, and consider the homomorphism

$$\varphi\colon B \to A\,, \qquad X \mapsto (c, 0)\,.$$

It is easily verified that φ is surjective. However $\tilde{\varphi}\colon \tilde{k}[X] \to \tilde{k} \oplus \tilde{k}$ cannot be surjective, since $|\varphi(X)|_{\sup} = |c| < 1$ and hence $\tilde{\varphi}(X) = 0$.

6.3.2. Finiteness of homomorphisms. — In this section we derive a criterion for the finiteness of a homomorphism of k-affinoid algebras $\varphi\colon B \to A$.

Proposition 1. *If $\tilde{\varphi}\colon \tilde{B} \to \tilde{A}$ is integral, then $\varphi\colon B \to A$ is finite.*

Obviously the proposition is a special case of the following

Theorem 2. *Let $\varphi\colon B \to A$ be a homomorphism of k-affinoid algebras. Assume that there exist affinoid generators $f_1, \ldots, f_n \in \mathring{A}$ of A such that $\tilde{f}_1, \ldots, \tilde{f}_n \in \tilde{A}$ are integral over $\tilde{B}$ (with respect to $\tilde{\varphi}\colon \tilde{B} \to \tilde{A}$). Then φ is finite.*

Proof. First we reduce the assertion to the case, where B is a free Tate algebra. By assumption, there exist integers $s_\nu > 0$ and elements $b_{\nu\mu} \in \mathring{B}$, $\mu = 1, \ldots, s_\nu$, such that

$$\tilde{f}_\nu^{s_\nu} + \tilde{\varphi}(\tilde{b}_{\nu 1})\,\tilde{f}_\nu^{s_\nu - 1} + \cdots + \tilde{\varphi}(\tilde{b}_{\nu s_\nu}) = 0\,, \qquad \nu = 1, \ldots, n\,.$$

We set $s := \sum_{\nu=1}^{n} s_\nu$ and consider the homomorphism $\psi\colon T_s \to B$ defined by

$$\psi(X_{\nu\mu}) := b_{\nu\mu}\,, \qquad \nu = 1, \ldots, n\,, \qquad \mu = 1, \ldots, s_\nu\,,$$

where $\{X_{\nu\mu};\ \nu = 1, \ldots, n,\ \mu = 1, \ldots, s_\nu\}$ is a chart of $T := T_s$. Obviously $\tilde{f}_1, \ldots, \tilde{f}_n \in \tilde{A}$ are integral over $\tilde{T}$ with respect to the map $\widetilde{\varphi \circ \psi}\colon \tilde{T} \to \tilde{A}$. Therefore if the theorem is known in the case where B is a free Tate algebra, we know that $\varphi \circ \psi\colon T \to A$ is finite. This implies the finiteness of φ.

We now assume $B = T$. We extend $\varphi\colon T \to A$ to an epimorphism $\varphi'\colon T\langle Y_1, \ldots, Y_n\rangle \to A$ by setting $\varphi'(Y_\nu) := f_\nu$, $\nu = 1, \ldots, n$, where $Y_1, \ldots, Y_n$ are new indeterminates. By assumption, $\widetilde{\varphi'(Y_\nu)}$ is integral over $\tilde{T}$. Hence it is enough to prove the following

Lemma 3. *Let T be a free Tate algebra, and let $\varphi\colon T\langle Y_1, \ldots, Y_n\rangle \to A$ be a finite homomorphism such that $\widetilde{\varphi(Y_1)}, \ldots, \widetilde{\varphi(Y_n)}$ are integral over $\tilde{T}$. Then $\varphi|_T\colon T \to A$ is finite.*

Proof. We proceed by induction on n. The case $n = 0$ is trivial. Let $n \geqq 1$. We set $T' := T\langle Y_1, \ldots, Y_n\rangle$. By assumption there is a Weierstrass polynomial

$$\omega := Y_n^m + a_1 Y_n^{m-1} + \cdots + a_m \in \mathring{T}[Y_n] \subset T\langle Y_1, \ldots, Y_{n-1}\rangle [Y_n]$$

such that $\tilde{\omega} \in \ker \tilde{\varphi}$. Since φ is strict (Proposition 6.1.3/4), Observation 6.3.1/4 shows $\omega^q \in \check{T}' + \ker \check{\varphi}$ for some $q \in \mathbb{N}$. In particular, there is a series $r \in \check{T}'$ such that $g := \omega^q - r \in \ker \check{\varphi}$. Furthermore, g is Y_n-distinguished of order mq, because $\tilde{g}$ is equal to $\tilde{\omega}^q$. Hence g is associated to a Weierstrass polynomial in Y_n (Weierstrass Preparation Theorem 5.2.2/1), and the Weierstrass Finiteness Theorem 5.2.3/4 implies that the restriction of φ to $T\langle Y_1, \ldots, Y_{n-1}\rangle$ is finite. Then the restriction of φ to T is finite by the induction hypothesis. □

In order to get the conclusion of Theorem 2, it does not suffice to assume that all elements of an affinoid generating system of A are integral over B. Indeed, choose $A = B = T_1 = k\langle X\rangle$ and define $\varphi\colon B \to A$ by $\varphi(X) := cX$, where $c \in k$, $0 < |c| < 1$. Then $\{X\}$ is an affinoid generating system of A, and $X \in A$ is obviously integral over B with respect to φ. Nevertheless, A is not integral over B. Assume the contrary. Then the inverse image $1 - X \in B$ of the unit $1 - cX \in A$ must be a unit in B, which is absurd. In particular, there cannot exist affinoid generators $f_1, \ldots, f_n \in \mathring{A}$ of A such that $\tilde{f}_1, \ldots, \tilde{f}_n$ are integral over $\tilde{B}$. On the other hand, if all elements of A are integral over B, then all elements of $\tilde{A}$ are integral over $\tilde{B}$ (cf. Proposition 6.3.4/1 in a later section).

6.3.3. Applications to group operations. — In the proof of Theorem 6.3.2/2. we never used the fact that B is k-affinoid. The construction of the map $\psi\colon T_s \to B$ only needs a k-Banach algebra B. This is why we can obtain the following criterion for k-affinoid algebras, which is a counterpart to Proposition 6.1.3/4.

Proposition 1. *Let B be a k-Banach algebra, and let A be k-affinoid. Let $\varphi\colon B \to A$ be a k-algebra homomorphism such that $\tilde{\varphi}\colon \tilde{B} \to \tilde{A}$ is integral. Then φ is finite, and $\varphi(B)$ is a closed subalgebra of A which is k-affinoid.*

Proof. If $f_1, \ldots, f_n \in \mathring{A}$ are affinoid generators of A, then $\tilde{f}_1, \ldots, \tilde{f}_n \in \tilde{A}$ are integral over $\tilde{B}$. As in the proof of Theorem 6.3.2/2, choose a k-algebra homomorphism $\psi\colon T_s \to B$ such that $\tilde{f}_1, \ldots, \tilde{f}_n$ are integral over $\tilde{T}_s$ with respect to the map $\tilde{\varphi} \circ \tilde{\psi}$. Then Theorem 6.3.2/2 implies that A is a finite T_s-module via $\varphi \circ \psi$ (and hence a finite B-module via φ). Viewing $\varphi(B)$ as a finite T_s-submodule of A, we see by Proposition 3.7.3/1 that $\varphi(B)$ is closed in A. Furthermore, $\varphi(B)$ is k-affinoid by Proposition 6.1.1/5. □

The proposition has the following immediate consequence:

Corollary 2. *Let B be a closed k-subalgebra of a k-affinoid algebra A such that $\mathring{A}$ is integral over $\mathring{B}$. Then B is k-affinoid, and A is finite over B.*

Using the corollary, one can construct k-affinoid algebras as follows. Let A be a given k-affinoid algebra, and consider a group G of k-algebra automorphisms of A. Since all automorphisms of A are continuous, it follows that

$$A^G := \{f \in A ; \gamma(f) = f \text{ for all } \gamma \in G\}$$

is a closed k-subalgebra of A.

We want to show that the algebra is even k-affinoid if the group G is finite. Namely, for each $f \in A$, we can consider the monic polynomial

$$p_f := \prod_{\gamma \in G} \bigl(Y - \gamma(f)\bigr) \in A[Y]$$

which is annihilated by f. Obviously p_f is invariant under G so that $p_f \in A^G[Y]$. Furthermore, each γ maps $\mathring{A}$ into $\mathring{A}$ (due to continuity); hence p_f has coefficients in $A^G \cap \mathring{A}$ if $f \in \mathring{A}$. It follows that $\mathring{A}$ is integral over $A^G \cap \mathring{A}$, and we deduce from Corollary 2 that

Proposition 3. *If G is a finite group of automorphisms of the k-affinoid algebra A, then A^G is a closed subalgebra of A, which is k-affinoid. Furthermore, A is finite over A^G.*

6.3.4. Finiteness of the reduction functor $A \rightsquigarrow \tilde{A}$.

— In this section we prove a converse of Proposition 6.3.2/1. First we state

Proposition 1. *If $\varphi\colon B \to A$ is an integral homomorphism of k-affinoid algebras, then $\mathring{\varphi}\colon \mathring{B} \to \mathring{A}$ and hence $\tilde{\varphi}\colon \tilde{B} \to \tilde{A}$ are integral.*

Proof. We know by Proposition 6.2.2/4 that, for each $f \in A$, there exists an integral equation

$$f^n + \varphi(b_1) f^{n-1} + \cdots + \varphi(b_n) = 0$$

of f over B such that $|f|_{\sup} = \max\limits_{1 \leqq i \leqq n} |b_i|_{\sup}^{1/i}$. If $f \in \mathring{A}$, i.e., if $|f|_{\sup} \leqq 1$, then $|b_i|_{\sup} \leqq 1$ and hence $b_i \in \mathring{B}$ for $i = 1, \ldots, n$. □

Remark. Applying Lemma 6.2.2/3, one can improve the above result: not only is $\mathring{A}$ integral over $\varphi(\mathring{B})$ if φ is integral, but in fact, $\mathring{A}$ equals the integral closure of $\varphi(\mathring{B})$ in A.

Theorem 2. *If $\varphi\colon B \to A$ is finite, then $\tilde{\varphi}\colon \tilde{B} \to \tilde{A}$ is finite.*

Proof. First we consider the special case where A is an integral domain. By Theorem 6.1.2/1, there exists a homomorphism $\psi\colon T_d \to B$ for some $d \geqq 0$

such that $\varphi \circ \psi\colon T_d \to A$ is finite and injective. If we know that $\widetilde{\varphi \circ \psi} = \tilde{\varphi} \circ \tilde{\psi}$ is finite, then $\tilde{\varphi}$ must be finite. Hence we may assume $B = T_d$ and $\ker \varphi = 0$. We want to show that Proposition 3.1.5/4 can be applied to A, viewed as a T_d-algebra. Namely, A (provided with $|\ |_{\sup}$) is a faithfully normed T_d-algebra by Proposition 6.2.2/2. Furthermore, A is of finite rank over T_d, since φ is finite. The map $\mathring{\varphi}\colon \mathring{T}_d \to \mathring{A}$ is integral by Proposition 1. The polynomial ring $\tilde{T}_d = \tilde{k}[X_1, \ldots, X_d]$ is Noetherian and Japanese (Proposition 4.4/4). Thus Proposition 3.1.5/4 implies that $\tilde{A}$ is finite over $\tilde{T}_d$.

Next we consider the general case. Similarly as before, we may assume that B is a free TATE algebra T_d. Denote by $\mathfrak{p}_1, \ldots, \mathfrak{p}_r$ the minimal prime ideals in A and by $\pi_i\colon A \to A/\mathfrak{p}_i$, $i = 1, \ldots, r$, the canonical residue epimorphisms. Furthermore, consider the composed map

$$T_d \xrightarrow{\varphi} A \xrightarrow{\pi} \bigoplus_{i=1}^{r} A/\mathfrak{p}_i,$$

where $\pi(f) := (\pi_1(f), \ldots, \pi_r(f))$ for all $f \in A$. Since π is an isometry with respect to $|\ |_{\sup}$ (see Lemma 6.2.1/3; view $\bigoplus_{i=1}^{r} A/\mathfrak{p}_i$ as a normed direct sum), the corresponding residue map

$$\tilde{\pi}\colon \tilde{A} \to \bigoplus_{i=1}^{r} \widetilde{A/\mathfrak{p}_i}, \qquad \tilde{f} \mapsto (\tilde{\pi}_1(\tilde{f}), \ldots, \tilde{\pi}_r(\tilde{f}))$$

is injective. Since all maps $\widetilde{\pi_i \circ \varphi} = \tilde{\pi}_i \circ \tilde{\varphi}$, $i = 1, \ldots, r$, are finite by what has been proved above, it follows that $\bigoplus_{i=1}^{r} \widetilde{A/\mathfrak{p}_i}$ is a finite $\tilde{T}_d$-module with respect to $\tilde{\pi} \circ \tilde{\varphi}$. Then $\tilde{A}$, as a submodule of a finite module over the Noetherian ring $\tilde{T}_d$, is finite too. Thus $\tilde{\varphi}$ is finite. □

Corollary 3. *For each k-affinoid algebra A, the residue algebra $\tilde{A}$ is a finitely generated $\tilde{k}$-algebra (i.e., a quotient of some polynomial algebra $\tilde{k}[X_1, \ldots, X_n]$).*

Proof. Choose a finite monomorphism $\varphi\colon T_d \hookrightarrow A$ (Corollary 6.1.2/2). The theorem tells us that $\tilde{\varphi}\colon \tilde{T}_d \to \tilde{A}$ is finite too. Hence $\tilde{A}$ is a finitely generated $\tilde{k}$-algebra. □

Remark. In the situation of the above proof, the map $\tilde{\varphi}\colon \tilde{T}_d \to \tilde{A}$ is injective by Proposition 6.2.2/1. Thus, by general facts of dimension theory (for example, use NAGATA [28], Corollary 10.10), we see that $\dim \tilde{A} = \dim \tilde{T}_d$ for the Krull dimensions of $\tilde{A}$ and $\tilde{T}_d$. Similarly, $\dim A = \dim T_d$. Since $\dim \tilde{T}_d = d$, and since $\dim T_d = d$ (see the remark following Corollary 6.1.2/2), it follows that A and $\tilde{A}$ have the same Krull dimension.

6.3.5. Summary. — The main results of the preceding sections can be summarized in the following way.

Theorem 1. *Let $\varphi \colon B \to A$ be a homomorphism of k-affinoid algebras. The following statements are equivalent:*

(i) *φ is finite.*

(ii) *φ is integral.*

(iii) *$\mathring{\varphi}$ is integral.*

(iv) *$\tilde{\varphi}$ is integral.*

(v) *$\tilde{\varphi}$ is finite.*

Proof. The implications (i) $\Rightarrow$ (ii) $\Rightarrow$ (iii) $\Rightarrow$ (iv) $\Rightarrow$ (i) are either trivial or just the statement of Propositions 6.3.4/1 and 6.3.2/1. Since (v) $\Rightarrow$ (iv) is trivial and since (i) $\Rightarrow$ (v) is the content of Theorem 6.3.4/2, all five statements are equivalent. □

The equivalence of (i) and (ii) is a Japaneseness statement in the category of k-affinoid algebras. For the proof, we did not use the Japaneseness of T_d, but that of $\tilde{T}_d$.

The statement "$\mathring{\varphi}$ is finite" does not occur in Theorem 1. Whether or not it is equivalent to the statements (i) to (v) depends on the ground field k. We will consider this question in the following sections.

6.4. The functor $A \rightsquigarrow \mathring{A}$

6.4.1. Finiteness Theorems. — Let $\varphi \colon B \to A$ be a homomorphism of k-affinoid algebras, and consider the induced homomorphisms $\mathring{\varphi} \colon \mathring{B} \to \mathring{A}$ and $\tilde{\varphi} \colon \tilde{B} \to \tilde{A}$. In (6.3) we investigated the relationship between φ and $\tilde{\varphi}$. In particular, we proved that φ is finite if and only if $\tilde{\varphi}$ is finite and, as a by-product, that both conditions are equivalent to the fact that $\mathring{\varphi}$ is integral. In this section we want to look for conditions assuring the finiteness of $\mathring{\varphi}$. This question was first studied in Grauert Remmert [16] for algebraically closed ground field k, whereas the general case was dealt with in Gruson [17]. Our considerations are based on the results of (5.2.7) and (5.3).

In order to see what happens in simple situations, we first look at the case, where B equals the ground field k, and where A is a finite field extension of k.

Proposition 1. *Let L be a finite field extension of k. Assume that the valuation on k is not discrete. Then $\mathring{L}$ is a finite $\mathring{k}$-module if and only if $f(L/k) = [L : k]$.*

Proof. If $f(L/k) = [L : k]$, then L is strictly k-cartesian by Corollary 2.5.1/6, and any orthonormal basis of L generates $\mathring{L}$ over $\mathring{k}$. Conversely, assume that $\mathring{L}$ is finite over $\mathring{k}$. Let $\{x_1, \ldots, x_n\} \subset \mathring{L}$ be a minimal set generating $\mathring{L}$ over $\mathring{k}$. We claim that $\{x_1, \ldots, x_n\}$ is an orthonormal basis of L over k. In order to verify this, consider a linear combination $x = \sum_{i=1}^{n} c_i x_i \in L$ with coefficients $c_1, \ldots, c_n \in k$. We have to show $|x| = \max_{1 \leq i \leq n} |c_i|$. Since $|x| \leq \max_{1 \leq i \leq n} |c_i|$ is clear by trivial reasons, it is enough to show that $|x| < \max_{1 \le i \le n} |c_i|$ is impossible. Proceeding indirectly, we assume that the latter estimate is true and that $\max_{1 \le i \le n} |c_i| = 1$. Then $|x| < 1$, and one can find a constant $c \in k$ such that $|x| < |c| < 1$ (because $|k^*|$ is dense in $\mathbb{R}_+ - \{0\}$). Since $|c^{-1}x| < 1$, we can write $c^{-1}x = \sum_{i=1}^{n} d_i x_i$ with coefficients $d_1, \ldots, d_n \in \mathring{k}$. This gives

$$\sum_{i=1}^{n} (c_i - cd_i)\, x_i = 0,$$

where $\max_{1 \le i \le n} |c_i - cd_i| = 1$, since $\max_{1 \le i \le n} |cd_i| \le |c| < 1 = \max_{1 \le i \le n} |c_i|$. If for example $|c_1 - cd_1| = 1$, then $x_1 \in \sum_{i=2}^{n} \mathring{k} x_i$, and $\{x_1, \ldots, x_n\}$ cannot be a minimal set of generators for $\mathring{L}$ over $\mathring{k}$. This contradiction shows that $\{x_1, \ldots, x_n\}$ is an orthonormal basis of L over k and that

$$f(L/k) = [\tilde{L} : \tilde{k}] = n = [L : k]. \qquad \square$$

If the valuation on k is discrete, the assertion of Proposition 1 does not remain true, because in this case $\mathring{L}$ is always a finite $\mathring{k}$-module (Proposition 2.4.2/3 and Lemma 5.2.7/6). Therefore the criterion given in Proposition 1 implies

Proposition 2. *The valuation ring $\mathring{L}$ of each finite extension L of k is finite over $\mathring{k}$ if and only if one of the following two conditions is satisfied:*

(i) *The valuation on k is discrete.*

(ii) *k is stable, and $|k^*|$ is divisible.*

Proof. We have only to consider the case where the valuation on k is not discrete. The condition $f(L/k) = [L : k]$ of Proposition 1 is equivalent to $e(L/k) = 1$ and $e(L/k)\, f(L/k) = [L : k]$ (since $e(L/k)\, f(L/k) \le [L : k]$ by Proposition 3.6.2/5). The ramification index $e(L/k)$ is 1 for all finite extensions L/k if and only if the value group of k equals the value group of its algebraic closure, i.e., if and only if $|k^*|$ is divisible (Observation 3.6.2/10). Furthermore the equation $e(L/k)\, f(L/k) = [L : k]$ holds for all finite extensions L/k if and only if k is stable (Proposition 3.6.2/6). Thus the assertion of the proposition is clear. $\square$

In order to give an example of a finite extension L/k such that $\mathring{L}$ is not finite over $\mathring{k}$, we take for k the field K constructed in (3.6.1). Recall that K was defined as the completion of the field $\mathbb{Q}_2(\alpha_0, \alpha_1, \ldots)$, where $\alpha_0 = 2$ and where α_{i+1} is a square root of α_i for all i. Then $L := K(\sqrt[3]{2})$ satisfies $e(L/K) = 3$ (since $|2|^{1/3} \notin |K|$), and we see that $\mathring{L}$ cannot be finite over $\mathring{K}$ (since $ef \le 3$ implies $f = 1 < 3$). Alternatively, one can consider the extension $K(\sqrt{3})$, which also has residue degree 1 over K (see the example in (3.6.1)). In particular, these examples show that a finite homomorphism of k-affinoid algebras $\varphi: B \to A$

does not, in general, give rise to a finite homomorphism $\mathring{\varphi}: \mathring{B} \to \mathring{A}$. However, we want to show that $\mathring{\varphi}: \mathring{B} \to \mathring{A}$ is always finite if A is reduced (i.e., if rad $A = 0$) and if k satisfies one of the conditions given in Proposition 2. That A must be reduced is plausible. Namely if $\mathring{\varphi}: \mathring{B} \to \mathring{A}$ is finite and if $\mathring{B}$ is reduced, any Banach norm on A must be bounded on $\mathring{A}$. However this cannot be the case if A contains a nilpotent element $a \neq 0$, since then $k \cong ka \subset \mathring{A}$.

Theorem 3. *Assume that k is stable. Let $\varphi: B \to A$ be a finite homomorphism of k-affinoid algebras where A is reduced. Then A is a pseudo-cartesian $B/\text{rad}\, B$-module (with respect to the supremum norm). More precisely, there exist elements $x_1, \ldots, x_s \in A$ such that every $a \in A$ admits a representation $a = \sum_{i=1}^{s} b_i x_i$ with coefficients $b_1, \ldots, b_s \in B$ satisfying $|a|_{\text{sup}} = \max_{1 \leq i \leq s} |b_i|_{\text{sup}} |x_i|_{\text{sup}}$.*

Before proving the theorem we want to derive some immediate consequences from it. We begin by clarifying on how to view A as a normed $B/\text{rad}\, B$-module. Recall that $|\ \ |_{\text{sup}}$ is a norm on each reduced k-affinoid algebra (Proposition 6.2.1/4 (iii)). In the situation of the theorem, we have $\text{rad}\, B \subset \ker \varphi$, and φ is contractive with respect to $|\ \ |_{\text{sup}}$ (Proposition 6.2.2/1). Then it is clear that A is a normed $B/\text{rad}\, B$-module with respect to $|\ \ |_{\text{sup}}$. Since the canonical map $\mathring{B} \to \overset{\circ}{\widehat{B/\text{rad}\, B}}$ is surjective (use Lemma 6.2.1/3), we see that $\mathring{\varphi}: \mathring{B} \to \mathring{A}$ is finite if and only if the corresponding map $\overset{\circ}{\widehat{B/\text{rad}\, B}} \to \mathring{A}$ is finite. Therefore Lemma 5.2.7/5 implies the following two statements.

Corollary 4. *Assume that k is stable. Let $\varphi: B \to A$ be a finite homomorphism of k-affinoid algebras where A is a reduced algebra satisfying $|A|_{\text{sup}} = |k|$. Then $\mathring{\varphi}: \mathring{B} \to \mathring{A}$ is finite.*

Corollary 5. *If k is stable and $|k^*|$ is divisible (e.g., if k is algebraically closed), then for every finite homomorphism of k-affinoid algebras $\varphi: B \to A$ where A is reduced, also $\mathring{\varphi}: \mathring{B} \to \mathring{A}$ is finite.*

Also the following statement is a corollary to Theorem 3.

Corollary 6. *If the valuation on k is discrete, then for every finite homomorphism of k-affinoid algebras $\varphi: B \to A$ where A is reduced, also $\mathring{\varphi}: \mathring{B} \to \mathring{A}$ is finite.*

Proof. The field k is stable by Proposition 3.6.2/1. Choosing an epimorphism $\psi: T_n \to B$, we see that $\mathring{\varphi}: \mathring{B} \to \mathring{A}$ is finite if we can show that $\mathring{\varphi} \circ \mathring{\psi}: \mathring{T}_n \to \mathring{A}$ is finite. Thus replacing B by T_n, we may assume $|B|_{\text{sup}} = |k|$. Then Lemma 5.2.7/6 yields the assertion of the corollary. □

Now we come to the *proof* of Theorem 3. First we consider the special case where A is an integral domain. By Theorem 6.1.2/1, there exists a homomorphism $\psi: T_d \to B$ for some $d \geq 0$ such that $\varphi \circ \psi: T_d \to A$ is finite and injective. If we can show that A is a pseudo-cartesian T_d-module via $\varphi \circ \psi$, then it is easily verified that A is a pseudo-cartesian $B/\text{rad}\, B$-module (since ψ is contractive with respect to $| \ |_{\sup}$). Hence we may assume $B = T_d$ and $\ker \varphi = 0$. Then the field of fractions $Q(A)$ is a finite extension of $Q(T_d)$, and $Q(T_d)$ is stable by Theorem 5.3.2/1. Therefore $Q(A)$ provided with the spectral norm $| \ |_{sp}$ over $Q(T_d)$ is a $Q(T_d)$-cartesian vector space. Because $| \ |_{sp}$ restricts to the supremum norm $| \ |_{\sup}$ on A (cf. Proposition 3.8.1/7 and the remark following Lemma 3.8.1/9), we see that A (provided with $| \ |_{\sup}$) is a finite T_d-submodule of a $Q(T_d)$-cartesian vector space. Then A is contained in a cartesian T_d-module, and Theorem 5.2.7/7 implies that A is a pseudo-cartesian T_d-module.

Next we consider the general case. Similarly as before, we may assume that B is a free TATE algebra T_d. Denote by $\mathfrak{p}_1, \ldots, \mathfrak{p}_r$ the minimal prime ideals in A and by $\pi_i\colon A \to A/\mathfrak{p}_i$, $i = 1, \ldots, r$, the canonical residue epimorphisms. By what we have proved above, each $A/\mathfrak{p}_i$ is a pseudo-cartesian T_d-module via $\pi_i \circ \varphi$. Hence the normed direct sum $\bigoplus_{i=1}^{r} A/\mathfrak{p}_i$ is a pseudo-cartesian T_d-module. Since A is reduced, we may view A as a normed submodule of $\bigoplus_{i=1}^{r} A/\mathfrak{p}_i$ (use Lemma 6.2.1/3). Then A is pseudo-cartesian by Corollary 5.2.7/11. □

6.4.2. Epimorphisms and isomorphisms. — In (6.3.1) we investigated how the injectivity of a homomorphism of k-affinoid algebras $\varphi\colon B \to A$ is related to the injectivity of the corresponding map $\tilde{\varphi}\colon \tilde{B} \to \tilde{A}$. Furthermore, we saw that, in the general situation studied in (6.3.1), not much could be said about the behavior of surjectivity. Under the additional assumptions that k is stable and $|k^*|$ is divisible, this can be remedied.

Proposition 1. *If k is stable and $|k^*|$ is divisible, then the following statements for a homomorphism of k-affinoid algebras $\varphi\colon B \to A$ where A is reduced are equivalent:*

(i) *$\mathring{\varphi}\colon \mathring{B} \to \mathring{A}$ is surjective.*

(ii) *$\tilde{\varphi}\colon \tilde{B} \to \tilde{A}$ is surjective.*

Each one of these conditions implies the surjectivity of φ.

Proof. Assume that $\tilde{\varphi}$ is surjective. Then $\tilde{\varphi}$ is finite. According to Theorem 6.3.5/1, also φ is finite. Hence $\mathring{\varphi}$ is finite by Corollary 6.4.1/5, and there are elements $x_1, \ldots, x_s \in \mathring{A}$ such that $\mathring{A} = \sum_{i=1}^{s} \mathring{B} x_i$. Since $|k^*|$ is divisible, $|k^*|$ is dense in $\mathbb{R}_+ - \{0\}$, and an easy consideration shows $\check{A} = \sum_{i=1}^{s} \check{B} x_i$. Because $\tilde{\varphi}$ is surjective, we know $\mathring{A} \subset \varphi(\mathring{B}) + \check{A}$, and thus

$$\mathring{A} \subset \varphi(\mathring{B}) + \sum_{i=1}^{s} \check{B} x_i .$$

Then $\mathring{A} = \varphi(\mathring{B})$ by the NAKAYAMA Lemma 1.2.4/6, and we see that $\mathring{\varphi}$ is surjective. Hence statement (ii) implies statement (i). The converse is obvious, as well as the fact that φ is surjective if $\mathring{\varphi}$ is surjective. □

Corollary 2. *If k is stable and $|k^*|$ is divisible, then the following statements for a homomorphism of reduced k-affinoid algebras are equivalent:*

(i) *$\varphi: B \to A$ is bijective.*

(ii) *$\mathring{\varphi}: \mathring{B} \to \mathring{A}$ is bijective.*

(iii) *$\tilde{\varphi}: \tilde{B} \to \tilde{A}$ is bijective.*

Proof. The implications (i) $\Leftrightarrow$ (ii) $\Rightarrow$ (iii) are trivial. In order to verify the remaining implication (iii) $\Rightarrow$ (i), assume that $\tilde{\varphi}$ is bijective. Then φ is surjective by Proposition 1 and injective by Theorem 6.3.1/6. □

The preceding result gives us a criterion to decide whether a given k-affinoid algebra is isomorphic to a free Tate algebra T_n.

Corollary 3. *Assume that k is stable and that $|k^*|$ is divisible. If A is a reduced k-affinoid algebra such that $\tilde{A}$ is a (free) polynomial ring $\tilde{k}[X_1, \ldots, X_n]$, then A is isomorphic to T_n.*

Proof. Choose elements $a_1, \ldots, a_n \in \mathring{A}$ which represent the indeterminates $X_1, \ldots, X_n \in \tilde{A}$, and consider the homomorphism

$$\varphi: k\langle X_1, \ldots, X_n\rangle \to A, \qquad X_i \mapsto a_i.$$

Then a straightforward verification shows that $\tilde{\varphi}$ is bijective. Hence φ is bijective by Corollary 2. □

6.4.3. Residue norm and supremum norm. Distinguished k-affinoid algebras and epimorphisms. — If a k-affinoid algebra A has nilpotent elements $\neq 0$, then $|\ |_{\sup}$ cannot be equal to any residue norm $|\ |_\alpha$ for whatever epimorphism $\alpha: T_n \to A$ one might choose. But, if A is reduced, we know (cf. Theorem 6.2.4/1) that $|\ |_{\sup}$ and $|\ |_\alpha$ are at least equivalent (i.e., induce the same topology) for all epimorphisms α. The question is: Can one determine α in such a way that $|\ |_\alpha$ becomes equal to $|\ |_{\sup}$? The following theorem provides a criterion for the answer to be affirmative.

Theorem 1. *Assume that k is stable, and let A be a k-affinoid algebra. Then the following two statements are equivalent:*

(i) *There is an epimorphism $\alpha: T_n \to A$ for some $n \geqq 0$ such that the residue norm $|\ |_\alpha$ coincides with the supremum semi-norm $|\ |_{\sup}$ on A.*

(ii) *A is reduced and $|A|_{\sup} = |k|$.*

Proof. Because $|A|_\alpha = |k|$ (cf. Proposition 6.1.1/2), obviously (ii) follows from (i). To show the converse, we shall use Corollary 6.4.1/4. Let $\varphi: T_m \to A$ be an epimorphism for some $m \geqq 0$. Then $\mathring{\varphi}$ is finite; i.e., there are elements $a_1, \ldots, a_s \in \mathring{A}$ such that $\mathring{A} = \sum_{i=1}^{s} \mathring{T}_m a_i$. Using Proposition 6.1.1/4, we can extend the epimorphism φ to an epimorphism

$$\alpha: T_n := T_m\langle Y_1, \ldots, Y_s\rangle \to A$$

such that $\alpha(Y_i) = a_i$ for $i = 1, \ldots, s$. Then $\mathring{A} = \sum_{i=1}^{s} \mathring{T}_m a_i \subset \alpha(\mathring{T}_n)$, and $\mathring{\alpha}$ must be surjective. We want to show $|\ \ |_\alpha = |\ \ |_{\text{sup}}$. Since $|\ \ |_{\text{sup}} \leqq |\ \ |_\alpha$ is true for all epimorphisms α (Corollary 3.8.2/2), we have only to verify $|\ \ |_\alpha \leqq |\ \ |_{\text{sup}}$. It follows from the surjectivity of $\mathring{\alpha}$ that $|f|_{\text{sup}} \leqq 1$ implies $|f|_\alpha \leqq 1$ for all $f \in A$. Since the supremum norm of any non-zero element $f \in A$ can be adjusted to 1 by scalar multiplication (use $|A|_{\text{sup}} = |k|$), one easily derives $|\ \ |_\alpha \leqq |\ \ |_{\text{sup}}$. □

Epimorphisms fulfilling property (i) of this theorem shall be given a special name.

Definition 2. *An epimorphism $\alpha\colon T_n \to A$ is called distinguished if the residue norm $|\ \ |_\alpha$ coincides with the supremum semi-norm $|\ \ |_{\text{sup}}$ on A. A k-affinoid algebra A is called distinguished if it admits a distinguished epimorphism $\alpha\colon T_n \to A$ for some $n \geqq 0$.*

If $\alpha\colon T_n \to A$ is an epimorphism, then for each $f \in A$, there exists an element $t \in T_n$ such that $\alpha(t) = f$ and $|t| = |f|_\alpha$ (because all ideals are strictly closed in T_n; see Corollary 5.2.7/8). Therefore the following properties of distinguished epimorphisms are obvious.

Proposition 3. *Each distinguished epimorphism $\alpha\colon T_n \to A$ satisfies the following properties:*

(i) $\alpha(\check{T}_n) = \check{A}$ *and* $\alpha(\mathring{T}_n) = \mathring{A}$. *In particular,* $\mathring{\alpha}\colon \mathring{T}_n \to \mathring{A}$ *is surjective.*

(ii) $\tilde{\alpha}\colon \tilde{T}_n \to \tilde{A}$ *is surjective.*

(iii) $|A|_{\text{sup}} = |A|_\alpha = |k|$.

Of course, T_n is distinguished for all $n \geqq 0$. On the other hand, a finite field extension L of k with $e(L/k) > 1$ provides us with an example of a k-affinoid algebra which is not distinguished. Furthermore, Theorem 1 shows that a reduced k-affinoid algebra A is distinguished if, in the case of a discrete valuation on k, we have $|A|_{\text{sup}} = |k|$, and in the case of a non-discrete valuation on k, the field k is stable and the value group $|k^*|$ is divisible. (Recall that k is stable by Proposition 3.6.2/1 if the valuation on k is discrete.)

We want to characterize distinguished epimorphisms in more detail.

Proposition 4. *Let $\alpha\colon T_n \to A$ be an epimorphism. Then the following statements are equivalent:*

(i) α *is distinguished.*

(ii) $\alpha(\check{T}_n) = \check{A}$.

(iii) *The $\mathring{T}_n$-ideal* $\check{T}_n + \ker \mathring{\alpha}$ *is reduced.*

(iv) $\ker \tilde{\alpha} = \tau(\ker \mathring{\alpha})$ *where* $\tau\colon \mathring{T}_n \to \tilde{T}_n$ *is the residue epimorphism.*

We give a cyclic *proof*. The implication (i) ⇒ (ii) is clear by Proposition 3. Next we ehow that statement (ii) implies statement (iii). Consider an element

$t \in \mathring{T}_n$ such that $t^s \in \check{T}_n + \ker \mathring{\alpha}$ for some $s > 0$. Then $\alpha(t^s) \in \check{A}$ and hence $\alpha(t) \in \check{A}$. By our assumption, there exists an element $t' \in \check{T}_n$ such that $\alpha(t') = \alpha(t)$. This implies $t - t' \in \ker \mathring{\alpha}$ and hence $t \in \check{T}_n + \ker \mathring{\alpha}$. Therefore statement (iii) follows from (ii). Since the implication (iii) $\Rightarrow$ (iv) is a consequence of Observation 6.3.1/4, it remains only to show that statement (iv) implies (i). Consider an element $f \in A$. In order to show $|f|_{\sup} = |f|_\alpha$, we may assume $f \neq 0$. Furthermore, $|f|_\alpha$ can be adjusted to 1 by scalar multiplication. Then there exists a series $t \in T_n$ such that $\alpha(t) = f$ and $|t| = 1 = |f|_\alpha$. We have $t \notin \check{T}_n + \ker \mathring{\alpha}$, since otherwise we would have $f = \alpha(t) \in \alpha(\check{T}_n)$ and hence $|f|_\alpha < 1$. Therefore $\tilde{t} = \tau(t) \notin \tau(\ker \mathring{\alpha})$. Since $\tau(\ker \mathring{\alpha})$ equals $\ker \tilde{\alpha}$, we see that $\tilde{f} = \tilde{\alpha}(\tilde{t}) \neq 0$. Hence we have $|f|_{\sup} = 1 = |f|_\alpha$. □

Since $\tilde{\alpha}: \tilde{T}_n \to \tilde{A}$ is surjective if $\varphi: T_n \to A$ is distinguished, we see that the equivalence of statements (i) and (iv) in the above proposition implies that

Corollary 5. *If $\alpha: T_n \to A$ is a distinguished epimorphism, then the residue algebra $\tilde{A}$ is isomorphic to the quotient $\tilde{T}_n/\tau(\ker \mathring{\alpha})$.*

Under certain assumptions on the ground field k, the statement (ii) in Proposition 4 is equivalent to $\alpha(\mathring{T}_n) = \mathring{A}$. Namely,

Corollary 6. *Let $\alpha: T_n \to A$ be an epimorphism. Assume that $|A|_{\sup} = |k|$ or that the valuation on k is not discrete. Then α is distinguished if and only if $\mathring{\alpha}: \mathring{T}_n \to \mathring{A}$ is surjective.*

Proof. It suffices to show that the surjectivity of $\mathring{\alpha}$ implies $\alpha(\check{T}_n) = \check{A}$. Consider an element $f \in \check{A}$ so that $|f|_{\sup} < 1$. If $|A|_{\sup} = |k|$, one can find a constant $c \in k$, $|c| > 1$, such that $|cf|_{\sup} \leqq 1$. The same is possible if $|k^*|$ is dense in $\mathbb{R}_+ - \{0\}$. Since $\mathring{\alpha}$ is surjective, there is an element $t \in \mathring{T}_n$ such that $cf = \alpha(t)$. Then we get $f = \alpha(c^{-1}t)$ and $c^{-1}t \in c^{-1}\mathring{T}_n \subset \check{T}_n$ which shows that $\alpha(\check{T}_n) = \check{A}$. □

The surjectivity of $\mathring{\alpha}$ is not in all cases sufficient for an epimorphism $\alpha: T_n \to A$ to be distinguished. Namely, take the field $k := \mathbb{Q}_2$ of 2-adic numbers as ground field and consider the epimorphism $\alpha: k\langle X\rangle \to k(\sqrt{2})$, $X \mapsto \sqrt{2}$. Then $\mathring{\alpha}$ is surjective (since the elements $1, \sqrt{2}$ form an orthogonal basis of $k(\sqrt{2})$ over k). However α cannot be distinguished, since $|k| \subsetneqq |k(\sqrt{2})|$.

PART C
Rigid Analytic Geometry

CHAPTER 7

Local theory of affinoid varieties

In Part B the category $\mathfrak{A}$ of affinoid algebras over a field k has been studied. Now we want to turn to the category of affinoid varieties which is dual to $\mathfrak{A}$. We start by giving a geometric meaning to such varieties, and, as a first step leading to the discussion of global varieties, we will investigate the local structure of affinoid varieties. Here a key role is played by the affinoid subdomains, which are the analytic counterpart of the affine open subsets in algebraic geometry. Our main objective towards the end of the chapter is the proof of the GERRITZEN GRAUERT theorem on locally closed immersions [12], which, in particular, clarifies the structure of affinoid subdomains.

In the following k always denotes a field with a complete non-trivial non-Archimedean valuation, and affinoid always means k-affinoid. We change our habit of denoting indeterminates by X, Y, etc. Instead, we will use the greek letters ξ, η, ζ and reserve X, Y etc. for varieties.

7.1. Affinoid varieties

7.1.1. Max T_n and the unit ball $B^n(k_a)$. — The TATE algebra $T_n = T_n(k)$ is defined as the algebra of all power series in n variables over k whose coefficients form a zero sequence. Thus, for any finite algebraic extension k' of k (provided with the unique complete valuation extending the valuation on k; see (3.2.4)), the elements of $T_n(k)$ give rise to k'-valued functions on $B^n(k')$, and, in particular, they can be viewed as functions on $B^n(k_a)$ taking values in k_a (cf. (5.1.4)).

Alternatively, T_n can be viewed as an algebra of functions on Max T_n, the set of all maximal ideals in T_n (cf. (3.8)). Recall that, for $f \in T_n$ and $\mathfrak{m} \in \operatorname{Max} T_n$, one defines $f(\mathfrak{m})$ as the residue class of f in $T_n/\mathfrak{m}$. Since $T_n/\mathfrak{m}$ is a finite algebraic extension of k, it can be embedded as a subfield of k_a, and thus $f(\mathfrak{m})$ can be viewed as an element of k_a. However, the embedding of $T_n/\mathfrak{m}$ into k_a is not, in general, unique. This means that $f(\mathfrak{m})$ is only determined up to conjugation over k. Therefore, with Γ denoting the Galois group of k_a over k, any $f \in T_n$ can be viewed as a function on Max T_n taking values in k_a/Γ. Note also that Γ acts isometrically on k_a. Hence $|f(\mathfrak{m})|$ is a well-defined real number, and we have $|f(\mathfrak{m})| = 0$, if and only if $f(\mathfrak{m}) = 0$, i.e., $f \in \mathfrak{m}$.

We want to show that the interpretations given above for T_n are closely related. For $x \in B^n(k_a)$, denote by $\mathfrak{m}_x$ the ideal of all $f \in T_n$ vanishing at x. Let $k(x)$ be the finite algebraic extension field of k generated by all components of x. Then, considering the epimorphism

$$h_x: T_n \to k(x), \qquad f \mapsto f(x),$$

it is clear that $\mathfrak{m}_x = \ker h_x$ is a maximal ideal in T_n. Thus, there is a canonical map

$$\tau: B^n(k_a) \to \operatorname{Max} T_n, \qquad x \mapsto \mathfrak{m}_x,$$

which has the remarkable property that, for all $f \in T_n$, the following diagram is commutative:

$$\begin{array}{ccc} B^n(k_a) & \xrightarrow{f} & k_a \\ \downarrow{\scriptstyle\tau} & & \downarrow \\ \operatorname{Max} T_n & \xrightarrow{f} & k_a/\Gamma . \end{array}$$

In order to verify this, let $x \in B^n(k_a)$. Since the map $h_x: T_n \to k(x) \subset k_a$ factors through $T_n/\mathfrak{m}_x$, it gives rise to an injection $i: T_n/\mathfrak{m}_x \hookrightarrow k_a$, satisfying $f(x) = h_x(f) = i \circ f(\mathfrak{m}_x)$. But this means that our diagram must be commutative. The nature of the map τ will be clarified by the following proposition. Note that the group Γ operates, in particular, on $\mathring{k}_a = B^1(k_a)$ and thus componentwise on $B^n(k_a)$.

Proposition 1. *The map* τ: $B^n(k_a) \to \operatorname{Max} T_n$ *is surjective and has finite fibres. Let* $x, y \in B^n(k_a)$. *Then* $\tau(x) = \tau(y)$ *if and only if there is a* $\gamma \in \Gamma = \operatorname{Gal}(k_a/k)$ *such that* $\gamma(x) = y$. *Hence* τ *induces a bijection* $B^n(k_a)/\Gamma \to \operatorname{Max} T_n$. *In particular, if* k *is algebraically closed,* τ *itself is bijective.*

Proof. First we verify the surjectivity of τ. Let $\mathfrak{m} \in \operatorname{Max} T_n$ be a maximal ideal. Then $T_n/\mathfrak{m}$ is finite algebraic over k. Hence there is a continuous embedding $T_n/\mathfrak{m} \hookrightarrow k_a$, which gives rise to a continuous map $\varphi: T_n \to k_a$ with kernel $\mathfrak{m}$. If $x_1, \ldots, x_n$ denote the images of the indeterminates of T_n via φ, then, by continuity, φ coincides with $h_x: T_n \to k_a$, where $x = (x_1, \ldots, x_n)$. Therefore, $\tau(x) = \ker h_x = \mathfrak{m}$, and consequently, τ must be surjective.

Relying on what we have just proved, we see that, for any $\mathfrak{m} \in \operatorname{Max} T_n$, the points of the fibre $\tau^{-1}(\mathfrak{m})$ correspond one-to-one to the different embeddings $T_n/\mathfrak{m} \hookrightarrow k_a$. Since $T_n/\mathfrak{m}$ is finite over k, there are only finitely many of them. Furthermore, it follows that $\tau(x) = \mathfrak{m} = \tau(y)$ holds for points $x = (x_1, \ldots, x_n)$, $y = (y_1, \ldots, y_n) \in B^n(k_a)$ if and only if there is a k-isomorphism $k(x) \xrightarrow{\sim} k(y)$ such that $x_i \mapsto y_i$, $i = 1, \ldots, n$. Since any k-isomorphism of that type extends to a k-automorphism of k_a, it is clear that $\tau(x) = \tau(y)$ if and only if there is a $\gamma \in \Gamma$ such that $\gamma(x) = y$. □

Thus, viewing the elements of T_n as functions on $B^n(k_a)$ or on $\operatorname{Max} T_n$ is essentially the same. More precisely, for $f \in T_n$, the induced map $\operatorname{Max} T_n$

$\to k_a/\Gamma$ is obtained from the corresponding map $B^n(k_a) \to k_a$ by taking quotients via Γ (i.e., by identifying conjugate points). Note also that, due to the identity theorems (Corollary 5.1.4/5 and Proposition 5.1.3/3), T_n can be identified with the induced algebra of functions on $B^n(k_a)$ or Max T_n.

It follows from Proposition 1 that any $\mathfrak{m} \in \operatorname{Max} T_n$ is of the type $\mathfrak{m} = \mathfrak{m}_x$ with a suitable $x \in B^n(k_a)$. We want to determine generators for such a maximal ideal $\mathfrak{m}$. Let $\xi = (\xi_1, \ldots, \xi_n)$ denote a system of indeterminates.

Lemma 2. *Let $\mathfrak{m} \subset T_n = k\langle\xi\rangle$ be a maximal ideal and let $\mathfrak{m}' := \mathfrak{m} \cap k[\xi]$. Then $\mathfrak{m}'$ is maximal in $k[\xi]$ and $\mathfrak{m} = \mathfrak{m}'k\langle\xi\rangle$. Moreover, the injection $k[\xi] \hookrightarrow k\langle\xi\rangle$ induces a bijection $k[\xi]/\mathfrak{m}' \xrightarrow{\sim} k\langle\xi\rangle/\mathfrak{m}$.*

Proof. Let $\mathfrak{m} = \mathfrak{m}_x$ with $x \in B^n(k_a)$. Considering the evaluation map $h_x\colon k\langle\xi\rangle \to k(x)$, we have $h_x(k[\xi]) = k(x)$. Hence the injection $k[\xi] \hookrightarrow k\langle\xi\rangle$ induces a bijection $\sigma\colon k[\xi]/\mathfrak{m}' \to k\langle\xi\rangle/\mathfrak{m}$. Furthermore, it gives rise to a commutative diagram:

$$\begin{array}{ccc} k[\xi]/\mathfrak{m}' & \xrightarrow{\ i\ } & k\langle\xi\rangle/\mathfrak{m}'k\langle\xi\rangle \\ & {\scriptstyle\sigma}\searrow \quad \swarrow{\scriptstyle\pi} & \\ & k\langle\xi\rangle/\mathfrak{m} & \end{array},$$

where i is injective and π is surjective. The field $k\langle\xi\rangle/\mathfrak{m}$ is finite algebraic over k. Hence $i(k[\xi]/\mathfrak{m}')$ is, in particular, a finite-dimensional vector space over k which is dense in $k\langle\xi\rangle/\mathfrak{m}'k\langle\xi\rangle$, since $k[\xi]$ is dense in $k\langle\xi\rangle$. According to Proposition 2.3.3/4, the vector space $i(k[\xi]/\mathfrak{m}')$ is complete and hence closed in $k\langle\xi\rangle/\mathfrak{m}'k\langle\xi\rangle$. Therefore i must be surjective and hence bijective. But then π must also be bijective. Hence the assertions of the lemma are clear. □

Proposition 3. *Let $\mathfrak{m} \subset T_n = k\langle\xi\rangle$ be a maximal ideal. Then there are polynomials $p_i \in k[\xi_1, \ldots, \xi_i]$ which are monic in ξ_i, $i = 1, \ldots, n$, such that the ideals $\mathfrak{m}' := \mathfrak{m} \cap k[\xi]$ and $\mathfrak{m}$ are generated by $p_1, \ldots, p_n$. If, in addition, k is algebraically closed, one can take $p_i = \xi_i - x_i$ where $x = (x_1, \ldots, x_n) \in B^n(k)$ is the point corresponding to $\mathfrak{m}$ (i.e., such that $\mathfrak{m} = \mathfrak{m}_x$).*

Proof. Choose a point $x = (x_1, \ldots, x_n) \in B^n(k_a)$ such that $\mathfrak{m} = \mathfrak{m}_x$ and, for $i = 1, \ldots, n$, let $\overline{p}_i \in k(x_1, \ldots, x_{i-1})[\xi_i]$ denote the minimal polynomial of x_i over $k(x_1, \ldots, x_{i-1})$. If $p_i \in k[\xi_1, \ldots, \xi_i]$ is a polynomial monic in ξ_i and representing $\overline{p}_i$ with respect to the canonical projection $k[\xi_1, \ldots, \xi_i] \to k(x_1, \ldots, x_{i-1})[\xi_i]$, then, since all these polynomials vanish on x, we have $p_1, \ldots, p_n \in \mathfrak{m}'$. Now, by induction, one easily shows that $k[\xi_1, \ldots, \xi_n]/(p_1, \ldots, p_n) = k(x)$. Thus, $p_1, \ldots, p_n$ generate the maximal ideal $\mathfrak{m}'$ in $k[\xi]$ and, by Lemma 2, they also generate the maximal ideal $\mathfrak{m}$ in $k\langle\xi\rangle$. Finally, if k is algebraically closed, the polynomials $\overline{p}_i$ have the form $\xi_i - x_i$. Hence one can also take $p_i := \xi_i - x_i$. □

7.1.2. Affinoid sets. Hilbert's Nullstellensatz. — Let F be a subset of T_n. We denote by

$$V_a(F) := \{x \in B^n(k_a); f(x) = 0 \quad \text{for all} \quad f \in F\} \quad \text{and}$$
$$V(F) := \{\mathfrak{m} \in \operatorname{Max} T_n; f(\mathfrak{m}) = 0 \quad \text{for all} \quad f \in F\}$$

the zero sets of F, where the elements of F are viewed as functions on $B^n(k_a)$ and $\operatorname{Max} T_n$, respectively. For $f \in F$ and $x \in B^n(k_a)$, we have $f(x) = 0$ if and only if $f \in \mathfrak{m}_x$, i.e., $f(\mathfrak{m}_x) = 0$. Thus, it is clear that *$V_a(F)$ is just the inverse image of $V(F)$ with respect to the projection $\tau: B^n(k_a) \to \operatorname{Max} T_n$*. Using this fact, it follows from Proposition 7.1.1/1 that the operation of $\Gamma = \operatorname{Gal}(k_a/k)$ on $B^n(k_a)$ induces an operation on $V_a(F)$ such that $V(F)$ is the quotient of $V_a(F)$ by Γ (i.e., $V(F)$ is obtained from $V_a(F)$ by identifying conjugate points).

The main purpose of the present chapter is to study zero sets of the type $V_a(F)$ or $V(F)$. These are called *affinoid subsets* of $B^n(k_a)$ or $\operatorname{Max} T_n$, respectively. Of course, from a strictly geometrical point of view, there is no real need to consider affinoid subsets in $\operatorname{Max} T_n$, and hence one should rely exclusively on $B^n(k_a)$ as base space. However, as we have seen before and will see again, there is no essential difference between affinoid subsets of $B^n(k_a)$ and those of $\operatorname{Max} T_n$. Thus, we can just as well restrict ourselves to affinoid subsets of $\operatorname{Max} T_n$, an approach which is much more elegant. It should be noted that everything which is proved in the following for affinoid subsets of $\operatorname{Max} T_n$ is easily carried over to the corresponding affinoid subsets of $B^n(k_a)$.

In order to emphasize the geometric significance of our approach, we call the elements of $\operatorname{Max} T_n$ *points*, using the notation $x, y, z, \ldots \in \operatorname{Max} T_n$. However, when thinking of $x, y, z, \ldots$ as subsets, namely, ideals in T_n, we write $\mathfrak{m}_x, \mathfrak{m}_y, \mathfrak{m}_z, \ldots$, instead. Thus, for example, $\mathfrak{m}_x = \{f \in T_n; f(x) = 0\}$ and, for $F \subset T_n$ we have $V(F) = \{x \in \operatorname{Max} T_n; F \subset \mathfrak{m}_x\}$.

If $\mathfrak{a} \subset T_n$ denotes the ideal generated by the subset $F \subset T_n$, then obviously $V(F) = V(\mathfrak{a}) = V(\operatorname{rad} \mathfrak{a})$. Since T_n is Noetherian (see Theorem 5.2.6/1), we have the following:

Proposition 1. *Let $Y \subset \operatorname{Max} T_n$ be an affinoid subset. Then there are finitely many functions $f_1, \ldots, f_r \in T_n$ such that $Y = V(f_1, \ldots, f_r)$.*

To any subset $E \subset \operatorname{Max} T_n$ (or $E \subset B^n(k_a)$), one associates the reduced T_n-ideal

$$\operatorname{id}(E) = \{f \in T_n; f \mid E = 0\} = \bigcap_{x \in E} \mathfrak{m}_x$$

of all functions vanishing on E. Note that $\operatorname{id}(E) = \operatorname{id}(\tau(E))$ when $E \subset B^n(k_a)$.

Proposition 2. *Let $Y \subset \operatorname{Max} T_n$ be an affinoid subset. Then $V(\operatorname{id}(Y)) = Y$.*

Proof. Obviously, $Y \subset V(\operatorname{id}(Y))$. In order to prove the opposite inclusion, let $x \in V(\operatorname{id}(Y))$, and assume $Y = V(F)$ for some subset $F \subset T_n$. Then all functions in $\operatorname{id}(Y)$ vanish at x. Since $F \subset \operatorname{id}(Y)$, it follows that $x \in Y$. □

A similar but non-trivial equality is derived from the fact that T_n is a Jacobson ring:

Theorem 3 (Hilbert's Nullstellensatz). *For any ideal* $\mathfrak{a} \subset T_n$, *we have* $\mathrm{id}(V(\mathfrak{a})) = \mathrm{rad}\,\mathfrak{a}$.

Proof. First recall that

$$V(\mathfrak{a}) = \{x \in \mathrm{Max}\, T_n;\, \mathfrak{a} \subset \mathfrak{m}_x\}$$

and

$$\mathrm{id}(V(\mathfrak{a})) = \bigcap_{x \in V(\mathfrak{a})} \mathfrak{m}_x.$$

Hence $\mathrm{id}(V(\mathfrak{a}))$ equals the intersection of all maximal ideals $\mathfrak{m} \subset T_n$ containing $\mathfrak{a}$ or, what amounts to the same thing, those containing $\mathrm{rad}\,\mathfrak{a}$. Since T_n is a Jacobson ring (see Theorem 5.2.6/3), we know that $\mathrm{rad}\,\mathfrak{a}$ is an intersection of maximal ideals in T_n. Thus $\mathrm{id}(V(\mathfrak{a})) = \mathrm{rad}\,\mathfrak{a}$. □

The following corollaries are direct consequences of Proposition 2 and Theorem 3.

Corollary 4. *The mappings* V *and* id *determine a one-to-one correspondence between reduced ideals in* T_n *and affinoid subsets in* $\mathrm{Max}\, T_n$.

Corollary 5. *Any functions* $f_1, \ldots, f_r \in T_n$ *having no common zeros generate the unit ideal in* T_n.

The mappings V and id satisfy various rules, some of which are listed below.

Proposition 6. (i) *Let* $\mathfrak{a} \subset \mathfrak{b} \subset T_n$ *be ideals. Then* $V(\mathfrak{a}) \supset V(\mathfrak{b})$.

(ii) *Let* $E \subset E' \subset \mathrm{Max}\, T_n$ *be subsets. Then* $\mathrm{id}(E) \supset \mathrm{id}(E')$.

(iii) *Let* $(\mathfrak{a}_i)_{i \in I}$ *be a family of ideals in* T_n. *Then* $V(\sum_{i\in I} \mathfrak{a}_i) = \bigcap_{i \in I} V(\mathfrak{a}_i)$.

(iv) *Let* $\mathfrak{a}, \mathfrak{b} \subset T_n$ *be ideals. Then* $V(\mathfrak{a}\mathfrak{b}) = V(\mathfrak{a} \cap \mathfrak{b}) = V(\mathfrak{a}) \cup V(\mathfrak{b})$.

Only (iv) requires a *proof*. Since $\mathfrak{a}\mathfrak{b} \subset \mathfrak{a} \cap \mathfrak{b} \subset \mathfrak{a}, \mathfrak{b}$, it follows from (i) that $V(\mathfrak{a}) \cup V(\mathfrak{b}) \subset V(\mathfrak{a} \cap \mathfrak{b}) \subset V(\mathfrak{a}\mathfrak{b})$. Furthermore, if $x \in \mathrm{Max}\, T_n$ is a point not belonging to $V(\mathfrak{a}) \cup V(\mathfrak{b})$, then there must be functions $f \in \mathfrak{a}$, $g \in \mathfrak{b}$ such that $f(x) \neq 0 \neq g(x)$. Hence $(f \cdot g)(x) \neq 0$. But $f \cdot g \in \mathfrak{a}\mathfrak{b}$. Thus, $x \notin V(\mathfrak{a}\mathfrak{b})$, and therefore $V(\mathfrak{a}\mathfrak{b}) \subset V(\mathfrak{a}) \cup V(\mathfrak{b})$. □

As a conclusion we see that the affinoid subsets of $\mathrm{Max}\, T_n$ form the closed sets of a topology on $\mathrm{Max}\, T_n$. This is the so-called *Zariski topology*, and the affinoid subsets of $\mathrm{Max}\, T_n$ will also be referred to as *Zariski-closed subsets*. Furthermore, an affinoid subset $Y \subset \mathrm{Max}\, T_n$ is called *reducible* if there are affinoid subsets $Y_1, Y_2 \subset \mathrm{Max}\, T_n$ such that $Y_1 \neq Y \neq Y_2$ and $Y = Y_1 \cup Y_2$. Otherwise, Y will be called *irreducible*.

Proposition 7. *A non-empty affinoid subset $Y \subset \operatorname{Max} T_n$ is irreducible if and only if* $\operatorname{id}(Y) \subset T_n$ *is a prime ideal.*

Proof. First assume that $Y \subset \operatorname{Max} T_n$ is irreducible. We have $\operatorname{id}(Y) \neq T_n$, since $Y \neq \emptyset$. In order to see that $\operatorname{id}(Y)$ is prime, we consider arbitrary functions $f, g \in T_n$ such that $f \cdot g \in \operatorname{id}(Y)$. Then $(fT_n + \operatorname{id}(Y))\,(gT_n + \operatorname{id}(Y)) \subset \operatorname{id}(Y)$. Hence, by Propositions 2 and 6, we get

$$Y = \big(V(f) \cap Y\big) \cup \big(V(g) \cap Y\big).$$

Since Y is irreducible, we may assume $Y = V(f) \cap Y$ and hence $Y \subset V(f)$. Then $f \in \operatorname{id}(Y)$, and $\operatorname{id}(Y)$ must be prime.

Conversely, assume that $\operatorname{id}(Y)$ is prime. Then, for affinoid subsets $Y_1, Y_2 \subset \operatorname{Max} T_n$ satisfying $Y = Y_1 \cup Y_2$, it follows from Propositions 2 and 6 that

$$Y = V\big(\operatorname{id}(Y_1)\big) \cup V\big(\operatorname{id}(Y_2)\big) = V\big(\operatorname{id}(Y_1) \cap \operatorname{id}(Y_2)\big).$$

Since $\operatorname{id}(Y_1) \cap \operatorname{id}(Y_2)$ is reduced, Hilbert's Nullstellensatz gives $\operatorname{id}(Y) = \operatorname{id}(Y_1) \cap \operatorname{id}(Y_2)$, and, since $\operatorname{id}(Y)$ is prime, we may say $\operatorname{id}(Y_1) \subset \operatorname{id}(Y)$. Again applying Propositions 2 and 6 gives $Y \subset Y_1$, and thus $Y = Y_1$. Therefore, Y must be irreducible. □

As a consequence of Proposition 7 and T_n being Noetherian, we have

Corollary 8. *Each affinoid subset $Y \subset \operatorname{Max} T_n$ admits a unique minimal decomposition $Y = Y_1 \cup \ldots \cup Y_r$ into irreducible affinoid subsets $Y_1, \ldots, Y_r \subset \operatorname{Max} T_n$.*

Proof. The assertion can be proved directly, or it can be deduced from the decomposition theorem for ideals in Noetherian rings. We follow the latter procedure here. Let

$$\operatorname{id}(Y) = \mathfrak{p}_1 \cap \ldots \cap \mathfrak{p}_r$$

be a minimal primary decomposition of $\operatorname{id}(Y)$. Note that, since $\operatorname{id}(Y)$ is reduced, all ideals $\mathfrak{p}_i$ are prime. Then

$$Y = V(\mathfrak{p}_1) \cup \ldots \cup V(\mathfrak{p}_r)$$

is a decomposition of Y into irreducible affinoid subsets of $\operatorname{Max} T_n$. Furthermore if $Y = Y_1 \cup \ldots \cup Y_s$ is an arbitrary minimal decomposition of Y into irreducible affinoid sets, then obviously

$$\operatorname{id}(Y) = \operatorname{id}(Y_1) \cap \ldots \cap \operatorname{id}(Y_s)$$

is a minimal primary decomposition of $\operatorname{id}(Y)$. Since the prime ideals occurring in such a decomposition are unique, it follows that $r = s$, and, without loss of generality, we may assume $\operatorname{id}(Y_i) = \mathfrak{p}_i$ for $i = 1, \ldots, r$. Thus, $Y_i = V(\mathfrak{p}_i)$ and the decomposition $Y = V(\mathfrak{p}_1) \cup \ldots \cup V(\mathfrak{p}_r)$ is minimal and unique. □

7.1.3. Closed subspaces of Max T_n. — In the previous sections the elements of T_n have been considered as functions on Max T_n and $B^n(k_a)$. A similar procedure is possible for affinoid subsets of these spaces. For example, let $\mathfrak{a} \subset T_n$ be an ideal and denote by $X = V(\mathfrak{a}) \subset \operatorname{Max} T_n$ the corresponding affinoid subset. Then, by restriction, the elements of T_n give rise to functions on X. Two elements $f, g \in T_n$ restrict to the same function on X if and only if $f \equiv g$ mod $\operatorname{id}(X)$, where $\operatorname{id}(X) = \operatorname{rad} \mathfrak{a}$ by HILBERT's Nullstellensatz (Theorem 7.1.2/3). Hence these functions are represented by the elements of $T_n/\mathfrak{a}$ and, furthermore, can be identified with the elements of $T_n/\operatorname{id}(X)$. By abuse of language, we call also the elements of $T_n/\mathfrak{a}$ *functions* on X. Note that a "function" $h \in T_n/\mathfrak{a}$ vanishes identically on X if and only if h is nilpotent.

There is another possibility of viewing the elements of $A := T_n/\mathfrak{a}$ as functions, namely as functions on the set Max A of maximal ideals in A. The procedure is the same as in (7.1.1) and (7.1.2). For $f \in A$ and $x \in \operatorname{Max} A$, define $f(x)$ as the residue class of f in $A/\mathfrak{m}_x$ with $\mathfrak{m}_x$ being another notation instead of x (analogously to the convention introduced for Max T_n). Furthermore, for subsets $F \subset A$ and $E \subset \operatorname{Max} A$, we set

$$V(F) := \{x \in \operatorname{Max} A\,; f(x) = 0 \quad \text{for all} \quad f \in F\} \quad \text{and}$$
$$\operatorname{id}(E) := \{f \in A\,; f(x) = 0 \quad \text{for all} \quad x \in E\}.$$

The sets of type $V(F)$ are called the *affinoid subsets* of Max A. In the special case where $A = T_n$, we know that V and id satisfy several properties (cf. (7.1.2)). However, it is easily checked that all results of (7.1.2) (including proofs) remain valid when T_n is replaced by A. This is true because in (7.1.2) we only used the fact that T_n is a Noetherian Jacobson ring which, of course, is also true for A. As an example, we state the analogue of HILBERT's Nullstellensatz.

Proposition 1. *For any ideal* $\mathfrak{b} \subset A$ *one has* $\operatorname{id}(V(\mathfrak{b})) = \operatorname{rad} \mathfrak{b}$.

We want to show that Max A and the affinoid subset $X = V(\mathfrak{a}) \subset \operatorname{Max} T_n$ are closely related. The canonical epimorphism $\sigma\colon T_n \to A$ induces an injection

$$^{a}\sigma\colon \operatorname{Max} A \to \operatorname{Max} T_n, \quad x \mapsto \sigma^{-1}(\mathfrak{m}_x).$$

In fact, $^{a}\sigma$ is actually a bijection between Max A and X, since the maximal ideals of A correspond bijectively to the maximal ideals in T_n containing $\ker \sigma = \mathfrak{a}$. Identifying Max A with X via $^{a}\sigma$ yields that the restriction of a function $f \in T_n$ to X is the same as considering $\sigma(f)$ as a function on Max A. Thus, it follows that the affinoid subsets of Max A are just the affinoid subsets of Max T_n which are contained in X, and the T_n-ideal of all functions vanishing on such a subset equals the inverse image via σ of the corresponding ideal in A.

For arbitrary affinoid algebras A, we will call Max A together with A an *affinoid variety*. In the next section we will study morphisms between affinoid varieties, and we will see that the map ${}^a\sigma$ just constructed is a special case of such a morphism, namely, a *closed immersion*. Note that any affinoid algebra is a quotient of some TATE algebra T_n, and hence any affinoid variety admits a closed immersion into Max T_n for some $n \geqq 0$.

7.1.4. Affinoid maps. The category of affinoid varieties. — We want to define morphisms between affinoid varieties. For the purpose of motivation, let $X \subset B^m(k_a)$ denote an affinoid subset and let

$$\varphi: X \to B^n(k_a)$$

be an arbitrary map. Then φ is called an *affinoid map* if it is given by affinoid functions on X, i.e., if there are functions $f_1, \ldots, f_n \in T_m/\mathrm{id}(X)$ (or, which amounts to the same thing, $f_1, \ldots, f_n \in T_m$) such that $\varphi(x) = \big(f_1(x), \ldots, f_n(x)\big)$ for all $x \in X$. Of course, $f_1, \ldots, f_n$ must have supremum norm $\leqq 1$ on X, which means that they must be power-bounded in $T_m/\mathrm{id}(X)$ (Proposition 6.2.3/1). Thus, by Proposition 6.1.1/4, the affinoid map $\varphi: X \to B^n(k_a)$ induces a unique continuous k-algebra homomorphism

$$\varphi^*: T_n \to T_m/\mathrm{id}(X)$$
$$\xi_i \mapsto f_i, \qquad i = 1, \ldots, n,$$

which, equivalently, can be described by

$$T_n \ni g \mapsto g \circ \varphi \in T_m/\mathrm{id}(X).$$

Obviously, φ can be recovered from φ^*, and it follows that $\varphi \mapsto \varphi^*$ defines a one-to-one correspondence between affinoid maps $X \to B^n(k_a)$ and k-algebra homomorphisms $T_n \to T_m/\mathrm{id}(X)$ (since these homomorphisms are continuous by Theorem 6.1.3/1). Furthermore, if $Y \subset B^n(k_a)$ denotes an affinoid subset in the image space, a given affinoid map $\varphi: X \to B^n(k_a)$ satisfies $\varphi(X) \subset Y$ if and only if $\mathrm{id}(Y) \subset \ker \varphi^*$. Thus, we can say more generally that the affinoid maps $X \to Y$ correspond bijectively to the k-algebra homomorphisms $T_n/\mathrm{id}(Y) \to T_m/\mathrm{id}(X)$.

Now let us go back to the viewpoint of maximal ideals. Denote by $\sigma: B \to A$ an arbitrary k-algebra homomorphism between k-affinoid algebras A and B. Then, for any $x \in \mathrm{Max}\, A$, the map σ induces an injection

$$B/\sigma^{-1}(\mathfrak{m}_x) \hookrightarrow A/\mathfrak{m}_x,$$

and it follows that $\sigma^{-1}(\mathfrak{m}_x)$ is maximal in B, since $A/\mathfrak{m}_x$ is finite algebraic over k. Thus, σ gives rise to a map

$${}^a\sigma: \mathrm{Max}\, A \to \mathrm{Max}\, B, \qquad x \mapsto \sigma^{-1}(\mathfrak{m}_x),$$

which can be interpreted as follows. Let $A := T_m/\mathrm{id}(X)$ and $B := T_n/\mathrm{id}(Y)$, where $X \subset B^m(k_a)$ and $Y \subset B^n(k_a)$ are affinoid subsets as above. The ideal $\mathrm{id}(X) \subset T_m$ defines, in particular, an affinoid subset in Max T_m which is obtained from X by identifying conjugate points (cf. (7.1.2)) and which can be canonically identified with $\operatorname{Max} A$ (cf. (7.1.3)). The resulting projection $\tau: X \to \operatorname{Max} A$ is given by $x \mapsto \operatorname{id}(x)$, i.e., by associating with $x \in X$ the maximal ideal of all functions in A vanishing at x. Similarly, there is a projection $\tau': Y \to \operatorname{Max} B$. We denote by $\varphi: X \to Y$ the affinoid map corresponding to the homomorphism $\sigma: B \to A$. Then we get a diagram

$$\begin{array}{ccc} X & \xrightarrow{\varphi} & Y \\ \downarrow{\scriptstyle\tau} & & \downarrow{\scriptstyle\tau'} \\ \operatorname{Max} A & \xrightarrow{{}^a\sigma} & \operatorname{Max} B \end{array}$$

which we show is commutative. Namely, let $x \in X$. By the definition of φ, we have $\sigma(f) = f \circ \varphi$ for all $f \in B$. Therefore, the ideal $\operatorname{id}(\varphi(x)) \subset B$ of all functions vanishing at $\varphi(x)$ equals the ideal of all $f \in B$ such that $f \circ \varphi$ vanishes at x, i.e., such that $\sigma(f) \in \operatorname{id}(x)$. Thus,

$$\operatorname{id}(\varphi(x)) = \sigma^{-1}(\operatorname{id}(x)),$$

which means the above diagram is commutative. Hence ${}^a\sigma$ is *derived from* φ *by identifying conjugate points.* Based on these considerations, we define affinoid varieties and morphisms between affinoid varieties as follows:

Definition 1. *A k-affinoid variety is a pair* $\operatorname{Sp} A = (\operatorname{Max} A, A)$, *where* A *is a k-affinoid algebra. A morphism* φ: $\operatorname{Sp} A \to \operatorname{Sp} B$ *of k-affinoid varieties (also called a k-affinoid morphism or a k-affinoid map) is a pair* $\varphi = ({}^a\sigma, \sigma)$ *where* $\sigma: B \to A$ *is a k-algebra homomorphism and* ${}^a\sigma$: $\operatorname{Max} A \to \operatorname{Max} B$ *is the map derived from* σ.

The case where $A = 0$ is not excluded in our definition; $\operatorname{Sp} 0 = (\emptyset, 0)$ is called the *empty k-affinoid variety.* For any affinoid variety $\operatorname{Sp} A$, the elements of the affinoid algebra A will be called the *affinoid functions* on the variety $\operatorname{Sp} A$. Just as we did here, we will in most cases use $\operatorname{Sp} A$ also in the sense of $\operatorname{Max} A$. Similarly, for any affinoid map $\varphi = ({}^a\sigma, \sigma)$: $\operatorname{Sp} A \to \operatorname{Sp} B$, we write φ instead of ${}^a\sigma$ and use the notation φ^* for σ.

The following formula shows the connection between φ and φ^*.

Lemma 2. *Let* φ: $\operatorname{Sp} A \to \operatorname{Sp} B$ *be an affinoid map. Then for any* $x \in \operatorname{Sp} A$ *and any* $g \in B$, *we have*

$$g(\varphi(x)) = \varphi^*(g)\,(x).$$

Proof. The map φ^*: $B \to A$ gives rise to a commutative diagram

$$\begin{array}{ccc} B & \xrightarrow{\varphi^*} & A \\ \downarrow & & \downarrow \\ B/\mathfrak{m}_{\varphi(x)} & \hookrightarrow & A/\mathfrak{m}_x \end{array}$$

which implies the desired equation. □

Definition 3. *An affinoid map φ: $\mathrm{Sp}\, A \to \mathrm{Sp}\, B$ is called a closed immersion if the inherent map $\varphi^*: B \to A$ is an epimorphism. The map φ is called finite if A is a finite B-module via φ^*.*

It is easy to see that a closed immersion φ: $\mathrm{Sp}\, A \to \mathrm{Sp}\, B$ is injective and identifies $\mathrm{Max}\, A$ with an affinoid subset of $\mathrm{Max}\, B$. Therefore, we say that in this case the map φ defines $\mathrm{Sp}\, A$ as a *closed subvariety* of $\mathrm{Sp}\, B$. Furthermore, by Lemma 2, all affinoid functions on $\mathrm{Sp}\, A$ may be viewed as "restrictions" of affinoid functions on $\mathrm{Sp}\, B$.

Of particular interest are the affinoid varieties $\mathbb{B}^n := \mathrm{Sp}\, T_n$, $n \geqq 0$, which we call "*unit balls*". Any affinoid variety X can be viewed as a closed subvariety of some unit ball $\mathbb{B}^n$ (cf. (7.1.3)). Also, due to the NOETHER Normalization Lemma (Corollary 6.1.2/2), one can find a finite surjective morphism $X \to \mathbb{B}^d$ for a suitable $d \geqq 0$ if $X \neq \emptyset$. Note that finite morphisms have finite fibres (e.g., use the Lying-over Theorem 10.8 of NAGATA [28] or Proposition 3.1.4/1 which deals with the special case where the image space consists of a single point).

Affinoid maps can be composed in a natural way, and it is clear that the k-affinoid varieties form a category which can be viewed as the opposite of the category of k-affinoid algebras. Thus, properties can be transformed between the two categories by dualizing. For example, it follows from Proposition 6.1.1/10:

Proposition 4. *In the category of k affinoid varieties the fibre product of two varieties $\mathrm{Sp}\, B_1$, $\mathrm{Sp}\, B_2$ over a variety $\mathrm{Sp}\, A$ can be constructed, namely,*

$$\mathrm{Sp}\, B_1 \times_{\mathrm{Sp} A} \mathrm{Sp}\, B_2 = \mathrm{Sp}\, (B_1 \mathbin{\widehat{\otimes}}_A B_2).$$

For completeness we add here a few words on fibre products. Let $\mathfrak{C}$ denote an arbitrary category. For two objects X, $Y \in \mathfrak{C}$, a *direct product* of X and Y in $\mathfrak{C}$ is an object $W \in \mathfrak{C}$ together with two $\mathfrak{C}$-morphisms $p_1: W \to X$ and $p_2: W \to Y$, called projections, such that the following universal property holds:

To any pair of $\mathfrak{C}$-morphisms $U \to X$, $U \to Y$, there corresponds a unique $\mathfrak{C}$-morphism $U \to W$ making the diagram

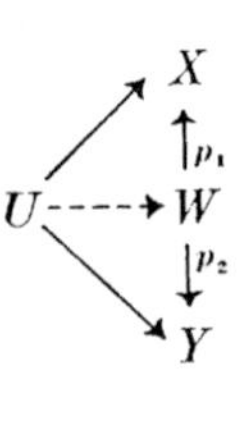

commutative.

It is clear that the direct product, if it exists, is unique up to natural isomorphism. In analogy with the category of sets, one uses the notation $W = X \times Y$.

To construct fibre products over a fixed object $Z \in \mathfrak{C}$, one proceeds as follows. One passes from $\mathfrak{C}$ to the category $\mathfrak{C}_Z$ of all objects over Z. The objects of $\mathfrak{C}_Z$ consist of all $\mathfrak{C}$-morphisms $X \to Z$ aiming at Z, where X varies over $\mathfrak{C}$, and a $\mathfrak{C}_Z$-morphism between two such objects $X \to Z$ and $Y \to Z$ is a $\mathfrak{C}$-morphism $X \to Y$ over Z, i.e., a $\mathfrak{C}$-morphism making the diagram

$$\begin{array}{ccc} X & \longrightarrow & Y \\ & \searrow \quad \swarrow & \\ & Z & \end{array}$$

commutative. Our definition of direct products applies, in particular, to the category $\mathfrak{C}_Z$; the direct product of two objects $X \to Z$ and $Y \to Z$ is denoted by $X \times_Z Y \to Z$. In order to simplify the notation, it is common not to mention the category $\mathfrak{C}_Z$ explicitly. One writes X and Y instead of $X \to Z$ and $Y \to Z$, and one calls $X \times_Z Y$ the *fibre product* in $\mathfrak{C}$ of the objects X, Y over Z.

For the *proof* of Proposition 4, it should be noted that the universal property defining fibre products is dual to the universal property characterizing (complete) tensor products of (complete) algebras, (cf. e.g., Proposition 3.1.1/2). Thus, in the situation of Proposition 4, it follows that the projections

$$\operatorname{Sp} B_1 \xleftarrow{p_1} \operatorname{Sp} B_1 \times_{\operatorname{Sp} A} \operatorname{Sp} B_2 \xrightarrow{p_2} \operatorname{Sp} B_2$$

are derived from the canonical A-algebra homomorphisms

$$B_1 \xrightarrow{\tau_1} B_1 \mathbin{\widehat{\otimes}_A} B_2 \xleftarrow{\tau_2} B_2 .$$

7.1.5. The reduction functor. — We have investigated in (6.3) the functor $\sim$ which associates to a k-affinoid algebra A its "reduction" $\tilde{A} = \mathring{A}/\check{A}$, where $\mathring{A}$ is the $\mathring{k}$-algebra of all power-bounded elements in A and $\check{A}$ is the $\mathring{A}$-ideal of all topologically nilpotent elements in A. Recall that

$$\mathring{A} = \{f \in A\,;\, |f|_{\sup} \leqq 1\},$$

$$\check{A} = \{f \in A\,;\, |f|_{\sup} < 1\},$$

(cf. Propositions 6.2.3/1 and 6.2.3/2) and $\tilde{A}$ is a finitely generated $\tilde{k}$-algebra (cf. Corollary 6.3.4/3). In the present section, we replace algebras by their corresponding varieties and study the resulting functor $\operatorname{Sp} A \rightsquigarrow \operatorname{Sp} \tilde{A}$, $\varphi \rightsquigarrow \tilde{\varphi}$, called the *reduction functor*, which goes from the category of k-affinoid varieties into the category of affine algebraic varieties over $\tilde{k}$ (or, more precisely, affine schemes of finite type over $\tilde{k}$).

For convenience we give a simplified definition of these varieties which is adequate for our purposes. Analogous to Definition 7.1.4/1, we understand by an *affine algebraic variety over* $\tilde{k}$ a pair $\operatorname{Sp} R := (\operatorname{Max} R, R)$ where R is a finitely generated $\tilde{k}$-algebra. Similarly, a *morphism* $\varphi\colon \operatorname{Sp} R \to \operatorname{Sp} R'$ of such varieties is a pair $({}^a\sigma, \sigma)$ where $\sigma\colon R' \to R$ is a $\tilde{k}$-algebra homomorphism and ${}^a\sigma\colon \operatorname{Max} R$

$\to \operatorname{Max} R'$ is derived from σ. In particular, if $\varphi\colon \operatorname{Sp} A \to \operatorname{Sp} B$ is a morphism of k-affinoid varieties given by the k-algebra homomorphism $\varphi^*\colon B \to A$, then the corresponding morphism $\tilde{\varphi}\colon \operatorname{Sp} \tilde{A} \to \operatorname{Sp} \tilde{B}$ obtained by applying the reduction functor is defined by the homomorphism $\widetilde{\varphi^*}\colon \tilde{B} \to \tilde{A}$. Note that we are considering only global functions on affine varieties. Furthermore, we are restricting our attention to "closed" points, which is, of course, permissible since we are dealing with finitely generated $\tilde{k}$-algebras (which are Jacobson rings; see Nagata [28], Theorem 14.9).

In the following let A denote an arbitrary k-affinoid algebra. We want to define a functorial map $\pi\colon \operatorname{Max} A \to \operatorname{Max} \tilde{A}$ which will prove to be surjective. Namely, for $x \in \operatorname{Max} A$, consider the canonical epimorphism $\sigma\colon A \to A/\mathfrak{m}_x$, as well as the corresponding map $\tilde{\sigma}\colon \tilde{A} \to \widetilde{A/\mathfrak{m}_x}$. Since $A/\mathfrak{m}_x$ is a finite algebraic extension of k, the residue field $\widetilde{A/\mathfrak{m}_x}$ must be finite algebraic over $\tilde{k}$ and thus $\tilde{\sigma}(\tilde{A})$ also is a finite algebraic extension of $\tilde{k}$. In particular, ker $\tilde{\sigma}$ is a maximal ideal in $\tilde{A}$. Therefore, we can define $\mathfrak{m}_{\pi(x)} := \ker \tilde{\sigma}$, and hence we get a map $\pi\colon \operatorname{Max} A \to \operatorname{Max} \tilde{A}$, also denoted by $x \mapsto \tilde{x}$.

Note that, in general, $\mathfrak{m}_{\pi(x)}$ differs from the $\tilde{A}$-ideal $\tilde{\mathfrak{m}}_x$, which is the image of $\mathfrak{m}_x \cap \mathring{A}$ via the projection $\mathring{A} \to \tilde{A}$.

Proposition 1. *For any $x \in \operatorname{Max} A$, one has $\mathfrak{m}_{\pi(x)} = \operatorname{rad} \tilde{\mathfrak{m}}_x$. Furthermore, if x is a rational point, i.e., if $[A/\mathfrak{m}_x : k] = 1$, then $\mathfrak{m}_{\pi(x)} = \tilde{\mathfrak{m}}_x$.*

Proof. Obviously, $\tilde{\mathfrak{m}}_x \subset \mathfrak{m}_{\pi(x)}$. Moreover, the first part of the assertion is a consequence of Observation 6.3.1/4, since the epimorphism $\sigma\colon A \to A/\mathfrak{m}_x$ is strict. Now assume that x is a rational point, i.e., $A/\mathfrak{m}_x = k$. We want to show $\mathfrak{m}_{\pi(x)} \subset \tilde{\mathfrak{m}}_x$. Let $\tilde{f} \in \mathfrak{m}_{\pi(x)} = \ker \tilde{\sigma}$ and denote by $f \in \mathring{A}$ a representative of $\tilde{f}$. Then $\sigma(f) \in \mathring{k}$ and $f - \sigma(f) \in \ker \sigma = \mathfrak{m}_x$. But this means $\tilde{f} = \widetilde{f - \sigma(f)} \in \tilde{\mathfrak{m}}_x$. □

Proposition 2. *Let $f \in \mathring{A}$ and $x \in \operatorname{Max} A$. Then $|f(x)| < 1$ if and only if $\tilde{f}(\pi(x)) = 0$.*

This follows immediately from the definition of π. More generally, the map $\pi_A = \pi\colon \operatorname{Max} A \to \operatorname{Max} \tilde{A}$ is easily checked to be functorial in the following sense:

Proposition 3. *Let $\varphi\colon \operatorname{Sp} A \to \operatorname{Sp} B$ be a morphism of k-affinoid varieties. Then the diagram*

$$\begin{array}{ccc} \operatorname{Max} A & \xrightarrow{\varphi} & \operatorname{Max} B \\ \downarrow{\scriptstyle \pi_A} & & \downarrow{\scriptstyle \pi_B} \\ \operatorname{Max} \tilde{A} & \xrightarrow{\tilde{\varphi}} & \operatorname{Max} \tilde{B} \end{array}$$

is commutative.

As indicated above, we want to show

Theorem 4. *The map $\pi\colon \operatorname{Max} A \to \operatorname{Max} \tilde{A}$ is surjective.*

The *proof* will be carried out in several steps. First we consider the case where $A = T_n$, i.e., A is the free TATE algebra in n variables. Let $\tilde{x} \in \operatorname{Max} \tilde{T}_n$ and let $\tilde{\sigma}\colon \tilde{T}_n \to \tilde{k}_a$ be a $\tilde{k}$-algebra homomorphism having kernel $\mathfrak{m}_{\tilde{x}}$. Applying Proposition 6.1.1/4, the map $\tilde{\sigma}$ can be lifted to a k-algebra homomorphism $\sigma\colon T_n \to k_a$. Then $\ker \sigma$ is a maximal ideal in T_n; hence there is an $x \in \operatorname{Max} T_n$ satisfying $\mathfrak{m}_x = \ker \sigma$. It follows readily from our construction that $\pi(x) = \tilde{x}$. Thus, π is surjective.

Next we assume that A is an arbitrary affinoid algebra without zero divisors. By the NOETHER Normalization Lemma (Corollary 6.1.2/2), there is a finite injection $\iota\colon T_n \hookrightarrow A$ for some $n \geqq 0$. Taking fields of fractions, $Q(A)$ is a finite algebraic extension of $Q(T_n)$. We denote by K the smallest extension field of $Q(T_n)$ which is quasi-Galois over $Q(T_n)$ and which contains A. Let G be the group of all $Q(T_n)$-automorphisms of K and let $B \subset K$ be the smallest k-subalgebra containing all algebras $\gamma(A)$, $\gamma \in G$, i.e., $B := T_n[\gamma(A);\, \gamma \in G]$. Then B is finite over T_n, and hence canonically an affinoid algebra (cf. Propositions 6.1.1/6 and 6.1.3/2). Thus, the injections $T_n \hookrightarrow A \hookrightarrow B$ give rise to a commutative diagram

$$\begin{array}{ccccc} \operatorname{Max} B & \to & \operatorname{Max} A & \to & \operatorname{Max} T_n \\ \downarrow{\scriptstyle \pi_B} & & \downarrow{\scriptstyle \pi_A} & & \downarrow{\scriptstyle \pi_{T_n}} \\ \operatorname{Max} \tilde{B} & \to & \operatorname{Max} \tilde{A} & \to & \operatorname{Max} \tilde{T}_n\,. \end{array}$$

Since the injections $T_n \hookrightarrow A \hookrightarrow B$ and $\tilde{T}_n \hookrightarrow \tilde{A} \hookrightarrow \tilde{B}$ are finite and hence integral, we can apply the Going-up Theorem to conclude that all horizontal maps are surjective. Thus, in order to get the surjectivity of π_A, we need only show that π_B is surjective.

The group G acts on K and, in particular, on B, by construction of B. For $\gamma \in G$, the corresponding automorphism $B \to B$ will also be denoted by γ. Since any such map $\gamma\colon B \to B$ leaves T_n, and thus k, fixed, it induces a $\tilde{k}$-algebra homomorphism $\tilde{\gamma}\colon \tilde{B} \to \tilde{B}$. Hence G acts on $\tilde{B}$. Furthermore, switching to the corresponding varieties, G acts on $\operatorname{Max} B$ and $\operatorname{Max} \tilde{B}$. Clearly, this action is compatible with the diagram

$$\begin{array}{ccc} \operatorname{Max} B & \to & \operatorname{Max} T_n \\ \downarrow{\scriptstyle \pi_B} & & \downarrow{\scriptstyle \pi_{T_n}} \\ \operatorname{Max} \tilde{B} & \to & \operatorname{Max} \tilde{T}_n\,, \end{array}$$

where all maps are surjective with the possible exception of π_B. Therefore, each fibre of $\operatorname{Max} \tilde{B} \to \operatorname{Max} \tilde{T}_n$ must contain a point of $\pi_B(\operatorname{Max} B)$, and the surjectivity of π_B will follow if we can show that G acts transitively on each fibre of $\operatorname{Max} \tilde{B} \to \operatorname{Max} \tilde{T}_n$.

So let's prove this and thereby conclude our second step. Let $\tilde{x} \in \operatorname{Max} \tilde{T}_n$ and let $\tilde{y}, \tilde{y}' \in \operatorname{Max} \tilde{B}$ be points lying over $\tilde{x}$. Assuming that there is no element

$\gamma \in G$ transforming $\tilde{y}$ into $\tilde{y}'$, we have $\mathfrak{m}_{\tilde{y}'} \not\subset \mathfrak{m}_{^a\tilde{\gamma}(\tilde{y})}$ for all $\gamma \in G$ and hence

$$\mathfrak{m}_{\tilde{y}'} \not\subset \bigcup_{\gamma \in G} \mathfrak{m}_{^a\tilde{\gamma}(\tilde{y})}.$$

Therefore, we can find an element $f \in \mathring{B}$ such that $\tilde{f} \in \mathfrak{m}_{\tilde{y}'}$ and, for all $\gamma \in G$,

$$\tilde{f} \notin \mathfrak{m}_{^a\tilde{\gamma}(\tilde{y})} = \tilde{\gamma}^{-1}(\mathfrak{m}_{\tilde{y}})$$

or, equivalently, $\tilde{\gamma}(\tilde{f}) \notin \mathfrak{m}_{\tilde{y}}$. If

$$f^r + a_1 f^{r-1} + \cdots + a_r = 0$$

is the equation of f over $Q(T_n)$, then, up to sign, a_r is a power of the product of all conjugates of f. Since the set of conjugates of f equals $\{\gamma(f); \gamma \in G\}$, we have

$$\tilde{a}_r \in \mathfrak{m}_{\tilde{y}'}, \qquad \tilde{a}_r \notin \mathfrak{m}_{\tilde{y}}.$$

Note that a_r must belong to T_n, since it is integral over T_n. Furthermore, $f \in \mathring{B}$ implies $a_r \in \mathring{T}_n$ (use Proposition 3.8.1/7 or 6.2.2/2). Thus, we have

$$\tilde{a}_r \in \mathfrak{m}_{\tilde{y}'} \cap \tilde{T}_n = \mathfrak{m}_{\tilde{x}}$$

which contradicts $\tilde{a}_r \notin \mathfrak{m}_{\tilde{y}}$ since $\tilde{y}$ also lies over $\tilde{x}$, i.e., $\mathfrak{m}_{\tilde{y}} \cap \tilde{T}_n = \mathfrak{m}_{\tilde{x}}$. Therefore, it follows that $\tilde{y}$ can be transformed into $\tilde{y}'$ by G, and hence G acts transitively on each fibre of $\operatorname{Max} \tilde{B} \to \operatorname{Max} \tilde{T}_n$. Thus, in the special case where A is an integral domain, our proof is finished.

Turning to the general case, let $\mathfrak{p}_1, \ldots, \mathfrak{p}_s$ denote the minimal prime ideals of A. Then the projections $\sigma_i\colon A \to A/\mathfrak{p}_i$, $i = 1, \ldots, s$, define a finite homomorphism

$$\sigma\colon A \to A' := \bigoplus_1^s A/\mathfrak{p}_i$$

having the nilradical rad A as kernel and hence yielding a surjective map

$$^a\sigma\colon \operatorname{Max} A' \to \operatorname{Max} A.$$

Applying the functor $\tilde{\ }$, we get a finite injection

$$\tilde{\sigma}\colon \tilde{A} \to \tilde{A}' = \bigoplus_1^s \widetilde{A/\mathfrak{p}_i}$$

(use Lemma 6.2.1/3 and Theorem 6.3.4/2), and therefore a surjection

$$^a\tilde{\sigma}\colon \operatorname{Max} \tilde{A}' \to \operatorname{Max} \tilde{A}.$$

Thus, the commutative diagram

$$\begin{array}{ccc} \operatorname{Max} A' & \xrightarrow{^a\sigma} & \operatorname{Max} A \\ \downarrow{\scriptstyle \pi_{A'}} & & \downarrow{\scriptstyle \pi_A} \\ \operatorname{Max} \tilde{A}' & \xrightarrow{^a\tilde{\sigma}} & \operatorname{Max} \tilde{A} \end{array}$$

shows that π_A is surjective if we can show that $\pi_{A'}$ is surjective.

Writing A_i instead of $A/\mathfrak{p}_i$, we have proved above that all maps $\pi_{A_i}\colon \operatorname{Max} A_i \to \operatorname{Max} \tilde{A}_i$ are surjective. It is easily seen that $\operatorname{Max} A'$ can be viewed as the disjoint union of all $\operatorname{Max} A_i$, similarly with $\operatorname{Max} \widetilde{A'}$, and that $\pi_{A'}$ coincides with π_{A_i} in $\operatorname{Max} A_i$. As a consequence, $\pi_{A'}$ must be surjective. □

7.2. Affinoid subdomains

7.2.1. The canonical topology on Sp A. — We consider the unit ball $B^n(k_a)$ with the topology induced by the natural topology of the affine space $(k_a)^n$. For an element $z = (z_1, \ldots, z_n) \in (k_a)^n$, we define the notation

$$|z| := \max_{1 \leq i \leq n} |z_i|.$$

Note that $|\ \ |$ is a norm on $(k_a)^n$ defining the topology of $(k_a)^n$ and hence that of $B^n(k_a)$. We start with a result which may be viewed as an improvement of Proposition 5.1.4/2.

Proposition 1. *Let $f \in T_n$ and let $x, y \in B^n(k_a)$. Then*

$$|f(x) - f(y)| \leq |f|\, |x - y|.$$

In particular, f gives rise to a continuous function on $B^n(k_a)$.

Proof. First let us assume that the above estimate is true for some power series g, h in T_n. We then show that this estimate must also hold for their product $f := gh$. This proceeds as usual:

$$\begin{aligned} |f(x) - f(y)| &= |g(x)\, h(x) - g(x)\, h(y) + g(x)\, h(y) - g(y)\, h(y)| \\ &\leq \max\{|g(x)|\, |h(x) - h(y)|, |h(y)|\, |g(x) - g(y)|\} \\ &\leq |g|\, |h|\, |x - y| = |f|\, |x - y|. \end{aligned}$$

Next we realize that our estimate is trivially true for the constant functions and the coordinate functions $\xi_1, \ldots, \xi_n \in T_n$; thus, it must hold for all monomials $\xi^\nu \in T_n$. Therefore, in the case of an arbitrary element $f = \sum a_\nu \xi^\nu \in T_n$, it follows that

$$|f(x) - f(y)| = |\sum a_\nu (x^\nu - y^\nu)| \leq \max_\nu |a_\nu|\, |x^\nu - y^\nu| \leq |f|\, |x - y|. \quad \square$$

We denote by $\tau\colon B^n(k_a) \to \operatorname{Max} T_n$ the surjection discussed in (7.1.1) and consider $\operatorname{Max} T_n$ with the quotient topology via τ, which will be called the *canonical topology* of $\operatorname{Max} T_n$ (or $\operatorname{Sp} T_n$). As an immediate consequence of Proposition 1, we have

Corollary 2. *For any $f \in T_n$ and any real number $\alpha > 0$, the subsets*

$$\{x \in \mathrm{Sp}\, T_n; f(x) \neq 0\},$$
$$\{x \in \mathrm{Sp}\, T_n; |f(x)| \leqq \alpha\},$$
$$\{x \in \mathrm{Sp}\, T_n; |f(x)| = \alpha\},$$
$$\{x \in \mathrm{Sp}\, T_n; |f(x)| \geqq \alpha\}$$

are open in Sp T_n *with respect to the canonical topology.*

In particular, it follows that the canonical topology is finer than the Zariski topology. In the following sections, we will further investigate the canonical topology. Unless otherwise stated, topological terms will always refer to this topology. For elements $f_1, \ldots, f_r \in T_n$, we introduce the notation

$$(\mathrm{Sp}\, T_n)\,(f_1, \ldots, f_r) := \{x \in \mathrm{Sp}\, T_n; |f_1(x)| \leqq 1, \ldots, |f_r(x)| \leqq 1\}.$$

Note that this subset of Sp T_n equals the intersection of the sets $(\mathrm{Sp}\, T_n)\,(f_i)$, $i = 1, \ldots, r$, and hence is open by Corollary 2.

Proposition 3. *The canonical topology on* Sp T_n *satisfies the following properties:*

(i) *The open sets* $(\mathrm{Sp}\, T_n)\,(f_1, \ldots, f_r)$ *for arbitrary* $f_1, \ldots, f_r \in T_n$ *form a basis.*

(ii) *For any* $x \in \mathrm{Sp}\, T_n$, *the open sets* $(\mathrm{Sp}\, T_n)\,(f_1, \ldots, f_r)$ *for arbitrary* $f_1, \ldots, f_r \in \mathfrak{m}_x$ *form a fundamental system of neighborhoods at the point* x.

Proof. Let U be an open subset of Sp T_n and let x be a point in U. It is only necessary to show that there are finitely many elements $f_1, \ldots, f_r \in \mathfrak{m}_x$ satisfying $U \supset (\mathrm{Sp}\, T_n)\,(f_1, \ldots, f_r)$. Considering the surjection $\tau: B^n(k_a) \to \mathrm{Sp}\, T_n$, we denote by $y_1, \ldots, y_s$ the points of the fibre $\tau^{-1}(x)$. Note that $\{y_1, \ldots, y_s\} \subset B^n(k_a)$ is precisely the zero set of the ideal $\mathfrak{m}_x$. The set $\tau^{-1}(U)$ is open in $B^n(k_a)$; hence, for each $\sigma = 1, \ldots, s$, it contains a ball

$$B^+(y_\sigma, \varepsilon_\sigma) := \{z \in B^n(k_a); |z - y_\sigma| \leqq \varepsilon_\sigma\}$$

with center y_σ and radius $\varepsilon_\sigma > 0$. Choosing $\varepsilon > 0$ small enough, we may assume

$$\tau^{-1}(U) \supset \bigcup_{\sigma=1}^{s} B^+(y_\sigma, \varepsilon).$$

By Proposition 7.1.1/3, there are polynomials $p_i \in k[\xi_1, \ldots, \xi_i]$, $i = 1, \ldots, n$, such that p_i is monic in ξ_i and such that $\mathfrak{m}_x$ is generated by $p_1, \ldots, p_n$. In order to conclude our proof, we need the following statement which amounts to an n-dimensional version of the lemma on continuity of roots:

Any sequence $z_j \in B^n(k_a)$ *such that* $\lim\limits_{j\to\infty} p_i(z_j) = 0$ *for* $i = 1, \ldots, n$ *contains a subsequence converging to a root of the polynomials* $p_1, \ldots, p_n$.

The case $n = 1$ follows easily from Corollary 3.4.1/2, since the polynomial $f_j(\xi_1) := p_1(\xi_1) - p_1(z_j)$ has z_j as a root for $j = 1, 2, \ldots$, and the sequence f_j clearly converges to p_1. Now let $n \geqq 2$. For any $z \in B^n(k_a)$, we denote by $z' \in B^{n-1}(k_a)$ the $(n-1)$-tuple of the first $n - 1$ components of z and by z'' its last component. Proceeding inductively, we suppose that our statement has already been verified for dimensions less than n. Thus, we can assume that (z_j) is a sequence having the additional property that $y' := \lim_{j\to\infty} z'_j$ exists and is a root of the polynomials $p_1, \ldots, p_{n-1}$, with y' necessarily belonging to $B^{n-1}(k_a)$. Using the fact that $\lim_{j\to\infty} p_n(z_j) = 0$, it follows from Proposition 1 that $\lim_{j\to\infty} p_n(y', z''_j) = 0$ Therefore, applying to the polynomial $p(\xi_n) := p_n(y', \xi_n)$ what we have already proved in the 1-dimensional case, the sequence z_j must contain a subsequence z_{ν_j} such that $y'' := \lim_{j\to\infty} z''_{\nu_j}$ exists and is a root of $p(\xi_n)$. But then, by construction, $y := (y', y'')$ is a root of $p_1, \ldots, p_n$ satisfying $y = \lim_{j\to\infty} z_{\nu_j}$. This verifies our statement concerning the continuity of roots.

It is clear that the given subsequence must converge to one of the points $y_1, \ldots, y_s$, since these are the only zeros of the ideal $\mathfrak{m}_x$ and hence the only roots in $B^n(k_a)$ of the generators $p_1, \ldots, p_n$. Consequently there must exist some element $c \in k$, $c \neq 0$, such that

$$\bigcup_{\sigma=1}^{s} B^+(y_\sigma, \varepsilon) \supset \{z \in B^n(k_a);\, |p_1(z)| \leqq |c|, \ldots, |p_n(z)| \leqq |c|\},$$

for otherwise our statement could not be true. Finally, applying the map τ gives

$$U \supset (\mathrm{Sp}\, T_n)\,(c^{-1}p_1, \ldots, c^{-1}p_n). \qquad \square$$

Now, for an arbitrary k-affinoid algebra A, we can define a topology on $\mathrm{Sp}\, A$ in the following way. Choosing a closed immersion $\mathrm{Sp}\, A \hookrightarrow \mathrm{Sp}\, T_n$ (cf. (7.1.3)), we assign to $\mathrm{Sp}\, A$ the topology induced by the canonical topology of $\mathrm{Sp}\, T_n$. Then it follows immediately from Proposition 3 that the subsets of type

$$(\mathrm{Sp}\, A)\,(f_1, \ldots, f_r) := \{x \in \mathrm{Sp}\, A;\, |f_1(x)| \leqq 1, \ldots, |f_r(x)| \leqq 1\}$$

for arbitrary $f_1, \ldots, f_r \in A$ form a basis for the topology of $\mathrm{Sp}\, A$. Consequently, this topology is independent of the chosen closed immersion and we call it the *canonical topology* of $\mathrm{Sp}\, A$. Considering the surjection $\tau\colon B^n(k_a) \to \mathrm{Sp}\, T_n$ and thinking of $\mathrm{Sp}\, A$ as of a Zariski-closed subset in $\mathrm{Sp}\, T_n$, it is an easy exercise to verify that the canonical topology on $\mathrm{Sp}\, A$ equals the quotient topology via the surjection $\tau^{-1}(\mathrm{Sp}\, A) \to \mathrm{Sp}\, A$. Here, of course, we assign to the affinoid subset $\tau^{-1}(\mathrm{Sp}\, A) \subset B^n(k_a)$ the topology induced by that of $B^n(k_a)$.

We have already used the fact that Proposition 3 carries over to the general case when T_n is replaced by an arbitrary k-affinoid algebra. The same is true for Corollary 2. For reference we state this explicitly.

Proposition 4. *For an arbitrary k-affinoid algebra A, the canonical topology on* Sp A *satisfies the properties of Corollary 2 and Proposition 3, with T_n being replaced by A.*

Furthermore, it is useful to know:

Proposition 5. *Let φ:* Sp $B \to$ Sp A *be a k-affinoid map, defined by a homomorphism of k-affinoid algebras $\varphi^*\colon A \to B$. Then φ is continuous with respect to the canonical topology.*

Proof. For arbitrary elements $f_1, \ldots, f_r \in A$, it follows from Lemma 7.1.4/2 that

$$\varphi^{-1}\big((\mathrm{Sp}\,A)\,(f_1, \ldots, f_r)\big) = (\mathrm{Sp}\,B)\,\big(\varphi^*(f_1), \ldots, \varphi^*(f_r)\big).$$

Hence the inverse image via φ of any basic open set in Sp A is a basic open set in Sp B. □

If, in addition, φ: Sp $B \to$ Sp A is a closed immersion, the map φ^* is surjective and hence the proof shows, in particular, that the canonical topology on Sp B equals the restriction of the canonical topology on Sp A to Sp B.

7.2.2. The universal property defining affinoid subdomains. — Let Sp A be an affinoid variety and U be a subset of Sp A. Then a given affinoid map φ: Sp $A' \to$ Sp A is said *to represent all affinoid maps into U* if φ maps Sp A' into U and if φ satisfies the following universal property:

Given any affinoid map ψ: Sp $B \to$ Sp A such that $\psi(\mathrm{Sp}\,B) \subset U$, there exists a unique affinoid map ψ': Sp $B \to$ Sp A' such that $\psi = \varphi \circ \psi'$, i.e., such that

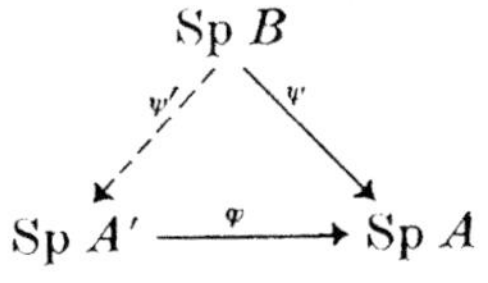

is commutative.

This universal property has some interesting consequences.

Proposition 1. *Let U be a subset of* Sp A *and let φ:* Sp $A' \to$ Sp A *be an affinoid map representing all affinoid maps into U. Then*

(i) *φ is injective and satisfies $\varphi(\mathrm{Sp}\,A') = U$.*

(ii) *For $x \in$ Sp A' and $n \in \mathbb{N}$, the map $\varphi^*\colon A \to A'$ induces an isomorphism $A/\mathfrak{m}^n_{\varphi(x)} \xrightarrow{\sim} A'/\mathfrak{m}^n_x$.*

(iii) *For $x \in$ Sp A', we have $\mathfrak{m}_x = \varphi^*(\mathfrak{m}_{\varphi(x)})\,A'$.*

Proof. For an arbitrary element $y \in U$, we consider the following commutative diagram

$$\begin{array}{ccc} A & \xrightarrow{\varphi^*} & A' \\ \downarrow{\scriptstyle \pi} & \swarrow{\scriptstyle \alpha} & \downarrow{\scriptstyle \pi'} \\ A/\mathfrak{m}_y^n & \xrightarrow{\sigma} & A'/\mathfrak{m}_y^n A', \end{array}$$

where π and π' denote the canonical projections and σ is induced by φ^*. (We use $\mathfrak{m}_y^n A'$ as a short notation for the ideal generated by $\varphi^*(\mathfrak{m}_y^n)$ in A'.) Since φ represents all affinoid maps into U, there exists a unique homomorphism $\alpha: A' \to A/\mathfrak{m}_y^n$ making the upper triangle commutative. Then it follows that both of the maps π' and $\sigma \circ \alpha$ make the diagram

$$\begin{array}{ccc} A & \xrightarrow{\varphi^*} & A' \\ & {\scriptstyle \sigma \circ \pi} \searrow \quad \swarrow & \\ & A'/\mathfrak{m}_y^n A' & \end{array}$$

commutative. Hence, again due to the universal property of φ, they must be equal. Thus, the lower triangle in the above diagram also is commutative.

The map σ is surjective, since π' is surjective. Furthermore, α is surjective, and we have $\ker \pi' = \mathfrak{m}_y^n A' \subset \ker \alpha$. Hence σ must, in fact, be bijective. Taking $n = 1$, we see then that $\mathfrak{m}_y A'$ is a maximal ideal in A'. Thus, the fibre $\varphi^{-1}(y)$ consists of precisely one element $x \in \operatorname{Sp} A'$, namely, $\mathfrak{m}_x = \mathfrak{m}_y A'$. This proves (i) and (iii). Moreover, (ii) must hold, since $\mathfrak{m}_x^n = \mathfrak{m}_y^n A'$ for all $n \in \mathbb{N}$ and since σ is bijective. □

Let $\varphi: \operatorname{Sp} A' \to \operatorname{Sp} A$ be as in Proposition 1; i.e., φ represents all affinoid maps into a subset $U \subset \operatorname{Sp} A$. Then we can identify the underlying set of $\operatorname{Sp} A'$ with its image U. Thus, U is equipped with the structure of an affinoid variety, and the homomorphism $\varphi^*: A \to A'$ can be interpreted as the process of restricting affinoid functions from $\operatorname{Sp} A$ to U. Note that the affinoid structure on U is unique (up to isomorphism), due to the universal property satisfied by φ.

Definition 2. *Let* $\operatorname{Sp} A$ *be an affinoid variety and let* U *be a subset of* $\operatorname{Sp} A$. *Then* U *is called an affinoid subdomain of* $\operatorname{Sp} A$ *if there exists an affinoid map* $\varphi: \operatorname{Sp} A' \to \operatorname{Sp} A$ *representing all affinoid maps into* U.

To give some trivial examples, we mention that $\emptyset$ and $\operatorname{Sp} A$ are affinoid subdomains of $\operatorname{Sp} A$. Namely, the affinoid maps $\operatorname{Sp} 0 \to \operatorname{Sp} A$ (corresponding to the zero homomorphism $A \to 0$) and $\mathrm{id}: \operatorname{Sp} A \to \operatorname{Sp} A$ represent all affinoid maps into $\emptyset$ and $\operatorname{Sp} A$, respectively.

In the situation of the definition, we also say that $\operatorname{Sp} A'$ is an affinoid subdomain of $\operatorname{Sp} A$ via φ, and we write $\operatorname{Sp} A' \hookrightarrow \operatorname{Sp} A$ or $\operatorname{Sp} A' \subset \operatorname{Sp} A$. When no confusion is possible, U is also used in the sense of $\operatorname{Sp} A'$. Furthermore, U or $\operatorname{Sp} A'$ is called an *open* subdomain of $\operatorname{Sp} A$ if U is an open subset in $\operatorname{Sp} A$. Open affinoid subdomains are introduced only for technical reasons. We will see later (in (7.2.5)) that all affinoid subdomains are open and that the canonical topology on an affinoid subdomain $\operatorname{Sp} A' \subset \operatorname{Sp} A$ is induced by the canonical topology of $\operatorname{Sp} A$.

In the following propositions, we list some elementary properties of affinoid subdomains which are formal consequences of the defining universal property.

Proposition 3. *Let* $\psi\colon \operatorname{Sp} A'' \hookrightarrow \operatorname{Sp} A'$ *and* $\varphi\colon \operatorname{Sp} A' \hookrightarrow \operatorname{Sp} A$ *denote affinoid subdomains. Then* $\operatorname{Sp} A''$ *is an affinoid subdomain of* $\operatorname{Sp} A$ *via the composition* $\varphi \circ \psi\colon \operatorname{Sp} A'' \hookrightarrow \operatorname{Sp} A' \hookrightarrow \operatorname{Sp} A$.

Proof. This is a direct consequence of the universal properties satisfied by φ and ψ. □

Proposition 4. *Let* $\varphi\colon \operatorname{Sp} B \to \operatorname{Sp} A$ *be an affinoid map and let* $\operatorname{Sp} A' \subset \operatorname{Sp} A$ *denote an affinoid subdomain of* $\operatorname{Sp} A$. *Then* $\varphi^{-1}(\operatorname{Sp} A')$ *is an affinoid subdomain of* $\operatorname{Sp} B$.

Considering $\operatorname{Sp} B$ *and* $\operatorname{Sp} A'$ *as varieties over* $\operatorname{Sp} A$, *the projection* $p_1 : \operatorname{Sp} B \times_{\operatorname{Sp} A} \operatorname{Sp} A' \to \operatorname{Sp} B$ *represents all affinoid maps into* $\varphi^{-1}(\operatorname{Sp} A')$. *Furthermore, the projection* p_2: $\operatorname{Sp} B \times_{\operatorname{Sp} A} \operatorname{Sp} A' \to \operatorname{Sp} A'$ *is the unique affinoid map making the following diagram commutative:*

$$\begin{array}{ccc} \operatorname{Sp} B & \xrightarrow{\ \varphi\ } & \operatorname{Sp} A \\ \uparrow{\scriptstyle p_1} & & \uparrow \\ \operatorname{Sp} B \times_{\operatorname{Sp} A} \operatorname{Sp} A' & \xrightarrow{\ p_2\ } & \operatorname{Sp} A'. \end{array}$$

Proof. The commutativity of the above diagram is obvious from the definition of the fibre product (see (7.1.4) and, in particular, Proposition 7.1.4/4). Thus, we get $\operatorname{im} p_1 \subset \varphi^{-1}(\operatorname{Sp} A')$. Furthermore, the uniqueness of p_2 is clear, since $\operatorname{Sp} A' \hookrightarrow \operatorname{Sp} A$ is an affinoid subdomain.

It only remains to show that p_1 represents all affinoid maps into $\varphi^{-1}(\operatorname{Sp} A')$. Therefore, let us consider an affinoid map $\psi\colon \operatorname{Sp} C \to \operatorname{Sp} B$ with $\operatorname{im} \psi \subset \varphi^{-1}(\operatorname{Sp} A')$ or, equivalently, $\operatorname{im} \varphi \circ \psi \subset \operatorname{Sp} A'$. Then $\varphi \circ \psi$ is factored through $\operatorname{Sp} A'$ by an affinoid map $\psi''\colon \operatorname{Sp} C \to \operatorname{Sp} A'$. The maps ψ, ψ'' can be considered as maps over $\operatorname{Sp} A$; thus, they define an affinoid map $\psi' : \operatorname{Sp} C \to \operatorname{Sp} B \times_{\operatorname{Sp} A} \operatorname{Sp} A'$ satisfying, in particular, $p_1 \circ \psi' = \psi$. This settles the existence part of the universal property for p_1. Just as easily, the uniqueness assertion can be derived from the uniqueness parts of the universal properties for $\operatorname{Sp} B \times_{\operatorname{Sp} A} \operatorname{Sp} A'$ and $\operatorname{Sp} A' \hookrightarrow \operatorname{Sp} A$. □

The proposition applies, in particular, to the case where $\operatorname{Sp} B$ is an affinoid subdomain of $\operatorname{Sp} A$ via φ; thus, we get

Corollary 5. *Let* $\operatorname{Sp} A' \subset \operatorname{Sp} A$ *and* $\operatorname{Sp} A'' \subset \operatorname{Sp} A$ *denote affinoid subdomains. Then* $\operatorname{Sp} A' \cap \operatorname{Sp} A''$ *is an affinoid subdomain of* $\operatorname{Sp} A$.

In the situation of Proposition 4 and in accordance with Definition 2, we will usually write $\varphi^{-1}(\mathrm{Sp}\, A')$ instead of $\mathrm{Sp}\, B \times_{\mathrm{Sp}A} \mathrm{Sp}\, A'$; thus, $\varphi^{-1}(\mathrm{Sp}\, A')$ will also denote the affinoid variety it induces. The affinoid map $p_2\colon \varphi^{-1}(\mathrm{Sp}\, A') \to \mathrm{Sp}\, A'$ will be referred to as the affinoid map *induced by* φ. We proceed similarly in the situation of Corollary 5 and write $\mathrm{Sp}\, A' \cap \mathrm{Sp}\, A''$ instead of $\mathrm{Sp}\, A' \times_{\mathrm{Sp}A} \mathrm{Sp}\, A''$.

Corollary 6. *Let* $\mathfrak{a} \subset A$ *be an ideal and consider the closed immersion* $\varphi\colon \mathrm{Sp}\, A/\mathfrak{a} \to \mathrm{Sp}\, A$. *Then for any affinoid subdomain* $\mathrm{Sp}\, A' \subset \mathrm{Sp}\, A$ *the induced affinoid map* $\varphi'\colon \varphi^{-1}(\mathrm{Sp}\, A') \to \mathrm{Sp}\, A'$ *coincides canonically with the closed immersion* $\mathrm{Sp}\, A'/\mathfrak{a}A' \to \mathrm{Sp}\, A'$.

Proof. According to Proposition 4, the algebra homomorphism corresponding to $\varphi'\colon \varphi^{-1}(\mathrm{Sp}\, A') \to \mathrm{Sp}\, A'$ is derived from $\varphi^*\colon A \to A/\mathfrak{a}$ by tensoring with A' over A. Furthermore, by Proposition 6.1.1/11 the map $\varphi'^*\colon A' = A \mathbin{\hat{\otimes}}_A A' \to A/\mathfrak{a} \mathbin{\hat{\otimes}}_A A'$ is surjective and has kernel $\mathfrak{a}A'$. This proves our assertion. □

It is not in general true that any union or even any finite union of affinoid subdomains of an affinoid variety is again an affinoid subdomain of this variety. Only for 1-dimensional varieties, one can show that finite unions of affinoid subdomains are affinoid subdomains again (see FIESELER [8], Satz 2.1). The picture changes if finite *disjoint* unions of affinoid subdomains are considered. In order to deal with this case, we need some preparations.

Lemma 7. *Let* $\mathrm{Sp}\, A_1$ *and* $\mathrm{Sp}\, A_2$ *be affinoid varieties. Then, for* $i = 1, 2$, *the canonical map* $\varphi_i^*\colon A_1 \oplus A_2 \to A_i$ *defines* $\mathrm{Sp}\, A_i$ *as a (Zariski-closed) affinoid subdomain of* $\mathrm{Sp}\, A_1 \oplus A_2$, *and* $\mathrm{Sp}\, A_1 \oplus A_2$ *is the disjoint union of the subdomains* $\mathrm{Sp}\, A_1$ *and* $\mathrm{Sp}\, A_2$.

Proof. First recall that $A_1 \oplus A_2$ is an affinoid algebra (see (6.1.1)). Fot $i = 1, 2$, let $\varphi_i\colon \mathrm{Sp}\, A_i \to \mathrm{Sp}\, A_1 \oplus A_2$ be the affinoid map associated to $\varphi_i^*\colon A_1 \oplus A_2 \to A_i$, and consider $e_1 := (0, 1)$ and $e_2 := (1, 0)$ as elements in $A_1 \oplus A_2$. Then φ_i identifies the points of $\mathrm{Sp}\, A_i$ with the Zariski-closed subser $V(e_i)$ of $\mathrm{Sp}\, A_1 \oplus A_2$, and we have only to show that φ_i defines $\mathrm{Sp}\, A_i$ as an affinoid subdomain of $\mathrm{Sp}\, A_1 \oplus A_2$. Let $\psi\colon \mathrm{Sp}\, B \to \mathrm{Sp}\, A_1 \oplus A_2$ be an affinoid map such that $\mathrm{im}\, \psi \subset \mathrm{im}\, \varphi_i$. Then $e_i(\psi(y)) = 0$ for all points $y \in \mathrm{Sp}\, B$. Hence, considering the associated homomorphism $\psi^*\colon A_1 \oplus A_2 \to B$, we see that $\psi^*(e_i)$ is nilpotent in B, say $(\psi^*(e_i))^n = 0$ (use HILBERT's Nullstellensatz 7.1.3/1). Since $e_i^n = e_i$, we must have $\psi^*(e_i) = 0$. Thus $\ker \varphi_i^* = (e_i) \subset \ker \psi^*$, and the map ψ^* factors uniquely through A_i. Hence, there is a unique affinoid map $\psi'\colon \mathrm{Sp}\, B \to \mathrm{Sp}\, A_i$ such that $\psi = \varphi_i \circ \psi'$. □

In the situation of Lemma 7, the variety $\mathrm{Sp}\, A_1 \oplus A_2$ is called the *disjoint union of the varieties* $\mathrm{Sp}\, A_1$ and $\mathrm{Sp}\, A_2$. We add a converse of Lemma 7 which is a direct consequence of Corollary 8.2.1/2 of TATE's Acyclicity Theorem (to be proved in Chapter 8).

Lemma 8. *Let* Sp A *be an affinoid variety and let* φ_i: Sp $A_i \hookrightarrow$ Sp A, $i = 1, 2$, *be disjoint affinoid subdomains of* Sp A *such that* Sp A = Sp $A_1 \cup$ Sp A_2. *Then the associated homomorphisms* $\varphi_i^*: A \to A_i$ *define an isomorphism* $A \xrightarrow{\sim} A_1 \oplus A_2$, *and* Sp A *is isomorphic to* Sp $A_1 \oplus A_2$.

Proof. Apply Corollary 8.2.1/2 to the affinoid covering {Sp A_1, Sp A_2} of Sp A. □

Proposition 9. *Let* U_1 *and* U_2 *be disjoint affinoid subdomains of the affinoid variety* Sp A. *Then* $U_1 \cup U_2$ *is an affinoid subdomain of* Sp A.

Proof. Let φ_i: Sp $A_i \to$ Sp A represent all affinoid maps into U_i, $i = 1, 2$. We want to show that the affinoid map φ: Sp $A_1 \oplus A_2 \to$ Sp A corresponding to the algebra homomorphism $A \to A_1 \oplus A_2$, $a \mapsto \varphi_1^*(a) \oplus \varphi_2^*(a)$, represents all affinoid maps into $U_1 \cup U_2$. Let ψ: Sp $B \to$ Sp A be an affinoid map such that $\operatorname{im}\psi \subset \operatorname{im}\varphi = U_1 \cup U_2$. Then $\psi^{-1}(U_1)$ and $\psi^{-1}(U_2)$ are disjoint affinoid subdomains of Sp B which cover Sp B. Thus, if $\psi^{-1}(U_i) =$ Sp B_i for $i = 1, 2$, we have Sp B = Sp $B_1 \oplus B_2$ by Lemma 8. The map Sp $B_i \to$ Sp A induced by ψ, factors uniquely through Sp A_i for $i = 1, 2$, and it is easily seen that ψ factors uniquely through Sp $A_1 \oplus A_2$. □

7.2.3. Examples of open affinoid subdomains. — In the following we make extensive use of generalized rings of fractions, which were introduced in (6.1.4). Let X = Sp A be an affinoid variety and let $f = (f_1, \ldots, f_m)$, $g = (g_1, \ldots, g_n)$ denote systems of functions in A. Then we set

$$X(f, g^{-1}) := X(f_1, \ldots, f_m, g_1^{-1}, \ldots, g_n^{-1})$$
$$:= \{x \in X;\ |f_i(x)| \leqq 1,\ |g_j(x)| \geqq 1,\ i = 1, \ldots, m,\ j = 1, \ldots, n\}.$$

Proposition 1. $X(f, g^{-1})$ *is an open affinoid subdomain of* X = Sp A, *and the affinoid map* Sp $A\langle f, g^{-1}\rangle \to$ Sp A *corresponding to the canonical homomorphism* $A \to A\langle f, g^{-1}\rangle$ *represents all affinoid maps into* $X(f, g^{-1})$.

Proof. $X(f, g^{-1})$ is open in X by Corollary 7.2.1/2 and Proposition 7.2.1/4. Furthermore, an arbitrary affinoid map φ: Sp $B \to$ Sp A satisfies $\operatorname{im}\varphi \subset X(f, g^{-1})$ if and only if for all $y \in$ Sp B

$$|\varphi^*(f_i)\,(y)| = \big|f_i(\varphi(y))\big| \leqq 1, \qquad i = 1, \ldots, m, \qquad \text{and}$$
$$|\varphi^*(g_j)\,(y)| = \big|g_j(\varphi(y))\big| \geqq 1, \qquad j = 1, \ldots, n,$$

(cf. Lemma 7.1.4/2). This is equivalent to the fact that the elements $\varphi^*(g_j)$ are units and the elements $\varphi^*(f_i)$, $\varphi^*(g_j)^{-1}$ are power-bounded in B (cf. Propositions 7.1.3/1 and 6.2.3/1). Thus, by Proposition 6.1.4/1, we see that the map Sp $A\langle f, g^{-1}\rangle \to$ Sp A represents all affinoid maps into $X(f, g^{-1})$. □

Similar to (6.1.4), we write $X(f)$ or $X(g^{-1})$ instead of $X(f, g^{-1})$, in the cases $n = 1$ and $g = 1 \in k$, or $m = 1$ and $f = 1 \in k$, respectively. Note also that $X(h^{-1})$ is well-defined if $h \in A$ is a unit.

Definition 2. *The open affinoid subdomains of type $X(f)$ and $X(f, g^{-1})$ are called Weierstrass domains in X and Laurent domains in X, respectively.*

It follows from Propositions 7.2.1/3 and 7.2.1/4 that the Weierstrass domains form a basis for the canonical topology of X. Furthermore, Weierstrass and Laurent domains are rigid in the following sense:

Proposition 3. *Let $X(f, g^{-1})$ and $X(f', g'^{-1})$ denote Laurent domains in X with f, f' being m-tuples and g, g' being n-tuples of affinoid functions on X. Then we have $X(f, g^{-1}) = X(f', g'^{-1})$ if $|f_i - f'_i|_{\sup} \leq 1$ and $|g_j - g'_j|_{\sup} < 1$ for all i and j.*

Proof. The assertion is a direct consequence of the non-Archimedean triangle inequality and the fact that it reads $|a + b| = \max\{|a|, |b|\}$ if $|a| \neq |b|$. □

Next we want to consider open affinoid subdomains in $X = \operatorname{Sp} A$ of a more general type. Let $g, f_1, \ldots, f_n \in A$ denote functions without common zeros, i.e., generating the unit ideal (see Proposition 7.1.3/1). Writing $\frac{f}{g} := \left(\frac{f_1}{g}, \ldots, \frac{f_n}{g}\right)$, we define

$$X\left(\frac{f}{g}\right) := X\left(\frac{f_1}{g}, \ldots, \frac{f_n}{g}\right)$$
$$:= \{x \in X;\ |f_i(x)| \leq |g(x)|,\ i = 1, \ldots, n\}.$$

Proposition 4. *The set $X\left(\frac{f}{g}\right)$ is an open affinoid subdomain of $X = \operatorname{Sp} A$, and the affinoid map $\operatorname{Sp} A\left\langle\frac{f}{g}\right\rangle \to \operatorname{Sp} A$ corresponding to the canonical homomorphism $A \to A\left\langle\frac{f}{g}\right\rangle$ represents all affinoid maps into $X\left(\frac{f}{g}\right)$.*

Proof. Let $x_0 \in X\left(\frac{f}{g}\right)$. Then $g, f_1, \ldots, f_n$ having no common zeros implies $g(x_0) \neq 0$. Thus, by Corollary 7.2.1/2 and Proposition 7.2.1/4,

$$\{x \in X;\ |g(x)| \geq |g(x_0)|,\ |f_i(x)| \leq |g(x_0)|,\ i = 1, \ldots, n\}$$

is an open neighborhood of x_0 contained in $X\left(\frac{f}{g}\right)$. Hence $X\left(\frac{f}{g}\right)$ is open in X.

In order to verify that the map $\operatorname{Sp} A\left\langle\frac{f}{g}\right\rangle \to \operatorname{Sp} A$ represents all affinoid maps into $X\left(\frac{f}{g}\right)$, we use Proposition 6.1.4/3. Then we only have to show that an arbitrary affinoid map $\varphi\colon \operatorname{Sp} B \to \operatorname{Sp} A$ satisfies $\operatorname{im} \varphi \subset X\left(\frac{f}{g}\right)$ *if and only if* $\varphi^*(g)$ is a unit and the elements $\varphi^*(f_i) \cdot \varphi^*(g)^{-1}$, $i = 1, \ldots, n$, are power-bounded in B. So let us assume that $\operatorname{im} \varphi \subset X\left(\frac{f}{g}\right)$. Then for all $y \in \operatorname{Sp} B$ and

$i = 1, \ldots, n,$ we have

$$|\varphi^*(f_i)\,(y)| = \big|f_i(\varphi(y))\big| \leqq \big|g(\varphi(y))\big| = |\varphi^*(g)\,(y)|\,.$$

The elements $g, f_1, \ldots, f_n$ generate the unit ideal in A, as do the images $\varphi^*(g)$, $\varphi^*(f_1), \ldots, \varphi^*(f_n)$ in B. Thus, by the above inequality, $\varphi^*(g)$ has no zero and, consequently, must be a unit in B. Therefore, the above inequality reads

$$\big|\big(\varphi^*(f_i)\cdot\varphi^*(g)^{-1}\big)\,(y)\big| \leqq 1, \qquad y \in \operatorname{Sp} B, \quad i = 1, \ldots, n,$$

and it follows from Proposition 6.2.3/1 that the elements $\varphi^*(f_i)\cdot\varphi^*(g)^{-1}$ are power-bounded. This proves the only if part. Going backwards the if part is easily obtained. □

Definition 5. *The open affinoid subdomains of type* $X\left(\frac{f}{g}\right)$ *are called rational domains in* X.

It is easily checked that the notation $X\left(\frac{f}{g}\right)$ is compatible with the notation introduced for Weierstrass domains and Laurent domains if some of the elements $g, f_1, \ldots, f_n$ are units in A. Furthermore, we point out that the condition $g, f_1, \ldots, f_n$ having no common zeros in $\operatorname{Sp} A$ is essential for $X\left(\frac{f}{g}\right)$ to be an affinoid subdomain of X. For example, let us consider the case where $A = T_1 = k\langle\xi\rangle$ and $f := \xi$, $g := c\xi$, with $c \in k$, $0 < |c| < 1$. Then

$$U := \{x \in \operatorname{Sp} A\,;\ |f(x)| \leqq |g(x)|\}$$

consists of one single point x_0, namely, the common zero of f and g. Hence if there is an affinoid map $\operatorname{Sp} A' \to \operatorname{Sp} A$ representing all affinoid maps into U, then A' contains exactly one maximal ideal $\mathfrak{m}$, and this ideal must coincide with the radical of A' by Hilbert's Nullstellensatz (Proposition 7.1.3/1). Therefore, Corollary 6.1.2/3 applies with $\mathfrak{q} := 0$ and yields that A' is a finite-dimensional vector space over k. By Proposition 7.2.2/1, we have the isomorphisms

$$A/\mathfrak{m}_{x_0}^n \xrightarrow{\sim} A'/\mathfrak{m}^n, \qquad n \in \mathbb{N}\,.$$

Considering k-vector space dimensions, the dimension on the right-hand side is bounded by the dimension of A'. However, the dimension of $A/\mathfrak{m}_{x_0}^n = k\langle\xi\rangle/(\xi^n)$ is n and thus not bounded. This contradiction shows that U cannot be an affinoid subdomain in $\operatorname{Sp} A$, which is, of course, due to the fact that f and g have a common zero.

We conclude this section by listing some elementary properties of the affinoid subdomains just introduced.

Proposition 6. *Let $\varphi\colon Y \to X$ denote an affinoid map and $X' \subset X$ an affinoid subdomain. If X' is a Weierstrass (Laurent or rational) domain in X, then $\varphi^{-1}(X')$ is a Weierstrass (Laurent or rational, respectively) domain in Y.*

Proof. We prove the assertion when X' is a Weierstrass domain. Using Lemma 7.1.4/2, we get

$$\varphi^{-1}(X(f)) = Y(\varphi^*(f)).$$

Similar equalities hold in the Laurent and rational cases. □

Proposition 7. *Let $X' \subset X$ and $X'' \subset X$ denote affinoid subdomains of the affinoid variety X. If X' and X'' are Weierstrass (Laurent or rational) domains in X, then $X' \cap X''$ also is a Weierstrass (Laurent or rational, respectively) subdomain in X.*

Only the rational case needs a *proof*. Let

$$X' := X\left(\frac{f_i}{f_m};\, i = 1, \ldots, m\right),$$

$$X'' := X\left(\frac{g_j}{g_n};\, j = 1, \ldots, n\right),$$

with $f_1, \ldots, f_m$ as well as $g_1, \ldots, g_n$ generating the unit ideal in A. Then the functions $f_i g_j$, $i = 1, \ldots, m$, $j = 1, \ldots, n$, generate the product ideal which is also the unit ideal. Thus, we can consider the rational domain

$$Y := X\left(\frac{f_i g_j}{f_m g_n};\, i = 1, \ldots, m,\, j = 1, \ldots, n\right).$$

Clearly, we have $X' \cap X'' \subset Y$. We show that the opposite inclusion also holds. Namely, let $x \in Y$. Then we have

$$|f_i(x)|\, |g_n(x)| \leqq |f_m(x)|\, |g_n(x)|, \qquad i = 1, \ldots, m.$$

Since $(f_m g_n)(x) \neq 0$ and thus $|g_n(x)| \neq 0$, we can cancel $|g_n(x)|$ on both sides which yields that $x \in X'$. Similarly, one shows $x \in X''$. Hence $X' \cap X'' = Y$ is a rational domain in X. □

Corollary 8. *Any Weierstrass domain in X is also Laurent, any Laurent domain in X is also rational.*

Proof. The first statement is trivial. Furthermore, an arbitrary Laurent domain $X(f_1, \ldots, f_m, g_1^{-1}, \ldots, g_n^{-1})$ in X equals the intersection of the rational domains $X\left(\frac{f_i}{1}\right)$, $X\left(\frac{1}{g_j}\right)$, $i = 1, \ldots, m$, $j = 1, \ldots, n$, and thus is rational by Proposition 7. □

Finally, it should be mentioned that the given approach to Weierstrass, Laurent and rational domains can be modified slightly. As before, let $X = \operatorname{Sp} A$ be an affinoid variety and denote by $f = (f_1, \ldots, f_m)$ a system of affinoid functions in A. Then, for any system $\varepsilon = (\varepsilon_1, \ldots, \varepsilon_m)$ of elements in $\sqrt{|k^*|}$ (the value group of the algebraic closure of k), we can consider the subset

$$\{x \in X;\ |f_i(x)| \leqq \varepsilon_i,\quad i = 1, \ldots, m\} \subset X,$$

which will suggestively be denoted by $X(\varepsilon^{-1}f)$ or $X(\varepsilon_1^{-1}f_1, \ldots, \varepsilon_m^{-1}f_m)$. Choosing an exponent $r \in \mathbb{N}$ such that $\varepsilon_1^r, \ldots, \varepsilon_m^r \in |k^*|$ and choosing elements $c_1, \ldots, c_m \in k$ with $|c_i| = \varepsilon_i^r$, $i = 1, \ldots, m$, we get

$$X(\varepsilon^{-1}f) = X(\varepsilon_1^{-r}f_1^r, \ldots, \varepsilon_m^{-r}f_m^r) = X(c_1^{-1}f_1^r, \ldots, c_m^{-1}f_m^r).$$

Thus, $X(\varepsilon^{-1}f)$ is a Weierstrass domain in X. Similarly, if $g = (g_1, \ldots, g_n)$ is another system of affinoid functions in A and $\delta = (\delta_1, \ldots, \delta_n)$ another system of elements in $\sqrt{|k^*|}$, one verifies that

$$\begin{aligned} X(\varepsilon^{-1}f, \delta g^{-1}) &:= X(\varepsilon_1^{-1}f_1, \ldots, \varepsilon_m^{-1}f_m, \delta_1 g_1^{-1}, \ldots, \delta_n g_n^{-1}) \\ &:= \{x \in X;\ |f_i(x)| \leqq \varepsilon_i, |g_j(x)| \geqq \delta_j, i = 1, \ldots, m, j = 1, \ldots, n\} \end{aligned}$$

is a Laurent domain in X. This procedure also extends to the rational case. Namely, let g be a function in A such that $g, f_1, \ldots, f_m$ are without common zeros. Then clearly,

$$\begin{aligned} X\left(\varepsilon^{-1}\frac{f}{g}\right) &:= X\left(\varepsilon_1^{-1}\frac{f_1}{g}, \ldots, \varepsilon_m^{-1}\frac{f_m}{g}\right) \\ &:= \{x \in X;\ |f_i(x)| \leqq \varepsilon_i\,|g(x)|,\quad i = 1, \ldots, m\} \end{aligned}$$

is a rational domain in X.

7.2.4. Transitivity properties. — In Proposition 7.2.2/3, we showed that the transitivity property holds for affinoid subdomains of general type. The purpose of this section is to derive the same result for Weierstrass, rational and open subdomains. For Laurent domains we will give a counter-example. In the following, $X = \operatorname{Sp} A$ always denotes an affinoid variety.

Proposition 1. *Let* $X\left(\frac{f}{g}\right) \subset X$ *be a rational domain, where* $\frac{f}{g} := \left(\frac{f_1}{g}, \ldots, \frac{f_n}{g}\right)$. *Then there exists a Laurent domain* X' *in* X *such that* $X\left(\frac{f}{g}\right) \subset X' \subset X$ *and* $X\left(\frac{f}{g}\right)$ *is a Weierstrass domain in* X'. *More precisely, there is an* $\varepsilon \in |k^*|$ *such that* $X' := X(\varepsilon g^{-1})$ *has the desired property and such that, in addition,* $X\left(\frac{f}{g}\right) \cap X(\varepsilon^{-1}g) = \emptyset$.

Proof. We denote by g'' the restriction of g to $X\left(\frac{f}{g}\right)$. Then g'' has no zero in $X\left(\frac{f}{g}\right)$ and hence is a unit in $A\left\langle\frac{f}{g}\right\rangle$. Choosing $\varepsilon \in |k^*|$ such that $|g''^{-1}|_{\sup} < \varepsilon^{-1}$, we get $|g(x)| > \varepsilon$ for all $x \in X\left(\frac{f}{g}\right)$. Thus $X\left(\frac{f}{g}\right) \subset X(\varepsilon g^{-1})$ and $X\left(\frac{f}{g}\right) \cap X(\varepsilon^{-1} g) = \emptyset$. Let $f', g' \in A\langle \varepsilon^{-1} g\rangle$ denote the restrictions of f, g to $X(\varepsilon g^{-1})$. Then g' is a unit, and we can write

$$X\left(\frac{f}{g}\right) = \left(X(\varepsilon g^{-1})\right)(g'^{-1}f'),$$

which shows that $X\left(\frac{f}{g}\right)$ is a Weierstrass domain in $X(\varepsilon g^{-1})$. □

Theorem 2. *Let $X'' \subset X' \subset X = \operatorname{Sp} A$ denote affinoid subdomains. If X'' is a Weierstrass domain in X' and furthermore X' is a Weierstrass domain in X, then X'' is a Weierstrass domain in X. The corresponding result is true for rational and for open subdomains.*

Proof. First we consider the Weierstrass case. We have $X' = X(f) = \operatorname{Sp} A\langle f\rangle$ and $X'' = X'(g)$, where f and g are tuples with entries in A or $A\langle f\rangle$, respectively. It follows from the definition of the ring $A\langle f\rangle$ (see (6.1.4)) that the image of A is dense in $A\langle f\rangle$. Thus, by the rigidity of Weierstrass domains (Proposition 7.2.3/3), we may assume that g admits an extension to X, which also will be denoted by g. But then $X'' = X(f) \cap X(g)$; hence X'' is a Weierstrass domain in X by Proposition 7.2.3/7.

Next we deal with the open case. Remembering that the Weierstrass domains form a basis for the canonical topology, we only have to show that, for any $x \in X''$, there is a Weierstrass domain $W \hookrightarrow X$ with $x \in W \subset X''$. Now let $x \in X''$. Since X' is open in X, there is a Weierstrass domain $W' \hookrightarrow X$ such that $x \in W' \subset X'$. It follows from Proposition 7.2.3/6 that W' is also open in X'. Furthermore, since X'' is open in X', there exists a Weierstrass domain $W \hookrightarrow X'$ such that $x \in W \subset X'' \cap W'$. Then W is, in particular, a Weierstrass domain in W' (use again Proposition 7.2.3/6) and thus a Weierstrass domain in X by what we have proved above. Hence the open case also is finished.

Finally, we come to the rational case. Let $X' = X\left(\frac{f}{g}\right) = \operatorname{Sp} A\left\langle\frac{f}{g}\right\rangle$ with $g \in A$ and f a tuple of elements in A. Using Proposition 7.2.3/7 and Proposition 1, it is enough to consider only the cases where $X'' = X'(h)$ or $X'' = X'(h^{-1})$, for a single function $h \in A\left\langle\frac{f}{g}\right\rangle$. By the definition of $A\left\langle\frac{f}{g}\right\rangle$ (see (6.1.4)), the image of $A[g^{-1}]$ is dense in $A\left\langle\frac{f}{g}\right\rangle$. Hence, due to the rigidity of Weierstrass and Laurent domains (cf. Proposition 7.2.3/3), we may assume that the function $g^n h$ extends to a function h' on X, for a suitable $n \in \mathbb{N}$. Then we have

$$X'(h) = X\left(\frac{f}{g}\right) \cap \{x \in X;\ |h'(x)| \leqq |g^n(x)|\} \quad \text{and}$$

$$X'(h^{-1}) = X\left(\frac{f}{g}\right) \cap \{x \in X;\ |g^n(x)| \leqq |h'(x)|\}.$$

By Proposition 1, there exsits a $c \in k^*$ such that $|g(x)|^n > |c|$ for all $x \in X\left(\frac{f}{g}\right)$. Then clearly $|c| \leqq |g^n(x)|$ for all $x \in X'(h)$ and $|c| \leqq |h'(x)|$ for all $x \in X'(h^{-1})$. Thus, the above equations can be rewritten as

$$X'(h) = X\left(\frac{f}{g}\right) \cap X\left(\frac{h'}{g^n}, \frac{c}{g^n}\right) \quad \text{and}$$

$$X'(h^{-1}) = X\left(\frac{f}{g}\right) \cap X\left(\frac{g^n}{h'}, \frac{c}{h'}\right),$$

which yields by Proposition 7.2.3/7 that $X'(h)$ and $X'(h^{-1})$ are rational domains in X. □

Corollary 3. *Let X' be an affinoid variety and assume that X' is an open affinoid subdomain of X via the affinoid map $\varphi: X' \to X$. Then, with respect to canonical topologies, φ defines a homeomorphism between X' and its image in X.*

Proof. The map φ is continuous by Proposition 7.2.1/5 and open by Theorem 2. □

Finally, we want to give a counter-example which shows that Theorem 2 is not true for Laurent domains. Using Proposition 1, it is clearly enough to construct a rational domain X' in $X = \operatorname{Sp} A$ which is not Laurent. This can be done as follows. We let $A := T_1 = k\langle\zeta\rangle$ and consider the rational subdomain

$$X' := X\left(\frac{\zeta}{(\zeta - c)^2}\right) \subset X = \operatorname{Sp} T_1,$$

for some $c \in k$, $0 < |c| < 1$. Then we introduce the subdomains

$$X_1 := X(\zeta^{-1}), \qquad X_2 := X(c^{-2}\zeta),$$

and show that $X' = X_1 \cup X_2$. Namely, if $x \notin X_1 \cup X_2$, i.e., $|c|^2 < |\zeta(x)| < 1$, then

$$|\zeta(x) - c|^2 = |\zeta(x)^2 - 2c\zeta(x) + c^2| < |\zeta(x)|,$$

and thus $x \notin X'$. This shows $X' \subset X_1 \cup X_2$. The reverse inclusion is obvious.

Now let us assume that X' is a Laurent domain in X; i.e.,

$$X' = X(f_1, \ldots, f_m, g_1^{-1}, \ldots, g_n^{-1})$$

for some $f_i, g_j \in T_1$. Then by Proposition 5.1.4/4, the functions $f_1, \ldots, f_m$ assume their maximum on X_1 and thus, in particular, on X'. Therefore, $f_1, \ldots, f_m$ must have a supremum norm $\leqq 1$ and, as a result, are superfluous; i.e.,

$$X' = X(g_1^{-1}, \ldots, g_n^{-1}).$$

Next let us look at the subdomain $X_2 = X(c^{-2}\zeta)$. The corresponding affinoid algebra is (cf. Proposition 6.1.4/2)

$$\begin{aligned} T_1 \langle c^{-2}\zeta\rangle &\cong k\langle\zeta\rangle \langle\eta\rangle/(\eta - c^{-2}\zeta) \\ &\cong k\langle\zeta, \eta\rangle/(\zeta - c^2\eta) \\ &\cong k\langle\zeta, \eta\rangle/(\zeta) \cong k\langle\eta\rangle, \end{aligned}$$

and the restriction homomorphism $T_1 \to T_1\langle c^{-2}\zeta\rangle$ can be identified with the map

$$\sigma\colon k\langle\zeta\rangle \to k\langle\eta\rangle, \qquad \zeta \mapsto c^2\eta .$$

Since X_2 is contained in X', the elements $\sigma(g_j)$, $j = 1, \ldots, n$, are units in $k\langle\eta\rangle$ and have supremum norm $\geqq 1$. Applying Proposition 5.1.3/1 in this situation yields that each one of the series $\sigma(g_j)$ has a dominating constant term of value $\geqq 1$; that is, if $g_j = \sum_{\nu=0}^{\infty} a_{j\nu}\zeta^\nu$, then

$$|a_{j0}| \geqq 1, \quad |a_{j\nu}|\,|c|^{2\nu} < |a_{j0}|, \qquad j = 1, \ldots, n, \quad \nu \in \mathbb{N}.$$

Thus, the constant term a_{j0} of g_j is "dominant on the disc X_2", and it is an easy consideration to see that then this term must also be dominant on a slightly bigger disc $X(\varepsilon^{-1}\zeta)$, with a suitable $\varepsilon \in \sqrt{|k^*|}$, $|c|^2 < \varepsilon < 1$. As a result, g_j assumes only values $\geqq 1$ on $X(\varepsilon^{-1}\zeta)$. If ε is chosen small enough to work for all $j = 1,\ldots,n$, we obtain $X(\varepsilon^{-1}\zeta) \subset X'$. But this contradicts $X(\varepsilon^{-1}\zeta) \not\subset X_1 \cup X_2$; hence X' cannot be a Laurent domain in X.

Thus, we have given an example of a rational domain which is not Laurent. It is easier to obtain examples of Laurent domains which are not Weierstrass. As before, let $X = \operatorname{Sp} T_1 = \operatorname{Sp} k\langle\zeta\rangle$. Then $X' := X(\zeta^{-1})$ is a Laurent domain in X which is not Weierstrass. For otherwise, $k\langle\zeta\rangle$ would be dense in the affinoid algebra of X', which is the algebra $k\langle\zeta, \zeta^{-1}\rangle$ of strictly convergent Laurent series (cf. (6.1.4)).

7.2.5. The Openness Theorem. — As previously indicated, we want to show that all affinoid subdomains are open. The proof of this fact requires some preparation.

In the following the map $\sigma\colon A \to B$ denotes a homomorphism between affinoid algebras. A system $b = (b_1, \ldots, b_n)$ of power-bounded elements in B is called an *affinoid generating system of B over A* if the continuous homomorphism

$$\sigma_1\colon A\langle\zeta_1, \ldots, \zeta_n\rangle \to B$$

extending σ and mapping ζ_i onto b_i, for $i = 1, \ldots, n$, (cf. Proposition 6.1.1/4) is surjective. Recall that affinoid generating systems (over k) were already introduced in (6.1.1). Each affinoid generating system of B over k is an affinoid generating system of B over A. In particular, affinoid generating systems of the considered type do exist in general.

Lemma 1. *Let* $\operatorname{Sp} A' \subset \operatorname{Sp} A$ *denote an affinoid subdomain and let* $\sigma' \colon A' \to A' \hat{\otimes}_A B$ *be the homomorphism induced by* $\sigma \colon A \to B$. *Then, for any affinoid generating system* $b = (b_1, \ldots, b_n)$ *of* B *over* A, *the image* $1 \hat{\otimes} b := (1 \hat{\otimes} b_1, \ldots, 1 \hat{\otimes} b_n)$ *in* $A' \hat{\otimes}_A B$ *is an affinoid generating system of* $A' \hat{\otimes}_A B$ *over* A'.

Proof. Denote by

$$\sigma_1 \colon A\langle \zeta_1, \ldots, \zeta_n \rangle \to B, \ \zeta_i \mapsto b_i, \quad \text{and}$$
$$\sigma_1' \colon A'\langle \zeta_1, \ldots, \zeta_n \rangle \to A' \hat{\otimes}_A B, \ \zeta_i \mapsto 1 \hat{\otimes} b_i,$$

the continuous homomorphisms extending σ and σ', respectively. Using standard properties of the tensor product and the fact that $A\langle\zeta\rangle \cong A \hat{\otimes}_k T_n$ and $A'\langle\zeta\rangle \cong A' \hat{\otimes}_k T_n$ by Proposition 6.1.1/7, it is easy to see that σ_1' is derived from σ_1 by tensoring with A' over A. But then, since σ_1 is surjective, the surjectivity of σ_1' follows from Proposition 2.1.8/6. □

Lemma 2. *Let* x *be a point of* $\operatorname{Sp} A$ *and denote by* $\mathfrak{m}_x$ *the corresponding maximal ideal in* A. *Assume that* $\sigma \colon A \to B$ *induces*

(a) *a finite homomorphism* $A/\mathfrak{m}_x \to B/\mathfrak{m}_x B$, *or*

(b) *an epimorphism* $A/\mathfrak{m}_x \to B/\mathfrak{m}_x B$, *or*

(c) *isomorphisms* $A/\mathfrak{m}_x^n \to B/\mathfrak{m}_x^n B$, *for all* $n \in \mathbb{N}$.

Then there exists an open affinoid subdomain $\operatorname{Sp} A' \subset \operatorname{Sp} A$ *containing* x *such that the map* $\sigma' \colon A' \to A' \hat{\otimes}_A B$ *induced by* σ *is finite, epimorphic, or isomorphic, respectively.*

Proof. Starting with case (a), we choose an affinoid generating system $b = (b_1, \ldots, b_n)$ of B over A. The elements $b_1, \ldots, b_n$ are power-bounded in B, and so are their images in $B/\mathfrak{m}_x B$. Hence, by Theorem 6.3.5/1, the residue classes of $b_1, \ldots, b_n$ in $B/\mathfrak{m}_x B$ are integral over $\overset{\circ}{\widehat{A/\mathfrak{m}_x}}$. Thus, we get equations

$$(*) \qquad b_i^{s_i} + a_{i1} b_i^{s_i - 1} + \cdots + a_{is_i} = d_i \in \mathfrak{m}_x B, \quad i = 1, \ldots, n,$$

with elements $a_{ij} \in A$, satisfying $|a_{ij}(x)| \leq 1$. Let $f_1, \ldots, f_r \in A$ denote generators for $\mathfrak{m}_x$, and let $d_{i\varrho} \in B$ be elements such that

$$d_i = \sum_{\varrho=1}^{r} d_{i\varrho} f_\varrho, \qquad i = 1, \ldots, n.$$

Choosing $c \in k$ with $|c| > \max_{i,\varrho} |d_{i\varrho}|_{\sup}$, we define

$$A' := A\langle a_{ij}, c f_\varrho; \, i = 1, \ldots, n, \, j = 1, \ldots, s_i, \, \varrho = 1, \ldots, r \rangle \quad \text{and}$$
$$B' := A' \hat{\otimes}_A B.$$

Then $\operatorname{Sp} A'$ is a Weierstrass domain in $\operatorname{Sp} A$ containing x. Furthermore, $\operatorname{Sp} B'$ is an affinoid subdomain of $\operatorname{Sp} B$, namely, the inverse image of $\operatorname{Sp} A'$ via ${}^a\sigma$: $\operatorname{Sp} B \to \operatorname{Sp} A$ (cf. Proposition 7.2.2/4). We denote by a'_{ij} the restriction of a_{ij} to $\operatorname{Sp} A'$ and similarly by b'_i, d'_i the restrictions of b_i, d_i to $\operatorname{Sp} B'$. By construction the following estimates hold:

$$(**) \qquad |a'_{ij}|_{\sup} \leqq 1, \quad |d'_i|_{\sup} < 1.$$

We now want to show that the canonical map $\sigma' \colon A' \to A' \mathbin{\widehat{\otimes}_A} B$ is finite. This will settle case (a), since $\operatorname{Sp} A'$, as a Weierstrass domain, is open in $\operatorname{Sp} A$. By Lemma 1, the system $b' := (b'_1, \ldots, b'_n)$ is an affinoid generating system of $A' \mathbin{\widehat{\otimes}_A} B$ over A'. Thus, the continuous homomorphism

$$\sigma'_1 \colon A'\langle \zeta_1, \ldots, \zeta_n \rangle \to A' \mathbin{\widehat{\otimes}_A} B$$

extending σ' and mapping ζ_i onto b'_i is surjective. The corresponding "reduced" homomorphism

$$\tilde{\sigma}'_1 \colon \tilde{A}'[\zeta_1, \ldots, \zeta_n] \to \widetilde{A' \mathbin{\widehat{\otimes}_A} B}$$

extends the map $\tilde{\sigma}' \colon \tilde{A}' \to \widetilde{A' \mathbin{\widehat{\otimes}_A} B}$, maps ζ_i onto $\tilde{b}'_i$, and is finite by Theorem 6.3.5/1. Therefore, $\tilde{\sigma}'$ is finite because we can show that $\tilde{b}'_1, \ldots, \tilde{b}'_n$ are integral over $\tilde{A}'$. Namely, by restriction, the equations $(*)$ become

$$b_i'^{s_i} + a'_{i1} b_i'^{s_i-1} + \cdots + a'_{is_i} = d'_i, \qquad i = 1, \ldots, n,$$

and, using the estimates $(**)$, we get

$$\tilde{b}_i'^{s_i} + \tilde{a}'_{i1} \tilde{b}_i'^{s_i-1} + \cdots + \tilde{a}'_{is_i} = 0, \qquad i = 1, \ldots, n.$$

Thus $\tilde{\sigma}'$ is finite, and, hence again by Theorem 6.3.5/1, σ' also is finite. This concludes the proof in case (a).

Next, consider case (b). Using Propositions 7.2.2/1 and 7.2.2/4, and some standard facts about tensor products, one easily verifies that assumption (b) is not affected, when $\operatorname{Sp} A$ is replaced by an affinoid subdomain $\operatorname{Sp} A'' \subset \operatorname{Sp} A$ containing x and σ is replaced by the map $\sigma'' \colon A'' \to A'' \mathbin{\widehat{\otimes}_A} B$ obtained from σ by tensoring with A''. Since case (b) is a specialization of case (a), we can find an open affinoid subdomain $\operatorname{Sp} A'' \subset \operatorname{Sp} A$ containing x such that $\sigma'' \colon A'' \to A'' \mathbin{\widehat{\otimes}_A} B$ is finite. If the assertion of the lemma (as far as case (b) is concerned) is known for $x \in \operatorname{Sp} A''$ and the map σ'', one can use Theorem 7.2.4/2 and easily obtain the corresponding assertion for $x \in \operatorname{Sp} A$ and the map σ. Thus, there is no restriction in assuming that σ is finite.

If σ is finite, we can choose elements $b_1, \ldots, b_n \in B$ generating B as an A-module. Since $B = \sigma(A) + \mathfrak{m}_x B$ by assumption (b), we get equations

$$b_i = \sigma(a_i) + \sum_{j=1}^{n} m_{ij} b_j, \qquad i = 1, \ldots, n,$$

for suitable elements $a_i \in A$, $m_{ij} \in \mathfrak{m}_x$. These equations can be rewritten as

$$\sum_{j=1}^{n} (\delta_{ij} - m_{ij})\, b_j = \sigma(a_i), \qquad i = 1, \ldots, n.$$

Hence, setting $d := \det(\delta_{ij} - m_{ij})$, we have $db_1, \ldots, db_n \in \sigma(A)$ by CRAMER's rule. We want to show that the assertion of the lemma is true for $A' := A\langle d^{-1}\rangle$. First, $\operatorname{Sp} A'$ is a Laurent domain in $\operatorname{Sp} A$ which contains x, since clearly $d(x) = 1$. We denote by $b_1', \ldots, b_n'$ the images of $b_1, \ldots, b_n$ in $A' \hat{\otimes}_A B$. Since d restricts to a unit in A', the relations $db_i \in \sigma(A)$ imply $b_i' \in \sigma'(A')$, for $i = 1, \ldots, n$. Furthermore, it follows easily from the bijectivity of the canonical map $A' \otimes_A B \to A' \hat{\otimes}_A B$ (cf. Proposition 3.7.3/6) that $b_1', \ldots, b_n'$ generate $A' \hat{\otimes}_A B$ as an A'-module. Thus, $\sigma'\colon A' \to A' \hat{\otimes}_A B$ must be surjective, which concludes the proof in case (b).

Finally, we consider case (c). Since this case is a specialization of case (b), it follows again from Theorem 7.2.4/2 and what we have proved above that $\sigma\colon A \to B$ can be assumed to be surjective. Furthermore, (c) implies

$$\ker \sigma \subset \bigcap_{n \in \mathbb{N}} \mathfrak{m}_x^n,$$

and thus, by KRULL's Intersection Theorem, there exists an element $m \in \mathfrak{m}_x$ such that $a := 1 - m$ annihilates all elements in $\ker \sigma$. Let $A' := A\langle a^{-1}\rangle$. Then $\operatorname{Sp} A'$ is a Laurent domain in $\operatorname{Sp} A$ which contains x, since $a(x) = 1$. Furthermore, the restriction map $\iota\colon A \to A'$ satisfies $\ker \sigma \subset \ker \iota$, since the element a, which annihilates $\ker \sigma$, is mapped onto a unit in A'. Therefore, the map ι factors through B, i.e., there is a commutative diagram

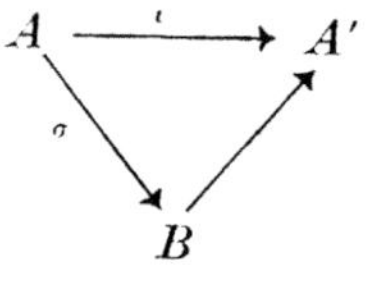

It follows from Proposition 7.2.2/4 that tensoring $\iota\colon A \to A'$ with A' over A leads to the identity map $A' \to A'$. Hence tensoring the above diagram with A' over A gives

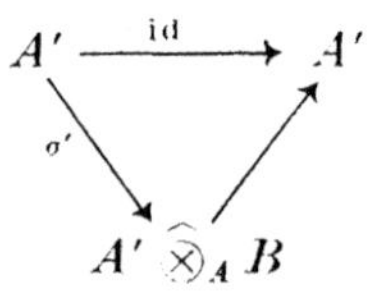

Consequently, σ' must be injective. But σ' is also surjective by Proposition 2.1.8/6, since σ was assumed to be surjective. Hence σ' is bijective. □

Now we are able to show

Theorem 3 (Openness Theorem). *All affinoid subdomains of* Sp *A are open.*

Proof. Let $\operatorname{Sp} A' \subset \operatorname{Sp} A$ denote an affinoid subdomain and let x be a point in $\operatorname{Sp} A'$. Then it is only necessary to show that there exists an open affinoid subdomain $\operatorname{Sp} A''$ in $\operatorname{Sp} A$ such that $x \in \operatorname{Sp} A'' \subset \operatorname{Sp} A'$. By Proposition 7.2.2/1, the restriction map $\sigma: A \to A'$ satisfies assumption (c) of Lemma **2**. Hence, there exists an open affinoid subdomain $\operatorname{Sp} A'' \subset \operatorname{Sp} A$ containing x such that the canonical map $\sigma': A'' \to A'' \mathbin{\widehat{\otimes}_A} A'$ is bijective. By Proposition 7.2.2/4 and Corollary 7.2.2/5, the affinoid variety

$$\operatorname{Sp}(A'' \mathbin{\widehat{\otimes}_A} A') = \operatorname{Sp} A'' \times_{\operatorname{Sp} A} \operatorname{Sp} A'$$

equals the intersection of $\operatorname{Sp} A'$ and $\operatorname{Sp} A''$, and the bijectivity of σ' implies $\operatorname{Sp} A'' = \operatorname{Sp} A'' \cap \operatorname{Sp} A'$. Thus, we get $x \in \operatorname{Sp} A'' \subset \operatorname{Sp} A'$. □

Corollary 4. *The affinoid subdomains of* $\operatorname{Sp} A$ *form a basis for the canonical topology of* $\operatorname{Sp} A$.

Furthermore, one obtains from Corollary 7.2.4/3

Corollary 5. *If X' is an affinoid subdomain of $X = \operatorname{Sp} A$, then the canonical topology on X restricts to the canonical topology on X'.*

7.2.6. Affinoid subdomains and reduction. — Let $X = \operatorname{Sp} A$ be an affinoid variety and denote by $\varphi: X' \hookrightarrow X$ an affinoid subdomain of X. In this general situation, not much can be said about the "reduced" map $\tilde{\varphi}: \tilde{X}' \to \tilde{X}$. For example, it can happen that $\tilde{\varphi}$ is surjective, even when X' is proper in X, or it is possible that $\tilde{\varphi}$ maps $\tilde{X}'$ onto a single point of $\tilde{X}$. Since the diagram

$$\begin{array}{ccc} X' & \xrightarrow{\ \varphi\ } & X \\ \downarrow{\scriptstyle \pi'} & & \downarrow{\scriptstyle \pi} \\ \tilde{X}' & \xrightarrow{\ \tilde{\varphi}\ } & \tilde{X} \end{array}$$

commutes, the latter phenomenon occurs when X' is contained in one fibre of the reduction map $\pi: X \to \tilde{X}$, for example, when $X = \mathbb{B}^1 = \operatorname{Sp} k\langle\zeta\rangle$ and $X' = X(c\zeta)$ with $c \in k$, $|c| > 1$. However, there is one interesting case which will be considered in this section, namely, when $X' = X(f^{-1})$ with $f \in A$, $|f|_{\sup} = 1$. We want to show that in this situation $\tilde{\varphi}: \tilde{X}' \to \tilde{X}$ defines $\tilde{X}'$ as a Zariski-open subvariety of $\tilde{X}$.

Lemma 1. *Let A be a k-affinoid algebra. Then for any elements $a, f \in A$, the set of non-zero values $|f^n a|_{\sup}$, $n \in \mathbb{N}$, is discrete in the set of positive real numbers.*

Proof. Considering the irreducible components of $\operatorname{Max} A$, we may assume that A is an integral domain. Then, by Corollary 6.1.2/2 and Proposition 6.2.2/2, the field of fractions $Q(A)$ is a finite extension of some field $Q(T_d)$, and the supremum norm on A is induced by the spectral norm of $Q(A)$ over $Q(T_d)$. Since, by Proposition 3.3.3/1, this norm equals the maximum of finitely many valuations on $Q(A)$, the assertion of the lemma is obvious. □

Lemma 2. *Let A be a k-affinoid algebra and let $a, f \in A$ with $|f|_{\sup} = 1$. Denote by $\sigma: A \to A\langle f^{-1}\rangle$ the canonical map. Then* $\lim\limits_{n\to\infty} |f^n a|_{\sup} = |\sigma(a)|_{\sup}$. *If $|\sigma(a)|_{\sup} > 0$, there exists an $n_0 \in \mathbb{N}$ such that $|f^n a|_{\sup} = |\sigma(a)|_{\sup}$ for all $n \geqq n_0$.*

Proof. The map σ is the restriction map corresponding to the affinoid subdomain $\operatorname{Sp} A\langle f^{-1}\rangle \hookrightarrow \operatorname{Sp} A$. Since $|f(x)| = 1$ for all $x \in \operatorname{Sp} A\langle f^{-1}\rangle$, we get

$$|f^n a|_{\sup} \geqq |f^{n+1} a|_{\sup} \geqq |a|_{\operatorname{Sp} A'} = |\sigma(a)|_{\sup}, \qquad n \in \mathbb{N},$$

where A' stands for $A\langle f^{-1}\rangle$ and $|a|_{\operatorname{Sp} A'}$ for $\sup\limits_{x \in \operatorname{Sp} A'} |a(x)|$. By Lemma 1, the sequence $|f^n a|_{\sup}$ approaches zero or becomes constant, say $|f^n a|_{\sup} = \alpha$ for $n \geqq n_0$. Therefore, we only have to show in the latter case that

$$\alpha \leqq |\sigma(a)|_{\sup}.$$

We assume the contrary. All functions $f^n a$ satisfy the Maximum Modulus Principle (Proposition 6.2.1/4). Since $|f^n(x)\, a(x)| \leqq |a(x)|$ for all $x \in \operatorname{Sp} A$, the functions $f^n a$ must assume the maximum on the affinoid subdomain

$$U := \{x \in \operatorname{Sp} A;\ |a(x)| \geqq \alpha\}.$$

Thus we see (since $\alpha > |\sigma(a)|_{\sup} = |a|_{\operatorname{Sp} A'}$) that U is disjoint from $\operatorname{Sp} A'$, and it follows again from the Maximum Modulus Principle that $\beta := |f|_U < 1$. Then we get

$$|f^n a|_U \leqq \beta^n\, |a|_U, \qquad n \in \mathbb{N},$$

and hence $\lim\limits_{n\to\infty} |f^n a|_U = 0$. But this is a contradiction to

$$|f^n a|_U = |f^n a|_{\sup} \geqq \alpha > 0. \qquad \square$$

Proposition 3. *Let $\operatorname{Sp} A$ be an affinoid variety and, for some $f \in A$ with $|f|_{\sup} = 1$, consider the Laurent domain $\varphi: \operatorname{Sp} A\langle f^{-1}\rangle \hookrightarrow \operatorname{Sp} A$. Then $\operatorname{Sp} \widetilde{A\langle f^{-1}\rangle} = \operatorname{Sp} \tilde{A}[\tilde{f}^{-1}]$, and the map $\tilde{\varphi}: \operatorname{Sp} \tilde{A}[\tilde{f}^{-1}] \to \operatorname{Sp} \tilde{A}$ is induced by the canonical homomorphism $\tilde{A} \to \tilde{A}[\tilde{f}^{-1}]$. Thus, $\tilde{\varphi}$ is the canonical embedding of the Zariski-open subvariety $\operatorname{Sp} \tilde{A}[\tilde{f}^{-1}]$ into $\operatorname{Sp} \tilde{A}$.*

Proof. We consider the homomorphism $\tilde{\varphi}^*: \tilde{A} \to \widetilde{A\langle f^{-1}\rangle}$. Since $\tilde{f}$ is a unit in $\widetilde{A\langle f^{-1}\rangle}$, we get the following factorization

$$\begin{array}{ccc} \tilde{A} & \xrightarrow{\ \tilde{\varphi}^*\ } & \widetilde{A\langle f^{-1}\rangle} \\ & {\scriptstyle \varrho}\searrow \quad \nearrow{\scriptstyle \iota} & \\ & \tilde{A}[\tilde{f}^{-1}] & \end{array},$$

and it is only necessary to show that ι is bijective. Let $a \in \mathring{A}$ be an element such that $\tilde{a} \in \ker \tilde{\varphi}^*$. Then $|\varphi^*(a)|_{\sup} < 1$; hence, by Lemma 2, there exists an

$n \in \mathbb{N}$ such that $|f^n a|_{\sup} < 1$. But then $\tilde{f}^n \tilde{a} = 0$ and therefore $\tilde{a} \in \ker \varrho$. Thus ι must be injective. Furthermore, ι is also surjective. Namely, $\varphi^*(A)[f^{-1}]$ is dense in $A\langle f^{-1}\rangle$ and, by Lemma 2, any element $b \in \varphi^*(A)[f^{-1}]$ can be written in the form $b = f^{-n}\varphi^*(a)$, with $n \in \mathbb{N}$, $a \in A$, and $|b|_{\sup} = |a|_{\sup}$, provided $|b|_{\sup} > 0$. □

7.3. Immersions of affinoid varieties

7.3.1. Ideal-adic topologies. — For convenience we gather here a few standard facts about ideal-adic topologies, which will be used during the next sections. In the following let R denote a ring (commutative with identity) and $\mathfrak{m} \subset R$ a proper ideal. The $\mathfrak{m}$-adic filtration

$$R \supset \mathfrak{m} \supset \mathfrak{m}^2 \supset \ldots$$

defines a semi-norm on R ($\mathfrak{m}$-adic semi-norm, see Proposition 1.3.3/1), and hence a topology, the so-called $\mathfrak{m}$-adic topology on R. Recall that the $\mathfrak{m}$-adic semi-norm on R is given by

$$|a| := \begin{cases} 0 & \text{if } a \in \bigcap_{n=0}^{\infty} \mathfrak{m}^n, \\ \alpha^{\max\{n; a \in \mathfrak{m}^n\}}, & \text{otherwise,} \end{cases}$$

where α is a constant in $\mathbb{R}$ with $0 < \alpha < 1$. It is easily seen from the definition that the ideal $\mathfrak{m}$ is open and closed in R.

Similarly, for any R-module M, one defines the $\mathfrak{m}$-adic topology of M using the filtration

$$M \supset \mathfrak{m}M \supset \mathfrak{m}^2M \supset \ldots .$$

Introducing a semi-norm on M as above, M becomes a "semi-normed" R-module. One can then apply the completion process of (1.1.7) and obtain the $\mathfrak{m}$-adic completions $\hat{R}$ and $\hat{M}$ of R and M, respectively. Note that $\hat{M}$ is canonically an $\hat{R}$-module. Furthermore if R is complete, an easy consideration shows that finite R-modules are always complete with respect to the $\mathfrak{m}$-adic topology.

Example. Let $\zeta_1, \ldots, \zeta_n$ be indeterminates and consider the polynomial ring $A := R[\zeta_1, \ldots, \zeta_n]$. Denote by $\mathfrak{a} \subset A$ the ideal generated by all indeterminates $\zeta_1, \ldots, \zeta_n$. Then the $\mathfrak{a}$-adic completion $\hat{A}$ of A is the ring $R[[\zeta_1, \ldots, \zeta_n]]$ of formal power series over R.

Proposition 1. *Let R be Noetherian. Then the $\mathfrak{m}$-adic completion $\hat{R}$ of R is also Noetherian.*

Proof. Let $f_1, \ldots, f_n \in R$ generate the ideal $\mathfrak{m}$. As in the example, consider the polynomial ring $A = R[\zeta_1, \ldots, \zeta_n]$ and define a surjective ring homomorphism $\varphi: A \to R$ by $\varphi|_R :=$ id and $\varphi(\zeta_i) := f_i$, $i = 1, \ldots, n$. Then $\mathfrak{a} := (\zeta_1, \ldots, \zeta_n)$, is mapped onto $\mathfrak{m}$ by φ. Therefore, φ is continuous and open and hence strict, when we consider the $\mathfrak{a}$-adic topology on A and the $\mathfrak{m}$-adic topology on R. According to Corollary 1.1.9/6, the map φ induces a surjection $\hat{\varphi}: \hat{A} \to \hat{R}$ between the completions of A and R, which, of course, is a ring homomorphism. Thus, $\hat{R}$ is seen to be Noetherian, if we can show that $\hat{A} = R[\![\zeta_1, \ldots, \zeta_n]\!]$ is Noetherian. But this follows by an induction argument from

Lemma 2. *Let R be Noetherian. Then the formal power series ring in one variable $R[\![\zeta]\!]$ is Noetherian.*

Proof. Let $\mathfrak{q}$ be an ideal in $R[\![\zeta]\!]$. Similarly as in the classical proof of Hilbert's Basis Theorem, we denote by $\mathfrak{q}_i \subset R$ the ideal generated by all $a \in R$ such that there exists an element

$$f = a\zeta^i + \text{higher terms}$$

in $\mathfrak{q}$. The ideals $\mathfrak{q}_i$ form an increasing sequence of ideals in R, which must become stationary, i.e., there exists an $s \in \mathbb{N}$ with $\mathfrak{q}_s = \mathfrak{q}_i$ for all $i \geqq s$. Now, for $i \leqq s$, it is possible to choose elements

$$f_{ij} = a_{ij}\zeta^i + \text{higher terms}, \qquad j = 1, \ldots, r_i,$$

in $\mathfrak{q}$ such that $\mathfrak{q}_i$ is generated by $a_{i1}, \ldots, a_{ir_i}$, since R is Noetherian. We claim that $\mathfrak{q}'$, the ideal generated by all elements f_{ij}, equals $\mathfrak{q}$.

Let $f \in \mathfrak{q}$. If i is a non-negative integer with $f \in \mathfrak{q} \cap (\zeta^i)$, then, by our construction, there exist $b_1, \ldots, b_{r_i} \in R$ such that

$$f - \sum_{j=1}^{r_i} b_j f_{ij} \in \mathfrak{q} \cap (\zeta^{i+1}) \quad \text{if} \quad i \leqq s,$$

or

$$f - \zeta^{i-s} \sum_{j=1}^{r_s} b_j f_{sj} \in \mathfrak{q} \cap (\zeta^{i+1}) \quad \text{if } i \geqq s.$$

Thus, by induction, one gets

$$\mathfrak{q} \subset \mathfrak{q}' + \mathfrak{q} \cap (\zeta^s)$$

and, of course,

$$\mathfrak{q} \cap (\zeta^i) \subset \zeta^{i-s}\mathfrak{q}' + \mathfrak{q} \cap (\zeta^{i+1})$$

for all $i \geqq s$. Since $\mathfrak{q}'$, as a finite $R[\![\zeta]\!]$-module, is complete in its (ζ)-adictopology, the last inclusion implies

$$\mathfrak{q} \cap (\zeta^s) \subset \mathfrak{q}'.$$

Therefore, we have

$$\mathfrak{q}' \subset \mathfrak{q} \subset \mathfrak{q}' + \mathfrak{q} \cap (\zeta^s) \subset \mathfrak{q}'$$

which means $\mathfrak{q}' = \mathfrak{q}$. □

Lemma 3. *Let R be Noetherian and let $\varphi: M \to N$ be an R-homomorphism of finite R-modules. Then φ is continuous and strict with respect to the $\mathfrak{m}$-adic topologies on M and N.*

Proof. Because $\varphi(\mathfrak{m}^n M) \subset \mathfrak{m}^n N$, the map φ is continuous. Since φ factors

$$\varphi: M \to \varphi(M) \hookrightarrow N,$$

we may assume that φ is surjective or injective in order to show that φ is strict. If φ is surjective, we have $\varphi(\mathfrak{m}^n M) = \mathfrak{m}^n N$; thus, φ is open and hence strict in this case. If $\varphi: M \hookrightarrow N$ is an injection, strictness follows from the Lemma of ARTIN-REES: There exists an integer n_0 such that

$$(\mathfrak{m}^n N) \cap M = \mathfrak{m}^{n-n_0}\big((\mathfrak{m}^{n_0} N) \cap M\big) \subset \mathfrak{m}^{n-n_0} M$$

or all $n \geqq n_0$. Consequently, the $\mathfrak{m}$-adic topology on N induces the $\mathfrak{m}$-adic topology on M, and φ is strict. □

Proposition 4. *Let R be Noetherian. If $M_1 \xrightarrow{\varphi} M_2 \xrightarrow{\psi} M_3$ is an exact sequence of finite R-modules, then the corresponding sequence of $\mathfrak{m}$-adic completions $\hat{M}_1 \xrightarrow{\hat{\varphi}} \hat{M}_2 \xrightarrow{\hat{\psi}} \hat{M}_3$ is exact.*

Proof. By Lemma 3, the homomorphisms φ and ψ are strict. Thus, our assertion follows from Corollary 1.1.9/6. □

Proposition 5. *Let R be Noetherian and M a finite R-module. Then the canonical map $\hat{R} \otimes_R M \to \hat{M}$ is bijective.*

Proof. Clearly, $\hat{R} \otimes_R M \to \hat{M}$ is bijective, when M is a finite free R-module. If M is not free, choose an exact sequence $F_2 \to F_1 \to M \to 0$ with finite free R-modules F_1, F_2, and consider the commutative diagram

$$\begin{array}{ccccccc}
\hat{R} \otimes_R F_2 & \to & \hat{R} \otimes_R F_1 & \to & \hat{R} \otimes_R M & \to & 0 \\
\downarrow{\scriptstyle \alpha_2} & & \downarrow{\scriptstyle \alpha_1} & & \downarrow{\scriptstyle \alpha_0} & & \\
\hat{F}_2 & \longrightarrow & \hat{F}_1 & \longrightarrow & \hat{M} & \longrightarrow & 0 .
\end{array}$$

The first row is exact, since the tensor product is right exact. The second row is exact by Proposition 4. Furthermore, α_1 and α_2 are isomorphisms, since F_1 and F_2 are free. But then, by diagram chasing (or by the Five Lemma), α_0 also must be an isomorphism. □

Corollary 6. *Let R be Noetherian. Then $\hat{R}$ is a flat R-module; i.e., $\hat{R} \otimes_R -$ is an exact functor on the category of R-modules.*

Proof. Since the tensor product is right exact, it is enough to show that, for any injection $M \hookrightarrow N$ of R-modules, $\hat{R} \otimes_R M \to \hat{R} \otimes_R N$ is injective. Using equations for elements in the kernel of this map, we may restrict ourselves to the case where M and N are finite R-modules. But then Proposition 4 and Proposition 5 imply the assertion. □

Corollary 7. *Let R be Noetherian and M a finite R-module. Then, for all $n \in \mathbb{N}$, the closure of $\mathfrak{m}^n M$ in $\hat{M}$ is $\hat{R}\mathfrak{m}^n M$, and the topology on $\hat{M}$ coincides with the $(\mathfrak{m}\hat{R})$-adic topology.*

Proof. The topology on $\hat{M}$ is given by the filtration

$$\hat{M} \supset \widehat{\mathfrak{m}M} \supset \widehat{\mathfrak{m}^2 M} \supset \ldots,$$

and the canonical map $\hat{R} \otimes_R \mathfrak{m}^n M \to \widehat{\mathfrak{m}^n M}$ is bijective by Proposition 5, i.e., $\widehat{\mathfrak{m}^n M} = \hat{R}\mathfrak{m}^n M$. □

Corollary 8. *Let R be Noetherian and let $\mathfrak{n} \subset R$ be an ideal such that* $\operatorname{rad} \mathfrak{m} = \operatorname{rad} \mathfrak{n}$. *Then $R \to \hat{R}$ induces a bijection $R/\mathfrak{n} \xrightarrow{\sim} \hat{R}/\mathfrak{n}\hat{R}$.*

Proof. Since $\operatorname{rad} \mathfrak{m} = \operatorname{rad} \mathfrak{n}$ the $\mathfrak{m}$-adic topology equals the $\mathfrak{n}$-adic topology on R. Thus, $R/\mathfrak{n} \to \hat{R}/\mathfrak{n}\hat{R}$ is surjective, since R is dense in $\hat{R}$. But the map is also injective, since $\mathfrak{n}$ is closed in R and since, by Corollary 7, the closure of $\mathfrak{n}$ in $\hat{R}$ is $\hat{R}\mathfrak{n}$. □

Corollary 9. *Let R be a local Noetherian ring with maximal ideal $\mathfrak{m}$. Then the $\mathfrak{m}$-adic completion $\hat{R}$ is also local Noetherian with maximal ideal $\hat{R}\mathfrak{m}$.*

Proof. The ring $\hat{R}$ is Noetherian by Proposition 1, and $R \to \hat{R}$ induces a bijection $R/\mathfrak{m} \to \hat{R}/\mathfrak{m}\hat{R}$ by Corollary 8. Thus, $\mathfrak{m}\hat{R}$ is maximal in $\hat{R}$. Furthermore, the above bijection shows that any element $u \in \hat{R} - \mathfrak{m}\hat{R}$ can be written as $u = u_1 - u_2$ with $u_2 \in \mathfrak{m}\hat{R}$ and u_1 the image of a unit in R. Using the geometric series for $(1 - u_1^{-1}u_2)^{-1}$, we get

$$u^{-1} = u_1^{-1} \sum_{\nu=0}^{\infty}{}' (u_1^{-1}u_2)^{\nu},$$

which is a well-defined element in $\hat{R}$. Hence $\mathfrak{m}\hat{R}$ is the unique maximal ideal in $\hat{R}$. □

7.3.2. Germs of affinoid functions. — Let X be an affinoid variety. We consider the contravariant functor $\mathcal{O}_X$, which assigns to each affinoid subdomain $\operatorname{Sp} A \subset X$ its affinoid algebra A and to each inclusion $\operatorname{Sp} A' \hookrightarrow \operatorname{Sp} A$ of affinoid subdomains in X the corresponding restriction homomorphism $A \to A'$. Thus, we can call $\mathcal{O}_X$ a presheaf on X, or, more precisely, on the basis for the canonical topology of X, which consists of all affinoid subdomains of X.

The presheaf $\mathcal{O}_X$ will be discussed at length in Chapter 8. Here we are only concerned about its stalks. For any $x \in X$, we define

$$\mathcal{O}_{X,x} := \varinjlim_{U \ni x} \mathcal{O}_X(U),$$

where the direct limit is taken over all affinoid subdomains U of X containing x. If U is such a domain, the canonical map $\mathcal{O}_X(U) \to \mathcal{O}_{X,x}$ is denoted by $f \mapsto f_x$, and we say that $f_x \in \mathcal{O}_{X,x}$ is the *germ of f at the point x* or simply that f_x is represented by f. Due to the definition of direct limits, the ring $\mathcal{O}_{X,x}$ is characterized by the facts that (i) any element in $\mathcal{O}_{X,x}$ is represented by some $f \in \mathcal{O}_X(U)$ for a suitable U and (ii) $f_x = g_x$ holds for affinoid functions $f \in \mathcal{O}_X(U)$ and

$g \in \mathcal{O}_X(V)$ if and only if there exists an affinoid subdomain $W \subset U \cap V$ containing x such that f and g restrict to the same function in $\mathcal{O}_X(W)$. Therefore, we call $\mathcal{O}_{X,x}$ the algebra of *germs of affinoid functions on X at the point x*.

Let $\mathfrak{m} \subset \mathcal{O}_X(X)$ denote the maximal ideal given by x. Then, for any affinoid subdomain $U \subset X$ with $x \in U$, the maximal ideal in $\mathcal{O}_X(U)$ corresponding to x is $\mathfrak{m}\mathcal{O}_X(U)$ as in Proposition 7.2.2/1. This proposition also shows that the evaluation maps

$$\mathcal{O}_X(U) \to k_a/\Gamma, \qquad f \mapsto f(x),$$

as discussed in (7.1.1) and (7.1.3) are compatible with the restriction maps. Thus, one obtains an evaluation map $\mathcal{O}_{X,x} \to k_a/\Gamma$, also denoted by $h \mapsto h(x)$.

Proposition 1. *$\mathcal{O}_{X,x}$ is a local ring with maximal ideal*

$$\mathfrak{m}\mathcal{O}_{X,x} = \{h \in \mathcal{O}_{X,x};\, h(x) = 0\}.$$

Proof. Let $h \in \mathcal{O}_{X,x}$, i.e., $h = f_x$ for some $f \in \mathcal{O}_X(U)$. Then $h(x) = 0$ if and only if $f(x) = 0$, i.e., if and only if $f \in \mathfrak{m}\mathcal{O}_X(U)$. Thus, $h(x) = 0$ implies $h \in \mathfrak{m}\mathcal{O}_{X,x}$. This verifies

$$\mathfrak{m}\mathcal{O}_{X,x} \supset \{h \in \mathcal{O}_{X,x};\, h(x) = 0\}.$$

The opposite inclusion is trivial. Since $\mathfrak{m}\mathcal{O}_{X,x}$ is a proper ideal in $\mathcal{O}_{X,x}$, it only remains to show that any $h \in \mathcal{O}_{X,x}$ with $h(x) \neq 0$ is a unit. Let h be the germ of some affinoid function $f \in \mathcal{O}_X(U)$. If $h(x) \neq 0$, we may assume $|h(x)| = |f(x)| \geqq 1$. Then $x \in U(f^{-1})$ and the restriction f' of f to $U(f^{-1})$ is a unit in $\mathcal{O}_X\big(U(f^{-1})\big)$. Hence $f'_x = h$ is also a unit in $\mathcal{O}_{X,x}$. □

Now let $\varphi\colon X \to Y$ be a morphism of affinoid varieties; choose $x \in X$ and set $y := \varphi(x)$. We consider pairs of affinoid subdomains $U \subset X$, $V \subset Y$ with $x \in U$ and $\varphi(U) \subset V$. For any such pair, φ induces an affinoid map

$$\varphi_{U,V}\colon U \to V,$$

which in turn corresponds to an algebra homomorphism

$$\varphi^*_{U,V}\colon \mathcal{O}_Y(V) \to \mathcal{O}_X(U).$$

The maps $\varphi^*_{U,V}$ are compatible with restriction homomorphisms; thus, taking direct limits on both sides, the maps $\varphi^*_{U,V}$ give rise to a homomorphism

$$\varphi^*_x\colon \mathcal{O}_{Y,y} \to \mathcal{O}_{X,x}$$

which is local, i.e., it maps the maximal ideal of $\mathcal{O}_{Y,y}$ into the maximal ideal of $\mathcal{O}_{X,x}$ (cf. Lemma 7.1.4/2).

Proposition 2. *Let $X = \operatorname{Sp} A$ be an affinoid variety and let $\mathfrak{a} \subset A$ be an ideal, giving rise to the closed immersion $\varphi\colon \operatorname{Sp} A/\mathfrak{a} \hookrightarrow X$.*

(i) *For any affinoid subdomain* $U \subset X$, *the map* $\varphi_U : \varphi^{-1}(U) \to U$ *obtained by restriction of* φ *is the closed immersion corresponding to the canonical epimorphism*

$$\varphi_U^* : \mathcal{O}_X(U) \to \mathcal{O}_X(U)/\mathfrak{a}\mathcal{O}_X(U).$$

(ii) *For any* $x \in \operatorname{Sp} A/\mathfrak{a}$, *the map*

$$\varphi_x^* : \mathcal{O}_{X,x} \to \mathcal{O}_{\operatorname{Sp} A/\mathfrak{a},x}$$

is obtained by taking the direct limit of all maps φ_U^* *such that* $x \in U$. *Furthermore,* φ_x^* *is surjective and* $\ker \varphi_x^* = \mathfrak{a}\mathcal{O}_{X,x}$.

Proof. Assertion (i) is just a reformulation of Corollary 7.2.2/6. Assertion (ii) follows from the facts that the canonical topology on X restricts to the canonical topology on $\operatorname{Sp} A/\mathfrak{a}$ and that $\varinjlim$ is exact. □

Proposition 3. *Let* $X = \operatorname{Sp} A$ *be an affinoid variety and let* x *be a point in* X. *Denote by* $\mathfrak{m} \subset A$ *the ideal corresponding to* x *and by* $A_{\mathfrak{m}}$ *the localization of* A *with respect to* $\mathfrak{m}$. *Then the canonical map* $A \to \mathcal{O}_{X,x}$ *is factored through an injective map* $A_{\mathfrak{m}} \to \mathcal{O}_{X,x}$, *and, for all* $n \in \mathbb{N}$, *it induces bijections*

$$A/\mathfrak{m}^n \xrightarrow{\sim} A_{\mathfrak{m}}/\mathfrak{m}^n A_{\mathfrak{m}} \xrightarrow{\sim} \mathcal{O}_{X,x}/\mathfrak{m}^n\mathcal{O}_{X,x}.$$

In particular, we get isomorphisms

$$\hat{A} \xrightarrow{\sim} \hat{A}_{\mathfrak{m}} \xrightarrow{\sim} \hat{\mathcal{O}}_{X,x}$$

between the $\mathfrak{m}$*-adic completion of* A *and the maximal-adic completions of* $A_{\mathfrak{m}}$ *and* $\mathcal{O}_{X,x}$.

Proof. For an arbitrary $n \in \mathbb{N}$, we consider the closed immersion $\varphi: \operatorname{Sp} A/\mathfrak{m}^n \hookrightarrow \operatorname{Sp} A$. Clearly, x is the only point in $\operatorname{Sp} A/\mathfrak{m}^n$. It follows from Proposition 2 and Proposition 7.2.2/1 that the map $A/\mathfrak{m}^n \to \mathcal{O}_{X,x}/\mathfrak{m}^n\mathcal{O}_{X,x}$ induced by $A \to \mathcal{O}_{X,x}$ is bijective. Since, by Proposition 1, all elements of $A - \mathfrak{m}$ are mapped onto units in $\mathcal{O}_{X,x}$, the map $A \to \mathcal{O}_{X,x}$ must factor through $A_{\mathfrak{m}}$. In particular, we get bijections

$$A/\mathfrak{m}^n \xrightarrow{\sim} A_{\mathfrak{m}}/\mathfrak{m}^n A_{\mathfrak{m}} \xrightarrow{\sim} \mathcal{O}_{X,x}/\mathfrak{m}^n\mathcal{O}_{X,x}$$

for all $n \in \mathbb{N}$, and thus bijections

$$\hat{A} \xrightarrow{\sim} \hat{A}_{\mathfrak{m}} \xrightarrow{\sim} \hat{\mathcal{O}}_{X,x}.$$

Due to Krull's Intersection Theorem, the maximal-adic topology on $A_{\mathfrak{m}}$ is Hausdorff. Therefore, the composition

$$A_{\mathfrak{m}} \to \mathcal{O}_{X,x} \to \hat{\mathcal{O}}_{X,x} = \hat{A}_{\mathfrak{m}}$$

is injective, and hence also $A_{\mathfrak{m}} \to \mathcal{O}_{X,x}$ is injective. □

We derive some immediate consequences from Proposition 3.

Corollary 4. *Let f be an affinoid function on an affinoid variety $X = \operatorname{Sp} A$. Then $f = 0$ if and only if all its germs f_x, $x \in X$, are zero.*

Proof. The assertion is clear by the following injections

$$A \hookrightarrow \prod_{\mathfrak{m} \in \operatorname{Max} A} A_{\mathfrak{m}} \hookrightarrow \prod_{x \in X} \mathcal{O}_{X,x}. \qquad \square$$

Corollary 5. *Let X be an affinoid variety and let $X = \bigcup_{i \in I} U_i$ be a covering consisting of affinoid subdomains $U_i \subset X$. Then the restriction maps $\mathcal{O}_X(X) \to \mathcal{O}_X(U_i)$ define an injection*

$$\mathcal{O}_X(X) \hookrightarrow \prod_{i \in I} \mathcal{O}_X(U_i).$$

Corollary 6. *If $X' = \operatorname{Sp} A'$ is an affinoid subdomain of $X = \operatorname{Sp} A$, then A' is a flat A-module via the restriction map $A \to A'$.*

Proof. Let $x \in X'$ be a point and denote by $\mathfrak{m} \subset A$ and $\mathfrak{m}' \subset A'$ the corresponding maximal ideals. Then, by Proposition 3, the map $A \to A'$ induces a bijection $\hat{A} \to \hat{A}'$ between the $\mathfrak{m}$-adic completion of A and the $\mathfrak{m}'$-adic completion of A'. Thus, by general facts of commutative algebra, A' is flat over A.

For completeness we give a second proof, which is more elementary. Let M be a finite A-module. Then there are canonical injections

$$M \hookrightarrow \prod_{\mathfrak{m} \in \operatorname{Max} A} M \otimes_A A_{\mathfrak{m}} \hookrightarrow \prod_{\mathfrak{m} \in \operatorname{Max} A} M \otimes_A \hat{A}_{\mathfrak{m}}.$$

(The first map is injective, since $M \otimes_A A_{\mathfrak{m}}$ is the localization of M with respect to the maximal ideal $\mathfrak{m} \subset A$; the second map is injective by KRULL's Intersection Theorem for modules.) Thus, using Proposition 3, the canonical map

$$M \to \prod_{x \in X} M \otimes_A \hat{\mathcal{O}}_{X,x}$$

is injective.

Now consider an injective A-homomorphism $M \hookrightarrow N$ of finite A-modules. Then, by Corollary 7.3.1/6 and Proposition 3, for all $x \in X$, the maps

$$M \otimes_A \hat{\mathcal{O}}_{X,x} \to N \otimes_A \hat{\mathcal{O}}_{X,x}$$

are injective, since $\hat{\mathcal{O}}_{X,x}$ may be viewed as the $\mathfrak{m}_x$-adic completion of A (where $\mathfrak{m}_x$ is the maximal ideal in A corresponding to x). We have the following commutative diagram

$$\begin{array}{ccc} M \otimes_A A' & \longrightarrow & N \otimes_A A' \\ \downarrow & & \downarrow \\ \prod_{x \in X'} M \otimes_A \hat{\mathcal{O}}_{X,x} & \hookrightarrow & \prod_{x \in X'} N \otimes_A \hat{\mathcal{O}}_{X,x} . \end{array}$$

Since for all $x \in X'$

$$(M \otimes_A A') \otimes_{A'} \hat{\mathcal{O}}_{X,x} = M \otimes_A \hat{\mathcal{O}}_{X,x},$$

the consideration above (applied to the A'-modules $M \otimes_A A'$ and $N \otimes_A A'$) shows that the vertical maps are injective. Hence the map in the first row is injective, and A' is flat over A. □

Proposition 7. *For any point x of an affinoid variety X, the local ring $\mathcal{O}_{X,x}$ is Noetherian.*

Proof. Let $X = \operatorname{Sp} A$, and let $\mathfrak{m} \subset A$ denote the maximal ideal corresponding to x. We begin by showing that $\mathcal{O}_{X,x}$ is separated, i.e., that KRULL's Intersection Theorem holds for $\mathcal{O}_{X,x}$. Any $h \in \bigcap_{n=1}^{\infty} \mathfrak{m}^n \mathcal{O}_{X,x}$ is the germ of some affinoid function f; we may assume $f \in A$. As $A/\mathfrak{m}^n$ is isomorphic to $\mathcal{O}_{X,x}/\mathfrak{m}^n\mathcal{O}_{X,x}$, we see that $f \in \bigcap_{n=1}^{\infty} \mathfrak{m}^n$. Hence the image of f in $A_\mathfrak{m}$ must be zero, because $A_\mathfrak{m}$, as a Noetherian local ring, is separated. Thus $h = f_x = 0$, proving that $\mathcal{O}_{X,x}$ is separated.

Next we want to see that any finitely generated ideal in $\mathcal{O}_{X,x}$ is closed in the maximal-adic topology. Choosing finitely many affinoid functions representing a set of generators, we only have to consider ideals of type $\mathfrak{a}\mathcal{O}_{X,x}$ where $\mathfrak{a}$ is an ideal in some affinoid algebra A'; we may assume $A = A'$. Due to Proposition 2, the quotient $\mathcal{O}_{X,x}/\mathfrak{a}\mathcal{O}_{X,x}$ can be interpreted as the algebra of germs of affinoid functions on the closed subvariety $\operatorname{Sp} A/\mathfrak{a} \subset \operatorname{Sp} A$ at the point x. Thus, $\mathcal{O}_{X,x}/\mathfrak{a}\mathcal{O}_{X,x}$ is separated, and hence $\mathfrak{a}\mathcal{O}_{X,x}$ is closed.

Now let us consider an increasing sequence of finitely generated ideals

$$\mathfrak{a}_1 \subset \mathfrak{a}_2 \subset \cdots \subset \mathcal{O}_{X,x}.$$

By completion we get a sequence of ideals

$$\hat{\mathfrak{a}}_1 \subset \hat{\mathfrak{a}}_2 \subset \cdots \subset \hat{\mathcal{O}}_{X,x}$$

which must become stationary, since $\hat{\mathcal{O}}_{X,x}$ is Noetherian by Proposition 7.3.1/1 and Proposition 3. Since $\mathcal{O}_{X,x}$ is separated, it maps injectively into its completion $\hat{\mathcal{O}}_{X,x}$. Thus, using the fact that the ideals $\mathfrak{a}_i$ are closed, the sequence of the $\mathfrak{a}_i$ also must become stationary, and hence $\mathcal{O}_{X,x}$ is Noetherian. □

We conclude this section with some remarks about local properties of affinoid varieties. An affinoid variety X is called *reduced, normal* or *smooth at a point* $x \in X$, if the local ring $\mathcal{O}_{X,x}$ is reduced, normal or regular, respectively. The *dimension* $\dim_x X$ *of X at the point x* is defined as the Krull dimension of $\mathcal{O}_{X,x}$. Note that these definitions are local in the following sense: If $X' \subset X$ is an affinoid subdomain containing x, then X is reduced, normal or smooth at x if and only if X' satisfies the corresponding property at x. Also $\dim_x X = \dim_x X'$.

The ring $\mathcal{O}_{X,x}$, which may look somewhat inaccessible, is of essential value when defining local properties of X at x. To make things easier, one is, of course, interested in local rings which may substitute for $\mathcal{O}_{X,x}$.

Proposition 8. *Let* $X = \operatorname{Sp} A$ *be an affinoid variety and let* $x \in X$ *be a point corresponding to the maximal ideal* $\mathfrak{m} \subset A$.

(i) *If one of the rings* $A_{\mathfrak{m}}$, $\mathcal{O}_{X,x}$, $\hat{A}_{\mathfrak{m}} = \hat{\mathcal{O}}_{X,x}$ *is reduced, normal or regular, then all three rings satisfy this property.*

(ii) $A_{\mathfrak{m}}$, $\mathcal{O}_{X,x}$ *and* $\hat{A}_{\mathfrak{m}} = \hat{\mathcal{O}}_{X,x}$ *have the same Krull dimension.*

The *proof* is mainly an application of facts from commutative algebra. We give only some references. Affinoid algebras without zero-divisors are Japanese by Proposition 6.1.2/4. Therefore, all affinoid algebras are pseudo-geometric rings in the terminology of NAGATA, § 36 [28], and hence Theorem 36.4 [28] settles the "reduced" case of (i). The "normal" case is dealt with in KIEHL [25], and the "regular" case is a consequence of Proposition 3, using the equality of Krull dimensions in (ii). Finally, (ii) is merely the standard fact that the dimension of a Noetherian local ring is equal to the dimension of its maximal-adic completion. □

Since a ring A is reduced or normal (i.e., integrally closed in its total ring of fractions) if and only if all localizations $A_{\mathfrak{m}}$, $\mathfrak{m} \in \operatorname{Max} A$, are reduced or normal, respectively, one derives from Proposition 8:

Corollary 9. *An affinoid variety* $X = \operatorname{Sp} A$ *is reduced or normal (at all of its points) if and only if* A *is reduced or normal, respectively.*

Corollary 10. *Let* $\operatorname{Sp} A'$ *be an affinoid subdomain of* $X = \operatorname{Sp} A$. *Then if* A *is reduced or normal,* A' *is reduced or normal, respectively.*

7.3.3. Locally closed immersions. — Let $\varphi: X \to Y$ be a morphism of affinoid varieties. Let $x \in X$ be a point such that φ is injective at x, i.e., $\{x\} = \varphi^{-1}(\varphi(x))$. Then φ is said to be a *locally closed immersion at* x if the homomorphism $\varphi_x^*: \mathcal{O}_{Y,\varphi(x)} \to \mathcal{O}_{X,x}$ is surjective, and φ is called an *open immersion at* x if φ_x^* is bijective. If φ is injective and satisfies the above conditions at all points $x \in X$, then φ is called a locally closed or open immersion, respectively. We state some simple properties of immersions, which are obvious from the definitions.

Proposition 1. *Open and closed immersions are locally closed. Any inclusion map* $\varphi: X' \hookrightarrow X$ *defining* X' *as an affinoid subdomain of an affinoid variety* X *is an open immersion.*

Proof. Closed immersions are locally closed by Proposition 7.3.2/2. □

Proposition 2. *Any composition of open (closed or locally closed) immersions, is again an open (closed or locally closed, respectively) immersion.*

Proposition 3. *Let* $\varphi: X \to Y$ *be a morphism of affinoid varieties and let* Y' *be an affinoid subdomain of* Y. *If* φ *is a closed, locally closed, or open immersion, then also the induced morphism* $\varphi_{Y'}: \varphi^{-1}(Y') \to Y'$ *is a closed, locally closed, or open immersion, respectively.*

Proof. If φ is a closed immersion, the assertion follows from Corollary 7.2.2/6. □

In the following we prove some characterizing properties of locally closed and open immersions, which, in particular, justify our terminology.

Proposition 4. *Let $\varphi\colon X \to Y$ be a morphism of affinoid varieties and fix a point $x \in X$. Assume that φ is injective at x. Then the following are equivalent:*

(i) $\varphi_x^*\colon \mathcal{O}_{Y,\varphi(x)} \to \mathcal{O}_{X,x}$ *is surjective; i.e., φ is a locally closed immersion at x.*

(ii) $\hat{\varphi}_x^*\colon \hat{\mathcal{O}}_{Y,\varphi(x)} \to \hat{\mathcal{O}}_{X,x}$, *the "completion" of φ_x^*, is surjective.*

(iii) *There exists an affinoid subdomain $Y' \subset Y$ containing $\varphi(x)$ such that the induced morphism $\varphi_{Y'}\colon \varphi^{-1}(Y') \to Y'$ is a closed immersion.*

Proof. The implication (i) ⇒ (ii) is a consequence of the exactness of the completion functor (cf. Proposition 7.3.1/4).

Now assume (ii). Let $X = \operatorname{Sp} A$, $Y = \operatorname{Sp} B$, and denote by $\mathfrak{m} \subset B$ the maximal ideal corresponding to $\varphi(x)$. Then the algebra homomorphism $\varphi^*\colon B \to A$ associated to φ gives rise to the following commutative diagram:

$$\begin{array}{ccc} B/\mathfrak{m} & \xrightarrow{\ \sigma\ } & A/\mathfrak{m}A \\ \downarrow{\scriptstyle\tau} & & \downarrow{\scriptstyle\tau'} \\ \hat{\mathcal{O}}_{Y,\varphi(x)}/\mathfrak{m}\hat{\mathcal{O}}_{Y,\varphi(x)} & \xrightarrow{\ \sigma'\ } & \hat{\mathcal{O}}_{X,x}/\mathfrak{m}\hat{\mathcal{O}}_{X,x}. \end{array}$$

The homomorphism σ' is surjective by our assumption. The ring $\hat{\mathcal{O}}_{Y,\varphi(x)}$ is the $\mathfrak{m}$-adic completion of B; hence τ is bijective by Corollary 7.3.1/8. In the same way, we get that τ' is bijective, since $\operatorname{rad}(\mathfrak{m}A)$ is the maximal ideal corresponding to x due to the injectivity of φ at x. Thus, σ must be surjective, and (iii) follows from assertion (b) in Lemma 7.2.5/2.

Finally (iii) ⇒ (i) by Proposition 1. □

Proposition 5. *Let $\varphi\colon X \to Y$ and $x \in X$ be as in Proposition 4. Then the following are equivalent:*

(i) $\varphi_x^*\colon \mathcal{O}_{Y,\varphi(x)} \to \mathcal{O}_{X,x}$ *is bijective; i.e., φ is an open immersion at x.*

(ii) $\hat{\varphi}_x^*\colon \hat{\mathcal{O}}_{Y,\varphi(x)} \to \hat{\mathcal{O}}_{X,x}$ *is bijective.*

(iii) *There exists an affinoid subdomain $Y' \subset Y$ containing $\varphi(x)$ such that $\varphi_{Y'}\colon \varphi^{-1}(Y') \to Y'$ is an isomorphism of affinoid varieties.*

The *proof* is only a slight modification of the preceding one. If (ii) is assumed, we consider, for $n \in \mathbb{N}$, the commutative diagram

$$\begin{array}{ccc} B/\mathfrak{m}^n & \xrightarrow{\ \sigma_n\ } & A/\mathfrak{m}^n A \\ \downarrow{\scriptstyle\tau_n} & & \downarrow{\scriptstyle\tau_n'} \\ \hat{\mathcal{O}}_{Y,\varphi(x)}/\mathfrak{m}^n\hat{\mathcal{O}}_{Y,\varphi(x)} & \xrightarrow{\ \sigma_n'\ } & \hat{\mathcal{O}}_{X,x}/\mathfrak{m}^n\hat{\mathcal{O}}_{X,x}\,. \end{array}$$

Then σ'_n is bijective, and as before also τ_n and τ'_n are bijective. Thus, σ_n is bijective and assertion (c) of Lemma 7.2.5/2 implies (iii). $\square$

Corollary 6. *Let $\varphi: X \to Y$ be a locally closed or open immersion at $x \in X$. Then there is an affinoid subdomain $X' \subset X$ containing x such that φ is respectively a locally closed or open immersion at all points $x' \in X'$.*

Furthermore, if $\varphi: X \to Y$ is an open immersion, it follows from Proposition 5 that φ is open with respect to the canonical topology. We proved this already in (7.2.5) in the case where φ is the inclusion of an affinoid subdomain into Y. But it should be mentioned at this point that open immersions and inclusions of affinoid subdomains are the same thing, a consequence which will be derived from the main theorem for locally closed immersions in (7.3.5) and TATE's Acyclicity Theorem in Chapter 8.

Proposition 7. *Let $\varphi: X \to Y$ be an open and closed immersion of affinoid varieties. Then X is a Weierstrass domain in Y via φ. Furthermore, X is Zariski-open and -closed in Y.*

Proof. Let $X = \operatorname{Sp} A$, $Y = \operatorname{Sp} B$, and denote by $\varphi^*: B \to A$ the homomorphism corresponding to φ. For any $x \in X$, the surjection φ^* induces a surjection $B_{\mathfrak{m}_{\varphi(x)}} \to A_{\mathfrak{m}_x}$ between the localizations at $\varphi(x)$ and x. Furthermore, this map is injective and hence bijective, since $\varphi_x^*: \mathcal{O}_{Y,\varphi(x)} \to \mathcal{O}_{X,x}$ is bijective (use Proposition 7.3.2/3). Thus, for any $x \in X$, there is an $f \in B$ with $f(\varphi(x)) \neq 0$ such that φ^* induces a bijection $B[f^{-1}] \to A[f^{-1}]$. This shows that $\varphi(X)$ is Zariski-open in Y.

Since $\varphi(X)$ is also Zariski-closed in Y, it has the natural structure of a Weierstrass domain in Y. Namely, $Y_1 := \varphi(X)$ and its complement $Y_2 := Y - Y_1$ are disjoint Zariski-closed subsets of Y. Thus, by the Chinese Remainder Theorem, one has

$$B/\operatorname{rad} B \cong B/\operatorname{id}(Y_1) \oplus B/\operatorname{id}(Y_2),$$

and there exists an affinoid function $h \in B$ such that $h \equiv 0 \bmod \operatorname{id}(Y_1)$ and $h \equiv 1 \bmod \operatorname{id}(Y_2)$. Since $Y_1 = Y(ch)$ for any constant $c \in k$, $|c| > 1$, we see that $\varphi(X) = Y_1$ is a Weierstrass domain in Y. It remains to be shown that the map $\varphi_1: X \to Y_1$ induced by φ is an isomorphism. As a map between sets, φ_1 is surjective and hence bijective. Thus, the isomorphisms

$$\mathcal{O}_Y(Y_1)_{\mathfrak{m}_{\varphi(x)}} = B_{\mathfrak{m}_{\varphi(x)}} \to A_{\mathfrak{m}_x}, \qquad x \in X,$$

show that $\varphi_1^*: \mathcal{O}_Y(Y_1) \to A$, and hence $\varphi_1: X \to Y_1$, are isomorphisms. $\square$

Proposition 8. *Let $\varphi: X \to Y$ be a finite morphism of affinoid varieties.*

(i) *There exists a (possibly empty) Zariski-open subset $Y' \subset Y$ such that φ is a locally closed immersion at a point $x \in X$ if and only if $x \in \varphi^{-1}(Y')$.*

(ii) *If φ is a locally closed immersion, then φ is a closed immersion.*

Proof. Let $X = \operatorname{Sp} A$, $Y = \operatorname{Sp} B$, and assume that φ is injective at a point $x \in X$. Then $\varphi^*: B \to A$ induces the following commutative diagram

$$\begin{array}{ccccccc} B & \to & B_{\mathfrak{m}_{\varphi(x)}} & \hookrightarrow & \mathcal{O}_{Y,\varphi(x)} & \hookrightarrow & \hat{\mathcal{O}}_{Y,\varphi(x)} \\ \downarrow{\scriptstyle\varphi^*} & & \downarrow{\scriptstyle\varphi^*_{\mathfrak{m}_x}} & & \downarrow{\scriptstyle\varphi^*_x} & & \downarrow{\scriptstyle\hat{\varphi}^*_x} \\ A & \to & A_{\mathfrak{m}_x} & \hookrightarrow & \mathcal{O}_{X,x} & \hookrightarrow & \hat{\mathcal{O}}_{X,x} \end{array}$$

with $\mathfrak{m}_x \subset A$ and $\mathfrak{m}_{\varphi(x)} \subset B$ denoting the maximal ideals corresponding to x and $\varphi(x)$, respectively; $\varphi^*_{\mathfrak{m}_x}$ is the canonical extension of φ^*.

By our assumptions, φ^* is finite. We want to show that $\varphi^*_{\mathfrak{m}_x}$ is finite also. Setting $S := B - \mathfrak{m}_{\varphi(x)}$ and $S' := \varphi^*(S)$, we consider the finite map $\varphi^*_S: B_S \to A_{S'}$ derived from φ^*. It is enough to show that φ^*_S coincides with $\varphi^*_{\mathfrak{m}_x}$, i.e., that $A_{S'} = A_{\mathfrak{m}_x}$. Clearly, $\mathfrak{m}_x A_{S'}$ is a maximal ideal in $A_{S'}$ and, in fact, is the only such ideal in $A_{S'}$. This is true because any maximal ideal of $A_{S'}$ restricts to a maximal ideal of $B_S = B_{\mathfrak{m}_{\varphi(x)}}$ and because φ is injective at x, so that $\mathfrak{m}_x A_{S'} = \operatorname{rad}(\mathfrak{m}_{\varphi(x)} A_{S'})$. Consequently, the canonical map $A_{S'} \to A_{\mathfrak{m}_x}$ must be bijective, and hence $\varphi^*_{\mathfrak{m}_x}$ is finite.

Next we want to show that $\varphi^*_{\mathfrak{m}_x}$ is surjective if and only if φ is a locally closed immersion at x. As above, it follows from the injectivity of φ at x that $\operatorname{rad}\left(\varphi^*(\mathfrak{m}_{\varphi(x)})\, A\right) = \mathfrak{m}_x$. Thus, the $\mathfrak{m}_x$-adic topology on $A_{\mathfrak{m}_x}$ coincides with the $\mathfrak{m}_{\varphi(x)}$-adic topology, when $A_{\mathfrak{m}_x}$ is viewed as a finite $B_{\mathfrak{m}_{\varphi(x)}}$-module via $\varphi^*_{\mathfrak{m}_x}$. Due to KRULL's Intersection Theorem for modules, $\operatorname{im} \varphi^*_{\mathfrak{m}_x}$ is a closed submodule of $A_{\mathfrak{m}_x}$. Now, if φ is a locally closed immersion at x, then $\hat{\varphi}^*_x$ is surjective, and as a result $\operatorname{im} \varphi^*_{\mathfrak{m}_x}$ is dense in $A_{\mathfrak{m}_x}$. Thus, $\varphi^*_{\mathfrak{m}_x}$ is surjective. Conversely, the surjectivity of $\varphi^*_{\mathfrak{m}_x}$ implies the surjectivity of $\hat{\varphi}^*_x$ by the exactness of the completion functor (Proposition 7.3.1/4), and hence the surjectivity of φ^*_x by Proposition 4.

Finally, let us verify the assertions of the proposition. If φ is a locally closed immersion at $x \in X$, then $\varphi^*_{\mathfrak{m}_x}: B_S \to A_{S'}$ (in the above notation) is surjective. The map φ^* being finite, we can find an $f \in S$ such that φ^* induces a surjection $B[f^{-1}] \to A[f^{-1}]$. Then, clearly, φ is injective at all points $x' \in U := X - \varphi^{-1}(V(f))$. Also $\varphi^*_{\mathfrak{m}_{x'}}$ is surjective at all points $x' \in U$. Therefore, φ is a locally closed immersion at all points $x' \in U$, and since U is the inverse image of the Zariski-open subset $Y - V(f) \subset Y$ and contains x, assertion (i) is clear.

If φ is a locally closed immersion at all points $x \in X$, then all maps $\varphi^*_{\mathfrak{m}_x}$ are surjective. Thus, viewing A as a B-module via $\varphi^*: B \to A$, all localizations of $\varphi^*(B)$ and A with respect to maximal ideals in B coincide. Therefore, by standard arguments, $\varphi^*(B) = A$; i.e., φ^* is surjective. This verifies (ii). □

7.3.4. Runge immersions. — Here we consider a special class of locally closed immersions which plays an essential role in the next section, where we prove the main structure theorem for locally closed immersions.

Definition 1. *A morphism of affinoid varieties $\varphi\colon X \to Y$ is called a Runge immersion if φ admits a factorization*

$$\begin{array}{ccc} X & \xrightarrow{\ \varphi\ } & Y \\ & {\scriptstyle\varphi'}\searrow \quad \nearrow{\scriptstyle\varphi''} & \\ & Y' & \end{array},$$

where φ' is a closed immersion and Y' is a Weierstrass domain in Y via φ''.

Any closed immersion is Runge; also any inclusion $X' \hookrightarrow X$ of a Weierstrass domain X' into an affinoid variety X is Runge. Furthermore, Runge immersions can be characterized in the following way:

Proposition 2. *Let $\varphi\colon \operatorname{Sp} B \to \operatorname{Sp} A$ be a morphism of affinoid varieties corresponding to the algebra homomorphism $\varphi^*\colon A \to B$. Then the following are equivalent:*

(i) *φ is a Runge immersion.*
(ii) *$\varphi^*(A)$ is dense in B.*
(iii) *$\varphi^*(A)$ contains an affinoid generating system of B over A.*

Proof. The implication (i) $\Rightarrow$ (ii) follows from the definitions. Also the implication (iii) $\Rightarrow$ (i) is easily verified. Namely, let $a_1, \ldots, a_n \in A$ be elements such that $\big(\varphi^*(a_1), \ldots, \varphi^*(a_n)\big)$ is an affinoid generating system of B over A. Since all $\varphi^*(a_i)$ are power-bounded in B, it follows from Proposition 6.1.4/1 that φ^* factors through $A\langle a_1, \ldots, a_n\rangle$, i.e., there is a commutative diagram

$$\begin{array}{ccc} A & \xrightarrow{\ \varphi^*\ } & B \\ & {\scriptstyle\sigma}\searrow \quad \nearrow{\scriptstyle\sigma'} & \\ & A\langle a_1, \ldots, a_n\rangle & \end{array}.$$

The map σ' is surjective, since $\sigma'(a_i) = \varphi^*(a_i)$ and since the elements $\varphi^*(a_1), \ldots, \varphi^*(a_n)$ generate B over A; hence φ is a Runge immersion.

Finally, using the fact that there always exist affinoid generating systems of B over A, the implication (ii) $\Rightarrow$ (iii) follows from

Lemma 3. *Let $\sigma\colon A \to B$ be a homomorphism of affinoid algebras and $b = (b_1, \ldots, b_n)$ an affinoid generating system of B over A. Then for any norm defining the topology of B, there exists an $\varepsilon > 0$ such that the following holds:*

If $d = (d_1, \ldots, d_n)$ is a system of elements in B satisfying $|b_i - d_i| \leqq \varepsilon$ for $i = 1, \ldots, n$, then d is an affinoid generating system of B over A.

Proof. Fix an algebra norm on A defining its topology. We consider on $A\langle \zeta_1, \ldots, \zeta_n\rangle$ the Gauss norm (cf. (1.4.1)) and on B the residue norm via the epimorphism $\sigma_1\colon A\langle \zeta_1, \ldots, \zeta_n\rangle \to B$ extending σ and mapping ζ_i onto b_i. By Banach's Theorem, it is enough to verify our assertion for this norm on B.

Then for a fixed $\alpha > 1$, any $f \in B$ has an inverse image $\sum a_\nu \zeta^\nu \in A\langle\zeta_1, \ldots, \zeta_n\rangle$ such that $|a_\nu| \leqq \alpha\,|f|$. Choosing an ε, $0 < \varepsilon < \alpha^{-1}$, and a system $d = (d_1, \ldots, d_n)$ of elements in B with $|b_i - d_i| \leqq \varepsilon$ for all i, the d_i are power-bounded. We can write

$$f = \sum a_\nu b^\nu = \sum a_\nu d^\nu + f_1\,,$$

and conclude just as in Proposition 7.2.1/1 that

$$|f_1| = |\sum a_\nu(b^\nu - d^\nu)| \leqq \alpha\varepsilon\,|f|\,.$$

Denote by $\sigma_2\colon A\langle\zeta_1, \ldots, \zeta_n\rangle \to B$ the homomorphism extending σ and mapping ζ_i onto d_i. Then by induction, there are elements $g_0, g_1, \ldots \in A\langle\zeta_1, \ldots, \zeta_n\rangle$ and $f_1, f_2, \ldots \in B$ satisfying

$$f = \sigma_2(g_0 + g_1 + \cdots + g_{r-1}) + f_r\,,$$
$$|g_i| \leqq (\varepsilon\alpha)^i\,\alpha\,|f|\,, \qquad |f_r| \leqq (\varepsilon\alpha)^r\,|f|\,,$$

for all $i, r \in \mathbb{N}$. Since $\varepsilon\alpha < 1$, we can take limits as $r \to \infty$. It follows that σ_2 is surjective and d is an affinoid generating system of B over A. □

Corollary 4. *Any composition of Runge immersions is again a Runge immersion.*

Proof. Use the equivalence of (i) and (ii) in Proposition 2. □

Proposition 5. *Let $\varphi\colon X \to Y$ be a Runge immersion and let $Y' \subset Y$ be an affinoid subdomain. Then also the induced morphism $\varphi^{-1}(Y') \to Y$ is a Runge immersion.*

Proof. Use Proposition 7.3.3/3 and Proposition 7.2.3/6. □

Proposition 6. *Let $\varphi\colon X \to Y$ be an open immersion of affinoid varieties which is also a Runge immersion. Then X is a Weierstrass domain in Y via φ.*

Proof. Since φ is a Runge immersion, there exists a commutative diagram

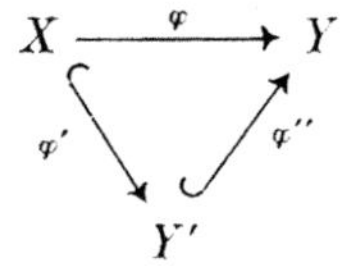

such that φ' is a closed immersion and φ'' defines Y' as a Weierstrass domain in Y. Hence φ'' is an open immersion. Furthermore, φ is an open immersion; thus φ' is an open immersion. Therefore, X is a Weierstrass domain in Y' via φ' by Proposition 7.3.3/7, and X is Weierstrass in Y via φ by Theorem 7.2.4/2. □

The remainder of this section will be concerned with an extension lemma for Runge immersions. Some preparation is necessary. We begin with a variation of the Maximum Modulus Principle for affinoid functions.

Lemma 7. *Let $X = \operatorname{Sp} A$ be an affinoid variety and let $f_1, \ldots, f_r \in A$ be affinoid functions on X. Then*

$$\alpha(x) := \max_{1 \leqq i \leqq n} |f_i(x)|$$

assumes its minimum on X.

Proof. The assertion is trivial when $f_1, \ldots, f_r$ have a common zero on X. Thus, we may assume that $f_1, \ldots, f_r$ generate the unit ideal in A. We consider the covering

$$X = \bigcup_{i=1}^{r} X_i$$

consisting of the rational subdomains

$$X_i := X\left(\frac{f_1}{f_i}, \ldots, \frac{f_r}{f_i}\right) = \{x \in X; \alpha(x) = |f_i(x)|\}.$$

By Proposition 6.2.1/4, the restriction of f_i to X_i assumes its minimum on X_i; thus, $\alpha(x)$ assumes its minimum on X. □

Now let $X = \operatorname{Sp} A$ be an affinoid variety and consider affinoid functions $f_1, \ldots, f_r, g \in A$ generating the unit ideal. For any $\varepsilon \in \sqrt{|k^*|}$, we define the rational domain

$$X_\varepsilon := X\left(\varepsilon^{-1}\frac{f_1}{g}, \ldots, \varepsilon^{-1}\frac{f_r}{g}\right) \subset X.$$

Note that $X_\varepsilon \subset X_{\varepsilon'}$ if $\varepsilon \leqq \varepsilon'$.

Proposition 8. *Let $Y \subset X$ be a finite union of closed subvarieties and of affinoid subdomains in X. Assume that $X_{\varepsilon_0} \cap Y = \emptyset$ for some $\varepsilon_0 \in \sqrt{|k^*|}$. Then there is an $\varepsilon > \varepsilon_0$ such that $X_\varepsilon \cap Y = \emptyset$.*

Proof. We may replace X by $X_{\varepsilon'}$ for some $\varepsilon' > \varepsilon_0$, and Y by $Y \cap X_{\varepsilon'}$ (cf. Corollary 7.2.2/5 and Corollary 7.2.2/6). Thus, we may assume that g is a unit in A. Then by Lemma 7, the function

$$\alpha(x) = \max_{1 \leqq i \leqq n} |(f_i g^{-1})(x)|$$

assumes its minimum on Y. Furthermore, since

$$X_\varepsilon = \{x \in X; \alpha(x) \leqq \varepsilon\},$$

we see that

$$\varepsilon_1 := \min_{x \in Y} \alpha(x) > \varepsilon_0.$$

Thus, for all $\varepsilon \in \sqrt{|k^*|}$, $\varepsilon_0 < \varepsilon < \varepsilon_1$, we have $X_\varepsilon \cap Y = \emptyset$. □

Corollary 9. *Let $h \in A$ and $\varepsilon_0 \in \sqrt{|k^*|}$ such that $|h(x)| < 1$ for all $x \in X_{\varepsilon_0}$. Then there exists an $\varepsilon > \varepsilon_0$ such that $|h(x)| < 1$ also for all $x \in X_\varepsilon$.*

Proof. Apply Proposition 8 to $Y := X(h^{-1})$. $\square$

Now let $\varphi\colon Y \to X$ be a morphism of affinoid varieties. The domain X_ε being defined as before, we set $Y_\varepsilon := \varphi^{-1}(X_\varepsilon)$ and denote by $\varphi_\varepsilon\colon Y_\varepsilon \to X_\varepsilon$ the morphism induced by φ. Also let $X = \operatorname{Sp} A$, $X_\varepsilon = \operatorname{Sp} A_\varepsilon$, $Y = \operatorname{Sp} B$, and $Y_\varepsilon = \operatorname{Sp} B_\varepsilon$.

Extension Lemma 10. *Assume that $\varphi_{\varepsilon_0}\colon Y_{\varepsilon_0} \to X_{\varepsilon_0}$ is a Runge immersion for some $\varepsilon_0 \in \sqrt{|k^*|}$. Then there exists an $\varepsilon > \varepsilon_0$ such that $\varphi_\varepsilon\colon Y_\varepsilon \to X_\varepsilon$ is a Runge immersion also.*

Proof. Replacing X by $X_{\varepsilon'}$ and Y by $Y_{\varepsilon'}$ for some $\varepsilon' > \varepsilon_0$, we may assume that all X_ε and Y_ε are Weierstrass domains in X and Y, respectively. For any $\varepsilon \geqq \varepsilon_0$, we have a commutative diagram

$$\begin{array}{ccc} A & \xrightarrow{\varphi^*} & B \\ \downarrow & & \downarrow \\ A_\varepsilon & \xrightarrow{\varphi_\varepsilon^*} & B_\varepsilon \\ \downarrow & & \downarrow \\ A_{\varepsilon_0} & \xrightarrow{\varphi_{\varepsilon_0}^*} & B_{\varepsilon_0} \end{array},$$

the vertical maps being the canonical restriction homomorphisms. Let $b = (b_1, \ldots, b_n)$ be an affinoid generating system of B over A. Denoting by $|\ \ |_\varepsilon$ the supremum norm on Y_ε, we can find elements $a_1, \ldots, a_n \in A$ such that, for any i, the images of $\varphi^*(a_i)$ and b_i in B_{ε_0} are arbitrarily close to each other (with respect to a given norm defining the topology of B_{ε_0}); in particular, we may assume that

$$|\varphi^*(a_i) - b_i|_{\varepsilon_0} < 1, \qquad i = 1, \ldots, n.$$

Here we use, of course, that X_{ε_0} is Weierstrass in X and that φ_{ε_0} is a Runge immersion, i.e., that by Proposition 2, the image of A is dense in B_{ε_0}. Furthermore, since the restriction of b to Y_{ε_0} is an affinoid generating system of B_{ε_0} over A_{ε_0} (cf. Lemma 7.2.5/1), we may assume by Lemma 3 that $(\varphi^*(a_1), \ldots, \varphi^*(a_n))$ restricts to an affinoid generating system of B_{ε_0} over A_{ε_0}. According to Corollary 9, we can find an $\varepsilon_1 > \varepsilon_0$ such that

$$(*) \qquad |\varphi^*(a_i) - b_i|_{\varepsilon_1} < 1, \qquad i = 1, \ldots, n.$$

Thus, for any ε, $\varepsilon_0 \leqq \varepsilon \leqq \varepsilon_1$, the elements $\varphi^*(a_i)|_{Y_\varepsilon}$ have supremum norm $\leqq 1$, i.e., are power-bounded in B_ε. Hence φ_ε^* must factor in the following way (cf. Proposition 6.1.4/1):

$$\begin{array}{ccc} A_\varepsilon & \xrightarrow{\varphi_\varepsilon^*} & B_\varepsilon \\ & \searrow & \nearrow{\scriptstyle \psi_\varepsilon^*} \\ & A_\varepsilon\langle a_1, \ldots, a_n\rangle & \end{array}.$$

The map ψ_ε^* is finite by Theorem 6.3.2/2. Namely, according to the theorem, it is enough to verify that the elements $\widetilde{b_i|_{Y_\varepsilon}}$ are integral over $\widetilde{A_\varepsilon\langle a_1, \ldots, a_n\rangle}$, and this follows trivially from (*).

Finally, denote by $\psi_\varepsilon\colon Y_\varepsilon \to X_\varepsilon(a_1, \ldots, a_n)$ the finite morphism corresponding to ψ_ε^*. We want to show that ψ_ε is a closed immersion if we take ε close enough to ε_0. By Proposition 7.3.3/8, there exists a Zariski-closed subset $X' \subset X_{\varepsilon_1}(a_1, \ldots, a_n)$ such that ψ_ε is a locally closed immersion at a point $y \in Y_\varepsilon$ if and only if $\varphi(y) \notin X'$. Since φ_{ε_0} is a Runge immersion, we see that $X' \cap X_{\varepsilon_0}(a_1, \ldots, a_n) = \emptyset$. But then, by Proposition 8 (applied to $X(a_1, \ldots, a_n)$ instead of X), one can find an ε, $\varepsilon_0 < \varepsilon \leqq \varepsilon_1$, such that

$$X' \cap X_\varepsilon(a_1, \ldots, a_n) = \emptyset.$$

This means, again by Proposition 7.3.3/8, that ψ_ε is a closed immersion for such an ε, and hence φ_ε is a Runge immersion. □

7.3.5. Main theorem for locally closed immersions. — We have reserved this section for the proof of the following theorem which characterizes locally closed immersions.

Theorem 1 (Gerritzen Grauert). *Let $\varphi\colon Y \to X$ be a locally closed immersion of affinoid varieties. Then there exists a covering $X = \bigcup_{i=1}^{r} X_i$ consisting of finitely many rational subdomains $X_i \subset X$ such that φ induces Runge immersions $\varphi_i\colon \varphi^{-1}(X_i) \to X_i$, $i = 1, \ldots, r$.*

Using Proposition 7.3.4/6, we get

Corollary 2. *If $\varphi\colon Y \to X$ is an open immersion, the maps φ_i define $\varphi^{-1}(X_i)$ as a Weierstrass domain of X_i, $i = 1, \ldots, r$.*

Corollary 3. *Let $X' \subset X$ be an affinoid subdomain. Then there exists a covering $X = \bigcup_{i=1}^{r} X_i$ consisting of finitely many rational subdomains $X_i \subset X$ such that $X_i \cap X'$ is a Weierstrass domain in X_i, $i = 1, \ldots, r$. In particular, X' is a finite union of rational subdomains in X.*

Proof. The inclusion $X' \hookrightarrow X$ is an open immersion, and the intersections $X_i \cap X'$ are rational subdomains of X by Theorem 7.2.4/2. □

In the following we gather some facts which will be needed for the proof of Theorem 1. We begin with a covering lemma for affinoid varieties which looks somewhat technical but is more or less trivial. Let X denote an affinoid variety. Using the notation introduced in the previous section, we consider, for $\varepsilon \in \sqrt{|k^*|}$, rational subdomains

$$X_\varepsilon^i := X\left(\varepsilon^{-1}\frac{g_1^{(i)}}{g_{n_i}^{(i)}}, \ldots, \varepsilon^{-1}\frac{g_{n_i-1}^{(i)}}{g_{n_i}^{(i)}}\right), \qquad i = 1, \ldots, r,$$

in X. Also we set $X_\varepsilon := \bigcup_{i=1}^{r} X_\varepsilon^i$.

Lemma 4. *Given any $\varepsilon \in \sqrt{|k^*|}$, $\varepsilon > 1$, there exist finitely many rational subdomains $V^1, \ldots, V^l \subset X$ such that*

$$\bigcup_{j=1}^{l} V^j \subset X - X_1$$

and

$$X = X_\varepsilon \cup \bigcup_{j=1}^{l} V^j.$$

Proof. The general case is easily reduced to the case $r = 1$; thus, we may assume $r = 1$ and forget about the index i. Set

$$V^j := X\left(\frac{g_1}{g_j}, \ldots, \frac{g_{n-1}}{g_j}, \frac{\varepsilon g_n}{g_j}\right), \qquad j = 1, \ldots, n-1.$$

Then, clearly, $V^1, \ldots, V^{n-1}$ satisfy the assertions. □

The main tool for the proof of Theorem 1 will be a Weierstrass preparation type argument. However, we have to consider a more general situation than that discussed in Chapter 5. Let A denote an affinoid algebra and $\zeta = (\zeta_1, \ldots, \zeta_n)$ a (non-empty) system of indeterminates. For each $x \in \operatorname{Sp} A$, we consider the canonical projection

$$\pi_x \colon A\langle\zeta\rangle \to A/\mathfrak{m}_x\langle\zeta\rangle, \qquad \sum a_\nu \zeta^\nu \mapsto \sum a_\nu(x)\, \zeta^\nu,$$

with $\mathfrak{m}_x \subset A$ denoting the maximal ideal given by x. Note that, in terms of varieties, π_x is the algebra homomorphism giving rise to the closed immersion $\{x\} \times \mathbb{B}^n \to \operatorname{Sp} A \times \mathbb{B}^n$.

An element $f \in A\langle\zeta\rangle$ is called *ζ_n-distinguished of degree s at the point $x \in \operatorname{Sp} A$* if $\pi_x(f) \in A/\mathfrak{m}_x\langle\zeta\rangle$ is ζ_n-distinguished of degree s in the sense of Definition 5.2.1/1. If f is ζ_n-distinguished of some degree $\leqq s$ at each point $x \in \operatorname{Sp} A$, then f is said to be *ζ_n-distinguished of degree $\leqq s$ on* $\operatorname{Sp} A$.

Proposition 5. *An element $f \in A\langle\zeta\rangle$ is a unit if and only if f is ζ_n-distinguished of degree $s = 0$ at all points $x \in \operatorname{Sp} A$.*

Proof. If f is a unit in $A\langle\zeta\rangle$, then $\pi_x(f)$ is a unit in $A/\mathfrak{m}_x\langle\zeta\rangle$ for all $x \in \operatorname{Sp} A$. By Proposition 5.1.3/1, this is equivalent to saying that $\pi_x(f)$ is ζ_n-distinguished of degree $s = 0$. Conversely, if f is ζ_n-distinguished of degree $s = 0$ at all points $x \in \operatorname{Sp} A$, then all elements $\pi_x(f)$ are units. Consequently, f cannot have any zero on $\operatorname{Sp} A \times \mathbb{B}^n$ and hence is a unit. □

Proposition 6. *Let $A \to A'$ be a homomorphism of affinoid algebras, defining $\operatorname{Sp} A'$ as an affinoid subdomain of $\operatorname{Sp} A$. Then an element $f \in A\langle\zeta\rangle$ is ζ_n-distinguished of degree s at a point $x \in \operatorname{Sp} A'$ if and only if its image $f' \in A'\langle\zeta\rangle$ is ζ_n-distinguished of degree s at x.*

Proof. Use Proposition 7.2.2/1. □

Lemma 7. *Let $f \in A\langle\zeta\rangle$ be ζ_n-distinguished of degree $\leqq s$ on* Sp A. *Then the set*

$$\{x \in \operatorname{Sp} A\,; f \text{ is } \zeta_n\text{-distinguished of degree } s \text{ at } x\}$$

is a rational domain in Sp A.

Proof. Write $f = \sum f_\nu \zeta_n^\nu$ with elements $f_\nu \in A\langle\zeta_1, \ldots, \zeta_{n-1}\rangle$ and let $a_\nu \in A$ denote the constant term of f_ν. Then f being ζ_n-distinguished of some degree $\sigma \leqq s$ at a point $x \in \operatorname{Sp} A$ means that

(i) $|\pi_x(f_\nu)| \leqq |\pi_x(f_\sigma)|$ for $\nu \leqq \sigma$,
(ii) $|\pi_x(f_\nu)| < |\pi_x(f_\sigma)|$ for $\nu > \sigma$, and
(iii) $\pi_x(f_\sigma)$ is a unit in $A/\mathfrak{m}_x\langle\zeta_1, \ldots, \zeta_{n-1}\rangle$.

Since $a_\nu(x) = \pi_x(a_\nu)$ is the constant term of the series $\pi_x(f_\nu)$, we get $|a_\nu(x)| \leqq |\pi_x(f_\nu)|$ which is, in fact, an equality for $\nu = \sigma$ by Proposition 5.1.3/1. Thus, we derive the following estimates

$$|a_\nu(x)| \leqq |\pi_x(f_\nu)| \leqq |\pi_x(f_\sigma)| = |a_\sigma(x)| \quad \text{for} \quad \nu \leqq \sigma \quad \text{and}$$
$$|a_\nu(x)| \leqq |\pi_x(f_\nu)| < |\pi_x(f_\sigma)| = |a_\sigma(x)| \quad \text{for} \quad \nu > \sigma.$$

Hence, in particular, $a_\sigma(x) \neq 0$. Since f is ζ_n-distinguished of some degree $\sigma \leqq s$ at each point $x \in \operatorname{Sp} A$, we see that $a_0, \ldots, a_s \in A$ have no common zero in Sp A. Therefore,

$$U := \{x \in \operatorname{Sp} A\,; |a_\nu(x)| \leqq |a_s(x)|, \quad \nu = 0, \ldots, s\}$$

is a rational subdomain in Sp A. Furthermore, it follows from the above estimates that f is ζ_n-distinguished of degree s at a point $x \in \operatorname{Sp} A$ if and only if $x \in U$. □

Next we prove a relative version of Theorem 5.2.3/4.

Proposition 8. *Let $f \in A\langle\zeta\rangle$ be ζ_n-distinguished of order s at all points $x \in$* Sp A. *Then the canonical map*

$$\iota\colon A\langle\zeta_1, \ldots, \zeta_{n-1}\rangle \to A\langle\zeta\rangle/(f)$$

is finite.

Proof. Write $f = \sum f_\nu \zeta_n^\nu$ with elements $f_\nu \in A\langle\zeta_1, \ldots, \zeta_{n-1}\rangle$. Then $\pi_x(f_s)$ is a unit and has no zero for all $x \in \operatorname{Sp} A$. As a result, f_s has no zeros and hence is a unit in $A\langle\zeta_1, \ldots, \zeta_{n-1}\rangle$. Replacing f by $f_s^{-1}f$, we may assume that $f_s = 1$, and furthermore

$$|f_\nu|_{\sup} \leqq 1 \quad \text{for} \quad \nu \leqq s \quad \text{and}$$
$$|f_\nu|_{\sup} < 1 \quad \text{for} \quad \nu > s.$$

Therefore, denoting by b the residue class of ζ_n in $A\langle\zeta\rangle/(f)$, we get

$$|b^s + \iota(f_{s-1})\, b^{s-1} + \cdots + \iota(f_0)|_{\sup} < 1.$$

Since b is an affinoid generating system of $A\langle\zeta\rangle/(f)$ over $A\langle\zeta_1, \ldots, \zeta_{n-1}\rangle$, the finiteness of ι follows from Theorem 6.3.2/2. □

In order to apply the above proposition, we must know how to generate ζ_n-distinguished elements. Therefore, we generalize the results of (5.2.4) to the relative case.

Proposition 9. *Let $f = \sum a_\nu \zeta^\nu$ be a series in $A\langle\zeta\rangle$ such that the coefficients $a_\nu \in A$ have no common zero. Then there is an A-algebra automorphism $\tau: A\langle\zeta\rangle \to A\langle\zeta\rangle$ such that, for some $s \in \mathbb{N}$, the series $\tau(f)$ is ζ_n-distinguished of degree $\leqq s$ on* Sp A.

Proof. For $x \in \operatorname{Sp} A$, denote by t_x the least upper bound of all natural numbers which occur in multi-indices ν satisfying $|a_\nu(x)| = |\pi_x(f)|$. We want to show that

$$t := \sup_{x \in \operatorname{Sp} A} t_x$$

is finite. As the coefficients a_ν generate the unit ideal, there must exist finitely many indices $\nu(1), \ldots, \nu(r)$ such that $a_{\nu(1)}, \ldots, a_{\nu(r)}$ generate the unit ideal in A. Hence there is an $\alpha > 0$ such that

$$\max_{1 \leqq i \leqq r} |a_{\nu(i)}(x)| \geqq \alpha$$

for all $x \in \operatorname{Sp} A$. Since $|a_\nu|_{\sup} < \alpha$ for almost all indices ν, we see that the t_x are bounded and hence that t is finite as claimed.

Now define natural numbers $c_1, \ldots, c_n \in \mathbb{N}$ inductively by

$$c_n := 1$$

$$c_{n-j} := 1 + t \sum_{d=0}^{j-1} c_{n-d}, \qquad j = 1, \ldots, n-1,$$

and consider the A-algebra automorphism

$$\tau: A\langle\zeta\rangle \to A\langle\zeta\rangle,$$
$$\zeta_i \mapsto \zeta_i + \zeta_n^{c_i}, \qquad i = 1, \ldots, n-1,$$
$$\zeta_n \mapsto \zeta_n.$$

It is clear that, for any $x \in \operatorname{Sp} A$, the map τ induces an $A/\mathfrak{m}_x$-algebra automorphism

$$\tau_x: A/\mathfrak{m}_x\langle\zeta\rangle \to A/\mathfrak{m}_x\langle\zeta\rangle.$$

Since $t \geqq t_x$, it follows from Proposition 5.2.4/2 that

$$\tau_x\big(\pi_x(f)\big) = \pi_x\big(\tau(f)\big)$$

is ζ_n-distinguished of some degree s_x satisfying

$$s_x \leqq t_x \sum_{i=1}^{n} c_i \leqq t \sum_{i=1}^{n} c_i.$$

Thus, $\tau(f)$ is ζ_n-distinguished of degree $\leqq t \sum_{i=1}^{n} c_i$ on Sp A. □

Lemma 10. *Let $\lambda\colon A \to B$ be a homomorphism of affinoid algebras and let $\Lambda\colon A\langle\zeta\rangle \to B$ be an epimorphism extending λ. Assume that for all $x \in \operatorname{Sp} A$ the induced epimorphisms $\Lambda_x\colon A/\mathfrak{m}_x\langle\zeta\rangle \to B/\mathfrak{m}_x B$ have non-trivial kernel. Then* $\ker \Lambda$ *contains a series $f = \sum a_\nu \zeta^\nu$ with coefficients $a_\nu \in A$ having no common zero on* $\operatorname{Sp} A$.

Proof. Given any $x \in \operatorname{Sp} A$, there exists a series $g \in \ker \Lambda$ such that $\pi_x(g) \neq 0$. Namely, since Λ_x is not injective, we can find a series $g_1 \in A\langle\zeta\rangle$ satisfying $0 \neq \pi_x(g_1) \in \ker \Lambda_x$. Then $\Lambda(g_1) \in \mathfrak{m}_x B$ and we can find an inverse image g_2 of $\Lambda(g_1)$ in $\mathfrak{m}_x A\langle\zeta\rangle$; hence, $g := g_1 - g_2$ is a series having the desired properties.

Now, for an arbitrary element $f_1 = \sum a_\nu^{(1)} \zeta^\nu \in \ker \Lambda$, denote by $X_1 \subset \operatorname{Sp} A$ the Zariski-closed subset defined by the coefficients $a_\nu^{(1)}$ of f_1. Then X_1 is defined by finitely many of the $a_\nu^{(1)}$, say by all coefficients $a_\nu^{(1)}$ having a multi-index ν with components $\nu_i < d$. If X_1 is not empty, we can find a $g \in \ker \Lambda$ such that $\pi_x(g) \neq 0$, for some $x \in X_1$. Therefore, the Zariski-closed subset $X_2 \subset \operatorname{Sp} A$ defined by the coefficients of

$$f_2 := f_1 + \zeta_1^d \ldots \zeta_n^d g$$

is strictly contained in X_1. If $X_2 \neq \emptyset$, we can repeat the above procedure for f_2. Continuing in this way, we arrive at a function $f_r \in \ker \Lambda$ whose corresponding Zariski-closed subset $X_r \subset \operatorname{Sp} A$ is empty, since any decreasing sequence of Zariski-closed subsets in $\operatorname{Sp} A$ becomes stationary due to the fact that A is Noetherian. Then $f := f_r$ satisfies the assertion of the lemma. □

Remark 11. *In the situation of Lemma* 10, *suppose that the map $\lambda\colon A \to B$ gives rise to an injective map $\varphi\colon \operatorname{Sp} B \to \operatorname{Sp} A$. Then, assuming $A \neq 0$, all epimorphisms $\Lambda_x\colon A/\mathfrak{m}_x\langle\zeta\rangle \to B/\mathfrak{m}_x B$ have a non-trivial kernel.*

Proof. If $\varphi\colon \operatorname{Sp} B \to \operatorname{Sp} A$ is injective, then, for any $x \in \operatorname{Sp} A$, the ideal $\mathfrak{m}_x B$ is contained in at most one maximal ideal of B. Thus, $B/\mathfrak{m}_x B$ is a local ring if it does not vanish. Therefore, Λ_x, which is surjective by our assumption, cannot be injective, since $A/\mathfrak{m}_x\langle\zeta\rangle$ is not local. □

For any homomorphism $\lambda\colon A \to B$ of affinoid algebras, we denote by $\langle B : A\rangle$ the minimal number n such that there exists an affinoid generating system of B over A consisting of n elements, i.e., such that there exists an epimorphism $\Lambda\colon A\langle\zeta_1, \ldots, \zeta_n\rangle \to B$ extending λ. In particular, $\langle B : A\rangle = 0$ characterizes the case where λ is surjective. The number $\langle B : A\rangle$ can be interpreted as the least integer n such that the map $\operatorname{Sp} B \to \operatorname{Sp} A$ induced by λ extends to a closed immersion $\operatorname{Sp} B \hookrightarrow \mathbb{B}^n \times \operatorname{Sp} A$, i.e., as the embedding dimension of $\operatorname{Sp} B$ over $\operatorname{Sp} A$. We now formulate the main lemma to be used in the proof of Theorem 1.

Lemma 12. *Let* $\operatorname{Sp} B \to \operatorname{Sp} A$ *be an injective morphism of affinoid varieties, corresponding to the algebra homomorphism $\lambda\colon A \to B$. Assume that $n := \langle B : A\rangle \geq 1$. Then there exists an epimorphism $\Lambda\colon A\langle\zeta_1, \ldots, \zeta_n\rangle \to B$ extending λ such* *that* $\ker \Lambda$ *contains an element* f *which, for some* $s \geqq 0$, *is* ζ_n*-distinguished of degree* $\leqq s$ *on* $\operatorname{Sp} A$.

Proof. First, it follows from $\langle B : A \rangle \geqq 1$ that $B \neq 0 \neq A$. Choosing an epimorphism $\Lambda\colon A\langle \zeta_1, \ldots, \zeta_n \rangle \to B$ extending λ, we see, by Lemma 10 and Remark 11, that $\ker \Lambda$ contains an element f such that $\pi_x(f) \neq 0$ for all $x \in X$. Then, by Proposition 9, we may apply an A-algebra automorphism to $A\langle \zeta_1, \ldots, \zeta_n \rangle$ which carries f into a ζ_n-distinguished series. □

After these preparations, we are ready to *prove* Theorem 1. Set $X = \operatorname{Sp} A$, $Y = \operatorname{Sp} B$, and consider the homomorphism $\lambda = \varphi^*\colon A \to B$ corresponding to $\varphi\colon Y \to X$. We use induction on $n := \langle B : A \rangle$. The case $n = 0$ is trivial since φ is a closed immersion and hence, in particular, a Runge immersion.

If $n \geqq 1$, we can apply Lemma 12. Hence there exists an epimorphism $\Lambda\colon A\langle \zeta_1, \ldots, \zeta_n \rangle \to B$ extending λ such that $\ker \Lambda$ contains an element f which, for some $s \geqq 0$, is ζ_n-distinguished of degree $\leqq s$ on $\operatorname{Sp} A$. Denote by $X' = \operatorname{Sp} A' \subset X$ the rational subdomain of all points $x \in X$, where f is ζ_n-distinguished of degree s (Lemma 7). We will now show that we can apply the induction hypothesis to the restricted map $\varphi'\colon \varphi^{-1}(X') \to X'$, which corresponds to the homomorphism

$$\lambda'\colon A' \to B \mathbin{\hat{\otimes}}_A A'$$

obtained from λ by tensoring with A' over A. Tensoring Λ in the same way, we get a homomorphism

$$\Lambda'\colon A'\langle \zeta_1, \ldots, \zeta_n \rangle \to B \mathbin{\hat{\otimes}}_A A'$$

which extends λ' and which is surjective by Corollary 7.2.2/6. If f' denotes the image of f in $A'\langle \zeta_1, \ldots, \zeta_n \rangle$, then $f' \in \ker \Lambda'$ and, due to Proposition 6, it is ζ_n-distinguished of degree s at all points $x \in \operatorname{Sp} A'$. Therefore, by Proposition 8, the epimorphism Λ' induces a finite homomorphism

$$\Lambda''\colon A'\langle \zeta_1, \ldots, \zeta_{n-1} \rangle \to B \mathbin{\hat{\otimes}}_A A'$$

extending λ'. Since λ' corresponds to a locally closed immersion of affinoid varieties, the same is true for Λ''. But then Λ'' must be surjective by Proposition 7.3.3/8, and hence $\langle B \mathbin{\hat{\otimes}}_A A' : A' \rangle \leqq n - 1$. Therefore, we can apply the induction hypothesis to $\varphi'\colon \varphi^{-1}(X') \to X'$. Hence there exists a covering $X' = \bigcup_{i=1}^{r} X^i$ consisting of rational subdomains $X^i \subset X'$ such that the induced maps $\varphi^{-1}(X^i) \to X^i$ are Runge immersions.

The X^i are also rational in X by Theorem 7.2.4/2. Fixing functions in A which describe the X^i we define rational domains X^i_ε for $\varepsilon \in \sqrt{|k^*|}$ as in (7.3.4) such that $X^i = X^i_1$. Then, by the Extension Lemma 7.3.4/10, there exists an $\varepsilon > 1$ such that the maps $\varphi^{-1}(X^i_\varepsilon) \to X^i_\varepsilon$ are still Runge immersions. Using

Lemma 4, we can find rational subdomains $V^1, \ldots, V^l \subset X$ which don't meet $X' = \bigcup_{i=1}^{r} X^i$ and such that

$$X = \bigcup_{i=1}^{r} X_\varepsilon^i \cup \bigcup_{j=1}^{l} V^j .$$

Let $V^j = \operatorname{Sp} A_j$. To conclude the proof it is enough to prove the assertion of Theorem 1 for the locally closed immersions $\varphi^{-1}(V^j) \to V^j$ which correspond to the homomorphisms $\lambda_j \colon A_j \to B \mathbin{\widehat{\otimes}}_A A_j$ induced by λ. Since Λ induces epimorphisms (cf. Corollary 7.2.2/6)

$$\Lambda_j \colon A_j\langle \zeta_1, \ldots, \zeta_n\rangle \to B \mathbin{\widehat{\otimes}}_A A_j ,$$

we see that $\langle B \mathbin{\widehat{\otimes}}_A A_j : A_j\rangle \leq n$ for all j. Furthermore, since none of the V^j intersects X', it follows from Proposition 6 that the image f_j of f in $A_j\langle \zeta_1, \ldots, \zeta_n\rangle$ is ζ_n-distinguished of degree $\leq s - 1$ on $V^j = \operatorname{Sp} A_j$. Because $f_j \in \ker \Lambda_j$, we can proceed with the maps $\varphi^{-1}(V^j) \to V^j$ as we did before with φ. Using an induction argument on s, we may assume that the f_j are ζ_n-distinguished of degree 0 on V^j. But then all f_j are units by Proposition 5, showing that $B \mathbin{\widehat{\otimes}}_A A_j = 0$ and $\varphi^{-1}(V^j) = \emptyset$. Thus, in this case, the maps $\varphi^{-1}(V^j) \to V^j$ are Runge immersions for trivial reasons. □

CHAPTER 8

Čech cohomology of affinoid varieties

This chapter is devoted to the presentation of TATE's Acyclicity Theorem which, for any affinoid variety X, gives precise information about its ČECH cohomology with values in the presheaf $\mathcal{O}_X$. A consequence of the theorem is that $\mathcal{O}_X$ (which, in general, is not a sheaf) satisfies sheaf properties when we consider finite coverings of X by affinoid subdomains. The definition of global varieties will be based on this fact.

The proof of TATE's Acyclicity Theorem requires a thorough knowledge of ČECH cohomology (as contained, for example in [36]). Therefore, we have included a detailed discussion of the necessary facts, including LERAY type comparison theorems for ČECH cohomology. These comparison theorems are used in order to reduce the general case of the Acyclicity Theorem to a special case, which can be attacked by explicit computation.

8.1. Čech cohomology with values in a presheaf

8.1.1. Cohomology of complexes. — Let R be a commutative ring. A *complex* $K^{\cdot}$ of R-modules consists of a collection of R-modules K^q, $q \in \mathbb{Z}$, and of R-module homomorphisms $d^q\colon K^q \to K^{q+1}$, called *coboundary maps*, such that $d^{q+1} \circ d^q = 0$ for all $q \in \mathbb{Z}$. Thus, $K^{\cdot}$ can be viewed as a sequence of R-modules

$$\cdots \to K^{q-1} \xrightarrow{d^{q-1}} K^q \xrightarrow{d^q} K^{q+1} \to \cdots$$

satisfying $\operatorname{im} d^{q-1} \subset \ker d^q$ for all q. The quotient

$$H^q(K^{\cdot}) := \ker d^q / \operatorname{im} d^{q-1}$$

is called the *q-th cohomology module* of the complex $K^{\cdot}$. Its size is a measure of how far the above sequence is from being exact at K^q.

A homomorphism $f\colon K^{\cdot} \to K'^{\cdot}$ between two complexes $K^{\cdot}$ and $K'^{\cdot}$ is a collection of R-module homomorphisms $f^q\colon K^q \to K'^q$, $q \in \mathbb{Z}$, commuting with coboundary maps. This means that $f^{q+1} \circ d^q = d'^q \circ f^q$ for all $q \in \mathbb{Z}$, where d^q, d'^q denote the coboundary maps of $K^{\cdot}$ and $K'^{\cdot}$, respectively. Therefore, for any homomorphism of complexes f, we have

$$f^q(\ker d^q) \subset \ker d'^q \qquad \text{and} \qquad f^q(\operatorname{im} d^{q-1}) \subset \operatorname{im} d'^{q-1}.$$

Consequently, f induces R-module homomorphisms

$$H^q(f)\colon H^q(K^\cdot) \to H^q(K'^\cdot), \qquad q \in \mathbb{Z},$$

and it is seen that H^q is a functor on the category of complexes of R-modules.

Now consider two particular homomorphisms of complexes $f, g\colon K^\cdot \to K'^\cdot$. If one wants to show that $H^q(f) = H^q(g)$ for all q, one can try to construct a *homotopy* h between f and g. This is a collection of R-module homomorphisms $h^q\colon K^q \to K'^{q-1}$ satisfying

$$f^q - g^q = d'^{q-1} \circ h^q + h^{q+1} \circ d^q$$

for all $q \in \mathbb{Z}$. Then $f^q - g^q$ must map $\ker d^q$ into $\operatorname{im} d'^{q-1}$. So it is obvious that $H^q(f) = H^q(g)$ if there exists such a homotopy h.

In order to compute cohomology modules, it is sometimes useful to apply some general techniques involving exact sequences. Consider a short exact sequence of complexes

$$0 \to K'^\cdot \xrightarrow{f} K^\cdot \xrightarrow{g} K''^\cdot \to 0.$$

Here 0 stands for the complex consisting only of zero modules, and the sequence being exact means that all sequences

$$0 \to K'^q \xrightarrow{f^q} K^q \xrightarrow{g^q} K''^q \to 0$$

are assumed to be exact. Such a short exact sequence gives rise to a so-called *long cohomology sequence* which is exact:

$$\cdots \to H^q(K'^\cdot) \xrightarrow{H^q(f)} H^q(K^\cdot) \xrightarrow{H^q(g)} H^q(K''^\cdot) \xrightarrow{\delta^q} H^{q+1}(K'^\cdot) \to \cdots$$

The exactness follows from a straightforward, but tedious, verification. Therefore we give only some hints on how to define the maps δ^q. Denote by d'^q, d^q, and d''^q the coboundary homomorphisms of $K'^\cdot$, $K^\cdot$, and $K''^\cdot$, respectively. Start with an element $a \in H^q(K''^\cdot)$. If $x \in \ker d''^q \subset K''^q$ is an element representing a, we can find an inverse image $y \in K^q$ of x with respect to the surjection $g^q\colon K^q \to K''^q$. As $d^q(y) \in \ker g^{q+1} = \operatorname{im} f^{q+1}$, there exists an element $z \in K'^{q+1}$ which is mapped onto $d^q(y)$ via the injection $f^{q+1}\colon K'^{q+1} \to K^{q+1}$. Then $d'^{q+1}(z) = 0$, since $d^q(y) \in \ker d^{q+1}$; hence z represents an element $b \in H^{q+1}(K'^\cdot)$, and we can define δ^q by setting $\delta^q(a) := b$. It is easily checked that δ^q is well-defined, since the element $z \in \ker d'^{q+1} \subset K'^{q+1}$ constructed above is uniquely determined by a, up to an element in $\operatorname{im} d'^q$.

It is also easily shown that the long cohomology sequence depends naturally on the corresponding exact sequence; i.e., any commutative diagram of complexes

$$\begin{array}{ccccccccc} 0 & \to & K'^\cdot & \to & K^\cdot & \to & K''^\cdot & \to & 0 \\ & & \downarrow & & \downarrow & & \downarrow & & \\ 0 & \to & L'^\cdot & \to & L^\cdot & \to & L''^\cdot & \to & 0 \end{array}$$

with exact rows gives rise to a commutative diagram

$$\begin{array}{ccccccccc} \cdots \to & H^q(K'^{\cdot}) & \to & H^q(K^{\cdot}) & \to & H^q(K''^{\cdot}) & \to & H^{q+1}(K'^{\cdot}) & \to \cdots \\ & \downarrow & & \downarrow & & \downarrow & & \downarrow & \\ \cdots \to & H^q(L'^{\cdot}) & \to & H^q(L^{\cdot}) & \to & H^q(L''^{\cdot}) & \to & H^{q+1}(L'^{\cdot}) & \to \cdots . \end{array}$$

Very often, short exact sequences arise in the following way. Let $K^{\cdot}$, $K'^{\cdot}$ be complexes. Assume that $K'^{\cdot}$ is a subcomplex of $K^{\cdot}$, i.e., that K'^q is a submodule of K^q, for all q, and that the injections $K'^q \to K^q$ define a homomorphism of complexes $K'^{\cdot} \to K^{\cdot}$. In this situation the quotient complex $K^{\cdot}/K'^{\cdot}$ can be considered. It consists of the modules K^q/K'^q and has coboundary homomorphisms which are induced by the coboundary maps of $K^{\cdot}$. Then the exact sequences $0 \to K'^q \to K^q \to K^q/K'^q \to 0$ are compatible with all coboundary maps, and thus give rise to a short exact sequence

$$0 \to K'^{\cdot} \to K^{\cdot} \to K^{\cdot}/K'^{\cdot} \to 0.$$

8.1.2. Cohomology of double complexes. — As before, let R be a commutative ring. A *double complex* $K^{\cdot\cdot}$ of R-modules consists of a collection of R-modules $K^{p,q}$, $p, q \in \mathbb{Z}$, and of R-module homomorphisms

$$'d^{p,q}\colon K^{p,q} \to K^{p+1,q} \qquad \text{and}$$
$$''d^{p,q}\colon K^{p,q} \to K^{p,q+1},$$

satisfying

$$'d^{p+1,q} \circ {'d^{p,q}} = 0,$$
$$''d^{p,q+1} \circ {''d^{p,q}} = 0, \qquad \text{and}$$
$$''d^{p+1,q} \circ {'d^{p,q}} + {'d^{p,q+1}} \circ {''d^{p,q}} = 0$$

for all $p, q \in \mathbb{Z}$. Thus, we may interpret $K^{\cdot\cdot}$ as a diagram

$$\begin{array}{ccccc} & \vdots & & \vdots & \\ & \downarrow & & \downarrow & \\ \cdots \to & K^{p,q} & \xrightarrow{''d^{p,q}} & K^{p,q+1} & \to \cdots \\ & \downarrow {'d^{p,q}} & & \downarrow {'d^{p,q+1}} & \\ \cdots \to & K^{p+1,q} & \xrightarrow{''d^{p+1,q}} & K^{p+1,q+1} & \to \cdots \\ & \downarrow & & \downarrow & \\ & \vdots & & \vdots & \end{array}$$

with coboundary homomorphisms $'d^{p,q}$ (for q fixed) as well as $''d^{p,q}$ (for p fixed) such that all squares are anticommutative.

It follows, in particular, that, for a fixed $q \in \mathbb{Z}$, the modules $K^{p,q}$ with coboundary homomorphisms $'d^{p,q}$ constitute a single complex as defined in (8.1.1). This complex is called the q-th column of $K^{\cdot\cdot}$, and it will be denoted by $'K^{\cdot q}$. Analogously, one defines the p-th row $''K^{p\cdot}$ of $K^{\cdot\cdot}$ by using the maps $''d^{p,q}$ as coboundary maps.

There is a third way of deriving a single complex from $K^{\cdot\cdot}$. For $r \in \mathbb{Z}$, set $K^r := \bigoplus_{p+q=r} K^{p,q}$ and define homomorphisms $d^r: K^r \to K^{r+1}$ by $d^r|K^{p,q} := {'d^{p,q}} + {''d^{p,q}}$ for all pairs (p, q) such that $p + q = r$. It is easily seen that $d^{r+1} \circ d^r = 0$ for all r; hence the modules K^r and the maps d^r constitute a complex $K^\cdot$ which is called the single complex associated to $K^{\cdot\cdot}$. The cohomology modules of $K^\cdot$ are also referred to as the cohomology modules of $K^{\cdot\cdot}$, i.e., we set

$$H^r(K^{\cdot\cdot}) := H^r(K^\cdot), \qquad r \in \mathbb{Z}.$$

In the following we are only interested in double complexes $K^{\cdot\cdot}$ which vanish at negative integers. Thus, *we always assume that* $K^{p,q} = 0$ *for* $p < 0$ *or* $q < 0$. Without referring to the machinery of spectral sequences, we want to prove some lemmata about the cohomology of such double complexes.

Lemma 1. *Let $K^{\cdot\cdot}$ be a double complex and consider the homomorphism of complexes ${'\pi}: K^\cdot \to {'K^{\cdot 0}}$ induced by the natural projections $K^r = \bigoplus_{p+q=r} K^{p,q} \to K^{r,0}$. Then, if $H^p({'K^{\cdot q}}) = 0$ for all $p \geqq 0$ and for all $q > 0$, the maps $H^r({'\pi}): H^r(K^\cdot) \to H^r({'K^{\cdot 0}})$ are bijective for all r.*

Proof. For $i \geqq 0$, we consider the subcomplex $K_i^\cdot$ of $K^\cdot$ which is defined by

$$K_i^r := \bigoplus_{\substack{p+q=r \\ q \geqq i}} K^{p,q}.$$

Note that $K_0^\cdot = K^\cdot$. Furthermore, there are natural isomorphisms $K_i^r/K_{i+1}^r \xrightarrow{\sim} K^{r-i,i}$ which constitute an isomorphism of complexes $K_i^\cdot/K_{i+1}^\cdot \xrightarrow{\sim} {'K^{\cdot i}}$ of degree $-i$ (i.e., one obtains an isomorphism in the usual sense if the indices of all modules in ${'K^{\cdot i}}$ are enlarged by i). Thus we have $H^r(K_i^\cdot/K_{i+1}^\cdot) = H^{r-i}({'K^{\cdot i}}) = 0$ for $i > 0$ and for all r by our assumption. Looking at the long cohomology sequence corresponding to

$$0 \to K_i^\cdot/K_{i+1}^\cdot \to K^\cdot/K_{i+1}^\cdot \to K^\cdot/K_i^\cdot \to 0,$$

we get bijections $H^r(K^\cdot/K_{i+1}^\cdot) \xrightarrow{\sim} H^r(K^\cdot/K_i^\cdot)$ for $i > 0$, and, by an induction argument also bijections

$$H^r(K^\cdot/K_i^\cdot) \xrightarrow{\sim} H^r(K^\cdot/K_1^\cdot) = H^r({'K^{\cdot 0}})$$

for $i > 0$. Since $H^r(K^\cdot)$ is canonically isomorphic to $H^r(K^\cdot/K_i^\cdot)$ for $i \geqq r + 2$, the assertion of the lemma follows. □

We use Lemma 1 in order to derive a key lemma which will be needed in (8.1.4) for the proof of a comparison theorem for Čech cohomology.

Lemma 2. *Let $K^{\cdot\cdot}$ be a double complex, $K^\cdot$ the associated single complex. Denote by $K''^\cdot$ the subcomplex of $K^\cdot$ defined by*

$$K''^q := \ker {'d^{0,q}}.$$

Then, if $H^p('K^{\cdot q}) = 0$ *for all* $p > 0$ *and for all* q, *the inclusion* $K''^{\cdot} \hookrightarrow K^{\cdot}$ *induces bijections*

$$H^r(K''^{\cdot}) \xrightarrow{\sim} H^r(K^{\cdot}),$$

for all r.

Proof. Considering the long cohomology sequence corresponding to the short exact sequence

$$0 \to K''^{\cdot} \to K^{\cdot} \to K^{\cdot}/K''^{\cdot} \to 0,$$

we have only to show that $H^r(K^{\cdot}/K''^{\cdot}) = 0$ for all r. For this purpose we introduce the double complex $L^{\cdot\cdot}$ defined by

$$L^{p,q} := \begin{cases} K^{p,q} & \text{if} \quad p \neq 0 \\ K^{0,q}/K''^q & \text{if} \quad p = 0 \end{cases}$$

with coboundary maps being induced by $K^{\cdot\cdot}$. Then, by our construction,

$$H^0('L^{\cdot q}) = 0$$

for all q. From the definition of $L^{\cdot\cdot}$ and the assumption on $H^p('K^{\cdot q})$, we have

$$H^p('L^{\cdot q}) = H^p('K^{\cdot q}) = 0$$

for all $p > 0$ and for all q. Since $K^{\cdot}/K''^{\cdot}$ is the single complex associated to $L^{\cdot\cdot}$, we get from Lemma 1

$$H^r(K^{\cdot}/K''^{\cdot}) = H^r('L^{\cdot 0}) = 0$$

for all r. □

8.1.3. Čech cohomology. — Let X be a set together with a system $\mathfrak{T}$ of subsets in X such that $U, V \in \mathfrak{T}$ implies $U \cap V \in \mathfrak{T}$. The elements of $\mathfrak{T}$ are referred to as the *open subsets* of X, analogous to the case where X is a topological space and $\mathfrak{T}$ is the system of all open subsets in X. Interpreting $\mathfrak{T}$ as a category with the morphisms being canonical inclusions of open sets, we call any contravariant functor from $\mathfrak{T}$ into the category of groups, rings, etc., a presheaf on X. If $\mathcal{F}$ is such a presheaf on X and if $U \subset V$ are open sets in X, then the corresponding homomorphism $\mathcal{F}(V) \to \mathcal{F}(U)$, called the restriction homomorphism, is denoted by $f \mapsto f|_U$. In the following we will define Čech cohomology attached to an open covering $\mathfrak{U}$ of X with values in a presheaf $\mathcal{F}$ on X. For our applications we are mainly interested in the case where X is an affinoid variety, $\mathfrak{T}$ is the system of affinoid subdomains of X, and $\mathcal{F}$ is the presheaf $\mathcal{O}_X$ defined in (7.3.2).

Now let us assume that X admits open coverings and let us consider such a covering $\mathfrak{U} = \{U_i\}_{i \in I}$, i.e., a covering $X = \bigcup_{i \in I} U_i$ with open subsets $U_i \in \mathfrak{T}$. Let $\mathcal{F}$ be a presheaf of abelian groups or rings on X. Setting

$$U_{i_0 \ldots i_q} := U_{i_0} \cap \cdots \cap U_{i_q}$$

for any indices $i_0, \ldots, i_q \in I$, we define, for $q \geqq 0$, the $\mathbb{Z}$-module of *q-cochains on* $\mathfrak{U}$ *with values in* $\mathscr{F}$ by

$$C^q(\mathfrak{U}, \mathscr{F}) := \prod_{(i_0, \ldots, i_q) \in I^{q+1}} \mathscr{F}(U_{i_0 \ldots i_q}).$$

Note that $C^q(\mathfrak{U}, \mathscr{F})$ is also an $\mathscr{F}(X)$-module in the case where X is an element of $\mathfrak{T}$ and $\mathscr{F}$ is a presheaf of rings. For any q-cochain $f \in C^q(\mathfrak{U}, \mathscr{F})$, we denote by $f_{i_0 \ldots i_q}$ its $(i_0, \ldots, i_q)$-component. We call f an *alternating q-cochain* if

$$f_{i_{\pi(0)} \ldots i_{\pi(q)}} = (\operatorname{sgn} \pi) f_{i_0 \ldots i_q}$$

for all permutations π of $\{0, \ldots, q\}$ and if, furthermore, $f_{i_0 \ldots i_q} = 0$ whenever the indices $i_0, \ldots, i_q$ are not pairwise distinct. The alternating q-cochains form a submodule $C_{\mathrm{a}}^q(\mathfrak{U}, \mathscr{F})$ of $C^q(\mathfrak{U}, \mathscr{F})$.

Setting $C^q(\mathfrak{U}, \mathscr{F}) = 0$ for $q < 0$, the modules $C^q(\mathfrak{U}, \mathscr{F})$ constitute a complex $C^{\cdot}(\mathfrak{U}, \mathscr{F})$ when we define coboundary homomorphisms $d^q: C^q(\mathfrak{U}, \mathscr{F}) \to C^{q+1}(\mathfrak{U}, \mathscr{F})$ by $d^q = 0$ for $q < 0$ and by

$$d^q(f)_{i_0 \ldots i_{q+1}} := \sum_{j=0}^{q+1} (-1)^j f_{i_0 \ldots \hat{\imath}_j \ldots i_{q+1}}|U_{i_0 \ldots i_{q+1}}$$

for $q \geqq 0$, where the notation $\hat{\imath}_j$ means omit i_j. As is easily checked, $d^{q+1} \circ d^q = 0$; thus, we get, in fact, a complex $C^{\cdot}(\mathfrak{U}, \mathscr{F})$, called the *Čech complex of cochains on* $\mathfrak{U}$ *with values in* $\mathscr{F}$. Its cohomology modules are denoted by

$$H^q(\mathfrak{U}, \mathscr{F}) := H^q\big(C^{\cdot}(\mathfrak{U}, \mathscr{F})\big).$$

As d^q maps alternating cochains into alternating cochains, the modules $C_{\mathrm{a}}^q(\mathfrak{U}, \mathscr{F})$ constitute a subcomplex $C_{\mathrm{a}}^{\cdot}(\mathfrak{U}, \mathscr{F})$ of $C^{\cdot}(\mathfrak{U}, \mathscr{F})$, called the *Čech complex of alternating cochains on* $\mathfrak{U}$ *with values in* $\mathscr{F}$. The corresponding cohomology modules are denoted by

$$H_{\mathrm{a}}^q(\mathfrak{U}, \mathscr{F}) := H^q\big(C_{\mathrm{a}}^{\cdot}(\mathfrak{U}, \mathscr{F})\big).$$

There is no essential difference between the complexes $C_{\mathrm{a}}^{\cdot}(\mathfrak{U}, \mathscr{F})$ and $C^{\cdot}(\mathfrak{U}, \mathscr{F})$, since, as we will see below, both yield the same cohomology. Thus, it is basically a matter of taste whether to work with all cochains or merely with alternating ones. However, it can be said that the complex $C^{\cdot}(\mathfrak{U}, \mathscr{F})$ is usually easier to handle in general considerations, whereas $C_{\mathrm{a}}^{\cdot}(\mathfrak{U}, \mathscr{F})$ is better adapted to explicit calculations. As an example, if $\mathfrak{U}$ is a finite covering consisting of n covering sets, we see immediately that $H_{\mathrm{a}}^q(\mathfrak{U}, \mathscr{F}) = 0$ for all $q \geqq n$, since $C_{\mathrm{a}}^q(\mathfrak{U}, \mathscr{F})$ vanishes for $q \geqq n$. The corresponding fact for the cohomology of $C^{\cdot}(\mathfrak{U}, \mathscr{F})$ is not obvious.

Proposition 1. *The injection* $\iota: C_{\mathrm{a}}^{\cdot}(\mathfrak{U}, \mathscr{F}) \to C^{\cdot}(\mathfrak{U}, \mathscr{F})$ *induces bijections* $H^q(\iota): H_{\mathrm{a}}^q(\mathfrak{U}, \mathscr{F}) \xrightarrow{\sim} H^q(\mathfrak{U}, \mathscr{F})$, *for all* q.

Proof. We will construct a homomorphism of complexes $p: C^{\cdot}(\mathfrak{U}, \mathscr{F}) \to C_{\mathrm{a}}^{\cdot}(\mathfrak{U}, \mathscr{F})$ such that $p \circ \iota$ is the identity on $C_{\mathrm{a}}^{\cdot}(\mathfrak{U}, \mathscr{F})$ and $\iota \circ p$ is homotopic to the identity on $C^{\cdot}(\mathfrak{U}, \mathscr{F})$. This will imply the assertion.

Digressing for a moment, we consider the sequence

$$0 \xleftarrow{d_0} F_0 \xleftarrow{d_1} F_1 \xleftarrow{d_2} F_2 \xleftarrow{d_3} \cdots$$

where F_q is the free $\mathbb{Z}$-module generated by I^{q+1}, i.e.,

$$F_q := \bigoplus_{(i_0,\ldots,i_q)\in I^{q+1}} \mathbb{Z}\,(i_0, \ldots, i_q),$$

and where $d_q\colon F_q \to F_{q-1}$ (for $q > 0$) is the linear map given by

$$d_q(i_0, \ldots, i_q) = \sum_{j=0}^{q} (-1)^j\,(i_0, \ldots, \hat{i}_j, \ldots, i_q).$$

Since $d_q \circ d_{q+1} = 0$ for $q \geqq 0$, the modules F_q and maps d_q constitute a complex $F.$ of $\mathbb{Z}$-modules. (To be consistent with the definition given in (8.1.1), we would have to set $F_q = 0$, $d_q = 0$ for $q < 0$ and consider the complex $K^{\cdot}$ with $K^q := F_{-q}$, $d^q := d_{-q}$. However, for convenience, we will use the above notations.) Calling a homomorphism $l\colon F_r \to F_s$ between two modules of $F.$ *simplicial* if, for all tuples $(i_0, \ldots, i_r) \in I^{r+1}$,

$$l(i_0, \ldots, i_r) \in \sum_{j_0,\ldots,j_s\in\{i_0,\ldots,i_r\}} \mathbb{Z}\,(j_0, \ldots, j_s),$$

we see that all maps d_q are simplicial.

We now fix a total ordering on I and define simplicial homomorphisms $g_q\colon F_q \to F_q$ in the following way. Let $(i_0, \ldots, i_q) \in I^{q+1}$. If two of the indices $i_0, \ldots, i_q$ coincide, we set $g_q(i_0, \ldots, i_q) = 0$. If all $i_0, \ldots, i_q$ are distinct, there exists a unique permutation π of $\{0, \ldots, q\}$ such that $i_{\pi(0)} < i_{\pi(1)} < \cdots < i_{\pi(q)}$, and we set $g_q(i_0, \ldots, i_q) := (\operatorname{sgn} \pi)\,(i_{\pi(0)}, \ldots, i_{\pi(q)})$. It is easily seen that the g_q constitute a homomorphism of complexes $g\colon F. \to F.$. We want to show that there is a simplicial homotopy between g and the identity map id.

To achieve this, we construct by induction simplicial homomorphisms $h_q\colon F_q \to F_{q+1}$ such that

$$g_q - \mathrm{id}_q = h_{q-1} \circ d_q + d_{q+1} \circ h_q, \qquad q \geqq 0.$$

Since $g_0 - \mathrm{id}_0$ is the zero map, we can start with $h_{-1} = 0$ and $h_0 = 0$. If h_{q-1} is already constructed for some $q \geqq 1$, a standard calculation shows that

$$g_q - \mathrm{id}_q - h_{q-1} \circ d_q\colon F_q \to F_q$$

maps F_q into $\ker d_q$. Furthermore, this map is simplicial. Thus, for an arbitrary $(i_0, \ldots, i_q) \in I^{q+1}$, its image in $\ker d_q$ has a representation

$$c := (g_q - \mathrm{id}_q - h_{q-1} \circ d_q)\,(i_0, \ldots, i_q) = \sum_{j_0,\ldots,j_q\in\{i_0,\ldots,i_q\}} c_{j_0\ldots j_q}(j_0, \ldots, j_q).$$

Also,

$$c' := \sum_{j_0,\ldots,j_q\in\{i_0,\ldots,i_q\}} c_{j_0\ldots j_q}(i_0, j_0, \ldots, j_q)$$

is seen to be an inverse of c with respect to the map d_{q+1}. (The complex $F_{\cdot}$ is exact at all F_q, $q > 0$.) Defining $h_q\colon F_q \to F_{q+1}$ by associating to any $(i_0, \ldots, i_q) \in I^{q+1}$ the corresponding element c' constructed above, we see that h_q is a homomorphism satisfying the desired property.

Turning back to Čech complexes and the proof of the proposition, we show that any simplicial homomorphism $l\colon F_r \to F_s$ between two modules of $F_{\cdot}$ induces a homomorphism $l^*\colon C^s(\mathfrak{U}, \mathscr{F}) \to C^r(\mathfrak{U}, \mathscr{F})$. Namely, if l is determined by the equations

$$l(i_0, \ldots, i_r) = \sum_{j_0,\ldots,j_s \in \{i_0,\ldots,i_r\}} c^{i_0\ldots i_r}_{j_0\ldots j_s}(j_0, \ldots, j_s),$$

we set, for any $f \in C^s(\mathfrak{U}, \mathscr{F})$,

$$\big(l^*(f)\big)_{i_0\ldots i_r} := \sum_{j_0,\ldots,j_s \in \{i_0,\ldots,i_r\}} c^{i_0\ldots i_r}_{j_0\ldots j_s} f_{j_0\ldots j_s}\big|_{U_{i_0\ldots i_r}}.$$

The correspondence $l \rightsquigarrow l^*$ is additive and functorial (contravariant); in particular, d_q^* equals the coboundary homomorphism $d^{q-1}\colon C^{q-1}(\mathfrak{U}, \mathscr{F}) \to C^q(\mathfrak{U}, \mathscr{F})$. Thus, the maps g_q^*, which are derived from the maps g_q considered above, constitute a complex homomorphism $g^*\colon C^{\cdot}(\mathfrak{U}, \mathscr{F}) \to C^{\cdot}(\mathfrak{U}, \mathscr{F})$ which is homotopic to the identity via the maps $h_{q-1}^*\colon C^q(\mathfrak{U}, \mathscr{F}) \to C^{q-1}(\mathfrak{U}, \mathscr{F})$. Since g^* maps $C^{\cdot}(\mathfrak{U}, \mathscr{F})$ onto $C^{\cdot}_{\mathrm{a}}(\mathfrak{U}, \mathscr{F})$ and restricts to the identity on $C_{\mathrm{a}}(\mathfrak{U}, \mathscr{F})$, it induces a complex homomorphism $p\colon C^{\cdot}(\mathfrak{U}, \mathscr{F}) \to C^{\cdot}_{\mathrm{a}}(\mathfrak{U}, \mathscr{F})$ as required in the beginning of the proof. □

Corollary 2. *Let $\mathfrak{U}$ be a finite covering consisting of n covering sets. Then $H^q(\mathfrak{U}, \mathscr{F}) = H^q_{\mathrm{a}}(\mathfrak{U}, \mathscr{F}) = 0$ for all $q \geqq n$.*

Proof. $C^q_{\mathrm{a}}(\mathfrak{U}, \mathscr{F}) = 0$ for $q \geqq n$. □

Now let us consider two open coverings $\mathfrak{U} = \{U_i\}_{i\in I}$ and $\mathfrak{V} = \{V_j\}_{j\in J}$ of X and assume that $\mathfrak{V}$ is a refinement of $\mathfrak{U}$, i.e., that there exists a map $\tau\colon J \to I$ satisfying $V_j \subset U_{\tau(j)}$, for all $j \in J$. Any such map τ induces homomorphisms

$$\tau^q\colon C^q(\mathfrak{U}, \mathscr{F}) \to C^q(\mathfrak{V}, \mathscr{F})$$

where a cochain $f \in C^q(\mathfrak{U}, \mathscr{F})$ is mapped onto the cochain $\tau^q(f)$ with components

$$\big(\tau^q(f)\big)_{j_0\ldots j_q} := f_{\tau(j_0)\ldots\tau(j_q)}\big|_{V_{j_0\ldots j_q}}.$$

The maps τ^q constitute a homomorphism of complexes

$$\tau^{\cdot}\colon C^{\cdot}(\mathfrak{U}, \mathscr{F}) \to C^{\cdot}(\mathfrak{V}, \mathscr{F})$$

as is easily checked, and it is also clear that $\tau^{\cdot}$ maps alternating cochains into alternating cochains.

Although the map $\tau\colon J \to I$ is not uniquely determined by the coverings $\mathfrak{U}$ and $\mathfrak{V}$, we can show that the induced maps

$$H^q(\tau^{\cdot})\colon H^q(\mathfrak{U}, \mathscr{F}) \to H^q(\mathfrak{V}, \mathscr{F})$$

are independent of τ. Namely, let $\tau': J \to I$ be a second map satisfying $V_j \subset U_{\tau'(j)}$ for all $j \in J$. Then one verifies that the homomorphisms

$$h^q: C^q(\mathfrak{U}, \mathscr{F}) \to C^{q-1}(\mathfrak{V}, \mathscr{F})$$

given by

$$\big(h^q(f)\big)_{j_0 \ldots j_{q-1}} = \sum_{\nu=0}^{q-1} (-1)^\nu f_{\tau(j_0) \ldots \tau(j_\nu), \tau'(j_\nu) \ldots \tau'(j_{q-1})}|_{V_{j_0 \ldots j_{q-1}}}$$

define a homotopy between $\tau^{\cdot}$ and $\tau'^{\cdot}$; thus, the maps $H^q(\tau^{\cdot})$ and $H^q(\tau'^{\cdot})$ must coincide for all q. We use the notation $\varrho^q(\mathfrak{V}, \mathfrak{U})$ instead of $H^q(\tau^{\cdot})$ or $H^q(\tau'^{\cdot})$. Note that $\varrho^q(\mathfrak{U}, \mathfrak{U}) = \mathrm{id}$ and that $\varrho^q(\mathfrak{W}, \mathfrak{U}) = \varrho^q(\mathfrak{W}, \mathfrak{V}) \circ \varrho^q(\mathfrak{V}, \mathfrak{U})$ if $\mathfrak{W}$ is a refinement of $\mathfrak{V}$ and $\mathfrak{V}$ is a refinement of $\mathfrak{U}$. In particular, we have

Proposition 3. *Assume that the coverings $\mathfrak{U}$ and $\mathfrak{V}$ are refinements of each other. Then $\varrho^q(\mathfrak{V}, \mathfrak{U}): H^q(\mathfrak{U}, \mathscr{F}) \to H^q(\mathfrak{V}, \mathscr{F})$ is bijective and its inverse is $\varrho^q(\mathfrak{U}, \mathfrak{V})$ for all q.*

In order to get cohomology modules on X which do not depend on a certain covering, one can take direct limits of the modules $H^q(\mathfrak{U}, \mathscr{F})$ for $\mathfrak{U}$ varying over the set of all open coverings of X and with the homomorphisms $\varrho^q(\mathfrak{V}, \mathfrak{U})$ serving as connecting homomorphisms. The resulting modules

$$\check{H}^q(X, \mathscr{F}) := \varinjlim_{\mathfrak{U}} H^q(\mathfrak{U}, \mathscr{F})$$

are called *Čech cohomology modules of X with values in $\mathscr{F}$*. (The direct limit is defined in the usual way, although the relation "being finer" does not in general constitute a partial ordering on the set of all open coverings of X. Two such open coverings need not be equal if they are refinements of each other.) All the above constructions can be carried out in the same way for alternating cochains, thus leading to the *alternating Čech cohomology modules*

$$\check{H}^q_{\mathrm{a}}(X, \mathscr{F}) := \varinjlim_{\mathfrak{U}} H^q_{\mathrm{a}}(\mathfrak{U}, \mathscr{F}).$$

If X belongs to the system $\mathfrak{T}$ of open subsets of X, then, for any open covering $\mathfrak{U} = \{U_i\}_{i\in I}$ of X, one defines an augmentation homomorphism

$$\varepsilon: \mathscr{F}(X) \to C^0(\mathfrak{U}, \mathscr{F})$$
$$f \mapsto (f|_{U_i})$$

mapping $\mathscr{F}(X)$ into $\ker d^0$. The map ε is used to construct the so-called *augmented Čech complex* $C^{\cdot}_{\mathrm{aug}}(\mathfrak{U}, \mathscr{F})$, which is given by $C^q_{\mathrm{aug}}(\mathfrak{U}, \mathscr{F}) := C^q(\mathfrak{U}, \mathscr{F})$ for $q \neq -1$ and $C^{-1}_{\mathrm{aug}}(\mathfrak{U}, \mathscr{F}) := \mathscr{F}(X)$ with coboundary homomorphisms d^q_{aug} as in $C^{\cdot}(\mathfrak{U}, \mathscr{F})$ except that $d^{-1}_{\mathrm{aug}} := \varepsilon$. The associated cohomology modules $H^q_{\mathrm{aug}}(\mathfrak{U}, \mathscr{F})$ coincide with the modules $H^q(\mathfrak{U}, \mathscr{F})$ for $q \geq 1$.

If all cohomology modules $H^q_{\mathrm{aug}}(\mathfrak{U}, \mathscr{F})$ vanish, i.e., if the sequence

$$0 \to \mathscr{F}(X) \xrightarrow{\varepsilon} C^0(\mathfrak{U}, \mathscr{F}) \xrightarrow{d^0} C^1(\mathfrak{U}, \mathscr{F}) \to \cdots$$

is exact, the covering $\mathfrak{U}$ is called *$\mathcal{F}$-acyclic*. This condition is equivalent to the fact that $H^q(\mathfrak{U}, \mathcal{F}) = 0$ for all $q \neq 0$ and that ε induces a bijection $\mathcal{F}(X) \xrightarrow{\sim} H^0(\mathfrak{U}, \mathcal{F})$. To give another characterization of acyclicity, we consider the trivial covering $\mathfrak{U}_0 := \{X\}$ of X. This covering is $\mathcal{F}$-acyclic by Corollary 2, since the augmentation is the identity map in this case. Therefore, an arbitrary open covering $\mathfrak{U}$ is $\mathcal{F}$-acyclic if and only if all homomorphisms

$$\varrho^q(\mathfrak{U}, \mathfrak{U}_0): H^q(\mathfrak{U}_0, \mathcal{F}) \to H^q(\mathfrak{U}, \mathcal{F})$$

are bijective. Namely, we have $H^q(\mathfrak{U}_0, \mathcal{F}) = 0$ for $q \geqq 1$. Furthermore, if $\tau^\cdot: C^\cdot(\mathfrak{U}_0, \mathcal{F}) \to C^\cdot(\mathfrak{U}, \mathcal{F})$ is the complex homomorphism associated to the refinement $\mathfrak{U}$ of $\mathfrak{U}_0$, then $\tau^0: C^0(\mathfrak{U}_0, \mathcal{F}) \to C^0(\mathfrak{U}, \mathcal{F})$ coincides with the augmentation $\varepsilon: \mathcal{F}(X) \to C^0(\mathfrak{U}, \mathcal{F})$. Using this characterization of acyclicity, one derives from Proposition 3:

Proposition 4. *Let $\mathfrak{U}$ and $\mathfrak{V}$ be open coverings of X which are refinements of each other. Then, assuming $X \in \mathfrak{X}$, the covering $\mathfrak{U}$ is $\mathcal{F}$-acyclic if and only if $\mathfrak{V}$ is $\mathcal{F}$-acyclic.*

8.1.4. A Comparison Theorem for Čech cohomology. — As before, X is a set together with a system $\mathfrak{X}$ of open subsets of X, and $\mathcal{F}$ is a presheaf on X. We want to define a double Čech complex $C^{\cdot\cdot}(\mathfrak{U}, \mathfrak{V}; \mathcal{F})$ depending on two open coverings $\mathfrak{U} = \{U_i\}_{i\in I}$ and $\mathfrak{V} = \{V_j\}_{j\in J}$ of X. For $p, q \geqq 0$, we set (using the same notations as before)

$$C^{p,q}(\mathfrak{U}, \mathfrak{V}; \mathcal{F}) := \prod_{\substack{(i_0,\ldots,i_p)\in I^{p+1}\\(j_0,\ldots,j_q)\in J^{q+1}}} \mathcal{F}(U_{i_0\ldots i_p} \cap V_{j_0\ldots j_q})$$

and define homomorphisms

$$'d^{p,q}: C^{p,q}(\mathfrak{U}, \mathfrak{V}; \mathcal{F}) \to C^{p+1,q}(\mathfrak{U}, \mathfrak{V}; \mathcal{F}) \quad \text{and}$$
$$''d^{p,q}: C^{p,q}(\mathfrak{U}, \mathfrak{V}; \mathcal{F}) \to C^{p,q+1}(\mathfrak{U}, \mathfrak{V}; \mathcal{F}),$$

where, for any $f \in C^{p,q}(\mathfrak{U}, \mathfrak{V}; \mathcal{F})$, the $(i_0, \ldots, i_{p+1}, j_0, \ldots, j_q)$-component of $'d^{p,q}(f)$ is given by

$$\sum_{\nu=0}^{p+1} (-1)^{\nu+q} f_{i_0\ldots\hat{\imath}_\nu\ldots i_{p+1}, j_0\ldots j_q}|_{U_{i_0\ldots i_{p+1}} \cap V_{j_0\ldots j_q}}$$

and the $(i_0, \ldots, i_p, j_0, \ldots, j_{q+1})$-component of $''d^{p,q}(f)$ is given by

$$\sum_{\mu=0}^{q+1} (-1)^{\mu} f_{i_0\ldots i_p, j_0\ldots\hat{\jmath}_\mu\ldots j_q}|_{U_{i_0\ldots i_p} \cap V_{j_0\ldots j_{q+1}}}.$$

It is immediate that the modules $C^{p,q}(\mathfrak{U}, \mathfrak{V}; \mathcal{F})$ together with the maps $'d^{p,q}$ and $''d^{p,q}$ constitute a double complex $C^{\cdot\cdot}(\mathfrak{U}, \mathfrak{V}; \mathcal{F})$. From this double complex, one can derive single complexes as outlined in (8.1.2). The q-th column and the p-th row of $C^{\cdot\cdot}(\mathfrak{U}, \mathfrak{V}; \mathscr{F})$ are described as follows:

$$'C^{\cdot q}(\mathfrak{U}, \mathfrak{V}; \mathscr{F}) = \prod_{(j_0,\ldots,j_q)\in J^{q+1}} C^{\cdot}_{(-1)^q}(\mathfrak{U}|_{V_{j_0\ldots j_q}}, \mathscr{F}) \quad \text{and}$$

$$''C^{p\cdot}(\mathfrak{U}, \mathfrak{V}; \mathscr{F}) = \prod_{(i_0,\ldots,i_p)\in I^{p+1}} C^{\cdot}(\mathfrak{V}|_{U_{i_0\ldots i_p}}, \mathscr{F}).$$

Here the product of complexes is understood in the obvious way. Furthermore, $C^{\cdot}_{(-1)^q}$ is the complex obtained from $C^{\cdot}$ by multiplying boundary operators with $(-1)^q$, and for any open $V \subset X$, the covering $\mathfrak{U}|_V := \{U_i \cap V\}_{i\in I}$ is the restriction of $\mathfrak{U}$ to V (likewise for $\mathfrak{V}$). Also, on the right-hand sides, $\mathscr{F}$ must be interpreted as its restriction to $V_{j_0\ldots j_q}$ or $U_{i_0\ldots i_p}$.

The augmentations

$$\mathscr{F}(V_{j_0\ldots j_q}) \to C^0(\mathfrak{U}|_{V_{j_0\ldots j_q}}, \mathscr{F})$$

induce homomorphisms

$$C^q(\mathfrak{V}, \mathscr{F}) \to \ker {}'d^{0,q} \subset C^{0,q}(\mathfrak{U}, \mathfrak{V}; \mathscr{F}),$$

which, in turn, can be interpreted as a homomorphism

$$\iota'' \colon C^{\cdot}(\mathfrak{V}, \mathscr{F}) \to C^{\cdot}(\mathfrak{U}, \mathfrak{V}; \mathscr{F})$$

into the single complex associated to $C^{\cdot\cdot}(\mathfrak{U}, \mathfrak{V}; \mathscr{F})$. Furthermore, ι'' maps $C^{\cdot}(\mathfrak{V}, \mathscr{F})$ into the subcomplex $C''^{\cdot}(\mathfrak{U}, \mathfrak{V}; \mathscr{F})$ of $C^{\cdot}(\mathfrak{U}, \mathfrak{V}; \mathscr{F})$ which is given by

$$C''^q(\mathfrak{U}, \mathfrak{V}; \mathscr{F}) = \ker {}'d^{0,q}$$

as in Lemma 8.1.2/2. Now assuming that the covering $\mathfrak{U}|_{V_{j_0\ldots j_q}}$ is $\mathscr{F}$-acyclic for all indices $j_0, \ldots, j_q \in J$ and for all q, we see that ι'' maps $C^{\cdot}(\mathfrak{V}, \mathscr{F})$ isomorphically onto $C''^{\cdot}(\mathfrak{U}, \mathfrak{V}; \mathscr{F})$. In addition, it follows from our description of the q-th row that

$$H^p({}'C^{\cdot q}(\mathfrak{U}, \mathfrak{V}; \mathscr{F})) = 0$$

for all $p > 0$ and for all q. Thus, Lemma 8.1.2/2 can be applied and we get

Lemma 1. *If the covering $\mathfrak{U}|_{V_{j_0\ldots j_q}}$ is $\mathscr{F}$-acyclic for all indices $j_0, \ldots, j_q \in J$ and for all q, then the homomorphism*

$$\iota'' \colon C^{\cdot}(\mathfrak{V}, \mathscr{F}) \to C^{\cdot}(\mathfrak{U}, \mathfrak{V}; \mathscr{F})$$

induces bijections

$$H^r(\iota'') \colon H^r(\mathfrak{V}, \mathscr{F}) \xrightarrow{\sim} H^r(C^{\cdot}(\mathfrak{U}, \mathfrak{V}; \mathscr{F}))$$

for all r.

Of course, there is an analogue of Lemma 1 for the homomorphism

$$\iota' \colon C^{\cdot}(\mathfrak{U}, \mathscr{F}) \to C^{\cdot}(\mathfrak{U}, \mathfrak{V}; \mathscr{F})$$

which is derived from the augmentations

$$\mathcal{F}(U_{i_0\ldots i_p}) \to C^0(\mathfrak{B}|_{U_{i_0\ldots i_p}}, \mathcal{F}).$$

Thus, we obtain the following result:

Comparison Theorem 2. *Assume that all coverings $\mathfrak{U}|_{V_{j_0\ldots j_q}}$ and $\mathfrak{B}|_{U_{i_0\ldots i_p}}$ are $\mathcal{F}$-acyclic. Then one gets bijections*

$$H^r(\mathfrak{U}, \mathcal{F}) \xrightarrow[\sim]{H^r(\iota')} H^r(C^\cdot(\mathfrak{U}, \mathfrak{B}; \mathcal{F})) \xleftarrow[\sim]{H^r(\iota'')} H^r(\mathfrak{B}, \mathcal{F})$$

for all r. In particular, assuming $X \in \mathfrak{T}$, the covering $\mathfrak{U}$ is $\mathcal{F}$-acyclic if and only if $\mathfrak{B}$ is $\mathcal{F}$-acyclic.

Proof. The acyclicity statement follows from Lemma 1 if one realizes that there is a canonical augmentation

$$\mathcal{F}(X) \to C^0(\mathfrak{U}, \mathfrak{B}; \mathcal{F}) = C^{0,0}(\mathfrak{U}, \mathfrak{B}; \mathcal{F})$$

which is compatible with the augmentations $\mathcal{F}(X) \to C^0(\mathfrak{U}, \mathcal{F})$ and $\mathcal{F}(X) \to C^0(\mathfrak{B}, \mathcal{F})$ via ι' and ι''. □

Corollary 3. *Assume that $\mathfrak{B}$ is a refinement of $\mathfrak{U}$ and that $\mathfrak{B}|_{U_{i_0\ldots i_p}}$ is $\mathcal{F}$-acyclic for all $i_0, \ldots, i_p \in I$ and for all p. Then, if $X \in \mathfrak{T}$, the covering $\mathfrak{U}$ is $\mathcal{F}$-acyclic if and only if $\mathfrak{B}$ is $\mathcal{F}$-acyclic.*

Proof. We have only to show that all coverings $\mathfrak{U}|_{V_{j_0\ldots j_q}}$ are $\mathcal{F}$-acyclic. However, this follows from Proposition 8.1.3/4, since $\mathfrak{U}|_{V_{j_0\ldots j_q}}$ and the trivial covering of $V_{j_0\ldots j_q}$ are refinements of each other. □

Corollary 4. *Assume that the covering $\mathfrak{B}|_{U_{i_0\ldots i_p}}$ is $\mathcal{F}$-acyclic for all indices $i_0, \ldots, i_p \in I$ and for all p. Then, if $X \in \mathfrak{T}$, the covering $\mathfrak{U} \times \mathfrak{B} := \{U_i \cap V_j\}_{i \in I, j \in J}$ of X is $\mathcal{F}$-acyclic if and only if $\mathfrak{U}$ is $\mathcal{F}$-acyclic.*

Proof. The coverings $\mathfrak{B}|_{U_{i_0\ldots i_p}}$ and $\mathfrak{U} \times \mathfrak{B}|_{U_{i_0\ldots i_p}}$ are refinements of each other; thus by our assumption and by Proposition 8.1.3/4, they are both $\mathcal{F}$-acyclic. Furthermore, $\mathfrak{U} \times \mathfrak{B}$ is a refinement of $\mathfrak{U}$. Therefore, Corollary 3 is applicable to the coverings $\mathfrak{U} \times \mathfrak{B}$ and $\mathfrak{U}$. □

8.2. Tate's Acyclicity Theorem

8.2.1. Statement of the theorem. — Let X be an affinoid variety and let $\mathfrak{T}$ be the system of all affinoid subdomains in X. As in (7.3.2), we denote by $\mathcal{O}_X$ the presheaf on X which associates to any affinoid subdomain $U \subset X$ its affinoid algebra $\mathcal{O}_X(U)$. Then, for any covering $\mathfrak{U} = \{U_i\}_{i \in I}$ of X by affinoid subdomains $U_i \subset X$, the Čech complex $C^\cdot(\mathfrak{U}, \mathcal{O}_X)$ is defined.

Theorem 1 (Tate [37]). *Let X be an affinoid variety and let $\mathfrak{U} = \{U_i\}_{i \in I}$ be a finite covering of X by affinoid subdomains $U_i \subset X$. Then $\mathfrak{U}$ is $\mathcal{O}_X$-acyclic.*

This is TATE's Acyclicity Theorem. The assumption of $\mathfrak{U}$ being a finite covering is essential, as we will see at the end of this section. Before we turn to the proof of Theorem 1, which will be carried out in the next two sections, we want to derive some consequences. To simplify our notation, finite coverings $\mathfrak{U}$ of X consisting of affinoid subdomains in X will be referred to as *affinoid coverings*. As a first consequence, we will show that $\mathcal{O}_X$ satisfies "sheaf properties" as far as affinoid coverings are concerned. This will be the starting point for the definition of global varieties in Chapter 9.

Corollary 2. *Let $\mathfrak{U} = \{U_i\}_{i\in I}$ be an affinoid covering of the affinoid variety X. Then the presheaf $\mathcal{O}_X$ satisfies the following properties:*

(i) *If $f, g \in \mathcal{O}_X(X)$ are affinoid functions such that $f|_{U_i} = g|_{U_i}$ for all $i \in I$, then $f = g$.*

(ii) *If $f_i \in \mathcal{O}_X(U_i)$, $i \in I$, are affinoid functions such that $f_i|_{U_i\cap U_j} = f_j|_{U_i\cap U_j}$ for all $i, j \in I$, then there exists a function $f \in \mathcal{O}_X(X)$ with $f|_{U_i} = f_i$ for all $i \in I$.*

Proof. By the theorem the sequence

$$0 \to \mathcal{O}_X(X) \xrightarrow{\varepsilon} C^0(\mathfrak{U}, \mathcal{O}_X) \xrightarrow{d^0} C^1(\mathfrak{U}, \mathcal{O}_X)$$

is exact. Thus, (i) follows from the injectivity of the augmentation ε, and (ii) is a consequence of $\operatorname{im} \varepsilon = \ker d^0$. □

There is a generalization of Corollary 2 for morphisms which reads as follows:

Corollary 3. *Let X, Y be affinoid varieties and $\mathfrak{U} = \{U_i\}_{i\in I}$ an affinoid covering of X.*

(i) *If $\varphi, \psi\colon X \to Y$ are affinoid morphisms such that $\varphi|_{U_i} = \psi|_{U_i}$ for all $i \in I$, then $\varphi = \psi$.*

(ii) *If $\varphi_i\colon U_i \to Y$ are affinoid morphisms, $i \in I$, such that $\varphi_i|_{U_i\cap U_j} = \varphi_j|_{U_i\cap U_j}$ for all $i, j \in I$, then there exists an affinoid morphism $\varphi\colon X \to Y$ with $\varphi|_{U_i} = \varphi_i$ for all $i \in I$.*

Proof. The assertions can be deduced from Corollary 2. A direct verification goes as follows. Any system of morphisms $\varphi_i\colon U_i \to Y$ as in (ii) induces a commutative diagram

$$\begin{array}{ccccccc} & & & \mathcal{O}_Y(Y) & & & \\ & & \lambda' \swarrow & \downarrow \lambda & & & \\ 0 \to & \mathcal{O}_X(X) & \xrightarrow{\varepsilon} & C^0(\mathfrak{U}, \mathcal{O}_X) & \xrightarrow{d^0} & C^1(\mathfrak{U}, \mathcal{O}_X) & \end{array}$$

where ε is the augmentation and λ is given by $\lambda(f) = (\varphi_i^*(f))$. By the supposition in (ii), we have $\operatorname{im} \lambda \subset \ker d^0$; thus, since ε is injective, there exists a unique λ' making the above diagram commutative. The map λ' is an algebra homomorphism; hence it induces a morphism $\varphi\colon X \to Y$ which, by construction, extends all φ_i. This proves the assertion of (ii). If $\psi\colon X \to Y$ is a second morphism such that $\psi|_{U_i} = \varphi_i$ for all $i \in I$, then also the corresponding algebra homomorphism $\psi^*\colon \mathcal{O}_Y(Y) \to \mathcal{O}_X(X)$ renders the above diagram commutative. As a consequence, the injectivity of ε implies $\psi^* = \lambda'$ and thus $\varphi = \psi$, which verifies (i). □

The Acyclicity Theorem together with the main theorem for locally closed immersions of Chapter 7 gives the following characterization of affinoid subdomains:

Corollary 4. *Let $\varphi\colon U \to X$ be a morphism of affinoid varieties. Then the following are equivalent:*

(i) *U is an affinoid subdomain of X via φ.*
(ii) *φ is an open immersion.*

In particular, if φ is a surjective open immersion, then φ is an isomorphism.

Proof. The implication (i) ⇒ (ii) follows from Proposition 7.3.3/1. Now assume that φ is an open immersion. By Corollary 7.3.5/2 and Proposition 7.2.2/3, there exists an affinoid covering $\mathfrak{U} = \{U_i\}_{i\in I}$ of U such that each U_i is an affinoid subdomain of X via $\varphi|_{U_i}$. We want to show that φ represents all affinoid morphisms into $\varphi(U)$. If $\psi\colon Y \to X$ is a morphism of affinoid varieties with $\psi(Y) \subset \varphi(U)$, then $V_i := \psi^{-1}(\varphi(U_i))$ is an affinoid subdomain of Y for all $i \in I$, and $\mathfrak{B} = \{V_i\}_{i\in I}$ is an affinoid covering of Y. For any affinoid subdomain $V \subset Y$ contained in some V_i, the map $V \hookrightarrow Y \xrightarrow{\psi} X$ factors uniquely through U_i and hence uniquely through U. In particular, for each $i \in I$, we have a commutative diagram

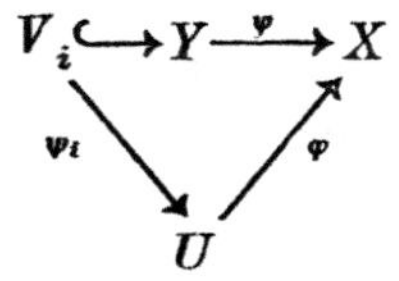

with a unique ψ_i. It follows then from Corollary 3 that the ψ_i extend to a unique morphism $\psi'\colon Y \to U$ such that

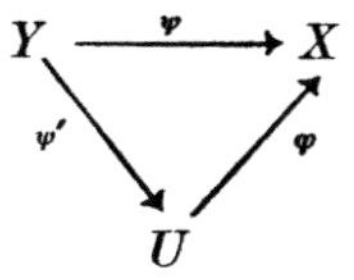

commutes. Consequently, φ represents all affinoid maps into $\varphi(U)$, and thus U is an affinoid subdomain of X via φ. □

Let $X = \operatorname{Sp} A$ be an affinoid variety. By an *$\mathcal{O}_X$-module* we understand a presheaf of abelian groups $\mathcal{F}$ on X, where $\mathcal{F}(U)$, for all affinoid subdomains $U \subset X$, is endowed with an $\mathcal{O}_X(U)$-module structure such that all these module structures are compatible with restriction homomorphisms. As an example we construct the $\mathcal{O}_X$-module associated to an A-module M. This is the presheaf $\mathcal{F}$ which assigns to any affinoid subdomain $U \subset X$ the $\mathcal{O}_X(U)$-module $M \otimes_A \mathcal{O}_X(U)$ with restriction homomorphisms being induced by $\mathcal{O}_X$. Then $\mathcal{F}$ is an $\mathcal{O}_X$-module, and we suggestively write $M \otimes \mathcal{O}_X$ instead of $\mathcal{F}$. In particular, note that $A \otimes \mathcal{O}_X = \mathcal{O}_X$. We want to generalize Theorem 1 to the case where $\mathcal{O}_X$ is replaced by an $\mathcal{O}_X$-module $M \otimes \mathcal{O}_X$ associated to some A-module M.

Corollary 5. *Let $X = \operatorname{Sp} A$ be an affinoid variety and M an A-module. Then every affinoid covering $\mathfrak{U}$ of X is $M \otimes \mathcal{O}_X$-acyclic.*

Proof. The assertion is a direct consequence of Theorem 1, when M is a free A-module, e.g., $M = A^{(\Lambda)}$ with some index set Λ. Namely, in this case the sequence

$$0 \to M \xrightarrow{\varepsilon} C^0(\mathfrak{U}, M \otimes \mathcal{O}_X) \xrightarrow{d^0} C^1(\mathfrak{U}, M \otimes \mathcal{O}_X) \to \cdots$$

is the Λ-fold direct sum of the sequence

$$0 \to A \to C^0(\mathfrak{U}, \mathcal{O}_X) \to C^1(\mathfrak{U}, \mathcal{O}_X) \to \cdots$$

with itself. If M is not free, choose a short exact sequence of A-modules

$$0 \to M' \to F \to M \to 0$$

with a free A-module F. Considering augmented Čech complexes, we get a canonical sequence

$$0 \to C^{\cdot}_{\text{aug}}(\mathfrak{U}, M' \otimes \mathcal{O}_X) \to C^{\cdot}_{\text{aug}}(\mathfrak{U}, F \otimes \mathcal{O}_X) \to C^{\cdot}_{\text{aug}}(\mathfrak{U}, M \otimes \mathcal{O}_X) \to 0$$

which is exact, since for any affinoid subdomain $\operatorname{Sp} A' \subset X = \operatorname{Sp} A$ the functor $- \otimes_A A'$ is exact on the category of A-modules (cf. Corollary 7.3.2/6). The associated long cohomology sequence contains isomorphisms

$$H^{q-1}_{\text{aug}}(\mathfrak{U}, M \otimes \mathcal{O}_X) \xrightarrow{\sim} H^q_{\text{aug}}(\mathfrak{U}, M' \otimes \mathcal{O}_X),$$

due to the fact that $C^{\cdot}_{\text{aug}}(\mathfrak{U}, F \otimes \mathcal{O}_X)$ has zero cohomology (F is a free A-module). Since the covering $\mathfrak{U}$ is finite, there exists an integer n such that $H^q_{\text{aug}}(\mathfrak{U}, N \otimes \mathcal{O}_X) = H^q(\mathfrak{U}, N \otimes \mathcal{O}_X) = 0$ for all $q \geq n$ and all A-modules N (cf. Corollary 8.1.3/2). Using an induction argument, the above isomorphisms show $H^q_{\text{aug}}(\mathfrak{U}, M \otimes \mathcal{O}_X) = 0$ for all q; hence $\mathfrak{U}$ is $M \otimes \mathcal{O}_X$-acyclic. □

Finally, we want to give an example of an infinite covering $\mathfrak{U}$ of an affinoid variety X such that $\mathfrak{U}$ is not $\mathcal{O}_X$-acyclic. Let A be an affinoid algebra which is an integral domain, but not a field, e.g., $A = T_1$, and let us consider the affinoid variety $X := \operatorname{Sp} A$. There is a non-unit $f \neq 0$ in A, and we can find a strictly decreasing sequence $\alpha_1 := |f|_{\sup} > \alpha_2 > \alpha_3 > \ldots$ of elements in $\sqrt{|k^*|}$ such that $\alpha := \lim_{i\to\infty} \alpha_i$ also is an element of $\sqrt{|k^*|}$. Then we introduce the affinoid subdomains $U_0 := X(\alpha^{-1} f)$ and $U_i := X(\alpha_i f^{-1})$, $i = 1, 2, \ldots$, and consider the covering $\mathfrak{U} := \{U_i\}_{i=0,1,2,\ldots}$ of X. Note that $U_0 \cap U_i = \emptyset$ for $i \neq 0$ and that all U_i are non-empty. We want to show that $\mathfrak{U}$ cannot be $\mathcal{O}_X$-acyclic. For this purpose we look at the 0-cochain $g \in C^0(\mathfrak{U}, \mathcal{O}_X)$ which is given by

$$g_i := \begin{cases} \text{unit element in } \mathcal{O}_X(U_0) & \text{if } i = 0, \\ 0 \quad \text{if} \quad i \neq 0. \end{cases}$$

Clearly, $g \in \ker d^0$. If $\mathfrak{U}$ were $\mathcal{O}_X$-acyclic, the augmentation homomorphism $\varepsilon: A \to C^0(\mathfrak{U}, \mathcal{O}_X)$ would be injective and g would have an inverse $g' \in A$ with respect to ε. We would have $\varepsilon\big(g'(1 - g')\big) = 0$ and hence $g'(1 - g') = 0$ without g' or $1 - g'$ being zero, contradicting the fact that A was assumed to be an integral domain. Thus $\mathfrak{U}$ cannot be $\mathcal{O}_X$-acyclic. The example relies, of course, on the fact that, with respect to the canonical topology, X is not connected.

A similar argument shows that any affinoid variety X of positive dimension (i.e., having a component not reduced to a single point) admits open coverings which are not $\mathcal{O}_X$-acyclic.

8.2.2. Affinoid coverings. — Our first step in the proof of TATE's Acyclicity Theorem is to show that the acyclicity of general affinoid coverings (which are finite by our terminology) can be derived from the acyclicity of a particular type of coverings, namely, Laurent coverings. In the next section, TATE's Theorem will then be proved for this particular type of coverings.

Let X be an affinoid variety, and consider finitely many affinoid functions $f_1, \ldots, f_n \in \mathcal{O}_X(X)$. Then, for each $j = 1, \ldots, n$,

$$\mathfrak{U}^j := \{X(f_j), X(f_j^{-1})\}$$

is an affinoid covering of X. The product covering

$$\mathfrak{U} := \mathfrak{U}^1 \times \cdots \times \mathfrak{U}^n$$

consisting of all intersections $U^1 \cap \ldots \cap U^n$ with $U^j \in \mathfrak{U}^j$ is called a *Laurent covering* of X. More precisely, $\mathfrak{U}$ is the *Laurent covering of X generated by* $f_1, \ldots, f_n$. Its covering sets are the Laurent domains $X(f_1^{\alpha_1}, \ldots, f_n^{\alpha_n})$ with $\alpha_1, \ldots, \alpha_n \in \{+1, -1\}$.

There is another type of coverings which has to be considered. Assume that the functions $f_1, \ldots, f_n \in \mathcal{O}_X(X)$ have no common zero on X. Then

$$\mathfrak{U} := \left\{X\left(\frac{f_1}{f_i}, \ldots, \frac{f_n}{f_i}\right)\right\}_{i=1,\ldots,n}$$

is an affinoid covering of X by rational subdomains. We refer to $\mathfrak{U}$ as a *rational covering*, or more precisely, as the *rational covering of X generated by* $f_1, \ldots, f_n$.

Proposition 1. *Let $\mathfrak{U}$ be an affinoid covering of the affinoid variety X, and let $X' \subset X$ be an affinoid subdomain. Then $\mathfrak{U}|_{X'}$ is an affinoid covering of X'. If $\mathfrak{U}$ is a rational or Laurent covering, $\mathfrak{U}|_{X'}$ is a covering of the same type.*

Proof. Use Corollary 7.2.2/5 if $\mathfrak{U}$ is a general affinoid covering. If $\mathfrak{U}$ is rational or Laurent, the assertion is obvious. □

In order to reduce the proof of TATE's Acyclicity Theorem to the case of Laurent coverings, we have to clarify the interdependence of the different types of affinoid coverings.

Lemma 2. *Let* $\mathfrak{U}$ *be an affinoid covering of* X. *Then there exists a rational covering* $\mathfrak{V}$ *of* X, *which is a refinement of* $\mathfrak{U}$.

Proof. Let $\mathfrak{U} = \{U_1, \ldots, U_r\}$. By Corollary 7.3.5/3, we may assume that all U_i are rational subdomains in X, say

$$U_i = X\left(\frac{g_1^{(i)}}{g_{n_i}^{(i)}}, \ldots, \frac{g_{n_i}^{(i)}}{g_{n_i}^{(i)}}\right).$$

For each $i = 1, \ldots, r$, the functions $g_1^{(i)}, \ldots, g_{n_i}^{(i)} \in \mathcal{O}_X(X)$ generate a rational covering

$$\mathfrak{V}^i = \{V_1^i, \ldots, V_{n_i}^i\}$$

of X, where

$$V_j^i := X\left(\frac{g_1^{(i)}}{g_j^{(i)}}, \ldots, \frac{g_{n_i}^{(i)}}{g_j^{(i)}}\right),$$

and where $V_{n_i}^i = U_i$. Set

$$I := \{(\nu_1, \ldots, \nu_r) \in \mathbb{N}^r;\ 1 \leqq \nu_i \leqq n_i,\ i = 1, \ldots, r\}$$

and

$$I' := \{(\nu_1, \ldots, \nu_r) \in I;\ \nu_i = n_i \quad \text{for} \quad \text{some } i \in \{1, \ldots, n\}\}.$$

Writing $V_{j_1\ldots j_r} := V_{j_1}^1 \cap \ldots \cap V_{j_r}^r$, the proof of Proposition 7.2.3/7 shows for all tuples $(j_1, \ldots, j_r) \in I$ that

$$V_{j_1\ldots j_r} = X\left(\frac{g_{\nu_1}^{(1)} \cdots g_{\nu_r}^{(r)}}{g_{j_1}^{(1)} \cdots g_{j_r}^{(r)}};\ (\nu_1, \ldots, \nu_r) \in I\right).$$

Furthermore, if $(j_1, \ldots, j_r) \in I'$, we claim that

$$V_{j_1\ldots j_r} = X\left(\frac{g_{\nu_1}^{(1)} \cdots g_{\nu_r}^{(r)}}{g_{j_1}^{(1)} \cdots g_{j_r}^{(r)}};\ (\nu_1, \ldots, \nu_r) \in I'\right).$$

To verify this, consider an element $x \in X$ satisfying

$$\max_{(\nu_1,\ldots,\nu_r)\in I'} |g_{\nu_1}^{(1)} \cdots g_{\nu_r}^{(r)}(x)| = |g_{j_1}^{(1)} \cdots g_{j_r}^{(r)}(x)|.$$

Let $(\mu_1, \ldots, \mu_r)$ be a tuple in $I - I'$. Then x belongs to some set U_i, say $x \in U_1$, and we have

$$|g_{\mu_1}^{(1)} \cdots g_{\mu_r}^{(r)}(x)| \leqq |g_{n_1}^{(1)} g_{\mu_2}^{(2)} \cdots g_{\mu_r}^{(r)}(x)| \leqq |g_{j_1}^{(1)} \cdots g_{j_r}^{(r)}(x)|,$$

since $(n_1, \mu_2, \ldots, \mu_r) \in I'$. This shows that the functions $g_{j_1}^{(1)} \ldots g_{j_r}^{(r)}$, $(j_1, \ldots, j_r) \in I'$, cannot have a common zero on X and that our claim is justified. Now

$$\mathfrak{V} := \{V_{j_1 \ldots j_r}\}_{(j_1 \ldots j_r) \in I'}$$

obviously is an affinoid covering of X, namely the rational covering generated by the functions $g_{j_1}^{(1)} \ldots g_{j_r}^{(r)}$, $(j_1, \ldots, j_r) \in I'$. By construction, $\mathfrak{V}$ is finer than $\mathfrak{U}$. □

Lemma 3. *Let $\mathfrak{U}$ be a rational covering of X. Then there exists a Laurent covering $\mathfrak{V}$ of X such that, for any $V \in \mathfrak{V}$, the covering $\mathfrak{U}|_V$ is a rational covering of V, which is generated by units in $\mathcal{O}_X(V)$.*

Proof. Let the rational covering $\mathfrak{U}$ be generated by the elements $f_1, \ldots, f_n \in \mathcal{O}_X(X)$, and choose a constant c in k^* such that

$$|c|^{-1} < \inf_{x \in X} \left(\max_{1 \leq i \leq n} |f_i(x)| \right)$$

(use Lemma 7.3.4/7). Let $\mathfrak{V}$ be the Laurent covering generated by the elements $cf_1, \ldots, cf_n$. We claim that $\mathfrak{V}$ is as desired. Consider a set

$$V = X\big((cf_1)^{\alpha_1}, \ldots, (cf_n)^{\alpha_n}\big) \in \mathfrak{V},$$

where $\alpha_1, \ldots, \alpha_n \in \{+1, -1\}$. We may assume that there is an $s \in \{0, 1, \ldots, n\}$ such that $\alpha_1 = \cdots = \alpha_s = +1$ and $\alpha_{s+1} = \cdots = \alpha_n = -1$. Then

$$X\left(\frac{f_1}{f_i}, \ldots, \frac{f_n}{f_i}\right) \cap V = \emptyset$$

for $i = 1, \ldots, s$, since

$$\max_{1 \leq i \leq s} |f_i(x)| \leq |c|^{-1} < \max_{1 \leq i \leq n} |f_i(x)|$$

if $x \in V$. In particular, we have

$$\max_{1 \leq i \leq n} |f_i(x)| = \max_{s+1 \leq i \leq n} |f_i(x)|$$

for all $x \in V$, and it follows that $\mathfrak{U}|_V$ is the rational covering of V generated by $f_{s+1}|_V, \ldots, f_n|_V$. By construction, these elements are units in $\mathcal{O}_X(V)$. □

Lemma 4. *Let $\mathfrak{U}$ be a rational covering of X, which is generated by units $f_1, \ldots, f_n \in \mathcal{O}_X(X)$. Then there exists a Laurent covering $\mathfrak{V}$ of X, which is a refinement of $\mathfrak{U}$.*

Proof. We claim that the Laurent covering $\mathfrak{V}$ of X generated by all products $f_i f_j^{-1}$ with $1 \leq i < j \leq n$ is as desired. Consider a set $V \in \mathfrak{V}$, say

$$V = \bigcap_{(i,j) \in I_1} X(f_i f_j^{-1}) \cap \bigcap_{(i,j) \in I_2} X(f_j f_i^{-1}),$$

where $I = I_1 \dot{\cup} I_2$ is a partition of the index set $I := \{(i, j) \in \mathbb{N}^2;\ 1 \leq i < j \leq n\}$. Writing $i \lll j$ if $(i, j) \in I_1$ and $j \lll i$ if $(i, j) \in I_2$, we get a relation $\lll$ on the set $S := \{1, \ldots, n\}$. For each pair of different elements $i, j \in S$, we have $i \lll j$ or $j \lll i$. Consider a maximal chain

$$i_1 \lll i_2 \lll \cdots \lll i_r$$

of elements in S. Then i_r is maximal in S, i.e., we have $i \lll i_r$ for all elements $i \in S$ which are different from i_r. Since $i \lll j$ implies $|f_i(x)| \leq |f_j(x)|$ for all $x \in V$, we see that

$$V \subset X\left(\frac{f_1}{f_{i_r}}, \ldots, \frac{f_n}{f_{i_r}}\right).$$

Hence $\mathfrak{V}$ is a refinement of $\mathfrak{U}$. □

As a consequence, we obtain the following result about the acyclicity of affinoid coverings:

Proposition 5. *Let $\mathcal{F}$ be a presheaf on the affinoid variety X. Assume that Laurent coverings are universally $\mathcal{F}$-acyclic on X, i.e., that for each affinoid subdomain $X' \subset X$, all Laurent coverings of X' are $\mathcal{F}$-acyclic. Then all affinoid coverings of X are $\mathcal{F}$-acyclic.*

Proof. Let $\mathfrak{U}$ be an affinoid covering of X. By Lemma 2, there exists a rational covering $\mathfrak{V}$ of X which refines $\mathfrak{U}$. Applying Corollary 8.1.4/3 (and Proposition 1), we see that $\mathfrak{U}$ is $\mathcal{F}$-acyclic if all rational coverings are universally $\mathcal{F}$-acyclic on X. Similarly, one can use Lemma 3, the Comparison Theorem 8.1.4/2, and our assumption on the Laurent coverings in order to show that rational coverings are universally $\mathcal{F}$-acyclic on X if the same is true for rational coverings which are generated by invertible functions. That the latter coverings are universally $\mathcal{F}$-acyclic on X, follows from Corollary 8.1.4/3, since any such covering can be refined by a Laurent covering (Lemma 4). □

In particular, TATE's Acyclicity Theorem needs a proof only for Laurent coverings.

8.2.3. Proof of the Acyclicity Theorem for Laurent coverings. — Let $\mathfrak{U}$ be a Laurent covering of the affinoid variety X. Using Corollary 8.1.4/4 and an induction argument, the acyclicity of $\mathfrak{U}$ can be reduced to the case where $\mathfrak{U}$ is generated by a single function $f \in \mathcal{O}_X(X)$, i.e., where $\mathfrak{U} = \{X(f), X(f^{-1})\}$. Thus, considering alternating cochains, we have to show that the sequence

$$0 \to \mathcal{O}_X(X) \xrightarrow{\varepsilon} C^0(\mathfrak{U}, \mathcal{O}_X) \xrightarrow{d^0} C^1_{\mathrm{a}}(\mathfrak{U}, \mathcal{O}_X) \to 0$$

is exact or, if $A := \mathcal{O}_X(X)$, that

$$0 \to A \xrightarrow{\varepsilon} A\langle f\rangle \times A\langle f^{-1}\rangle \xrightarrow{d^0} A\langle f, f^{-1}\rangle \to 0$$

is exact. Here we have identified $C^1_{\mathrm{a}}(\mathfrak{U}, \mathcal{O}_X)$ with $\mathcal{O}_X\big(X(f) \cap X(f^{-1})\big)$ in such a way that d^0: $A\langle f\rangle \times A\langle f^{-1}\rangle \to A\langle f, f^{-1}\rangle$ is the map $(g_1, g_2) \mapsto g_1 - g_2$. The above sequence is part of the following commutative diagram of A-module homomorphisms

$$\begin{array}{ccccccc}
 & & & 0 & & 0 & \\
 & & & \downarrow & & \downarrow & \\
 & & & (\zeta - f)\, A\langle\zeta\rangle \times (1 - f\eta)\, A\langle\eta\rangle & \xrightarrow{\lambda'} & (\zeta - f)\, A\langle\zeta, \zeta^{-1}\rangle & \to 0 \\
 & & & \hookuparrow & & \hookuparrow & \\
0 \to & A & \xrightarrow{\iota} & A\langle\zeta\rangle \times A\langle\eta\rangle & \xrightarrow{\lambda} & A\langle\zeta, \zeta^{-1}\rangle & \to 0 \\
 & \| & & \downarrow & & \downarrow & \\
0 \to & A & \xrightarrow{\varepsilon} & A\langle f\rangle \times A\langle f^{-1}\rangle & \xrightarrow{d^0} & A\langle f, f^{-1}\rangle & \to 0 \\
 & & & \downarrow & & \downarrow & \\
 & & & 0 & & 0 &
\end{array}$$

The symbols ζ and η denote indeterminates, ι is the canonical injection, λ is the map $\big(h_1(\zeta), h_2(\eta)\big) \mapsto h_1(\zeta) - h_2(\zeta^{-1})$, and λ' is induced by λ. Furthermore, the vertical maps in question are characterized by $\zeta \mapsto f$ and $\eta \mapsto f^{-1}$. The first column of the diagram is exact due to the characterization of generalized rings of fractions given in (6.1.4). Also the second column is exact since

$$A\langle f, f^{-1}\rangle = A\langle\zeta, \eta\rangle/(\zeta - f, 1 - f\eta)$$

and since the ideal $(\zeta - f, 1 - f\eta)$ coincides with $(\zeta - f, 1 - \zeta\eta)$. This implies

$$A\langle f, f^{-1}\rangle = A\langle\zeta, \zeta^{-1}\rangle/(\zeta - f).$$

The equations

$$A\langle\zeta, \zeta^{-1}\rangle = A\langle\zeta\rangle + \zeta^{-1}A\langle\zeta^{-1}\rangle \quad \text{and}$$

$$(\zeta - f)\, A\langle\zeta, \zeta^{-1}\rangle = (\zeta - f)\, A\langle\zeta\rangle + (1 - f\zeta^{-1})\, A\langle\zeta^{-1}\rangle$$

show the surjectivity of λ and λ'; thus, in particular, the first row is exact. But also the second row is exact, since

$$0 = \lambda\left(\sum_{\nu\geq 0} a_\nu\zeta^\nu, \sum_{\nu\geq 0} b_\nu\eta^\nu\right) = \sum_{\nu\geq 0} a_\nu\zeta^\nu - \sum_{\nu\geq 0} b_\nu\zeta^{-\nu}$$

implies $a_\nu = 0 = b_\nu$ for $\nu > 0$ and $a_0 - b_0 = 0$. Finally, looking at the third row, the injectivity of ε follows from Corollary 7.3.2/5. But then, by some diagram chasing, also the third row is exact. Hence the covering $\mathfrak{U}$ is $\mathcal{O}_X$-acyclic. This concludes the proof of TATE's Acyclicity Theorem. □

CHAPTER 9

Rigid analytic varieties

The present chapter deals with the basic theory of global rigid analytic varieties. Most of the material is scattered through the literature, and the formal machinery we use may be well-known to the reader; however for reasons of easy accessibility, we have chosen to include all necessary details.

We start by discussing GROTHENDIECK topologies and sheaf theory with respect to GROTHENDIECK topologies; see [1]. Then, using TATE's Acyclicity Theorem, affinoid varieties can be viewed as locally G-ringed spaces so that the definition of global rigid analytic varieties becomes straightforward. We pay special attention to explicit construction principles and to examples.

The second part of the chapter deals with KIEHL's work on coherent modules. Although we do not prove the Direct Image Theorem, we derive some consequences from it, namely the Proper Mapping Theorem and the STEIN Factorization.

It seems that not all possible implications of these results have been explored yet. For example, studying meromorphic functions, one can prove the same theorems on the analytic and algebraic dependence of such functions which were obtained by the third author of this book in the classical complex case for compact analytic spaces.

At the end of the chapter we look at elliptic curves with bad reduction. A glueing lemma for annuli and the classification of affinoid subdomains of the unit disc lead to the construction of the "universal covering" of such curves. Thereby we obtain TATE's result, which says that these curves correspond bijectively to the 1-dimensional analytic tori.

9.1. Grothendieck topologies

9.1.1. G-topological spaces. — The theory of Grothendieck topologies will not be presented here in full generality. Instead we give a treatment which is more suited to our needs. To fix notation, let X be a set. By a covering of some subset $U \subset X$, we mean a family $\{U_i\}_{i \in I}$ of subsets in X satisfying $U = \bigcup_{i \in I} U_i$. Given two coverings $\mathfrak{U} = \{U_i\}_{i \in I}$ and $\mathfrak{V} = \{V_j\}_{j \in J}$ of V we say that $\mathfrak{V}$ refines $\mathfrak{U}$ or that $\mathfrak{V}$ is a refinement of $\mathfrak{U}$ if there exists a map $\tau\colon J \to I$ such that $V_j \subset U_{\tau(j)}$ for all $j \in J$.

Definition 1. *A Grothendieck topology $\mathfrak{T}$ on X (also called G-topology) consists of*

(a) *a system S of subsets in X, called admissible open or $\mathfrak{T}$-open subsets of X, and*

(b) *a family $\{\operatorname{Cov} U\}_{U\in S}$ of systems of coverings, called admissible or $\mathfrak{T}$-coverings, where $\operatorname{Cov} U$ for $U \in S$ contains coverings $\{U_i\}_{i\in I}$ of U by sets $U_i \in S$.*

The system S and the family $\{\operatorname{Cov} U\}_{U\in S}$ are subject to the following conditions:

(i) $U, V \in S \Rightarrow U \cap V \in S$.

(ii) $U \in S \Rightarrow \{U\} \in \operatorname{Cov} U$.

(iii) *If* $U \in S$, $\{U_i\}_{i\in I} \in \operatorname{Cov} U$, *and* $\{V_{ij}\}_{j\in J_i} \in \operatorname{Cov} U_i$ *for* $i \in I$, *then* $\{V_{ij}\}_{i\in I, j\in J_i} \in \operatorname{Cov} U$.

(iv) *If* $U, V \in S$ *with* $V \subset U$ *and if* $\{U_i\}_{i\in I} \in \operatorname{Cov} U$, *then* $\{V \cap U_i\}_{i\in I} \in \operatorname{Cov} V$.

We call X a *G-topological space* and write more explicitly $X_{\mathfrak{T}}$, when the G-topology on X has to be specified. In order to see how G-topologies are related to ordinary topologies, we mention first that arbitrary unions of admissible open subsets in X need not again be admissible open in X. Therefore, the admissible open subsets of X cannot, in general, be viewed as the open sets of a topology on X. However, one can associate a topology with any G-topology $\mathfrak{T}$ on X. Simply take the topology which has as a basis the system of all $\mathfrak{T}$-open subsets in X (enlarged by the set X, if X is not covered by its $\mathfrak{T}$-open subsets). On the other hand, starting with an ordinary topological space X and fixing a basis B of open sets with the additional assumption that $U, V \in B$ implies $U \cap V \in B$, a G-topology can be defined on X as follows: take $S := B$ as the system of admissible open sets, and for $U \in S$, let $\operatorname{Cov} U$ be the system of all coverings $\{U_i\}_{i\in I}$ of U where $U_i \in S$. Alternatively, $\operatorname{Cov} U$ could be taken as the system of all *finite* coverings $\{U_i\}_{i\in I}$ with $U_i \in S$, which, in general, leads to a different G-topology on X. Thus, the concept of ordinary topologies is generalized by that of G-topologies. As will be seen later, the definition of a G-topology provides only a minimal amount of topological structure, so that sheaf theory and Čech cohomology can be done at a reasonable level.

Many topological terms have G-topological analogues. We give some examples. A G-topological space X admitting itself as an admissible open subset is called *quasi-compact* if every admissible covering of X has a finite admissible refinement. Similarly, X is called *connected* if there is no admissible covering $\{U_i\}_{i\in I}$ of X such that

$$\bigcup_{i\in I_1} U_i \cap \bigcup_{i\in I_2} U_i = \emptyset \qquad \text{and} \qquad \bigcup_{i\in I_1} U_i \neq \emptyset \neq \bigcup_{i\in I_2} U_i$$

for subsets $I_1, I_2 \subset I$ with $I = I_1 \mathbin{\dot\cup} I_2$.

A map $\varphi\colon X \to Y$ between G-topological spaces is called *continuous* if the following conditions are satisfied:

(i) For any admissible open set $V \subset Y$, its inverse image $\varphi^{-1}(V)$ is admissible open in X.

(ii) For any admissible covering $\{V_j\}_{j\in J}$ of an admissible open set $V \subset Y$, the system $\{\varphi^{-1}(V_j)\}_{j\in J}$ is an admissible covering of $\varphi^{-1}(V)$.

Considering the induced (ordinary) topologies on X and Y, it is clear that condition (i) implies continuity of φ in the ordinary sense. If $\mathfrak{T}$ and $\mathfrak{T}'$ are two G-topologies on a set X, we call $\mathfrak{T}'$ *finer* than $\mathfrak{T}$ or $\mathfrak{T}$ *weaker* than $\mathfrak{T}'$ if the identity map $X_{\mathfrak{T}'} \to X_{\mathfrak{T}}$ is continuous, i.e., if each $\mathfrak{T}$-open subset of X is also $\mathfrak{T}'$-open and each $\mathfrak{T}$-covering is also a $\mathfrak{T}'$-covering.

If $\varphi: Y \to X$ is a map between a set Y and a G-topological space X, there exists a unique weakest G-topology on Y making φ continuous. This G-topology is called the *inverse image* or *induced G-topology* via φ. Its admissible open sets and coverings are just the inverse images via φ of the admissible open sets and coverings belonging to the G-topology of X.

Let X be a G-topological space with G-topology $\mathfrak{T}$ and system of $\mathfrak{T}$-open subsets S. Then a subsystem $B \subset S$ is called a *basis* for $\mathfrak{T}$ if each $\mathfrak{T}$-open subset in X admits a $\mathfrak{T}$-covering by sets belonging to B. In any case, the system S of all $\mathfrak{T}$-open subsets of X is a basis for $\mathfrak{T}$; however, note that two G-topologies on the same set X do not necessarily coincide when they have a common basis (cf. our examples of G-topologies given above).

It is not surprising that all the above definitions coincide with the ordinary ones for topological spaces, when these spaces are viewed as G-topological spaces by calling all open sets and all open coverings admissible.

9.1.2. Enhancing procedures for *G*-topologies. — We start with a definition which has no non-trivial counterpart for topological spaces.

Definition 1. *Let X be a set admitting $\mathfrak{T}$ and $\mathfrak{T}'$ as G-topologies. $\mathfrak{T}'$ is called slightly finer than $\mathfrak{T}$, if the following conditions are satisfied:*

(i) *$\mathfrak{T}'$ is finer than $\mathfrak{T}$.*

(ii) *The $\mathfrak{T}$-open subsets of X form a basis for $\mathfrak{T}'$.*

(iii) *For each $\mathfrak{T}'$-covering $\mathfrak{U}$ of a $\mathfrak{T}$-open subset $U \subset X$, there exists a $\mathfrak{T}$-covering which refines $\mathfrak{U}$.*

In (9.2.3), we will see that sheaf theory for a G-topological space X remains essentially the same when the G-topology $\mathfrak{T}$ of X is replaced by a slightly finer one. Therefore, it is natural to ask if there is a finest G-topology among all the G-topologies slightly finer than $\mathfrak{T}$. In the following, we will pursue this question in a somewhat more general setting.

Let $\mathfrak{C}$ be a subcategory of the category of sets and suppose that each $X \in \mathfrak{C}$ carries a G-topology $\mathfrak{T}_X$ such that all morphisms of $\mathfrak{C}$ are continuous. Then $\mathfrak{T}$, the collection of all $\mathfrak{T}_X$, is called a G-topology on $\mathfrak{C}$ and may be regarded as a functor from $\mathfrak{C}$ into the category of G-topological spaces. If $\mathfrak{T}$ and $\mathfrak{T}'$ are G-topologies on $\mathfrak{C}$, we call $\mathfrak{T}'$ finer (respectively slightly finer) than $\mathfrak{T}$ if $\mathfrak{T}'_X$ is finer (respectively slightly finer) than $\mathfrak{T}_X$ for all $X \in \mathfrak{C}$. Similary, if P is a general property applicable to G-topologies, a G-topology $\mathfrak{T}$ on $\mathfrak{C}$ is said to satisfy P if $\mathfrak{T}_X$ satisfies P for all $X \in \mathfrak{C}$. Of particular interest will be the following properties:

(G_0) $\emptyset$ and X are admissible open subsets in X.

(G_1) Let $U \subset X$ be an admissible open set and let $V \subset U$ be a subset. Assume that there exists an admissible covering $\{U_i\}_{i\in I}$ of U such that $V \cap U_i$ is admissible open in X for all $i \in I$. Then V is admissible open in X.

(G_2) Let $\mathfrak{U} = \{U_i\}_{i\in I}$ be a covering of an admissible open set $U \subset X$ such that U_i is admissible open in X for all $i \in I$. Assume further that $\mathfrak{U}$ has a refinement which is admissible. Then $\mathfrak{U}$ is an admissible covering of U.

Proposition 2. *Let $\mathfrak{T}$ be a G-topology on a category of sets $\mathfrak{C}$ as above. Then there exists a unique finest G-topology $\mathfrak{T}'$ on $\mathfrak{C}$ among all the G-topologies slightly finer than $\mathfrak{T}$. The G-topology $\mathfrak{T}'$ satisfies conditions* (G_1) *and* (G_2). *If condition* (G_0) *is satisfied by $\mathfrak{T}$, then it is also satisfied by $\mathfrak{T}'$.*

Before giving the proof, we want to describe the G-topology $\mathfrak{T}'$ explicitly. Let $X \in \mathfrak{C}$ and $U \subset X$. An arbitrary covering $\{U_i\}_{i\in I}$ of U is called *compatible with* $\mathfrak{T}$ if for any $\mathfrak{C}$-morphism $\varphi\colon Y \to X$ and any $\mathfrak{T}_Y$-open subset $Y' \subset Y$ satisfying $\varphi(Y') \subset U$, the covering $\{\varphi^{-1}(U_i) \cap Y'\}_{i\in I}$ of Y' has a $\mathfrak{T}_Y$-covering which refines it. Using this terminology we can state

Lemma 3. (i) *A G-topology $\mathfrak{T}'$ on $\mathfrak{C}$ can be defined as follows: For $X \in \mathfrak{C}$, a subset $U \subset X$ is called $\mathfrak{T}'_X$-open if it admits a covering $\{U_i\}_{i\in I}$ by $\mathfrak{T}_X$-open sets U_i, which is compatible with $\mathfrak{T}$. Similarly, a covering $\{V_j\}_{j\in J}$ of any $\mathfrak{T}'_X$-open subset $V \subset X$ by $\mathfrak{T}'_X$-open sets V_j is called a $\mathfrak{T}'_X$-covering if $\{V_j\}_{j\in J}$ is compatible with $\mathfrak{T}$.*

(ii) *The G-topology $\mathfrak{T}'$ of* (i) *coincides with the G-topology $\mathfrak{T}'$ mentioned in Proposition* 2.

Once assertion (i) has been proved, Proposition 2, as well as assertion (ii) of Lemma 3, is an easy consequence. Namely, the G-topology $\mathfrak{T}'$ of assertion (i) is by definition slightly finer than $\mathfrak{T}$ and indeed finer than any G-topology on $\mathfrak{C}$ which is slightly finer than $\mathfrak{T}$. Thus, apart from conditions (G_0), (G_1), (G_2), the proofs of Proposition 2 and Lemma 3 reduce to the verification of assertion (i) in Lemma 3, which in turn will be based on some facts about coverings compatible with $\mathfrak{T}$.

Lemma 4. (i) *Let U, V be subsets of some $X \in \mathfrak{C}$. If $\{U_i\}_{i\in I}$ is a covering of U compatible with $\mathfrak{T}$, then also the covering $\{U_i \cap V\}_{i\in I}$ is compatible with $\mathfrak{T}$.*

(ii) *Let U be a subset of some $X \in \mathfrak{C}$. If the coverings $\{U_i\}_{i\in I}$ of U and $\{U_{ij}\}_{j\in J_i}$ of U_i, $i \in I$, are compatible with $\mathfrak{T}$, then also $\{U_{ij}\}_{i\in I, j\in J_i}$ is a covering of U compatible with $\mathfrak{T}$.*

Proof. Assertion (i) is obvious. To verify assertion (ii), consider an arbitrary $\mathfrak{C}$-morphism $\varphi\colon Y \to X$ and assume that $Y' \subset Y$ is a $\mathfrak{T}_Y$-open subset satisfying $\varphi(Y') \subset U$. Since $\{U_i\}_{i\in I}$ is compatible with $\mathfrak{T}$, the covering $\{\varphi^{-1}(U_i) \cap Y'\}_{i\in I}$ has a $\mathfrak{T}_Y$-covering $\{Y_\lambda\}_{\lambda\in L}$ which refines it. Let $\tau\colon L \to I$ be a corresponding map such that $Y_\lambda \subset \varphi^{-1}(U_{\tau(\lambda)})$ or, equivalently, $\varphi(Y_\lambda) \subset U_{\tau(\lambda)}$ for all $\lambda \in L$. Since also the coverings $\{U_{ij}\}_{j\in J_i}$, $i \in I$, are compatible with $\mathfrak{T}$, it follows, in particular, that, for every $\lambda \in L$, the covering $\{\varphi^{-1}(U_{\tau(\lambda)j}) \cap Y_\lambda\}_{j \in J_{\tau(\lambda)}}$ of Y_λ has a $\mathfrak{T}_Y$-covering $\{Y_{\lambda\mu}\}_{\mu \in M_\lambda}$ which refines it. Then $\{Y_{\lambda\mu}\}_{\lambda \in L, \mu \in M_\lambda}$ is a $\mathfrak{T}_Y$-covering of Y', and by construction it is a refinement of $\{\varphi^{-1}(U_{ij}) \cap Y'\}_{i \in I, j \in J_i}$. Thus, it is clear that $\{U_{ij}\}_{i \in I, j \in J_i}$ is compatible with $\mathfrak{T}$. □

Now turning to the *proof* of assertion (i) in Lemma 3, we fix an $X \in \mathfrak{C}$ and show that the $\mathfrak{T}'_X$-open sets and the $\mathfrak{T}'_X$-coverings, as defined in Lemma 3, constitute a G-topology on X, i.e., satisfy conditions (i)—(iv) of Definition 9.1.1/1. Let $U, V \subset X$ be $\mathfrak{T}'_X$-open subsets and let $\{U_i\}_{i \in I}$, $\{V_j\}_{j \in J}$ denote coverings of U, V, compatible with $\mathfrak{T}$ and consisting of $\mathfrak{T}_X$-open sets. By (i) of Lemma 4, the covering $\{U_i \cap V\}_{i \in I}$ of $U \cap V$ is compatible with $\mathfrak{T}$, and for any $i \in I$, the same is true for the covering $\{U_i \cap V_j\}_{j \in J}$ of $U_i \cap V$. Thus, by (ii) of Lemma 4, also the covering $\{U_i \cap V_j\}_{i \in I, j \in J}$ of $U \cap V$ is compatible with $\mathfrak{T}$. Since all sets $U_i \cap V_j$ are $\mathfrak{T}_X$-open, $U \cap V$ must be $\mathfrak{T}'_X$-open. Thus, condition (i) of Definition 9.1.1/1 is satisfied. Furthermore, condition (ii) is obvious. Conditions (iii) and (iv) are trivial consequences of Lemma 4. Hence, $\mathfrak{T}'_X$, for any $X \in \mathfrak{C}$, is a G-topology on X.

To conclude the proof of Lemma 3, one still has to check that all $\mathfrak{C}$-morphisms are continuous with respect to the G-topologies $\mathfrak{T}'_X$, i.e., that the $\mathfrak{T}'_X$ constitute a G-topology on $\mathfrak{C}$. But this is straightforward, since $\mathfrak{T}$ is a G-topology on $\mathfrak{C}$ and since the inverse image by a $\mathfrak{C}$-morphism of any covering compatible with $\mathfrak{T}$ is again compatible with $\mathfrak{T}$. □

Finally, knowing the existence and the construction of the G-topology $\mathfrak{T}'$, the *proof* of Proposition 2 reduces to the verification of conditions (G_0), (G_1) and (G_2) for $\mathfrak{T}'$. This is a triviality, except possibly for condition (G_1). Therefore, we restrict ourselves to showing that $\mathfrak{T}'$ satisfies (G_1). For $X \in \mathfrak{C}$, let $U \subset X$ be a $\mathfrak{T}'_X$-open subset admitting $\{U_i\}_{i \in I}$ as a $\mathfrak{T}'_X$-covering. Furthermore, let $V \subset U$ be a subset such that $V \cap U_i$ is $\mathfrak{T}'_X$-open in X for all $i \in I$. Then, due to (i) of Lemma 4, the covering $\{V \cap U_i\}_{i \in I}$ is compatible with $\mathfrak{T}$ since $\{U_i\}_{i \in I}$ is. Choosing a $\mathfrak{T}'_X$-covering $\{V_{ij}\}_{j \in J_i}$ of $V \cap U_i$ consisting of $\mathfrak{T}_X$-open sets for all $i \in I$, it follows from (ii) of Lemma 4 that $\{V_{ij}\}_{i \in I, j \in J_i}$ is a covering of V compatible with $\mathfrak{T}$. Then V is $\mathfrak{T}'_X$-open by Lemma 3. □

We want to apply Proposition 2 to the special situation where the category $\mathfrak{C}$ consists of a single set X and where the identity map id: $X \to X$ is the only morphism. Any G-topology on X can then be viewed as a G-topology on $\mathfrak{C}$, and we get from Proposition 2

Corollary 5. *Let X be a set and $\mathfrak{T}$ a G-topology on X. Then there exists a unique finest G-topology $\mathfrak{T}'$ on X among all the G-topologies slightly finer than $\mathfrak{T}$. The G-topology $\mathfrak{T}'$ satisfies conditions (G_1) and (G_2). In addition, if $\mathfrak{T}$ satisfies condition (G_0), so does $\mathfrak{T}'$.*

In the situation of Proposition 2, where $\mathfrak{T}$ is a G-topology on a category of sets $\mathfrak{C}$, we constructed the finest G-topology $\mathfrak{T}'$ on $\mathfrak{C}$ among all the G-topologies slightly finer than $\mathfrak{T}$. According to Corollary 5, one can also construct for each $X \in \mathfrak{C}$ the finest G-topology $\mathfrak{T}''_X$ on X among all the G-topologies slightly finer than $\mathfrak{T}_X$. As a result, $\mathfrak{T}''_X$ is finer than $\mathfrak{T}'_X$ for each $X \in \mathfrak{C}$; however, the $\mathfrak{T}''_X$ do not necessarily constitute a G-topology on $\mathfrak{C}$, meaning that $\mathfrak{T}''_X$ will be different from $\mathfrak{T}'_X$ in general. Namely, as simple examples show, the continuity of a morphism $\varphi\colon Y \to X$ between G-topological spaces can be lost, when the G-topologies of X and Y are replaced by the slightly finer ones of Corollary 5.

9.1.3. Pasting of G-topological spaces. — We consider the following situation: X is a set, $\{X_i\}_{i \in I}$ is a covering of X, and each X_i carries a G-topology $\mathfrak{T}_i$ such that for all $i, j \in I$,

(a) $X_i \cap X_j$ is $\mathfrak{T}_i$-open in X_i and $\mathfrak{T}_j$-open in X_j and
(b) $\mathfrak{T}_i$ and $\mathfrak{T}_j$ induce the same G-topology on $X_i \cap X_j$.

Among all G-topologies on X admitting the sets X_i as admissible open sets and inducing on X_i the G-topology $\mathfrak{T}_i$, for $i \in I$, there are two canonical ones, a weakest and a finest, denoted by $\mathfrak{T}^{\mathrm{w}}$ and $\mathfrak{T}$, respectively. The $\mathfrak{T}^{\mathrm{w}}$-open sets are just the subsets $U \subset X$ which are admissible open with respect to some $\mathfrak{T}_i$ (in particular, $U \subset X_i$ for some $i \in I$). Accordingly, the $\mathfrak{T}^{\mathrm{w}}$-coverings consist of all $\mathfrak{T}_i$-coverings, $i \in I$. The G-topology $\mathfrak{T}$ is described as follows: a subset $U \subset X$ is $\mathfrak{T}$-open if and only if $U \cap X_i$ is $\mathfrak{T}_i$-open for all $i \in I$. A covering $\{U_j\}_{j \in J}$ is a $\mathfrak{T}$-covering if and only if $\{U_j \cap X_i\}_{j \in J}$ is a $\mathfrak{T}_i$-covering for all $i \in I$. One shows by direct verification that $\mathfrak{T}^{\mathrm{w}}$ and $\mathfrak{T}$ are G-topologies and that they satisfy the stated properties. Since there are no difficulties, we leave the details to the reader.

Proposition 1. *The G-topology $\mathfrak{T}$ is slightly finer than $\mathfrak{T}^{\mathrm{w}}$ if all G-topologies $\mathfrak{T}_i$, $i \in I$, satisfy condition (G_2).*

Proof. It is only necessary to show that the $\mathfrak{T}^{\mathrm{w}}$-open sets form a basis for $\mathfrak{T}$; the rest is trivial. Let $U \subset X$ be $\mathfrak{T}$-open. Then $\{U \cap X_i\}_{i \in I}$ is a covering of U by $\mathfrak{T}^{\mathrm{w}}$-open sets. Restricting to an arbitrary X_j, the covering $\{U \cap X_i \cap X_j\}_{i \in I}$ of $U \cap X_j$ is a $\mathfrak{T}_j$-covering due to (G_2), since the trivial covering $\{U \cap X_j\}$ is a refinement. Thus, $\{U \cap X_i\}_{i \in I}$ is a $\mathfrak{T}$-covering of U by $\mathfrak{T}^{\mathrm{w}}$-open sets showing that the $\mathfrak{T}^{\mathrm{w}}$-open sets form a basis for $\mathfrak{T}$. □

Proposition 2. *Assume that all G-topologies $\mathfrak{T}_i$, $i \in I$, satisfy conditions (G_0), (G_1) and (G_2). Then the G-topology $\mathfrak{T}$ is uniquely characterized on X by the following properties:*

(i) *X_i is $\mathfrak{T}$-open in X and $\mathfrak{T}$ induces the G-topology $\mathfrak{T}_i$ on X_i, for all $i \in I$.*
(ii) *$\mathfrak{T}$ satisfies conditions (G_0), (G_1) and (G_2).*
(iii) *$\{X_i\}_{i \in I}$ is a $\mathfrak{T}$-covering of X.*

Proof. The G-topology $\mathfrak{T}$ satisfies condition (i) and, by a straightforward verification also conditions (ii) and (iii). Thus, if $\mathfrak{T}'$ is a second G-topology on X with properties (i), (ii) and (iii), $\mathfrak{T}'$ will be weaker than $\mathfrak{T}$ by definition of $\mathfrak{T}$. To see that $\mathfrak{T}'$, in fact, equals $\mathfrak{T}$, let $U \subset X$ be $\mathfrak{T}$-open. Then $U \cap X_i$ is $\mathfrak{T}_i$-open in X_i for all $i \in I$ and hence also $\mathfrak{T}'$-open in X since $\mathfrak{T}'$ satisfies (i). Therefore, applying condition (G_1) to the $\mathfrak{T}'$-covering $\{X_i\}_{i\in I}$ and the subset $U \subset X$, one recognizes U as $\mathfrak{T}'$-open and $\{X_i \cap U\}_{i\in I}$ as a $\mathfrak{T}'$-covering of U. Similarly, if $\{U_j\}_{j\in J}$ is a $\mathfrak{T}$-covering of U, its restriction $\{X_i \cap U_j\}_{j\in J}$ to X_i is a $\mathfrak{T}_i$-covering of $X_i \cap U$ for all $i \in I$ and hence also a $\mathfrak{T}'$-covering by (i). Using the "transitivity" of admissible coverings, $\{X_i \cap U_j\}_{i\in I, j\in J}$ is seen to be a $\mathfrak{T}'$-covering of U which is a refinement of the covering $\{U_j\}_{j\in J}$. Thus, $\{U_j\}_{j\in J}$ is a $\mathfrak{T}'$-covering due to condition (G_2), and $\mathfrak{T}'$ must equal $\mathfrak{T}$. □

Proposition 2 states, in particular, that the structure of a G-topological space X satisfying (G_0), (G_1) and (G_2) can be recovered from an admissible covering $\{X_i\}_{i\in I}$ of X, when all induced topologies on the sets X_i, $i \in I$, are known.

9.1.4. *G*-topologies on affinoid varieties. — In (7.2.1) we defined the canonical topology on an affinoid variety X, and we saw by Corollary 7.2.5/4 that the affinoid subdomains of X form a basis for this topology. Since the intersection of two affinoid subdomains in X is again an affinoid subdomain of X, we can introduce a G-topology on X as follows: we call all affinoid subdomains of X admissible open and all affinoid coverings of such sets admissible. Recall that an affinoid covering is a *finite* covering by affinoid subdomains. This G-topology induces the canonical topology on X and is called the *weak G-topology of X*. Any affinoid morphism $Y \to X$ is continuous with respect to the weak G-topology on X and Y by Proposition 7.2.2/4. Therefore, considering the category of affinoid varieties, the collection of all weak G-topologies constitutes a G-topology $\mathfrak{T}$ on this category. Proposition 9.1.2/2 can be applied to conclude that there exists a unique finest G-topology $\mathfrak{T}'$, being slightly finer than $\mathfrak{T}$. For any affinoid variety X, the resulting G-topology $\mathfrak{T}'_X$ is called the *strong G-topology of X*. By construction, all affinoid morphisms are continuous with respect to the strong G-topology. Also the following is obvious from our definitions:

Proposition 1. *The weak G-topology on an affinoid variety X satisfies condition (G_0); the strong G-topology, conditions (G_0), (G_1) and (G_2). The affinoid subdomains of X form a basis for both G-topologies.*

An explicit characterization of the strong G-topology has been given in Lemma 9.1.2/3. Since affinoid subdomains carry the structure of an affinoid variety, the lemma simplifies in our situation to

Proposition 2. *The strong G-topology on an affinoid variety X is characterized by the following properties:*

(i) *A subset $U \subset X$ is admissible open if and only if it admits a covering $\{U_i\}_{i\in I}$ by affinoid subdomains $U_i \subset X$ such that, for any affinoid morphism $\varphi: Y \to X$ with $\varphi(Y) \subset U$, the covering $\{\varphi^{-1}(U_i)\}_{i\in I}$ of Y has an affinoid covering which refines it.*

(ii) *A covering $\{V_j\}_{j\in J}$ by admissible open sets V_j of an admissible open set $V \subset X$ is admissible if and only if, for any affinoid morphism $\varphi: Y \to X$ with*

$\varphi(Y) \subset U$, the covering $\{\varphi^{-1}(V_j)\}_{j\in J}$ of Y has an affinoid covering which refines it.

As direct consequences we get

Corollary 3. *Let X be an affinoid variety and $U \subset X$ an affinoid subdomain. Then the strong G-topology of X restricts on U to the strong G-topology of the affinoid variety U.*

Corollary 4. *Let X be an affinoid variety and consider on X the strong G-topology. Then any finite union of affinoid subdomains in X is admissible open, and any finite covering of such a set by affinoid subdomains of X is admissible.*

That the strong G-topology is, in general, different from the weak G-topology is a consequence of

Proposition 5. *Let $X = \operatorname{Sp} A$ be an affinoid variety. For $f \in A$ consider the following sets:*

$$U \ := \{x \in X;\ |f(x)| < 1\},$$
$$U' \ := \{x \in X;\ |f(x)| > 1\},$$
$$U'' := \{x \in X;\ |f(x)| > 0\}.$$

Any finite union of sets of this type is admissible open, and any finite covering by sets of this type is admissible with respect to the strong G-topology on X.

Proof. Choosing a strictly increasing sequence $\varepsilon_\nu \in \sqrt{|k^*|}$ satisfying $\lim\limits_{\nu\to\infty} \varepsilon_\nu = 1$, we get

$$U = \bigcup_{\nu=1}^{\infty} X(\varepsilon_\nu^{-1} f)$$

(using notations for affinoid subdomains of X as in (7.2.3)). To see that U is admissible open and that the above covering of U is admissible, we consider an arbitrary affinoid morphism $\varphi\colon \operatorname{Sp} B \to \operatorname{Sp} A$ satisfying $\varphi(\operatorname{Sp} B) \subset U$. If $\varphi^*\colon A \to B$ denotes the corresponding algebra homomorphism, the Maximum Modulus Principle (Proposition 6.2.1/4) for $\operatorname{Sp} B$ yields $|\varphi^*(f)|_{\sup} < 1$, since $|f(x)| < 1$ for all $x \in U$. Thus, for almost all indices ν, we have $\varphi^{-1}\big(X(\varepsilon_\nu^{-1} f)\big) = \operatorname{Sp} B$, implying that U is admissible open and that the covering $\{X(\varepsilon_\nu^{-1} f)\}_{\nu\in\mathbf{N}}$ is admissible.

Similarly, using strictly decreasing sequences $\varepsilon_\nu \in \sqrt{|k^*|}$ having 1 or 0 as limit, one shows the corresponding results for U' and U''. Of course, in these cases, one has to rely on the fact that $\varphi^*(f)$ assumes its minimum on $\operatorname{Sp} B$. The situation is a little bit more complicated when finite unions of sets of type U, U' and U'' are considered. Here one proceeds basically in the same way as above; however, the Maximum Modulus Principle has to be replaced by the following more general version.

Lemma 6. *Let B be an affinoid algebra and let $f = (f_1, \ldots, f_l)$, $g = (g_1, \ldots, g_m)$, $h = (h_1, \ldots, h_n)$ be systems of functions in B such that each $x \in \operatorname{Sp} B$ satisfies at* *least one of the equations* $|f_\lambda(x)| < 1$, $|g_\mu(x)| > 1$, *or* $|h_\nu(x)| > 0$, $\lambda = 1, \ldots, l$, $\mu = 1, \ldots, m$, $\nu = 1, \ldots, n$. *Then there exist constants* $\alpha, \beta, \gamma \in \sqrt{|k^*|}$, $\alpha < 1$, $\beta > 1$ *such that each* $x \in \operatorname{Sp} B$ *also satisfies one of the equations* $|f_\lambda(x)| \leqq \alpha$, $|g_\mu(x)| \geqq \beta$, *or* $|h_\nu(x)| \geqq \gamma$.

Proof. If $X = \operatorname{Sp} B$ admits a covering $\{U_1, \ldots, U_r\}$ by affinoid subdomains and if for each $\varrho \in \{1, \ldots, r\}$ constants $\alpha_\varrho < 1$, $\beta_\varrho > 1$, $\gamma_\varrho > 0$ can be found such that the assertion of the lemma is true for all $x \in U_\varrho$ with respect to these constants, then clearly the lemma will be proved. In particular, if we choose an $\varepsilon \in \sqrt{|k^*|}$, $\varepsilon < 1$, and consider the covering

$$X = X(\varepsilon f_1^{-1}, \ldots, \varepsilon f_l^{-1}) \cup \bigcup_{\lambda=1}^{l} X(\varepsilon^{-1} f_\lambda),$$

the lemma only has to be verified for $X(\varepsilon f_1^{-1}, \ldots, \varepsilon f_l^{-1})$ instead of X, and we may assume that $f_1, \ldots, f_l$ are units in B. Then we can replace the system g by $(f_1^{-1}, \ldots, f_l^{-1}, g_1, \ldots, g_m)$, and thereby simplify our problem to the case where f has been dropped and only the systems g and h are of interest.

In this situation, the functions $h_1, \ldots, h_n$ cannot have a common zero on $X(g_1, \ldots, g_m)$ by assumption. Thus, by Lemma 7.3.4/7, there exists a $\delta \in \sqrt{|k^*|}$ such that $\max_{1 \leqq \nu \leqq n} |h_\nu(x)| > \delta$ for all $x \in X(g_1, \ldots, g_m)$, or, equivalently, for any $x \in X(\delta^{-1}h_1, \ldots, \delta^{-1}h_n)$ there must exist an index $\mu \in \{1, \ldots, m\}$ satisfying $|g_\mu(x)| > 1$. Hence, considering the covering

$$X = X(\delta^{-1}h_1, \ldots, \delta^{-1}h_n) \cup \bigcup_{\nu=1}^{n} X(\delta h_\nu^{-1}),$$

we can replace X by $X(\delta^{-1}h_1, \ldots, \delta^{-1}h_n)$, thereby reducing our problem to the case where only g is of interest and the systems f and h have been dropped.

Turning to this special case, the functions $g_1, \ldots, g_m$ cannot have a common zero on X. Therefore,

$$X = \bigcup_{\mu=1}^{m} X\left(\frac{g_1}{g_\mu}, \ldots, \frac{g_m}{g_\mu}\right)$$

is a well-defined rational covering of X such that

$$\max_{1 \leqq i \leqq m} |g_i(x)| = |g_\mu(x)| > 1$$

for all $x \in X\left(\frac{g_1}{g_\mu}, \ldots, \frac{g_m}{g_\mu}\right)$. Since, by the Maximum Modulus Principle, g_μ assumes its minimum on the affinoid variety $X\left(\frac{g_1}{g_\mu}, \ldots, \frac{g_m}{g_\mu}\right)$, the assertion of the lemma is clear, and Proposition 5 follows. □

Corollary 7. *All Zariski-open subsets of an affinoid variety X are admissible open with respect to the strong G-topology on X. All coverings of X by Zariski-open subsets are admissible with respect to the strong G-topology on X.*

Proof. Each Zariski-open covering of X contains a finite subcovering. Therefore the assertions are clear by Proposition 5. □

In particular, for any affinoid variety X, the strong G-topology is finer than the Zariski topology (viewed as a G-topology in the obvious way). This fact can be used in order to characterize the connectedness of affinoid varieties.

Proposition 8. *Let X be an affinoid variety. Then the following are equivalent:*

(i) *X is connected with respect to the Zariski topology.*
(ii) *X is connected with respect to the weak G-topology.*
(iii) *X is connected with respect to the strong G-topology.*

Proof. If X is not connected with respect to the Zariski topology, we see by Corollary 7 (or by a simple direct argument) that it is not connected with respect to the strong G-topology. Furthermore, the latter implies that X cannot be connected with respect to the weak G-topology. Namely, any covering of X, which is admissible with respect to the strong G-topology, has a refinement which is admissible with respect to the weak G-topology. Thus it remains only to show that X is not Zariski-connected if it is not connected with respect to the weak G-topology. Let

$$X = U_1 \cup \ldots \cup U_m \cup V_1 \cup \ldots \cup V_n$$

be a covering of X by non-empty affinoid subdomains, and assume that

$$U := \bigcup_{i=1}^{m} U_i \qquad \text{and} \qquad V := \bigcup_{j=1}^{n} V_j$$

are disjoint. By Corollary 8.2.1/2 of TATE's Acyclicity Theorem, there exists an affinoid function $f \in \mathcal{O}_X(X)$ such that

$$f|_{U_i} = 1 \ (= \text{unit element in } \mathcal{O}_X(U_i)), \qquad i = 1, \ldots, m,$$

and

$$f|_{V_j} = 0 \ (= \text{zero element in } \mathcal{O}_X(V_j)), \qquad j = 1, \ldots, n.$$

Then

$$U = \{x \in X; f(x) \neq 0\},$$

and

$$V = \{x \in X; (1 - f)(x) \neq 0\}.$$

Hence U and V are Zariski-open in X, and we see that X is not Zariski-connected. □

In order to see whether or not an affinoid variety X is connected (with respect to the Zariski topology), one considers its unique minimal decomposition

$$X = X_1 \cup \ldots \cup X_r$$

into irreducible Zariski-closed subsets (cf. Corollary 7.1.2/8). If X is not connected, say $X = X' \mathbin{\dot\cup} X''$ with non-empty Zariski-open (and hence Zariski-closed) subsets X' and X'', one can decompose X' and X'' and thereby obtain a new minimal decomposition of X into irreducible Zariski-closed subsets. Then, by the uniqueness of such a decomposition, X' and X'' are unions of some of the sets $X_1, \ldots, X_r$. Thus X is connected if and only if there is no partition $\{1, \ldots, r\} = I_1 \,\dot\cup\, I_2$ such that $I_1 \neq \emptyset \neq I_2$ and $X_i \cap X_j = \emptyset$ for all $i \in I_1, j \in I_2$.

Let $Z_1, \ldots, Z_s$ denote the connected components of X with respect to the Zariski topology (i.e., the maximal Zariski-connected subsets of X). Each Z_j is a union of some of the sets $X_1, \ldots, X_r$, and $X = \bigcup_{j=1}^{s} Z_j$ is a covering of X by disjoint Zariski-closed (and hence Zariski-open) subsets of X. By the Chinese Remainder Theorem, one knows that, for each $j = 1, \ldots, s$, there is a function $g_j \in \mathcal{O}_X(X)$ such that g_j is identically 1 on Z_j and identically zero on $X - Z_j$ so that $Z_j = \{x \in X; |g_j(x)| \geqq 1\}$. Hence each Z_j is a connected affinoid subdomain of X. We call $Z_1, \ldots, Z_s$ the *connected components* of the affinoid variety X.

In particular, if X is irreducible (which is the case if $\mathcal{O}_X(X)/\mathrm{rad}\, \mathcal{O}_X(X)$ is an integral domain, see Proposition 7.1.2/7), there is only one connected component and X is connected itself.

9.2. Sheaf theory

9.2.1. Presheaves and sheaves on G-topological spaces. — We have already dealt with presheaves before, in particular, with the presheaf $\mathcal{O}_X$ belonging to an affinoid variety X. In the following we want to investigate more systematically sheaf theory on G-topological spaces. It should be mentioned at this point that ordinary sheaf theory on topological spaces is a special case of what is presented below. One simply has to view topological spaces as G-topological spaces by calling all open sets and all open coverings admissible.

Definition 1. *A presheaf of abelian groups or rings, etc., on a G-topological space X is a contravariant functor $\mathcal{F}$ from the category of all admissible open subsets of X, with inclusions as morphisms, into the category of abelian groups or rings, etc.*

Hence, a presheaf $\mathcal{F}$, say of abelian groups on X, assigns to each admissible open subset $U \subset X$ an abelian group $\mathcal{F}(U)$ and to each inclusion $U \hookrightarrow V$ of admissible open subsets in X a so-called restriction homomorphism

$$\mathcal{F}(V) \to \mathcal{F}(U)$$
$$f \mapsto f|_U, \qquad f \in \mathcal{F}(V),$$

such that the following properties are fulfilled: the trivial inclusion $U \subset U$ induces the identity map $\mathcal{F}(U) \to \mathcal{F}(U)$, and whenever $U \subset V \subset W$ are admissible open subsets of X, the restriction homomorphism $\mathcal{F}(W) \to \mathcal{F}(U)$ is the composition of the restrictions $\mathcal{F}(W) \to \mathcal{F}(V)$ and $\mathcal{F}(V) \to \mathcal{F}(U)$.

Definition 2. *A presheaf $\mathcal{F}$ of abelian groups or rings, etc., on X is called a sheaf if, for all admissible open subsets $U \subset X$ and all admissible coverings $\{U_i\}_{i\in I}$ of U, the following conditions are satisfied:*

(i) *If $f, g \in \mathcal{F}(U)$ are elements such that $f|_{U_i} = g|_{U_i}$ for all $i \in I$, then $f = g$.*

(ii) *If $f_i \in \mathcal{F}(U_i)$, $i \in I$, is a family of elements such that $f_i|_{U_i\cap U_j} = f_j|_{U_i\cap U_j}$ for all $i, j \in I$, there exists an $f \in \mathcal{F}(U)$ with $f|_{U_i} = f_i$ for all $i \in I$.*

Note that by (i) the element f in (ii) is uniquely determined. Conditions (i) and (ii) can be phrased more elegantly by requiring the diagram

$$\mathcal{F}(U) \xrightarrow{\sigma} \prod_{i\in I} \mathcal{F}(U_i) \underset{\sigma''}{\overset{\sigma'}{\rightrightarrows}} \prod_{i,j\in I} \mathcal{F}(U_i \cap U_j)$$

to be exact. Here σ is induced by the restrictions $\mathcal{F}(U) \to \mathcal{F}(U_i)$; whereas σ' and σ'' are induced by the restrictions $\mathcal{F}(U_i) \to \mathcal{F}(U_i \cap U_j)$ and $\mathcal{F}(U_j) \to \mathcal{F}(U_i \cap U_j)$, respectively. As usual, a diagram

$$A \xrightarrow{\sigma} B \underset{\sigma''}{\overset{\sigma'}{\rightrightarrows}} C$$

is called exact if A is mapped bijectively by σ onto the set of all $b \in B$ satisfying $\sigma'(b) = \sigma''(b)$.

If the empty set $\emptyset$ is admissible open in X, it is convenient to require $\mathcal{F}(\emptyset) = 0$ for any sheaf $\mathcal{F}$ on X. This property can be formally deduced from condition (i) if the empty covering (with index set $I = \emptyset$) is accepted as an admissible covering of $\emptyset$.

Example. Let X be an affinoid variety. Then the presheaf $\mathcal{O}_X$ of (7.3.2), which associates to each affinoid subdomain of X its corresponding affinoid algebra, is, in fact, a sheaf with respect to the weak G-topology on X. This follows by Corollary 8.2.1/2 from Tate's Acyclicity Theorem.

Many concepts of ordinary sheaf theory carry over to the G-topological situation; here are some examples. Let $\mathcal{F}$ be a presheaf, say of abelian groups on a G-topological space X. If $U \subset X$ is an admissible open subset, $\mathcal{F}$ restricts to a presheaf on U, denoted by $\mathcal{F}|_U$ and called the *restriction of $\mathcal{F}$ to U*. As in (7.3.2), one defines the *stalk of $\mathcal{F}$ at a point $x \in X$* by

$$\mathcal{F}_x := \varinjlim \mathcal{F}(U),$$

where the direct limit is taken over all admissible open subsets $U \subset X$ containing x. In particular, if the system of these sets is empty, one sets $\mathcal{F}_x = 0$.

A *homomorphism $\varphi\colon \mathcal{F} \to \mathcal{F}'$ of presheaves or sheaves on X* is a collection of homomorphisms $\varphi_U\colon \mathcal{F}(U) \to \mathcal{F}'(U)$ compatible with restriction homomorphisms, where U runs over all admissible open subsets of X. Any such homomorphism φ induces homomorphisms between the stalks

$$\varphi_x\colon \mathcal{F}_x \to \mathcal{F}'_x, \qquad x \in X.$$

If $\mathcal{O}$ is a presheaf of rings on X, an *$\mathcal{O}$-module* is a presheaf of abelian groups $\mathcal{F}$ on X together with an $\mathcal{O}(U)$-module structure on $\mathcal{F}(U)$ for each admissible open set $U \subset X$ (likewise for sheaves). Of course, these module structures have to be compatible with all restriction homomorphisms, and it follows that the stalk $\mathcal{F}_x$ is an $\mathcal{O}_x$-module for all $x \in X$.

Finally we want to mention a few things about Čech cohomology on a G-topological space X with values in a presheaf $\mathcal{F}$. If $\mathfrak{U}$ is an admissible covering of some admissible open subset $U \subset X$, the Čech complex $C^{\cdot}(\mathfrak{U}, \mathcal{F})$ and the cohomology groups $H^p(\mathfrak{U}, \mathcal{F})$ are defined as in (8.1.3), although strictly speaking $\mathcal{F}$ would have to be replaced by its restriction to U. Considering only admissible coverings, the Čech cohomology theory developed in (8.1) is completely applicable. In particular, the Čech cohomology groups

$$\check{H}^p(U, \mathcal{F}) = \varinjlim_{\mathfrak{U} \in \operatorname{Cov} U} H^p(\mathfrak{U}, \mathcal{F})$$

can be constructed, the direct limit being taken over all admissible coverings $\mathfrak{U}$ of U. Note that the direct limit does exist, since any two admissible coverings $\mathfrak{U} = \{U_i\}_{i \in I}$ and $\mathfrak{V} = \{V_j\}_{j \in J}$ of U admit a common admissible refinement, for example, the covering $\mathfrak{U} \times \mathfrak{V} = \{U_i \cap V_j\}_{i \in I, \in J}$.

It is not really necessary to extend the above direct limit over the system $\operatorname{Cov} U$ of *all* admissible coverings of U; any cofinal subsystem $\operatorname{Cov}' U$ (where $\operatorname{Cov}' U$ is called cofinal in $\operatorname{Cov} U$ if each covering in $\operatorname{Cov} U$ has a refinement in $\operatorname{Cov}' U$) will lead to the same Čech cohomology groups. For example, if B is a basis for the G-topology of X, one could take the system of all admissible coverings $\{U_i\}_{i \in I}$ of U with sets $U_i \in B$. Thus, it is seen that the Čech cohomology groups $\check{H}^p(U, \mathcal{F})$ depend only on the "restriction" of $\mathcal{F}$ to B, or, more generally, they can be defined for any "presheaf" on the basis B.

9.2.2. Sheafification of presheaves. — Starting with a presheaf $\mathcal{F}$ of abelian groups or rings, etc., on a G-topological space X, we want to obtain a sheaf which best approximates $\mathcal{F}$. Such a procedure will be of general interest for the construction of sheaves.

Definition 1. *A sheafification of the presheaf $\mathcal{F}$ is a homomorphism $\mathcal{F} \to \mathcal{F}'$ into a sheaf $\mathcal{F}'$ having the following universal property: given any homomorphism $\mathcal{F} \to \mathcal{G}$ into a sheaf $\mathcal{G}$, there exists a unique homomorphism $\mathcal{F}' \to \mathcal{G}$ making the diagram*

$$\begin{array}{ccc} \mathcal{F} & \longrightarrow & \mathcal{F}' \\ & \searrow \quad \swarrow & \\ & \mathcal{G} & \end{array}$$

commutative.

The sheaf $\mathcal{F}'$ is uniquely determined by $\mathcal{F}$ up to isomorphism; it is called the *sheaf associated with* $\mathcal{F}$. Note that the identity map $\mathcal{F} \to \mathcal{F}$ is a sheafification of $\mathcal{F}$ if $\mathcal{F}$ is already a sheaf. In the following we will prove the existence of sheafifications by using Čech cohomology. The classical construction relying on stalks is not applicable here, since there are non-zero sheaves having only zero stalks. (In order to obtain simple examples of such sheaves, consider a

G-topological space X such that, for each admissible open subset $U \subset X$, the system Cov U of admissible coverings of U contains only the trivial covering $\{U\}$. For such spaces X, the notions of presheaves and sheaves coincide.)

For any presheaf $\mathcal{F}$ on X, one constructs a presheaf $\check{H}^0(X, \mathcal{F})$ which is given by

$$U \rightsquigarrow \check{H}^0(U, \mathcal{F}), \qquad U \subset X \text{ admissible open.}$$

The restrictions are canonically defined; namely let $U \subset V$ be admissible open subsets of X and let $\mathfrak{V} = \{V_i\}_{i \in I}$ be an admissible covering of V. The induced covering $\mathfrak{V}|_U = \{V_i \cap U\}_{i \in I}$ of U is also admissible, and there is a natural homomorphism of Čech complexes

$$C^\cdot(\mathfrak{V}, \mathcal{F}) \to C^\cdot(\mathfrak{V}|_U, \mathcal{F}).$$

This homomorphism induces a homomorphism

$$H^0(\mathfrak{V}, \mathcal{F}) \to H^0(\mathfrak{V}|_U, \mathcal{F})$$

and hence, by taking direct limits, a homomorphism

$$\check{H}^0(V, \mathcal{F}) \to \check{H}^0(U, \mathcal{F}),$$

which is the restriction homomorphism corresponding to the inclusion $U \subset V$. The properties of a presheaf are easily verified.

For any admissible open subset $U \subset X$ and any admissible covering $\mathfrak{U}$ of U, the augmentation homomorphism $\mathcal{F}(U) \to C^0(\mathfrak{U}, \mathcal{F})$ of (8.1.3) induces a homomorphism $\mathcal{F}(U) \to H^0(\mathfrak{U}, \mathcal{F})$. Taking direct limits over admissible coverings, one obtains homomorphisms $\mathcal{F}(U) \to \check{H}^0(U, \mathcal{F})$ which are compatible with restriction homomorphisms and thus constitute a homomorphism of presheaves $\mathcal{F} \to \check{\mathcal{H}}^0(X, \mathcal{F})$. Clearly, $\check{\mathcal{H}}^0(X, \cdot)$ is a functor, and each homomorphism of presheaves $\mathcal{F} \to \mathcal{G}$ induces a commutative diagram

$$\begin{array}{ccc} \mathcal{F} & \longrightarrow & \check{\mathcal{H}}^0(X, \mathcal{F}) \\ \downarrow & & \downarrow \\ \mathcal{G} & \longrightarrow & \check{\mathcal{H}}^0(X, \mathcal{G}). \end{array}$$

Note that $\mathcal{F} \to \check{\mathcal{H}}^0(X, \mathcal{F})$ is an isomorphism when $\mathcal{F}$ is a sheaf, since all homomorphisms $\mathcal{F}(U) \to H^0(\mathfrak{U}, \mathcal{F})$ are then isomorphisms. — For later reference we state

Lemma 2. *Let $U \subset X$ be admissible open and let g be an element of $\check{H}^0(U, \mathcal{F})$. Then there exists an admissible covering $\{U_i\}_{i \in I}$ of U such that, for any $i \in I$, the element $g|_{U_i}$ belongs to the image of the augmentation homomorphism $\mathcal{F}(U_i) \to \check{H}^0(U_i, \mathcal{F})$.*

Proof. The result is obvious, since g must be represented by some element in $H^0(\mathfrak{U}, \mathcal{F})$ for a suitable admissible covering $\mathfrak{U}$ of U. □

The functor $\check{\mathcal{H}}^0(X, \cdot)$ does not, in general, turn presheaves into sheaves;

however, the surprising fact is that, if applied twice, $\check{\mathscr{H}}^0(X,\cdot)$ does yield sheaves!

Lemma 3. *Let $\mathscr{F}$ be a presheaf on X. Then $\check{\mathscr{H}}^0(X,\mathscr{F})$ satisfies sheaf condition* (i) *of Definition* 9.2.1/2. *Furthermore, if $\mathscr{F}$ already fulfills this condition, $\check{\mathscr{H}}^0(X,\mathscr{F})$ also satisfies sheaf condition* (ii) *of Definition* 9.2.1/2 *and thus is a sheaf.*

Proof. Let $\mathfrak{U} = \{U_i\}_{i\in I}$ be an admissible covering of some admissible open subset $U \subset X$. We consider the diagram

$$\check{H}^0(U,\mathscr{F}) \xrightarrow{\ \sigma\ } \prod_{i\in I} \check{H}^0(U_i,\mathscr{F}) \overset{\sigma'}{\underset{\sigma''}{\rightrightarrows}} \prod_{i,j\in I} \check{H}^0(U_i \cap U_j,\mathscr{F}),$$

with maps σ, σ', σ'' as explained in the paragraph following Definition 9.2.1/2. To verify the first part of our assertion, we have to show that σ is injective. The second part will follow if the diagram is exact.

For each $i \in I$, we choose an admissible covering $\mathfrak{V}_i = \{V_{i\nu}\}_{\nu\in J_i}$ of U_i. Then $\mathfrak{V} := \{V_{i\nu}\}_{i\in I,\nu\in J_i}$ is an admissible covering of U, and $\mathfrak{V}_{ij} := \{V_{i\nu} \cap V_{j\mu}\}_{\nu\in J_i,\mu\in J}$ an admissible covering of $U_i \cap U_j$. Using these coverings, we extend the above diagram to the following:

$$\begin{array}{ccccc}
\prod\limits_{i\in I,\nu\in J_i} \mathscr{F}(V_{i\nu}) & \xrightarrow{\ \varrho\ } & \prod\limits_{i\in I}\prod\limits_{\nu\in J_i} \mathscr{F}(V_{i\nu}) & \overset{\varrho'}{\underset{\varrho''}{\rightrightarrows}} & \prod\limits_{i,j\in I}\prod\limits_{\substack{\nu\in J_i\\ \mu\in J_j}} \mathscr{F}(V_{i\nu} \cap V_{j\mu}) \\
\uparrow & & \uparrow & & \uparrow \\
H^0(\mathfrak{V},\mathscr{F}) & \xrightarrow{\ \tau\ } & \prod\limits_{i\in I} H^0(\mathfrak{V}_i,\mathscr{F}) & \overset{\tau'}{\underset{\tau''}{\rightrightarrows}} & \prod\limits_{i,j\in I} H^0(\mathfrak{V}_{ij},\mathscr{F}) \\
\downarrow & & \downarrow & & \downarrow \\
\check{H}^0(U,\mathscr{F}) & \xrightarrow{\ \sigma\ } & \prod\limits_{i\in I} \check{H}^0(U_i,\mathscr{F}) & \overset{\sigma'}{\underset{\sigma''}{\rightrightarrows}} & \prod\limits_{i,j\in I} \check{H}^0(U_i \cap U_j,\mathscr{F})
\end{array}$$

By the vertical arrows, we mean the canonical maps; ϱ is the canonical bijection. Furthermore, for

$$f = (f_{i\nu}) \in \prod_{i\in I}\prod_{\nu\in J_i} \mathscr{F}(V_{i\nu}),$$

$\varrho'(f)$ and $\varrho''(f)$ are defined by

$$\varrho'(f)_{ij\nu\mu} := f_{i\nu}|_{V_{i\nu}\cap V_{j\mu}} \quad \text{and}$$
$$\varrho''(f)_{ij\nu\mu} := f_{j\mu}|_{V_{i\nu}\cap V_{j\mu}}, \quad i,j \in I, \quad \nu \in J_i, \quad \mu \in J_j.$$

The maps in the second row are induced by those in the first row. The diagram is commutative when either the $'$-mappings or the $''$-mappings are considered.

Let Cov$'$ U denote the system of all admissible coverings of U which are obtained from the covering $\mathfrak{U}$ in the same way as $\mathfrak{V}$ was, namely, by "inserting" an admissible covering $\mathfrak{V}_i$ of U_i for each $U_i \in \mathfrak{U}$. Then Cov$'$ U is cofinal in the system Cov U of all admissible coverings of U, and as a result, σ is the direct limit of all maps τ for $\mathfrak{V}$ ranging in Cov$'$ U. Since ϱ is bijective, τ is injective.

Thus, σ, as a direct limit of injections, must also be injective, which proves the first part of our assertion.

Now let us assume that $\mathscr{F}$ satisfies sheaf condition (i) of Definition 9.2.1/2. Then it is not hard to see that all vertical maps in the lower part of the diagram are injective (use cofinal systems of admissible coverings as before). Considering an element

$$g = (g_i) \in \prod_{i\in I} \check{H}^0(U_i, \mathscr{F})$$

with $\sigma'(g) = \sigma''(g)$, we may assume by Lemma 2 that the coverings $\mathfrak{V}_i$ are fine enough so that each component g_i is represented by some element in $H^0(\mathfrak{V}_i, \mathscr{F})$. Then g is represented by an element $f \in \prod_{i\in I} H^0(\mathfrak{V}_i, \mathscr{F})$ which satisfies $\tau'(f) = \tau''(f)$, since all vertical maps are injective. The second row of the diagram is exact for trivial reasons. Therefore, we must have $f \in \operatorname{im} \tau$ and, in particular, $g \in \operatorname{im} \sigma$. Thus, the second part of the assertion is clear. □

Proposition 4. *Let $\mathscr{F}$ be a presheaf on X. Then the composition of the canonical maps*

$$\mathscr{F} \to \check{\mathscr{H}}^0(X, \mathscr{F}) \to \check{\mathscr{H}}^0\big(X, \check{\mathscr{H}}^0(X, \mathscr{F})\big)$$

is a sheafification of $\mathscr{F}$.

Proof. Any homomorphism of presheaves $\psi\colon \mathscr{F} \to \mathscr{G}$ induces a commutative diagram

$$\begin{array}{ccccc}
\mathscr{F} & \longrightarrow & \check{\mathscr{H}}^0(X, \mathscr{F}) & \longrightarrow & \check{\mathscr{H}}^0\big(X, \check{\mathscr{H}}^0(X, \mathscr{F})\big) \\
\downarrow{\scriptstyle\psi} & & \downarrow & & \downarrow \\
\mathscr{G} & \longrightarrow & \check{\mathscr{H}}^0(X, \mathscr{G}) & \longrightarrow & \check{\mathscr{H}}^0\big(X, \check{\mathscr{H}}^0(X, \mathscr{G})\big).
\end{array}$$

If, in addition, $\mathscr{G}$ is a sheaf, then the maps in the lower row are isomorphisms. Hence there exists a factorization

$$\begin{array}{ccc}
\mathscr{F} & \longrightarrow & \check{\mathscr{H}}^0\big(X, \check{\mathscr{H}}^0(X, \mathscr{F})\big) \\
& {\scriptstyle\psi}\searrow \quad \swarrow{\scriptstyle\chi} & \\
& \mathscr{G} &
\end{array}$$

of ψ through $\check{\mathscr{H}}^0\big(X, \check{\mathscr{H}}^0(X, \mathscr{F})\big)$ in this case.

To see that χ is uniquely determined, let $U \subset X$ be admissible open, and let $g \in \check{H}^0\big(U, \check{\mathscr{H}}^0(X, \mathscr{F})\big)$. Applying Lemma 2 twice, there exists an admissible covering $\{U_i\}_{i\in I}$ of U such that, for each $i \in I$, the element $g|_{U_i}$ has an inverse image $g_i \in \mathscr{F}(U_i)$ with respect to the map $\mathscr{F}(U_i) \to \check{H}^0\big(U_i, \check{\mathscr{H}}^0(X, \mathscr{F})\big)$. Then we must have

$$\chi_U(g)|_{U_i} = \psi_{U_i}(g_i)$$

for all $i \in I$, and therefore $\chi_U(g) \in \mathscr{G}(U)$ is uniquely determined due to the fact that $\mathscr{G}$ is a sheaf. Thus, χ is unique. This concludes our proof, since $\check{\mathscr{H}}^0\big(X, \check{\mathscr{H}}^0(X, \mathscr{F})\big)$ is a sheaf by Lemma 3. □

We want to derive some consequences from the explicit description of the sheafification given in Proposition 4. In the following let $\mathcal{F} \to \mathcal{F}'$ be a sheafification of the presheaf $\mathcal{F}$.

Corollary 5. *For any admissible open set $U \subset X$, the induced homomorphism $\mathcal{F}|_U \to \mathcal{F}'|_U$ is a sheafification of $\mathcal{F}|_U$.*

Corollary 6. *For any admissible open set $U \subset X$ and any $g \in \mathcal{F}'(U)$, there exists an admissible covering $\{U_i\}_{i\in I}$ of U such that, for each $i \in I$, the element $g|_{U_i}$ is in the image of the homomorphism $\mathcal{F}(U_i) \to \mathcal{F}'(U_i)$.*

Corollary 7. *For any $x \in X$, the induced homomorphism between the stalks $\mathcal{F}_x \to \mathcal{F}'_x$ is bijective.*

Due to Lemma 2 and Proposition 4, only Corollary 7 requires *proof*. The surjectivity of $\mathcal{F}_x \to \mathcal{F}'_x$ follows from Corollary 6. In order to see the injectivity, we fix an $x \in X$ and define a sheaf $\mathcal{G}$ by

$$\mathcal{G}(U) := \begin{cases} \mathcal{F}_x & \text{if } x \in U \\ 0 & \text{if } x \notin U \end{cases}$$

for $U \subset X$ admissible open with restrictions being the identity or zero map. The canonical homomorphism $\mathcal{F} \to \mathcal{G}$ factors through $\mathcal{F}'$ and induces a bijection $\mathcal{F}_x \xrightarrow{\sim} \mathcal{G}_x$. Then, in particular, $\mathcal{F}_x \to \mathcal{F}'_x$ must be injective. □

Since the process of sheafification of presheaves has been established, it is now possible to carry out many of the usual operations on sheaves. For some examples concerning modules over a sheaf of rings, we refer to (9.4).

9.2.3. Extension of sheaves. — In the following consider on a base space X two G-topologies $\mathfrak{T}$ and $\mathfrak{T}'$ where $\mathfrak{T}'$ is finer than $\mathfrak{T}$. When dealing with sheaves on X, we use the terminology $\mathfrak{T}$-sheaves and $\mathfrak{T}'$-sheaves in order to specify the corresponding G-topology. It is clear that each $\mathfrak{T}'$-sheaf $\mathcal{F}'$ induces a $\mathfrak{T}$-sheaf $\mathcal{F}$ simply by restricting $\mathcal{F}'$ to the $\mathfrak{T}$-open subsets of X. In this situation we call $\mathcal{F}'$ a *$\mathfrak{T}'$-extension* of $\mathcal{F}$. Of course, the same applies to presheaves.

Proposition 1. *Let $\mathfrak{T}'$ be slightly finer than $\mathfrak{T}$. Then any $\mathfrak{T}$-sheaf $\mathcal{F}$ extends to a $\mathfrak{T}'$-sheaf $\mathcal{F}'$ which is uniquely determined up to isomorphism. Furthermore, any homomorphism of $\mathfrak{T}$-sheaves $\mathcal{F}_1 \to \mathcal{F}_2$ extends uniquely to a homomorphism $\mathcal{F}'_1 \to \mathcal{F}'_2$ between the corresponding $\mathfrak{T}'$-extensions.*

Proof. The system B of all $\mathfrak{T}$-open subsets in X is a basis for $\mathfrak{T}'$; hence any $\mathfrak{T}'$-open subset $U \subset X$ admits a $\mathfrak{T}'$-covering $\{U_i\}_{i\in I}$ by sets $U_i \in B$. The system of these coverings is cofinal in the system of all $\mathfrak{T}'$-coverings of U. Therefore, considering $\mathcal{F}$ as a $\mathfrak{T}'$-presheaf on the basis B, one can construct $\check{\mathcal{H}}^0(X_{\mathfrak{T}'}, \mathcal{F})$ as a $\mathfrak{T}'$-presheaf on X in the same way as in (9.2.2). We have

$$\check{\mathcal{H}}^0(X_{\mathfrak{T}'}, \mathcal{F}) = \check{\mathcal{H}}^0(X_{\mathfrak{T}'}, \mathcal{F}')$$

for any $\mathfrak{T}'$-presheaf $\mathcal{F}'$ extending $\mathcal{F}$. Restricting $\check{\mathcal{H}}^0(X_{\mathfrak{T}'}, \mathcal{F})$ with respect to the G-topology $\mathfrak{T}$, one obtains the $\mathfrak{T}$-presheaf $\check{\mathcal{H}}^0(X_{\mathfrak{T}}, \mathcal{F})$, since the system of all $\mathfrak{T}$-coverings of a $\mathfrak{T}$-open set $U \subset X$ is cofinal in the system of all $\mathfrak{T}'$-coverings of U. Furthermore, the canonical homomorphism $\mathcal{F} \to \check{\mathcal{H}}^0(X_{\mathfrak{T}}, \mathcal{F})$ is an isomorphism of $\mathfrak{T}$-sheaves; thus, $\mathcal{F}' := \check{\mathcal{H}}^0(X_{\mathfrak{T}'}, \mathcal{F})$ may be viewed as a $\mathfrak{T}'$-presheaf extending $\mathcal{F}$.

The sheaf $\mathcal{F}'$ is, in fact, a $\mathfrak{T}'$-sheaf. This can be seen by examining the proof of Lemma 9.2.2/3. Alternatively, we could have set $\mathcal{F}' := \check{\mathcal{H}}^0\big(X_{\mathfrak{T}'}, \check{\mathcal{H}}^0(X_{\mathfrak{T}'}, \check{\mathcal{H}}^0(X_{\mathfrak{T}'}, \mathcal{F}))\big)$. In this case, Lemma 9.2.2/3 would apply as it stands. If $\mathcal{F}''$ is another $\mathfrak{T}'$-sheaf extending $\mathcal{F}$, there are isomorphisms

$$\mathcal{F}'' \xrightarrow{\sim} \check{\mathcal{H}}^0(X_{\mathfrak{T}'}, \mathcal{F}'') = \check{\mathcal{H}}^0(X_{\mathfrak{T}'}, \mathcal{F}),$$

showing that $\mathcal{F}'$ is unique up to isomorphism.

Finally, if $\varphi\colon \mathcal{F}_1 \to \mathcal{F}_2$ is a homomorphism of $\mathfrak{T}$-sheaves, one can easily construct an extension $\varphi'\colon \mathcal{F}_1' \to \mathcal{F}_2'$ between the corresponding $\mathfrak{T}'$-extensions $\mathcal{F}_1'$ and $\mathcal{F}_2'$ by using the functorial property of $\check{\mathcal{H}}^0$. Applying properties of sheaves, φ' is seen to be unique. □

Proposition 2. *Let $\mathfrak{T}$ and $\mathfrak{T}'$ be G-topologies on X with $\mathfrak{T}'$ being slightly finer than $\mathfrak{T}$. Consider on X a $\mathfrak{T}$-presheaf $\mathcal{F}$ and a $\mathfrak{T}'$-presheaf $\mathcal{F}'$ extending $\mathcal{F}$. Then if $\mathcal{F}' \to \mathcal{G}'$ is a sheafification of $\mathcal{F}'$, the restriction of this homomorphism with respect to the G-topology $\mathfrak{T}$ is a sheafification of $\mathcal{F}$.*

Proof. Due to Proposition 9.2.2/4, we may assume that the sheafification of $\mathcal{F}'$ is given by the canonical homomorphism

$$\mathcal{F}' \to \check{\mathcal{H}}^0\big(X_{\mathfrak{T}'}, \check{\mathcal{H}}^0(X_{\mathfrak{T}'}, \mathcal{F}')\big).$$

By the same argument as presented in the beginning of the preceding proof, one sees that the $\mathfrak{T}'$-presheaf $\check{\mathcal{H}}^0(X_{\mathfrak{T}'}, \mathcal{F}')$ restricts to the $\mathfrak{T}$-presheaf $\check{\mathcal{H}}^0(X_{\mathfrak{T}}, \mathcal{F})$. Applying this process twice, the above sheafification restricts to the canonical homomorphism

$$\mathcal{F} \to \check{\mathcal{H}}^0\big(X_{\mathfrak{T}}, \check{\mathcal{H}}^0(X_{\mathfrak{T}}, \mathcal{F})\big)$$

which is a sheafification of $\mathcal{F}$. □

9.3. Analytic varieties. Definitions and constructions

9.3.1. Locally G-ringed spaces and analytic varieties. — We will introduce analytic varieties as global spaces which are "locally" affinoid in some sense. Our considerations are based on the notion of G-ringed spaces, which is analogous to that of ringed spaces. A *G-ringed space* is a pair $(X, \mathcal{O}_X)$ where X is a G-topological space and $\mathcal{O}_X$ is a sheaf of rings on X, called the structure sheaf of X. If all stalks $\mathcal{O}_{X,x}$, $x \in X$, are local rings, $(X, \mathcal{O}_X)$ is called a *locally G-ringed space*. Of particular interest will be the case where $\mathcal{O}_X$ is a sheaf of algebras over some fixed ring R. Then $(X, \mathcal{O}_X)$ is called a *G-ringed space over R*.

For such spaces we need appropriate morphisms. A *morphism of G-ringed spaces* $(X, \mathcal{O}_X) \to (Y, \mathcal{O}_Y)$ is a pair (ψ, ψ^*) where $\psi: X \to Y$ is a continuous map and ψ^* is a collection of ring homomorphisms

$$\psi_V^*: \mathcal{O}_Y(V) \to \mathcal{O}_X(\psi^{-1}(V)),$$

with V ranging over the system of all admissible open subsets of Y, such that all ψ_V^* are compatible with restriction homomorphisms. Taking direct limits one gets, for each $x \in X$, a ring homomorphism

$$\psi_x^*: \mathcal{O}_{Y,\psi(x)} \to \mathcal{O}_{X,x}$$

between corresponding stalks. The pair (ψ, ψ^*) is called a *morphism of locally G-ringed spaces* if $(X, \mathcal{O}_X)$ and $(Y, \mathcal{O}_Y)$ are locally G-ringed spaces and if all homomorphisms ψ_x^* are local, i.e., map the maximal ideal of $\mathcal{O}_{Y,\psi(x)}$ into the maximal ideal of $\mathcal{O}_{X,x}$. When G-ringed spaces over some ring R are considered, we require, of course, that all homomorphisms ψ_V^* are R-algebra homomorphisms. The pair (ψ, ψ^*) is then called an *R-morphism* or a *morphism over R*. For simplicity we often write X instead of $(X, \mathcal{O}_X)$ and ψ instead of (ψ, ψ^*).

Let $(X, \mathcal{O}_X)$ be a G-ringed space and $U \subset X$ an admissible open subset. Then U is a G-topological space with the G-topology being induced by X, and furthermore, $\mathcal{O}_X|_U$, the restriction of $\mathcal{O}_X$ to U, is a sheaf of rings on U. The resulting G-ringed space $(U, \mathcal{O}_X|_U)$ is called an *open subspace of* $(X, \mathcal{O}_X)$, and the inclusion $U \hookrightarrow X$ induces a canonical morphism of G-ringed spaces $(U, \mathcal{O}_X|_U) \to (X, \mathcal{O}_X)$. Such a morphism and, more generally, any morphism $(U, \mathcal{O}_U) \to (X, \mathcal{O}_X)$ identifying $(U, \mathcal{O}_U)$ with an open subspace of $(X, \mathcal{O}_X)$ will be called an *open immersion*.

In the following let k be a fixed ground field with a non-trivial complete non-Archimedean valuation. If $X = \operatorname{Sp} A$ is a k-affinoid variety, the presheaf $\mathcal{O}_X$ of (7.3.2) which associates to each affinoid subdomain of X its corresponding k-affinoid algebra, is a sheaf on X with respect to the weak G-topology (cf. Corollary 8.2.1/2). Now considering the strong G-topology on X, which is slightly finer than the weak one, the sheaf $\mathcal{O}_X$ can be extended uniquely to a sheaf of k-algebras $\mathcal{O}'_X$ with respect to the strong G-topology on X (cf. Proposition 9.2.3/1).

From now on the weak G-topology will be of minor importance. We will always associate with affinoid varieties X the *strong G-topology* which satisfies conditions (G_0), (G_1) and (G_2) and is characterized by Proposition 9.1.4/2. Then $(X, \mathcal{O}'_X)$ is a G-ringed space and locally G-ringed by Proposition 7.3.2/1. It is called the *locally G-ringed space associated with* X. In the following, by abuse of notation, we again write $\mathcal{O}_X$, instead of $\mathcal{O}'_X$, for the structure sheaf of this ringed space.

We want to establish a relationship between k-affinoid maps and k-morphisms of the associated locally G-ringed spaces. Let $\varphi: X \to Y$ be a k-affinoid map between affinoid varieties $X = \operatorname{Sp} A$ and $Y = \operatorname{Sp} B$ given by the k-algebra homomorphism $\varphi^*: B \to A$. Then one derives from φ a k-morphism of locally G-ringed spaces $(\psi, \psi^*): (X, \mathcal{O}_X) \to (Y, \mathcal{O}_Y)$ as follows. Let $\psi: X \to Y$

be the map of G-topological spaces, underlying φ. Then ψ is continuous by the construction of the strong G-topology on affinoid varieties. To define the collection ψ^* of k-algebra homomorphisms, we first consider an arbitrary affinoid subdomain $V \subset Y$. Then $\varphi^{-1}(V)$ is an affinoid subdomain in X by Proposition 7.2.2/4, and φ induces a unique k-affinoid map $\varphi_V\colon \varphi^{-1}(V) \to V$. Denoting by $\psi_V^*\colon \mathcal{O}_Y(V) \to \mathcal{O}_X(\varphi^{-1}(V))$ the k-algebra homomorphism corresponding to φ_V, we get a collection of homomorphisms ψ_V^* with V ranging in the system of all affinoid subdomains in Y, which is compatible with restriction homomorphisms.

It is now clear how to define ψ_V^* for arbitrary admissible open subsets $V \subset Y$, since the affinoid subdomains of Y form a basis for the (strong) G-topology of Y and since we are dealing with sheaves. Namely, choose an admissible covering $\mathfrak{V} = \{V_i\}_{i\in I}$ of V by affinoid subdomains $V_i \subset Y$. Then, for $g \in \mathcal{O}_Y(V)$, define $\psi_V^*(g) \in \mathcal{O}_X(\psi^{-1}(V))$ by

$$\psi_V^*(g)|_{\psi^{-1}(V_i)} := \psi_{V_i}^*(g|_{V_i}), \quad i \in I.$$

Since $\{\psi^{-1}(V_i)\}_{i\in I}$ is an admissible covering of $\psi^{-1}(V)$, it can be verified without difficulties that ψ_V^* is a well-defined k-algebra homomorphism, independent of the covering $\mathfrak{V}$, and that all ψ_V^* are compatible with restriction homomorphisms. Thus, we have derived from φ a k-morphism $(\psi, \psi^*)\colon (X, \mathcal{O}_X) \to (Y, \mathcal{O}_Y)$ which is a morphism of locally G-ringed spaces (cf. (7.3.2), especially the paragraph following Proposition 7.3.2/1). Note also that $\psi_Y^* = \varphi^*$.

Proposition 1. *Let X and Y be k-affinoid varieties. Then assigning to each k-affinoid map $X \to Y$ the associated k-morphism $(X, \mathcal{O}_X) \to (Y, \mathcal{O}_Y)$ (as constructed above) sets up a one-to-one correspondence between k-affinoid maps $X \to Y$ and morphisms of locally G-ringed spaces $(X, \mathcal{O}_X) \to (Y, \mathcal{O}_Y)$.*

Since k-affinoid maps $\operatorname{Sp} A \to \operatorname{Sp} B$ by definition correspond bijectively to k-algebra homomorphisms $B \to A$, we need only show the following:

Lemma 2. *Let $X = \operatorname{Sp} A$ and $Y = \operatorname{Sp} B$ be k-affinoid varieties. For each k-algebra homomorphism $\varphi^*\colon B \to A$, there exists a unique morphism of locally G-ringed spaces $(\psi, \psi^*)\colon (X, \mathcal{O}_X) \to (Y, \mathcal{O}_Y)$ satisfying $\psi_Y^* = \varphi^*$.*

Proof. Only the uniqueness assertion has to be verified. Let $\varphi^*\colon B \to A$ and $(\psi, \psi^*)\colon (X, \mathcal{O}_X) \to (Y, \mathcal{O}_Y)$ satisfy $\psi_Y^* = \varphi^*$. Then for each $x \in X$, there is a commutative diagram

$$\begin{array}{ccc} B & \xrightarrow{\varphi^* = \psi_Y^*} & A \\ \downarrow & & \downarrow \\ \mathcal{O}_{Y,\psi(x)} & \xrightarrow{\psi_x^*} & \mathcal{O}_{X,x}\,. \end{array}$$

Since ψ_x^* is local, φ^* must map the maximal ideal $\mathfrak{m}_{\psi(x)} \subset B$ corresponding to $\psi(x)$ into the maximal ideal $\mathfrak{m}_x \subset A$ corresponding to x. Thus, $\mathfrak{m}_{\psi(x)} = (\varphi^*)^{-1}(\mathfrak{m}_x)$, and as a result, the map $\psi\colon X \to Y$ equals the underlying map of

G-topological spaces which is obtained from the k-affinoid map $\varphi\colon X \to Y$ associated with φ^*.

In order to see that all maps $\psi_V^*\colon \mathcal{O}_Y(V) \to \mathcal{O}_X(\psi^{-1}(V))$ are uniquely determined by φ^*, it is obviously enough to consider only affinoid subdomains $V \subset Y$. For any such V, its inverse image $\psi^{-1}(V) = \varphi^{-1}(V)$ is an affinoid subdomain in X by Proposition 7.2.2/4, and there is a commutative diagram

$$\begin{array}{ccc} B & \xrightarrow{\varphi^* = \psi_Y^*} & A \\ \downarrow & & \downarrow \\ \mathcal{O}_Y(V) & \xrightarrow{\psi_V^*} & \mathcal{O}_X(\psi^{-1}(V)), \end{array}$$

the vertical maps being restriction homomorphisms. Since, as a k-affinoid map, the inclusion $V \hookrightarrow Y = \operatorname{Sp} B$ corresponds to the restriction homomorphism $B \to \mathcal{O}_Y(V)$, it follows from the defining universal property of affinoid subdomains or from Proposition 7.2.2/4 that ψ_V^* is uniquely determined by φ^*. □

Summarizing what has been proved so far, we can say that associating with each k-affinoid variety its corresponding locally G-ringed space and with each k-affinoid map its corresponding k-morphism of locally G-ringed spaces constitutes a fully faithful functor from the category of k-affinoid varieties into the category of locally G-ringed spaces over k. We want to show that this functor respects open immersions.

Proposition 3. *Let $\varphi\colon U \to X$ be a k-affinoid map between k-affinoid varieties U and X. Then φ is an open immersion if and only if the associated k-morphism of locally G-ringed spaces $(\psi, \psi^*)\colon (U, \mathcal{O}_U) \to (X, \mathcal{O}_X)$ is an open immersion. In particular, U is an affinoid subdomain of X via φ if and only if $(U, \mathcal{O}_U)$ is an open subspace of $(X, \mathcal{O}_X)$ via (ψ, ψ^*).*

Proof. Let φ be an open immersion, or what amounts to the same thing by Corollary 8.2.1/4, let U be an affinoid subdomain of X via φ. Interpreting φ as an inclusion, the strong G-topology of X induces on U the strong G-topology of U due to Corollary 9.1.4/3. Furthermore, it follows from Proposition 9.2.3/1 that ψ^* identifies $\mathcal{O}_U$ with $\mathcal{O}_X|_U$, since both sheaves restrict to the same sheaf when the weak G-topology is considered on U. Thus, (ψ, ψ^*) must be an open immersion.

Conversely, if (ψ, ψ^*) is an open immersion, ψ is, in particular, injective and all homomorphisms $\psi_x^*\colon \mathcal{O}_{X,\psi(x)} \to \mathcal{O}_{U,x}$ for $x \in U$ must be isomorphisms. Consequently, φ is an open immersion. □

From now on k-affinoid varieties will always be viewed as locally G-ringed spaces over k. Any locally G-ringed space over k, isomorphic to such a space, will be called a k-affinoid variety. We use notations $(X, \mathcal{O}_X)$ or even X and $\operatorname{Sp} A$. Accordingly, k-affinoid maps will be thought of as k-morphisms of locally G-ringed spaces; we will write (φ, φ^*) or φ. Now we are in a position to introduce global analytic varieties.

Definition 4. *A rigid analytic variety over k (also called k-analytic variety) is a locally G-ringed space $(X, \mathcal{O}_X)$ over k (also denoted by X) such that the following conditions are satisfied:*

(i) *The G-topology of X fulfills conditions (G_0), (G_1) and (G_2) of (9.1.2).*

(ii) *X admits an admissible affinoid covering, i.e., an admissible covering $\{X_i\}_{i\in I}$ with $(X_i, \mathcal{O}_X|_{X_i})$ being a k-affinoid variety for all $i \in I$.*

A morphism of rigid analytic varieties $(X, \mathcal{O}_X) \to (Y, \mathcal{O}_Y)$ over k (also called k-analytic map) is a k-morphism of locally G-ringed spaces (φ, φ^*): $(X, \mathcal{O}_X) \to (Y, \mathcal{O}_Y)$ (also denoted by $\varphi: X \to Y$).

The reference to the ground field k will be suppressed in general, when no confusion is possible. As we have seen, affinoid varieties are examples of analytic varieties, since the strong G-topology on an affinoid variety satisfies conditions (G_0), (G_1), and (G_2). Furthermore, by Proposition 1, the analytic maps between affinoid varieties are just the affinoid ones. We mention also that for any affinoid variety $(X, \mathcal{O}_X)$ a covering $\mathfrak{X} = \{X_i\}_{i\in I}$ of X is an admissible affinoid covering in the sense of Definition 4 if and only if each X_i is an affinoid subdomain of X and $\mathfrak{X}$ contains a finite subcovering. This follows from Proposition 3 and the fact that the strong G-topology on X satisfies condition (G_2) and is slightly finer than the weak G-topology on X. Hence, for affinoid varieties, admissible affinoid coverings and (finite) affinoid coverings are essentially the same.

Proposition 5. *Let $(X, \mathcal{O}_X)$ be an analytic variety. Then the G-topology of X admits a basis consisting of admissible open sets $V \subset X$ such that $(V, \mathcal{O}_X|_V)$ is always an affinoid variety.*

Proof. Let $\{X_i\}_{i\in I}$ be an admissible affinoid covering of X. Then, for any admissible open subset $U \subset X$, the covering $\{U \cap X_i\}_{i\in I}$ is an admissible covering of U. Since $U \cap X_i$ is an admissible open subset of the "affinoid" set X_i, it follows from Proposition 9.1.4/1 and Proposition 3 that each $U \cap X_i$ has an admissible affinoid covering. Hence U has an admissible affinoid covering. □

It is a consequence of Proposition 5 that any open subspace $(U, \mathcal{O}_X|_U)$ of an analytic variety $(X, \mathcal{O}_X)$ is again an analytic variety. We call such a variety an *open analytic subvariety of* $(X, \mathcal{O}_X)$ or even an *open affinoid subvariety* if $(U, \mathcal{O}_X|_U)$ is affinoid.

Let X be a k-affinoid variety, denote by k_a the algebraic closure of k, and let Γ be the Galois group of k_a over k. We saw in (7.1.1) and (7.1.3) that any $x \in X$ induces an evaluation map $h_x: \mathcal{O}_X(X) \to k_a/\Gamma$ which by Proposition 7.2.2/1 is compatible with restriction homomorphisms. More precisely, if $X' \subset X$ is an affinoid subdomain of X containing x and if $h'_x: \mathcal{O}_X(X') \to k_a/\Gamma$ denotes the corresponding evaluation map, then h_x equals the composition $h'_x \circ \text{res}$, where $\text{res}: \mathcal{O}_X(X) \to \mathcal{O}_X(X')$ is the restriction homomorphism. Using this fact, we also get evaluation maps in the more general case where X is a

k-analytic variety. If $x \in X$ is given, we choose an open affinoid subvariety $X' \subset X$ containing x and thus get an evaluation map at x by the composition

$$\mathcal{O}_X(X) \xrightarrow{\text{res}} \mathcal{O}_X(X') \xrightarrow{\text{eval}} k_a/\Gamma,$$

which is independent of X' by Proposition 5 and the considerations above. The same applies, of course, to any analytic subvariety containing x, and in fact any evaluation at x is factored through a unique evaluation map $\mathcal{O}_{X,x} \to k_a/\Gamma$ (cf. also (7.3.2)). Thus, as in the case of affinoid varieties, the elements of $\mathcal{O}_X(U)$ for any admissible open $U \subset X$ induce functions $U \to k_a/\Gamma$. In particular, we can talk about a zero of some "function" $f \in \mathcal{O}_X(U)$ and about the absolute value $|f(x)|$ it assumes at a point $x \in U$.

9.3.2. Pasting of analytic varieties. — In the following we will present a general method for constructing analytic varieties, which strongly relies on the conditions (G_0), (G_1) and (G_2) for G-topologies. Let $(X, \mathcal{O}_X)$ be an analytic variety and consider an admissible covering $\mathfrak{X} = \{X_i\}_{i \in I}$ of X. As a first step, we want to show that the structure of $(X, \mathcal{O}_X)$ can be recovered from the admissible covering $\mathfrak{X}$, if all subvarieties $(X_i, \mathcal{O}_X|_{X_i})$ are known. Since the G-topology of X satisfies conditions (G_0), (G_1) and (G_2), it is uniquely determined by the G-topologies induced on the X_i by Proposition 9.1.3/2. Moreover, the structure sheaf $\mathcal{O}_X$ is uniquely determined (up to isomorphism) by the sheaves $\mathcal{O}_X|_{X_i}$, $i \in I$, since for any admissible open $U \subset X$ the covering $\mathfrak{U} := \{U \cap X_i\}_{i \in I}$ of U is admissible, hence implying

$$\mathcal{O}_X(U) = H^0(\mathfrak{U}, \mathcal{O}_X).$$

(We also could have applied Proposition 9.2.3/1 to the G-topologies $\mathfrak{T}$ and $\mathfrak{T}^w$ of (9.1.3).) In any case, it is seen that, within the framework of admissible coverings, the structure of $(X, \mathcal{O}_X)$ depends only on "local" data. Next we want to construct analytic varieties by pasting together "local" varieties via prescribed "intersections".

Proposition 1. *Consider the following data:*

(i) *analytic varieties X_i, where i runs in some index set I,*
(ii) *open subvarieties $X_{ij} \subset X_i$ and isomorphisms $\varphi_{ij}: X_{ij} \xrightarrow{\sim} X_{ji}$ for all $i, j \in I$.*

Assume that the following conditions are satisfied:

(a) $\varphi_{ij} \circ \varphi_{ji} = \mathrm{id}$, $X_{ii} = X_i$, $\varphi_{ii} = \mathrm{id}$ *for all $i, j \in I$.*
(b) *The map φ_{ij} induces isomorphisms $\varphi_{ijl}: X_{ij} \cap X_{il} \to X_{ji} \cap X_{jl}$ such that $\varphi_{ijl} = \varphi_{lji} \circ \varphi_{ilj}$ for all $i, j, l \in I$.*

Then it is possible to construct an analytic variety X by pasting together the varieties X_i, with X_{ij} and X_{ji} being identified via φ_{ij} and serving as the intersection $X_i \cap X_j$, such that $\{X_i\}_{i \in I}$ is an admissible covering of X. More precisely:

There exists an analytic variety X with an admissible covering $\{X'_i\}_{i\in I}$, and there are isomorphisms $\psi_i\colon X_i \xrightarrow{\sim} X'_i$ giving rise to isomorphisms $\psi_{ij}\colon X_{ij} \xrightarrow{\sim} X'_i \cap X'_j$ such that

$$\begin{array}{ccc} X_{ij} & & \\ \Big\downarrow{\scriptstyle \varphi_{ij}} & \searrow^{\psi_{ij}} & \\ & & X'_i \cap X'_j \\ & \nearrow_{\psi_{ji}} & \\ X_{ji} & & \end{array}$$

is commutative for all $i, j \in I$. The analytic variety X is unique up to natural isomorphism.

Proof. The uniqueness assertion follows from our considerations above; hence we only have to construct X. Let $\hat{X}$ be the disjoint union of the "sets" X_i. We define an equivalence relation $\sim$ on $\hat{X}$ by calling $x \in X_i$ and $y \in X_j$ equivalent if and only if $x \in X_{ij}$, $y \in X_{ji}$ and $\varphi_{ij}(x) = y$. Then $\sim$ is reflexive and symmetric due to condition (a). Furthermore, condition (b) implies the transitivity: namely, let $x \in X_i$, $y \in X_l$, and $z \in X_j$ such that $x \sim y$ and $y \sim z$. Then we have $x \in X_{il}$, $y \in X_{li}$ and $y = \varphi_{il}(x)$ as well as $y \in X_{lj}$, $z \in X_{jl}$ and $z = \varphi_{lj}(y)$. Hence y belongs to the intersection $X_{li} \cap X_{lj}$, which implies $x \in X_{il} \cap X_{ij}$ by condition (b). We get

$$z = \varphi_{lj}\big(\varphi_{il}(x)\big) = \varphi_{lji}\big(\varphi_{ilj}(x)\big) = \varphi_{ijl}(x) = \varphi_{ij}(x),$$

and therefore $x \sim z$. Thus, $\sim$ is an equivalence relation.

Now define $X := \hat{X}/\sim$. Then the canonical maps $X_i \to X$ are injective and, for all $i \in I$, we identify X_i with its image in X. Under this process, the subset $X_{ij} \subset X_i$ is identified with $X_{ji} \subset X_j$ via the map $\varphi_{ij}\colon X_{ij} \to X_{ji}$, and it becomes the intersection $X_i \cap X_j$. (In order to keep notations simple we avoid introducing the maps ψ_i and ψ_{ij} stated in the proposition.) Since, for any $i, j \in I$, the map φ_{ij} is a homeomorphism with respect to the induced G-topologies on X_{ij} and X_{ji}, it is seen immediately that the situation of (9.1.3) applies to the covering $\{X_i\}_{i\in I}$ of X. Hence, by Proposition 9.1.3/2, there exists a G-topology $\mathfrak{T}$ on X which satisfies (G_0), (G_1) and (G_2), admits the X_i as admissible open subsets, induces on X_i the given G-topology, and renders $\{X_i\}_{i\in I}$ an admissible covering of X.

In order to define the structure sheaf on X, we consider in addition to $\mathfrak{T}$ the G-topology $\mathfrak{T}^{\mathrm{w}}$ on X, as defined in (9.1.3). Recall that a subset $U \subset X$ is $\mathfrak{T}^{\mathrm{w}}$-open if and only if it is admissible open in X_i, for some $i \in I$. For a $\mathfrak{T}^{\mathrm{w}}$-open subset $U \subset X$, we denote by I_U the set of all indices $i \in I$ such that U is contained in X_i. Then U is admissible open in X_i for all $i \in I_U$, and I_U is not empty. Further, if $\mathcal{O}_{X_i}$ denotes the structure sheaf of X_i, the isomorphisms $\varphi_{ij}\colon X_{ij} \to X_{ji}$ (as morphisms of ringed spaces now) give rise to isomorphisms

$$(\varphi^*_{ij})_U\colon \mathcal{O}_{X_j}(U) \to \mathcal{O}_{X_i}(U), \qquad i, j \in I_U.$$

For each $\mathfrak{T}^{\mathrm{w}}$-open set $U \subset X$, we identify all rings $\mathcal{O}_{X_i}(U)$ via the maps $(\varphi^*_{ij})_U$, $i, j \in I_U$. The procedure is similar to the one used above; here, of course, one has to consider the equivalence relation on $\dot{\bigcup}_{i\in I_U} \mathcal{O}_{X_i}(U)$ which is induced by the maps $(\varphi_{ij}^*)_U$. Since the identification process is compatible with restriction homomorphisms, one sees that, for any $i, j \in I$, the isomorphism

$$\mathcal{O}_{X_j}|_{X_i\cap X_j} \xrightarrow{\sim} \mathcal{O}_{X_i}|_{X_i\cap X_j}$$

furnished by φ_{ij}^* becomes an identity. Hence we can define a $\mathfrak{T}^{\mathrm{w}}$-sheaf of rings $\mathcal{O}^{\mathrm{w}}$ on X simply by

$$\mathcal{O}^{\mathrm{w}}|_{X_i} := \mathcal{O}_{X_i}, \quad i \in I.$$

Due to Proposition 9.1.3/1, the G-topology $\mathfrak{T}$ is slightly finer than $\mathfrak{T}^{\mathrm{w}}$; therefore we can apply Proposition 9.2.3/1. Thus, $\mathcal{O}^{\mathrm{w}}$ extends to a $\mathfrak{T}$-sheaf $\mathcal{O}_X$, which is essentially unique. It is seen without difficulties now that $(X, \mathcal{O}_X)$ is an analytic variety having the required properties. □

If analytic varieties are to be constructed explicitly with the help of Proposition 1, it is of course most effective to take all varieties X_i and X_{ij}, $i, j \in I$, as affinoid varieties. Namely, affinoid varieties and maps can be defined in terms of affinoid algebras and algebra homomorphisms. Thus, in this case the assumptions of Proposition 1 can be stated without referring to the notion of locally G-ringed spaces. We will present some examples in (9.3.4).

9.3.3. Pasting of analytic maps. — Another impact of conditions (G_0), (G_1), and (G_2) appearing in the definition of analytic varieties is that analytic maps can be defined "locally".

Proposition 1. *Let X and Y be analytic varieties admitting admissible coverings $\{X_i\}_{i\in I}$ and $\{Y_i\}_{i\in I}$, respectively. Let $\varphi_i: X_i \to Y_i$, $i \in I$, be analytic maps such that, for any $i, j \in I$, both φ_i and φ_j restrict to the same analytic map $\varphi_{ij}: X_i \cap X_j \to Y_i \cap Y_j$. Then there exists a unique analytic map $\varphi: X \to Y$ extending all φ_i, $i \in I$.*

Proof. Of course the φ_i define a unique map $\varphi: X \to Y$ between the "sets" X and Y. We want to show that φ is continuous with respect to the G-topologies on X and Y. If $V \subset Y$ is admissible open, then also $V_i := V \cap Y_i$ is admissible open in Y for all $i \in I$. Since the maps φ_i are continuous, $\varphi_i^{-1}(V_i) = \varphi^{-1}(V) \cap X_i$ is admissible open in X for all $i \in I$. Hence $\varphi^{-1}(V)$ is admissible open in X by condition (G_1). Furthermore, let $\mathfrak{V} = \{V_j\}_{j\in J}$ be an admissible covering of V. Then $\mathfrak{V}_i := \{V_j \cap Y_i\}_{j\in J}$ is an admissible covering of $V \cap Y_i$, and since the maps φ_i are continuous, $\{\varphi_i^{-1}(V_j \cap Y_i)\}_{j\in J}$ is an admissible covering of $\varphi_i^{-1}(V \cap Y_i) = \varphi^{-1}(V) \cap X_i$ for all $i \in I$. Now obviously $\{\varphi^{-1}(V) \cap X_i\}_{i\in I}$ is an admissible covering of $\varphi^{-1}(V)$; hence $\{\varphi_i^{-1}(V_j \cap Y_i)\}_{i\in I, j\in J}$ is an admissible covering of $\varphi^{-1}(V)$. But this covering is a refinement of $\{\varphi^{-1}(V_j)\}_{j\in J}$. Hence, by condition (G_2), also $\{\varphi^{-1}(V_j)\}_{j\in J}$ is an admissible covering of $\varphi^{-1}(V)$. As a result, $\varphi: X \to Y$ is continuous.

Now it is clear how to define φ as a morphism of locally G-ringed spaces, i.e., how to define the collection φ^* of homomorphisms

$$\varphi_V^*: \mathcal{O}_Y(V) \to \mathcal{O}_X(\varphi^{-1}(V))$$

for $V \subset Y$ admissible open. If we require the homomorphisms φ_V^* to be compatible with restrictions and with all homomorphisms of the collections φ_i^*, $i \in I$, the only possible way to define $\varphi_V^*(g)$ for $g \in \mathcal{O}_Y(V)$ is by

$$\varphi_V^*(g)|_{\varphi^{-1}(V)\cap X_i} := (\varphi_i^*)_{V\cap Y_i}(g|_{V\cap Y_i}), \qquad i \in I.$$

Due to the fact that $\{V \cap Y_i\}_{i\in I}$ and $\{\varphi^{-1}(V) \cap X_i\}_{i\in I}$ are admissible coverings of V and $\varphi^{-1}(V)$, respectively, this is a well-defined homomorphism, and it is easily checked that $(\varphi, \varphi^*): (X, \mathcal{O}_X) \to (Y, \mathcal{O}_Y)$ is the unique analytic map extending all φ_i. □

The above proposition enables us to generalize Lemma 9.3.1/2 in the following way:

Corollary 2. *Let X and Y be analytic varieties and assume that Y is affinoid. Then, for any algebra homomorphism $\sigma: \mathcal{O}_Y(Y) \to \mathcal{O}_X(X)$, there exists a unique analytic map $\varphi: X \to Y$ such that $\varphi_Y^*: \mathcal{O}_Y(Y) \to \mathcal{O}_X(X)$ equals σ.*

Proof. Let $\{X_i\}_{i\in I}$ denote an admissible affinoid covering of X. Composing σ with the restriction homomorphisms $\mathcal{O}_X(X) \to \mathcal{O}_X(X_i)$, we get algebra homomorphisms $\sigma_i: \mathcal{O}_Y(Y) \to \mathcal{O}_X(X_i)$, $i \in I$. Then, according to Lemma 9.3.1/2, each σ_i corresponds to an affinoid map $\varphi_i: X_i \to Y$. The φ_i clearly satisfy the condition of Proposition 1 with $Y_i := Y$ for all $i \in I$. Hence they define an analytic map $\varphi: X \to Y$ which must satisfy $\varphi_Y^* = \sigma$, since $(\varphi_i^*)_Y = \sigma_i$.

If $\psi: X \to Y$ is any other analytic map satisfying $\psi_Y^* = \sigma$, then for the restricted maps $\psi_i: X_i \to Y$, we must have $(\psi_i^*)_Y = \sigma_i$; hence $\psi_i = \varphi_i$ by Lemma 9.3.1/2 and $\varphi = \psi$ by Proposition 1. □

9.3.4. Some basic examples. — In order to show how the pasting procedures of (9.3.2) and (9.3.3) apply in concrete situations, we want to explicitly construct some analytic varieties which are of general interest.

Example 1. The affine n-space $\mathbb{A}_k^n$. — Let $\zeta = (\zeta_1, \ldots, \zeta_n)$ be a system of indeterminates and choose a constant $c \in k$, $|c| > 1$. As always, k is our fixed ground field. Let k_a be the algebraic closure of k. For $i \in \mathbb{N} \cup \{0\}$, we denote by A_i the k-algebra of all power series $\sum a_\nu \zeta^\nu \in k[[\zeta]]$ converging on the ball of radius $|c|^i$ and center 0 in k_a^n, i.e., $A_i = T_{n,\varrho}$ with $\varrho = (|c|^i, \ldots, |c|^i)$ in the terminology of (6.1.5). Then the A_i occur in the decreasing sequence

$$k\langle\zeta\rangle = A_0 \supset A_1 \supset A_2 \supset \ldots \supset k[\zeta],$$

and each A_i can be viewed as a free TATE algebra T_n with $c^{-i}\zeta_1, \ldots, c^{-i}\zeta_n$ serving as indeterminates. In particular, all A_i are affinoid. Since there is a canonical isomorphism $A_i \cong A_{i+1}\langle c^{-i}\zeta\rangle$, the inclusion $A_{i+1} \hookrightarrow A_i$ corresponds to an affinoid map $\operatorname{Sp} A_i \to \operatorname{Sp} A_{i+1}$ which identifies $\operatorname{Sp} A_i$ with an affinoid subdomain of $\operatorname{Sp} A_{i+1}$. Hence the above decreasing sequence of algebras yields an increasing sequence of affinoid subdomains

$$\mathbb{B}^n = \operatorname{Sp} A_0 \overset{\varphi_0}{\hookrightarrow} \operatorname{Sp} A_1 \overset{\varphi_1}{\hookrightarrow} \operatorname{Sp} A_2 \overset{\varphi_2}{\hookrightarrow} \ldots.$$

Setting $X_i := \operatorname{Sp} A_i$, $X_{ij} := X_{\min\{i,j\}}$ and defining $\varphi_{ij}: X_{ij} \to X_{ji}$ to be the identity map for $i, j \geqq 0$, we can apply Proposition 9.3.2/1 in order to construct an analytic variety by pasting the affinoid varieties $\operatorname{Sp} A_i$. The resulting analytic variety is called the affine n-space and is denoted by $\mathbb{A}_k^n$ or $\mathbb{A}^n$. It is characterized by the fact that it contains the above sequence of affinoid varieties $\operatorname{Sp} A_i$ as an increasing sequence of open affinoid subvarieties and that $\{\operatorname{Sp} A_i\}_{i \geqq 0}$ is an admissible covering of $\mathbb{A}^n$. Note also that our construction of the affine n-space is independent of the chosen constant $c \in k$. Namely, if c is replaced by some other constant $c' \in k$, $|c'| > 1$, one can use Proposition 9.3.3/1 and show that the resulting analytic variety is isomorphic to $\mathbb{A}^n$.

We want to look a little more closely at the structure of $\mathbb{A}^n$. Since the polynomial ring $k[\zeta]$ is contained in all the algebras A_i, we see that $k[\zeta] \subset \mathcal{O}_{\mathbb{A}^n}(\mathbb{A}^n)$. In particular, the functions $\zeta_1, \ldots, \zeta_n \in \mathcal{O}_{\mathbb{A}^n}(\mathbb{A}^n)$ are referred to as a set of coordinates for $\mathbb{A}^n$. The unique common zero of $\zeta_1, \ldots, \zeta_n$ is called the origin of $\mathbb{A}^n$. Going back once more to the above decreasing sequence of algebras $A_0 \supset A_1 \supset \ldots$ and looking at the spectra of maximal ideals, we get with Lemma 7.1.1/2 a sequence

$$\operatorname{Max} A_0 \subset \operatorname{Max} A_1 \subset \ldots \subset \operatorname{Max} k[\zeta].$$

By the analogue of Corollary 6.1.2/3 for finitely generated k-algebras, any $x \in \operatorname{Max} k[\zeta]$ occurs as the kernel of some k-homomorphism $\sigma: k[\zeta] \to k_a$ into the algebraic closure k_a of k. If $i \geqq 0$ is given, then $|\sigma(\zeta_j)| \leqq |c|^i$ for $j = 1, \ldots, n$ is a necessary and sufficient condition for σ extending to a k-homomorphism $\sigma': A_i \to k_a$ or, equivalently, x belonging to $\operatorname{Max} A_i$. Therefore, $\operatorname{Max} k[\zeta]$ is covered by the $\operatorname{Max} A_i$ and, in particular,

$$\operatorname{Max} A_i = \{x \in \operatorname{Max} k[\zeta];\ |\zeta_j(x)| \leqq |c|^i,\ j = 1, \ldots, n\}.$$

As a result, we can view the affine n-space $\mathbb{A}^n$ as an analytic version of the affine algebraic n-space over k. The open affinoid subvariety $\operatorname{Sp} A_i \subset \mathbb{A}^n$ corresponds then to the "ball" of radius $|c|^i$, centered at the origin of the affine algebraic n-space, but has, of course, no counterpart in terms of algebraic geometry. We use the notation $\mathbb{B}^n(\alpha)$ with $\alpha \in \sqrt{|k^*|}$ for the open affinoid subvariety in $\mathbb{A}^n$ consisting of those points where $\zeta_1, \ldots, \zeta_n$ have absolute value $\leqq \alpha$.

Example 2. Algebraic varieties. — We just saw how the affine algebraic n-space (i.e. the affine scheme $\operatorname{Spec} k[\zeta_1, \ldots, \zeta_n]$) can be viewed as an analytic variety. Here we want to give a generalization for arbitrary algebraic varieties.

Let B be a finitely generated k-algebra; i.e., $B = k[\zeta_1, \ldots, \zeta_n]/\mathfrak{a}$ with $\mathfrak{a}$ being an ideal of the polynomial ring $k[\zeta_1, \ldots, \zeta_n]$. Using the notations defined in Example 1, we get a sequence of k-algebra homomorphisms

$$A_0/\mathfrak{a}A_0 \leftarrow A_1/\mathfrak{a}A_1 \leftarrow \cdots \leftarrow B$$

from the decreasing sequence $A_0 \supset A_1 \supset \ldots$. By Proposition 7.2.2/4 and Corollary 7.2.2/6, the above sequence corresponds to an increasing sequence of affinoid subdomains

$$\mathrm{Sp}\, A_0/\mathfrak{a}A_0 \hookrightarrow \mathrm{Sp}\, A_1/\mathfrak{a}A_1 \hookrightarrow \ldots,$$

and, as in Example 1, one constructs an analytic variety admitting $\{\mathrm{Sp}\, A_i/\mathfrak{a}A_i\}_{i \geq 0}$ as an admissible affinoid covering. This is the analytic variety X^{an} associated with the affine scheme $X := \mathrm{Spec}\, B$. Just as in the case of the affine n-space $\mathbb{A}^n$, one shows that the spectrum $\mathrm{Max}\, B$ of all maximal ideals in B is identified via the above sequence of k-algebra homomorphisms with the underlying set of X^{an}.

Now let $U \subset \mathrm{Spec}\, B$ be a Zariski-open subset. Thinking only in terms of maximal ideals (which is allowed since B as a finitely generated k-algebra is a Jacobson ring), we see by Corollary 9.1.4/7 that all intersections $U \cap \mathrm{Sp}\, A_i/\mathfrak{a}A_i$ are admissible open in X^{an}. Then by condition (G_1), which is satisfied by the G-topology of any analytic variety, U itself is admissible open in X^{an}. Therefore, the open subschemes of $X = \mathrm{Spec}\, B$ correspond to open analytic subvarieties of the associated analytic variety X^{an}. Using this fact, one can associate an analytic variety X^{an} with an arbitrary scheme X of locally finite type over k. Namely such a scheme X is obtained by pasting together affine schemes X_j of finitely generated k-algebras. Then using Proposition 9.3.2/1, one can construct X^{an} in the same way by pasting together the associated analytic varieties X_j^{an}.

To assure uniqueness of the construction, one should look at morphisms. Namely, any morphism of schemes of locally finite type over k induces an analytic map between the associated analytic varieties in such a way that the correspondence $X \rightsquigarrow X^{\mathrm{an}}$ is a functor. To verify this, one applies Proposition 9.3.3/1 and reduces the problem to the case of affine algebraic maps. Here only algebraic maps between affine m- and n-spaces are of interest. We leave the details to the reader and show only the following:

Let $\zeta_1, \ldots, \zeta_n$ and $\eta_1, \ldots, \eta_m$ be coordinate functions for $\mathbb{A}^n$ and $\mathbb{A}^m$, respectively, and consider a k-homomorphism $\sigma: k[\zeta_1, \ldots, \zeta_n] \to k[\eta_1, \ldots, \eta_m]$. Then there is a unique analytic map $\varphi: \mathbb{A}^m \to \mathbb{A}^n$ such that $\varphi^*_{\mathbb{A}^n}: \mathcal{O}_{\mathbb{A}^n}(\mathbb{A}^n) \to \mathcal{O}_{\mathbb{A}^m}(\mathbb{A}^m)$ is a homomorphism extending σ.

Let $c \in k^*$ be a constant and choose $d \in k^*$ with absolute value not smaller than the sup norms which the functions $\sigma(\zeta_1), \ldots, \sigma(\zeta_n)$ take on the affinoid variety $\mathbb{B}^m(|c|) \subset \mathbb{A}^m$. Then σ extends uniquely to a k-homomorphism

$$\sigma': k\langle d^{-1}\zeta_1, \ldots, d^{-1}\zeta_n\rangle \to k\langle c^{-1}\eta_1, \ldots, c^{-1}\eta_m\rangle$$

belonging to an affinoid map

$$\mathbb{A}^m \supset \mathbb{B}^m(|c|) \to \mathbb{B}^n(|d|) \subset \mathbb{A}^n.$$

Considering a standard admissible affinoid covering of $\mathbb{A}^m$, one can apply Proposition 9.3.3/1 and obtain an analytic map $\mathbb{A}^m \to \mathbb{A}^n$ which is unique, since any extension σ' of σ, as used above, is unique.

Example 3. The projective n-space $\mathbb{P}_k^n$. — Having discussed how to associate an analytic variety with any algebraic variety over k, the projective n-space $\mathbb{P}_k^n$, thought of as an analytic variety, can be obtained by pasting together $n+1$ copies of the affine n-space $\mathbb{A}_k^n$ in the usual way. Then one realizes that $\mathbb{P}_k^n$ is already covered by $n+1$ copies of the unit ball $\mathbb{B}_k^n$. This corresponds to the fact that (for algebraically closed k) any point $x \in \mathbb{P}^n$ can be expressed in homogeneous coordinates $x = (x_0, \ldots, x_n)$ satisfying $\max\limits_{0 \leq j \leq n} |x_j| = 1$. Therefore, one can proceed with the construction of $\mathbb{P}_k^n$ as follows.

Let $X_0, \ldots, X_n$ be affinoid varieties isomorphic to $\mathbb{B}_k^n$. We write symbolically

$$X_i = \operatorname{Sp} k\left\langle \frac{\zeta_0}{\zeta_i}, \ldots, \frac{\zeta_n}{\zeta_i}\right\rangle, \quad i = 0, \ldots, n,$$

thinking of the $\dfrac{\zeta_j}{\zeta_i}$ as indeterminates for $i \neq j$ and identifying $\dfrac{\zeta_i}{\zeta_i}$ with the constant 1. For $i, j = 0, \ldots, n$, we define affinoid subdomains

$$X_{ij} = X_i\left(\left(\frac{\zeta_j}{\zeta_i}\right)^{-1}\right) \subset X_i$$

and isomorphisms $\varphi_{ij}\colon X_{ij} \to X_{ji}$ via the algebra homomorphisms

$$\varphi_{ij}^*\colon k\left\langle \frac{\zeta_0}{\zeta_j}, \ldots, \frac{\zeta_n}{\zeta_j}, \left(\frac{\zeta_i}{\zeta_j}\right)^{-1}\right\rangle \to k\left\langle \frac{\zeta_0}{\zeta_i}, \ldots, \frac{\zeta_n}{\zeta_i}, \left(\frac{\zeta_j}{\zeta_i}\right)^{-1}\right\rangle$$

$$\frac{\zeta_\nu}{\zeta_j} \mapsto \left(\frac{\zeta_j}{\zeta_i}\right)^{-1} \frac{\zeta_\nu}{\zeta_i}, \quad \nu = 0, \ldots, n.$$

An easy computation shows that the varieties X_i, X_{ij} and the maps φ_{ij} satisfy the conditions of Proposition 9.3.2/1; hence, the varieties X_i can be pasted together via the maps φ_{ij}. The resulting variety is called the projective n-space, denoted by $\mathbb{P}_k^n$ or $\mathbb{P}^n$.

Example 4. Analytic tori $\mathbb{A}_k^*/(q)$. — For $q \in k$, $0 < |q| < 1$, we consider the following affinoid subdomains of the unit disc $\mathbb{B}_k^1 = \operatorname{Sp} k\langle\zeta\rangle$:

$$\begin{aligned}
U &:= \mathbb{B}_k^1(q^{-1}\zeta, q\zeta^{-1}),\\
X_1 &:= \mathbb{B}_k^1(|q|^{-1/2}\zeta, q\zeta^{-1}),\\
X_2 &:= \mathbb{B}_k^1(|q|^{1/2}\zeta^{-1}), \quad \text{and}\\
V &:= \mathbb{B}_k^1(\zeta^{-1}).
\end{aligned}$$

Then U, X_1, X_2, V are the annuli centered at zero with radii $|q|$, $|q|$ and $|q|^{1/2}$, $|q|^{1/2}$ and 1, and 1, respectively. The affinoid algebra corresponding to U is

$k\langle q^{-1}\zeta, q\zeta^{-1}\rangle$, i.e., the free algebra of strictly convergent Laurent series in the variable $q^{-1}\zeta$. Hence we have a canonical isomorphism

$$\varphi\colon U \to V = \mathrm{Sp}\, k\langle \zeta, \zeta^{-1}\rangle$$

given by

$$\varphi^*\colon k\langle\zeta, \zeta^{-1}\rangle \to k\langle q^{-1}\zeta, q\zeta^{-1}\rangle$$
$$\zeta \mapsto q^{-1}\zeta.$$

Now set $X_{12} := U \cup (X_1 \cap X_2)$ and $X_{21} := V \cup (X_1 \cap X_2)$. Then X_{12} and X_{21}, as disjoint unions of affinoid subdomains in $\mathbb{B}_k^1$, are again affinoid, and we get isomorphisms

$$\varphi_{12}\colon X_{12} \to X_{21}, \qquad \varphi_{21}\colon X_{21} \to X_{12}$$

by defining $\varphi_{12}|_U := \varphi$, $\varphi_{12}|_{X_1\cap X_2} := \mathrm{id}|_{X_1\cap X_2}$, and $\varphi_{21} := \varphi_{12}^{-1}$. It is clear that the varieties X_i, X_{ij} and isomorphisms φ_{ij} with $i, j = 1, 2$ satisfy the conditions of Proposition 9.3.2/1. Therefore, one derives from $X_1 \cup X_2$, which is the annulus with radii $|q|$ and 1 in $\mathbb{B}_k^1$, an analytic variety X by pasting together $U \subset X_1$ with $V \subset X_2$ via φ, i.e., by identifying the circumferences of the annulus $X_1 \cup X_2$ via "multiplication by q". The variety X is called the analytic torus associated with q and is denoted by $\mathbb{A}_k^*/(q)$ or $\mathbb{A}^*/(q)$.

The notation is chosen because analytic tori are, in fact, quotients. Denote by $\mathbb{A}^*$ the Zariski-open subvariety in $\mathbb{A}^1$ obtained by removing the origin. Then $\mathbb{A}^*$ is associated to the affine algebraic variety $\mathrm{Spec}\, k[\zeta, \zeta^{-1}]$. For $n \in \mathbb{Z}$, the homomorphism

$$k[\zeta, \zeta^{-1}] \to k[\zeta, \zeta^{-1}]$$
$$\zeta \mapsto q^n\zeta$$

corresponds to an analytic map $\mathbb{A}^* \to \mathbb{A}^*$, which can be interpreted as "multiplication by q^n", and one sees that the subgroup $(q) \subset k^*$, generated by q, acts on $\mathbb{A}^*$. Since $\{q^nX_1, q^nX_2\}_{n\in\mathbb{Z}}$ is an admissible affinoid covering of $\mathbb{A}^*$, it is clear that the analytic torus associated with $q \in k$ can be viewed as the quotient $\mathbb{A}^*/(q)$.

Likewise, higher-dimensional tori can be constructed: take the quotient of a direct product $(\mathbb{A}^*)^r$ by a "discrete" subgroup Γ of rank r. Note also that unlike the complex analytic case, analytic tori cannot be obtained as quotients $\mathbb{A}_k^1/\Gamma$ with Γ denoting a "discrete" additive subgroup of k, since there is no exponential map $\mathbb{A}^1 \to \mathbb{A}^*$.

9.3.5. Fibre products. — In (7.1.4), we saw that fibre products can always be constructed in the category of affinoid varieties. Here we want to generalize this result to analytic varieties. Let X and Y be analytic varieties over a fixed analytic variety S. Recall that a *fibre product* of X and Y over S (in the category of analytic varieties) is an analytic variety W over S together with two S-morphisms called projections $p_1\colon W \to X$, $p_2\colon W \to Y$ such that the following universal property holds:

Given any two S-morphisms $Z \to X$ and $Z \to Y$, there is a unique S-morphism $Z \to W$ making the diagram

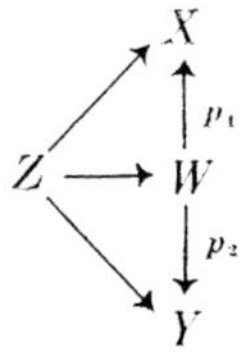

commutative.

The analytic variety W is unique up to canonical isomorphism, if it exists. As usual (cf. our remarks at the end of (7.1.4)), the fibre product of X and Y over S is denoted by $X \times_S Y$. In particular, if $S = \operatorname{Sp} k$ with k being the ground field, $X \times Y = X \times_{\operatorname{Sp} k} Y$ is called the *direct product* of X and Y. As a first step towards the construction of general fibre products, we show that a fibre product in the category of affinoid varieties also satisfies the properties of a fibre product in the bigger category of analytic varieties.

Proposition 1. *Let X and Y be affinoid varieties over S and assume that S also is affinoid. Let W with projections $p_1: W \to X$ and $p_2: W \to Y$ be a fibre product of X and Y over S in the category of affinoid varieties. Then it is also a fibre product in the category of analytic varieties.*

Proof. Let Z be an arbitrary analytic variety over S and consider two S-morphisms $\varphi: Z \to X$ and $\psi: Z \to Y$. Then for any admissible affinoid covering $\{Z_i\}_{i \in I}$ of Z, the maps φ and ψ induce S-morphisms $\varphi_i: Z_i \to X$ and $\psi_i: Z_i \to Y$, $i \in I$. Since the Z_i are affinoid varieties and W is a fibre product of X and Y over S in the category of affinoid varieties, we get unique S-morphisms $\varrho_i: Z_i \to W$ such that all diagrams

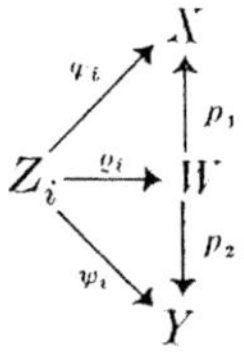

are commutative. Furthermore, for any indices $i, j \in I$, and any open affinoid subvariety $Z' \subset Z_i \cap Z_j$, it follows by the same reason that ϱ_i and ϱ_j both restrict to the same map $Z' \to W$. Thus, by Proposition 9.3.3/1, the maps ϱ_i and ϱ_j must coincide on $Z_i \cap Z_j$ and therefore yield an analytic map $\varrho: Z \to W$ which is easily seen to be the unique S-morphism making the diagram

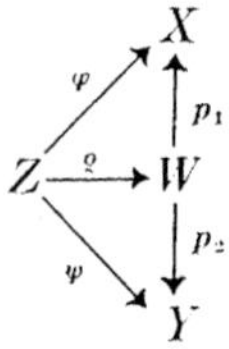

commutative. □

Theorem 2. *Let X and Y be analytic varieties over some analytic variety S. Then the fibre product $X \times_S Y$ of X and Y over S can be constructed in the category of analytic varieties. Furthermore, if X, Y and S are affinoid, also $X \times_S Y$ is affinoid.*

Proof. It follows from Proposition 7.1.4/4 and Proposition 1 that $X \times_S Y$ exists and is affinoid if X, Y and S are affinoid. In the general case, the fibre product $X \times_S Y$ will be constructed by pasting together affinoid fibre products of type $X_0 \times_{S_0} Y_0$ with X_0, Y_0, S_0 being open affinoid subvarieties of X, Y, S, respectively. The following property of fibre products, which is obvious from the definition, is fundamental:

Lemma 3. *Assume that the fibre product of X and Y over S exists and is given by the commutative diagram*

$$\begin{array}{ccc} X & & \\ \uparrow{\scriptstyle p_1} & \searrow{\scriptstyle \sigma_1} & \\ X \times_S Y & \xrightarrow{\sigma} & S \\ \downarrow{\scriptstyle p_2} & \nearrow{\scriptstyle \sigma_2} & \\ Y & & \end{array}$$

Let $X_0 \subset X$, $Y_0 \subset Y$, and $S_0 \subset S$ be open analytic subvarieties such that $\sigma_1(X_0) \subset S_0$ and $\sigma_2(Y_0) \subset S_0$. Then X_0 and Y_0 are analytic varieties over S_0, and $p_1^{-1}(X_0) \cap p_2^{-1}(Y_0)$ with projections induced by p_1 and p_2 is a fibre product of X_0 and Y_0 over S_0.

Continuing with the proof of the theorem, we assume for a moment that the fibre product $X \times_S Y$ exists and is given by a diagram as in the lemma. Let $\{S_i\}_{i \in I}$ be an admissible affinoid covering of S. For each $i \in I$, let $\{X_{ij}\}_{j \in J_i}$ and $\{Y_{il}\}_{l \in L_i}$ be admissible affinoid coverings of $\sigma_1^{-1}(S_i)$ and $\sigma_2^{-1}(S_i)$, respectively. Then

$$\{p_1^{-1}(X_{ij})\}_{j \in J_i} \quad \text{and} \quad \{p_2^{-1}(Y_{il})\}_{l \in L_i}$$

are admissible coverings of $\sigma^{-1}(S_i)$, and hence

$$\{p_1^{-1}(X_{ij}) \cap p_2^{-1}(Y_{il})\}_{j \in J_i, l \in L_i}$$

is an admissible covering of $\sigma^{-1}(S_i)$. Furthermore, since $\{\sigma^{-1}(S_i)\}_{i \in I}$ is an admissible covering of $X \times_S Y$,

$$\{p_1^{-1}(X_{ij}) \cap p_2^{-1}(Y_{il})\}_{i \in I, j \in J_i, l \in L_i}$$

must be an admissible covering of $X \times_S Y$. It follows from Lemma 3 and what we already know about affinoid fibre products that the varieties $p_1^{-1}(X_{ij}) \cap p_2^{-1}(Y_{il})$ are affinoid, namely, they are just the affinoid fibre products $W_{ijl} := X_{ij} \times_{S_i} Y_{il}$. Therefore, if the existence is known, the fibre product $X \times_S Y$ can be constructed by pasting together the affinoid varieties W_{ijl}. Note also that again by Lemma 3 the intersections $W_{ijl} \cap W_{i'j'l'}$ are intrinsically characterized as the fibre products

$$W_{ijl}^{i'j'l'} := (X_{ij} \cap X_{i'j'}) \times_{(S_i \cap S_{i'})} (Y_{il} \cap Y_{i'l'}).$$

Now it is clear how to construct the fibre product $X \times_S Y$ in general. We give only a sketch, since the procedure is straightforward and completely trivial, although tedious. We have data as in Proposition 9.3.2/1, namely, affinoid varieties W_{ijl} and, for any pair W_{ijl}, $W_{i'j'l'}$, isomorphic open subvarieties

$$W_{ijl}^{i'j'l'} \subset W_{ijl}, \qquad W_{i'j'l'}^{ijl} \subset W_{i'j'l'}.$$

Note that the existence of the fibre products $W_{ijl}^{i'j'l'}$ is guaranteed by Lemma 3 due to the existence of the affinoid fibre products W_{ijl}. It is then a matter of repeatedly using Lemma 3 and the universal property of fibre products in order to see that these varieties satisfy conditions (a) and (b) of Proposition 9.3.2/1. Hence one can construct an analytic variety W by pasting together the varieties W_{ijl} in the prescribed way. The variety W is canonically a variety over S, and there are projections p_1: $W \to X$ and p_2: $W \to Y$ over S (use Proposition 9.3.3/1). Finally, W is a fibre product of X and Y over S, since it is "locally" a fibre product of affinoid parts of X and Y. □

We conclude this section by listing some canonical isomorphisms between fibre products which are formal consequences of the definition and therefore are valid in any category.

(i) $X \times_S Y = Y \times_S X$.
(ii) $(X \times_S Y) \times_S Z = X \times_S (Y \times_S Z)$.
(iii) $X \times_S S = X$.
(iv) $(X \times_S S') \times_{S'} S'' = X \times_S S''$.
(v) $(X \times_Z Y) \times_S S' = (X \times_S S') \times_{(Z \times_S S')} (Y \times_S S')$.

Formulas (iv) and (v) apply in particular to the so-called process of *base change*. If S' is a fixed analytic variety over S, then for any analytic variety X over S, the fibre product $X \times_S S'$ can be viewed as a variety over S'. Since any S-morphism $\varphi: X \to Y$ induces an S'-morphism $\varphi \times \mathrm{id}$: $X \times_S S' \to Y \times_S S'$, one sees that $X \rightsquigarrow X \times_S S'$ is a functor, called base change with respect to S'. The case where $S = \mathrm{Sp}\, k$ and $S' = \mathrm{Sp}\, k'$ with k' denoting a finite algebraic extension of the ground field k is of particular interest. Any base change of this type is more specifically called *extension of the ground field*, a process which will be defined more generally in the next section.

9.3.6. Extension of the ground field. — Let k' be a complete field which is an extension of the ground field k. Then for any k-affinoid algebra A, the Banach algebra $A \hat{\otimes}_k k'$ is k'-affinoid by Corollary 6.1.1/9. As a consequence, $A \rightsquigarrow A \hat{\otimes}_k k'$ defines a functor from the category of k-affinoid algebras into the category of k'-affinoid algebras. Dualizing we get a field extension functor $X = \mathrm{Sp}\, A \rightsquigarrow X \hat{\otimes} k' := \mathrm{Sp}\, A \hat{\otimes}_k k'$ from the category of k-affinoid varieties into the category of k'-affinoid varieties. This functor fits into the pattern of fibre products only if k' is finite algebraic over k. For, according to Corollary 6.1.2/3, only in this case can k' be viewed as a k-affinoid algebra which allows one to write $X \hat{\otimes} k' = X \times_{\mathrm{Sp}\,k} \mathrm{Sp}\, k'$. Therefore, if k' is not finite over k, the field extension functor $X \rightsquigarrow X \hat{\otimes} k'$ cannot be defined for general analytic varieties simply by referring to fibre products. Instead a direct construction principle has to be applied.

Proposition 1. *Let $X = \mathrm{Sp}\, A$ be a k-affinoid variety and let k' be a complete field containing k.*

(i) *If $\varphi \colon U \to X$ is a k-affinoid map defining U as an affinoid subdomain of X, then $U \hat{\otimes} k'$ is an affinoid subdomain of $X \hat{\otimes} k'$ via the k-affinoid map $\varphi \hat{\otimes} k' \colon U \hat{\otimes} k' \to X \hat{\otimes} k'$. Any functions in A describing U as a Weierstrass, Laurent, or rational subdomain of X also characterize $U \hat{\otimes} k'$ as a subdomain of $X \hat{\otimes} k'$.*

(ii) *If $\mathfrak{U} = \{U_1, \ldots, U_n\}$ is an affinoid covering of X, then $\mathfrak{U} \hat{\otimes} k' = \{U_1 \hat{\otimes} k', \ldots, U_n \hat{\otimes} k'\}$ is an affinoid covering of $X \hat{\otimes} k'$.*

Proof. Let us first assume that the affinoid subdomain $U \subset X$ of (i) is rational, e.g., $U = X\left(\frac{f_1}{g}, \ldots, \frac{f_n}{g}\right)$ with functions $f_1, \ldots, f_n, g \in A$ generating the unit ideal in A. Then the map φ is given by the canonical k-algebra homomorphism

$$\varphi^* \colon A \to A\langle \zeta_1, \ldots, \zeta_n\rangle/(g\zeta_i - f_i;\, i = 1, \ldots, n)$$

which by tensorizing with k' yields the canonical map

$$\varphi^* \hat{\otimes} k' \colon A \hat{\otimes}_k k' \to (A \hat{\otimes}_k k')\langle \zeta_1, \ldots, \zeta_n\rangle/(g\zeta_i - f_i;\, i = 1, \ldots, n)$$

(use Propositions 6.1.1/7 and 6.1.1/12). Identifying A with its canonical image in $A \hat{\otimes}_k k'$ (Corollary 6.1.1/9), the functions $f_1, \ldots, f_n, g$ must also generate the unit ideal in $A \hat{\otimes}_k k'$. Therefore, $\varphi \hat{\otimes} k'$, which is the map associated with the homomorphism $\varphi^* \hat{\otimes} k'$, is just the injection of the rational subdomain

$$(X \hat{\otimes} k')\left(\frac{f_1}{g}, \ldots, \frac{f_n}{g}\right) \hookrightarrow X \hat{\otimes} k'.$$

This proves (i) for rational subdomains. The Weierstrass and Laurent cases are proved similarly. Furthermore, assertion (ii) is easily verified for rational coverings.

To conclude the proof of (i), we consider an arbitrary affinoid subdomain $U \hookrightarrow X$. By Corollary 7.3.5/3, there exist finitely many rational subdomains $V_1, \ldots, V_n$ in X such that $\mathfrak{V} = \{V_1, \ldots, V_n\}$ is a covering of U. The covering

$\mathfrak{V}$ is refined by some rational covering $\mathfrak{W} = \{W_j\}_{j\in J}$ of U (Lemma 8.2.2/2), and each $W_j \in \mathfrak{W}$ is a rational subdomain in X by Theorem 7.2.4/2. Then $\mathfrak{W} \hat{\otimes} k' = \{W_j \hat{\otimes} k'\}_{j\in J}$ is a rational covering of $U \hat{\otimes} k'$, and, for each $j \in J$, the k'-morphisms

$$W_j \hat{\otimes} k' \hookrightarrow U \hat{\otimes} k' \to X \hat{\otimes} k'$$

derived from the inclusions $W_j \hookrightarrow U \hookrightarrow X$ define $W_j \hat{\otimes} k'$ as a rational subdomain in $U \hat{\otimes} k'$ and $X \hat{\otimes} k'$, respectively. Thus, $U \hat{\otimes} k' \to X \hat{\otimes} k'$ is an open immersion by Proposition 7.3.3/1, and it defines $U \hat{\otimes} k'$ as an affinoid subdomain in $X \hat{\otimes} k'$ by Corollary 8.2.1/4.

It remains to verify assertion (ii) for general affinoid coverings. However, each affinoid covering of X can be refined by a rational covering (Lemma 8.2.2/2), and we know already that assertion (ii) holds for rational coverings. Thus it must be true for all affinoid coverings. □

Corollary 2. *Let U and V be affinoid subdomains of the k-affinoid variety X. Then for any complete field k' containing k, the affinoid subdomains $U \hat{\otimes} k$, $V \hat{\otimes} k'$, and $(U \cap V) \hat{\otimes} k'$ are related by the equation*

$$(U \hat{\otimes} k') \cap (V \hat{\otimes} k') = (U \cap V) \hat{\otimes} k' .$$

Proof. One can either rely on Proposition 7.2.2/4 and use the canonical isomorphism

$$(B_1 \hat{\otimes}_k k') \hat{\otimes}_{A \hat{\otimes}_k k'} (B_2 \hat{\otimes}_k k') \cong (B_1 \hat{\otimes}_k B_2) \hat{\otimes}_k k' ,$$

where $X = \operatorname{Sp} A$, $U = \operatorname{Sp} B_1$, and $V = \operatorname{Sp} B_2$, or observe that due to Proposition 1 the assertion is trivially true if U and V are rational subdomains of X or, more generally, finite unions of rational subdomains of X. The latter, however, is always the case by Corollary 7.3.5/3. □

Roughly speaking, Proposition 1 and Corollary 2 mean that the field extension functor $X \rightsquigarrow X \hat{\otimes} k'$ preserves affinoid subdomains and intersections of affinoid subdomains. This is what is needed in order to extend the functor to the special class of those *k-analytic varieties X where the intersection of any two open affinoid subvarieties is again affinoid or, more generally, a finite union of open affinoid subvarieties of X*. If such a variety X is given, one can choose an admissible affinoid covering $\{X_i\}_{i\in I}$ of X and define $X \hat{\otimes} k'$ by pasting together the affinoid varieties $X_i \hat{\otimes} k'$ via the "intersections" $(X_i \cap X_j) \hat{\otimes} k'$. The conditions of Proposition 9.3.2/1 are easily verified; for instance, condition (b) is a consequence of Corollary 2. The resulting k'-analytic variety $X \hat{\otimes} k'$ is independent of the chosen covering $\{X_i\}_{i\in I}$ of X, and furthermore one shows by using Proposition 9.3.3/1 that $X \rightsquigarrow X \hat{\otimes} k'$ is a functor. We leave the details to the reader.

9.4. Coherent modules

9.4.1. $\mathcal{O}$-modules. — Let X be a G-topological space and $\mathcal{O}$ a sheaf of rings on X. Recall that, in terms of sheaves, an *$\mathcal{O}$-module* is a sheaf of abelian groups $\mathcal{F}$ on X together with an $\mathcal{O}(U)$-module structure on $\mathcal{F}(U)$ for each admissible open $U \subset X$ such that all these module structures are compatible with restriction homomorphisms. In the following we will study some general properties of such $\mathcal{O}$-modules; however, our main interest lies in the case where X is an analytic variety and $\mathcal{O}$ is the structure sheaf of X.

To begin, let us consider an *$\mathcal{O}$-module homomorphism* φ: $\mathcal{F} \to \mathcal{G}$ between $\mathcal{O}$-modules $\mathcal{F}$ and $\mathcal{G}$ on X, i.e., a collection of homomorphisms φ_U: $\mathcal{F}(U) \to \mathcal{G}(U)$ with U ranging over the system of all admissible open subsets of X such that each φ_U is an $\mathcal{O}(U)$-module homomorphism and all φ_U are compatible with the restriction homomorphisms of $\mathcal{F}$ and $\mathcal{G}$. Given such a φ, one constructs the following presheaves on X:

$$\mathcal{F}' : U \mapsto \ker \varphi_U ,$$

$$\mathcal{G}' : U \mapsto \operatorname{im} \varphi_U ,$$

with the restriction homomorphisms being induced by the corresponding homomorphisms of $\mathcal{F}$ and $\mathcal{G}$. It is easily seen that $\mathcal{F}'$ is again a sheaf, which canonically carries the structure of an $\mathcal{O}$-module. We use the notation $\mathcal{F}' = \ker \varphi$. Obviously, $\ker \varphi$ is a *submodule* of $\mathcal{F}$, i.e., for each $U \subset X$ admissible open, the $\mathcal{O}(U)$-module $(\ker \varphi)(U) = \ker \varphi_U$ is a submodule of $\mathcal{F}(U)$. As a sheaf homomorphism, φ induces for each $x \in X$ a map φ_x: $\mathcal{F}_x \to \mathcal{G}_x$ between the corresponding stalks which in this case is an $\mathcal{O}_x$-module homomorphism. Due to the exactness of the direct limit, the stalks of the submodule $\ker \varphi$ satisfy

$$(\ker \varphi)_x = \ker \varphi_x .$$

The presheaf $\mathcal{G}'$ is not, in general, a sheaf; however, we can apply Proposition 9.2.2/4 and consider the sheafification of $\mathcal{G}'$, which will be denoted by $\operatorname{im} \varphi$. It is easily checked that $\operatorname{im} \varphi$ is an $\mathcal{O}$-module and, in fact, a submodule of $\mathcal{G}$. Namely, due to the universal property of the sheafification, φ factors through $\operatorname{im} \varphi$ and the homomorphism $\operatorname{im} \varphi \to \mathcal{G}$ clearly has trivial kernel (use, for example, Corollary 9.2.2/6). Looking at stalks, the exactness of the direct limit implies $\mathcal{G}'_x = \operatorname{im} \varphi_x$ for all $x \in X$. Thus,

$$(\operatorname{im} \varphi)_x = \operatorname{im} \varphi_x ,$$

since the stalks of $\mathcal{G}'$ equal the stalks of its sheafification by Corollary 9.2.2/7.

For any $\mathcal{O}$-module $\mathcal{F}$ and any $\mathcal{O}$-submodule $\mathcal{F}'$, one can construct the *quotient module* $\mathcal{F}/\mathcal{F}'$. This is the sheaf associated with the presheaf

$$U \mapsto \mathcal{F}(U)/\mathcal{F}'(U) .$$

The canonical homomorphism φ: $\mathcal{F} \to \mathcal{F}/\mathcal{F}'$ satisfies $\ker \varphi = \mathcal{F}'$ and $\operatorname{im} \varphi = \mathcal{F}/\mathcal{F}'$. Furthermore, all $\mathcal{O}$-module homomorphisms $\mathcal{F} \to \mathcal{G}$ with kernel containing $\mathcal{F}'$ must factor through $\mathcal{F}/\mathcal{F}'$. Using Corollary 9.2.2/7, the exactness of the direct limit implies

$$(\mathcal{F}/\mathcal{F}')_x = \mathcal{F}_x/\mathcal{F}'_x$$

for all $x \in X$.

With the kernel and image of an $\mathcal{O}$-module homomorphism being defined, the notion of *exact sequences* can be introduced. A sequence of $\mathcal{O}$-module homomorphisms

$$\cdots \to \mathcal{F}_{i-1} \xrightarrow{\varphi_{i-1}} \mathcal{F}_i \xrightarrow{\varphi_i} \mathcal{F}_{i+1} \to \cdots$$

is called exact at $\mathcal{F}_i$ if it satisfies $\operatorname{im} \varphi_{i-1} = \ker \varphi_i$. In particular,

$$0 \to \mathcal{F}' \to \mathcal{F} \to \mathcal{F}/\mathcal{F}' \to 0$$

is an exact sequence when $\mathcal{F}'$ is a submodule of $\mathcal{F}$.

Proposition 1. *Let $\mathcal{F}' \xrightarrow{\varphi} \mathcal{F} \xrightarrow{\psi} \mathcal{F}''$ be an exact sequence of $\mathcal{O}$-module homomorphisms. Then for any admissible open $U \subset X$, the restricted sequence $\mathcal{F}'|_U \to \mathcal{F}|_U \to \mathcal{F}''|_U$ is exact. Furthermore, the sequence $\mathcal{F}'_x \xrightarrow{\varphi_x} \mathcal{F}_x \xrightarrow{\psi_x} \mathcal{F}''_x$ is exact for all $x \in X$.*

Proof. Due to Corollary 9.2.2/5, the construction of $\operatorname{im} \varphi$ is compatible with the restriction of φ to any admissible open subset $U \subset X$. Since the corresponding fact for $\ker \varphi$ is trivially true, the first part of the assertion follows. The second part is obvious, since $\operatorname{im} \varphi = \ker \psi$ implies

$$\operatorname{im} \varphi_x = (\operatorname{im} \varphi)_x = (\ker \psi)_x = \ker \psi_x$$

for all $x \in X$. □

Proposition 2. *Let $\mathcal{F}' \xrightarrow{\varphi} \mathcal{F} \xrightarrow{\psi} \mathcal{F}''$ be $\mathcal{O}$-module homomorphisms and consider an admissible covering $\{U_i\}_{i \in I}$ of X. Then $\mathcal{F}' \to \mathcal{F} \to \mathcal{F}''$ is exact if and only if the restricted sequences $\mathcal{F}'|_{U_i} \to \mathcal{F}|_{U_i} \to \mathcal{F}''|_{U_i}$ are exact for all $i \in I$.*

Proof. We consider the submodules $\operatorname{im} \varphi$ and $\ker \psi$ of $\mathcal{F}$ and assume that both modules restrict to the same module over any U_i. Then necessarily $\operatorname{im} \varphi = \ker \psi$, and the if part of the assertion is clear. The only if part follows from Proposition 1. □

If $\{\mathcal{F}_i\}_{i \in I}$ is a family of $\mathcal{O}$-modules, the *direct sum* $\bigoplus_{i \in I} \mathcal{F}_i$ is the $\mathcal{O}$-module defined by

$$U \mapsto \bigoplus_{i \in I} \mathcal{F}_i(U).$$

In particular, if all $\mathcal{F}_i$ are isomorphic to $\mathcal{O}$ and if I is a finite index set consisting of n elements, we simply write $\mathcal{O}^n$ instead of $\bigoplus_{i \in I} \mathcal{F}_i$. It is clear that the formation of stalks commutes with the direct sum.

The *tensor product* $\mathcal{F} \otimes_{\mathcal{O}} \mathcal{G}$ of two $\mathcal{O}$-modules $\mathcal{F}$ and $\mathcal{G}$ is defined as the sheaf associated with the presheaf

$$U \mapsto \mathcal{F}(U) \otimes_{\mathcal{O}(U)} \mathcal{G}(U),$$

where the restriction homomorphisms are the canonical ones. It is not hard to see that the tensor product of $\mathcal{O}$-modules is characterized by a universal property similar to the one used for ordinary tensor products. Since the direct limit commutes with tensor products, one obtains from Corollary 9.2.2/7

$$(\mathscr{F} \otimes_{\mathcal{O}} \mathscr{G})_x = \mathscr{F}_x \otimes_{\mathcal{O}_x} \mathscr{G}_x$$

for all $x \in X$.

An $\mathcal{O}$-submodule $\mathscr{I}$ of $\mathcal{O}$ (considered as a module over itself) is called an *$\mathcal{O}$-ideal* or an *ideal of $\mathcal{O}$*. Furthermore, if $\mathscr{F}$ is an arbitrary $\mathcal{O}$-module, one defines the *product* $\mathscr{I}\mathscr{F}$ as the sheaf associated with the presheaf

$$U \mapsto \mathscr{I}(U)\,\mathscr{F}(U).$$

Then $\mathscr{I}\mathscr{F}$ is a submodule of $\mathscr{F}$, namely, just the image of the canonical $\mathcal{O}$-homomorphism $\mathscr{I} \otimes_{\mathcal{O}} \mathscr{F} \to \mathscr{F}$. In particular, one obtains $(\mathscr{I}\mathscr{F})_x = \mathscr{I}_x\mathscr{F}_x$ for all $x \in X$.

For any morphism of G-ringed spaces $\varrho\colon (X, \mathcal{O}_X) \to (Y, \mathcal{O}_Y)$ and any $\mathcal{O}_X$-module $\mathscr{F}$, one defines an $\mathcal{O}_Y$-module $\varrho_*(\mathscr{F})$ which is given by the sheaf

$$V \mapsto \mathscr{F}\big(\varrho^{-1}(V)\big), \; V \subset Y \text{ admissible open.}$$

The module $\varrho_*(\mathscr{F})$ is called the *direct image* of $\mathscr{F}$ via ϱ. In particular, $\varrho_*(\mathcal{O}_X)$ is an $\mathcal{O}_Y$-module, and the homomorphisms $\varrho_V^*\colon \mathcal{O}_Y(V) \to \mathcal{O}_X\big(\varrho^{-1}(V)\big)$ can be interpreted as an $\mathcal{O}_Y$-module homomorphism $\varrho^{\#}\colon \mathcal{O}_Y \to \varrho_*(\mathcal{O}_X)$.

Finally we want to point out that all the module constructions presented above are compatible with the process of restricting $\mathcal{O}$-modules to admissible open subsets $U \subset X$ (or $V \subset Y$ and $\varrho^{-1}(V) \subset X$ in the case of the direct image). This is clear either by definition or from Corollary 9.2.2/5. We could proceed now with the general theory of $\mathcal{O}$-modules and consider $\mathcal{O}$-modules of finite type and of finite presentation as well as coherent $\mathcal{O}$-modules. However, this would only be of minor importance for applications to analytic varieties, since this case needs a special treatment anyway. Therefore, we will restrict ourselves from now on to the situation we are really interested in, namely, where X is an analytic variety and $\mathcal{O}$ is the structure sheaf of X.

9.4.2. Associated modules. — In the following let $X = \operatorname{Sp} A$ be an affinoid variety. Its structure sheaf $\mathcal{O}_X$ is simply denoted by $\mathcal{O}$, and we write $\mathcal{O}^{\mathrm{w}}$ for the restriction of $\mathcal{O}$ with respect to the weak G-topology of X. There is a natural functor from the category of A-modules into the category of $\mathcal{O}$-modules, which is defined as follows. Given any A-module M, associate to each affinoid subdomain $\operatorname{Sp} A' \subset X$ the A'-module $M \otimes_A A'$. Introducing canonical restriction homomorphisms, we get a presheaf with respect to the weak G-topology on X, which is denoted by $M \otimes \mathcal{O}^{\mathrm{w}}$ and which is, in fact, a sheaf by Corollary 8.2.1/5. (Note that the "presheaf" $\mathcal{O}_X$ of (8.2.1) equals our sheaf $\mathcal{O}^{\mathrm{w}}$.) Then one can apply Proposition 9.2.3/1 and extend $M \otimes \mathcal{O}^{\mathrm{w}}$ to a sheaf $M \otimes \mathcal{O}$ with respect to the (strong) G-topology of X, which is slightly finer than the weak one. Since $M \otimes \mathcal{O}^{\mathrm{w}}$ is an $\mathcal{O}^{\mathrm{w}}$-module, $M \otimes \mathcal{O}$ is seen to be an $\mathcal{O}$-module and is called the *$\mathcal{O}$-module associated to M*. The construction is compatible with restriction homomorphisms in the following sense:

Proposition 1. *Let M be an A-module and denote by $\operatorname{Sp} A' \subset X = \operatorname{Sp} A$ an affinoid subdomain. Then the restriction of $M \otimes \mathcal{O}$ to $\operatorname{Sp} A'$ is the $\mathcal{O}|_{\operatorname{Sp} A'}$-module associated to the A'-module $M \otimes_A A'$.*

Since any homomorphism of A-modules $\varphi: M' \to M$ induces an $\mathcal{O}^{\mathrm{w}}$-module homomorphism $\varphi \otimes \mathcal{O}^{\mathrm{w}}: M' \otimes \mathcal{O}^{\mathrm{w}} \to M \otimes \mathcal{O}^{\mathrm{w}}$ and thus by Proposition 9.2.3/1 an $\mathcal{O}$-module homomorphism $\varphi \otimes \mathcal{O}: M' \otimes \mathcal{O} \to M \otimes \mathcal{O}$, one easily verifies that

$$M \rightsquigarrow M \otimes \mathcal{O}^{\mathrm{w}} \qquad \text{and} \qquad M \rightsquigarrow M \otimes \mathcal{O}$$

constitute functors F^{w} and F.

Proposition 2. *The functor F which assigns to each A-module M the associated $\mathcal{O}$-module $M \otimes \mathcal{O}$ has the following properties:*

(i) *If M' and M are arbitrary A-modules, F defines a bijection $\operatorname{Hom}_A(M', M) \xrightarrow{\sim} \operatorname{Hom}_{\mathcal{O}}(M' \otimes \mathcal{O}, M \otimes \mathcal{O})$ between the set of all A-homomorphisms $M' \to M$ and the set of all $\mathcal{O}$-homomorphisms $M' \otimes \mathcal{O} \to M \otimes \mathcal{O}$.*

(ii) *F is exact.*

(iii) *F commutes with the formation of quotient modules, with finite direct sums, and with tensor products.*

Proof. Due to Proposition 9.2.3/1, it is enough to verify assertion (i) for the functor F^{w} instead of F. But that F^{w} satisfies (i) is trivial, since for any A-module homomorphism $M' \to M$ and any affinoid subdomain $\operatorname{Sp} A' \subset \operatorname{Sp} A$, there exists a unique A'-module homomorphism $M' \otimes_A A' \to M \otimes_A A'$ making the diagram

$$\begin{array}{ccc} M' & \longrightarrow & M \\ \downarrow & & \downarrow \\ M' \otimes_A A' & \longrightarrow & M \otimes_A A' \end{array}$$

commutative.

Furthermore, for any affinoid subdomain $\operatorname{Sp} A' \subset \operatorname{Sp} A$, the algebra A' is a flat module over A by Corollary 7.3.2/6. As a consequence, any exact sequence of A-modules $M' \xrightarrow{\varphi} M \xrightarrow{\psi} M''$ induces an exact sequence of $\mathcal{O}^{\mathrm{w}}$-modules

$$(*) \qquad M' \otimes \mathcal{O}^{\mathrm{w}} \xrightarrow{\varphi \otimes \mathcal{O}^{\mathrm{w}}} M \otimes \mathcal{O}^{\mathrm{w}} \xrightarrow{\psi \otimes \mathcal{O}^{\mathrm{w}}} M'' \otimes \mathcal{O}^{\mathrm{w}}.$$

Considering also the sequence

$$(**) \qquad M' \otimes \mathcal{O} \xrightarrow{\varphi \otimes \mathcal{O}} M \otimes \mathcal{O} \xrightarrow{\psi \otimes \mathcal{O}} M'' \otimes \mathcal{O},$$

the $\mathcal{O}$-module $\ker(\psi \otimes \mathcal{O})$ is an extension of the $\mathcal{O}^{\mathrm{w}}$-module $\ker(\psi \otimes \mathcal{O}^{\mathrm{w}})$. The same is true by Proposition 9.2.3/2 and the definition of the image for the modules $\operatorname{im}(\varphi \otimes \mathcal{O})$ and $\operatorname{im}(\varphi \otimes \mathcal{O}^{\mathrm{w}})$. Therefore, Proposition 9.2.3/1 shows that the exactness of $(*)$ implies the exactness of $(**)$. This verifies assertion (ii).

Finally turning to assertion (iii), due to the exactness of F, we only have to show that F commutes with tensor products. The corresponding fact is, of course, true for the functor F^w, since

$$(M \otimes_A M') \otimes_A A' = (M \otimes_A A') \otimes_{A'} (M' \otimes_A A')$$

for any A-modules M', M and any algebra A' over A. But then, just as before, one applies Proposition 9.2.3/2 and obtains that also F commutes with tensor products. □

If M is an A-module and $N \subset M$ is a submodule, one derives from the exact sequence $0 \to N \to M$ an exact sequence $0 \to N \otimes \mathcal{O} \to M \otimes \mathcal{O}$. Thus, $N \otimes \mathcal{O}$ can canonically be viewed as a submodule of $M \otimes \mathcal{O}$. We consider the special case where N is the kernel or the image of an A-module homomorphism.

Corollary 3. *Let $\varphi\colon M' \to M$ be a homomorphism of A-modules and $\varphi \otimes \mathcal{O}\colon M' \otimes \mathcal{O} \to M \otimes \mathcal{O}$ the associated $\mathcal{O}$-module homomorphism. Then* $\ker(\varphi \otimes \mathcal{O}) = (\ker \varphi) \otimes \mathcal{O}$ *and* $\operatorname{im}(\varphi \otimes \mathcal{O}) = (\operatorname{im} \varphi) \otimes \mathcal{O}$.

Proof. The exact sequence $0 \to \ker \varphi \to M' \xrightarrow{\varphi} M$ leads to the exact sequence

$$0 \to (\ker \varphi) \otimes \mathcal{O} \to M' \otimes \mathcal{O} \xrightarrow{\varphi \otimes \mathcal{O}} M \otimes \mathcal{O},$$

which implies the first equation. Similarly, the exact sequence $M' \xrightarrow{\varphi} M \xrightarrow{\psi} M/\operatorname{im} \varphi$ leads to the exact sequence

$$M' \otimes \mathcal{O} \xrightarrow{\varphi \otimes \mathcal{O}} M \otimes \mathcal{O} \xrightarrow{\psi \otimes \mathcal{O}} (M/\operatorname{im} \varphi) \otimes \mathcal{O},$$

showing that

$$\operatorname{im}(\varphi \otimes \mathcal{O}) = \ker(\psi \otimes \mathcal{O}) = (\ker \psi) \otimes \mathcal{O} = (\operatorname{im} \varphi) \otimes \mathcal{O}. \qquad \square$$

As an immediate consequence, one gets from Proposition 2 and Corollary 3

Corollary 4. *A sequence of A-module homomorphisms $M' \to M \to M''$ is exact if and only if the associated sequence of $\mathcal{O}$-modules $M' \otimes \mathcal{O} \to M \otimes \mathcal{O} \to M'' \otimes \mathcal{O}$ is exact.*

Proposition 5. *Let $\varrho\colon Y \to X$ be a morphism between affinoid varieties $Y = \operatorname{Sp} B$ and $X = \operatorname{Sp} A$, induced by an algebra homomorphism $\varrho^*\colon A \to B$, which is finite. Then viewing ϱ^* as a homomorphism of A-modules, the homomorphism of $\mathcal{O}_X$-modules $\varrho^{\#}\colon \mathcal{O}_X \to \varrho_*(\mathcal{O}_Y)$ is associated to ϱ^*. Furthermore, for any B-module M, the $\mathcal{O}_X$-module $\varrho_*(M \otimes \mathcal{O}_Y)$ is associated to the A-module M_A, which is derived from M by restricting scalars to A via ϱ^*.*

Proof. If $\operatorname{Sp} A' \subset X = \operatorname{Sp} A$ is an affinoid subdomain, we apply Proposition 7.2.2/4 to conclude that $\varrho^*_{\operatorname{Sp} A'}$ is the natural homomorphism $A' \to A' \hat{\otimes}_A B$. Since B is finite over A, the complete tensor product $A' \hat{\otimes}_A B$ coincides with the ordinary tensor product $A' \otimes_A B$ due to Proposition 3.7.3/6. Therefore, the first assertion follows. Writing $\mathscr{G} := \varrho_*(M \otimes \mathcal{O}_Y)$, we see that

$$\begin{aligned}\mathscr{G}(\operatorname{Sp} A') &= M \otimes_B (B \mathbin{\widehat{\otimes}}_A A') \\ &= M \otimes_B (B \otimes_A A') = M_A \otimes_A A'.\end{aligned}$$

Hence Proposition 9.2.3/1 shows $\mathscr{G} = M_A \otimes \mathcal{O}_X$. □

Finally, we want to derive some properties of the stalks of associated modules. Let $X = \operatorname{Sp} A$ be an affinoid variety, and let $\mathscr{F}$ be an $\mathcal{O}_X$-module which is associated to some A-module M. We choose a point $x \in X$ and denote by $\mathfrak{m}$ the maximal ideal in A corresponding to x, by A_x and $M_x = M \otimes_A A_x$ the localizations of A and M with respect to $\mathfrak{m}$, and by $\hat{\mathcal{O}}_{X,x}$ and $\hat{\mathscr{F}}_x$ the $\mathfrak{m}$-adic completions of the stalks $\mathcal{O}_{X,x}$ and $\mathscr{F}_x$. Then one has a canonical commutative diagram as follows:

$$\begin{array}{ccccc} M \otimes_A A_x & \longrightarrow & M \otimes_A \mathcal{O}_{X,x} & \longrightarrow & M \otimes_A \hat{\mathcal{O}}_{X,x} \\ \downarrow & & \downarrow & & \downarrow \\ M_x & \longrightarrow & \mathscr{F}_x & \longrightarrow & \hat{\mathscr{F}}_x \end{array} \quad .$$

Proposition 6. *Suppose that in the above situation M is a finite A-module. Then all vertical maps are isomorphisms and all horizontal maps are injections. The module $\hat{\mathscr{F}}_x$ which is the $\mathfrak{m}$-adic completion of $\mathscr{F}_x$ can also be viewed as the $\mathfrak{m}$-adic completion of M_x via the map $M_x \to \mathscr{F}_x$ of the second row. In particular, if one of the modules M_x, $\mathscr{F}_x$, or $\hat{\mathscr{F}}_x$ vanishes, all three of them must vanish.*

Proof. Tensor products commute with direct limits. Therefore, the map $M \otimes_A \mathcal{O}_{X,x} \to \mathscr{F}_x$ is bijective. The stalk $\mathcal{O}_{X,x}$ is a Noetherian local ring by Proposition 7.3.2/7, and $\mathscr{F}_x$ is a finite $\mathcal{O}_{X,x}$-module, since M is a finite A-module. Writing

$$M \otimes_A \hat{\mathcal{O}}_{X,x} = (M \otimes_A \mathcal{O}_{X,x}) \otimes_{\mathcal{O}_{X,x}} \hat{\mathcal{O}}_{X,x} = \mathscr{F}_x \otimes_{\mathcal{O}_{X,x}} \hat{\mathcal{O}}_{X,x},$$

we see by Proposition 7.3.1/5 that $M \otimes_A \hat{\mathcal{O}}_{X,x} \to \hat{\mathscr{F}}_x$ is bijective. The same proposition shows that $\hat{\mathscr{F}}_x$ can also be viewed as the $\mathfrak{m}$-adic completion of M_x, because $\hat{\mathcal{O}}_{X,x}$ is the $\mathfrak{m}$-adic completion of A_x (see Proposition 7.3.2/3). Finally, the injectivity of all horizontal maps follows from KRULL's Intersection Theorem (for finite modules over Noetherian local rings). □

Corollary 7. *As before, let $X = \operatorname{Sp} A$ be an affinoid variety, and let $\mathscr{F} = M \otimes \mathcal{O}_X$ be an $\mathcal{O}_X$-module which is associated to a finite A-module M. Then the canonical map $\mathscr{F}(X) \to \prod_{x \in X} \mathscr{F}_x$ is injective.*

Proof. The canonical map $\mathscr{F}(X) = M \to \prod_{x \in X} M_x$ is injective. Hence Proposition 6 implies the assertion. □

In particular, we see that the $\mathcal{O}_X$-module $\mathscr{F} = M \otimes \mathcal{O}_X$ vanishes if and only if all its stalks $\mathscr{F}_x$ vanish.

9.4.3. $\mathfrak{U}$-coherent modules. — If $(X, \mathcal{O}_X)$ is a G-ringed space and $\mathcal{F}$ is an $\mathcal{O}_X$-module, $\mathcal{F}$ can be called a *coherent module* if there exists an admissible covering $\mathfrak{U} = \{U_i\}_{i\in I}$ of X such that for each $i \in I$ the restriction $\mathcal{F}|_{U_i}$ equals the cokernel of some $\mathcal{O}_X|_{U_i}$-module homomorphism of type

$$(\mathcal{O}_X|_{U_i})^r \to (\mathcal{O}_X|_{U_i})^s.$$

Since for any affinoid variety $X = \operatorname{Sp} A$, the finite direct sum $\mathcal{O}_X^n$ is associated to the free A-module A^n, the notion of coherence in terms of analytic varieties amounts to the following:

Definition 1. *Let X denote an analytic variety. An $\mathcal{O}_X$-module $\mathcal{F}$ is called coherent if there exists an admissible affinoid covering $\mathfrak{U} = \{\operatorname{Sp} A_i\}_{i\in I}$ of X such that for each $i \in I$ the restricted module $\mathcal{F}|_{\operatorname{Sp} A_i}$ is associated to some finite A_i-module M_i, i.e., if*

$$\mathcal{F}|_{\operatorname{Sp} A_i} = M_i \otimes \mathcal{O}_{\operatorname{Sp} A_i}.$$

More precisely, $\mathcal{F}$ is called $\mathfrak{U}$-coherent if the covering $\mathfrak{U}$ is to be specified.

In particular, if $X = \operatorname{Sp} A$ is affinoid and $\mathcal{F}$ is associated to a finite A-module, $\mathcal{F}$ is coherent and, in fact, $\mathfrak{U}$-coherent for all affinoid coverings $\mathfrak{U}$ of X (cf. Proposition 9.4.2/1). The same proposition shows that for any coherent module $\mathcal{F}$ on an arbitrary analytic variety X and any open analytic subvariety $V \subset X$, the restriction $\mathcal{F}|_V$ is a coherent $\mathcal{O}_V$-module. Furthermore, if $\mathcal{F}$ is $\mathfrak{U}$-coherent for some admissible affinoid covering $\mathfrak{U}$ of X, then $\mathcal{F}$ is also $\mathfrak{V}$-coherent for any admissible affinoid covering $\mathfrak{V}$ of X which is finer than $\mathfrak{U}$. As a consequence, for finitely many coherent $\mathcal{O}_X$-modules $\mathcal{F}_1, \ldots, \mathcal{F}_n$, there is always an admissible affinoid covering U of X such that $\mathcal{F}_1, \ldots, \mathcal{F}_n$ are $\mathfrak{U}$-coherent simultaneously.

Proposition 2. *Let X be an analytic variety and let $\mathcal{F}, \mathcal{F}'$ be $\mathcal{O}_X$-modules which are $\mathfrak{U}$-coherent for some admissible affinoid covering $\mathfrak{U}$ of X.*

(i) *If $\mathcal{F}'$ is a submodule of $\mathcal{F}$, then $\mathcal{F}/\mathcal{F}'$ is $\mathfrak{U}$-coherent.*

(ii) *If $\varphi\colon \mathcal{F}' \to \mathcal{F}$ is an $\mathcal{O}_X$-module homomorphism, then* $\ker \varphi$ *and* $\operatorname{im} \varphi$ *are $\mathfrak{U}$-coherent.*

(iii) *$\mathcal{F}' \oplus \mathcal{F}$ and $\mathcal{F}' \otimes_{\mathcal{O}_X} \mathcal{F}$ are $\mathfrak{U}$-coherent.*

(iv) *If $\mathcal{I}$ is a $\mathfrak{U}$-coherent $\mathcal{O}_X$-ideal, then also the $\mathcal{O}_X$-module $\mathcal{I}\mathcal{F}$ is $\mathfrak{U}$-coherent.*

Proof. Assertions (i) and (iii) are immediate consequences of Proposition 9.4.2/2, and assertion (ii) follows from Corollary 9.4.2/3 due to the fact that affinoid algebras are Noetherian. Finally, assertion (iv) is clear from (ii) and (iii), since $\mathcal{I}\mathcal{F}$ equals the image of the canonical homomorphism $\mathcal{I} \otimes_{\mathcal{O}_X} \mathcal{F} \to \mathcal{F}$. □

The theory of coherent modules is simplified to a certain extent by the following

Theorem 3 (Kiehl). *Let $X = \operatorname{Sp} A$ be an affinoid variety and denote by $\mathscr{F}$ an $\mathcal{O}_X$-module. Then the following are equivalent:*

(i) *$\mathscr{F}$ is a coherent $\mathcal{O}_X$-module.*
(ii) *$\mathscr{F}$ is $\mathfrak{U}$-coherent for all affinoid coverings $\mathfrak{U}$ of X.*
(iii) *$\mathscr{F}$ is associated to a finite A-module.*

Applying this to analytic varieties, we get immediately

Corollary 4. *Let X be an analytic variety and let $\mathscr{F}$ be a coherent $\mathcal{O}_X$-module. Then $\mathscr{F}$ is $\mathfrak{U}$-coherent for all admissible affinoid coverings $\mathfrak{U}$ of X.*

To prove the theorem, we have only to show that $\mathscr{F}$ is associated to a finite A-module, provided $\mathscr{F}$ is $\mathfrak{U}$-coherent for some affinoid covering $\mathfrak{U}$ of X. According to Lemmata 8.2.2/2, 8.2.2/3, and 8.2.2/4, we may restrict ourselves to the case where $\mathfrak{U}$ is a Laurent covering. Furthermore, by an induction argument, we can always assume that the Laurent covering is generated by a single element $f \in A$. Thus, for the proof of the theorem, we have to verify the following two lemmata where $\mathfrak{U}$ is a Laurent covering generated by some element $f \in A$:

Lemma 5. *Let $\mathscr{F}$ be $\mathfrak{U}$-coherent. Then $H^1(\mathfrak{U}, \mathscr{F}) = 0$.*

Lemma 6. *Let $\mathscr{F}$ be $\mathfrak{U}$-coherent and assume $H^1(\mathfrak{U}, \mathscr{F}') = 0$ for each $\mathfrak{U}$-coherent $\mathcal{O}_X$-module $\mathscr{F}'$. Then $\mathscr{F}$ is associated to a finite A-module.*

Proof of Lemma 5. We have $\mathfrak{U} = \{U_1, U_2\}$ with $U_1 = X(f)$ and $U_2 = X(f^{-1})$. Let

$$M_1 := \mathscr{F}(U_1), \quad M_2 := \mathscr{F}(U_2), \quad \text{and} \quad M_{12} := \mathscr{F}(U_1 \cap U_2).$$

Then M_1, M_2, and M_{12} are finite modules over $A\langle f\rangle$, $A\langle f^{-1}\rangle$, and $A\langle f, f^{-1}\rangle$, respectively. The Čech complex of alternating cochains $C_a^{\cdot}(\mathfrak{U}, \mathscr{F})$ degenerates to

$$0 \to M_1 \times M_2 \xrightarrow{d^0} M_{12} \to 0,$$

and it is only necessary to show that $d^0\colon M_1 \times M_2 \to M_{12}$ is surjective, since $H^1(\mathfrak{U}, \mathscr{F})$ can be computed by using alternating cochains (cf. Proposition 8.1.3/1). We fix an arbitrary residue norm on A and consider $A\langle f\rangle$, $A\langle f^{-1}\rangle$, and $A\langle f, f^{-1}\rangle$ with the residue norms induced by the canonical epimorphisms

$$A\langle\zeta\rangle \to A\langle\zeta\rangle/(f - \zeta) = A\langle f\rangle,$$
$$A\langle\eta\rangle \to A\langle\eta\rangle/(1 - f\eta) = A\langle f^{-1}\rangle,$$
$$A\langle\zeta, \eta\rangle \to A\langle\zeta, \eta\rangle/(f - \zeta, 1 - f\eta) = A\langle f, f^{-1}\rangle.$$

Then all restriction homomorphisms of the commutative diagram

$$\begin{array}{ccccc} & & A\langle f\rangle & & \\ & \nearrow & & \searrow & \\ A & & & & A\langle f, f^{-1}\rangle \\ & \searrow & & \nearrow & \\ & & A\langle f^{-1}\rangle & & \end{array}$$

are contractive. Choosing an arbitrary constant $\beta > 1$, any $g \in A\langle f, f^{-1}\rangle$ can be represented by a series

$$g' = \sum c_{\mu\nu}\zeta^\mu\eta^\nu \in A\langle\zeta, \eta\rangle,$$

satisfying $|c_{\mu\nu}| \leqq \beta\,|g|$. Thus, we see that

(*) For any $g \in A\langle f, f^{-1}\rangle$, there exist elements $g^+ \in A\langle f\rangle$ and $g^- \in A\langle f^{-1}\rangle$ such that $|g^+| \leqq \beta\,|g|$, $|g^-| \leqq \beta\,|g|$, and $g = g^+|_{U_1\cap U_2} - g^-|_{U_1\cap U_2}$.

Let $\{v_1, \ldots, v_n\} \subset M_1$ and $\{w_1, \ldots, w_m\} \subset M_2$ be generator systems of M_1 and M_2, respectively. Then the restrictions $v_1', \ldots, v_n'$ of the v_i, as well as the restrictions $w_1', \ldots, w_m'$ of the w_j, to $U_1 \cap U_2$ generate the $A\langle f, f^{-1}\rangle$-module M_{12}, since $\mathcal{F}$ is $\mathfrak{U}$-coherent. According to Proposition 3.7.3/3, we can define complete module norms on M_1, M_2, M_{12} via the epimorphisms

$$(A\langle f\rangle)^n \to M_1, \quad (h_1, \ldots, h_n) \mapsto \sum_{i=1}^{n} h_i v_i,$$

$$(A\langle f^{-1}\rangle)^m \to M_2, \quad (h_1, \ldots, h_m) \mapsto \sum_{j=1}^{m} h_j w_j,$$

$$(A\langle f, f^{-1}\rangle)^n \to M_{12}, \quad (h_1, \ldots, h_n) \mapsto \sum_{i=1}^{n} h_i v_i'.$$

The surjectivity of the coboundary homomorphism $d^0\colon M_1 \times M_2 \to M_{12}$ will then be a consequence of the following assertion:

(**) Let $\varepsilon > 0$ be a constant. For each $u \in M_{12}$, there exist elements $u^+ \in M_1$ and $u^- \in M_2$ with

$$|u^+| \leqq \alpha\,|u|, \qquad |u^-| \leqq \alpha\,|u|,$$

$$|u - (u^+|_{U_1\cap U_2}) + (u^-|_{U_1\cap U_2})| \leqq \varepsilon\,|u|,$$

where $\alpha > 0$ is a suitable constant independent of u.

So let us verify assertion (**). Since the elements v_i', as well as the w_j', generate M_{12}, there are equations

$$v_i' = \sum_{j=1}^{m} c_{ij} w_j', \quad i = 1, \ldots, n, \quad c_{ij} \in A\langle f, f^{-1}\rangle,$$

$$w_j' = \sum_{l=1}^{n} d_{jl} v_l', \quad j = 1, \ldots, m, \quad d_{jl} \in A\langle f, f^{-1}\rangle.$$

The image of $A\langle f^{-1}\rangle$ is dense in $A\langle f, f^{-1}\rangle$. Therefore, we can find elements $c_{ij}' \in A\langle f^{-1}\rangle$ which satisfy

$$\max_{i,j,l} |c_{ij} - c_{ij}''|\,|d_{jl}| \leqq \beta^{-2}\varepsilon,$$

where $\beta > 1$ is a constant as in assertion (*) and c_{ij}'' is the image of c_{ij}' in $A\langle f, f^{-1}\rangle$. We want to show that assertion (**) is true for the constant

$$\alpha := \beta^2 \cdot \max_{i,j} (|c_{ij}'| + 1).$$

For any $u \in M_{12}$, the definition of the norm on M_{12} guarantees the existence of elements $a_1, \ldots, a_n \in A\langle f, f^{-1}\rangle$ such that

$$u = \sum_{i=1}^{n} a_i v'_i \quad \text{and} \quad |a_i| \leqq \beta\,|u|.$$

We write $a_i = a_i^+|_{U_1 \cap U_2} - a_i^-|_{U_1 \cap U_2}$ according to assertion (*) with elements $a_i^+ \in A\langle f\rangle$, $a_i^- \in A\langle f^{-1}\rangle$ satisfying $|a_i^+| \leqq \beta\,|a_i|$ and $|a_i^-| \leqq \beta\,|a_i|$. Setting

$$u^+ := \sum_{i=1}^{n} a_i^+ v_i \in M_1,$$

$$u^- := \sum_{i,j} a_i^- c'_{ij} w_j \in M_2,$$

we have

$$|u^+| \leqq \max_i |a_i^+| \leqq \beta \max_i |a_i| \leqq \beta^2\,|u| \leqq \alpha\,|u|,$$

$$|u^-| \leqq \max_{i,j} |a_i^-|\,|c'_{ij}| \leqq \beta^2\,|u| \max_{i,j} |c'_{ij}| \leqq \alpha\,|u|,$$

and, omitting restrictions on $U_1 \cap U_2$,

$$u = \sum_{i=1}^{n} (a_i^+ v'_i - a_i^- v'_i) = u^+ - \sum_{i,j} a_i^- c_{ij} w'_j$$

$$= u^+ - u^- - \sum_{i,j} a_i^- (c_{ij} - c''_{ij})\, w'_j.$$

Hence we get

$$|u - u^+ + u^-| = \left|\sum_{i,j,l} a_i^- (c_{ij} - c''_{ij})\, d_{jl} v'_l\right|$$

$$\leqq \max_{i,j,l} |a_i^-|\,|c_{ij} - c''_{ij}|\,|d_{jl}| \leqq \beta^2\,|u|\,\beta^{-2}\varepsilon = \varepsilon\,|u|,$$

thereby verifying assertion (**) and thus completing the proof of Lemma 5. □

Proof of Lemma 6. It is not necessary to make a difference between Laurent and general affinoid coverings. Therefore we consider the case where $\mathfrak{U} = \{U_1, \ldots, U_n\}$ with affinoid subdomains $U_i = \operatorname{Sp} A_i \subset X$. Since $\mathcal{F}$ is $\mathfrak{U}$-coherent, $\mathcal{F}|_{U_i}$ is associated to a finite A_i-module M_i, $i = 1, \ldots, n$. For $x \in X$, we denote by $\mathfrak{m}_x \subset A$ the corresponding maximal ideal of A and by $\mathfrak{m}_x\mathcal{O}_X$ the associated $\mathcal{O}_X$-ideal. Multiplying this ideal with $\mathcal{F}$ gives an $\mathcal{O}_X$-submodule $\mathfrak{m}_x\mathcal{F}$ of $\mathcal{F}$ which is $\mathfrak{U}$-coherent, since $\mathfrak{m}_x\mathcal{F}|_{U_i}$ is associated to the finite A_i-module $\mathfrak{m}_x M_i$, $i = 1, \ldots, n$. Then $\mathcal{F}/\mathfrak{m}_x\mathcal{F}$ is $\mathfrak{U}$-coherent by Proposition 2, and thus we have an exact sequence of $\mathfrak{U}$-coherent modules

$$0 \to \mathfrak{m}_x\mathcal{F} \to \mathcal{F} \to \mathcal{F}/\mathfrak{m}_x\mathcal{F} \to 0.$$

If $U' = \operatorname{Sp} A'$ is an affinoid subdomain of X, contained in some U_i, the above exact sequence restricts to an exact sequence on U' by Proposition 9.4.1/1. Furthermore, due to Proposition 9.4.2/1, this restricted sequence is associated to the sequence

$$0 \to \mathfrak{m}_x\mathcal{F}(U') \to \mathcal{F}(U') \to \mathcal{F}/\mathfrak{m}_x\mathcal{F}(U') \to 0,$$

which is exact, for example, by Corollary 9.4.2/4. Therefore, we get an exact sequence of Čech complexes

$$0 \to C^\cdot(\mathfrak{U}, \mathfrak{m}_x\mathscr{F}) \to C^\cdot(\mathfrak{U}, \mathscr{F}) \to C^\cdot(\mathfrak{U}, \mathscr{F}/\mathfrak{m}_x\mathscr{F}) \to 0$$

and hence an exact cohomology sequence

$$0 \to \mathfrak{m}_x\mathscr{F}(X) \to \mathscr{F}(X) \to \mathscr{F}/\mathfrak{m}_x\mathscr{F}(X) \to H^1(\mathfrak{U}, \mathfrak{m}_x\mathscr{F}) \to \dots.$$

Since $H^1(\mathfrak{U}, \mathfrak{m}_x\mathscr{F}) = 0$, this implies the first of the following two assertions:

(*) The canonical map $\mathscr{F}(X) \to \mathscr{F}/\mathfrak{m}_x\mathscr{F}(X)$ is surjective.

(**) The restriction homomorphism $\mathscr{F}/\mathfrak{m}_x\mathscr{F}(X) \to \mathscr{F}/\mathfrak{m}_x\mathscr{F}(U_j)$ is bijective, provided $x \in U_j$.

To verify (**), we consider again an affinoid subdomain $U' = \operatorname{Sp} A'$ of X contained in some U_i. Setting $A'_j := \mathcal{O}_X(U' \cap U_j)$ and assuming $x \in U_j$, the canonical homomorphism

$$A'/\mathfrak{m}_xA' \to A'_j/\mathfrak{m}_xA'_j$$

is bijective. This follows from Proposition 7.2.2/1 if $x \in U' \cap U_j$, and is trivial if $x \notin U'$, since then both algebras must vanish. Using elementary tensor calculus, one obtains the restriction homomorphism

$$\mathscr{F}/\mathfrak{m}_x\mathscr{F}(U') \to \mathscr{F}/\mathfrak{m}_x\mathscr{F}(U' \cap U_j)$$

by tensoring the above bijection with M_i over A_i; hence it is bijective. This argument shows that the canonical homomorphism of Čech complexes

$$C^\cdot(\mathfrak{U}, \mathscr{F}/\mathfrak{m}_x\mathscr{F}) \to C^\cdot(\mathfrak{U}|_{U_j}, \mathscr{F}/\mathfrak{m}_x\mathscr{F})$$

is bijective. In particular, we get an isomorphism in the zero-th cohomology, which proves assertion (**).

Now assertions (*) and (**) together say that for $x \in U_j$ the canonical homomorphism $\mathscr{F}(X) \to \mathscr{F}/\mathfrak{m}_x\mathscr{F}(U_j)$ is surjective. This homomorphism can also be interpreted as the composition

$$\begin{array}{ccccc} \mathscr{F}(X) & \longrightarrow & \mathscr{F}(U_j) & \longrightarrow & \mathscr{F}/\mathfrak{m}_x\mathscr{F}(U_j) \\ & & \| & & \| \\ & & M_j & \longrightarrow & M_j/\mathfrak{m}_xM_j\,. \end{array}$$

In particular, the image of $\mathscr{F}(X)$ in $\mathscr{F}/\mathfrak{m}_x\mathscr{F}(U_j)$ generates this module over A_j. By the classical Lemma of NAKAYAMA, $\mathscr{F}(X)$ must generate the A_j-module $\mathscr{F}(U_j) = M_j$ after localization by $\mathfrak{m}_xA_j$. This is true for all $x \in U_j$, and hence we see that, in fact, $\mathscr{F}(X)$ generates $\mathscr{F}(U_j)$ over A_j. Therefore, we can choose finitely many elements $f_1, \dots, f_s \in \mathscr{F}(X)$ such that their images generate all modules $\mathscr{F}(U_j)$ simultaneously for $j = 1, \dots, n$. The elements $f_1, \dots, f_s$ define an $\mathcal{O}_X$-module homomorphism $\varphi\colon \mathcal{O}_X^s \to \mathscr{F}$ which is clearly surjective by construction, i.e., satisfies $\operatorname{im} \varphi = \mathscr{F}$. The direct product $\mathcal{O}_X^s$ is, in particular, $\mathfrak{U}$-coherent so that φ is a homomorphism of $\mathfrak{U}$-coherent modules. Hence, by

Proposition 2, the module $\ker\varphi$ is $\mathfrak{U}$-coherent. Repeating for $\ker\varphi$ what we did with $\mathscr{F}$ gives an exact sequence

$$\mathcal{O}_X^r \to \mathcal{O}_X^s \to \mathscr{F} \to 0\,.$$

Since $\mathcal{O}_X^r$ and $\mathcal{O}_X^s$ are associated to A^r and A^s, we see by Proposition 9.4.2/2 that $\mathscr{F}$ is associated to a finite A-module. This concludes the proof of Lemma 6 and thus the proof of the theorem. □

9.4.4. Finite morphisms. — In this section, we give an application of Theorem 9.4.3/3. A morphism $\varrho\colon Y \to X$ of analytic varieties is called *finite* if there exists an admissible affinoid covering $\{U_i\}_{i\in I}$ of X such that for all $i \in I$ the induced map $\varrho^{-1}(U_i) \to U_i$ is a finite morphism of affinoid varieties in the sense of Definition 7.1.4/3 (meaning that the associated homomorphism of affinoid algebras $\varrho^*_{U_i}\colon \mathcal{O}_X(U_i) \to \mathcal{O}_Y\big(\varrho^{-1}(U_i)\big)$ is finite). One concludes from the following proposition that in the case where X and Y themselves are affinoid, this notion of finiteness is not different from the one given in Definition 7.1.4/3.

Proposition 1. *A morphism $\varrho\colon Y \to X$ of analytic varieties is finite if and only if the following two conditions are satisfied:*

(i) *The inverse image $\varrho^{-1}(U)$ of any open affinoid subvariety $U \subset X$ is an open affinoid subvariety in Y.*

(ii) *$\varrho^*(\mathcal{O}_Y)$ is a coherent $\mathcal{O}_X$-module.*

Proof. Due to Proposition 9.4.2/5, we have only to show that ϱ satisfies condition (i) if it is finite. The rest is clear. Therefore, we assume that ϱ is finite and consider an arbitrary open affinoid subvariety U in X.

By Theorem 9.4.3/3, the algebra $B := \mathcal{O}_Y\big(\varrho^{-1}(U)\big) = \varrho_*(\mathcal{O}_Y)\,(U)$ is a finite module over the affinoid algebra $A := \mathcal{O}_X(U)$ via the homomorphism

$$\varrho^*_U\colon \mathcal{O}_X(U) \to \mathcal{O}_Y\big(\varrho^{-1}(U)\big)\,.$$

Hence by Proposition 6.1.1/6, we can view B as an affinoid algebra. Applying Corollary 9.3.3/2, the canonical diagram

$$\begin{array}{ccc} A & \longrightarrow & \mathcal{O}_Y\big(\varrho^{-1}(U)\big) \\ & \searrow \quad \nearrow{\scriptstyle\sim} & \\ & B & \end{array}$$

leads to a commutative diagram of analytic maps

$$\begin{array}{ccc} \varrho^{-1}(U) & \xrightarrow{\ \varrho'\ } & U = \operatorname{Sp} A \\ & {\scriptstyle\chi}\searrow \quad \nearrow{\scriptstyle\psi} & \\ & \operatorname{Sp} B & \end{array}$$

with ϱ' being a restriction of ϱ. There is an admissible affinoid covering $\{U_i\}_{i \in I}$ of X such that all morphisms $\varrho^{-1}(U_i) \to U_i$ are affinoid. Therefore, for any open subvariety $V \subset X$ such that $V \subset U_i$ for some $i \in I$, the inverse image $\varrho^{-1}(V) \subset Y$ is affinoid. Hence there exists an admissible affinoid covering $\{V_j\}_{j \in J}$ of U such that $\varrho^{-1}(V_j)$ is affinoid for all $j \in J$. As a consequence, χ induces morphisms of affinoid varieties

$$\chi_j : \varrho^{-1}(V_j) \to \psi^{-1}(V_j), \qquad j \in J.$$

Since the $\mathcal{O}_U$-modules $\psi_*(\mathcal{O}_{\operatorname{Sp} B})$ and $\varrho'_*(\mathcal{O}_{\varrho^{-1}(U)})$ are both associated to the finite A-module B (use Proposition 9.4.2/5), we see that all maps χ_j are isomorphisms, because their corresponding homomorphisms of affinoid algebras must be isomorphisms. But then χ is an isomorphism by Corollary 8.2.1/4, and $\varrho^{-1}(U)$ is affinoid. □

Corollary 2. *A morphism $\varrho: Y \to X$ of analytic varieties is finite if and only if for any open affinoid subvariety $U \subset X$ the inverse image $\varrho^{-1}(U)$ is affinoid and the induced morphism $\varrho^{-1}(U) \to U$ is finite in the sense of Definition* 7.1.4/3, *i.e., the corresponding homomorphism of affinoid algebras is finite.*

As a consequence it is seen that a composition $X \xrightarrow{\varrho} Y \xrightarrow{\sigma} Z$ of analytic maps is finite if ϱ and σ are finite. Furthermore, finite morphisms are preserved by any base change functor, and the fibre product of finite morphisms over any analytic variety S is again a finite morphism.

Proposition 3. *Let $\varrho: Y \to X$ be a finite morphism of analytic varieties and let $\mathcal{F}$ be a coherent $\mathcal{O}_Y$-module. Then $\varrho_*(\mathcal{F})$ is a coherent $\mathcal{O}_X$-module.*

The *proof* is obvious from Proposition 9.4.2/5. □

9.5. Closed analytic subvarieties

9.5.1. Coherent ideals. The nilradical. — Let $X = \operatorname{Sp} A$ be an affinoid variety with structure sheaf $\mathcal{O}_X$. If $\mathfrak{a} \subset A$ denotes an ideal, we see by Corollary 9.4.2/4 that the associated $\mathcal{O}_X$-module $\mathfrak{a} \otimes \mathcal{O}_X$ is a submodule of $\mathcal{O}_X$ and thus an $\mathcal{O}_X$-ideal. Therefore, $\mathfrak{a} \otimes \mathcal{O}_X$ is called the *$\mathcal{O}_X$-ideal associated to* $\mathfrak{a}$. Furthermore, any $\mathcal{O}_X$-ideal $\mathcal{J}$ which is associated to some A-module M is, in fact, associated to an ideal in A. This also follows from Corollary 9.4.2/4, since by Proposition 9.4.2/2 the A-module homomorphisms $M \to A$ correspond bijectively to the $\mathcal{O}_X$-module homomorphisms $M \otimes \mathcal{O}_X \to \mathcal{O}_X$.

Now, more generally, let X be an arbitrary analytic variety. An $\mathcal{O}_X$-ideal $\mathcal{J}$ is called *coherent* if there exists an admissible affinoid covering $\{\operatorname{Sp} A_i\}_{i \in I}$ of X such that $\mathcal{J}|_{\operatorname{Sp} A_i}$ is associated to an ideal in A_i for all $i \in I$. By the above consideration, this notion of coherence coincides with the one given for $\mathcal{O}_X$-modules in Definition 9.4.3/1 when $\mathcal{J}$ is viewed as an $\mathcal{O}_X$-module. In addition, Theorem 9.4.3/3 implies that $\mathcal{J}$ is coherent if and only if $\mathcal{J}|_{\operatorname{Sp} A}$, for any open affinoid subvariety $\operatorname{Sp} A$ of X, is associated to an ideal in A.

For any $\mathcal{O}_X$-ideal $\mathcal{J}$, one defines the *nilradical* rad $\mathcal{J}$ as the sheaf associated to the presheaf

$$U \mapsto \operatorname{rad}\big(\mathcal{J}(U)\big),$$

where $\operatorname{rad}\big(\mathcal{J}(U)\big)$ denotes the nilradical of $\mathcal{J}(U)$, considered as an ideal in $\mathcal{O}_X(U)$. Using Corollary 9.2.2/6 it is easily checked that rad $\mathcal{J}$ which is an $\mathcal{O}_X$-module is, in fact, a submodule of $\mathcal{O}_X$ and thus an $\mathcal{O}_X$-ideal containing $\mathcal{J}$.

Proposition 1. *Let X be an analytic variety and $\mathcal{J}$ an $\mathcal{O}_X$-ideal. Then*

(i) $(\operatorname{rad}\mathcal{J})(U) = \operatorname{rad}\big(\mathcal{J}(U)\big)$ *for any open affinoid subvariety U of X.*
(ii) $(\operatorname{rad}\mathcal{J})_x = \operatorname{rad}(\mathcal{J}_x)$ *for all $x \in X$.*
(iii) $\operatorname{rad}(\operatorname{rad}\mathcal{J}) = \operatorname{rad}\mathcal{J}$.

Proof. If $U \subset X$ is an open affinoid subvariety of X, we see $\operatorname{rad}\big(\mathcal{J}(U)\big) \subset (\operatorname{rad}\mathcal{J})(U)$ by the definition of the nilradical of $\mathcal{J}$. To verify the opposite inclusion, let $f \in \mathcal{O}_X(U)$ be a function contained in $(\operatorname{rad}\mathcal{J})(U)$. By Corollary 9.2.2/6, there exists an admissible covering $\{U_i\}_{i\in I}$ of U such that

$$(f|_{U_i})^{r_i} \in \mathcal{J}(U_i)$$

for all $i \in I$ and suitable integers r_i. Since admissible coverings of affinoid varieties always admit finite subcoverings, the index set I can be assumed to be finite. Then we get $f^r \in \mathcal{J}(U)$ with $r := \max_{i\in I} r_i$ and hence $f \in \operatorname{rad}\big(\mathcal{J}(U)\big)$. This verifies (i). Assertion (ii) follows from Corollary 9.2.2/7 and the fact that direct limits commute with the formation of the nilradical. Finally, (iii) is a consequence of (i). □

Proposition 2. *Let $X = \operatorname{Sp} A$ be an affinoid variety and $\mathfrak{a} \subset A$ an ideal. Denoting by $\mathcal{J} := \mathfrak{a} \otimes \mathcal{O}_X$ the $\mathcal{O}_X$-ideal associated to $\mathfrak{a}$, the $\mathcal{O}_X$-ideal* rad $\mathcal{J}$ *is associated to the ideal* $\operatorname{rad}\mathfrak{a} \subset A$.

Proof. For any affinoid subdomain $U = \operatorname{Sp} A'$ in X, we have $\mathcal{J}(U) = \mathfrak{a}A'$, and therefore by Proposition 1

$$(\operatorname{rad}\mathcal{J})(U) = \operatorname{rad}(\mathfrak{a}A').$$

Hence by Proposition 9.2.3/1, the only thing to verify is

$$\operatorname{rad}(\mathfrak{a}A') = (\operatorname{rad}\mathfrak{a})\,A'.$$

Writing $\mathfrak{b} := \operatorname{rad}\mathfrak{a}$, we see by Proposition 7.2.2/4 and Corollary 7.2.2/6 that the map $A/\mathfrak{b} \to A'/\mathfrak{b}A'$, induced by the restriction homomorphism $A \to A'$, defines $\operatorname{Sp} A'/\mathfrak{b}A'$ as an affinoid subdomain of $\operatorname{Sp} A/\mathfrak{b}$. Therefore, by Corollary 7.3.2/10, the algebra $A'/\mathfrak{b}A'$ is reduced, since $A/\mathfrak{b}$ is reduced. This implies

$$\operatorname{rad}(\mathfrak{a}A') = \operatorname{rad}(\mathfrak{b}A') = \mathfrak{b}A' = (\operatorname{rad}\mathfrak{a})\,A'. \qquad \square$$

Corollary 3. *Let X be an analytic variety and $\mathcal{J}$ a coherent $\mathcal{O}_X$-ideal. Then also* rad $\mathcal{J}$ *is coherent.*

This is a direct consequence of Proposition 2. In particular, the zero ideal $0 \subset \mathcal{O}_X$ is always coherent. Hence denoting by rad $\mathcal{O}_X$ the nilradical of 0 in $\mathcal{O}_X$, we get

Corollary 4. *Let X be an analytic variety. Then* rad $\mathcal{O}_X$ *is a coherent $\mathcal{O}_X$-ideal.*

Proposition 5. *Let X be an analytic variety and $\mathcal{I}$ a coherent $\mathcal{O}_X$-ideal. Then $\mathcal{I} = \operatorname{rad} \mathcal{I}$ if and only if $\mathcal{I}_x = \operatorname{rad} \mathcal{I}_x$ for all $x \in X$.*

Proof. Since $(\operatorname{rad} \mathcal{I})_x = \operatorname{rad} \mathcal{I}_x$ by Proposition 1, the only if part of the assertion is clear. To verify the if part we may assume that X is affinoid, say $X = \operatorname{Sp} A$. Then $\mathcal{I}$ is associated to some ideal $\mathfrak{a} \subset A$, and by our assumption the ideal $\mathcal{I}_x = \mathfrak{a}\mathcal{O}_{X,x}$ is equal to its nilradical for all $x \in X$. In order to show $\mathfrak{a} = \operatorname{rad} \mathfrak{a}$, which implies $\mathcal{I} = \operatorname{rad} \mathcal{I}$, we consider the affinoid variety $Y := \operatorname{Sp} A/\mathfrak{a}$. Applying Proposition 7.3.2/2, it is seen that all stalks of $\mathcal{O}_Y$ are reduced. Hence by Corollary 7.3.2/4 or by Corollary 7.3.2/9, also the algebra $A/\mathfrak{a}$ is reduced. □

9.5.2. Analytic subsets. — In (7.1.2) and (7.1.3), we defined affinoid subsets of affinoid varieties. Here we want to generalize this concept to analytic varieties.

Definition 1. *Let X denote an analytic variety. A subset $Y \subset X$ is called a (closed) analytic subset of X if there exist an admissible covering $\{U_i\}_{i\in I}$ of X and, for each $i \in I$, finitely many functions $f_{i1}, \ldots, f_{in_i} \in \mathcal{O}_X(U_i)$ such that*

$$Y \cap U_i = \{x \in U_i;\, f_{i1}(x) = 0, \ldots, f_{in_i}(x) = 0\}.$$

If X is an affinoid variety and $Y \subset X$ is an affinoid subset, then clearly Y is also an analytic subset of X. Conversely, we will show below that any analytic subset of X is, in fact, affinoid.

In the following let X denote an arbitrary analytic variety. Defining the support of an $\mathcal{O}_X$-module $\mathcal{F}$ by

$$\operatorname{supp} \mathcal{F} := \{x \in X;\, \mathcal{F}_x \neq 0\},$$

we associate to each $\mathcal{O}_X$-ideal $\mathcal{I}$ the subset

$$V(\mathcal{I}) := \operatorname{supp} \mathcal{O}_X/\mathcal{I} \subset X.$$

Using Proposition 9.4.1/1, we can write

$$V(\mathcal{I}) = \{x \in X;\, \mathcal{O}_{X,x}/\mathcal{I}_x \neq 0\}.$$

It is easily checked that $V(\mathcal{I})$ satisfies the usual properties as stated in Proposition 7.1.2/6 for the case of ideals in T_n. Furthermore, the notions $V(\mathcal{I})$ for $\mathcal{O}_X$-ideals $\mathcal{I}$ and $V(\mathfrak{a})$ for ideals $\mathfrak{a}$ in affinoid algebras, (as defined in (7.1.3)) are compatible in the following sense:

Proposition 2. *Let $X = \operatorname{Sp} A$ be an affinoid variety and $\mathfrak{a} \subset A$ an ideal. If $\mathcal{I} := \mathfrak{a} \otimes \mathcal{O}_X$ denotes the $\mathcal{O}_X$-ideal associated to $\mathfrak{a}$, then $V(\mathcal{I}) = V(\mathfrak{a})$.*

Proof. By assumption we have

$$V(\mathcal{I}) = \operatorname{supp} \mathcal{O}_X/\mathfrak{a} \otimes \mathcal{O}_X = \{x \in X; \mathfrak{a}\mathcal{O}_{X,x} \subsetneqq \mathcal{O}_{X,x}\}.$$

If $x \in X$ is a point in $V(\mathfrak{a})$, then by Proposition 7.3.2/1, the ideal $\mathfrak{a}\mathcal{O}_{X,x}$ is contained in the maximal ideal of $\mathcal{O}_{X,x}$, showing $x \in V(\mathcal{I})$. Conversely, assume $x \notin V(\mathfrak{a})$. Then there is a function $f \in \mathfrak{a}$ such that $f(x) \neq 0$. By Proposition 7.3.2/1, the germ $f_x \in \mathcal{O}_{X,x}$ is a unit; hence we have $\mathfrak{a}\mathcal{O}_{X,x} = \mathcal{O}_{X,x}$ and thus $x \notin V(\mathcal{I})$. □

A direct consequence of Proposition 2 is

Corollary 3. *Let X be an analytic variety and let $\mathcal{I}$ be a coherent $\mathcal{O}_X$-ideal. Then $V(\mathcal{I})$ is an analytic subset of X.*

What we proved above can also be obtained in the more general context of coherent $\mathcal{O}_X$-modules, as the following proposition shows. The results of Proposition 2 and Corollary 3 are thus interpreted as properties of the coherent $\mathcal{O}_X$-module $\mathcal{O}_X/\mathcal{I}$.

Proposition 4. *Let X be an analytic variety and let $\mathcal{F}$ be a coherent $\mathcal{O}_X$-module. Then* supp $\mathcal{F}$ *is an analytic subset of X. More specifically, if $X = \operatorname{Sp} A$ is affinoid and $\mathcal{F}$ is associated to the finite A-module M, then* supp $\mathcal{F}$ *is equal to $V(\operatorname{Ann}(M))$, where $\operatorname{Ann}(M) := \{a \in A; aM = 0\}$ is the annihilator of M.*

Proof. It is enough to consider the affinoid case where $X = \operatorname{Sp} A$ and $\mathcal{F} = M \otimes \mathcal{O}_X$. For $x \in X$, we denote by M_x the localization of M with respect to the maximal ideal in A corresponding to x. Then we see by Proposition 9.4.2/6 that

$$\operatorname{supp} \mathcal{F} = \operatorname{supp} M := \{x \in X; M_x \neq 0\}.$$

Since M is a finite A-module, an easy verification shows $\operatorname{supp} M = V(\operatorname{Ann}(M))$. □

Just as in (7.1.2) and (7.1.3), we can associate to any subset Y of an analytic variety X an $\mathcal{O}_X$-ideal $\mathcal{I}$ defined by

$$\mathcal{I}(U) := \{f \in \mathcal{O}_X(U); f(x) = 0 \text{ for all } x \in U \cap Y\},$$

where $U \subset X$ is admissible open. We use the notations $\mathcal{I} = \mathfrak{id}(Y) = \mathfrak{id}_X(Y)$. Here we also have a certain compatibility with the process id of (7.1.2) and (7.1.3).

Proposition 5. *Let $X = \operatorname{Sp} A$ be an affinoid variety and let $Y \subset X$ denote an affinoid subset, i.e., $Y = V(\mathfrak{a})$ for some ideal $\mathfrak{a} \subset A$. Then the $\mathcal{O}_X$-ideal $\mathfrak{id}(Y)$ is associated to the A-ideal*

$$\operatorname{id}_A(Y) = \{f \in A; f(x) = 0 \text{ for all } x \in Y\}.$$

Proof. For any affinoid subdomain $X' = \operatorname{Sp} A' \subset X$, we have

$$Y \cap X' = V(\mathfrak{a}A'),$$

and hence by HILBERT's Nullstellensatz (Proposition 7.1.3/1)

$$\mathrm{id}_{A'}(Y \cap X') = \mathrm{rad}\,(\mathfrak{a}A').$$

Therefore $\mathfrak{id}(Y) = \mathrm{rad}\,(\mathfrak{a} \otimes \mathcal{O}_X)$, or $\mathfrak{id}(Y) = (\mathrm{rad}\,\mathfrak{a}) \otimes \mathcal{O}_X$ by Proposition 9.5.1/2. Since $\mathrm{rad}\,\mathfrak{a} = \mathrm{id}_A(Y)$ by HILBERT's Nullstellensatz, the assertion of the proposition follows. □

Corollary 6. *Let X be an analytic variety and let Y be an analytic subset of X. Then $\mathfrak{id}(Y)$ is a coherent $\mathcal{O}_X$-ideal satisfying $V(\mathfrak{id}(Y)) = Y$.*

Proof. By the definition of analytic sets, there exists an admissible covering $\{U_i\}_{i \in I}$, which we can assume is affinoid, such that $Y \cap U_i$ is an affinoid subset of U_i for all $i \in I$. Writing $U_i = \mathrm{Sp}\,A_i$ and $\mathfrak{a}_i := \mathrm{id}_{A_i}(Y \cap U_i)$, it follows from Proposition 5 that the $\mathcal{O}_{U_i}$-ideal $\mathfrak{id}_{U_i}(Y \cap U_i)$ is associated to $\mathfrak{a}_i$. But since $\mathfrak{id}_{U_i}(Y \cap U_i)$ is just the restriction of $\mathfrak{id}_X(Y)$ to U_i, it is clear that $\mathfrak{id}_X(Y)$ is a coherent $\mathcal{O}_X$-ideal. Furthermore, one gets

$$V(\mathfrak{id}_X(Y)) \cap U_i = V(\mathfrak{id}_{U_i}(Y \cap U_i)) = V(\mathfrak{a}_i)$$

by Proposition 2, and

$$V(\mathfrak{a}_i) = V(\mathrm{id}_{A_i}(Y \cap U_i)) = Y \cap U_i$$

by Proposition 7.1.2/2. Therefore, we must have $V(\mathfrak{id}_X(Y)) = Y$. □

Combining Corollaries 3 and 6, we get the following characterization of analytic subsets:

Corollary 7. *Let X be an analytic variety. Then a subset $Y \subset X$ is analytic if and only if there exists a coherent $\mathcal{O}_X$-ideal $\mathcal{I}$ satisfying $Y = V(\mathcal{I})$.*

Furthermore, applying Theorem 9.4.3/3 and Proposition 2, one obtains

Corollary 8. *Let X be an affinoid variety. Then any affinoid subset of X is analytic and vice versa.*

It follows, in particular, that an analytic subset Y of an analytic variety X is characterized by the following property which is stronger than the one actually required in Definition 1. Namely, for each open affinoid subvariety $U \subset X$, there are functions $f_1, \ldots, f_n \in \mathcal{O}_X(U)$ having $Y \cap U$ as common zero set. We conclude this section by proving an analogue of HILBERT's Nullstellensatz for coherent ideals.

Proposition 9. *Let X be an analytic variety and $\mathcal{I}$ a coherent $\mathcal{O}_X$-ideal. Then*

$$\mathfrak{id}(V(\mathcal{I})) = \mathrm{rad}\,\mathcal{I}.$$

Proof. The question being local, we may assume that X is affinoid. Then $\mathcal{I}$ is associated to some ideal $\mathfrak{a} \subset A := \mathcal{O}_X(X)$, and HILBERT's Nullstellensatz for ideals in A (Proposition 7.1.3/1) says

$$\mathrm{id}_A(V(\mathfrak{a})) = \mathrm{rad}\,\mathfrak{a}.$$

Applying Proposition 9.5.1/2, Proposition 2, and Proposition 5 and considering associated $\mathcal{O}_X$-ideals, we get

$$\mathfrak{id}(V(\mathcal{I})) = \operatorname{rad} \mathcal{I}. \qquad \square$$

9.5.3. Closed immersions of analytic varieties. — A morphism $\varrho : Y \to X$ of analytic varieties is called a *closed immersion* if there exists an admissible affinoid covering $\{U_i\}_{i \in I}$ of X such that, for all $i \in I$, the induced map $\varrho^{-1}(U_i) \to U_i$ is a closed immersion of affinoid varieties in the sense of Definition 7.1.4/3. Then, as a map between the underlying sets, ϱ is injective, and we call Y a *closed analytic subvariety of* X via ϱ. It should also be realized that the above notion of closed immersions coincides with the one given in Definition 7.1.4/3 if X and Y are affinoid. Namely, it is seen by the following proposition that, for an arbitrary closed immersion $\varrho: Y \to X$ of analytic varieties, the homomorphism $\varrho^{\#}: \mathcal{O}_X \to \varrho_*(\mathcal{O}_Y)$ is a surjection of coherent $\mathcal{O}_X$-modules. Hence in the affinoid case, $\varrho: Y \to X$ must correspond to an epimorphism of affinoid algebras $\varrho^*: \mathcal{O}_X(X) \to \mathcal{O}_Y(Y)$ by Corollary 9.4.2/4.

Proposition 1. *Let* $\varrho: Y \to X$ *be a closed immersion of analytic varieties. Then*

(i) $\varrho^{\#}: \mathcal{O}_X \to \varrho_*(\mathcal{O}_Y)$ *is a surjective homomorphism of coherent* $\mathcal{O}_X$*-modules.*

(ii) $\mathcal{I} := \ker \varrho^{\#}$ *is a coherent* $\mathcal{O}_X$*-ideal satisfying* $V(\mathcal{I}) = \varrho(Y)$.

(iii) *For all* $y \in Y$, *the homomorphism between corresponding stalks* $\varrho_y^*: \mathcal{O}_{X,\varrho(y)} \to \mathcal{O}_{Y,y}$ *is surjective and has kernel* $\mathcal{I}_{\varrho(y)}$; *hence it induces an isomorphism* $\mathcal{O}_{X,\varrho(y)}/\mathcal{I}_{\varrho(y)} \xrightarrow{\sim} \mathcal{O}_{Y,y}$.

Proof. Let $\{\operatorname{Sp} A_i\}_{i \in I}$ be an admissible affinoid covering of X such that, for all $i \in I$, the subvariety $\varrho^{-1}(\operatorname{Sp} A_i) \subset Y$ is affinoid, say $\varrho^{-1}(\operatorname{Sp} A_i) = \operatorname{Sp} B_i$, and the induced morphism $\varrho_i: \operatorname{Sp} B_i \to \operatorname{Sp} A_i$ corresponds to an epimorphism $\varrho_i^*: A_i \to B_i$. Viewing B_i as an A_i-module via ϱ_i^*, we see by Proposition 9.4.2/5 that the restriction of $\varrho^{\#}: \mathcal{O}_X \to \varrho_*(\mathcal{O}_Y)$ to $\operatorname{Sp} A_i$ is the $\mathcal{O}_{\operatorname{Sp} A_i}$-module homomorphism associated to $\varrho_i^*: A_i \to B_i$. Then one can apply Proposition 9.4.1/2, Corollary 9.4.2/3, and Proposition 9.5.2/2 in order to deduce assertions (i) and (ii). Furthermore, assertion (iii) follows from Proposition 9.4.1/1. $\square$

In the situation of Proposition 1, it is seen that Y is identified via ϱ with the analytic subset $V(\mathcal{I}) \subset X$, and, by abuse of language, the structure sheaf $\mathcal{O}_Y$ is obtained by "restricting" $\mathcal{O}_X/\mathcal{I}$ to $V(\mathcal{I})$. Moreover, it will be seen below that this process defines a one-to-one correspondence between closed analytic subvarieties of X and coherent $\mathcal{O}_X$-ideals. However, before we turn to this problem, we give a further characterization of closed immersions.

Proposition 2. *Let* $\varrho: Y \to X$ *be a morphism of analytic varieties. Then the following are equivalent:*

(i) ϱ *is a closed immersion.*

(ii) *For each open affinoid subvariety* $U \subset X$, *the inverse image* $V := \varrho^{-1}(U)$ *is affinoid in* Y *and the homomorphism of corresponding affinoid algebras* $\varrho_U^*: \mathcal{O}_X(U) \to \mathcal{O}_Y(V)$ *is surjective.*

(iii) ϱ *is a finite morphism and* $\varrho^{\#}: \mathcal{O}_X \to \varrho_*(\mathcal{O}_Y)$ *is surjective.*

Proof. That (ii) implies (i) is trivial, and that (i) implies (iii) follows from the definitions and Proposition 1. Thus, we have only to show that (iii) implies (ii). Therefore, assume ϱ is finite and $\varrho^\#$ is surjective. Then $\varrho_*(\mathcal{O}_Y)$ is a coherent $\mathcal{O}_X$-module via the surjection $\varrho^\#: \mathcal{O}_X \to \varrho_*(\mathcal{O}_Y)$ (cf. Proposition 9.4.4/1). The same proposition states that for any open affinoid subvariety $U \subset X$, also $V := \varrho^{-1}(U)$ is open affinoid in Y. Hence the restriction of $\varrho^\#: \mathcal{O}_X \to \varrho_*(\mathcal{O}_Y)$ to U is a surjection of $\mathcal{O}_U$-modules which is associated to the homomorphism of affinoid algebras $\varrho_U^*: \mathcal{O}_X(U) \to \mathcal{O}_Y(V)$. But then ϱ_U^* must be surjective by Corollary 9.4.2/4, and (ii) follows. □

As in the case of finite morphisms, we see that the composition of closed immersions yields again a closed immersion. Also, using Proposition 2.1.8/6, it is verified without difficulty that closed immersions are preserved by any base change functor and that the fibre product of closed immersions over any analytic variety S is again a closed immersion.

Proposition 3. *Let X be an analytic variety and $\mathcal{I}$ a coherent $\mathcal{O}_X$-ideal. Then there exists a closed immersion of analytic varieties $\varrho: Y \to X$ such that $\varrho_*(\mathcal{O}_Y) \simeq \mathcal{O}_X/\mathcal{I}$ and therefore $\varrho(Y) = V(\mathcal{I})$. Furthermore, Y and ϱ are unique up to natural isomorphism.*

Proof. By Proposition 9.4.2/2 and Proposition 9.4.2/5, the assertion is trivial if X is affinoid. If X is not affinoid, we denote by $\{X_i = \mathrm{Sp}\, A_i\}_{i\in I}$ the family of all open affinoid subvarieties in X. For $i, j \in I$, let $\mathcal{I}|_{X_i}$ be associated to the ideal $\mathfrak{a}_i \subset A_i$ and let Y_{ij} denote the inverse image of $X_i \cap X_j$ via the canonical immersion $\varrho_i: \mathrm{Sp}\, A/\mathfrak{a}_i \to \mathrm{Sp}\, A_i$. Using the uniqueness assertion in the affinoid case, one can apply Proposition 9.3.2/1 and paste the varieties $Y_i := \mathrm{Sp}\, A_i$ together via the "intersections" Y_{ij} in such a way that $\{Y_i\}_{i\in I}$ is an admissible covering of the obtained analytic variety Y. Furthermore, the maps $\varrho_i: Y_i \to X_i$ determine a morphism $\varrho: Y \to X$ according to Proposition 9.3.3/1, which, by construction, is a closed immersion. That Y and ϱ are essentially unique follows from the uniqueness assertion in the affinoid case. □

For any analytic variety X, the nilradical $\mathrm{rad}\, \mathcal{O}_X$ of the structure sheaf is a coherent $\mathcal{O}_X$-ideal by Corollary 9.5.1/4. Hence, according to Proposition 3, it defines a closed subvariety of X, which will be denoted by X_{red}. Since $V(\mathrm{rad}\, \mathcal{O}_X) = V(0) = X$, we see that X and X_{red} are locally G-ringed spaces over the same underlying G-topological space, the structure sheaf of X_{red} being $\mathcal{O}_X/\mathrm{rad}\, \mathcal{O}_X$. It is a direct consequence of assertion (ii) in Proposition 9.5.1/1 and assertion (iii) in Proposition 1 that X_{red} is a *reduced analytic variety*, i.e., all stalks of the structure sheaf $\mathcal{O}_{X_{\mathrm{red}}}$ are reduced local rings. For this reason, X_{red} is called the *nilreduction of X*.

Proposition 4. *Let X be an analytic variety and $Y \subset X$ an analytic subset. Then Y can be defined as a reduced closed analytic subvariety of X. The analytic structure on Y is unique up to natural isomorphism.*

Proof. By Corollary 9.5.2/6 there is a coherent $\mathcal{O}_X$-ideal $\mathcal{J}$ satisfying $V(\mathcal{J}) = Y$. Assuming that $\mathcal{J}$ defines a reduced closed analytic subvariety of X, we must have $\operatorname{rad}(\mathcal{O}_{X,x}/\mathcal{J}_x) = 0$, or equivalently $\operatorname{rad}\mathcal{J}_x = \mathcal{J}_x$ for all $x \in X$. Hence we get $\operatorname{rad}\mathcal{J} = \mathcal{J}$ by Proposition 9.5.1/5, and HILBERT's Nullstellensatz (Proposition 9.5.2/9) implies

$$\mathfrak{id}(Y) = \mathfrak{id}(V(\mathcal{J})) = \operatorname{rad}\mathcal{J} = \mathcal{J},$$

showing that $\mathcal{J}$ is uniquely determined by Y. Therefore, the uniqueness assertion follows from Proposition 3.

On the other hand, $\mathfrak{id}(Y)$ is, according to Corollary 9.5.2/6, a coherent $\mathcal{O}_X$-ideal satisfying $V(\mathfrak{id}(Y)) = Y$. Since $\operatorname{rad}\mathfrak{id}(Y)$ obviously equals $\mathfrak{id}(Y)$, it is seen that $\mathfrak{id}(Y)$ defines a reduced closed analytic subvariety of X. This settles the existence part of the assertion. □

Finally we want to characterize closed immersions in terms of locally closed immersions. As in the affinoid case, an injective morphism $\varrho: Y \to X$ is called a *locally closed immersion* if, for all $y \in Y$, the homomorphism of stalks $\varrho_y^*: \mathcal{O}_{X,\varrho(y)} \to \mathcal{O}_{Y,y}$ is surjective. Of course, closed immersions are locally closed by Proposition 1.

Proposition 5. *Assume that a morphism of analytic varieties $\varrho: Y \to X$ satisfies the following conditions:*

(i) *ϱ is a locally closed immersion,*

(ii) *$\varrho(Y)$ is an analytic subset of X,*

(iii) *there exists an admissible affinoid covering $\{U_i\}_{i\in I}$ of X and, for each $i \in I$, a finite admissible covering of $\varrho^{-1}(U_i)$ by open affinoid subvarieties of Y.*

Then ϱ is a closed immersion.

For the proof we need an auxiliary result concerning the canonical topology of an analytic variety X. Analogously to (7.2.1), this topology is defined as the (ordinary) topology which is generated by all open subvarieties of X.

Lemma 6. *Let $\varrho: Y \to X$ be a bijective morphism of analytic varieties which satisfies condition* (iii) *of Proposition* 5. *Then ϱ is a homeomorphism with respect to canonical topologies on X and Y.*

Proof. We may assume that X is affinoid. Then Y has an admissible covering consisting of finitely many open affinoid subvarieties of Y. Let $Y' = \operatorname{Sp} B$ be an arbitrary open affinoid subvariety of Y. We show that $\varrho(Y')$ is a closed subset of X. Namely, consider a point $x \in X - \varrho(Y')$ and choose generators $f_1, \ldots, f_n$ for the corresponding maximal ideal in $A := \mathcal{O}_X(X)$. Denoting by $\sigma: A \to B$ the algebra homomorphism of the affinoid map $Y' \to X$ induced by ϱ, the elements $\sigma(f_1), \ldots, \sigma(f_n)$ generate the unit ideal in B. Hence using Lemma 7.3.4/7, there is an $\varepsilon \in |k|$, $0 < \varepsilon < 1$, such that

$$\max_\nu |\sigma(f_\nu)(y)| > \varepsilon$$

for all $y \in Y'$. Then $X(\varepsilon^{-1}f_1, \ldots, \varepsilon^{-1}f_n)$ is a neighborhood of x in X which is disjoint from $\varrho(Y')$. Therefore $\varrho(Y')$ is closed. Next we fix a point $y \in Y$ and a neighborhood $V \subset Y$ of y. An easy argument shows that there are finitely many open affinoid subvarieties $V_1, \ldots, V_m \subset Y$, not containing y, such that

$$Y = V \cup V_1 \cup \ldots \cup V_m .$$

Since $\varrho(V_1 \cup \ldots \cup V_m)$ is closed in X — as we just showed — it follows that $\varrho(V)$ is a neighborhood of $\varrho(y)$. Thus, ϱ is a homeomorphism. □

Proof of Proposition 5. We have only to consider the case where X is affinoid, say $X = \operatorname{Sp} A$. Then Y admits a finite admissible covering $\{Y_1, \ldots, Y_r\}$ by open affinoid subvarieties of Y. Furthermore, we may assume that ϱ is surjective and hence defines a bijection between the points of Y and X. Namely, denote by $\mathfrak{a}$ the kernel of the homomorphism $\varrho_X^*: A \to \mathcal{O}_Y(Y)$. Then $\varrho: Y \to \operatorname{Sp} A$ can be replaced by the induced morphism $\varrho': Y \to \operatorname{Sp} A/\mathfrak{a}$. We claim that ϱ' is surjective — i.e. that $\varrho(Y)$, which by construction is contained in $V(\mathfrak{a})$, in fact equals $V(\mathfrak{a})$. Let $f \in \operatorname{id}_A(\varrho(Y))$ be a function vanishing on the affinoid subset $\varrho(Y) \subset \operatorname{Sp} A$. Then $f(\varrho(y)) = 0$ for all $y \in Y$, and hence the restriction of $\varrho_X^*(f)$ to Y_j is nilpotent in $\mathcal{O}_Y(Y_j)$ for all j. Therefore, $\varrho_X^*(f)$ is nilpotent in $\mathcal{O}_Y(Y)$, which means that $f \in \operatorname{rad} \mathfrak{a}$. Thus, we have $\operatorname{id}_A(\varrho(Y)) \subset \operatorname{rad} \mathfrak{a}$ and hence $\varrho(Y) \supset V(\mathfrak{a})$; i.e., $\varrho(Y)$ equals $V(\mathfrak{a})$ and ϱ' is surjective.

We assume now that X and the covering $\{Y_1, \ldots, Y_r\}$ of Y are as above and that $\varrho: Y \to X$ is surjective. Then ϱ is a homeomorphism with respect to canonical topologies by Lemma 6. The same is true for the induced map $\varrho_{\text{red}}: Y_{\text{red}} \to X_{\text{red}}$. In particular, the maps $(\varrho_j)_{\text{red}}: (Y_j)_{\text{red}} \to X_{\text{red}}$, regarded locally, are surjective closed immersions between reduced affinoid varieties and hence local isomorphisms, or in other words, open immersions. Therefore, by Corollary 8.2.1/4, each $(Y_j)_{\text{red}}$ is an affinoid subdomain of X_{red} via $(\varrho_j)_{\text{red}}$ or, by Corollary 7.3.5/3, a finite union of rational subdomains of X_{red}. Consequently, $\varrho(Y_j)$ is a finite union of rational subdomains in X. If $X' \subset X$ is such a rational subdomain, then $\varrho^{-1}(X')$ is an open affinoid subvariety of Y, and thereby we can reduce the proof of the proposition to the case where $\varrho: Y \to X$ is a surjective locally closed immersion between affinoid varieties. Applying Theorem 7.3.5/1, we may additionally assume that ϱ is a Runge immersion. But then ϱ must be a closed immersion. □

9.6. Separated and proper morphisms

9.6.1. Separated morphisms. — For ordinary topological spaces X, the Hausdorff property is equivalent to the fact that the diagonal in $X \times X$ is closed. We introduce a similar notion for analytic varieties.

Definition 1. *Let $\varrho: X \to Y$ be a morphism of analytic varieties, and denote by $\Delta: X \to X \times_Y X$ the diagonal morphism, i.e., the unique morphism whose composition with both projections $X \times_Y X \to X$ is the identity $X \to X$. Then ϱ*

is called separated if Δ is a closed immersion. Furthermore, an arbitrary analytic variety X is called separated (over the ground field k) if the canonical morphism $X \to \mathrm{Sp}\, k$ is separated.

Not all analytic varieties are separated. For example, take two copies X_1 and X_2 of the unit disc $\mathbb{B}^1$, and consider an open analytic subvariety $U \subset \mathbb{B}^1$ which is not affinoid, like any "open" disc in $\mathbb{B}^1$. Then one constructs an analytic variety X, having the admissible affinoid covering $X = X_1 \cup X_2$, by glueing X_1 and X_2 together via U. Since $U = X_1 \cap X_2$ is not affinoid, one easily verifies that X cannot be separated (see also Proposition 6 below).

Proposition 2. *If $\varrho: X \to Y$ is a morphism of affinoid varieties, then ϱ is separated. In particular, any affinoid variety is separated.*

Proof. Let $X = \mathrm{Sp}\, A$ and $Y = \mathrm{Sp}\, B$. The diagonal morphism $\Delta: \mathrm{Sp}\, A \to \mathrm{Sp}\, A \times_{\mathrm{Sp}\, B} \mathrm{Sp}\, A = \mathrm{Sp}\, A \mathbin{\hat{\otimes}}_B A$ corresponds to the canonical homomorphism $A \mathbin{\hat{\otimes}}_B A \to A$, which is of course surjective. □

Similarly, if X is an analytic variety with an admissible affinoid covering $\{U_i\}_{i\in\mathbb{N}}$ such that U_i is contained in U_{i+1}, then X is separated. Namely, in this case, $\{U_i \times U_i\}_{i\in\mathbb{N}}$ is an admissible covering of $X \times X$, and the diagonal morphism $X \to X \times X$ is a closed immersion, since all affinoid maps $U_i \to U_i \times U_i$ are closed immersions. This applies in particular to the affine n-space $\mathbb{A}_k^n$ constructed in (9.3.4), or more generally, to any analytic variety associated to an affine algebraic scheme over k. Furthermore, it is not difficult to show that, for any separated scheme of locally finite type over k, the associated analytic variety of (9.3.4) is also separated. Thereby, we see in particular that the projective n-space $\mathbb{P}_k^n$ is a separated analytic variety. This fact can also be obtained by direct reasoning as follows. Consider the defining admissible affinoid covering $\mathbb{P}_k^n = X_0 \cup \ldots \cup X_n$ of (9.3.4), where

$$X_i = \mathrm{Sp}\, k\left\langle \frac{\zeta_0}{\zeta_i}, \ldots, \frac{\zeta_n}{\zeta_i} \right\rangle \quad \text{and}$$

$$X_i \cap X_j = \mathrm{Sp}\, k\left\langle \frac{\zeta_0}{\zeta_i}, \ldots, \frac{\zeta_n}{\zeta_i}, \left(\frac{\zeta_j}{\zeta_i}\right)^{-1} \right\rangle = \mathrm{Sp}\, k\left\langle \frac{\zeta_0}{\zeta_j}, \ldots, \frac{\zeta_n}{\zeta_j}, \left(\frac{\zeta_i}{\zeta_j}\right)^{-1} \right\rangle.$$

Then $\{X_i \times X_j\}_{i,j=0,\ldots,n}$ is an admissible affinoid covering of $\mathbb{P}_k^n \times \mathbb{P}_k^n$, and the inverse of $X_i \times X_j$ with respect to the diagonal morphism $\Delta: \mathbb{P}_k^n \to \mathbb{P}_k^n \times \mathbb{P}_k^n$ is $X_i \cap X_j$. Since one verifies immediately that all maps $X_i \cap X_j \to X_i \times X_j$ are closed immersions of affinoid varieties, Δ is a closed immersion; i.e., $\mathbb{P}_k^n$ is separated. A similar procedure shows that analytic tori $\mathbb{A}_k^*/(q)$, as constructed in (9.3.4), are separated.

Proposition 3. *Let $\varrho: X \to Y$ be a morphism of analytic varieties.*

(i) *If ϱ is separated and $U \subset X$, $V \subset Y$ are open analytic subvarieties with $\varrho(U) \subset V$, then the induced map $U \to V$ is also separated.*

(ii) *Let $\{Y_i\}_{i\in I}$ be an admissible open covering of Y. Then ϱ is separated if and only if all induced maps $\varrho^{-1}(Y_i) \to Y_i$ are separated.*

Proof. Let $\Delta: X \to X \times_Y X$ be the diagonal morphism corresponding to ϱ, and let $U \subset X$, $V \subset Y$ be open analytic subvarieties with $\varrho(U) \subset V$. It follows from Lemma 9.3.5/3 that $U \times_V U$ is canonically an open analytic subvariety of $X \times_Y X$, having U as inverse image with respect to Δ. Thus, if Δ is a closed immersion, $U \to U \times_V U$ is also a closed immersion (use, for instance, the criterion of Proposition 9.5.3/2). This verifies assertion (i). By a similar argument, one obtains assertion (ii), since in that case $\{\varrho^{-1}(Y_i) \times_{Y_i} \varrho^{-1}(Y_i)\}_{i \in I}$ is an admissible open covering of $X \times_Y X$. □

Propositions 2 and 3 show in particular that all finite morphisms and all closed immersions of analytic varieties are separated. Also, if U is an open analytic subvariety of an analytic variety X, the inclusion morphism $U \to X$ is separated, since the morphism $U \to U \times_X U$ is an isomorphism in this case.

Lemma 4. *Let $Y \to Z$ be a morphism of affinoid varieties and let X_1, X_2 be analytic varieties over Y. Then the canonical morphism $X_1 \times_Y X_2 \to X_1 \times_Z X_2$ is a closed immersion.*

Proof. Considering admissible affinoid coverings of X_1 and X_2 and applying Lemma 9.3.5/3, the assertion is reduced to the case where X_1 and X_2 are affinoid. Then we see by Proposition 6.1.1/10 that $X_1 \times_Y X_2 \to X_1 \times_Z X_2$ is a closed immersion. □

Proposition 5. *Let $\varrho: X \to Y$ and $\sigma: Y \to Z$ be morphisms of analytic varieties. If $\sigma \circ \varrho$ is separated, then ϱ is also separated. In particular, if X is a separated variety, any morphism $\varrho: X \to Y$ is separated.*

Proof. Using Proposition 3, the assertion is easily reduced to the case where Y and Z are affinoid. We denote by $\Delta: X \to X \times_Y X$ and $\Delta': X \to X \times_Z X$ the diagonal morphisms corresponding to ϱ and $\sigma \circ \varrho$, and by $\tau: X \times_Y X \to X \times_Z X$ the canonical morphism, which is a closed immersion by Lemma 4. Since $\Delta' = \tau \circ \Delta$, one concludes without difficulty that Δ is a closed immersion. □

Proposition 6. *Let $\varrho: X \to Y$ be a separated morphism of analytic varieties, and suppose that Y is affinoid. Let U, V be open affinoid subvarieties of X. Then $U \cap V$ is also affinoid.*

Proof. We consider the diagonal morphism $\Delta: X \to X \times_Y X$. Then $U \times_Y V$ is an open affinoid subvariety of $X \times_Y X$ (see Lemma 9.3.5/3), whose inverse image in X is $U \cap V$. Thus, $U \cap V$ is affinoid by Proposition 9.5.3/2. □

We add here a criterion for an analytic morphism having an affinoid base Y to be separated. In view of Proposition 3, a generalization to the case where Y is an arbitrary analytic variety is straightforward.

Proposition 7. *Let $\varrho: X \to Y$ be a morphism of analytic varieties, and suppose that Y is affinoid. Then ϱ is separated if and only if the following conditions are satisfied:*

(i) *the image of the diagonal morphism $\Delta: X \to X \times_Y X$ is an analytic subset of $X \times_Y X$,*

(ii) *there exists an admissible affinoid covering* $\{X_i\}_{i\in I}$ *of* X *such that each intersection* $X_i \cap X_j$, $i, j \in I$, *has a finite admissible covering by open affinoid subvarieties.*

Proof. That any separated morphism $\varrho: X \to Y$ satisfies conditions (i) and (ii) follows from Proposition 9.5.3/1 and Proposition 6. Conversely, assume that ϱ satisfies conditions (i) and (ii). We will apply Proposition 9.5.3/5 in order to see that $\Delta: X \to X \times_Y X$ is a closed immersion. First, Δ is a locally closed immersion, since Δ composed with any projection $X \times_Y X \to X$ yields the identity. Thus, we have only to check condition (iii) of Proposition 9.5.3/5. The covering $\{X_i \times_Y X_j\}_{i,j\in I}$ is an admissible covering of $X \times_Y X$ such that $\Delta^{-1}(X_i \times_Y X_j) = X_i \cap X_j$. Since $X_i \cap X_j$ has by assumption a finite affinoid covering, everything is clear. □

We state without proof that separated morphisms are preserved by finite compositions, by any base change functor, or by fibre products over any analytic variety S.

9.6.2. Proper morphisms. — The definition of proper morphisms is a delicate point, if one wants these morphisms to satisfy properties similar to those of proper maps in complex analysis or algebraic geometry. In our approach, the following notion similar to the classical notion of relative compactness is basic. Let $X = \operatorname{Sp} A$ be an affinoid variety over some affinoid variety $Y = \operatorname{Sp} B$ via a morphism $X \to Y$, and let $U \subset X$ denote an affinoid subdomain. We write $U \Subset_Y X$ and say that U is *relatively compact in* X *over* Y if there exists an affinoid generating system $\{f_1, \ldots, f_r\}$ of A over B such that

$$U \subset \{x \in X;\ |f_1(x)| < 1, \ldots, |f_r(x)| < 1\},$$

or, due to the Maximum Modulus Principle (Proposition 6.2.1/4), if there exists an $\varepsilon \in \sqrt{|k^*|}$, $\varepsilon < 1$, such that $U \subset X(\varepsilon^{-1}f_1, \ldots, \varepsilon^{-1}f_r)$.

Lemma 1. *Let* X_1 *and* X_2 *be affinoid varieties over some affinoid variety* Y, *and, for* $i = 1, 2$, *let* $U_i \subset X_i$ *denote affinoid subdomains such that* $U_i \Subset_Y X_i$. *Then*

$$U_1 \times_Y X_2 \Subset_{X_2} X_1 \times_Y X_2 \quad \textit{and} \quad U_1 \times_Y U_2 \Subset_Y X_1 \times_Y X_2,$$

where, in the first case, $U_1 \times_Y X_2$ *and* $X_1 \times_Y X_2$ *are considered as varieties over* X_2 *via the canonical projection onto the second factor.*

Proof. Let $X_i = \operatorname{Sp} A_i$ and $U_i = \operatorname{Sp} A_i'$ for $i = 1, 2$, and let $Y = \operatorname{Sp} B$. Furthermore, we denote by $\{f_1, \ldots, f_r\}$ and $\{g_1, \ldots, g_s\}$ affinoid generating systems of A_1 and A_2 over B, respectively, such that, for a suitable $\varepsilon \in \sqrt{|k^*|}$, $\varepsilon < 1$, one has

$$U_1 \subset X_1(\varepsilon^{-1}f_1, \ldots, \varepsilon^{-1}f_r) \quad \text{and} \quad U_2 \subset X_2(\varepsilon^{-1}g_1, \ldots, \varepsilon^{-1}g_s).$$

Then one uses Proposition 6.1.1/10 and verifies that $\{f_1 \mathbin{\widehat{\otimes}} 1, \ldots, f_r \mathbin{\widehat{\otimes}} 1\}$ is an affinoid generating system of $A_1 \mathbin{\widehat{\otimes}}_B A_2$ over A_2 and that $\{f_1 \mathbin{\widehat{\otimes}} 1, \ldots, f_r \mathbin{\widehat{\otimes}} 1, 1 \mathbin{\widehat{\otimes}} g_1, \ldots, 1 \mathbin{\widehat{\otimes}} g_s\}$ is an affinoid generating system of $A_1 \mathbin{\widehat{\otimes}}_B A_2$ over B. Since

$$U_1 \times_Y X_2 \subset X_1(\varepsilon^{-1}f_1, \ldots, \varepsilon^{-1}f_r) \times_Y X_2$$

and

$$U_1 \times_Y U_2 \subset X_1(\varepsilon^{-1}f_1, \ldots, \varepsilon^{-1}f_r) \times_Y X_2(\varepsilon^{-1}g_1, \ldots, \varepsilon^{-1}g_s),$$

the assertions of the lemma are clear. □

The above lemma applies, in particular, to the case where X_2 is an affinoid subdomain of Y, say $X_2 = Y' \subset Y$. Then $U_1' := U_1 \times_Y Y'$ and $X_1' := X_1 \times_Y Y'$ are affinoid subdomains of $X_1 \times_Y Y = X_1$ — namely just the inverse images of Y' with respect to the canonical maps $U_1 \to Y$ and $X_1 \to Y$ (see Proposition 7.2.2/4) — and we have $U_1' \Subset_{Y'} X_1'$.

Definition 2. *A morphism of analytic varieties $\varrho: X \to Y$ is called proper if ϱ is separated and furthermore satisfies the following condition:*

There exists an admissible affinoid covering $\{Y_i\}_{i\in I}$ of Y, and, for each $i \in I$, there are two finite admissible affinoid coverings $\{X_{ij}\}_{j=1,\ldots,n_i}$ and $\{X'_{ij}\}_{j=1,\ldots,n_i}$ of $\varrho^{-1}(Y_i)$ such that $X_{ij} \Subset_{Y_i} X'_{ij}$ for all indices i and j.

First, we want to show that the properness of morphisms is a condition which is local in the following sense:

Proposition 3. *Let $\varrho: X \to Y$ be a morphism of analytic varieties, and let $\{Y_i\}_{i\in I}$ be an admissible open covering of Y. Setting $X_i := \varrho^{-1}(Y_i)$, the map $\varrho: X \to Y$ is proper if and only if all induced maps $\varrho_i: X_i \to Y_i$ are proper.*

Proof. The assertion with proper replaced by separated follows from Proposition 9.6.1/3. Thus, one verifies immediately that the properness of the maps ϱ_i implies the properness of ϱ. To prove the converse, let ϱ be proper. We consider the special case where Y is affinoid and X has finite admissible affinoid coverings $\{X_j\}_{j=1,\ldots,n}$ and $\{X'_j\}_{j=1,\ldots,n}$ with $X_j \Subset_Y X'_j$. Then, if Y' is an affinoid subdomain of Y, it follows from Lemma 1 (see the above discussion following the proof) that $X_j \cap \varrho^{-1}(Y') \Subset_{Y'} X'_j \cap \varrho^{-1}(Y')$ for $j = 1, \ldots, n$ and hence that $\varrho^{-1}(Y') \to Y'$ is proper. Therefore, in the general case, one can find admissible affinoid coverings of the varieties Y_i such that over each affinoid variety of such a covering the map ϱ is proper. Consequently, all maps ϱ_i must be proper. □

Proposition 4. *Let $\varrho: X \to Y$ and $\sigma: Y \to Z$ be morphisms of analytic varieties. If $\sigma \circ \varrho$ is proper and if σ is separated, then ϱ is proper.*

Proof. First, we use σ being separated to show that the morphism $\tau: X \to X \times_Z Y$ determined by the maps $\mathrm{id}: X \to X$ and $\varrho: X \to Y$ is a closed immersion.

Namely, the diagonal morphism $\Delta\colon Y \to Y \times_Z Y$ is a closed immersion, and τ is obtained from Δ by carrying out a base change with respect to X over Y. Hence τ is also a closed immersion.

Next we observe that ϱ is separated by Proposition 9.6.1/5 and that, due to Proposition 3, we have only to consider the case where Z is affinoid and where X has admissible affinoid coverings $\{X_j\}_{j=1,\ldots,n}$ and $\{X'_j\}_{j=1,\ldots,n}$ such that $X_j \Subset_Z X'_j$ for all j. Let $\{Y_i\}_{i\in I}$ be an admissible affinoid covering of Y. If $i \in I$ is fixed, then $\{\tau^{-1}(X_j \times_Z Y_i)\}_{j=1,\ldots,n}$ and $\{\tau^{-1}(X'_j \times_Z Y_i)\}_{j=1,\ldots,n}$ are admissible coverings of $\varrho^{-1}(Y_i)$ which are affinoid, since τ is a closed immersion. Furthermore, we have $X_j \times_Z Y_i \Subset_{Y_i} X'_j \times_Z Y_i$ by Lemma 1; hence, also $\tau^{-1}(X_j \times_Z Y_i) \Subset_{Y_i} \tau^{-1}(X'_j \times_Z Y_i)$. This shows that ϱ is proper. □

In particular, it follows that a morphism of analytic varieties $\varrho\colon X \to Y$ is proper if X is proper (i.e., proper over Sp k with k denoting the ground field) and Y is separated. To give some examples of proper analytic varieties, we consider the projective n-space $\mathbb{P}_k^n$ and the analytic torus $\mathbb{A}^*/(q)$. In both cases we have separatedness, as we have seen in (9.6.1). The analytic structure of $\mathbb{P}_k^n$ was defined in (9.3.4) by the admissible affinoid covering $\mathbb{P}_k^n = X_0 \cup \ldots \cup X_n$, where $X_i \cong \mathbb{B}_k^n$. Taking X'_i as a ball in $\mathbb{P}_k^n$ containing X_i and having a radius > 1, one obtains another admissible affinoid covering $\mathbb{P}_k^n = X'_0 \cup \ldots \cup X'_n$. Since then $X_i \Subset_{\mathrm{Sp}\,k} X'_i$ for all i, it is seen that $\mathbb{P}_k^n$ is proper. Similarly, starting with the defining covering of $\mathbb{A}^*/(q)$ by the two annuli X_1, X_2 as given in (9.3.4), one finds annuli X'_1, X'_2 in $\mathbb{A}^*/(q)$ such that $X_1 \Subset_{\mathrm{Sp}\,k} X'_1$ and $X_2 \Subset_{\mathrm{Sp}\,k} X'_2$, so that also $\mathbb{A}^*/(q)$ is proper.

Proposition 5. *Let $\varrho\colon X \to Y$ be a finite morphism of analytic varieties. Then ϱ is proper.*

Proof. We know already that finite morphisms are separated. Furthermore, due to Proposition 3, we may assume that X and Y are affinoid, say $X = \mathrm{Sp}\,A$ and $Y = \mathrm{Sp}\,B$. Then A is finite over B and hence, in particular, a finitely generated B-algebra. Therefore, A has an affinoid generating system $\{f_1, \ldots, f_r\}$ over B such that $|f_i|_{\sup} < 1$ for $i = 1, \ldots, r$. Consequently, $X \Subset_Y X$; i.e., $\varrho\colon X \to Y$ is proper. □

Since closed immersions are finite, we see in particular that all closed immersions are proper. Also, it is easily verfied that any composition of a closed immersion with a proper morphism is proper. We state without proof that proper morphisms are preserved by fibre products and any base change functor.

9.6.3. The Direct Image Theorem and the Theorem on Formal Functions. — Here we state without proof and in a simplified version two basic theorems for proper morphisms and derive some consequences. Proofs of the theorems can be found in Kiehl [23].

Theorem 1 (Direct Image Theorem). *Let $\varrho: X \to Y$ be a proper morphism of analytic varieties, and let $\mathscr{F}$ be a coherent $\mathcal{O}_X$-module. Then $\varrho_*(\mathscr{F})$ is a coherent $\mathcal{O}_Y$-module.*

For the second theorem, we need some preparation. Let $\varrho: X \to Y$ be a proper morphism with $Y = \operatorname{Sp} B$ being affinoid. We consider a coherent $\mathcal{O}_X$-module $\mathscr{F}$, a point $y \in Y$, and the corresponding maximal ideal $\mathfrak{m} \subset B$. If $U = \operatorname{Sp} A$ is an open affinoid subvariety of X, the $\mathcal{O}_U$-module $\mathscr{F}|_U$ is associated to a finite A-module M, and M may also be viewed as a B-module via the canonical algebra homomorphism $B \to A$. In particular, $\mathfrak{m}M$ makes sense as a B-submodule or, as we prefer, as an A-submodule of M. Thus, applying Proposition 9.2.3/1 (combined with the methods of (9.1.3)), one defines a coherent $\mathcal{O}_X$-module $\mathfrak{m}\mathscr{F}$ by setting $(\mathfrak{m}\mathscr{F})|_U := \mathfrak{m}(\mathscr{F}|_U)$. The same procedure is of course possible for any power $\mathfrak{m}^n$ of $\mathfrak{m}$, so that, for any $n \in \mathbb{N}$, one can consider the canonical map $\mathscr{F}(X) \to (\mathscr{F}/\mathfrak{m}^n\mathscr{F})(X)$ or the map

$$\phi_n: \mathscr{F}(X) \otimes_B \hat{\mathcal{O}}_{Y,y} \to (\mathscr{F}/\mathfrak{m}^n\mathscr{F})(X) \otimes_B \hat{\mathcal{O}}_{Y,y},$$

where $\hat{\mathcal{O}}_{Y,y}$ is the $\mathfrak{m}$-adic completion of $\mathcal{O}_{Y,y}$. We want to further interpret the maps ϕ_n. Since ϱ is proper, $\varrho_*(\mathscr{F})$ is a coherent $\mathcal{O}_Y$-module by Theorem 1. In particular, Y being affinoid, $\varrho_*(\mathscr{F})$ must be associated to the finite B-module $\mathscr{F}(X)$. Consequently, $\mathscr{F}(X) \otimes_B \hat{\mathcal{O}}_{Y,y}$ equals the $\mathfrak{m}$-adic completion of the stalk $\varrho_*(\mathscr{F})_y$ (see Proposition 9.4.2/6), which we denote by $\varrho_*(\mathscr{F})_y^\wedge$. Furthermore, the B-module $(\mathscr{F}/\mathfrak{m}^n\mathscr{F})(X)$ is annihilated by $\mathfrak{m}^n$, and $\hat{\mathcal{O}}_{Y,y}$ is the $\mathfrak{m}$-adic completion of B (cf. Proposition 7.3.2/3). Hence one obtains

$$(\mathscr{F}/\mathfrak{m}^n\mathscr{F})(X) \otimes_B \hat{\mathcal{O}}_{Y,y} = (\mathscr{F}/\mathfrak{m}^n\mathscr{F})(X),$$

and the maps ϕ_n can be rewritten as

$$\phi_n: \varrho_*(\mathscr{F})_y^\wedge \to (\mathscr{F}/\mathfrak{m}^n\mathscr{F})(X).$$

The homomorphism

$$\phi: \varrho_*(\mathscr{F})_y^\wedge \to \varprojlim (\mathscr{F}/\mathfrak{m}^n\mathscr{F})(X),$$

derived from the ϕ_n, is the subject of the following theorem. Notice that ϕ is also defined when Y is not affinoid. Namely, one chooses an open affinoid subvariety $V \subset Y$ containing y and constructs a map

$$\varrho_*(\mathscr{F})_y^\wedge \to \varprojlim (\mathscr{F}/\mathfrak{m}^n\mathscr{F})\big(\varrho^{-1}(V)\big)$$

as above, where now $\mathfrak{m} := \mathfrak{id}(\{y\})$ is the $\mathcal{O}_Y$-ideal of all functions vanishing at y. Since $\mathscr{F}/\mathfrak{m}^n\mathscr{F}$ vanishes outside the fibre $\varrho^{-1}(y)$ (which is contained in $\varrho^{-1}(V)$), one verifies that the restriction homomorphism

$$(\mathscr{F}/\mathfrak{m}^n\mathscr{F})(X) \to (\mathscr{F}/\mathfrak{m}^n\mathscr{F})\big(\varrho^{-1}(V)\big)$$

is an isomorphism.

Theorem 2 (Theorem on Formal Functions). *Let $\varrho: X \to Y$ be a proper morphism of analytic varieties, let $\mathscr{F}$ be a coherent $\mathcal{O}_X$-module, let y be a point of Y,* *and let* $\mathfrak{m} := \mathfrak{id}(\{y\})$ *be the corresponding* $\mathcal{O}_Y$*-ideal. Then the canonical homomorphism*

$$\phi : \varrho_*(\mathcal{F})_y^{\wedge} \to \varprojlim (\mathcal{F}/\mathfrak{m}^n\mathcal{F})(X)$$

is an isomorphism.

The proofs of Theorems 1 and 2 involve cohomological methods which we do not present here. As a simple application of both theorems, we show

Proposition 3 (Proper Mapping Theorem). *Let* $\varrho: X \to Y$ *be a proper morphism of analytic varieties, and let* M *be an analytic subset of* X. *Then* $\varrho(M)$ *is an analytic subset of* Y.

Proof. Due to Proposition 9.5.3/4, we can view M as a closed analytic subvariety of X. Restricting ϱ to this subvariety, we may assume $M = X$. Furthermore, we need only consider the case where Y is affinoid. Since $\varrho_*(\mathcal{O}_X)$, due to Theorem 1, is a coherent $\mathcal{O}_Y$-module, we see by Proposition 9.5.2/4 that $\operatorname{supp}(\varrho_*(\mathcal{O}_X))$ is an analytic subset of Y. Therefore it is enough to verify that $\varrho(X) = \operatorname{supp}(\varrho_*(\mathcal{O}_X))$. Let y be a point of Y, and let $\mathfrak{m} := \mathfrak{id}(\{y\})$ be the corresponding $\mathcal{O}_Y$-ideal. If y is contained in $\varrho(X)$, a local consideration shows that the canonical mappings $k \to (\mathcal{O}_X/\mathfrak{m}^n\mathcal{O}_X)(X)$, with k denoting the ground field, are non-zero. Therefore, $\varprojlim (\mathcal{O}_X/\mathfrak{m}^n\mathcal{O}_X)(X)$ is non-zero. Conversely, if $y \notin \varrho(X)$, all $\mathcal{O}_X$-modules $\mathcal{O}_X/\mathfrak{m}^n\mathcal{O}_X$ must vanish, so that $\varprojlim (\mathcal{O}_X/\mathfrak{m}^n\mathcal{O}_X)(X) = 0$ in this case. Using Theorem 2 and Proposition 9.4.2/6, the assertion follows. □

As a second application we give a proof for the Stein Factorization of proper morphisms. Let $\varrho: X \to Y$ be a morphism of analytic varieties. If M is an analytic subset of Y, it follows from the local description of M that $\varrho^{-1}(M)$ is an analytic subset of X. In particular, for any $y \in Y$, the fibre $\varrho^{-1}(y)$ is an analytic subset of X. Any analytic subset of X can be viewed as a closed subvariety of X (see Proposition 9.5.3/4). Thus, $\varrho^{-1}(y)$ carries a natural G-topology, and we call the fibre $\varrho^{-1}(y)$ *connected* if this G-topology is connected.

Lemma 4. *Let* $\varrho: X \to Y$ *be a proper morphism of analytic varieties such that the induced homomorphism of* $\mathcal{O}_Y$*-modules* $\varrho^{\#}: \mathcal{O}_Y \to \varrho_*(\mathcal{O}_X)$ *is an isomorphism. Then* ϱ *is surjective and has connected fibres. If all fibres of* ϱ *contain only a single point, then* ϱ *is an isomorphism.*

Proof. We may assume that Y is affinoid. The image $\varrho(X)$ is an analytic subset of Y by Proposition 3. Hence we must have $\varrho(X) = Y$, since $\varrho^{\#}: \mathcal{O}_Y \to \varrho_*(\mathcal{O}_X)$ is bijective. If y is a point in Y, then by our assumptions,

$$\varrho_*(\mathcal{O}_X)_y = \varinjlim \mathcal{O}_X(\varrho^{-1}(V)) = \varinjlim \mathcal{O}_Y(V) = \mathcal{O}_{Y,y},$$

where the direct limits are taken over all admissible open subsets V of Y containing y. Hence $\varrho_*(\mathcal{O}_X)_y^{\wedge}$ is isomorphic to the maximal-adic completion of $\mathcal{O}_{Y,y}$ and thus a local ring. On the other hand, by Theorem 2, we must have

$$\hat{\mathcal{O}}_{Y,y} = \varrho_*(\mathcal{O}_X)_y^{\wedge} = \varprojlim (\mathcal{O}_X/\mathfrak{m}^n\mathcal{O}_X)(X)$$

for the $\mathcal{O}_Y$-ideal $\mathfrak{m} := \mathfrak{id}(\{y\})$. A local consideration shows that, for each $n \in \mathbb{N}$, the $\mathcal{O}_X$-ideal $\mathfrak{m}^n\mathcal{O}_X$ is coherent and satisfies $V(\mathfrak{m}^n\mathcal{O}_X) = \varrho^{-1}(y)$. Therefore, we can apply Proposition 9.5.3/3 and define $\varrho^{-1}(y)$ as a closed analytic subvariety of X by taking as structure sheaf the "restriction" of $\mathcal{O}_X/\mathfrak{m}^n\mathcal{O}_X$ to $\varrho^{-1}(y)$. We denote this subvariety of X by X_n. All varieties X_n have the same nilreduction; therefore, the corresponding G-topologies on $\varrho^{-1}(y)$ coincide. If $\varrho^{-1}(y)$ is not connected, there are disjoint non-empty admissible open subsets $V_1, V_2 \subset \varrho^{-1}(y)$ such that $\{V_1, V_2\}$ is an admissible covering of $\varrho^{-1}(y)$. Consequently, we have for each $n \in \mathbb{N}$

$$(\mathcal{O}_X/\mathfrak{m}^n\mathcal{O}_X)(X) = \mathcal{O}_{X_n}\big(\varrho^{-1}(y)\big) = \mathcal{O}_{X_n}(V_1) \oplus \mathcal{O}_{X_n}(V_2),$$

and therefore

$$\varprojlim\, (\mathcal{O}_X/\mathfrak{m}^n\mathcal{O}_X)(X) = \varprojlim\, \mathcal{O}_{X_n}(V_1) \oplus \varprojlim\, \mathcal{O}_{X_n}(V_2).$$

Since both terms on the right-hand side do not vanish, we see that $\varprojlim\, (\mathcal{O}_X/\mathfrak{m}^n\mathcal{O}_X)(X)$ cannot be a local ring, contradicting the fact that this ring is isomorphic to $\hat{\mathcal{O}}_{Y,y}$. Therefore, $\varrho^{-1}(y)$ must be connected.

It remains to consider the case where ϱ, as a map of sets, is bijective. Because ϱ is proper and Y is affinoid, there exists a finite admissible affinoid covering of X. Hence, ϱ is a homeomorphism with respect to canonical topologies by Lemma 9.5.3/6, and the isomorphism $\varrho^{\#}: \mathcal{O}_Y \to \varrho_*(\mathcal{O}_X)$ shows that ϱ is a locally closed immersion. Thus, we see by Proposition 9.5.3/5 that ϱ, in fact, is a closed immersion. But $\varrho^{\#}: \mathcal{O}_Y \to \varrho_*(\mathcal{O}_X)$ being an isomorphism, ϱ must be an isomorphism. □

Proposition 5 (Stein Factorization). *Let $\varrho: X \to Y$ be a proper morphism of analytic varieties. Then there exists an analytic variety Z and a factorization of ϱ*

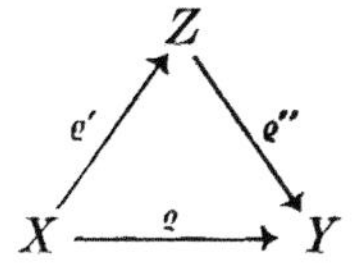

where $\varrho': X \to Z$ is a surjective proper morphism with connected fibres such that $\varrho'^{\#}: \mathcal{O}_Z \to \varrho_*(\mathcal{O}_X)$ is an isomorphism and where $\varrho'': Z \to X$ is finite.

Proof. Let $\{Y_i\}_{i\in I}$ be an admissible affinoid covering of Y and set $X_i := \varrho^{-1}(Y_i)$. The $\mathcal{O}_Y$-module $\varrho_*(\mathcal{O}_X)$ is coherent by Theorem 1. Therefore, all algebra homomorphisms $\varrho^*_{Y_i}: \mathcal{O}_Y(Y_i) \to \mathcal{O}_X(X_i)$ are finite. In particular, $\mathcal{O}_X(X_i)$ may be viewed as an affinoid algebra (cf. Proposition 6.1.1/6). According to Corollary 9.3.3/2, the maps $\varrho_i: X_i \to Y_i$ induced by ϱ have factorizations

$$X_i \xrightarrow{\varrho_i'} \operatorname{Sp} \mathcal{O}_X(X_i) \xrightarrow{\varrho_i''} Y_i.$$

One can now use the pasting techniques of (9.3.2) to construct an analytic variety Z by pasting together the affinoid varieties $\operatorname{Sp} \mathcal{O}_X(X_i)$ via the intersections $(\varrho_i'')^{-1}(Y_i \cap Y_j) \cong (\varrho_j'')^{-1}(Y_i \cap Y_j)$. Accordingly, one uses (9.3.3) to define morphisms $\varrho' : X \to Z$ and $\varrho'' : Z \to Y$ extending the maps ϱ_i' and ϱ_i''. By our construction, ϱ'' is finite, and hence in particular separated. Therefore ϱ' is proper (see Proposition 9.6.2/4). That ϱ' is surjective and has connected fibres follows from Lemma 4, since $(\varrho')^\# : \mathcal{O}_Z \to \varrho'_*(\mathcal{O}_X)$ is an isomorphism. □

Corollary 6. *Let $\varrho : X \to Y$ be a morphism of analytic varieties. Then ϱ is finite if and only if ϱ is proper and has finite fibres.*

Proof. Any finite morphism has finite fibres and is proper by Proposition 9.6.2/5. Conversely, if $\varrho : X \to Y$ is a proper morphism with finite fibres, we can consider the Stein Factorization $X \xrightarrow{\varrho'} Z \xrightarrow{\varrho''} Y$, where ϱ' is a bijection and $(\varrho')^\# : \mathcal{O}_Z \to \varrho'_*(\mathcal{O}_X)$ is an isomorphism. But then ϱ' is an isomorphism by Lemma 4. Consequently, ϱ is finite, since ϱ'' is finite. □

9.7. An application to elliptic curves

9.7.1. Families of annuli. — Throughout this section we consider affinoid and analytic varieties over a fixed affinoid variety $\operatorname{Sp} A$ which is assumed to be connected. Let ζ, η denote indeterminates, and let Λ be a unit in A with $|\Lambda|_{\sup} \leqq 1$. Then $\operatorname{Sp} B := \operatorname{Sp} A\langle\zeta, \eta\rangle/(\zeta\eta - \Lambda)$ is called a (*generalized*) *annulus over* $\operatorname{Sp} A$. We may view $\operatorname{Sp} B$ as the Laurent domain in $\operatorname{Sp} A\langle\zeta\rangle$ where the function $\Lambda^{-1}\zeta$ takes only values $\geqq 1$. Therefore, we can write $B = A\langle\zeta, \Lambda\zeta^{-1}\rangle$, and we claim that each $f \in B$ has a unique Laurent expansion $f = \sum_{\nu=-\infty}^{+\infty} a_\nu \zeta^\nu$ with coefficients $a_\nu \in A$ satisfying $\lim_{\nu\to+\infty} a_\nu = 0$ and $\lim_{\nu\to-\infty} a_\nu\Lambda^\nu = 0$. Namely, consider the set L of all these Laurent series. An easy computation shows that L is a k-affinoid algebra if the norm is defined by

$$\left\| \sum_{\nu=-\infty}^{+\infty} a_\nu\zeta^\nu \right\| := \max\left\{ \max_{\nu\geqq 0} |a_\nu|, \max_{\nu<0} |a_\nu\Lambda^\nu| \right\}$$

(where $|\ \ |$ is the Banach norm on A). Now let $A\langle\zeta\rangle \to C$ be a homomorphism of affinoid algebras mapping $\Lambda^{-1}\zeta$ onto a unit whose inverse is power-bounded in C. Then this homomorphism factors uniquely through L. Since the algebra B satisfies the same universal property, B is canonically isomorphic to L, and we see that each $f \in B$ has a unique Laurent expansion as claimed. The function ζ (or $\Lambda\zeta^{-1}$) is referred to as a coordinate function of $\operatorname{Sp} B$. Note that the generalized annulus $\operatorname{Sp} B$ consists of a family of annuli (in the ordinary sense), parametrized by the points of $\operatorname{Sp} A$.

Lemma 1. *Let $\operatorname{Sp} B = \operatorname{Sp} A\langle\zeta, \Lambda\zeta^{-1}\rangle$ be an annulus over $\operatorname{Sp} A$, and let $f = \sum_{\nu=-\infty}^{+\infty} a_\nu\zeta^\nu$ be an arbitrary function in B. Then the following are equivalent:*

(i) *The function f has no zero on $\operatorname{Sp} B$; i.e., f is a unit in B.*

(ii) *There exists a $\nu_0 \in \mathbb{Z}$ such that the term $a_{\nu_0}\zeta^{\nu_0}$ of the Laurent expansion is* *dominant on* Sp B; *i.e.*,

$$|(a_\nu\zeta^\nu)(y)| < |(a_{\nu_0}\zeta^{\nu_0})(y)| \quad \textit{for all} \quad \nu \neq \nu_0 \quad \textit{and all} \quad y \in \operatorname{Sp} B.$$

(iii) *There exists a* $\nu_0 \in \mathbb{Z}$ *such that* a_{ν_0} *is a unit in* A *and such that* $|a_{\nu_0}^{-1}a_\nu|_{\sup} < 1$ *for all* $\nu > \nu_0$ *as well as* $|\Lambda^{\nu-\nu_0}a_{\nu_0}^{-1}a_\nu|_{\sup} < 1$ *for all* $\nu < \nu_0$.

Proof. The assertion of the lemma is clear if A equals the ground field k and if $\Lambda = 1$. Namely, we have $B = k\langle\zeta, \zeta^{-1}\rangle$ in this case, and an element $f \in B$ with $|f|_{\sup} = 1$ is a unit if and only if its residue in the reduction $\tilde{k}[\zeta, \zeta^{-1}]$ is a unit. Considering the general case, let us assume that f is a unit. First we want to show that, for each $y \in \operatorname{Sp} B$, the series $\sum a_\nu\zeta^\nu$ has a term which is dominant at y. We denote by x the image of y with respect to the canonical map $\operatorname{Sp} B \to \operatorname{Sp} A$, by $\mathfrak{m}_x \subset A$ the maximal ideal corresponding to x, and by Y_x the fibre over x; i.e.,

$$Y_x = \operatorname{Sp} B/\mathfrak{m}_x B = \operatorname{Sp} A/\mathfrak{m}_x\langle\zeta, \Lambda(x)\,\zeta^{-1}\rangle.$$

Then $y \in Y_x$, and we may view $A/\mathfrak{m}_x$ as a new ground field. Thus it is only necessary to consider the special case where $A = k$ and $B = k\langle\zeta, \Lambda\zeta^{-1}\rangle$ with $\Lambda \in k$, $0 < |\Lambda| \leq 1$. We claim that, for each $y \in \operatorname{Sp} B$, the series $\sum a_\nu\zeta^\nu$ has a term which is, in fact, dominant on the affinoid subdomain

$$U_y := \operatorname{Sp} B\langle|\zeta(y)|^{-1}\zeta, |\zeta(y)|\,\zeta^{-1}\rangle$$

of Sp B, which contains y. This is clear if $|\zeta(y)| \in |k|$. Namely, then U_y is isomorphic to $\operatorname{Sp} k\langle\xi, \xi^{-1}\rangle$ (where ξ corresponds to the function $c^{-1}\zeta$ for some $c \in k^*$ satisfying $|c| = |\zeta(y)|$), and the assertion follows from what has been shown at the beginning. If $|\zeta(y)| \notin |k|$, we replace the ground field k by a finite extension k' such that $|\zeta(y)| \in |k'|$. The same reasoning as before can be applied to the variety $U_y \mathbin{\hat{\otimes}} k' \simeq \operatorname{Sp} k'\langle\xi, \xi^{-1}\rangle$. Thereby we see that our claim is justified.

Returning to the general case, we denote by $\nu(y)$, as a function of $y \in \operatorname{Sp} B$, the index of the term in $\sum a_\nu\zeta^\nu$ which is dominant at the point y. Then $\nu(y)$ can take at most finitely many values on Sp B, since the terms $a_\nu\zeta^\nu$ cannot have a common zero on Sp B; i.e., since finitely many of them must generate the unit ideal in B and since $\lim\limits_{\nu\to\pm\infty} |a_\nu\zeta^\nu|_{\sup} = 0$. For the same reason, we see that, for each $\nu_0 \in \mathbb{Z}$, the set

$$Y_{\nu_0} := \{y \in \operatorname{Sp} B;\ |(a_\nu\zeta^\nu)(y)| \leq |(a_{\nu_0}\zeta^{\nu_0})(y)| \quad \text{for all} \quad \nu \neq \nu_0\}$$

defines a rational subdomain of Sp B. Furthermore, by our considerations above, we can replace $\leq$ by $<$ and thus get $Y_{\nu_0} = \{y \in \operatorname{Sp} B;\ \nu(y) = \nu_0\}$. Then $Y_{\nu_0} = \emptyset$ for almost all indices ν_0, and

$$\operatorname{Sp} B = \bigcup_{\nu=-\infty}^{+\infty} Y_\nu$$

can be viewed as an admissible covering of Sp B by finitely many rational subdomains which are pairwise disjoint. However, $\operatorname{Sp} B = \operatorname{Sp} A\langle\zeta, \Lambda\zeta^{-1}\rangle$ is connected; namely, it contains $\operatorname{Sp} A\langle\zeta, \zeta^{-1}\rangle = \operatorname{Sp} A \times \operatorname{Sp} k\langle\zeta, \zeta^{-1}\rangle$ as a Zariski-dense affinoid subdomain which is connected because $\operatorname{Sp} A$ was assumed to be connected (use Proposition 9.1.4/8). Therefore, the function $\nu(y)$ must be constant, and we have verified that condition (i) of our lemma implies condition (ii). The converse is trivial.

It remains to show the equivalence of conditions (ii) and (iii). We may assume $A = k$; hence, $\Lambda \in k$ and $f = \sum a_\nu \zeta^\nu \in k\langle\zeta, \Lambda\zeta^{-1}\rangle$. Let $\nu_0 \in \mathbb{Z}$ be a fixed index such that $a_{\nu_0} \neq 0$. Since for any $\nu \in \mathbb{Z}$ the ratio

$$\frac{|a_\nu|\, r^\nu}{|a_{\nu_0}|\, r^{\nu_0}} = \frac{|a_\nu|}{|a_{\nu_0}|}\, r^{\nu-\nu_0}, \quad |\Lambda| \leqq r \leqq 1,$$

is maximal for $r = 1$ if $\nu > \nu_0$ and maximal for $r = |\Lambda|$ if $\nu < \nu_0$, we see that condition (iii) is merely a reformulation of condition (ii). □

In the situation of Lemma 1, we call ν_0 the order of the unit f with respect to the coordinate function ζ; i.e., we define $\operatorname{ord}_\zeta(f) := \nu_0$. We have $\operatorname{ord}_\zeta(f) = -\operatorname{ord}_{\Lambda\zeta^{-1}}(f)$ for all units $f \in B$, and the coordinate functions of $\operatorname{Sp} B$ can be characterized as follows:

Proposition 2. *Let* $\operatorname{Sp} B = \operatorname{Sp} A\langle\zeta, \Lambda\zeta^{-1}\rangle$ *be an annulus over* $\operatorname{Sp} A$, *and let* $f = \sum_{\nu=-\infty}^{+\infty} a_\nu \zeta^\nu$ *be an arbitrary function in* B. *Then* f *is a coordinate function of* $\operatorname{Sp} B$, *i.e., there exists a unit* $\Lambda' \in A$ *and an isomorphism* $A\langle\xi, \Lambda'\xi^{-1}\rangle \xrightarrow{\sim} B$ *over* A *mapping* ξ *to* f *if and only if* f *is a unit in* B *which satisfies one of the following conditions:*

(i) $\operatorname{ord}_\zeta(f) = 1$ *and* a_1 *is a unit in* $\mathring{A}$ *or*
(ii) $\operatorname{ord}_\zeta(f) = -1$ *and* $\Lambda^{-1}a_{-1}$ *is a unit in* $\mathring{A}$.

Furthermore, any Λ' *as above differs from* Λ *only by a unit in* $\mathring{A}$; *i.e.,* $\Lambda'\Lambda^{-1}$ *is a unit in* $\mathring{A}$.

Proof. Let $A\langle\xi, \Lambda'\xi^{-1}\rangle \xrightarrow{\sim} B$ be an A-isomorphism mapping ξ to f. Then f must be a unit, and, replacing ζ by the coordinate function $\Lambda\zeta^{-1}$ if necessary, we may assume that $\nu_0 := \operatorname{ord}_\zeta(f) \geqq 0$. If $Y_x \subset \operatorname{Sp} B$ is the fibre over a point $x \in \operatorname{Sp} A$, then ξ, and hence f, must have supremum norm equal to 1 on Y_x. Since $a_{\nu_0}\zeta^{\nu_0}$ is the dominant term of the Laurent expansion of f, this means that $|a_{\nu_0}(x)| = 1$ for all $x \in \operatorname{Sp} A$, i.e., that a_{ν_0} is a unit in $\mathring{A}$. Also it follows that the above isomorphism induces an A-isomorphism

$$A\langle\xi, \xi^{-1}\rangle \xrightarrow{\sim} A\langle\zeta, \zeta^{-1}\rangle, \quad \xi \mapsto \sum_{\nu=-\infty}^{+\infty} a_\nu \zeta^\nu.$$

Looking at the corresponding isomorphism between the reductions over $\tilde{k}$ (the residue field of the ground field k), we see that ν_0 necessarily equals 1. This verifies the only if part of our assertion and also the additional assertion about Λ'. Namely, if x is a point in $\operatorname{Sp} A$, then $|\Lambda(x)|^{-1}$ is the supremum norm of

ζ^{-1} or of f^{-1} on the fibre Y_x, and $|A'(x)|^{-1}$ is the corresponding supremum norm for ξ^{-1}. However, due to the given A-isomorphism, both norms must coincide.

To verify the if part of the assertion, we may assume that f is a unit satisfying condition (i). Otherwise one can replace the coordinate function ζ by $A\zeta^{-1}$. Then a_1 is a unit in $\mathring{A}$, and it is enough to consider the case where $a_1 = 1$. Since $\operatorname{ord}_\zeta(\zeta^{-1}f) = 0$, we can write

$$f = \zeta\left(1 + \sum_{\nu=1}^{\infty} b_\nu \zeta^\nu + \sum_{\nu=1}^{\infty} c_\nu (A\zeta^{-1})^\nu\right)$$

with elements $b_\nu, c_\nu \in A$ which form zero sequences and satisfy $|b_\nu|_{\sup} < 1$ and $|c_\nu|_{\sup} < 1$ (cf. Lemma 1). By a standard approximation argument, one can then show that the A-homomorphism $A\langle\zeta, \eta\rangle \to A\langle\zeta, \eta\rangle$ given by

$$\zeta \mapsto \zeta\left(1 + \sum_{\nu=1}^{\infty} b_\nu \zeta^\nu + \sum_{\nu=1}^{\infty} c_\nu \eta^\nu\right)$$

$$\eta \mapsto \eta\left(1 + \sum_{\nu=1}^{\infty} b_\nu \zeta^\nu + \sum_{\nu=1}^{\infty} c_\nu \eta^\nu\right)^{-1}$$

is an A-automorphism leaving the ideal $(\zeta\eta - A)$ fixed. Therefore, it induces an A-automorphism $A\langle\zeta, A\zeta^{-1}\rangle \to A\langle\zeta, A\zeta^{-1}\rangle$ which maps ζ to f. Consequently, f is a coordinate function of $\operatorname{Sp} B$. □

Next we will show how annuli can be pasted together in order to get "bigger" annuli. The properties of coordinate functions proved above are basic for such a procedure. In preparation we start with the following

Lemma 3. *Let f be a function in $A\langle\zeta, \zeta^{-1}\rangle$ such that $|\zeta^{-1} - f|_{\sup} < 1$. Then any $h \in A\langle\zeta, \zeta^{-1}\rangle$ can be uniquely written as $h = \sum_{\nu=0}^{\infty} a_\nu \zeta^\nu + \sum_{\mu=1}^{\infty} b_\mu f^\mu$, where $a_\nu, b_\mu \in A$ and $|a_\nu|_{\sup}, |b_\mu|_{\sup} \leq |h|_{\sup}$.*

Proof. We fix a residue norm on A via some epimorphism $T_n \to A$. Consider on $A\langle\zeta, \zeta^{-1}\rangle$ an A-algebra norm as in (6.1.4); i.e., for any $h = \sum_{\nu=-\infty}^{+\infty} a_\nu \zeta^\nu \in A\langle\zeta, \zeta^{-1}\rangle$, we set $|h| := \max |a_\nu|$. The norm on A may be chosen such that $|\zeta^{-1} - f| < 1$. Then any h as before can be written as

$$h = \sum_{\nu=-\infty}^{+\infty} a_\nu \zeta^\nu = \sum_{\nu=0}^{\infty} a_\nu \zeta^\nu + \sum_{\mu=1}^{\infty} a_{-\mu} f^\mu + h'$$

where

$$h' = \sum_{\mu=1}^{\infty} a_{-\mu}(\zeta^{-\mu} - f^\mu),$$

and hence $|h'| \leq |\zeta^{-1} - f|\,|h|$. Repeating the same process for h', a standard limit argument shows that h can be written as required. The uniqueness assertion is verified by a similar argument. □

Proposition 4. *For $i = 1, 2$, let $Y_i = \operatorname{Sp} A\langle\zeta_i, A_i\zeta_i^{-1}\rangle$ be an annulus over $\operatorname{Sp} A$ with $|A_i|_{\sup} < 1$, and consider the annulus $Y_i' = \operatorname{Sp} A\langle\zeta_i, \zeta_i^{-1}\rangle$ as an affinoid subdomain of Y_i. Let Y be a separated analytic variety which is obtained by pasting together Y_1 and Y_2 via some isomorphism $Y_1' \xrightarrow{\sim} Y_2'$. Then Y is an annulus over $\operatorname{Sp} A$ isomorphic to $\operatorname{Sp} A\langle\xi, A_1A_2\xi^{-1}\rangle$. More precisely, there exists a coordinate function f on Y such that $Y_1 = Y(A_1f^{-1})$, $Y_2 = Y(A_1^{-1}f)$ and such that $f|_{Y_1}$ is a coordinate function of Y_1 with $\operatorname{ord}_{\zeta_1}(f|_{Y_1}) = -1$ and $A_1^{-1}f|_{Y_2}$ is a coordinate function of Y_2 with $\operatorname{ord}_{\zeta_2}(A_1^{-1}f|_{Y_2}) = 1$.*

Proof. We identify $A\langle\zeta_1, \zeta_1^{-1}\rangle$ with $A\langle\zeta_2, \zeta_2^{-1}\rangle$ via the appropriate isomorphism and show first of all that $\operatorname{ord}_{\zeta_1}(\zeta_2) = -1$. Assuming $\operatorname{ord}_{\zeta_1}(\zeta_2) = +1$, which is the only other possibility (cf. Proposition 2), we look at the diagonal morphism $Y \to Y \times Y$. Since Y is separated, it induces a closed immersion $Y_1 \cap Y_2 \to Y_1 \times Y_2$, i.e., the canonical map

$$A\langle\zeta_1, A_1\zeta_1^{-1}\rangle \mathbin{\hat{\otimes}_k} A\langle\zeta_2, A_2\zeta_2^{-1}\rangle \to A\langle\zeta_1, \zeta_1^{-1}\rangle = A\langle\zeta_2, \zeta_2^{-1}\rangle$$

is surjective. Hence $\{\zeta_1, \zeta_2, A_1\zeta_1^{-1}, A_2\zeta_2^{-1}\}$ is an affinoid generating system of $A\langle\zeta_i, \zeta_i^{-1}\rangle$ over A, and, by Theorem 6.3.4/2, the reduction $\tilde{A}[\tilde{\zeta}_i, \tilde{\zeta}_i^{-1}]$ is finite over the subalgebra generated over $\tilde{A}$ by the residues $\tilde{\zeta}_1, \tilde{\zeta}_2, \tilde{A}_1\tilde{\zeta}_1^{-1}, \tilde{A}_2\tilde{\zeta}_2^{-1}$. However, since $|A_1|_{\sup}, |A_2|_{\sup} < 1$, and since, by Proposition 2 and our assumption, $\tilde{\zeta}_1$ differs from $\tilde{\zeta}_2$ only by a unit in $\tilde{A}$, we see that $\tilde{A}[\tilde{\zeta}_1, \tilde{\zeta}_1^{-1}]$ must be finite over $\tilde{A}[\tilde{\zeta}_1]$, which is of course not true. Therefore we must have $\operatorname{ord}_{\zeta_1}(\zeta_2) = -1$, or in other words, $\operatorname{ord}_{\zeta_1^{-1}}(\zeta_2) = 1$. Due to Proposition 2, we can multiply ζ_2 by a unit in $\mathring{A}$ and thus assume that $|\zeta_1^{-1} - \zeta_2|_{\sup} < 1$. According to Lemma 3, we write

$$\zeta_1^{-1} - \zeta_2 = \sum_{\nu=0}^{\infty} a_\nu\zeta_1^\nu + \sum_{\mu=1}^{\infty} b_\mu\zeta_2^\mu$$

with coefficients $a_\nu, b_\mu \in A$ satisfying $|a_\nu|_{\sup}, |b_\mu|_{\sup} < 1$. Then, using Lemma 1 and Proposition 2, we can replace ζ_2 by $\zeta_2 + \sum_{\mu=1}^{\infty} b_\mu\zeta_2^\mu$ as coordinate function of Y_2 and replace $A_1\zeta_1^{-1}$ by $A_1\zeta_1^{-1} - A_1\sum_{\nu=0}^{\infty} a_\nu\zeta_1^\nu$ as coordinate function of Y_1. Thereby we have reduced the problem to the case where $\zeta_1^{-1}|_{Y_1\cap Y_2} = \zeta_2|_{Y_1\cap Y_2}$. Now consider the commutative diagram of A-homomorphisms

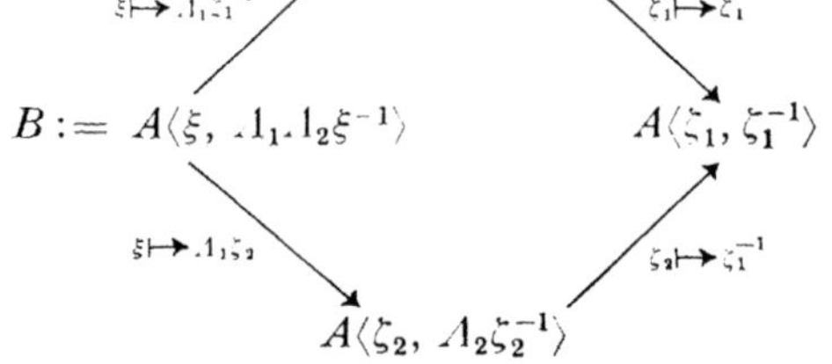

and the associated diagram of affinoid maps over Sp A

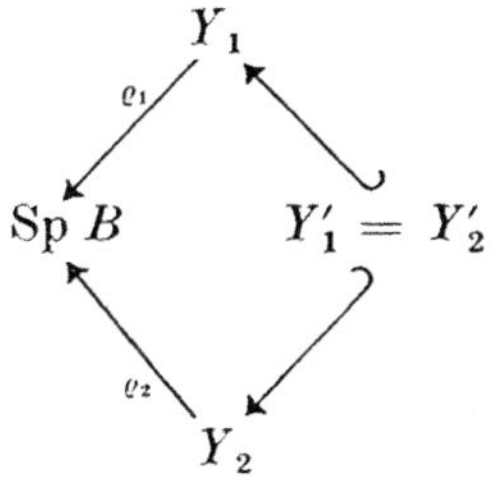

Then ϱ_1 identifies Y_1 with the affinoid subdomain Sp $B\langle A_1\xi^{-1}\rangle$ of Sp B, and ϱ_2 identifies Y_2 with the affinoid subdomain Sp $B\langle A_1^{-1}\xi\rangle$ of Sp B. Furthermore, $Y_1' = Y_2'$ is identified with $\varrho_1(Y_1) \cap \varrho_2(Y_2)$ by ϱ_1 and ϱ_2, so that in fact Sp B $= Y$. Thus, Y is an annulus and ξ is a coordinate function of Y satisfying the properties stated in the assertion. Of course, the same would be true for any coordinate function f of Y such that $\mathrm{ord}_\xi(f) = 1$. □

9.7.2. Affinoid subdomains of the unit disc. — In this section the ground field k is assumed to be algebraically closed, so that we can identify the points of the unit disc $\mathbb{B}^1 = \mathrm{Sp}\, k\langle\zeta\rangle$ with the corresponding points of the unit disc $\{x \in k;\, |x| \leqq 1\}$ in k (cf. Proposition 7.1.1/1). For $a \in \mathbb{B}^1$ and $r \in |k|$, $0 \leqq r \leqq 1$, we denote by

$$B^-(a, r) = \{x \in \mathbb{B}^1;\, |x - a| < r\} \quad \text{and}$$
$$B^+(a, r) = \{x \in \mathbb{B}^1;\, |x - a| \leqq r\}$$

the "open" and "closed" discs, respectively, of radius r, centered at a (see also (1.1.3)). A subset $X \subset \mathbb{B}^1$ is called *standard* if $X = \emptyset$ or if there exist points $a_0, \ldots, a_n \in \mathbb{B}^1$ and radii $r_0, \ldots, r_n \in |k^*|$, $r_i \leqq 1$, such that

$$X = B^+(a_0, r_0) - \bigcup_{i=1}^{n} B^-(a_i, r_i).$$

By definition, any standard subset is an affinoid subdomain of $\mathbb{B}^1$. Since each point of a disc may serve as its center, one verifies without difficulty

Lemma 1. *Let X and Y be standard subsets of $\mathbb{B}^1$ such that $X \cap Y \neq \emptyset$. Then $X \cup Y$ and $X \cap Y$ are standard subsets of $\mathbb{B}^1$.*

However, if $X_1, \ldots, X_s$, $s \geqq 2$, are pairwise disjoint non-empty standard subsets of $\mathbb{B}^1$, then $X_1 \cup \ldots \cup X_s$ is not standard. Furthermore, this union is an affinoid subdomain of $\mathbb{B}^1$ which is not connected (Proposition 7.2.2/9). The main result of this section is the following:

Theorem 2. *A subset $X \subset \mathbb{B}^1$ is standard if and only if X is a connected affinoid subdomain of $\mathbb{B}^1$. Furthermore, X is an affinoid subdomain of $\mathbb{B}^1$ if and only if X is a finite disjoint union of standard subsets in $\mathbb{B}^1$.*

Proof. Let X be an affinoid subdomain of $\mathbb{B}^1$. Starting with the second assertion, we need only show that X is a finite union of standard subsets of $\mathbb{B}^1$ (see Lemma 1). By Corollary 7.3.5/3, we may assume that X is a rational subdomain of $\mathbb{B}^1$. Therefore, let $f_1, \ldots, f_s, g$ be elements in $k\langle\zeta\rangle$ without a common zero such that

$$X = \mathbb{B}^1\left(\frac{f_1}{g}, \ldots, \frac{f_s}{g}\right).$$

Applying the WEIERSTRASS Preparation Theorem (Theorem 5.2.2/1), we can write each one of the elements $f_1, \ldots, f_s, g$ as a product of a unit in $k\langle\zeta\rangle$ and a polynomial in $k[\zeta]$. If $e \in k\langle\zeta\rangle$ is such a unit, then $|e(x)|$ is constant for all $x \in \mathbb{B}^1$ (cf. Proposition 5.1.3/1). Therefore we need only consider the case where $f_1, \ldots, f_s, g$ are polynomials, and, in fact, it is enough to look at the case where $s = 1$. Namely, let d_i be the greatest common divisor of the polynomials f_i and g. Then we have

$$X = \bigcap_{i=1}^{s} \left\{x \in \mathbb{B}^1; \left|\frac{f_i}{d_i}(x)\right| \leq \left|\frac{g}{d_i}(x)\right|\right\}$$

with rational subdomains of $\mathbb{B}^1$ on the right-hand side. If each one of these is a finite union of standard subsets of $\mathbb{B}^1$, then X has the same property by Lemma 1.

Thus we may assume that $X = \mathbb{B}^1\left(\frac{f}{g}\right)$ with polynomials

$$f = \prod_{i=1}^{n} (\zeta - \alpha_i) \quad \text{and} \quad g = c\prod_{j=1}^{m} (\zeta - \beta_j)$$

satisfying $c \neq 0$ and $\alpha_i \neq \beta_j$ for all i, j. For $r \in |\mathring{k}|$, we denote by

$$\Gamma_r := \{x \in \mathbb{B}^1; |x| = r\}$$

the circumference of the disc $B^+(0, r)$ and by $\| \;\|_r$ the supremum norm of functions on Γ_r. Then we have

$$\|f\|_r = \prod_{i=1}^{n} \|\zeta - \alpha_i\|_r = \prod_{i=1}^{n} \max(r, |\alpha_i|)$$

and a similar equality for g. Viewing $\log \|f\|_r$ as a function of $t := \log r$, we see that $\log \|f\|_r$ is the sum of the piecewise linear convex functions $\max(t, \log |\alpha_i|)$; hence it is piecewise linear and convex. Since the same is true for g, one sees that

$$R := \{r \in |\mathring{k}|; \|f\|_r \leq \|g\|_r\}$$

is a finite union of intervals in $|k|$, i.e., of sets of type

$$\{r \in |k|; r_1 \leq r \leq r_2\}$$

with $r_1, r_2 \in |k|$. Furthermore, it is easily seen that either $[0, \varepsilon] \cap |k| \subset R$ or $[0, \varepsilon] \cap R = \emptyset$ for some $\varepsilon > 0$. Therefore,

$$S := \bigcup_{r \in R} \Gamma_r = \{x \in \mathbb{B}^1; \|f\|_{|x|} \leq \|g\|_{|x|}\}$$

is the union of a "closed" disc (if $0 \in R$) and of finitely many "closed" annuli in $\mathbb{B}^1$, all centered at 0. Hence, in particular, S is a finite union of standard subsets of $\mathbb{B}^1$.

Let S' be the standard subset of $\mathbb{B}^1$ consisting of all points x such that $|g(x)| = \|g\|_{|x|}$; i.e.,

$$S' := \mathbb{B}^1 - \bigcup_{j=1}^{m} B^-(\beta_j, |\beta_j|).$$

Then $S \cap S'$ is a subset of X satisfying

$$S \cap S' \cap \Gamma_r = X \cap \Gamma_r$$

for all $r \notin \Delta := \{|\alpha_1|, \ldots, |\alpha_n|, |\beta_1|, \ldots, |\beta_m|\} - \{0\}$, since $|f(x)| = \|f\|_{|x|}$ and $|g(x)| = \|g\|_{|x|}$ if $|x| \notin \Delta$. Therefore, we can write

$$X = (S \cap S') \cup \bigcup_{r \in \Delta} X \cap \Gamma_r.$$

Hence in order to show that $X = \mathbb{B}^1\left(\frac{f}{g}\right)$ is a finite union of standard subsets of $\mathbb{B}^1$, it is enough to show the corresponding fact for each set $X \cap \Gamma_r$, where $r \in |\mathring{k}| - \{0\}$. Using induction on $n + m = \deg f + \deg g$, one proceeds as follows. The case $n = 0 = m$ is trivial. If $n + m > 0$, say $n > 0$ (the case $m > 0$ is analogous), we may apply a translation to $\mathbb{B}^1$ and thus assume $\alpha_1 = 0$. Then, for $r \in |\mathring{k}| - \{0\}$ and $\alpha \in \Gamma_r$, we can write

$$X \cap \Gamma_r = \left\{x \in \mathbb{B}^1; \left|\alpha \prod_{i=2}^{n} (x - \alpha_i)\right| \leq |g(x)|\right\} \cap \Gamma_r.$$

Hence by Lemma 1 and the induction hypothesis, $X \cap \Gamma_r$ is a finite union of standard subsets of $\mathbb{B}^1$. This concludes the verification of the second assertion of the theorem. Furthermore, the first assertion is also clear. Namely, if X is a standard subset of $\mathbb{B}^1$, its connected components consist of finitely many disjoint standard subsets of $\mathbb{B}^1$ by what has just been shown. Since a disjoint union of several non-empty standard subsets of $\mathbb{B}^1$ is not standard again, we see that X must be connected. The converse is obvious. □

Corollary 3. *Let X and Y be connected affinoid subdomains of $\mathbb{B}^1$. Then $X \cap Y$ is connected.*

Proof. Assuming $X \cap Y \neq \emptyset$, the assertion follows from Lemma 1 and Theorem 2. □

9.7.3. Tate's elliptic curves. — In this section we assume that the ground field k is algebraically closed and has residue characteristic $\neq 2$. For $\lambda \in k - \{0, 1\}$, we denote by C_λ the elliptic curve in $\mathbb{P}_k^2$ given by LEGENDRE's equation

$$Y^2Z - X(X - Z)(X - \lambda Z) = 0$$

(where X, Y, Z are coordinates of $\mathbb{P}_k^2$). Considering each C_λ as an analytic variety, we will show that the curves C_λ with $0 < |\lambda| < 1$ are just the analytic tori introduced in (9.3.4). This fact, first proved by TATE, was the starting point for the investigation of rigid analytic varieties. In contrast to TATE's approach, we will give a proof in terms of analytic varieties. In particular, the necessary facts about elliptic curves are kept to a minimum.

First we want to choose suitable coordinates of $\mathbb{P}_k^2$. For this purpose, let $i \in k$ be a square root of -1. Taking λZ as a new indeterminate instead of Z, the equation of C_λ reads

$$Y^2Z - X(X - Z)(\lambda X - Z) = 0.$$

For $|\lambda| < 1$, we look at the curve $\tilde{C}_\lambda$ in $\mathbb{P}_{\tilde{k}}^2$ which is obtained by "reducing coefficients". Then $\tilde{C}_\lambda$ is given by the equation $\left(Y^2 + X(X - Z)\right) Z = 0$; i.e., $\tilde{C}_\lambda$ is the union of the two curves with equations $Y^2 + X(X - Z) = 0$ and $Z = 0$, which intersect at the points $(1, \tilde{i}, 0)$ and $(1, -\tilde{i}, 0)$. The tangents of the curve $Y^2 + X(X - Z) = 0$ at these points have the equations $Z - 2\tilde{i}\,Y - 2X = 0$ and $Z + 2\tilde{i}\,Y - 2X = 0$, so that the picture is the following:

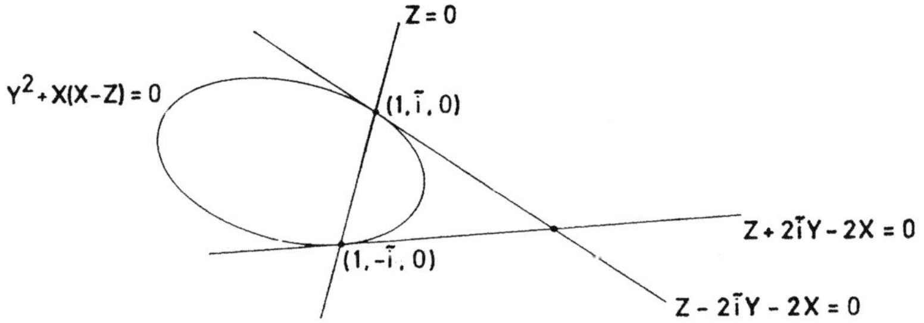

Switching back to $\mathbb{P}_k^2$, we take $Z - 2iY - 2X$ and $Z + 2iY - 2X$ as new coordinates instead of X and Y. Then the above equation of C_λ is transformed into

$$(XY - Z^2)\,Z + \lambda F(X + Y, Z) = 0,$$

where

$$F(\xi, \zeta) := \frac{1}{16}\,\xi^3 - \frac{1}{8}\,\xi^2\zeta - \frac{1}{4}\,\xi\zeta^2 + \frac{1}{2}\,\zeta^3.$$

Considering the standard covering $\mathbb{P}_k^2 = U_0 \cup U_1 \cup U_2$ where

$$U_0 := \operatorname{Sp} k\left\langle\frac{Y}{X}, \frac{Z}{X}\right\rangle, \quad U_1 := \operatorname{Sp} k\left\langle\frac{X}{Y}, \frac{Z}{Y}\right\rangle, \quad U_2 := \operatorname{Sp} k\left\langle\frac{X}{Z}, \frac{Y}{Z}\right\rangle,$$

the coordinates of $\mathbb{P}_k^2$ have been choosen in such a way that $C_\lambda \subset U_0 \cup U_1$ if $|\lambda| < 1$. This is clear from our construction (or can be verified by a simple computation).

In order to see how the corresponding analytic torus varies with the elliptic curve C_λ, we consider families of curves C_λ. Similarly as in (9.7.1), let $\operatorname{Sp} A$ be a connected and reduced affinoid variety, and let Λ be a unit in A

such that $\alpha := |A|_{\sup} < 1$. Then the homogeneous polynomial

$$(XY - Z^2)\,Z + AF(X + Y, Z) \in A[X, Y, Z]$$

defines a closed analytic subvariety C of the projective space $\mathbb{P}^2_A := \mathbb{P}^2_k \times \operatorname{Sp} A$ over $\operatorname{Sp} A$, and C may be viewed as a family of elliptic curves over $\operatorname{Sp} A$. In particular, if $\operatorname{Sp} A = \operatorname{Sp} k\langle \alpha^{-1}\zeta, \varepsilon\zeta^{-1}\rangle$ is an annulus in $\mathbb{B}^1 = \operatorname{Sp} k\langle\zeta\rangle$, with radii $\varepsilon, \alpha \in |k|$, $0 < \varepsilon \leqq \alpha < 1$, and if A denotes the restriction to $\operatorname{Sp} A$ of ζ, then C is just the family of all C_λ with $\varepsilon \leqq |\lambda| \leqq \alpha$. As for the curves C_λ with $|\lambda| < 1$, one has

$$C \subset (U_0 \times \operatorname{Sp} A) \cup (U_1 \times \operatorname{Sp} A),$$

so that $C = C^0 \cup C^1$ with $C^j := C \cap (U_j \times \operatorname{Sp} A)$, $j = 0, 1$, is an admissible affinoid covering of C.

Lemma 1. *C^0 is an annulus, isomorphic to* $\operatorname{Sp} A\langle\zeta, A\zeta^{-1}\rangle$ *over* $\operatorname{Sp} A$. *With respect to this isomorphism, the intersection $C^0 \cap C^1$ is the disjoint union of the affinoid subdomains* $\operatorname{Sp} A\langle\zeta, \zeta^{-1}\rangle$ *and* $\operatorname{Sp} A\langle A^{-1}\zeta, A\zeta^{-1}\rangle$. *The same is true for C^1 and the open subvariety $C^0 \cap C^1$ of C^1.*

Proof. The equation of C is symmetric in X and Y. Therefore it is enough to consider $C^0 \subset U_0 \times \operatorname{Sp} A$. Writing Y and Z instead of $\frac{Y}{X} - \left(\frac{Z}{X}\right)^2$ and $\frac{Z}{X}$, we have

$$C^0 = \operatorname{Sp} A\langle Y, Z\rangle / \big(YZ + AF(1 + Y + Z^2, Z)\big),$$

and $C^0 \cap C^1$ corresponds to the subdomain consisting of those points in C^0 where $Y + Z^2$ takes only absolute value 1. Since YZ has supremum norm < 1 on C^0, this is the disjoint union of the two subdomains in C^0, where Y takes absolute value 1 and where Z takes absolute value 1.

The polynomial $F(1 + Y + Z^2, Z) \in \mathring{k}[Y, Z]$, which has constant term $F(1, 0) = \frac{1}{16}$, can be written as

$$F(1 + Y + Z^2, Z) = \frac{1}{16} + YF_1 + ZF_2$$

with $F_1, F_2 \in \mathring{k}[Y, Z]$. Then

$$YZ + AF(1 + Y + Z^2, Z) = (Y + AF_2)\,(Z + AF_1) + A\left(\frac{1}{16} - AF_1F_2\right),$$

and $\frac{1}{16} - AF_1F_2$ is a unit in $\mathring{A}\langle Y, Z\rangle$, since $|AF_1F_2|_{\sup} < 1 = \left|\frac{1}{16}\right|$. Furthermore, we have $|AF_1|_{\sup} < 1$ and $|AF_2|_{\sup} < 1$. A standard approximation argument (see Corollary 5.1.3/8 and its proof) shows that $Y + AF_2$ and $-\left(\frac{1}{16} - AF_1F_2\right)^{-1}(Z + AF_1)$ can be used as new indeterminates ζ, η over A

in $A\langle Y, Z\rangle$. Hence we get

$$C^0 = \operatorname{Sp} A\langle \zeta, \eta\rangle/(\zeta\eta - \Lambda);$$

i.e., C^0 is an annulus over $\operatorname{Sp} A$. Since ζ differs from Y only by the term ΛF_2, which has supremum norm < 1, and since η — up to a unit in $\mathring{A}\langle Y, Z\rangle$ — differs from Z only by ΛF_1, which also has supremum norm < 1, we see that the intersection $C^0 \cap C^1$ is as claimed. □

In the situation of Lemma 1, we denote the two components of $C^0 \cap C^1$ by C^0_+ and C^0_- when thinking of them as affinoid subdomains of C^0. Viewing them as subdomains of C^1, we use the notation C^1_+ instead of C^0_-, and C^1_- instead of C^0_+. For $\mu \in \mathbb{Z}$, let $V^{2\mu}$ and $V^{2\mu+1}$ be different copies of C^0 and C^1, respectively. The affinoid subdomains of $V^{2\mu}$ and $V^{2\mu+1}$ corresponding to the subdomains $C^0_+, C^0_- \subset C^0$ and $C^1_+, C^1_- \subset C^1$ are denoted by $V^{2\mu}_+$, $V^{2\mu}_-$ and $V^{2\mu+1}_+$, $V^{2\mu+1}_-$, respectivtly. Then one can construct an analytic variety V over $\operatorname{Sp} A$, admitting $V = \bigcup_{\nu\in\mathbb{Z}} V^\nu$ as admissible affinoid covering, by pasting $V^{2\mu}$ to $V^{2\mu+1}$ via the canonical isomorphism $V^{2\mu}_+ \xrightarrow{\sim} V^{2\mu+1}_-$ and by pasting $V^{2\mu+1}$ to $V^{2\mu+2}$ via the canonical isomorphism $V^{2\mu+1}_+ \xrightarrow{\sim} V^{2\mu+2}_-$, where ν ranges in $\mathbb{Z}$. There is a canonical morphism $V \to C$, and one obtains C from V by identifying all subvarieties $V^{2\mu} \subset V$ as well as all subvarieties $V^{2\mu+1} \subset V$. We set

$W^n_m := \bigcup_{\nu=m+1}^{n} V^\nu$ for $m, n \in \mathbb{Z}$, $m < n$.

Lemma 2. *W^n_m is an annulus over* $\operatorname{Sp} A$ *isomorphic to* $\operatorname{Sp} A\langle \zeta, \Lambda^{n-m}\zeta^{-1}\rangle$. *More precisely, there exists an analytic function f on V, unique up to a unit in $\mathring{A}$, such that, for all $m, n \in \mathbb{Z}$ with $m < n$, the function $\Lambda^n f$ restricts to a coordinate function of W^n_m.*

Proof. First of all, one concludes from the separatedness of C that V is separated. Then one can apply Proposition 9.7.1/4 and show by induction that W^n_m is an annulus over $\operatorname{Sp} A$ isomorphic to $\operatorname{Sp} A\langle \zeta, \Lambda^{n-m}\zeta^{-1}\rangle$. Furthermore, the assertion about coordinate functions in Proposition 9.7.1/4 shows that, for $l \in \mathbb{N}$, there exists a function f_l on W^l_{-l} such that $\Lambda^n f_l|_{W^n_m}$ is a coordinate function of W^n_m for all $m, n \in \mathbb{Z}$ with $-l \leqq m < n \leqq l$. Therefore, we necessarily have

$$\operatorname{ord}_{\Lambda^l f_l}(\Lambda^l f_{l+1}|_{W^l_{-l}}) = 1,$$

and we may assume that the dominant term of the Laurent series

$$\Lambda^l f_{l+1}|_{W^l_{-l}} = \sum_{\nu=-\infty}^{+\infty} a^{(l)}_\nu (\Lambda^l f_l)^\nu$$

has coefficient $a^{(l)}_1 = 1$ (cf. Proposition 9.7.1/2). By assertion (iii) of Lemma 9.7.1/1, we have $|a^{(l)}_\nu|_{\sup} < 1$ for $\nu > 1$ and $|\Lambda^{2l(\nu-1)} a^{(l)}_\nu|_{\sup} < 1$ for $\nu < 1$. Hence, for $l \geqq j$, we can estimate the supremum norm of $f_l|_{W^j_{-j}} - f_{l+1}|_{W^j_{-j}}$ as follows:

$$|f_l - f_{l+1}|_{W^j_{-j}} = \left| \sum_{\nu>1} a_\nu^{(l)} \Lambda^{(l-j)\nu-l} (\Lambda^j f_l)^\nu + \sum_{\nu<1} a_\nu^{(l)} \Lambda^{(l+j)\nu-l} (\Lambda^{-j} f_l)^\nu \right|_{W^j_{-j}}$$

$$\leqq \max \left(\max_{\nu>1} |a_\nu^{(l)} \Lambda^{(l-j)\nu-l}|_{\sup}, \max_{\nu<1} |\Lambda^{2l(\nu-1)} a_\nu^{(l)} \Lambda^{-(l-j)\nu+l}|_{\sup} \right)$$

$$\leqq \max(|\Lambda^{l-2j}|_{\sup}, |\Lambda^l|_{\sup}) \leqq \alpha^{l-2j}$$

where $\alpha = |\Lambda|_{\sup} < 1$. Thus, for each $j \in \mathbb{N}$, the sequence $(f_l|_{W^j_{-j}})_{l\geqq j}$ converges on W^j_{-j}, and the sequence $(f_l)_{l\geqq 1}$ has a well-defined limit f on V. That f gives rise to coordinate functions of the W^n_m as required can be verified using the criterion of Proposition 9.7.1/2.

Finally we have to show that f is unique up to a unit in $\mathring{A}$. Let h be any other analytic function on V such that $\Lambda^n h$ always restricts to a coordinate function of W^n_m. We can write

$$h = \sum_{\nu=-\infty}^{+\infty} a_\nu f^\nu$$

with coefficients $a_\nu \in A$, where the series is convergent on each W^n_m, i.e., on the whole variety V. It follows from Lemma 9.7.1/1 that one term of this series, say of index ν_0, is dominant on each W^n_m and hence on V. By Proposition 9.7.1/2, we must have $\nu_0 = 1$, since $\nu_0 = -1$ can be ruled out easily. Also it follows that a_1 is a unit in $\mathring{A}$. Thus it remains to show that $a_\nu = 0$ for $\nu \neq 1$. Applying assertion (iii) of Lemma 9.7.1/1 to the variety W^l_{-l} for $l \in \mathbb{N}$ and to the series

$$h = \sum_{\nu=-\infty}^{+\infty} a_\nu \Lambda^{-l\nu} (\Lambda^l f)^\nu$$

yields the following estimates:

$$|a_\nu|_{\sup} \leqq |a_1 \Lambda^{l(\nu-1)}|_{\sup} \, |a_1^{-1} \Lambda^l a_\nu \Lambda^{-l\nu}|_{\sup}$$
$$\leqq |a_1|_{\sup} \, |\Lambda|_{\sup}^{l(\nu-1)} \leqq \alpha^l \quad \text{for} \quad \nu > 1, \quad \text{and}$$
$$|a_\nu|_{\sup} \leqq |a_1 \Lambda^{l(1-\nu)}|_{\sup} \, |a_1^{-1} \Lambda^l a_\nu \Lambda^{-l\nu} \Lambda^{2l(\nu-1)}|_{\sup}$$
$$\leqq |a_1|_{\sup} \, |\Lambda|_{\sup}^{l(1-\nu)} \leqq \alpha^l \quad \text{for} \quad \nu < 1,$$

where $\alpha = |\Lambda|_{\sup} < 1$. Taking the limit as $l \to \infty$, we see that $a_\nu = 0$ for $\nu \neq 1$. □

The function f can in fact be used in order to show that V is isomorphic over $\operatorname{Sp} A$ to the direct product $\mathbb{A}^* \times \operatorname{Sp} A$. For example, this assertion is immediately clear in the case where $A = k$. We consider now two automorphisms $\varphi, \psi \colon V \to V$ over $\operatorname{Sp} A$, where, for all $\nu \in \mathbb{Z}$, the restriction of φ to V^ν equals the canonical isomorphism $V^\nu \xrightarrow{\sim} V^{\nu-2}$. Similarly, ψ is obtained by pasting together the isomorphisms $V^\nu \xrightarrow{\sim} V^{\nu+2}$ given by the homomorphism

$$A\langle \Lambda^{\nu+2} f, \Lambda^{-\nu-1} f^{-1} \rangle \to A\langle \Lambda^\nu f, \Lambda^{-\nu+1} f^{-1} \rangle$$
$$f \mapsto \Lambda^{-2} f,$$

where A remains fixed. Thus ψ may be viewed as "multiplication" by Λ^{-2} and $\varphi \circ \psi\colon V \to V$ is an automorphism, leaving all subvarieties W_m^n invariant. Looking at the automorphisms $\varphi^*, \psi^*\colon \mathcal{O}(V) \to \mathcal{O}(V)$ of the ring of global analytic functions on V, we conclude from Lemma 2 that $\psi^* \circ \varphi^*(f) = af$ with a unit $a \in \mathring{A}$. Then

$$\varphi^*(f) = (\psi^*)^{-1}(af) = \Lambda^2 af.$$

Since C is obtained from V by identification via φ, we see that, in fact, C is obtained from $V^0 \cup V^1$ by identifying V^1_+ with V^0_- via φ, i.e., by "multiplication" with $\Lambda^2 a$. From this fact we derive the following result:

Theorem 3. *Each elliptic curve C_λ with $\lambda \in k$, $0 < |\lambda| < 1$, given by the equation $Y^2Z - X(X - Z)(X - \lambda Z) = 0$ in $\mathbb{P}^2$, is isomorphic to an analytic torus $\mathbb{A}^*/(q)$. More precisely, there exist cofficients $c_2, c_3, \ldots \in \mathring{k}$ with $|c_2| = 1$ such that $C_\lambda \cong \mathbb{A}^*/(q(\lambda))$, where*

$$q(\lambda) = \sum_{\nu=2}^{\infty} c_\nu \lambda^\nu .$$

Proof. Considering a fixed $\lambda \in k$, $0 < |\lambda| < 1$, we can set $A := k$ and $\Lambda := \lambda$. Then our considerations above show that $C_\lambda \cong \mathbb{A}^*/(q)$ for some $q \in k$ satisfying $|q| = |\lambda^2|$. In order to see how q depends on λ, we choose $\varepsilon, \alpha \in |k|$ with $0 < \varepsilon < \alpha < 1$ and set $A := k\langle \alpha^{-1}\Lambda, \varepsilon\Lambda^{-1}\rangle$, where Λ is viewed as an indeterminate over k. Then, pointwise, $\mathrm{Sp}\, A$ may be interpreted as the annulus $\{\lambda \in k; \varepsilon \leqq |\lambda| \leqq \alpha\}$, and C as the family of elliptic curves C_λ with $\lambda \in \mathrm{Sp}\, A$. By what we proved above, there exists a unit $a \in \mathring{A}$ such that C is the quotient of V by "multiplication" with $\Lambda^2 a$. Therefore, we get $C_\lambda \cong \mathbb{A}^*/(q(\lambda))$ with $q(\lambda) = \lambda^2 a(\lambda)$. If

$$a = \sum_{\nu=-\infty}^{+\infty} d_\nu \Lambda^\nu$$

is the Laurent expansion of a, the coefficients $d_\nu \in k$ are independent of the choice of ε and α as a review of our construction shows. Thus the above series converges on the set $\{\lambda \in k; 0 < |\lambda| < 1\}$. Since $|a(\lambda)| = 1$ always, we necessarily have $d_\nu = 0$ for $\nu < 0$, $|d_\nu| \leqq 1$ for $\nu \geqq 0$, and $|d_0| = 1$ (use, for example, Lemma 9.7.1/1). Therefore $q(\lambda) = \sum_{\nu=0}^{\infty} d_\nu \lambda^{\nu+2}$, as stated. □

The power series $q(\lambda)$ defines a surjective map from $B^* := \{\lambda \in k; 0 < |\lambda| < 1\}$ onto itself (for example, use Proposition 5.1.3/1). Therefore, each analytic torus $\mathbb{A}^*/(q)$ occurs as an elliptic curve C_λ with $\lambda \in B^*$. However, λ is not uniquely determined by q. Namely, an application of the Weierstrass Preparation Theorem shows that the series $q(\lambda)$ assumes all its non-zero values twice on B^*. Since two analytic tori $\mathbb{A}^*/(q)$ and $\mathbb{A}^*/(q')$ cannot be isomorphic unless $q = q'$ (this will be proved in the remainder of this section; see Proposition 5 below), it follows that each $q \in B^*$ uniquely determines two different numbers $\lambda_1, \lambda_2 \in B^*$ such that $C_{\lambda_1} \cong \mathbb{A}^*/(q) \cong C_{\lambda_2}$.

Lemma 4. *Let U be a connected affinoid subdomain of the unit disc $\mathbb{B}^1$, and let $\varphi\colon U \to \mathbb{A}^*/(q)$ be an analytic morphism defining U as an open subvariety of the analytic torus $\mathbb{A}^*/(q)$. If $p\colon \mathbb{A}^* \to \mathbb{A}^*/(q)$ denotes the canonical projection and if $x_0 \in U$ and $y_0 \in \mathbb{A}^*$ are points satisfying $\varphi(x_0) = p(y_0)$, then there exists a unique analytic morphism $\hat{\varphi}\colon U \to \mathbb{A}^*$ such that $\hat{\varphi}(x_0) = y_0$ and $p \circ \hat{\varphi} = \varphi$.*

Proof. Let ζ be the restriction to $\mathbb{A}^*$ of a coordinate function of $\mathbb{A}^1$. For $\alpha, \beta \in |k^*|$, $\alpha \leqq \beta$, we set

$$\hat{V}(\alpha, \beta) := \{y \in \mathbb{A}^*; \alpha \leqq |\zeta(y)| \leqq \beta\}$$

and consider $V(\alpha, \beta) := p\big(\hat{V}(\alpha, \beta)\big)$ as an open analytic subvariety of $\mathbb{A}^*/(q)$. In particular, we write

$$V_1 := V(|q|, |q|^{1/2}) \quad \text{and} \quad V_2 := V(|q|^{1/2}, 1).$$

Then $\{V_1, V_2\}$ is an admissible affinoid covering of $\mathbb{A}^*/(q)$ by annuli. We claim that there exists an $\alpha_0 \in |k^*|$ such that $\varphi(U) \cap V(\alpha_0, \alpha_0) = \emptyset$. Proceeding indirectly, we assume the contrary. For $i = 1, 2$, the intersection $\varphi(U) \cap V_i$ is an affinoid subdomain of the annulus V_i. Hence by Theorem 9.7.2/2, it is a finite disjoint union of standard subsets of V_i (where V_i is viewed as an annulus in $\mathbb{B}^1$ and where a standard subset of V_i is meant as a standard subset of $\mathbb{B}^1$ which is contained in V_i), say

$$\varphi(U) \cap V_i = U_{i1} \,\dot{\cup}\, \ldots \,\dot{\cup}\, U_{ir_i}.$$

Since $\varphi(U) \cap V(\alpha, \alpha) \neq \emptyset$ for all α, a simple consideration shows that one of the sets $U_{i1}, \ldots, U_{ir_i}$, say U_{i1}, must satisfy $U_{i1} \cap V(\alpha, \alpha) \neq \emptyset$ for all α such that $V(\alpha, \alpha)$ is contained in V_i. Furthermore, each U_{ij} for $j > 1$ must be a subset of some disc contained in $V(\alpha, \alpha)$ for some α. (In fact, one can show that $U_{ij} = \emptyset$ for $j > 1$ since U is connected.) Now $\varphi^{-1}(U_{11})$ and $\varphi^{-1}(U_{21})$ are connected affinoid subdomains of $U \subset \mathbb{B}^1$. Hence the intersection

$$\varphi^{-1}(U_{11}) \cap \varphi^{-1}(U_{21}) \cong U_{11} \cap U_{21}$$

is connected (Corollary 9.7.2/3). However, this is impossible, since

$$U_{11} \cap U_{21} \subset V(|q|^{1/2}, |q|^{1/2}) \,\dot{\cup}\, V(1, 1)$$

and since, by our construction, $U_{11} \cap U_{21} \cap V(\alpha, \alpha) \neq \emptyset$ for $\alpha = |q|^{1/2}$ and $\alpha = 1$. Consequently, there must exist an $\alpha_0 \in |k^*|$ such that $\varphi(U) \cap V(\alpha_0, \alpha_0) = \emptyset$, and, in fact, one can find $\alpha, \beta \in |k^*|$ with $\alpha \leqq \beta$ and $\alpha\beta^{-1} > |q|$ such that $\varphi(U) \subset V(\alpha, \beta)$. We may assume that $y_0 \in \hat{V}(\alpha, \beta)$. Then it is obvious that φ can be lifted as required, since the projection $p\colon \mathbb{A}^* \to \mathbb{A}^*/(q)$ restricts to an isomorphism $\hat{V}(\alpha, \beta) \xrightarrow{\sim} V(\alpha, \beta)$. That any lifting $\hat{\varphi}\colon U \to \mathbb{A}^*$ mapping x_0 to y_0 is unique follows from the connectedness of U and the fact that p is a local isomorphism. □

Proposition 5. *Let $\tau\colon \mathbb{A}^*/(q) \to \mathbb{A}^*/(q')$ be an analytic isomorphism of analytic tori. Then $q = q'$.*

Proof. We consider the diagram

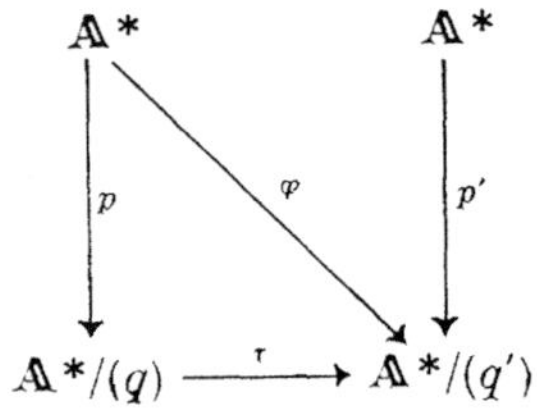

where p and p' denote the canonical projections and where $\varphi := \tau \circ p$. Choosing points $x, y \in \mathbb{A}^*$ with $\varphi(x) = p'(y)$, it follows from Lemma 4 that φ lifts uniquely to a map $\hat{\varphi}\colon \mathbb{A}^* \to \mathbb{A}^*$ satisfying $\hat{\varphi}(x) = y$. Namely, the corresponding fact is true for the restriction of φ to any annulus $\hat{V}\,(|q|^{\frac{n}{2}}, |q|^{\frac{n-1}{2}})$, where n ranges over $\mathbb{Z}$ and where the notation $\hat{V}$ is as in the proof of Lemma 4. Since $\tau^{-1} \circ p'$ can be lifted similarly, it follows that $\hat{\varphi}\colon \mathbb{A}^* \to \mathbb{A}^*$ is an analytic isomorphism.

Let ζ be the restriction to $\mathbb{A}^*$ of a coordinate function on $\mathbb{A}^1$. If $\hat{\varphi}^*\colon \mathcal{O}(\mathbb{A}^*) \to \mathcal{O}(\mathbb{A}^*)$ is the endomorphism of the ring of global analytic functions on $\mathbb{A}^*$ corresponding to $\hat{\varphi}$, it follows from Lemma 9.7.1/1 that $\hat{\varphi}^*(\zeta) = c\zeta^m$ for some constant $c \in k^*$ and for some $m \in \mathbb{Z}$. Since $\hat{\varphi}$ is an isomorphism, only the case $m = 1$ or $m = -1$ is possible. Thus the map $\hat{\varphi}\colon \mathbb{A}^* \to \mathbb{A}^*$ is described with respect to ζ by $x \mapsto cx^{\pm 1}$. Since $\hat{\varphi}$ is a lifting of τ, we have

$$cq^{\pm 1} = \hat{\varphi}(1 \cdot q) = \hat{\varphi}(1) \cdot q'^n = cq'^n$$

for some $n \in \mathbb{Z}$; i.e., q is a (necessarily positive) power of q'. Similarly, q' is a positive power of q. However, this can only be true when $q = q'$. □

Finally, we want to discuss elliptic curves in terms of their j-invariants, which classify these curves up to isomorphism. Since char $\tilde{k} \neq 2$ implies char $k \neq 2$, each elliptic curve over k can be written in Legendre form, i.e., it is isomorphic to some curve C_λ with $\lambda \in k - \{0, 1\}$. Furthermore, it is only necessary to consider the curves C_λ with $|\lambda| \leq 1$, as a change of coordinates in $\mathbb{P}^2$ shows. The j-invariant of C_λ is given by

$$j = 2^8 \frac{(1 - \lambda + \lambda^2)^3}{\lambda^2(1 - \lambda)^2}.$$

Theorem 6. *For each $j \in k$, $|j| > 1$, there exists a unique $q \in k$, $0 < |q| < 1$, such that the elliptic curve with j-invariant j is isomorphic to the analytic torus $\mathbb{A}^*/(q)$. The correspondence $j \mapsto q$ is given by a series $q = \sum_{\nu=1}^{\infty} d_\nu j^{-\nu}$ with coefficients $d_\nu \in \mathring{k}$, where $|d_1| = 1$. In particular, the elliptic curves with absolute j-invariant $|j| > 1$ correspond bijectively to the analytic tori $\mathbb{A}^*/(q)$ with $0 < |q| < 1$.*

Proof. The elliptic curve C_λ with $\lambda \in \mathring{k} - \{0, 1\}$ has absolute j-invariant $|j| > 1$ if and only if $0 < |\lambda| < 1$ or $0 < |1 - \lambda| < 1$. Since $C_\lambda \cong C_{1-\lambda}$ (as a change of coordinates in $\mathbb{P}^2$ or a computation of j-invariants shows), the elliptic curves with $|j| > 1$ are already represented by the curves C_λ with $\lambda \in B^* := \{a \in k; 0 < |a| < 1\}$. Furthermore, we know that each such curve C_λ is isomorphic to a unique analytic torus $\mathbb{A}^*/(q(\lambda))$, where $\lambda \mapsto q(\lambda)$ is a surjective map $B^* \to B^*$ which assumes all its values twice. The map $B^* \to B^*$, $\lambda \mapsto \frac{1}{j(\lambda)}$, is also surjective and assumes all its values twice, since, for $\lambda \in B^*$,

$$\frac{1}{j(\lambda)} = 2^{-8}\lambda^2(1 - \lambda)^2 (1 - \lambda + \lambda^2)^{-3}$$

can be written as a power series in λ over $\mathring{k}$, starting with a term $c\zeta^2$, where $|c| = 1$. Because $j(\lambda_1) = j(\lambda_2)$ implies $q(\lambda_1) = q(\lambda_2)$, it is plausible that the elliptic curves with absolute j-invariant $|j| > 1$ correspond bijectively to the analytic tori $\mathbb{A}^*/(q)$ with $q \in B^*$; i.e., that the map $\{a \in k; |a| > 1\} \to B^*$, $j \mapsto q(j)$, is bijective.

We will give a precise verification of this fact by showing that $q(j)$ is a power series in j^{-1} as stated. The polynomial $1 + \zeta$ has a square root in the formal power series ring $\mathring{k}[[\zeta]]$. Namely, the coefficients of a series $f \in \mathring{k}[[\zeta]]$ satisfying $f^2 = 1 + \zeta$ can be determined recursively (use the fact that char $\tilde{k} \neq 2$). Therefore we can find a square root of $1 - \lambda + \lambda^2$ in $\mathring{k}[[\lambda]]$ (where λ is viewed as an indeterminate), and we can consider

$$\lambda' := 2^{-4}\lambda(1 - \lambda)\left(\sqrt{1 - \lambda + \lambda^2}\right)^{-3}$$

as a new parameter on B^*. Then $\frac{1}{j(\lambda)} = \lambda'^2$, and the series $q(\lambda)$ changes into a series $q(\lambda')$ having coefficients in $\mathring{k}$ and starting with a term $c\lambda'^2$ of degree 2, where $|c| = 1$. However $q(\lambda') = q(-\lambda')$, since q depends only on j. Therefore all odd terms of $q(\lambda')$ vanish, so that $q(\lambda')$ is in fact a power series in j^{-1}, as stated. □

The elliptic curves with absolute j-invariant $|j| \leqq 1$ are referred to as the elliptic curves with *good reduction*. They are so named, because any such elliptic curve induces an elliptic curve $\tilde{C}$ over the residue field $\tilde{k}$ of k simply by reducing the coefficients of a suitable equation for C. Namely, all elliptic curves are represented by the curves C_λ, where $0 < |\lambda| \leqq 1$. Thus the elliptic curves with absolute j-invariant $|j| \leqq 1$ are represented by the curves C_λ, where $|\lambda| = 1$ and $|1 - \lambda| = 1$. Reducing the coefficients of the equation $Y^2Z - X(X - Z)(X - \lambda Z) = 0$ for C_λ leads to the curve $Y^2Z - X(X - Z)(X - \tilde{\lambda}Z) = 0$, which is an elliptic curve in $\mathbb{P}^2_{\tilde{k}}$, since $\tilde{\lambda} \neq 0,1$. However if $|\lambda| < 1$ (or $|1 - \lambda| < 1$), we saw at the beginning of this section that reducing the coefficients of the equation for C_λ leads to a singular curve $\tilde{C}_\lambda$, a phenomenon called *bad reduction*. Thus, we showed that the analytic tori are just the elliptic curves having bad reduction. See [4], § 1 for a geometric argument which shows that elliptic curves with good reduction cannot be isomorphic to analytic tori.

Bibliography

1. Artin, M.: Grothendieck topologies. Notes on a seminar by M. Artin, Harvard University (1962).
2. Bosch, S.: Orthonormalbasen in der nichtarchimedischen Funktionentheorie. Manuscripta math. **1**, 35–57 (1969).
3. Bosch, S.: Multiplikative Untergruppen in abeloiden Mannigfaltigkeiten. Math. Ann. **239**, 165–183 (1979).
4. Bosch, S.: Formelle Standardmodelle hyperelliptischer Kurven. Math. Ann. **251**, 19–42 (1980).
5. Bourbaki, N.: Eléments de mathématique. Espaces vectoriels topologiques, Chap. I. Hermann, Paris (1953).
6. Bourbaki, N.: Eléments de mathématique. Topologie générale, Chap. IX. Hermann, Paris (1958).
7. Deuring, M.: Die Typen der Multiplikatorenringe elliptischer Funktionenkörper. Abh. Math. Sem. Univ. Hamburg **14**, 197–272 (1941).
8. Fieseler, K.-H.: Zariski's Main Theorem für affinoide Kurven. Math. Ann. **251**, 97–110 (1980).
9. Gerritzen, L.: Erweiterungsendliche Ringe in der nichtarchimedischen Funktionentheorie. Inventiones math. **2**, 178–190 (1966).
10. Gerritzen, L.: Die Norm der gleichmäßigen Konvergenz auf reduzierten affinoiden Algebren. J. Reine Angew. Math. **231**, 114–120 (1968).
11. Gerritzen, L.: Zur nichtarchimedischen Uniformisierung von Kurven. Math. Ann. **210**, 321–337 (1974).
12. Gerritzen, L. and Grauert, H.: Die Azyklizität der affinoiden Überdeckungen. Global Analysis, Papers in Honor of K. Kodaira, 159–184. University of Tokyo Press, Princeton University Press (1969).
13. Gerritzen, L. and Güntzer, U.: Über Restklassennormen auf affinoiden Algebren. Inventiones math. **3**, 71–74 (1967).
14. Gerritzen, L. and van der Put, M.: Schottky groups and Mumford curves. Lecture Notes in Math. **817**. Springer, Berlin–Heidelberg–New York (1980).
15. Grauert, H. and Remmert, R.: Nichtarchimedische Funktionentheorie. Weierstraß-Festschrift, Wissenschaftl. Abh. Arbeitsgemeinschaft für Forschung des Landes Nordrhein-Westfalen **33**, 393–476 (1966).
16. Grauert, H. and Remmert, R.: Über die Methode der diskret bewerteten Ringe in der nicht-archimedischen Analysis. Inventiones math. **2**, 87–133 (1966).
17. Gruson, L.: Fibrés vectoriels sur un polydisque ultramétrique. Ann. scient. Ec. Norm. Sup., 4e série, **1**, 45–89 (1968).
18. Güntzer, U.: Modellringe in der nichtarchimedischen Funktionentheorie. Indag. math. **29**, 334–342 (1967).
19. Güntzer, U.: The norm of uniform convergence on the k-algebraic maximal spectrum of an algebra over a non-archimedean valuation field k. Bull. Soc. math. France, Mémoire **39/40**, 101–121 (1974).
20. Güntzer, U. and Remmert, R.: Bewertungstheorie und Spektralnorm. Abh. Math. Sem. Univ. Hamburg **38**, 32–48 (1972).

21. GÜNTZER, U. and REMMERT, R.: Kahl bewertete Ringe in der nicht-archimedischen Analysis. Manuscripta math. **8**, 33–38 (1973).
22. KAPLANSKY, I.: Maximal fields with valuations. Duke Math. J. **9**, 303–321 (1942).
23. KIEHL, R.: Der Endlichkeitssatz für eigentliche Abbildungen in der nichtarchimedischen Funktionentheorie. Inventiones math. **2**, 191–214 (1967).
24. KIEHL, R.: Theorem A und Theorem B in der nichtarchimedischen Funktionentheorie. Inventiones math. **2**, 256–273 (1967).
25. KIEHL, R.: Die analytische Normalität affinoider Ringe. Arch. Math. **18**, 479–484 (1967).
26. KÖPF, U.: Über eigentliche Familien algebraischer Varietäten über affinoiden Räumen. Schriftenreihe Math. Inst. Univ. Münster, 2. Serie. Heft **7** (1974).
27. MUMFORD, D.: An analytic construction of degenerating curves over complete local fields. Compositio Math. **24**, 129–174 (1972).
28. NAGATA, M.: Local rings. Interscience Publishers, New York (1962).
29. PUT, M. VAN DER: Espaces de Banach non archimédiens. Bull. Soc. math. France **97**, 309–320 (1969).
30. PUT, M. VAN DER: Non-archimedean function algebras. Indag. math. **33**, 60–77 (1971).
31. PUT, M. VAN DER and VAN TIEL, J.: Espaces nucléaires non archimédiens. Indag. math. **29**, 556–561 (1967).
32. RAYNAUD, M.: Variétés abéliennes et géométrie rigide. Actes, Congrès intern. math. Tome **1**, 473–477 (1970).
33. RAYNAUD, M.: Géométrie analytique rigide d'apres Tate, Kiehl, ... Bull. Soc. math. France, Mémoire **39/40**, 319–327 (1974).
34. ROQUETTE, P.: Analytic theory of elliptic functions over local fields. Hamburger Math. Einzelschriften. Neue Folge, Heft **1** (1970).
35. SCHMIDT, F. K.: Über die Erhaltung der Kettensätze der Idealtheorie bei beliebigen endlichen Körpererweiterungen. Math. Z. **41**, 443–450 (1936).
36. SERRE, J.-P.: Faisceaux algébriques cohérents. Ann. Math. **61**, 197–278 (1955).
37. TATE, J.: Rigid analytic spaces. Private notes (1962). Reprinted in Inventiones math. **12**, 257–289 (1971).
38. WAERDEN, B. VAN DER: Algebra II, 5th ed. Springer, Berlin–Göttingen–Heidelberg (1967).
39. ZARISKI, O. and SAMUEL, P.: Commutative algebra (Vol. II). Van Nostrand, Princeton (1960).

Further Publications

The following books all deal with non-Archimedean analysis. We have tried to classify them according to their main intentions. [D] means differential equations, [F] means foundations (p-adic numbers, valuation theory, etc.), [FA] means functional analysis, [N] means number theory, and [RG] means rigid analytic geometry.

[F] AMICE, Y.: Les nombres p-adiques. Presses Universitaires de France, Paris (1975).
[F] BACHMAN, G.: Introduction to p-adic numbers and valuation theory. Academic Press, New York–London (1964).
[RG] BERGER, R., KIEHL, R., KUNZ, E., and NASTOLD, H.-J.: Differentialrechnung in der analytischen Geometrie. Lecture Notes in Math. **38**. Springer, Berlin–Heidelberg–New York (1967).
[F] BOURBAKI, N.: Eléments de mathématique. Algèbre commutative, Chap. VI. Hermann, Paris (1964).

[D] Dwork, B.: Lectures on p-adic differential equations. Springer, New York—Heidelberg—Berlin (1982).
[F] Endler, O.: Valuation theory. Springer, Berlin—Heidelberg—New York (1972).
[RG] Fresnel, J. and van der Put, M.: Géométrie analytique rigide et applications. Birkhäuser, Boston—Basel—Stuttgart (1981).
[N] Iwasawa, K.: Lectures on p-adic L-functions. Princeton University Press, Princeton (1972).
[F] Koblitz, N.: p-adic numbers, p-adic analysis, and zeta-functions. Springer, New York—Heidelberg—Berlin (1977).
[N] Koblitz, N.: p-adic analysis: a short course on recent work. Cambridge University Press, Cambridge (1980).
[F] Mahler, K.: p-adic numbers and their functions, 2nd ed. Cambridge University Press, Cambridge (1981).
[FA] Monna, A. F.: Analyse non-archimédienne. Springer, Berlin—Heidelberg—New York (1970).
[FA] Narici, L., Beckenstein, E. and Bachman, G.: Functional analysis and valuation theory. Marcel Dekker Inc., New York (1971).
[F] Ribenboim, P.: Théorie des valuations. Les Presses de l'Université de Montréal. Montréal (1964/68).
[FA] Rooij, A. C. M. van: Non-Archimedean functional analysis. Marcel Dekker Inc., New York—Basel (1978).
[F] Schilling, O. F. G.: The theory of valuations. AMS, New York (1950).

In the literature, there is an abundance of articles which are related to non-Archimedean analysis, and it is nearly impossible to specify them all or even a few representative ones (see the bibliographies of the books mentioned above). Therefore we rsetrict ourselves to *rigid analytic geometry* and supplement our list of publications by some articles which, as main object of study, are dealing with special problems and methods in this field. These articles are not referred to elsewhere in the book.

Bartenwerfer, W.: Einige Fortsetzungssätze in der p-adischen Analysis. Math. Ann. **185**, 191—210 (1970).

Bartenwerfer, W.: Der allgemeine Kontinuitätssatz für k-meromorphe Funktionen im Dizylinder. Math. Ann. **191**, 196—234 (1971).

Bartenwerfer, W.: Der Kontinuitätssatz für reindimensionale k-affinoide Räume. Math. Ann. **193**, 139—170 (1971).

Bartenwerfer, W.: Ein nichtarchimedisches Analogon zum Kugelsatz für analytische Mengen. J. Reine Angew. Math. **255**, 44—59 (1972).

Bartenwerfer, W.: Die nichtarchimedische Version des Rothsteinschen Kontinuitätssatzes. Math. Ann. **202**, 89—134 (1973).

Bartenwerfer, W.: Die Fortsetzung holomorpher und meromorpher Funktionen in eine k-holomorphe Hyperfläche hinein. Math. Ann. **212**, 331—358 (1975).

Bartenwerfer, W.: Ein Konvergenzsatz für 0-Koketten affinoider Funktionen. Nachr. Akad. Wiss. Göttingen, 43—72 (1976).

Bartenwerfer, W.: Die erste „metrische" Kohomologiegruppe glatter affinoider Räume. Indag. math. **40**, 1—14 (1978).

Bartenwerfer, W.: Die höheren metrischen Kohomologiegruppen affinoider Räume. Math. Ann. **241**, 11—34 (1979).

Bartenwerfer, W.: Die Lösung des nichtarchimedischen Corona-Problems für beliebige Dimension. J. Reine Angew. Math. **319**, 133—141 (1980).

BARTENWERFER, W.: k-holomorphe Vektorraumbündel auf offenen Polyzylindern. J. Reine Angew. Math. **326**, 214—220 (1981).

BOSCH, S.: k-affinoide Gruppen. Inventiones math. **10**, 128—176 (1970).

BOSCH, S.: k-affinoide Tori. Math. Ann. **192**, 1—16 (1971).

BOSCH, S.: Homogene Räume k-affinoider Gruppen. Inventiones math. **19**, 165—218 (1973).

BOSCH, S.: Rigid analytische Gruppen mit guter Reduktion. Math. Ann. **223**, 193—205 (1976).

BOSCH, S.: Zur Kohomologietheorie rigid analytischer Räume. Manuscripta math. **20**, 1—27 (1977).

BOSCH, S., DWORK, B., and ROBBA, P.: Un théorème de prolongement pour des fonctions analytiques. Math. Ann. **252**, 165—173 (1980).

BOSCH, S.: A rigid analytic version of M. Artin's theorem on analytic equations. Math. Ann. **255**, 395—404 (1981).

BRÜSKE, R.: Saturierte Moduln und André-Kohomologie in der affinoiden Geometrie. Schriftenreihe Math. Inst. Univ. Münster, 2. Serie. Heft **11** (1976).

DENNEBERG, D.: Divisoren in eindimensionalen affinoiden und affinen Räumen. Math. Ann. **181**, 137—151 (1969).

GERRITZEN, L.: On one-dimensional affinoid domains and open immersions. Inventiones math. **5**, 106—119 (1968).

GERRITZEN, L.: Über Endomorphismen nichtarchimedischer holomorpher Tori. Inventiones math. **11**, 27—36 (1970).

GERRITZEN, L.: On multiplication algebras of Riemann matrices. Math. Ann. **194**, 109—122 (1971).

GERRITZEN, L.: On non-archimedean representations of abelian varieties. Math. Ann. **196**, 323—346 (1972).

GERRITZEN, L.: Periode und Index eines prinzipal-homogenen Raumes über gewissen abelschen Varietäten. Manuscripta math. 8, 131—142 (1973).

GERRITZEN, L.: Zerlegungen der Picard-Gruppe nichtarchimedischer holomorpher Räume. Compositio Math. **35**, 23—38 (1977).

GERRITZEN, L.: Unbeschränkte Steinsche Gebiete von $\mathbb{P}_1$ und nichtarchimedische automorphe Formen. J. Reine Angew. Math. **297**, 21—34 (1978).

GERRITZEN, L.: Zur analytischen Beschreibung des Raumes der Schottky-Mumford-Kurven. Math. Ann. **255**, 259—271 (1981).

GOSS, D.: The algebraist's upper half-plane. Bull. Am. Math. Soc. New Ser. **2**, 391—415 (1980).

GRAUERT, H.: Affinoide Überdeckungen eindimensionaler affinoider Räume. Publ. Math. IHES **34** (1968).

GÜNTZER, U.: Zur Funktionentheorie einer Veränderlichen über einem vollständigen nichtarchimedischen Grundkörper. Arch. Math. **17**, 415—431 (1966).

GÜNTZER, U.: Über den Zusammenhang zwischen affinoiden Mengen und ihren affinen Modellen. Math. Ann. **181**, 97—102 (1969).

GÜNTZER, U.: Strikt konvergente Laurent-Reihen über nicht-archimedisch normierten vollständigen Ringen. Compositio Math. **21**, 21—35 (1969).

GÜNTZER, U.: Valuation theory and spectral norm on non-Archimedean function algebras. Proceedings of the 4th Session of the Greek Math. Soc. 54—68 (1972).

GÜNTZER, U.: Transitivity theorems for the non-Archimedean spectral semi-norm of ring extensions. The Greek Math. Soc., C. Carathéodory symposium 1973, 173—195.

HEINRICH, E.: Die Picardgruppen affinoider Algebren und ihrer affinoiden Reduktionen. Arch. Math. **31**, 146—150 (1978).

HERRLICH, F.: Die Ordnung der Automorphismengruppe einer p-adischen Schottkykurve. Math. Ann. **246**, 125—130 (1980).

HERRLICH, F.: Endlich erzeugbare p-adische diskontinuierliche Gruppen. Arch. Math. **35**, 505—515 (1980).

KIEHL, R.: Die de Rham Kohomologie algebraischer Mannigfaltigkeiten über einem bewerteten Körper. Publ. Math. IHES **33** (1967).

KIEHL, R.: Analytische Familien affinoider Algebren. S.-Ber. Heidelberger Akad. Wiss., 25—49 (1968).

KIEHL, R.: Ausgezeichnete Ringe in der nichtarchimedischen analytischen Geometrie. J. Reine Angew. Math. **234**, 89—98 (1969).

LAZARD, M.: Les zéros des fonctions analytiques d'une variable sur un corps valué complet. Publ. Math. IHES **14** (1962).

LÜTKEBOHMERT, W.: Steinsche Räume in der nichtarchimedischen Funktionentheorie. Schriftenreihe Math. Inst. Univ. Münster, 2. Serie. Heft **6** (1973).

LÜTKEBOHMERT, W.: Der Satz von Remmert-Stein in der nichtarchimedischen Funktionentheorie. Math. Z. **139**, 69—84 (1974).

LÜTKEBOHMERT, W.: Fortsetzbarkeit k-analytischer Mengen. Math. Ann. **217**, 131—143 (1975).

LÜTKEBOHMERT, W.: Ein Kontinuitätssatz für k-holomorphe Mengen. Manuscripta math. **18**, 257—272 (1976).

LÜTKEBOHMERT, W.: Fortsetzbarkeit k-meromorpher Funktionen. Math. Ann. **220**, 273—284 (1976).

LÜTKEBOHMERT, W.: Vektorraumbündel über nichtarchimedischen holomorphen Räumen. Math. Z. **152**, 127—143 (1977).

LÜTKEBOHMERT, W.: Ein globaler Starrheitssatz für Mumfordkurven. J. Reine Angew. Math. **340**, 118—139 (1983).

MEHLMANN, F.: Ein Beweis für einen Satz von Raynaud über flache Homomorphismen affinoider Algebren. Schriftenreihe Math. Inst. Univ. Münster, 2. Serie. Heft **19** (1981).

MORITA, Y.: On the induced h-structure on an open subset of the rigid analytic space $\mathbb{P}^1(k)$. Math. Ann. **242**, 47—58 (1979).

NASTOLD, H.-J.: Lokale nichtarchimedische Funktionentheorie. Math. Ann. **164**, 213—218 (1966).

PUT, M. VAN DER: La conjecture de la Couronne en analyse complexe et p-adique. Sem. Theor. Nombres 1979—1980, Exposé No. 6 (1980).

PUT, M. VAN DER: The class group of a one-dimensional affinoid space. Ann. Inst. Fourier **30**, No. 4, 155—164 (1980).

PUT, M. VAN DER: Essential singularities of rigid analytic functions. Indag. Math. **43**, 423—429 (1981).

PUT, M. VAN DER: Cohomology on affinoid spaces. Compositio Math. **45**, 165—198 (1982).

REMMERT, R.: Algebraische Aspekte in der nichtarchimedischen Analysis. Proceedings of a conference on local fields, 86—117. Springer, Berlin—Heidelberg—New York (1967).

Glossary of Notations

$\mathbb{N}$	positive integers
$\mathbb{Q}$	rational numbers
$\mathbb{R}$	real numbers
$\mathbb{R}_+$	non-negative real numbers
k^*	multiplicative group of a field k
G	abelian group, 9
$\lvert\ \ \rvert$	ultrametric function on G, 9
$G^0(r)$	elements of G of value $\leqq r$, 10
$G^\vee(r)$	elements of G of value $< r$, 10
$G^\sim(r)$	residue group, 10
$\ker \lvert\ \ \rvert$	kernel of an ultrametric function $\lvert\ \ \rvert$, 10
$\lvert G \rvert$	value set of G under $\lvert\ \ \rvert$, 10
v	filtration on G, 11
$(G, \lvert\ \ \rvert), G$	semi-normed group, 11, 12
$d(x, y)$	distance between x and y, 12
$x \underset{r}{\sim} y$	x and y r-near, 12
$B^+(a, r)$	ball with center a and radius r, with circumference, 12
$B^-(a, r)$	ball with center a and radius r, without circumference, 12
$S(a, r)$	sphere with center a and radius r, 13
$\varphi^\sim(r)$	residue homomorphism induced by φ, 14
$\lvert a, H \rvert$	distance from a to H, 14
$\lvert\ \ \rvert_{\text{res}}$	residue semi-norm, 16
$\hat{G}$	completion of G, 17
$\hat{\varphi}$	"completion" of a homomorphism φ, 18
A	commutative ring with identity, 23
A^+	additive group of A, 23
$\lvert\ \ \rvert$	(semi-) norm on A, 23
$(A, \lvert\ \ \rvert), A$	(semi-) normed ring, 23
v	filtration on A, 25
$\operatorname{rad} A$	nilradical of A, 25
A_S	localization of A by a multiplicative system S, 26
$\mathfrak{N}$	category of semi-normed rings, 26
A°	"closed" unit ball in A, 26
$A^\vee$	"open" unit ball in A, 26
$A^\sim$	residue ring of A, 26
$\varphi^\sim$	residue homomorphism induced by φ, 26
$\check{A}$	topologically nilpotent elements of A, 26
$E(A)$	units of A, 27
$\mathfrak{m}$	maximal ideal in A, 27
$\mathring{A}$	power-bounded elements of A, 28
$\tilde{A}$	residue ring of A, 29
$\dot{A}$	largest subring of A containing $\check{A}$ as an ideal, 29
${}^\sim A$	invariant residue ring of A, 30
$\lvert\ \ \rvert_c$	power-multiplicative semi-norm, obtained by a smoothing procedure (multiplicative with respect to c), 33

$A[\![X]\!]$	formal power series in X over A, 34
$A[X]$	polynomials in X over A, 34
deg	degree function, 34
ord	order function, 34
$v_{\mathfrak{a}}$	$\mathfrak{a}$-adic filtration, 35
$\vert\ \vert_{\mathfrak{a}}$	$\mathfrak{a}$-adic semi-norm, 35
$A\langle X\rangle$	strictly convergent power series in X over A, 35
$\vert\ \vert'$, $\vert\ \vert$	Gauss semi-norm on $A\langle X\rangle$, 35, 37
$A\langle X_1, \ldots, X_n\rangle$	strictly convergent power series in $X_1, \ldots, X_n$ over A, 37
$M\langle X\rangle$	strictly convergent power series with coefficients in M, 37
$\mathrm{Hom}_{\mathrm{b}}(A, B)$	bounded homomorphisms from A to B, 39
$\mathrm{Der}_A(B)$	continuous A-derivations of B, 40
$\vert\ \vert$	valuation on A, 41
$(A, \vert\ \vert)$, A	valued ring, 41
$\sigma(p)$	spectral value of a monic polynomial p, 44
$A[\![Y_1, Y_2, \ldots]\!]$	formal power series in countably many variables $Y_1, Y_2, \ldots$ with coefficients in A, 46
∂A	elements of A of value 1, 53
A	normed ring, 63
$L, M, N, \ldots$	normed A-modules, 63
$e(M/A)$, e	ramification index, 66
$\mathrm{rk}_A\, M$	A-rank of M, 67
$\bigoplus_i M_i$	normed direct sum, 67
$\widehat{\bigoplus_i} M_i$	completion of $\bigoplus_i M_i$, 67
$b(\prod_i M_i)$	normed bounded direct product, 67
$c(\prod_i M_i)$	normed restricted direct product, 68
$\mathscr{L}(L, M)$	bounded A-linear maps from L to M, 69
$L \mathbin{\widehat{\otimes}_A} M$	complete tensor product of L and M, 71
$x \mathbin{\widehat{\otimes}} y$	image of $x \otimes y$ in $L \mathbin{\widehat{\otimes}_A} M$, 72
$\psi_1 \mathbin{\widehat{\otimes}} \psi_2$	complete tensor product of bounded A-linear maps ψ_1 and ψ_2, 74
M°	elements of M of value ≤ 1, 81
$M^\vee$	elements of M of value < 1, 81
$M^\sim$	residue module, 81
$f(M/A)$	residue degree of M over A, 82
A^n	normed n-fold direct sum of copies of A, 82
$e_1, \ldots, e_n$	canonical basis of A^n, 83
$A^{(I)}$	normed direct sum of copies of A, indexed by I, 83
$\{e_i\}_{i\in I}$	canonical basis of $A^{(I)}$, 84
$A^{(\infty)}$	normed direct sum of countably many copies of A, 84
$c_I(A)$	normed restricted direct product of copies of A, indexed by I, 84
$b_I(A)$	normed bounded direct product of copies of A, indexed by I, 84
$c(A)$	normed restricted direct product of countably many copies of A, 84
$b(A)$	normed bounded direct product of countably many copies of A, 84
$T_n(A)$, $T(A)$	strictly convergent power series in n variables over A, 87
$T_n(M)$, $T(M)$	module of strictly convergent power series with coefficients in M, 87
T	functor $M \rightsquigarrow T(M)$, 89
K	valued field, 89
V	K-vector space, 89
$\mathfrak{F}(V)$	family of all finite-dimensional K-subspaces of V, 90
$U^\perp$	norm-direct supplement of U, 95
A	normed ring, 125

$Q(A)$	field of fractions of A, 125
B	A-algebra, 125
$G(B/A)$	group of A-algebra automorphisms of B, 128
K	field with power-multiplicative norm, 134
L	K-algebra, 134
$\lvert\ \rvert_{\mathrm{sp}}$	spectral norm, 134, 137
$\mathrm{Tr}_{L/K}\, y$	trace of $y \in L$ over K, 139
$\lvert\ \rvert_i$	extension of a valuation, 141
$\lvert\ \rvert_{\mathfrak{S}}$	extension of a valuation, 142
K_{a}	algebraic closure of K, 145
K_{sep}	separable algebraic closure of K, 148
Δ	discriminant (of a polynomial), 148
$r(\alpha)$	distance between α and the set of its conjugates $\neq \alpha$, 148
$\lvert\ \rvert_p$	p-adic valuation on $\mathbb{Q}$, 149
$\mathbb{Q}_p$	p-adic completion of $\mathbb{Q}$, 150
$\mathbb{C}_p$	(algebraically closed) field of p-adic numbers, 150
$K^{p^{-1}}$	field of p-th roots of elements of K, 154
card S	cardinality of a set S, 160
$\mathfrak{M}_A$	category of finite normed A-modules, 164
$\mathrm{Max}_k\, A$	spectrum of k-algebraic maximal ideals of A, 168
$\lvert\ \rvert_{\mathrm{sup}}$	semi-norm of uniform convergence, supremum semi-norm, 169
red A	nilreduction of A, 172
A	integral domain, 183
$Q(M)$	tensor product of an A-module M with $Q(A)$, 183
A_{a}	integral closure of A, 186
$A^{p^{-1}}$	ring of p-th roots of elements of A, 186
k	complete valued field, 192
$k\langle X_1, \ldots, X_n\rangle$, $T_n(k)$, T_n	strictly convergent power series in n indeterminates with coefficients in k, 192
$\check{T}_n$	topologically nilpotent elements of T_n, 193
$\mathring{T}_n$	power-bounded elements of T_n, 193
$\tilde{T}_n$	residue algebra of T_n, 193
$\sim$	residue map, 193
Max T_n	maximal ideals of T_n, 194
$B^n(k)$	n-dimensional unit ball, 196
ϕ'	affinoid map, defined by a homomorphism ϕ between Tate algebras, 199
$j(\mathfrak{a})$	Jacobson radical of $\mathfrak{a}$, 206
$j(A)$	Jacobson radical of the zero-ideal in A, 206
$k(X)$	field of rational functions in X, 214
$\lvert\ \rvert_t$	valuation induced by the (total) degree on $k(X)$, 214
$\lvert\ \rvert_\alpha$	residue norm with respect to an epimorphism α, 222
$\mathfrak{A}$	category of k-affinoid algebras, 222
$k\langle f_1, \ldots, f_n\rangle$	affinoid algebra with affinoid generators $f_1, \ldots, f_n$, 223
$\mathfrak{M}_A$	category of finite complete normed A-modules, 227
$A\langle X, X^{-1}\rangle$	strictly convergent Laurent series in X over A, 230
$A\langle f, g^{-1}\rangle$	strictly convergent power series in f and g^{-1} over A, 231
$A\langle f\rangle$	algebra of strictly convergent power series in f over A, 232
$A\langle g^{-1}\rangle$	algebra of strictly convergent power series in g^{-1} over A, 232
$A\left\langle \frac{f}{g}\right\rangle$	algebra of strictly convergent power series in $\frac{f}{g}$ over A, 233
$P_\varrho(k)$, P_ϱ	polydisc with polyradius ϱ, 234
$T_{n,\varrho}(k)$, $T_{n,\varrho}$	power series converging on P_ϱ, 234
Max A	maximal ideals of A, 236

$\lvert\ \rvert_{\sup}$	supremum semi-norm on A, 236
red A	nilreduction of A, 237
red	reduction map modulo nilradical, 237
τ, τ_A, τ_B	reduction map modulo topologically nilpotent elements, 243
A^G	fix-algebra of a group action on A, 247
k	complete valued field, 259
ξ, η, ζ	indeterminates, 259
Γ	Galois group of k_a over k, 259
$k(x)$	extension of k generated by the components of $x = (x_1, \ldots, x_n)$, 260
τ	canonical map from points to maximal ideals, 260
$V_a(F)$, $V(F)$	zero set of F, 262
$\mathfrak{m}_x$	maximal ideal corresponding to a point x, 262
id(E)	ideal of functions vanishing on E, 262
φ^*	homomorphism of affinoid algebras corresponding to an affinoid map φ, 266
${}^a\sigma$	affinoid map corresponding to a homomorphism σ of affinoid algebras, 266
Sp A	affinoid variety, 267
$\mathbb{B}^n$	n-dimensional unit ball, as an affinoid variety, 268
$X \times_Z Y$	fibre product, 269
π: Max $A \to$ Max $\tilde{A}$	reduction map, 270
(Sp A) $(f_1, \ldots, f_r)$	subset of Sp A, 274, 275
$X(f, g^{-1})$	Laurent subdomain of X, 280
$X(f)$	Weierstrass subdomain of X, 280, 281
$X(g^{-1})$	Laurent subdomain of X, 280, 281
$X\left(\frac{f}{g}\right)$	rational subdomain of X, 281, 282
$\sqrt{\lvert k^*\rvert}$	value group of the algebraic closure of k, 283
$X(\varepsilon^{-1}f)$	Weierstrass subdomain of X, 284
$X(\varepsilon^{-1}f, \delta g^{-1})$	Laurent subdomain of X, 284
$X\left(\varepsilon^{-1}\frac{f}{g}\right)$	rational subdomain of X, 284
$\mathcal{O}_X$	presheaf of affinoid functions on X, 296
$\mathcal{O}_{X,x}$	stalk of $\mathcal{O}_X$ at a point x, 296
f_x	germ of an affinoid function f at a point x, 296
φ_x^*	homomorphism between stalks, induced by a morphism φ, 297
$\dim_x X$	dimension of X at a point x, 300
$\langle B : A\rangle$	minimal cardinality of affinoid generating systems of B over A, 313
$K^{\cdot}$	complex of modules, 316
d^q	coboundary map, 316
$H^q(K^{\cdot})$	q-th cohomology module of a complex $K^{\cdot}$, 316
$H^q(f)$	homomorphism between cohomology modules, 317
$K^{\cdot}/K'^{\cdot}$	quotient complex, 318
$K^{\cdot\cdot}$	double complex, 318
$'d^{p,q}$	coboundary map, 318
$''d^{p,q}$	coboundary map, 318
$'K^{\cdot q}$	q-th column of $K^{\cdot\cdot}$, 318
$''K^{p\cdot}$	p-th row of $K^{\cdot\cdot}$, 318
$H^q(K^{\cdot\cdot})$	q-th cohomology module of $K^{\cdot\cdot}$, 319
$\mathfrak{T}$	system of "open" subsets of a set X, 320
$\mathcal{F}$	presheaf on X, 320
$f\vert_U$	restriction of an element $f \in \mathcal{F}(V)$ to a subset $U \subset V$, 320
$U_{i_0\ldots i_q}$	intersection of the sets $U_{i_0}, \ldots, U_{i_q}$, 320
$C^q(\mathfrak{U}, \mathcal{F})$	module of q-cochains on $\mathfrak{U}$ with values in $\mathcal{F}$, 321

$C_a^q(\mathfrak{U}, \mathscr{F})$	module of alternating q-cochains on $\mathfrak{U}$ with values in $\mathscr{F}$, 321
$C^\cdot(\mathfrak{U}, \mathscr{F})$	Čech complex of cochains on $\mathfrak{U}$ with values in $\mathscr{F}$, 321
$H^q(\mathfrak{U}, \mathscr{F})$	q-th cohomology module of $C^\cdot(\mathfrak{U}, \mathscr{F})$, 321
$C_a^\cdot(\mathfrak{U}, \mathscr{F})$	Čech complex of alternating cochains on $\mathfrak{U}$ with values in $\mathscr{F}$, 321
$H_a^q(\mathfrak{U}, \mathscr{F})$	q-th cohomology module of $C_a^\cdot(\mathfrak{U}, \mathscr{F})$, 321
$\varrho^q(\mathfrak{V}, \mathfrak{U})$	homomorphism between cohomology modules, associated to a refinement of coverings, 324
$\check{H}^q(X, \mathscr{F})$	Čech cohomology module of X with values in $\mathscr{F}$, 324
$\check{H}_a^q(X, \mathscr{F})$	alternating Čech cohomology module of X with values in $\mathscr{F}$, 324
ε	augmentation homomorphism, 324
$C_{\text{aug}}^\cdot(\mathfrak{U}, \mathscr{F})$	augmented Čech complex of cochains on $\mathfrak{U}$ with values in $\mathscr{F}$, 324
$H_{\text{aug}}^q(\mathfrak{U}, \mathscr{F})$	q-th cohomology module of $C_{\text{aug}}^\cdot(\mathfrak{U}, \mathscr{F})$, 324
$C^{\cdot\cdot}(\mathfrak{U}, \mathfrak{V}; \mathscr{F})$	Čech complex depending on two coverings $\mathfrak{U}$ and $\mathfrak{V}$, 325
$\mathfrak{U} \times \mathfrak{V}$	product covering, 327
$\mathcal{O}_X$	presheaf of affinoid functions on X, 327
$M \otimes \mathcal{O}_X$	$\mathcal{O}_X$-module associated to M, 330
$\mathfrak{T}$	Grothendieck topology on a set X, 337
S	system of admissible open subsets of X, 337
Cov U	system of admissible open coverings of U, 337
(G_0), (G_1), (G_2)	properties of G-topologies, 339
$\mathfrak{T}^{\text{w}}$	weak G-topology, obtained by pasting G-topological spaces, 341
$\mathscr{F}$	presheaf on X, 346
$\mathscr{F}_x$	stalk of $\mathscr{F}$ at a point x, 347
$\check{\mathscr{H}}^0(X, \mathscr{F})$	zeroth Čech cohomology of $\mathscr{F}$, as a presheaf on X, 349
$(X, \mathcal{O}_X)$	(locally) G-ringed space, 353
(ψ, ψ^*)	morphism of (locally) G-ringed spaces, 354
ψ_V^*	homomorphisms defining ψ^*, 354
ψ_x^*	homomorphism between stalks, 354
$\mathcal{O}_X\|_U$	restriction of $\mathcal{O}_X$ to U, 354
$\mathbb{A}_k^n$, $\mathbb{A}^n$	affine n-space over k, 362
$\mathbb{B}^n(\alpha)$	"closed" ball of radius α in $\mathbb{A}^n$, centered at zero, 362
X^{an}	analytic variety associated to an affine algebraic scheme X, 363
$\mathbb{P}_k^n$, $\mathbb{P}^n$	projective n-space over k, 364
$\mathbb{A}^*/(q)$	analytic torus, 365
$\mathbb{A}^*$	affine 1-space with origin removed, 365
$X \hat{\otimes} k'$	k'-affinoid variety, obtained by extension of the ground field, 368
$\varphi \hat{\otimes} k'$	k'-affinoid map, obtained by extension of the ground field, 369
$\mathcal{O}$	sheaf of rings, 371
$\mathscr{F}, \mathscr{G}, \ldots$	$\mathcal{O}$-modules, 371
$\ker \varphi$	kernel of an $\mathcal{O}$-module homomorphism φ, 371
$\operatorname{im} \varphi$	image of an $\mathcal{O}$-module homomorphism φ, 371
$\mathscr{F}/\mathscr{F}'$	quotient of $\mathcal{O}$-modules, 371
$\bigoplus_{i \in I} \mathscr{F}_i$, $\mathcal{O}^n$	direct sum of $\mathcal{O}$-modules, 372
$\mathscr{F} \otimes_{\mathcal{O}} \mathscr{G}$	tensor product of $\mathcal{O}$-modules, 372
$\mathscr{I}\mathscr{F}$	product of an $\mathcal{O}$-ideal $\mathscr{I}$ with an $\mathcal{O}$-module $\mathscr{F}$, 373
$\varrho_*(\mathscr{F})$	direct image of $\mathscr{F}$ with respect to a morphism ϱ, 373
$\varrho^\#$	canonical homomorphism from $\mathcal{O}_Y$ to $\varrho_*(\mathcal{O}_X)$, associated to a morphism $\varrho: (X, \mathcal{O}_X) \to (Y, \mathcal{O}_Y)$, 373
$X = \operatorname{Sp} A$	affinoid variety, 373
$\mathcal{O}_X$, $\mathcal{O}$	structure sheaf of X, 373
$\mathcal{O}^{\text{w}}$	restriction of $\mathcal{O}$ with respect to the weak G-topology of X, 373
$M \otimes \mathcal{O}^{\text{w}}$	$\mathcal{O}^{\text{w}}$-module associated to an A-module M, 373
$M \otimes \mathcal{O}$	$\mathcal{O}$-module associated to an A-module M, 373

$\varphi \otimes \mathcal{O}^{\mathrm{w}}$	$\mathcal{O}^{\mathrm{w}}$-module homomorphism associated to an A-module homomorphism φ, 374
$\varphi \otimes \mathcal{O}$	$\mathcal{O}$-module homomorphism associated to an A-module homomorphism φ, 374
F^{w}	functor from the category of A-modules to the category of $\mathcal{O}^{\mathrm{w}}$-modules, 374
F	functor from the category of A-modules to the category of $\mathcal{O}$-modules, 374
$\operatorname{rad} \mathcal{I}$	nilradical of an $\mathcal{O}_X$-ideal $\mathcal{I}$, 384
$\operatorname{rad} \mathcal{O}_X$	nilradical of $\mathcal{O}_X$, 385
$\mathcal{O}_X$	structure sheaf of an analytic variety X, 385
$\operatorname{supp} \mathcal{F}$	support of an $\mathcal{O}_X$-module $\mathcal{F}$, 385
$V(\mathcal{I})$	zero set of an $\mathcal{O}_X$-ideal $\mathcal{I}$ in X, 385
$\operatorname{Ann}(M)$	annihilator of a module M, 386
$\mathfrak{id}(Y)$, $\mathfrak{id}_X(Y)$	$\mathcal{O}_X$-ideal of all elements of $\mathcal{O}_X$ vanishing on Y, 386
$\operatorname{id}_A(Y)$	A-ideal of all elements of A vanishing on Y, 386
X_{red}	nilreduction of an analytic variety X, 389
Δ	diagonal morphism, 391
$U \Subset_Y X$	U relatively compact in X over Y, 394
$\operatorname{ord}_\zeta(f)$	order of a unit f, defined on an annulus, 402
$\mathbb{P}^2_A$	2-dimensional projective space over A, 409
j	j-invariant of an elliptic curve, 414

Index

9 783540 125464

GPSR Compliance

The European Union's (EU) General Product Safety Regulation (GPSR) is a set of rules that requires consumer products to be safe and our obligations to ensure this.

If you have any concerns about our products, you can contact us on ProductSafety@springernature.com

In case Publisher is established outside the EU, the EU authorized representative is:

Springer Nature Customer Service Center GmbH
Europaplatz 3
69115 Heidelberg, Germany

Batch number: 10755284

Printed by Printforce, the Netherlands